Nucleic Acid Targeted Drug Design

On the cover: Three examples of DNA binding are illustrated in one chimeric complex. The long multicolored molecule is double-stranded DNA; carbon atoms are black, nitrogen is blue, oxygen is red, and phosphorus is yellow. The orange molecule is netropsin, a minor groove binder (Chapter 7). The green cluster below netropsin is cisplatin, a platinum coordination complex that covalently binds to the major groove edge of guanine (Chapter 4). Below this is triostin A, a bis-intercalator (Chapter 2), which approaches its binding site from the minor groove side of the bases. This imaginary structure was modeled and imaged by T. A. Larsen, coauthor of Chapter 7, using the computer graphic program RMS by D. S. Goodsell (Chapter 7, ref 200), which colors, shades, and casts shadows to create a realistic looking object. Published three-dimensional coordinates from x-ray diffraction studies and idealized DNA coordinates derived from fiber diffraction data were used to create the model.

Nucleic Acid Targeted Drug Design

edited by

C. L. Propst

Center for Biotechnology
Northwestern University
Evanston, Illinois

Thomas J. Perun

Abbott Laboratories
Abbott Park, Illinois

Marcel Dekker, Inc.   New York • Basel • Hong Kong

Library of Congress Cataloging-in-Publication Data

Nucleic acid targeted drug design / edited by C.L. Propst, Thomas J. Perun.
p. cm.
Includes bibliographical references and index.
ISBN 0-8247-8662-9
1. Drugs--Design. 2. DNA-drug interactions. I. Propst, C. L. (Catherine Lamb). II. Perun, Thomas J.
RS420.N83 1992
615'.19--dc20 92-17585
CIP

This book is printed on acid-free paper.

MARCEL DEKKER, INC.
270 Madison Avenue, New York, New York 10016

Current printing (last digit):
10 9 8 7 6 5 4 3 2 1

PRINTED IN THE UNITED STATES OF AMERICA

To
Julia Nuse Perun
and
Charles Hewlings Clark Mills

Preface

Several years ago, when computer-aided, structure-based drug design was less appreciated than it is today, we felt that a book was needed that would put this emerging technology into a readily learnable format for anyone wishing to understand this approach to drug design. Coming from backgrounds of molecular biology and organic chemistry, respectively, we originally decided that the book should cover the design of drugs targeted against nucleic acids (i.e., DNA and RNA) and against proteins (e.g., enzymes and receptors). In implementing our project, however, we found that technology had not progressed sufficiently for us to be comfortable discussing nucleic acids as drug targets. Thus we decided to limit our subject matter to the protein-based targets, namely enzymes and receptors, and to defer discussing the nucleic acids until a later date. The result of our effort on protein targets, *Computer-Aided Drug Design* (*CADD*), was published in 1989 and is currently in use as both a reference book and a graduate text.

Now the time is right to discuss the nucleic acids as targets for molecular design. Technology in this field has moved along rapidly and is generating exciting information that needs to be brought together in a common, instructive format. Much information in this field remains compartmentalized in its traditional disciplines, and nowhere is dialogue among chemists, biologists, and computer specialists more seriously needed than in this area. So it is with pleasure that we have put together chapters by some of the most knowledgeable, established and young investigators in these intersecting disciplines. The science presented in these chapters is state of the art and will be of interest to anyone who follows the field of nucleic acid/ligand interactions. In addition to providing an update for scientists on current research in this area, *Nucleic Acid Targeted Drug Design* (*NATD*) is intended to be of use to students who plan to enter the field of drug design and to individuals who are involved with drug research but may not be fully familiar with structure-based molecular design. Thus we have elected to continue using the two-section "Methods" and "Applications" format that was so successful for *CADD*.

Somewhat surprisingly, there is very little crossover of information between scientists who use this technology to design drugs for protein-based targets and those who use it to design drugs for nucleic acid targets. Scientists who specialize in designing molecules targeted for proteins will be fascinated by the differences, as well as the similarities, detailed in *NATD*. Conversely, scientists who specialize in targeting molecules for the nucleic acids can find the reciprocal information in the first book, *CADD*. In this respect, *CADD* represents Part I and *NATD* represents Part II in the continuum of strategies and specifics on mechanistic-based drug design.

C. L. Propst
Thomas J. Perun

Contents

APPLICATIONS

Contributors

Mavis Agbandje Department of Biological Sciences, Purdue University, West Lafayette, Indiana

Mary D. Barkley Department of Chemistry, Louisiana State University, Baton Rouge, Louisiana

Norbert Bischofberger Gilead Sciences, Inc., Foster City, California

Mario Gustavo Cardozo Department of Medicinal Chemistry and Pharmacognosy, College of Pharmacy, University of Illinois at Chicago, Chicago, Illinois

James C. Dabrowiak Department of Chemistry, Syracuse University, Syracuse, New York

Peter C. Dedon* Department of Biological Chemistry and Molecular Pharmacology, Harvard Medical School, Boston, Massachusetts

James O. Deshler Department of Molecular Genetics, Beckman Research Institute of the City of Hope, Duarte, California

Irving H. Goldberg Department of Biological Chemistry and Molecular Pharmacology, Harvard Medical School, Boston, Massachusetts

Jerry Goodisman Department of Chemistry, Syracuse University, Syracuse, New York

*Present affiliation: Division of Toxicology, Massachusetts Institute of Technology, Cambridge, Massachusetts.

A. J. Hopfinger Departments of Medicinal Chemistry, Chemistry, and Chemical Engineering, University of Illinois at Chicago, Chicago, Illinois

Laurence H. Hurley College of Pharmacy, Drug Dynamics Institute, The University of Texas at Austin, Austin, Texas

Yoshiyuki Kawakami Department of Drug Research IV, Tsukuba Research Laboratories, Eisai Company, Ltd., Tsukuba, Ibaraki, Japan

Thomas Kodadek Department of Chemistry, The University of Texas at Austin, Austin, Texas

Mary L. Kopka Molecular Biology Institute, University of California at Los Angeles, Los Angeles, California

Teresa A. Larsen Molecular Biology Institute, University of California at Los Angeles, Los Angeles, California

Robert McKenna Department of Biological Sciences, Purdue University, West Lafayette, Indiana

Lester A. Mitscher Department of Medicinal Chemistry, University of Kansas, Lawrence, Kansas

Thomas J. Perun Drug Design and Delivery Division, Abbott Laboratories, Abbott Park, Illinois

C. L. Propst Center for Biotechnology, Northwestern University, Evanston, Illinois

Shashidhar N. Rao Searle Research and Development, Division of G. D. Searle and Company, Skokie, Illinois

William A. Remers Department of Pharmaceutical Sciences, University of Arizona, Tucson, Arizona

Howard Robinson Department of Physiology and Biophysics, University of Illinois at Urbana–Champaign, Urbana, Illinois

John J. Rossi Department of Molecular Genetics, Beckman Research Institute of the City of Hope, Duarte, California

Regan G. Shea Gilead Sciences, Inc., Foster City, California

Linus L. Shen Anti-infective Research Division, Abbott Laboratories, Abbott Park, Illinois

Allison A. Stankus Department of Chemistry, Syracuse University, Syracuse, New York

Andrew H.-J. Wang Department of Physiology and Biophysics, University of Illinois at Urbana–Champaign, Urbana, Illinois

Nucleic Acid Targeted Drug Design

1

Designing Drugs that Interact with Nucleic Acids

Thomas J. Perun
Abbott Laboratories, Abbott Park, Illinois

C. L. Propst
Northwestern University, Evanston, Illinois

I. INTRODUCTION

Since the discovery of DNA as the major genetic material in 1944 (1) and the elucidation of its physical structure in 1953 (2), this molecule has held a central position in our understanding of biology. DNA and its companion molecule RNA provide the first steps in the production of proteins that are required for normal, and sometimes abnormal, physiological processes. Thus it is only natural that we should look to these molecules as potential targets in the design of drugs to prevent or eliminate diseases.

The association of the nucleic acids with so many important life processes, and the familiar chemical structures we associate with these molecules, however, often work against using the nucleic acids as targets for drug action. The basic structural motif of the double helix with its four bases, which was so instrumental in unraveling the genetic code, also confers an element of redundancy that must be overcome in the design and discovery of selective binding agents. Moreover, these nucleic acids usually exist within cells. Thus a drug acting at the nucleic acid level must traverse protective biological membranes to reach its target. DNA is sequestered in the nuclear region of a cell and is doubly protected by membranes. Eukaryotic DNA is further complicated by its chromosomal organization.

Because of its complexity, nucleic acid targeted drug design has not been a major goal in pharmaceutical discovery programs. Currently available drugs that target DNA have come out of traditional drug research, and frequently it has been found that these drugs interact with DNA only in retrospect. Moreover these drugs are often toxic and frequently are used only for the most life-threatening diseases for which no other treatments are available. Central among applications of these drugs are treatments for severe viral infections and cancers.

Nevertheless, the appeal to develop effective, low-toxicity drugs that stop diseases early in their progression is strong. Recent advances in structure-based molecular design of drugs targeted against proteins have clearly shown the power of this approach. With our current understanding of how various molecules interact with the nucleic acids and improvements in structure-based molecular design technology coming out of the protein area, it is now possible to seriously tackle the rational design of nucleic acid targeted drugs.

II. THE DRUG DESIGN PROCESS

There are five established techniques for discovering drugs. The first is serendipity, where a drug is discovered purely by accident, sometimes in the clinic where a side effect becomes a new therapeutic use. Probably the most famous instance of drug discovery by serendipity is Fleming's discovery of penicillin. Although drug discovery by pure accident does happen occasionally, for the most part drugs are discovered through an organized effort to find a way to treat a specific disease. Serendipity, although important to drug design, usually is most productive when it occurs as part of other drug design techniques such as those listed below.

Historically, most drugs have been discovered through a process called screening. Here in vitro or in vivo model systems are developed to try to mimic the disease state. Materials are then screened in these test systems in an attempt to find some molecule that will have a favorable impact on the disease. Some screens are simple assay systems, such as a liquid suspension of a pathogenic microorganism. Tests like these have led to the discovery of most antibiotics, including the tetracyclines and macrolides. Still other in vitro screens involve enzyme or receptor systems and have yielded such drugs as the antiulcer drug cimetidine (3). In vivo screening systems using animal models of disease are also very important in drug discovery. The Watanabe (WHHL) rabbit is but one example (4). This animal model is used extensively in the screening for antihyperlipidemic drugs. The advantage of screening for new drugs is that this technique requires a minimum of knowledge about the disease process, the drug molecule, or the biological target with which the drug interacts. Using this technique, it is possible to discover totally new structure types with the requisite biological activity from natural product or chemical libraries. The disadvantage is that screening is a more-or-less random process, involving trial and error. As such, it is unpredictable in results and time-consuming in labor.

Another major technique for drug discovery involves lead compound modification. Here a lead molecule that has some activity as a drug is modified by chemical or biological techniques in order to improve its characteristics. The lead molecule can come from any source, including serendipity, screening, or rational design using a variety of the techniques discussed in this section. The modification process uses this first molecule as a starting point in order to generate improved molecules. In point of fact, this process is so useful that virtually all drugs, and many other molecules with desirable characteristics, are treated in this way in an attempt to produce better products. Most second-, third-, and higher-generation drugs in a class are generated via this method. A classic example is the antibiotic series of cephalosporins. Here the lead molecule cephalothin, itself a useful first-generation antibiotic, was used as a basis for the design via modification of second-generation drugs such as cefotaxime and third-generation drugs such as ceftazidime (5). The advantage of modifying lead compounds is the high probability of being able to generate a better, or at least somewhat different, molecule. The disadvantage is that this approach does not allow for the discovery of new chemical entities (NCEs), completely new structures that provide the quantum leaps forward in pharmaceutical discovery.

The most powerful technique of the future in the discovery of pharmaceutics is structure-based drug design (SBDD), also known as computer-aided or mechanism-based drug design. As more is learned about the biology and chemistry of a disease, it becomes possible to select specific molecular *targets* at which to arrest the disease process. These targets may be proteins (e.g., enzymes or receptors) or nucleic acids (RNA or DNA). Once a target has been selected, a molecule that interacts with the target in such a way as to prevent or stop the disease can often be designed. In this book, molecules that interact with targets are called *effectors*. Targets and effectors interact with each other in a physically complementary way, which has been likened to a key fitting into a lock (6). Although the lock-and-key model of interactions is somewhat simplistic given the complexity of the interactions between target and effector, it is nonetheless a useful analogy. The computer is a necessary adjunct to this technology because of the massive amount of information that must be harnessed in order to develop drugs via this technique. There are two major advantages of this technology. First, when used in conjunction with the modification process, it has the capacity to speed up the entire drug design process by suggesting and eliminating potential types of modifications (7). More important, however, is its ability to suggest entirely new types of molecules that may be useful as drugs, especially against the diseases for which conventional drug discovery techniques have been relatively unsuccessful. The major disadvantage of SBDD is the vast amount of knowledge required to carry out the process. Structure-based drug design is also sometimes called rational drug design. Although structure-based design is clearly very rational, to say that this technique, alone, is rational does disservice to the substantial thought processes that routinely go into developing drugs using the other techniques of drug discovery.

The purpose of drug design is to prevent or eliminate a disease. Under the best of circumstances this is a difficult process. Therefore, most serious attempts to find a pharmaceutical to treat any specific disease will use a combination of the above techniques in attempting to solve the problem. Figure 1 presents schematically some of the various interactions that can occur in the discovery of a new drug. This book will concentrate on structure-based drug design techniques as they are applied to nucleic acid targets. The application of this technology to protein-based targets has been previously discussed in detail (8).

III. COMPUTER-AIDED DRUG DESIGN

It is important to recognize that the drug design process utilizes a number of different technologies to obtain the most precise information on how an effector interacts with its macromolecular target. The most common methods used are X-ray crystallography and nuclear magnetic resonance (NMR) spectroscopy in combination with computer-generated molecular models. X-ray crystallography and NMR are necessary to provide the experimental data used to develop or refine the three-dimensional structure of the target and its effectors, separately or when complexed. Computer modeling allows one to generate and test new models of the macromolecules and their complexes in greater number and with greater speed than can be accomplished with experimental techniques alone.

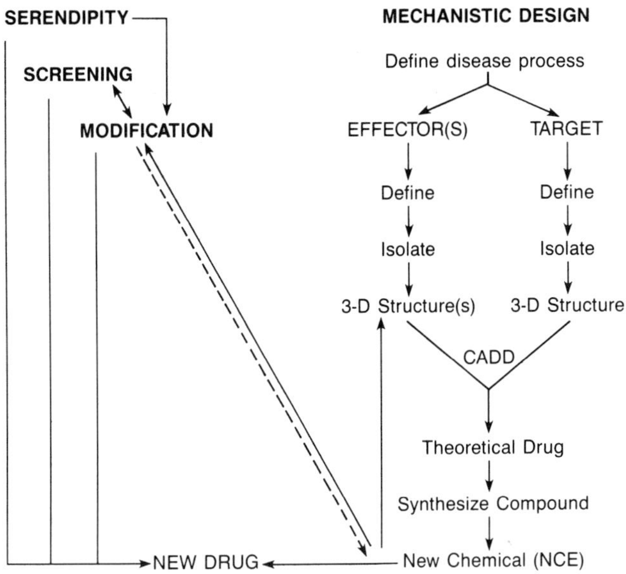

Figure 1 Schematic representation of some of the various interactions that occur in the discovery of a new drug. Mechanistic, structure-based drug design uses the known, or surmised, shape of the target and characterized effector molecules to develop drugs that will fit the target molecule. The process of interactive matching of three-dimensional structures is carried out using a computer, hence the name computer-aided drug design (CADD).

X-ray crystallography has become the method of choice when detailed structural data about a target and its effector are needed. It has been used with great success for proteins, as well as much larger structures such as whole viruses (9). Methods have been developed for increasing the speed with which initial structures are obtained and refinements in the structure are made. There has even been new technology applied to the crystallization procedure itself, using robotics to automate the selection and testing of the multiple variables necessary to obtain suitable crystals for X-ray analysis. Since the X-ray–determined structure is often the starting point for a drug design effort, it may become the rate-limiting step. As improvements have been made in this technology to speed up the obtaining of results, drug design has become a more iterative process. This has been demonstrated in a dramatic manner in the development of novel inhibitors of thymidylate synthase (10). In this example, the structure of the enzyme as determined by X-ray crystallographic analysis was used to construct a computer-generated model. Putative inhibitors were designed using the model and synthesized for testing. These compounds were then cocrystallized with the enzyme and their fit in the active site determined through another X-ray diffraction analysis. Using this iterative process, refinement of the inhibitor structure was possible with only a small number of compounds needing to be synthesized.

X-ray crystallographic analysis of drug-DNA complexes has not developed as rapidly as the corresponding drug-protein crystallography. Nevertheless, progress is being made in this area, and crystal structures of intercalating agents and groove binders have been obtained. Methods have had to be developed for preparation of the pure oligonucleotides needed for successful crystallization. Sometimes the oligonucleotide has had to be modified chemically to allow crystallization of complexes. Once suitable crystals have been obtained, other difficulties arise in solving the structure of the drug-DNA complex. How these problems have been overcome is discussed in Chapter 2 by Wang and Robinson.

Another technique used for obtaining experimental information about macromolecules and their complexes with effector molecules is NMR spectroscopy. The structural studies carried out using NMR complement those with X-ray crystallography as they are conducted primarily with solutions. This allows a more dynamic investigation of the changes that occur in effector-target complexes. For instance, a substrate for an enzyme could exist in a number of different conformations only one of which will be acted on by the enzyme. NMR techniques have been developed that can identify the active conformation in the presence of the enzyme (11). Often conformational changes in the macromolecular target or the effector molecule occur before a tight complex is obtained. This is the kind of information NMR studies can provide.

Like X-ray crystallography, NMR techniques have advanced dramatically over the last decade so that information can be obtained in a more timely manner. Since the size of the macromolecular target is more limited with NMR, newer spectrometers, along with more sophisticated multidimensional NMR techniques, have been developed that double or triple the size of the molecules that could be studied only a few years ago. Commercial 600-MHz spectrometers are now available, and three-dimensional NMR techniques make it possible to determine the structure of proteins with 150 amino acids or more. The use of isotopic labeling with either the macromolecule or the small molecule has given rise to four-dimensional NMR techniques for analyzing a complex. These isotope-editing studies help to overcome the size limitation for obtaining structural information by focusing on the binding region of the macromolecule with its effector (12).

As noted earlier with X-ray crystallography, NMR techniques are particularly useful in defining protein structures, especially for small and medium-sized proteins. The characteristic protons of specific amino acids can often be identified readily in a spectrum and provide a useful starting point in the complete assignment of spectral protons. It has been possible to determine the complete secondary and tertiary structures of some proteins by the use of intramolecular NOE interactions. DNA structures usually exist in extended polymeric structures and thus present fewer intramolecular interactions for analysis. On the other hand, the characteristic helical structures of oligonucleotides provide a useful starting point for examining the interactions of small molecules with these structures by intramolecular NOE assignments. Just as in the study of small molecule-protein interactions, the use of isotopic labeling of nucleotides and their binding agents will become an important technique for the study of these complexes. NMR is currently being used to determine the conformations of proteins that interact with RNA and DNA. In the future we should see the use of NMR in the investigation of the interactions themselves.

Molecular modeling and nucleic acid structures have been intertwined from the time of the discovery of the DNA double helix. In those early days, models for visualizing the interactions between two strands of DNA had to be constructed with rigid hand-held or framework-supported models. The development of computer techniques for construction and manipulation of models has revolutionized the ability to study the three-dimensional shape of macromolecules and the interaction of these structures with their partners. The computer graphics programs available enable one to examine more characteristics than a stick-figure model will allow. In addition to projecting the backbone structure, these techniques show the electron densities of atoms, the solvent-accessible surfaces, and other physicochemical properties that are useful in determining the probable binding modes of effector molecules. The application of computer modeling to the structural analysis of drug-DNA binding is discussed in Chapter 3 by Rao.

Another application of computer programs in the drug design process is the energy optimization of structures in their free and binding mode. A number of programs have been developed for the energy minimization of protein structures, and some of these have been applied to nucleic acid structures. Some examples are discussed in Chapter 3 and a number of the application chapters. The study of the energetics of binding between oligonucleotides and intercalating agents is more advanced than with other binding modes. Similarly, computational analyses of intercalating agents have begun to develop quantitative structure-activity relationships (QSAR), which are explored in Chapter 5 by Hopfinger et al.

IV. THE NUCLEIC ACIDS AS TARGETS

In a mechanism-based drug design effort, the three most common targets for interrupting a disease process are enzymes, receptors, and nucleic acids.

Enzymes are often the target of choice because of the growing number of them that have been identified in the study of disease processes and their availability through genetic engineering techniques. If they belong to a known class of enzymes, the design of inhibitors is frequently possible through a comparison of the active site with others of the class and through structural knowledge obtained from inhibitors of other members (13). For an enzyme inhibitor to be useful in an in vivo situation, it must inhibit a key step in a biochemical pathway where no alternate pathway to the product is available. The inhibitor should also have an affinity consistent with inhibitory concentrations at or near the nanomolar level to be effective in shutting down the process.

Receptors are also useful targets for developing new drugs since they are involved in the first step of the signaling pathway for hormones and other modulators of biological function. Since they exist primarily in a membrane-bound form, their structures have not been as well defined as enzymes. This is rapidly changing, and the structural information now being obtained will create useful targets for future drug design efforts.

The nucleic acids make logical targets for drug design, since all enzymes and receptor proteins are derived from DNA and RNA. Blocking at one of these early steps in a disease-related pathway would clearly be an effective means of shutting down the synthesis of offending proteins. Moreover, inhibiting synthesis at the nucleic acid stage has an appealing simplicity since amplification occurs when each DNA molecule is transcribed into many molecules of RNA, which are further translated into a multitude of protein products.

B-form DNA, usually considered to be the major biologically active DNA, twists in the form of a right-handed, double-stranded helix forming distinctive large and small grooves on its exterior surface. These grooves, termed respectively the major and minor grooves, are an important source of structural information for effector molecules interacting with native DNA. The bases that encode the sequence specificity are tucked away inside the double helix and must be read from the exterior. Access to the bases occurs via these grooves. Figure 2 shows the structure of B-DNA, with clearly visible major and minor grooves. Additional local information may be provided by a variety of DNA structural variations such as hairpins, kinks, and left-handed coils (14).

Although DNA is a remarkably well-defined macromolecule, it makes a complex drug design target. The major characteristics we associate with this nucleic acid—its consistent four bases, paired and stacked inside a double-stranded helix—work against it in developing new drugs. A major tenet of drug design is molecular specificity, which is usually required to reduce side effects. While protein variability is achieved via 20 different

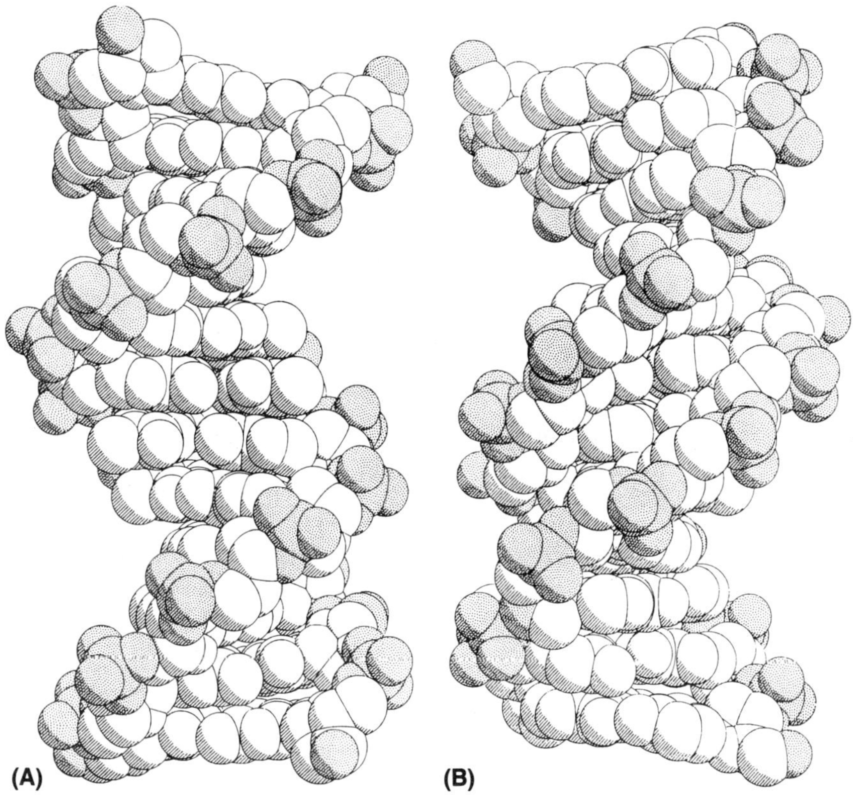

Figure 2 B-DNA showing major and minor grooves. Space-filling representations of the B-DNA dodecamer CGCGAATTCGCG clearly show the major (A) and minor (B) grooves. This DNA includes the *EcoR1* restriction site sequence, GAATTC. (Reprinted with permission from Ref. 18.)

amino acids, DNA has only four different bases. This means that it takes a span of about 15–16 bases to define a unique DNA sequence in humans via base specificity (15). Intact DNA must be read from the exterior of the helical structure. Since only four bases can be read from the major groove and two from the minor groove, a reading molecule with high specificity must be very large and of unique enough shape to read many bases. The major molecules that are able to read DNA with high specificity are large proteins and complementary strands of DNA or RNA. Smaller molecules with less complex shapes are able to interact with DNA, but these do not have much specificity. It is a testimony to the difficulty of designing these drugs that most DNA-targeted drugs are small molecules of limited specificity and high toxicity.

When RNA is the target, many of the same principles apply as with DNA. RNA, however, differs from DNA in a number of obvious ways. While DNA normally occurs as a double-stranded, B-form (10-fold) helix, RNA is basically single-stranded. Internal pairing of RNA bases results in double-stranded regions of A-form (11-fold) helices. The difference in the formation of B-form DNA and A-form RNA is structural, resulting from the extra oxygen in the sugar of RNA. This difference in helical form means drugs that target double-stranded DNA will not necessarily interact with similar sequences of double-stranded RNA for spatial reasons. On the other hand, drugs designed to target RNA selectively may be able to take advantage of its A-form, double-stranded regions or home in on its single-stranded or loop structures.

V. DRUGS THAT INTERACT WITH NUCLEIC ACIDS

Because DNA and RNA are so integral to the proper functioning of all living organisms, it is difficult to utilize these biological molecules as targets without the possibility of disrupting a normal function. For this reason, only serious diseases have been treated with drugs that interact with nucleic acids, where some degree of toxicity is permitted. A key to obtaining less toxic drugs is to discover techniques for binding molecules to DNA and RNA in a very specific manner. If this can be accomplished, new compounds can be designed that will utilize nucleic acid targets and will be used to treat many more diseases than presently possible.

Virtually all drugs currently available that interact with nucleic acids are small molecules and have limited specificity. These drugs can be categorized in a number of ways. Drugs, in general, however, interact with DNA in basically one of three ways (see Kopka and Larsen, Chapter 7):

1. Drugs bind in the minor or major grooves, via noncovalent interactions.
2. Drugs intercalate between base pairs.
3. Drugs form a covalent bond with DNA.

Some drugs, e.g., neocarzinostatin, are versatile and use more than one of these techniques in binding to DNA (see Chapters 4, 10).

Studying these drugs by classes and comparison of their different DNA interaction sequences is one of the first steps in improving specificity and allows general principles to be developed. Chapter 4, by Dabrowiak, Stankus, and Goodisman, discusses techniques that are available for determining binding specificity and reviews in detail how many of the currently available compounds interact with DNA.

The chapters in the applications section of this book discuss a number of examples where some degree of sequence specificity has been obtained and new compounds are being made to increase selectivity and produce better drugs. Chapters 7 and 8 deal with drugs that bind in the minor groove of DNA. Kopka and Larsen (Chapter 7) discuss the pioneering work with netropsin, a compound that binds selectively to AT sequences. While the original compound was found from natural product screening, determination of binding characteristics leads to the design of compounds with increased selectivity. Similarly, in Chapter 8 Remers, Barkley, and Hurley discuss a series of compounds based on tomaymycin, a GC-selective drug, which was first found in nature but has since led to the design of new structures that have better activity profiles.

Compounds that bind to double-stranded DNA by insertion between adjacent base pairs are known as intercalators. The effectiveness of intercalating agents in the treatment of cancer has been known for some time. In Chapter 11, Agbandje and McKenna discuss these drugs and their own attempts to design and synthesize new molecules with a reduction in nonspecific binding.

A unique class of DNA-binding compounds known as the enediynes are the subject of Chapter 10 by Dedon and Goldberg. Neocarzinostatin, the first antitumor antibiotic found in this class, is a versatile molecule combining groove binding and intercalation. Compounds in this class have a reactive moiety that is unleashed after binding occurs, and the radicals produced cause breaks in the DNA strand. Efforts are underway in a number of laboratories to combine the reactivity of this chemical structure with selective binding agents utilizing total synthesis approaches.

In Chapter 9 Mitscher and Shen show how binding to bacterial DNA by synthetic quinolone antibacterials can inhibit a critical nucleic acid regulatory enzyme, DNA gyrase, through a ternary complex. Because the enzyme involved is unique to prokaryotes, the drugs are safe and effective in human infections.

Proteins clearly have unique abilities to bind to nucleic acids. Knowledge about how large proteins bind to nucleic acid sequences with exquisite specificity and shut down or otherwise regulate important gene functions has come from research in many areas of genetic regulation. Frequently these DNA binding proteins are regulatory proteins that fit into the major groove, such as the lambda (16) and Lac (17) repressors. This area, which will have major implications in the future for the design of drugs, is reviewed by Kodadek in Chapter 6.

Recently the concept of using nucleic acid sequences as drugs, as well as drug targets, to shut down desired pathways with high specificity has been gaining momentum. Such oligonucleotides can be either DNA or RNA based and have the option of binding in the form of duplexes or triplexes.

Chapter 13, by Bischofberger and Shea, discusses drug design with these types of molecules. In addition, RNA molecules can be designed in such a way as to cause strand cleavage at specifically targeted sites. The potential usefulness of catalytic antisense RNAs (ribozymes) is discussed by in Chapter 12.

VI. FUTURE CONSIDERATIONS

As mentioned previously, current drugs that act on DNA do so with low selectivity. The problem of drug toxicity thus becomes an important factor in the use of such agents. Unlike enzymes and other protein structures that may be uniquely associated with a disease process, nucleic acid structures often resemble, and in many cases utilize, the basic framework of the structure involved in normal human function. Drugs that act on nucleic acids are therefore reserved for serious and life-threatening diseases where safer treatments are not available.

Much effort is currently being made to identify new techniques for modifying nucleic acid function. Future therapies may be available that combine the selectivity and binding affinity of modified polymeric structures with specifically designed chemically reactive moities. New modalities such as antisense binding by synthetically modified nucleic acid oligomers and the use of nucleic acid structures as enzymes (ribozymes) also offer high hopes of achieving some of the goals of nucleic acid targeted therapy. Successes in these areas should make it possible to utilize nucleic acid targets in many more diseases than currently feasible. When this is achieved, the limiting factors will be drug absorption, metabolism, and excretion, but this is the subject of yet another book.

ACKNOWLEDGMENTS

The authors thank Dr. M. L. Kopka for providing Figures 2A and B and for her constructive discussions and critiques.

REFERENCES

1. Avery, O.T., C.M. Macleod, and M. McCarty, Studies on the chemical nature of the substance inducing transformation of pneumococcal types. I. Induction of transformation by a desoxyribonucleic acid fraction isolated from pneumococcus type III. *J. Exp. Med. 79*: 137 (1944).
2. Watson, J.D., and F.H.C. Crick, Molecular structure of nucleic acids: a structure for deoxyribose nucleic acid. *Nature 171*: 737–738 (1953).
3. Ganellin, C., Discovery of cimetidine. In *Medicinal Chemistry—The Role of*

Organic Chemistry in Drug Research, S. Robert and B. Price, eds., Academic Press, London (1985), pp. 93–118.
4. Watanabe, Y., Serial inbreeding of rabbits with hereditary hyperlipidema (WHHL rabbit). *Atherosclerosis 36*: 261–268 (1980).
5. Sweetman, A.J., Ceftazidime. *Drugs of the Future 6*: 612–615 (1981).
6. Fischer, E., Einfluss der Configuration auf die Wirkung der Enzyme. *Ber. Deutsch. Chem. Ges. 27*: 2984–2993 (1894).
7. Propst, C.L., Structure-based drug design: biology, chemistry, and computers in concert. *SIM News 41*: 5–7 (1991).
8. Perun, T.J., and C. L. Propst, eds. *Computer-Aided Drug Design*, Marcel Dekker, New York (1989).
9. Rossmann, M.G., E. Arnold, J.W. Erickson, E.A. Frankenberger, J.P. Griffith, H.J. Hecht, J.R. Johnson, G. Kamer, M. Luo, A.G. Mosser, R.R. Rueckert, B. Sherry, and G. Vriend, Structure of a human common cold virus and functional relationship to other picornaviruses. *Nature 317*: 145–153 (1985).
10. Appelt, K., R.J. Bacquet, C.A. Bartlett, C.L.J. Booth, S.T. Freer, M.A.M. Fuhry, M.R. Gehring, S.M. Herrmann, E.F. Howland, C.A. Janson, T.R. Jones, C.C. Kan, V. Kathardekar, K.K. Lewis, G.P. Marzoni, D.A. Matthews, C. Mohr, E.W. Moomaw, C.A. Morse, S.J. Oatley, R.C. Ogden, M.R. Reddy, S.H. Reich, W.S. Schoettlin, W.W. Smith, M.D. Varney, J.E. Villafranca, R.W. Ward, S. Webber, S.E. Webber, K.M. Welsh, and J. White, Design of enzyme inhibitors using iterative protein crystallographic analysis. *J. Med. Chem. 34*: 1925–1934 (1991).
11. Fesik, S. Approaches to drug design using nuclear magnetic resonance spectroscopy. In *Computer-Aided Drug Design*, T.J. Perun and C.L. Propst, eds., Marcel Dekker, New York (1989), pp. 133–184.
12. Fesik, S.W., R.T. Gampe, Jr., H.L. Eaton, G. Gemmecker, E.T. Olejniczak, P. Neri, T.F. Holzman, D.A. Egan, R. Edalji, R. Simmer, R. Helfrich, J. Hochlowski, and M. Jackson, NMR studies of [U-^{13}C] cyclosporin A bound to cyclophilin: bound conformation and portions of cyclosporin involved in binding. *Biochemistry 30*: 6574–6583 (1991).
13. Hutchins, C., and J. Greer, Comparative modeling of proteins in the design of novel renin inhibitors. *Crit. Rev. Biochem. Mol. Biol. 26*: 77–127 (1991).
14. Barton, J.K., Recognizing DNA structures. *Chem. Engin. News*, Sept. 26, 1988, pp. 30–42.
15. Hurley, L.H., and F.L. Boyd. Chapter 26. Approaches toward the design of sequence-specific drugs for DNA. In *Annual Reports in Medicinal Chemistry*, Vol. 22. D. Bailey, ed. Academic Press, New York (1987), chapter 26, pp. 259–268.
16. Jordon, S.R., and C.O. Pabo, Structure of the lambda complex at 2.5 Å resolution: details of the repressor-operator interactions. *Science 242*: 893–899 (1988).
17. Boelens, R., R.M. Scheek, J.H. vanBoom, and R. Kaptein, Complex of lac repressor headpiece with a 14 base-pair lac operator fragment studied by two-dimensional nuclear magnetic resonance. *J. Mol. Biol. 193*: 213–216 (1987).
18. Wing, R., H. Drew, T. Takano, C. Broka, S. Tanaka, K. Itakura, and R.E. Dickerson. Crystal structure analysis of a complete turn of B-DNA. *Nature 287*: 755–758 (1980).

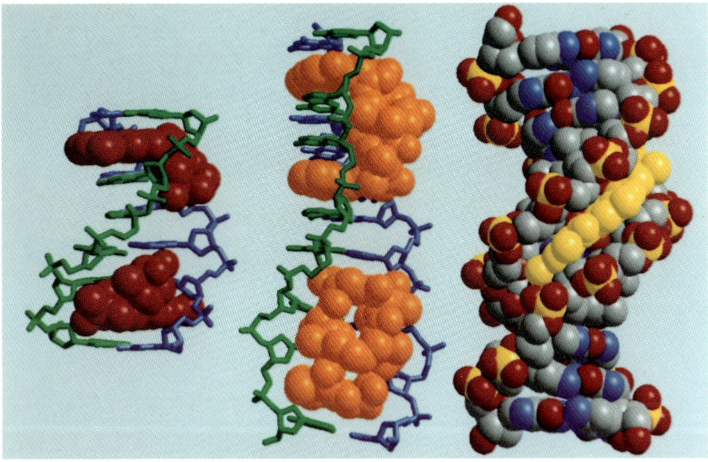

Figure 2.1 Three-dimensional structures of three different antitumor drug-DNA complexes solved by high resolution x-ray crystallography. (Left) Intercalator daunorubicin-d(CGTACG) complex (11), (middle) bis-intercalator triostine A-d(GCGTACGC) (12), and (right) minor-groove binder distamycin A-d(CGCAAATTTGCG) complex (13). The figure was produced by a molecular graphic program MARS (molecular architecture rendering system) developed for Silocon Graphics IRIS computer at UIUC.

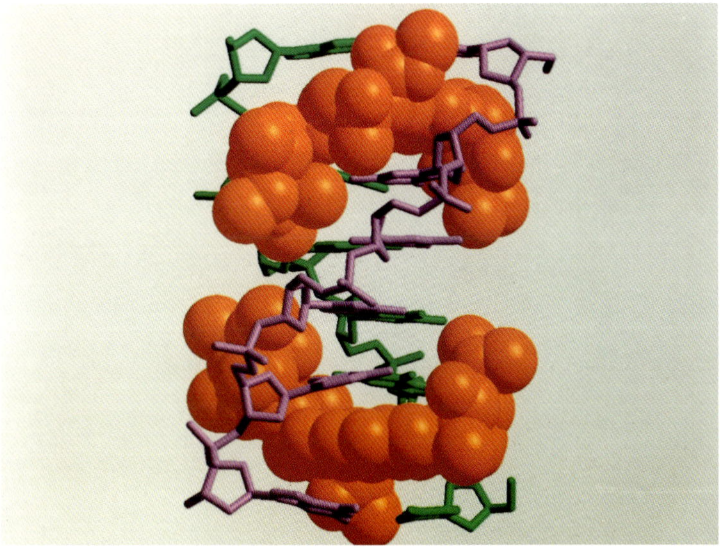

Figure 2.18 Molecular structure of the 2 : 1 complex between nogalamycin and a modified d(CGTACG) hexamer DNA at 1.3 Å resolution. Two nogalamycins are intercalated between the CpG steps at both ends of the distorted right-handed B-DNA hexamer duplex. The aglycone spans the base pairs such that the sugars are in both grooves (nogalose in the minor groove and the aminoglucose in the major groove) of the duplex. Specific hydrogen bonds were found between the drug and DNA.

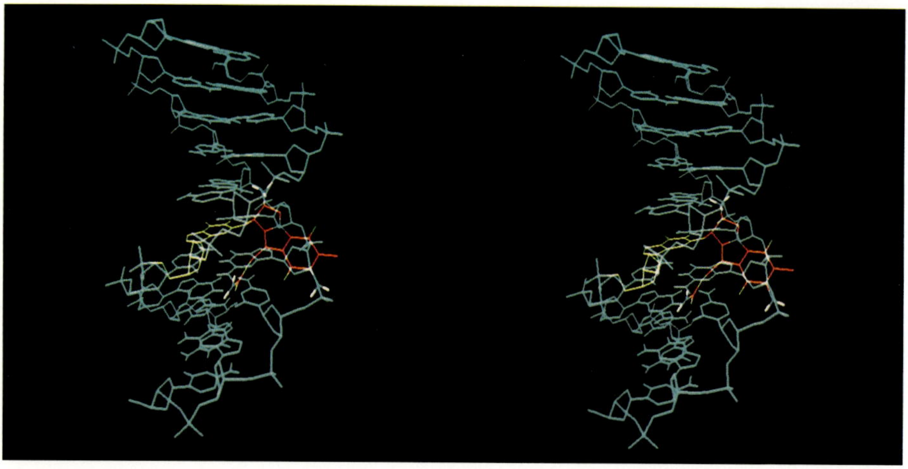

Plate 3.1 Stereo pair of the covalent complex between d(GCGCGCGCGC)$_2$ and mitomycin C. The DNA and the drug components are colored cyan and red, while the guanine nucleotide covalently linked to mitomycin C is shown in yellow.

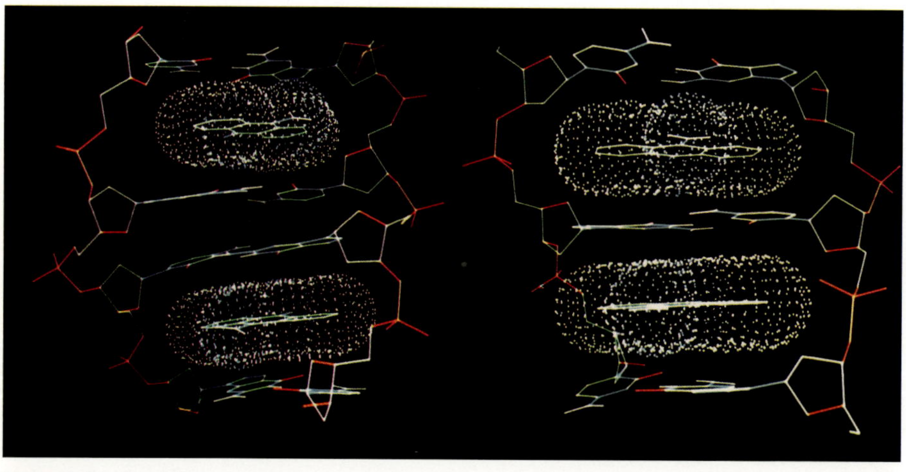

Plate 3.2 Double intercalation complexes between d(CGCGCGC)·d(GCGCGCG) and 9-aminoacridine with the neighbor exclusion rule obeyed (left) and with the neighbor exclusion rule violated (right). Base pairs not directly involved in intercalation with the 9-amino acridines are not shown.

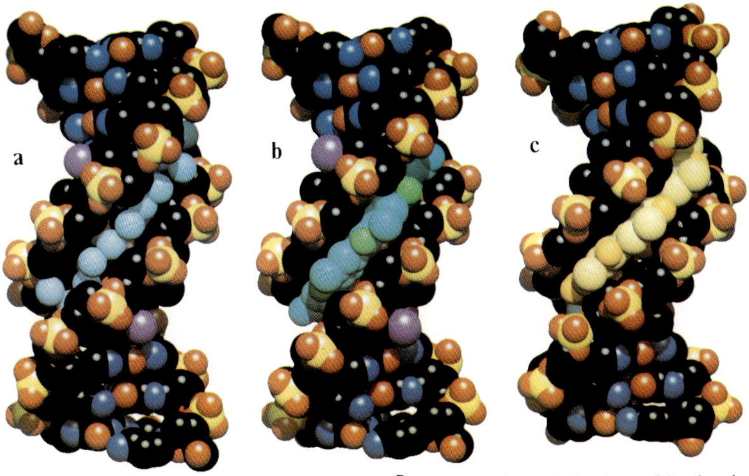

Plate 7.1 Minor groove view of CGCGAATT^{Br}CGCG with (a) "spine of hydration" filling the minor groove. Water molecules are pale blue. DNA carbon atoms are black, nitrogens—blue, oxygens—red, phosphorus—yellow, and bromine—violet. (b) With netropsin bound in shades of aqua for carbon and nitrogen atoms and oxygens in green. Note blue atom at bottom left of groove is a H_2O molecule. (c) CGCAAATTTGCG with distamycin bound in shades of yellow. Light yellow indicates carbon and nitrogen atoms with dark yellow for oxygen atoms. Note partially hidden H_2O molecule in groove at bottom left.

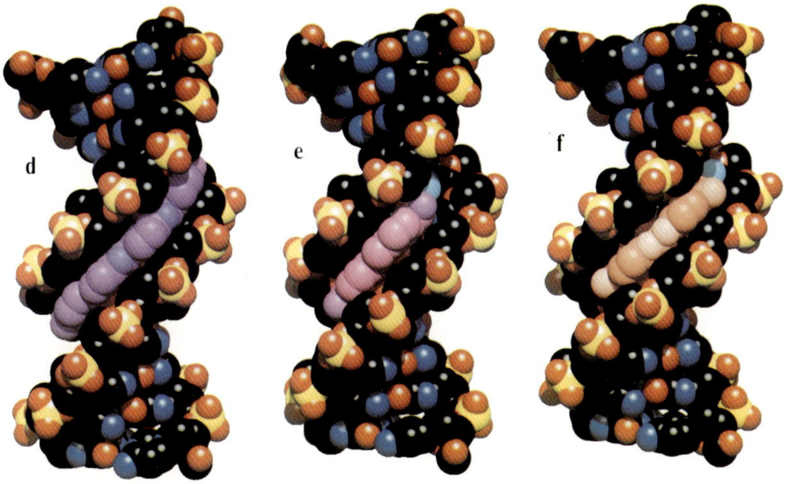

Plate 7.2 Minor groove view of CGCGAATTCGCG with (d) Hoechst 33258 bound in minor groove. DNA carbon atoms are black, nitrogens—blue, oxygens—red, and phosphorus—yellow. Hoechst molecule is shown in shades of violet with carbons in light violet and nitrogens in dark violet and phenol oxygen at right top of Hoechst in pink violet. (e) DAPI shown in shades of deep pink for carbons and lavender for nitrogens. Note bound H_2O molecule at top of minor groove. (f) Berenil in shades of salmon with carbon atoms dark salmon and nitrogens light salmon. Again H_2O molecule is at top of groove, but in a different position than in DAPI structure. Raster images calculated using RMS by D.S. Goodsell (200).

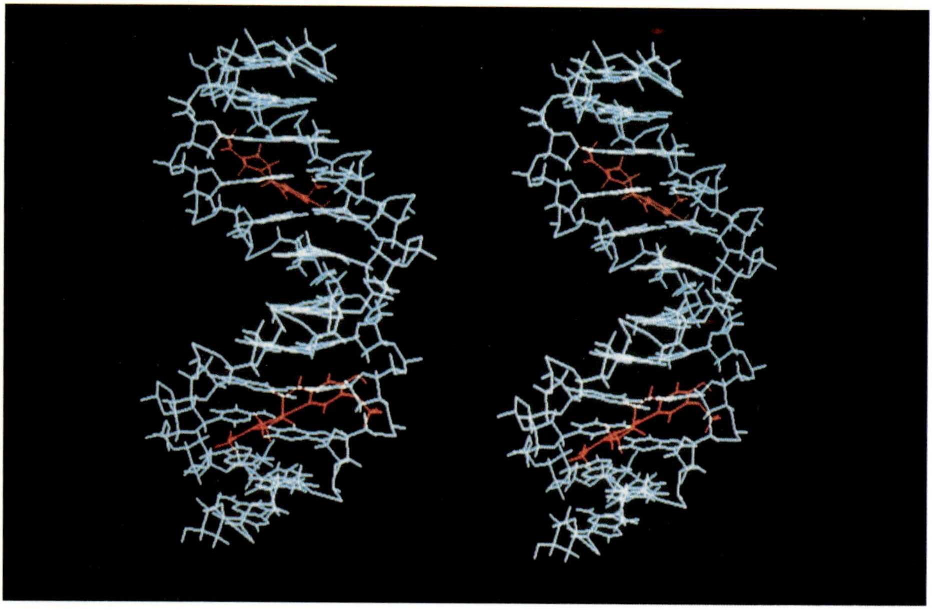

Figure 8.16 Stereo drawing of ^4G attached to tomaymycin showing the conformation obtained through minimization with AMBER (see Molecular Modeling).

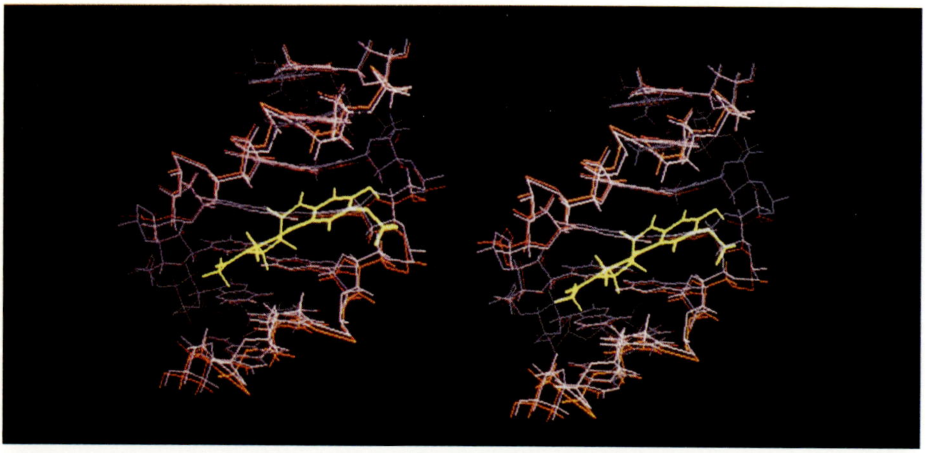

Figure 8.17 Stereo drawing of the superimposed d(CICGAATT·ATTCICG) in orange and tomaymycin–d(CICGAAT·ATTCICG) in magenta and yellow.

METHODS

2

Structural Studies of Drug-DNA Interactions by X-ray Crystallography and NMR Spectroscopy

Andrew H.-J. Wang and Howard Robinson
University of Illinois at Urbana–Champaign, Urbana, Illinois

I. INTRODUCTION

DNA and RNA play a central functional role in biological systems. Many compounds are known to interfere with the function of nucleic acids and proteins that interact with nucleic acids. Some of those compounds, natural and synthetic, have been found to possess cytotoxic activities, and they have been exploited as potential antitumor/anticancer agents. The mode of action of those compounds has been the subject of intense research for the past two decades from which a general picture has emerged. There are five major types of DNA binding antitumor drugs, namely, intercalator, noncovalent groove binder, covalent binding/cross-linking agent, DNA cleaving agent, and nucleoside-analog (1).

In order to fully understand how antitumor compounds interact with DNA at the molecular level, it would be desirable to have a detailed knowledge of the three-dimensional structure of DNA and its potentials in adopting various conformations. In addition to the well-known right handed doubled helical structures of B-DNA and A-DNA (and the similar RNA-11) conformations, nucleic acids can exist in other stable alternative conformations. For example, a stretch of alternating purine-pyrimidine sequences, such as $(C\text{-}G)_n$ or $(T\text{-}G)_n$, readily forms a left-handed Z-DNA conformation. This novel structure was discovered by a high resolution x-ray diffraction structural analysis of the DNA hexamer d(CGCGCG) (2,3). Other repetitive sequences like $(TTGGGG)_n$ occur in certain eucaryotic telomere DNA and they have been shown to have a tetra-stranded structure (4–7). Oligo(A) sequences, when suitably spaced along the DNA double helix, induce bending in DNA molecules, which has been suggested to have a significant implication in gene regulation (8). Further studies suggested that alternative DNA conformations may be associated with certain specific biological functions.

Questions related to the molecular aspects of the drug binding are of particular interest, as they may be relevant for the design of better drugs. The structure of the complexes between antitumor drugs and DNA oligonucleotides have been studied extensively using the two most powerful techniques for visualizing the three-dimensional structure of biological macromolecules, namely single-crystal X-ray crystallography and NMR spectroscopy. This has been made possible mainly due to the great advances in synthetic oligonucleotide chemistry which enable us to obtain pure oligonucleotide molecules of defined sequence in large amounts for the biophysical experiments. Their structures, both in crystals and in solution, can be solved, refined and analyzed to afford us an unprecedentedly clear view of the architecture of DNA molecules and drug-DNA complexes (9,10). Figure 1 shows three examples of the three-dimensional structures of drug-DNA complexes solved by high-resolution x-ray diffraction analysis (11–13).

In this chapter, we provide an overall view of these two important techniques. This is not meant to be a comprehensive discussion of the two techniques, nor is it an exhaustive review of the structural studies of antitumor drug-DNA complexes studied by these techniques. Many excellent texts are already available to which the readers are referred (14–19). Instead, we attempt to provide sufficient relevant background information to lay the

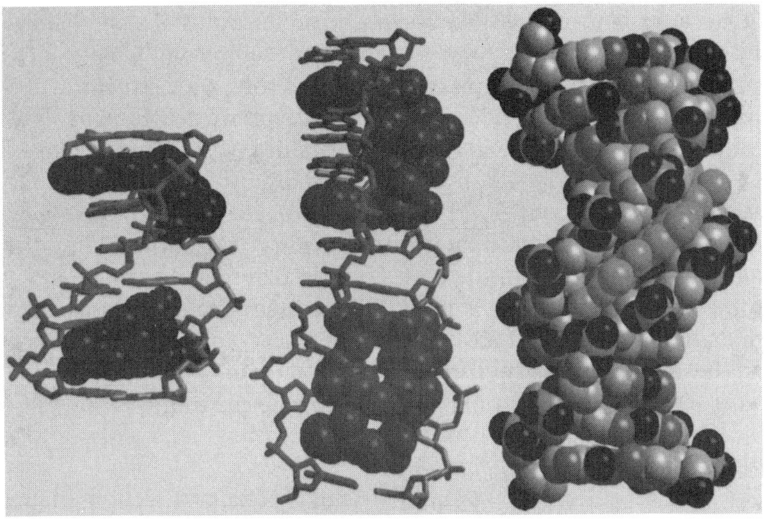

Figure 1 (see also Color Plate following Chapter 1) Three-dimensional structures of three different antitumor drug-DNA complexes solved by high resolution x-ray crystallography. (Left) Intercalator daunorubicin-d(CGTACG) complex (11), (middle) bis-intercalator triostine A-d(GCGTACGC) (12), and (right) minor-groove binder distamycin A-d(CGCAAATTTGCG) complex (13). The figure was produced by a molecular graphic program MARS (molecular architecture rendering system) developed for Silocon Graphics IRIS computer at UIUC.

groundwork for the subsequent chapters which deal with the results of various drug-DNA systems studied by x-ray diffraction and/or NMR spectroscopy.

II. X-RAY CRYSTALLOGRAPHY

X-ray crystallography has been recognized as the most powerful technique in elucidating the three-dimensional structure of molecules. Its influence in science, both in chemistry, physics, and medicine, has been amply demonstrated by 18 Nobel Prizes to scientists whose research had x-ray connections. While the method dates back more than 70 years, there has been a revolution in terms of the theory, instrumentation, and methodology during the past several years of this mature science. In the past, the fiber x-ray diffraction technique has been used to provide various models for nucleic acid polymers. However, this method has serious limitations because it cannot provide reliable fine-structure details. Recently, single-crystal x-ray diffraction analysis of DNA oligomers has changed the situation completely.

This is so because many of the DNA fragments produce crystals that diffract x-rays to high resolution. Now crystallographers are not only able to deal with molecules of very high complexity, but also to solve their structures at a high pace. Nevertheless, the underlying basic concepts of structural analysis by x-ray diffraction, which are illustrated schematically in Figure 2, remain essentially the same. There are five primary steps, which include (1) the preparation of suitable crystals, (2) collecting diffraction data, (3) solving the phase problem to calculate the electron density map, (4) building the molecular model, and (5) refinement of the molecular model.

The objective of the process is to obtain a correct electron density map from which the structure of a molecule can be traced, interpreted and then refined. The electron density distribution $\rho(x,y,z)$, of a molecule in the crystal lattice at a position (x,y,z) is expressed by the equation

$$\rho(x,y,z) = \frac{1}{V}\sum_h\sum_k\sum_l \quad |F(hkl)|e^{-i[2\pi(hx+ky+lz) \; - \; \Phi(hkl)]} \tag{1}$$

Here $|F(hkl)|$ and $\Phi(hkl)$ are the magnitude and the phase angle, respectively, of the *structure factor* of a diffracted spot (*reflection*). This reflection arises from the diffraction of x-rays by a series of planes in the crystal which are separated by the interplanar spacing of $d(hkl)$, where h,k,l are the Miller indices. The quantities $|F(hkl)|$ and $\Phi(hkl)$ need to be determined experimentally in order to calculate the electron density map.

X-rays are scattered by the electrons of atoms in a molecule. In a crystal, molecules are arranged in a highly regular and repetitive manner, and they scatter x-rays collectively. A reflection can be observed when the incident x-ray beam (with a wavelength λ) glances the Miller planes at an angle θ, resulting in a reflected beam exiting at the same angle θ due to the constructive interference of the x-rays in the path. This condition can be met when the *Bragg equation*

$$n\lambda = 2d_{hkl} \sin \theta \tag{2}$$

is satisfied.

In this chapter, the five steps for x-ray diffraction studies of drug-DNA interactions are concisely discussed. Many excellent comprehensive treatises (14–16) and review articles (20,21) have appeared recently to which the readers are referred for more specific information.

A. Structural Determination by X-ray Crystallography

1. *Crystallization of Drug-DNA Complexes*

The success of a structural work of biological macromolecules depends critically on whether useful crystals can be obtained or not (22). So far, the

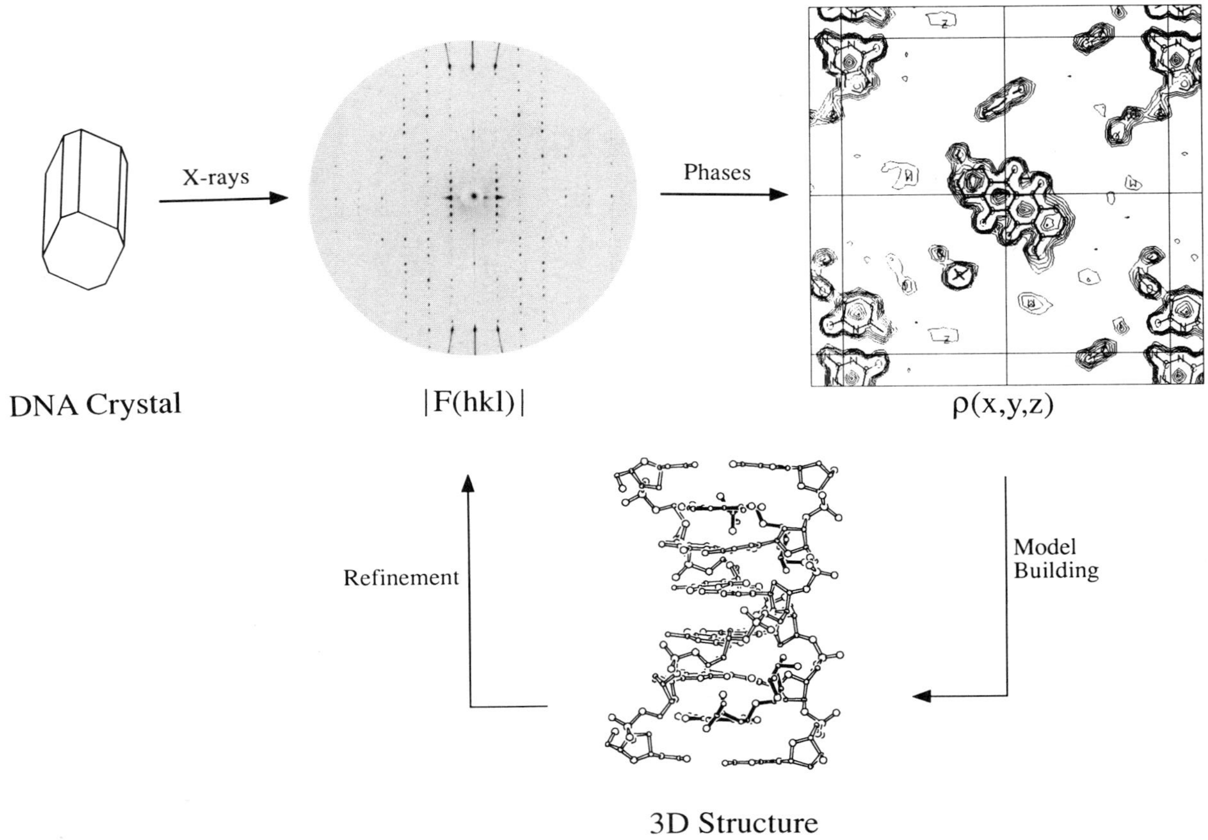

Figure 2 A schematic diagram showing the major steps in the structure determination of drug-DNA complex by x-ray crystallography.

crystallization of nucleic acid molecules and their complexes with antitumor/anticancer drugs remains at a somewhat empirical, trial-and-error stage. Nevertheless, some general information on crystallization conditions have been accumulated, as described below.

The development of automated chemical synthesis of oligonucleotides has provided the opportunity to prepare reliable, pure DNA molecules in large amounts for various biophysical studies. We now routinely prepare DNA molecules on an automated DNA synthesizer using a 10-micromole scale on a solid support resin column to yield 10 mg or more of material. After deblocking of all the protecting groups and cleavage from the resin by ammonia treatment, the final products may be purified by high-pressure liquid chromatography (HPLC), or simply by a gel filtration column (e.g., Sephadex G50). The purity of the final product should be better than 95%. After purification, the molecules are converted into the ammonium form and may be stored indefinitely as a white/lyophilyzed powder at $-20°C$. Most of the DNA oligonucleotide molecules that have been studied by x-ray crystallography or NMR have self-complementary sequences and range from 4 to 20 nucleotides long.

Specific chemical modifications have been useful either during crystallization or during subsequent structure determination. Molecules selectively substituted with thiophospho linkage in the sugar-phosphate backbone seem to have advantages in crystallization under certain conditions. For example, we were able to crystallize nogalamycin complexed to d[CGT(pS)ACG], where pS is a thiophospho linkage in the *R*-configuration, but not to normal d[CGTACG] (23). Sometimes modified oligonucleotides are needed for various purposes. For example, the isostructural 5-bromodeoxyuridine can be used to replace thymidine in DNA to serve as an effective heavy atom derivative to obtain multiple isomorphous replacement phase information (vide infra).

The pH of the solution plays important roles in the crystallization experiment. The most commonly used buffer solution is cacodylate [dimethyl arsinate] solution ($pK_a = 6.19$), which has a favorable buffering strength in the pH ranges of 5 to 7. By varying the pH one can often obtain different crystal forms. In addition, low pH conditions may be used to change the property of the DNA molecule. For example, cisplatin-d(pGpG) complex is negatively charged at neutral pH. However, at lower pH (3.5–4.0) the molecule is a neutral species and its solubility is reduced. We were able to crystallize this complex either from acetate or glycine buffers (24). Another example is that the complex of triostin A with d(GCGTACGC) can be crystallized more consistently from acetate buffer (pH = 4.5), due to the guanine-cytosine Hoogsteen base pairing (which requires a protonated cytosine) in the molecule (12). Finally, the pH appears also to influence crystal growth kinetics. In general, low pH conditions accelerate the crystallization process. If crystals appear too rapidly, the process can be slowed down by using a slightly higher pH conditions (e.g. pH = 7.5).

Both metal ions (e.g., sodium, magnesium) and organic spermine molecules were found to be essential ingredients for the crystallization of many nucleic acid molecules (25). Other heavy atoms like calcium, strontium, and barium can be used to replace magnesium. Sometimes, they facilitate crystal formation, or produce different lattices. Other heavy-metal ions, such as Zn^{+2}, Co^{+2} and Ni^{+2} cobalt(III) hexaammine are sometimes found to be useful in promoting crystallization as well. Organic ions like spermine and spermidine, because of their positive charges, strongly interact with nucleic acids. The flexible spermine molecules readily "glue" the molecules in the lattice together, using hydrogen bonds and electrostatic interactions (26).

The most commonly used precipitating agent in the crystallization of oligonucleotides is 2-methyl-2,4-pentanediol (2-MPD). This viscous organic solvent has a higher boiling point than water and is completely miscible with water. DNA has low solubility in this organic solvent. By putting higher concentrations of 2-MPD in the reservoir, water from the crystallization solution is drawn away into the reservoir due to the lower water vapor pressure of the reservoir. This procedure, termed *vapor diffusion*, increases the concentrations of the materials in the crystallization solution in a slow, nondisruptive way, eventually leading to crystals. Another organic precipitant that may be used is low molecular weight polyethylene glycol (PEG), such as PEG 400. Volatile organic solvents, such as isopropanol, have also occasionally been used. The method that we use most commonly is a sitting drop vapor diffusion method employing the plastic sandwich box and glass spot plate described by McPherson (22). Occasionally some special techniques are useful in increasing the size of the crystals. These include feeding and seeding. In addition, harvesting bad crystals and recycling the material often improves the quality of the crystals significantly.

An Example. Here as an example, we describe the detailed optimized conditions for the crystallization of the antitumor drug nogalamyccin complexed to DNA oligonucleotides (23). Many different DNA sequences have been tested without success, including d(CGTACG) which was successfully crystallized with daunomycin previously (11). Finally, two modified DNA hexamers, d[CGT(pS)ACG] and d[m^5CGT(pS)Am^5CG], which have the same basic sequence as d(CGTACG), were found to be useful. A typical crystallization mixture contained 0.5 mM oligonucleotide, 8 mM $MgCl_2$, 30 mM sodium cacodylate (pH 5), and 0.5 mM nogalamycin in 2% (w/v) PEG400. The solution was equilibrated with 30% PEG400 at room temperature by vapor diffusion. Bright orange-red crystals for both complexes appeared after one week.

2. Data Collection

When suitable crystals of the nucleic acid samples are obtained, a single crystal is selected and mounted in a capillary along with a droplet of the mother liquor for data collection. The capillary is sealed at both ends with dental wax and followed by a coat of fingernail polish. This procedure keeps the crystal in a moist state, which is important in maintaining the integrity of the nucleic acids crystals, since they always contain large solvent channels in the crystal lattices. If the crystals are left exposed to open air, they disintegrate almost instantaneously. The capillary containing the precious crystal is placed on a goniometer head which has translational and angular (arc) adjustment components. The crystal is centered in the x-ray beam on a data collection instrument, e.g., a diffractometer.

Several types of x-ray sources are available for diffraction work. While the sealed-tube x-ray generator remains as a useful source, it is gradually being replaced by the more powerful rotating anode x-ray generator in a regular laboratory environment. The most commonly used x-rays are that from a copper anode K_α (CuK_α) line with the wavelength of ~1.54 Å. More recently, synchrotron radiation provides a totally different dimension for a new kind of x-rays. This source is several orders of magnitude higher (10^8–10^9) in its intensity (brilliance, flux, etc.) when compared to the most powerful commercially available rotating anode source (18 kW). Specifically, the effective brightness of the x-ray radiation from the rotating anode source is about 10^8 photons sec^{-1} mm^{-2} $mrad^{-2}$ (27), while the respective maximum undulator brightness from the 7 GeV Advanced Photon Source at Argonne National Laboratory will be 10^{18} photons sec^{-1} mm^{-2} $mrad^{-2}$ (28). There is no doubt that synchrotron radiation will play a unique role in the structural studies of the problems of drug design related to nucleic acids and proteins in the future.

At the present time, the most widely used instrument for data collection is the fully automated computer-controlled four-circle diffractometer using a rotating-anode x-ray generator. This method offers a reliable and rapid way for collecting intensity data at a rate of about 1000 reflections per day. The resulting intensity data I_{hkl}, is related to the structure factor (F_{hkl}) in the following way:

$$I_{hkl} = k \cdot \mathrm{Lp} \cdot |F_{hkl}|^2 \tag{3}$$

Here k is a factor used to make corrections for absorption, crystal decay, etc., and Lp is a geometric/physical correction factor called the Lorentz-polarization factor. It can be seen that although one can obtain the amplitude of the structure factor $|F_{hkl}|$, the phase information ϕ_{hkl} associated with the reflection is lost during the data collection and can only be recovered via indirect means.

Recently, a number of electronic two-dimensional area detectors have been developed which have significantly improved the speed of data collection. These include the multiwire detector (29), TV detector based on vidicon SIT tube (30), imaging plate detector (31), and charged-coupled device (CCD) detector (32). All of them are capable of collecting thousands of reflections per day and have already reshaped the way we think about data collection. A benefit of the rapid data collection is that one can obtain highly redundant data due to the symmetry of the crystal lattice. These redundant data may be averaged to obtain a significantly more reliable data set.

Another interesting development is the Laue method for collecting data on a synchrotron source (33,34). The principle of Laue photography is to use a large continuous bandwidth (e.g., between 0.2 and 2.5 Å) synchrotron x-rays having relatively uniform high intensity. Because the Bragg condition is satisfied for every Miller plane by at least one wavelength of the polychromatic incident x-rays, a large number of lattice planes from a stationary crystal diffract simultaneously. It has been shown that extremely high data collection rates (~100,000 reflections/sec) can be achieved by this technique. This method holds a great potential for studying the real-time in situ, interactions between ligand and receptor in the crystals.

3. *Solving the Phase Problem*

The most challenging problem in x-ray crystallography has been how to recover the lost phase information of the diffraction data, i.e., the *phase problem*. For molecules with fewer than 100 nonhydrogen atoms, the powerful *direct method* has been developed and its use has become widespread (35,36). The importance of this method is reflected in the Nobel Prize in chemistry to J. Karle and H. Hauptman who developed the theory of direct method in the 1950s. The structures of many antitumor and anticancer compounds have been solved by this method. However, for most problems associated with the drug-DNA complexes, there are several hundred atoms in the asymmetric unit of the crystal lattice. They are clearly beyond the power of the direct method at present and can only be solved using the methods that were originally developed for protein crystallography as described below.

The classical method is the *multiple isomorphous replacement* (MIR) technique, which is schematically shown in Figure 3. In this method, one prepares a series of heavy-atom substituted crystals which are isomorphous to the native crystal. Those isomorphous heavy-atom derivative crystals have very similar unit cell dimensions and have the same space group, but their diffraction patterns are slightly perturbed (compared to that of the native crystal) due to the presence of heavy atoms in the lattice. The diffrac-

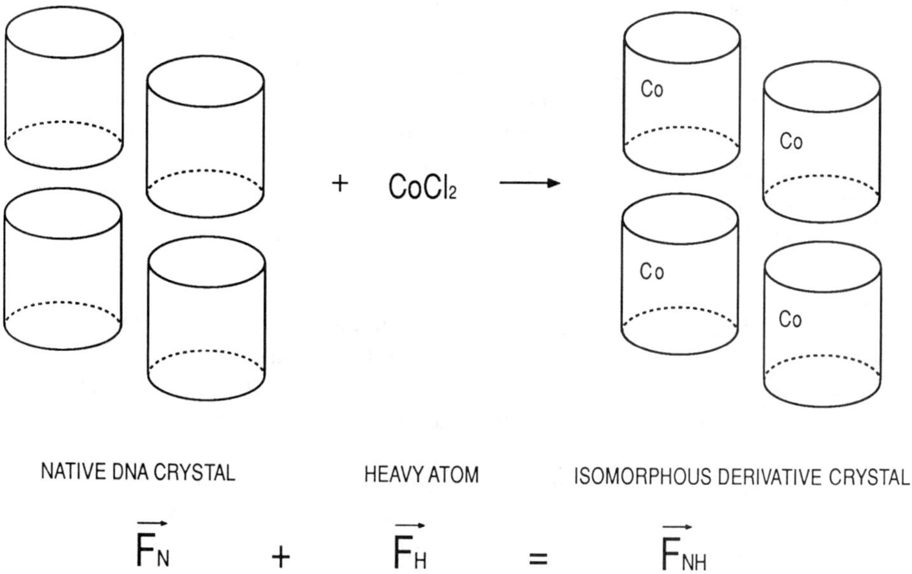

Figure 3 A schematic drawing illustrating the method of multiple isomorphous replacement for solving the phase problem. The cylindrical drums represent the packing of DNA double helixes in the crystal lattice. The diffraction data of the native crystal, $|F_N|$, are collected. Another native crystal is soaked in a solution containing low concentration of $CoCl_2$ which may interact with DNA duplex at a single site (e.g., N7 of a guanine). The diffraction data from the heavy-atom-modified crystal, $|F_{NH}|$, are then collected. The phase information can be calculated using $|F_N|$ and $|F_{NH}|$. Usually, more than one heavy atom derivative is prepared in order to obtain more reliable phase information.

tion data of both the native (N) and heavy-atom (H) derivative crystals are collected which provide their corresponding structure factor amplitudes, $|F_N|$, $|F_{NH1}|$, $|F_{NH2}|$, etc.

The structure factor of the heavy atom derivative 1 may be expressed vectorially as

$$\mathbf{F}_{NH1} = \mathbf{F}_N + \mathbf{F}_{H1} \tag{4}$$

or

$$|F_N| \exp(i\Phi_N) = |F_{NH1}| \exp(i\Phi_{NH1}) - \mathbf{F}_{H1} \tag{5}$$

where $\mathbf{F}_{H1}$ is calculated from the positions of the heavy atom H1 which can be located by the method of difference Patterson map (a special type of fourier map using $|F(hkl)|^2$, instead of $|F(hkl)|$, as the coefficient in Eq. (1.))

This allows two possible solutions for the phase angle of Φ_N for Eq. (5). Additional heavy atom derivatives, NH2, NH3, . . . , in combination with the first derivative (NH1) will afford an unique solution of Φ_N (with some uncertainties due to the experimental errors) so that an electron density Fourier map, termed MIR map, may be calculated.

In the work of DNA-drug complexes, the heavy atom derivatives may be prepared in two ways. The first choice is to substitute the thymine base with a 5-bromouracil base. These two bases are nearly isostructural, except that the bromine atom (35 electrons) is significantly heavier than the methyl group (~6 electrons). This method often produces excellent isomorphous crystals. Alternatively, a cytosine base may be replaced by a 5-bromocytosine base, if there is no thymine base in the DNA sequence or additional heavy atoms are needed. The second useful method is to soak the native crystals in solutions containing heavy-metal ions such as Co^{+2}, Ba^{+2}, etc., as described earlier. Sometimes, the crystals may be grown from a solution in which lighter-metal ions (Na^+, Mg^{+2}) are replaced by heavy-metal ions (Rb^+, Cs^+, Ba^{+2}, Sr^{+2}).

The MIR method requires no a priori assumption about the structure and is most suitable for solving complicated or unusual structures. We have successfully applied this technique in solving the structure of the unorthodox left-handed Z-DNA structure (2) and the structure of the complex between the bis-intercalator antitumor agent triostine A and DNA hexamer, d(CGTACG) (37).

The other widely used technique for solving the phase problem in nucleic acid crystallography is the *molecular replacement* (MR) method. The first step is to construct a model of the molecule which is assumed to be very similar to the real structure. The model is then rotated in the unit cell to determine the possible orientation of the molecule. Finally, the model in the preferred orientation is systemically translated along the three directions (x,y,z) of the unit cell in order to find the correct position of the molecule in the crystal. Hence, the method is often called the rotation-translation search method. Several useful computer program packages, including ULTIMA (38), MERLOT (39), X-PLOR (40), have been adapted for this method. The success of this method heavily depends on, among other things, the construction of a model that is sufficiently close to the real structure. This method is particularly suitable for nucleic acids alone, as they often adopt a structure in one of the three families, A-DNA, B-DNA, or Z-DNA. However, this is not always straightforward for drug-DNA complexes, as they often adopt rather unexpected structures which makes it difficult to build an initial model that is close to the real structure.

A powerful application of the MR method is in investigating the effect of chemical modification on DNA conformation or drug-DNA interaction.

When a good crystal lattice is available, one can often generate isomorphous crystals from similar, but different molecules. For example, when the structure of d(CGCGCG) Z-DNA was determined (2), we crystallized several related DNA sequences to understand the sequence-dependent effect on Z-DNA. The structure of d(m^5CGTAm^5CG) was determined at 1.2 Å resolution to address questions of the stability of TA base pair in Z-DNA (41). Similarly, d(br^5CGATbr^5CG) was crystallized in the Z-DNA conformation facilitated by the cobalt(III) hexaammine ion and its structural analysis helped us understand the influence of a sequence that differs from purine-pyrimidine alternation on the Z-DNA formation (42). We also crystallized the d(CGCGTG) hexamer to study the effect of G-T mismatched wobble base pair in Z-DNA (43). All these molecules crystallized in the isomorphous $P2_12_12_1$ lattice, which permitted straightforward structure determination by the molecular replacement method. For those studies the phases for the new crystals were derived directly from the original canonical d(CGCGCG) crystal.

A similar type of sequence engineering has been applied to the $P2_12_12_1$ lattice associated with the dodecamer d(CGCGAATTCGCG) in the B-DNA conformation (44). In fact, many complexes between a series of minor-groove binding agents and several related dodecamer sequences (e.g., d(CGCAAATTTGCG), d(CGCGATATCGCG)) have been crystallized in the isomorphous $P2_12_12_1$ lattice and their structure solved by the molecular replacement method (45–47).

A unique property associated with the diffraction of a crystal containing certain types of heavy atoms (metal ion like Fe, Co, Cd, or Hg in metalloproteins, selenium in Se-methionine, etc.) is the anomalous scattering due to the unusual absorption of x-rays by the heavy atoms. This happens when the wavelength of the x-rays is near the absorption edge of the heavy atoms (48). Normally, the diffracted intensity from the Bragg planes of (h,k,l) and $(\bar{h},\bar{k},\bar{l})$ are identical (called Friedel's law). However, $|F(h,k,l)|$ and $|F(\bar{h},\bar{k},\bar{l})|$ become nonequivalent when heavy atoms are present in the lattice. The difference between $|F(h,k,l)|$ and $|F(\bar{h},\bar{k},\bar{l})|$, arising from this anomalous scattering of x-rays, may be used to locate the positions of the heavy atoms in the unit cell and to provide valuable phase information. This technique has been extended by using the synchrotron radiation. The tunability of the synchrotron radiation source allows the anomalous data to be collected at several different wavelengths. The new method, termed *multiple anomalous dispersion* (MAD), has been successfully applied to several protein structures (49,50). Recently, the structure of the complex between chromomycin A_3 and a DNA octamer has been solved by the MAD technique using a bromosubstituted (thymine by 5-bromouracil) DNA molecule (51).

Two special methods, noncrystallographic symmetry averaging (52) and solvent flattening (53), for improving phase information have been shown to be quite useful in generating significantly better electron density maps. The former method takes advantage of the fact that subunits of a multisubunit protein have a similar conformation. The electron densities associated with independent subunits related by noncrystallographic symmetry (e.g., local twofold symmetry) are identified first and then averaged. The resulting new (averaged) electron density map is back-Fourier-transformed to obtain the new phase information which is used to calculate a new map. The procedure is iterated many cycles until no change between cycles is seen. The final map is often substantially "cleaner." This procedure also assumes that the solvent regions have a uniform level of electron density. Therefore, during the averaging steps the electron density of the solvent regions is "flattened." This "solvent flattening" procedure may be applied to any noisy electron density map, such as those from the *single isomorphous replacement* (SIR) calculation (54). While these two methods have not been used in solving the structure of drug-DNA complex as yet, their potential should not be ignored.

4. *Model Building*

After the phase information Φ_N is obtained for all the reflections, it is used to calculate the electron density map according to Eq. (1). The ease with which one can interpret the map depends on the noisiness of the map which is related to the quality of the phase information of the reflections. The better the phases are, the cleaner the map is. In general, a good MIR map requires two or more high quality isomorphous derivative crystals.

A convenient way to inspect the overall structure of the unit cell is a "minimap" at a scale of ~5 mm/Å. An example of one section of such minimaps of daunomycin-d(CGTACG) complex is shown in Figure 2. Many sections are stacked along a particular direction (usually along one of the three unit cell axes) to provide a three dimensional representation of the unit cell contents.

A very useful program, FRODO, for building a model fitted within the electron density map on a computer has been developed by Alwyn Jones (55). The program has been modified recently and ported onto a UNIX-based Silicon Graphics Iris computer by C. Cambillau, with a new name called TOM. The electron density maps are displayed as an interconnected net of isodensity points, often called chicken-wire density envelope. The map plus the model can be manipulated interactively, e.g., rotate, translate, zoom, etc., in the computer to optimize the fit of the model to the density. Figure 4 is an example of a ($2F_o$-F_c) electron density of the complex between daunorubicin and d(CGCGCG) cross-linked by formaldehyde (56).

5. *Refinement of the Structure*

The initial model structure, either traced from the MIR Fourier map or obtained as the result of a correct rotation-translation search, can be used to

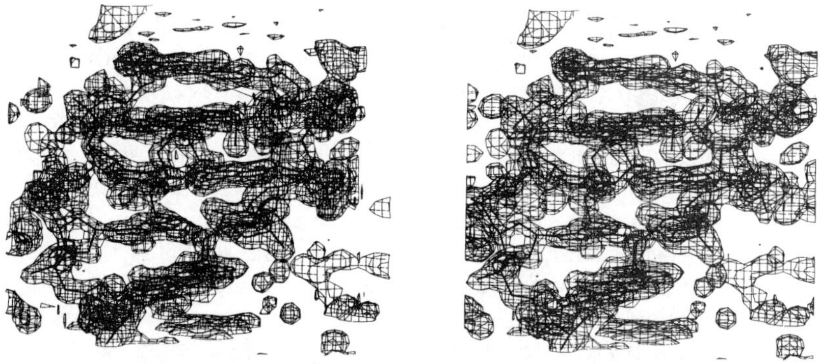

Figure 4 A stereo drawing of a portion of the electron density Fourier map using $2|F_o|-|F_c|$ as the coefficient for the Fourier calculation at 1.5 Å resolution. This type of map is often called sum map as it contains the sum of the normal $|F_o|$ Fourier map and the $|F_o|-|F_c|$ difference Fourier map. The molecule shown here is the adduct of daunorubicin and d(CGCGCG) cross-linked by formaldehyde (56). Many water molecules are clearly visible.

calculate the theoretical structure factor, $|F_c|$. Inevitably the $|F_c|$ based on the initial model is not fully in agreement with the observed experimental structure factor $|F_o|$. The initial model requires many adjustments so that $|F_c|$ and $|F_o|$ become close to each other in value. The process is called *refinement* and is highly iterative.

For small molecules (with fewer than 100 atoms), their crystals are often of high quality; i.e., the molecules are packed in the lattice in a highly regular manner. Those crystals diffract x-rays to high Bragg θ angle. For example, it is not uncommon that crystals of organic compounds diffracts CuK_α x-rays (λ = 1.54 Å) to θ greater than 60°. On the basis of Bragg equation (Eq. 2), this is equivalent to being able to resolve atoms that are separated by the *d* spacing, i.e., *resolution*, of 0.89 Å. The resolution of 0.89 Å, at the first glance, seems to be at odds with the common knowledge regarding the accuracy of small molecule structure which often can be determined with the estimated standard deviation (esd) of bond lengths of 0.002 Å. The very small esd's are the consequence of an important characteristic of the x-ray diffraction analysis, namely, the overdeterminancy.

We shall use the crystal structure of the d(CGCGCG) Z-DNA (2) to illustrate the concept of the over-determinancy. This hexamer crystallized in the orthorhombic space group $P2_12_12_1$ with the unit cell dimensions of a = 17.9 Å, b = 30.98 Å, and c = 44.57 Å (volume of the unit cell ~25,000 Å^3). There are two DNA hexamer strands, two spermines, one magnesium ion, 68 water molecules, totalling 340 nonhydrogen atoms, in the asymmetric unit of the unit cell. Diffraction data up to 0.9 Å resolution ($2\theta = 125°$ using CuK$_\alpha$ x-rays) were collected with 14,000 (out of 21,000) observed reflections above $2\sigma(F_o)$ of the background. In order to define the three-dimensional structure of a molecule fully, so that the $|F_c|$ of the model can match as close as possible with the $|F_o|$, three positional parameters (x,y,z) fixing the positions and six thermal parameters (temperature factors) describing the anisotropic motions of every individual atom in the molecule are needed. There are a total of $340 \times 9 = 3060$ independent parameters to be determined. Therefore, in this crystal structure there are 4.6 observations per variable parameter to be determined; in other words, a 4.6-fold overdeterminancy. In fact, the ratio is even higher (~8–10) for small molecules. In those cases, the individual variable parameters may be simultaneously adjusted by the process of least-squares fitting. One can often obtain positional accuracy of atoms down to a few thousandths of an Ångstrom by this least-square refinement procedure.

However, crystals of biological macromolecules tend to diffract x-rays to significantly lower Bragg angle, resulting in substantially fewer observed reflections. Consequently, the reflections/parameter ratio decreases to the point that the refinement procedures involving individual atoms cannot be used. An ingenious method was developed to overcome the problem of underdeterminancy. Since biological macromolecules (proteins and nucleic acids) are made of small building units (amino acids and nucleotides respectively), it can be safely assumed that their geometric parameters (bond distances and bond angles) are very similar to those of the building units. Instead of refining the individual atomic positions, torsional angles are varied to minimize the difference between $|F_o|$ and $|F_c|$, minding several other factors. A popular program (PROLSQ) developed originally for proteins by W. Hendrickson (57), later modified for nucleic acids (NUCLSQ) by G. Quigley and E. Westhof (58), incorporates the following function Ω_{total} to be minimized by a least-square procedure.

$$\Omega_{\text{total}} = \Omega_F + \Omega_{\text{distance}} + \Omega_{\text{vdW}} + \Omega_{\text{planar}} + \Omega_{\text{chiral}} + \Omega_B \quad (6)$$

Here the various terms to be minimized are associated with the structure factors (Ω_F), bonding distances (Ω_{distance}), nonbonded van der Waals contacts (Ω_{vdW}), planar groups such as the peptide bond or nucleic acid bases (Ω_{planar}), stereoconfiguration of chiral centers (Ω_{chiral}), and temperature factors (Ω_B). This type of stereochemically constrained refinement substantially reduces the number of parameters to be refined by a factor of about 5 and it becomes possible to refine a protein or DNA structure even at relatively low resolution (e.g., ~3 Å). Table 1 compares the overdeterminancy ratios for the refinement of a structure at different resolution limits.

More recently, a new refinement procedure incorporating molecular dy-

Table 1 Crystallographic Over-determinancy Ratios

		Reflections per variable parameter[a]		
Resolution (Å)	No. of independent reflections observed[b]	Positional and anisotropic thermal parameters (9 per atom)	Positional parameters only (3 per atom)	Conformational angles only (0.5 per atom)
3.0	600			3.5
2.0	1,800		1.8	11.0
1.5	4,000		4.0	25.0
1.2	8,000	2.5	8.0	50.0
0.9	14,000	5.0	15.0	80.0

[a]These numbers are calculated on the basis of the refinement of d(CGCGCG) Z-DNA crystal structure in which 240 DNA nonhydrogen atoms (MW ≈ 6,600) plus about 100 non-DNA atoms (water, ion, and spermine) have been located in the asymmetric unit of the unit cell (dimensions a = 18 Å, b = 31 Å, and c = 44 Å, space group $P2_12_12_1$) (2).
[b]Approximate number of reflections observed at 2.0 σ (F_{obs}) level.

namics in conjunction with the structure factor data has been developed by A. Brunger (40). The program X-PLOR, has been implemented on several computer systems. It employs a procedure called simulated annealing in the molecular dynamics simulation process by "heating" the molecular system to high temperature (~3000 K) and then gradually cooling back to low temperature (~100 K). At all times, the system is constrained by the observed structure factor $|F_o|$ to prevent the molecule from exploding. This process appears to be effective in getting the model out of the local energy minima and in finding the global minimum more rapidly, as shown in several protein structure refinements. Our experience in the refinements of the structure DNA and drug-DNA complexes is that both methods, NUCLSQ and X-PLOR, seem to work equally well, possibly due to the good quality of the starting models.

An important procedure in the refinement cycles is the difference Fourier method. The initial model of a macromolecular crystals inevitably does not account for everything in the unit cell. The R-factor at this stage is usually about 25–30%. The remaining material in the unit cell (e.g., solvent molecules, ions) can be located through the difference Fourier electron maps which are calculated based on the following equation.

$$\Delta\rho(x,y,z)=\frac{1}{V}\sum_h\sum_k\sum_l \left(\left|\,|F_o(hkl)|-|F_c(hkl)|\,\right|\right)e^{-i[2\pi(hx+ky+lz)-\Phi_c(hkl)]} \tag{7}$$

Here $|F_o|$ and $|F_c|$ are the observed experimental and the calculated (based on the current model) structure factors respectively, and Φ_c is the calculated phase angle. This method is particularly useful in locating the missing portion of the model. For example, many minor-groove binding drugs such as netrospin, distamycin, Hoechst 33258 have been identified from the difference Fourier maps of their respective crystals of drug-DNA dodecamer complexes (for a review, see Ref. 10). Figure 5 shows the difference Fourier map of the Hoechst 33258-d(CGCGAATTCGCG) complex (5A) and the daunorubicin-d(CGCGCG) adduct (5B) as displayed by the program FRODO/TOM (vide supra). In the former case the Hoechst drug was not included in model for the calculation of phase angles. It can be seen that a continuous electron density is clearly visible in the minor-groove of the dodecamer DNA double helix. In the latter case, the daunosamine is removed in the Fourier calculation and residual electron density agreed with the refined model extremely well. The high-resolution nature of the structure is also evident by the well-resolved density.

B. Assessing the Structure

It cannot be overemphasized that any refinement procedure has limitations. A structure, even properly refined, at low resolution simply does not have the kind of accuracy needed to address the fine details of the structure. Figure 6 shows the electron density maps of a guanine-cytosine base pair of the Z-DNA at various resolution. It can be seen that a great deal of structural information is available at 1.0 Å resolution. Not much ambiguity exists. On the other hand, a structure of 3.0 Å resolution can provide the overall shape of the molecule, the orientation of the base pairs, or the location of the phosphate and sugar groups, etc., but it cannot unambiguously define the backbone conformation, or reveal the detailed base pair geometry such as propeller twist, buckle, etc.

How does one monitor the progress of the refinement of a structure? The most often used indicator is the crystallographic residual factor (R-factor) which measures the agreement (or discrepancy, depending on which way you look at it) between the $|F_o|$ and $|F_c|$. The R-factor is defined as

$$\mathrm{R} = \frac{\Sigma \mid |F_o| - |F_c| \mid}{\Sigma |F_o|} \tag{8}$$

For small molecules, one can now routinely attain an R-factor of below 5%, due to the very reliable measurement of the intensity data. In contrast, the crystals of macromolecules contain large solvent channels which make it difficult to account for the presence of disordered solvent molecules. In addition, some regions of those large molecules are flexible, which contrib-

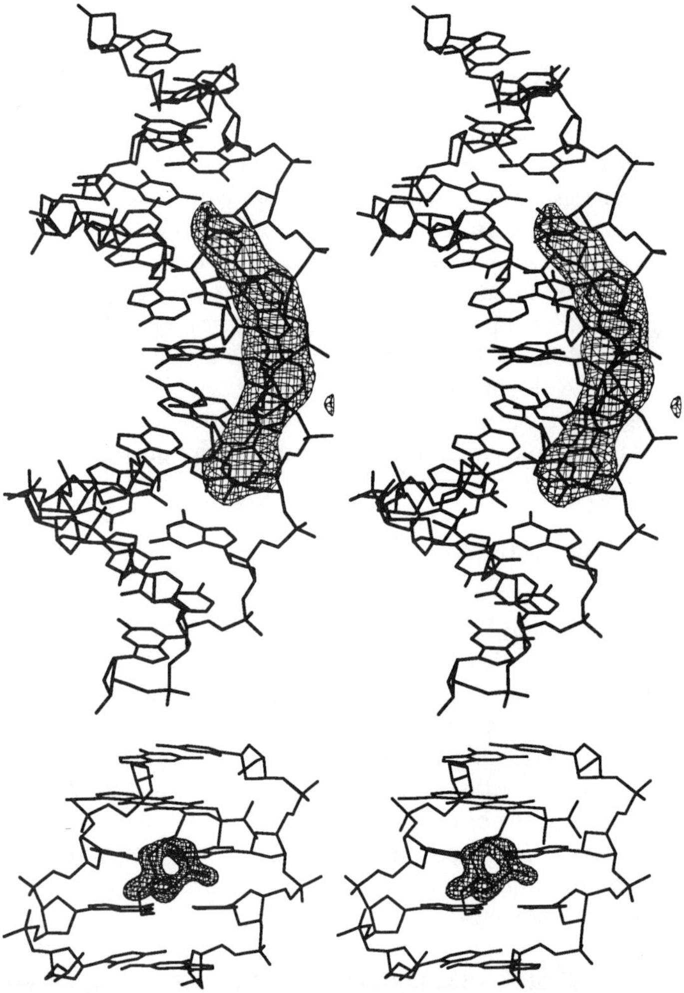

Figure 5 Two examples of difference Fourier maps. They are useful for locating missing atoms (e.g., drug, solvent molecules), adjusting misplaced atoms, etc. (A). The minor-groove binding drug Hoechst 33258 was located from the difference Fourier map of the crystal grown from a mixture of the Hoechst drug and the DNA dodecamer d(CGCGAATTCGCG) (109). The map shown here is based on the model that has already been refined, and the drug is subsequently removed for the Fourier calculation at 2.0 Å resolution. It shows that the Hoechst drug is definitively bound in the narrow minor groove of the B-DNA dodecamer helix near the AATT region. (B). The daunosamine moiety of the daunorubcin-d(CGCGCG) adduct (56) is removed from the model for the Fourier calculation. The daunosamine sugar is seen to fit the difference Fourier electron density very well, and it demonstrates the correctness of the model.

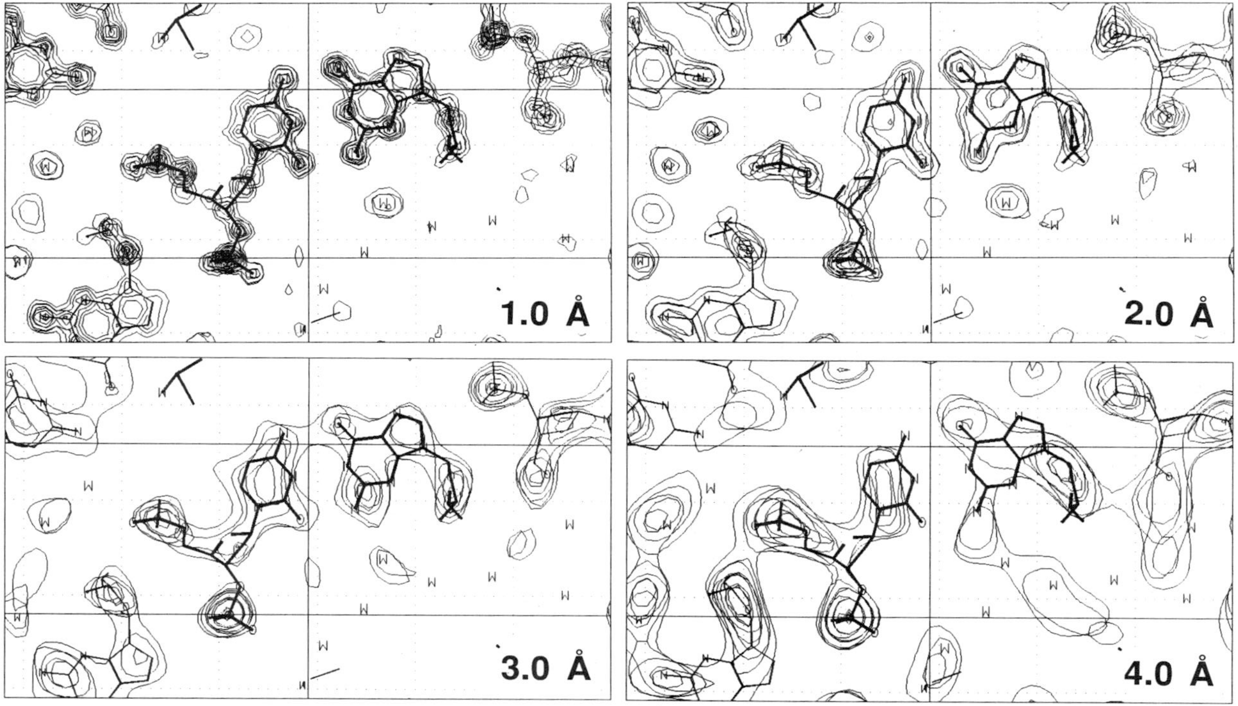

Figure 6 Electron density maps of the d(CGCGCG) hexamer in the Z-DNA conformation at different resolution (2). The map is shown as a composite of three sections, with 0.7 Å separation between sections, along the *c*-axis direction. At atomic (1.0 Å) resolution, the atoms from the DNA molecule as well as water molecules are all well-resolved. At 2.0 Å resolution, the shapes of the bases are still quite clear. However, the structure solved at a resolution lower than 3.0 Å does not have the reliability for detailed geometric and conformational analysis, as evident in the two lower panels.

ute to the uncertainty of the model. Nevertheless, we can improve the model by carefully adding first shell water molecules and hydrated metal ions, using the difference Fourier method. Usually a macromolecular structure which has been refined to an *R*-factor between 15% and 20% with good constraints can be considered satisfactory.

However, it is important to point out that the *R*-factor alone should not be the only criteria for judging the correctness of the structure. One should inspect the electron density map very carefully to see whether there is any obvious discrepancy between the model and the electron density in the map. We have had instances where the structure of a drug-DNA complex has been refined at 3.0 Å resolution to an *R*-factor of 12% with good constraints, but the model clearly did not fit well in the Fourier map (needless to say the results have not been published). It remains unclear why this has happened, since the correct structure is still yet to be solved. This issue of misinterpretation of structure is the topic of a recent commentary (59).

III. NMR SPECTROSCOPY

Nuclear magnetic resonance (NMR) spectroscopy has emerged as the most powerful technique in elucidating the three-dimensional structure of biological macromolecules *in solution* (17–19, 60). The phenomenal progress is mainly due to the great advances in two areas. First is the availability of the superconducting high-field NMR spectrometer with the state-of-the-art electronics and computer associated with the instrument. These new instruments, with magnet of field strength ranging from 9 to 14 T (translating into the resonance frequency of ca. 400 to 600 MHz for the proton nucleus), provide a very high sensitivity such that biological samples (proteins or nucleic acids) can be studied at relatively low concentration (0.1–10 mM). In parallel, these instruments employ the Fourier transform (FT) technology which allows a rapid data collection. The second important development is the application of multidimensional NMR spectroscopy (17–19). A number of useful pulse sequences have been developed to provide methods for obtaining information regarding the nuclear spin connectivity (specifically, through bond and through space connectivity) of nuclei (in particular protons). The connectivity through space, in the form of nuclear Overhauser enhancement (NOE), is related to the distance between protons which may be used to deduce a three-dimensional model of the molecule. A powerful method, the two-dimensional NOE spectroscopy (NOESY), has been developed. Similarly, the information regarding the connectivity through bonds can be obtained by the two-dimensional J-correlated spectroscopy (COSY). Both techniques, plus their subsequent more sophisticated and improved versions, have been widely applied to problems of biological macromolecular structure. We shall discuss those techniques later in this chapter.

Figure 7 depicts the pathway for solving the 3D molecular structure by NMR spectroscopy. No attempt is made to describe the basic principles of NMR here. The reader is referred to numerous excellent texts for this topic (17–19). Instead, we shall focus on the topics that are relevant to the problems associated with nucleic acids and their complexes with antitumor drugs.

A. Structural Determination by NMR Spectroscopy

There are five major steps depicted in Figure 7 of the structure determination process, as outlined below.

1. Sample Preparation

An example of preparing the NMR sample of DNA oligonucleotide is described below. The lyophilyzed powder of d(CGTACG) (7.8 mg), prepared by a DNA synthesizer (vide supra), was dissolved in 500 μL of 99.8% H_2O in 50 mM sodium phosphate buffer (pH = 7.0) and 150 mM NaCl. The solution was lyophilyzed and the material was dissolved in 500 μL of 99.8% D_2O

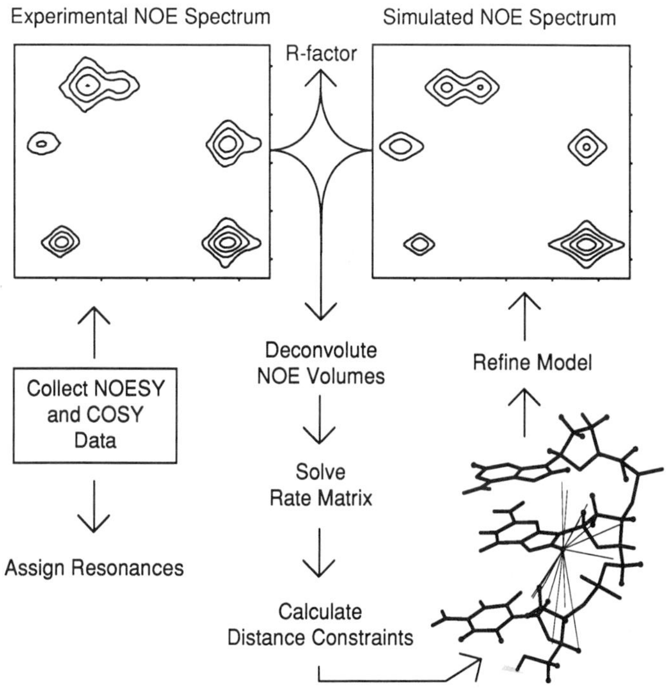

Figure 7 A schematic diagram showing the major steps in the structure determination of drug-DNA complex in solution with NMR spectroscopic data.

and lyophilyzed again. This step was repeated again. At this point, the sample was dissolved in 500 μL of 99.8% D_2O, centrifuged to remove any precipitate and the supernate loaded into an NMR tube (Wilmad #528PP). The sample in the NMR tube was dried under a slow stream of dry argon gas, delivered to 0.5 cm above the liquid level by a Teflon capillary tube, in a glove bag filled with argon gas. The final sample was prepared by adding 500 μL of 99.96% D_2O (Aldrich Chemical Co.) and the tube is capped and sealed.

For the preparation of the NMR sample of the drug-DNA complex, the same procedure is followed except the drug molecule was added to the DNA solution at the initial stage. Many antitumor drugs are not soluble in water, therefore they are dissolved in organic solvent as stock solution. For example, actinomycin D, nogalamycin, chromomycin A_3 are dissolved in methanol, while echinomycin, triostin A and luzopeptin are dissolved in methylene chloride/methanol mixture.

Some drugs (like mitomycin C, CC-1065, mithramycin, cisplatin) form covalent adduct with DNA (1). The adducts are prepared with DNA having specific sequence and purified by HPLC to obtain a homogeneous species for the NMR work.

2. *Data Collection of 1D and 2D NMR Spectra*

Figure 8 shows the 1D proton NMR spectrum (500 MHz) of the d(CGATCG) DNA hexamer in D_2O. The data were collected on a GE GN500 500-MHz NMR spectrometer and processed by the program FELIX

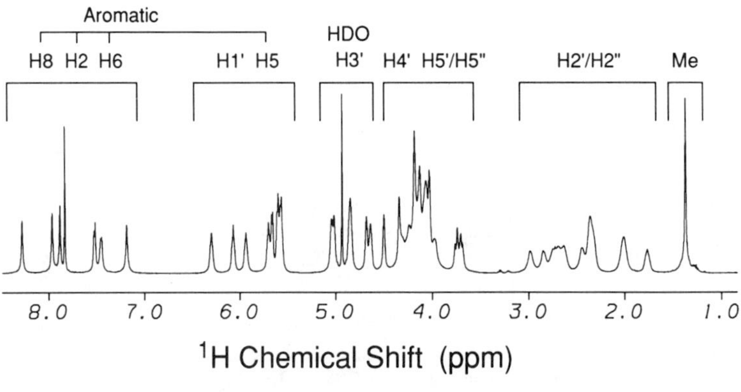

Figure 8 A typical one-dimensional NMR spectrum of a DNA oligonucleotide. The spectrum is that of the d(CGATCG) recorded in D_2O at neutral pH. The resonances associated with the nonexchangeable protons are dispersed into their characteristic regions of chemical shift.

(Hare Research). It can be seen that the spectrum is divided into three principal regions of chemical shifts: 7.2–8.6 ppm for most of the aromatic protons, 5.6–6.5 ppm for the H1′ protons and the cytosine H5 protons, and 1.0–5.6 ppm for the remaining deoxyribose protons and the methyl protons of thymine. In some instances, it is useful to record the spectrum in H_2O from which information regarding the exchangeable protons may be obtained. The huge proton signal from water (110 molar proton) peak at 4.83 ppm at 20°C has to be suppressed so that other resonances are visible. The exchangeable amino and imino proton resonances are located in 7–9 ppm and 12–14 ppm, respectively. Figure 9 summarizes the general locations of the chemical shifts of protons in nucleic acids.

Some protons are *J*-coupled through the chemical bonds, causing the resonance to split into multiplets. For example in DNA, H1′ and H2′ are coupled with the coupling constant of $J_{H1',H2'} \sim 10$ Hz, and H2′ is further coupled to H3′ with a coupling constant of $J_{H2',H3'} \sim 3.8$ Hz. It has been shown that the vicinal coupling constants $J_{a,b}$ may be used to estimate the $H_a{-}C{-}C{-}H_b$ torsion angle (θ) by the Karplus equation (61).

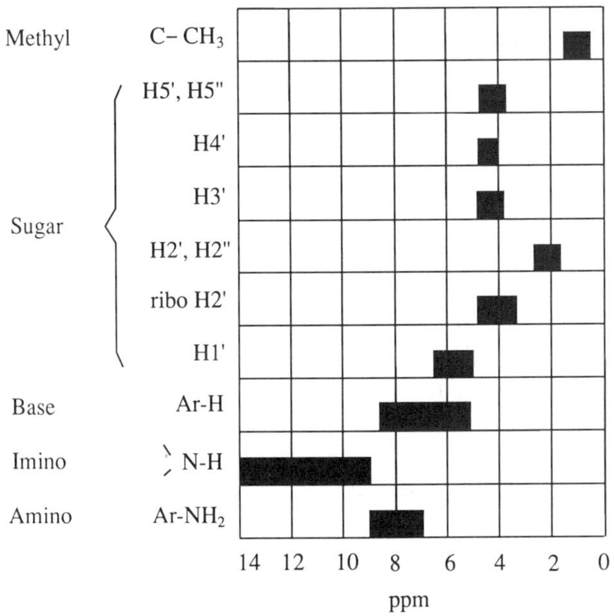

Figure 9 A schematic diagram showing the characteristic chemical shifts of the various protons in nucleic acids. Notice that the imino protons (from guanine and thymine bases) are well downfield from the rest of the protons.

$$J_{a,b} = (5.6 \cos 2\theta - \cos \theta + 7.8) \text{ Hz} \quad (9)$$

This relationship has been extended for nucleic acid system, taking into account the electronegativity of atoms attached to the carbon atoms (62). The modified Karplus equation is expressed as follow.

$$J_{a,b} = (P_1 \cos^2\theta + P_2 \cos \theta + P_3 + \Sigma\, \Delta x_i[P_4 + P_5 \cos^2(\zeta\theta + P_6|\Delta x_i|)]) \text{ Hz} \quad (10)$$

Here P_1 to P_6 are constants determined by a least-squares procedure using 315 model compounds, and Δx_i is the electronegativity of substituent i, ζ is either +1 or −1, depending on the orientation of the substituents (62).

Since those H—C—C—H torsion angles (e.g., H1′—C1′—C2′—H2′) are closely related to the sugar pucker conformation, they have been used to evaluate the conformational state of the sugar moiety (ribose in RNA or deoxyribose in DNA) of the nucleic acid backbone (63).

There are two major types of sugar puckers, C2′-endo and C3′-endo, as depicted in Figure 10. Various DNA conformations are associated with different sugar puckers. For example, B-DNA has C2′-*endo*-deoxyribose and A-DNA has C3′-endo-deoxyribose. Interestingly the left-handed Z-DNA of alternating dG-dC sequence adopts mixed sugar puckers in which the dC is in the C2′-endo and the dG is in the C3′-endo conformation. Another flexibility in nucleic acids is in the glycosyl torsional torsion (χ), which may exist in the anti or syn conformation. The nucleotide in A-DNA and B-DNA has the anti conformation, while the dG nucleotide in Z-DNA has the syn conformation.

Different sugar and glycosyl conformations in the nucleic acid molecule place their protons in a unique arrangement which are revealed in their respective COSY and NOESY spectra (to be described below). Figure 10 compares the interproton distances in the DNA nucleotide in the A- and B-DNA conformations. It can be seen that in B-DNA the H8 proton of a purine or H6 proton of a pyrimidine is not only close to its own H2′, H2″ protons, but also close to the H1′, H2″ protons of the 5′-side preceding deoxyribose. Indeed, strong NOEs are observed between those protons and they may be used to follow the connectivity along the sugar-phosphate backbone of DNA/RNA. This further suggests that different DNA/RNA conformations have their respective characteristic NOESY and COSY patterns.

By now, it should be clear that these 2-D COSY and NOESY techniques are indispensable tools for the structure determination of biological macromolecules in solution. The concept of the 2-D NMR experiment was originally formulated by Jeener (64) and extensively expanded by Ernst and, subsequently, others (18). Basically, the 2D methods incorporate an addi-

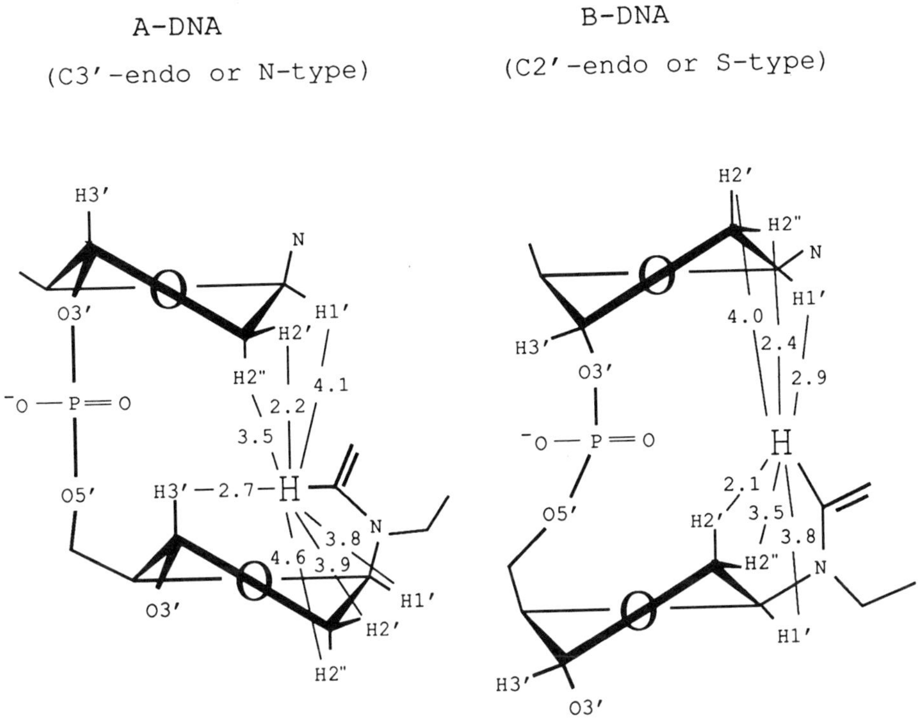

Figure 10 The effect of different sugar puckers on the inter-proton distances in nucleic acids. In B-DNA, the deoxyribose adopts a C2′-endo conformation in which the H6/H8 proton of the nucleotide is close to the H2′/H2″ protons of its own deoxyribose as well as those of the deoxyribose on the 5′-side, but far away from its own H3′ proton. In contrast, the H6/H8 proton is very close to the H3′ proton (2.7 Å) in A-DNA or RNA, which adopts a C3′-endo sugar information. In Z-DNA, the dG residue adopts a syn glycosyl conformation which swings the base around such that its H8 proton is away from both H2′/H2″ and H3′ protons, but very close to its own H1′ proton.

tional time domain (evolution domain, usually called the t_1 domain) between an initial pulse (normally a $\pi/2$ pulse) and the second pulse (another $\pi/2$ pulse) before the acquisition of the free induction decay (FID) in the second time domain (t_2 domain). These two time domains are converted by Fourier transforms into their respective frequency domains, F_1 and F_2, which are related to the chemical shifts of the protons. The pulse sequences that are used in a prototype of the COSY and NOESY experiment are shown schematically in Figure 11. It can be seen that there is an additional $\pi/2$

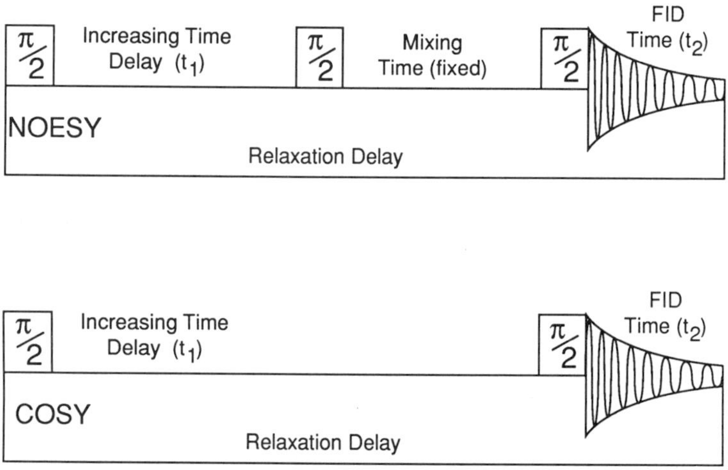

Figure 11 A simplified schematic drawing showing the pulse sequences for the two most commonly used two-dimensional NMR spectroscopy, *J*-correlated spectroscopy (COSY) and nuclear Overhauser effect spectroscopy (NOESY).

pulse and delay (τ_m), termed the mixing time, in the NOESY experiment. During the mixing time period, the protons in the molecule have an opportunity to cross relax amongst themselves, resulting in the exchange of spins. This cross-relaxation is known as the nuclear Overhauser enhancement and is observed as the off-diagonal cross-peaks in the 2D-NOESY spectrum. Figure 12 shows the complete 2D-NOESY spectrum of the nonexchangeable protons of d(CGATCG).

Interestingly, the rate (ρ_{ij}) of the cross-relaxation between protons *i* and *j* is inversely proportional to sixth power of the distance between the two protons, r_{ij}.

$$\rho_{ij} \propto r_{ij}^{-6} \cdot F(\tau_c) \tag{11}$$

Here $F(\tau_c)$ is a function of the rotational correlation time of the molecule.

This inverse sixth power of the distance dependence of these cross-relaxation rates immediately suggests that the measured NOE becomes very weak when the distance between two protons is longer than 6–7 Å. It should be clearly understood that the observed cross-peak intensity results from multiple pathways for relaxation within the mixing time period. At very short mixing time, the observed NOE cross-peak volume is dominated by the largest term, i.e., the direct relaxation between the two spins. However, as the mixing time becomes longer (>25 ms), many secondary relaxation pathways become significant (or even dominate) and a direct interpretation of

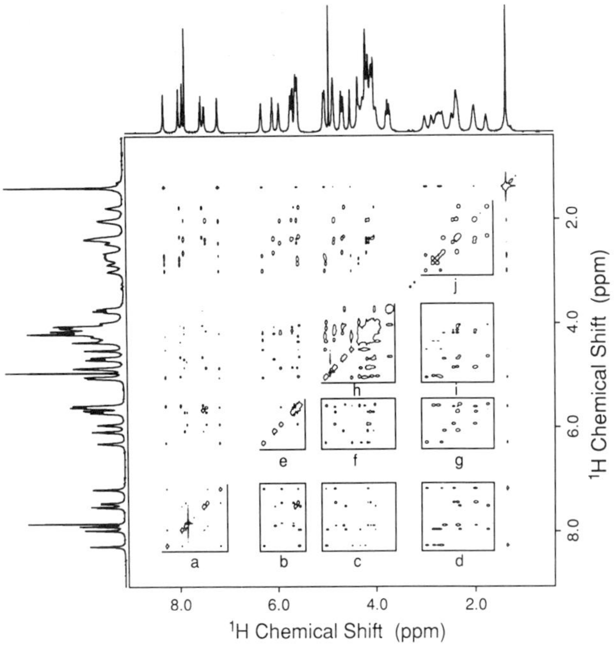

Figure 12 The phase-sensitive 2D-NOESY of d(CGATCG) in D_2O at 10°C. Different regions of the cross-peaks between protons are labeled. (a) aromatic-aromatic. (b). aromatic-H1′/H5. (c) aromatic-H3′, H4′, and H5′/H5″. (d) aromatic-H2′/H2″. (e) H1′/H5-H1′ H5. (f) H1′/H5C-H3′, H4′, and H5′/H5″. (g) H1′/H5-H2′/H2″. (h) H3′, H4′, and H5′/H5″-H3′, H4′, and H5′/H5″. (i) H3′, H4′, and H5′/H5″-H2′/H2″. (j) H2′/H2″-H2′/H2″. In addition, the methyl group of thymine (TMe) has crosspeaks to other protons. The spectrum was collected at a mixing time (τ_m) of 200 msec. Data points in 512 blocks were collected 32 times with 2048 complex points each. The entire process was run twice to allow extraction of phase information. There are roughly 400 to 700 NOE cross-peaks, and they are used for the structure refinement described in the text.

the cross-peak volume becomes unreliable. Thus, at longer mixing times the cross-peak volumes are interpreted within the context of a relaxation matrix, encompassing the rate of relaxation between all nuclei simultaneously.

3. *Resonance Assignments*

The next task is to identify all the proton resonances in the spectrum. There is no "definitive" way to follow, as each system is unique. Normally, one tries to find certain landmarks in the COSY and NOESY spectra to begin the tracing of the connectivity pathway. In the COSY spectrum, the most prominent cross-peaks are those between the H5 and H6 of the same cytosine base, and they can be easily identified and used to locate the starting point of the connectivity pathway in the corresponding NOESY spectrum. Figure 13 shows an expanded diagram of the cross-peak region (region (b) in Fig. 12) between the H1′ protons and the aromatic protons of d(CGATCG). In this case, one of the cytosines is a 5′ terminal base, C1, which does not have

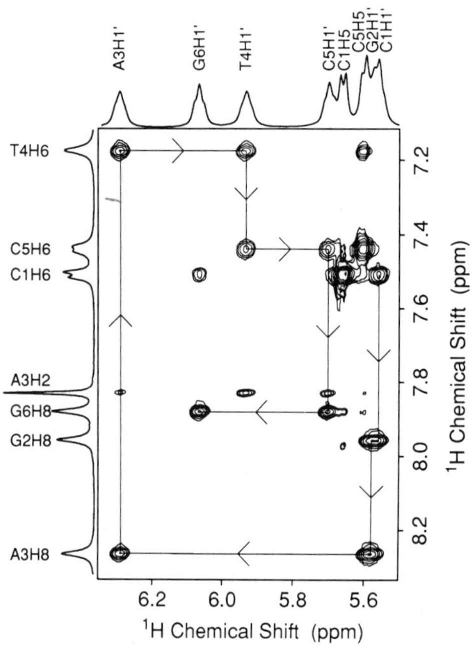

Figure 13 An expanded NOESY spectrum of the region (b) in Figure 12, showing the crosspeaks between the aromatic protons (H6 of cytosine/thymine, H8 of guanine/adenine, and H2 of adenine) and the H1′ and cytosine H5 protons. (The nucleotides in the d(CGATCG) hexamer are numbered as C1, G2, A3, T4, C5, and G6 respectively.) This region is particularly useful for starting the process of the resonance assignment. The two strong cross-peaks between the cytosine H5 (doublet) and the H6 (doublet), due to their fixed close distance (2.46 Å), are easily recognized landmarks. One of the cytosine H6 protons is from a terminal dC nucleotide, and it should have only one cross-peak to its own H1′ proton. For the other H6/H8 protons, each has two cross-peaks to to H1′ protons, one to its own H1′ and the other to the H1′ of the preceding 5′ residue. (This is illustrated in Fig. 14.) One can sequentially step along this H6/H8 to H1′ connectivity pathway, as shown by the arrows in the figure, to assign all the resonances in this region. There are several small cross-peaks between A3H2 and other protons. An interesting cross-peak, between C1H6 and G6H1′, is likely due to the transient stacking of two hexamer helixes which brings the terminal base pairs close to each other.

a neighbor on its 5′ side (see Fig. 14). Hence, the H6 of C1 has an NOE cross-peak to its own H1′ only, but the H6 of C5 has NOE cross-peaks both to its own H1′ and to the T4 H1′ protons. This analysis, termed sequential assignment (65), allows one to establish the connectivity pathway as indicated in the Figure 13.

The assignment of the aromatic protons and the H1′ protons is then expanded into other regions of the 2D spectra. For example, H2′/H2″ protons may be identified by their cross-connectivity to the aromatic protons as illustrated by the expanded region (region (d) of Fig. 12) of Figure 15. Using Figure 10 it is easy to understand the various rectangular relations between the crosspeaks. The NOE data here suggest that the sugar conformations in this hexamer at physiological condition (neutral pH, low salt) are all in the C2′-endo family.

In general, a complete assignment of the spectrum is possible for a system up to 20 nucleotides long. However, ambiguity often occurs, especially in the H5′ and H5″ regions, and to a lesser extent, in the H2′ and H2″ regions. This is due to the fact that the NOE strengths associated with the those

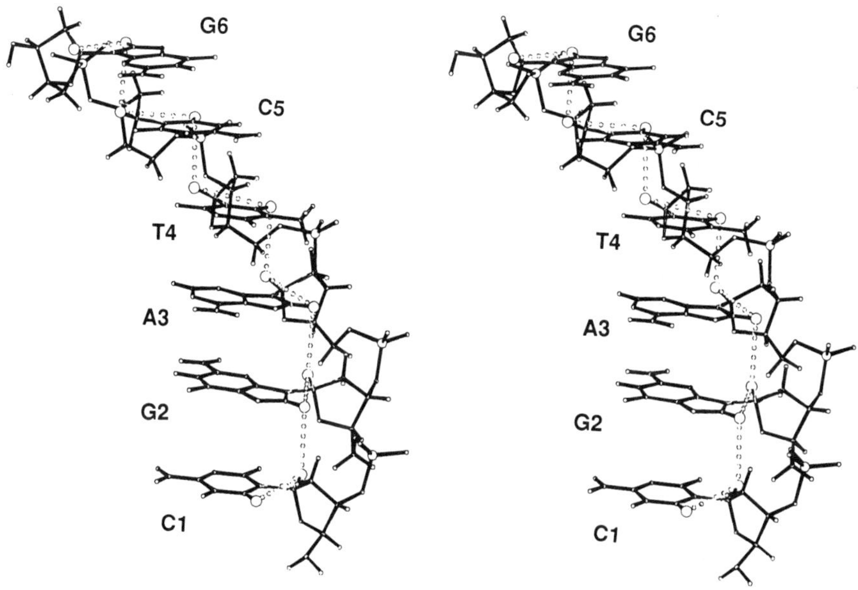

Figure 14 A stereodrawing of one strand of the d(CGATCG) duplex in the B-DNA conformation. The through-space connections between the H6/H8 and H1′ protons (both shown as large circles) are represented as strings of small circles. The distances between those protons range from 2.8 to 3.8 Å.

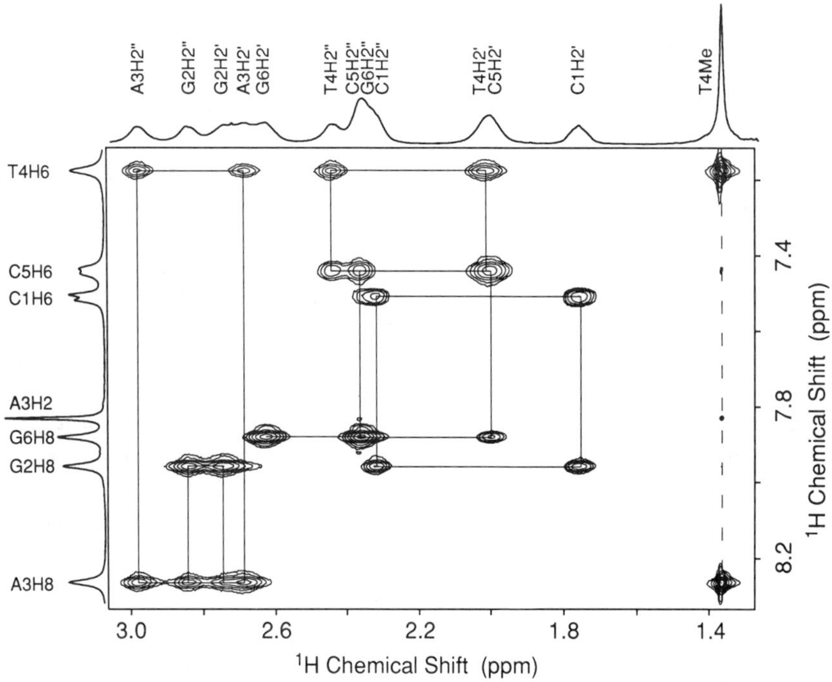

Figure 15 An expanded region (d) of the NOESY spectrum to illustrate the characteristic pattern of H6/H8 to H2′/H2″ cross-peak due to the C2′-endo sugars and anti conformation of bases in B-DNA. Each H6/H8 proton has four cross-peaks to the H2′/H2″ protons. The recognition of these patterns permits a qualitative identification of the type of backbone conformation of the nucleic acid being studied.

prochiral pairs (H5′/H5″ and H2′/H2″) of protons are so similar, it becomes very difficult to distinguish which is which. Recently, there have been efforts to introduce specifically deuterated deoxyribose into DNA fragment in order to break this prochirality, but no definitive result has yet been obtained (66).

The full resonance assignment of the NOESY spectrum provides the entry point for obtaining the entire dataset of the volumes of all NOE crosspeaks. However, a major problem in the measurement of the NOE volume associated with a given proton is that many crosspeaks are severely overlapped. What is a proper procedure to resolve those overlapped peaks remains an interesting open question and it will be discussed later. In general, for a DNA hexamer like d(CGATCG), about 400 to 700 independent NOE cross-peaks can be measured.

4. *Model Building*

The three dimensional structure of the molecule that is being studied is implicitly encoded in the several hundred NOEs that are observed. How does one generate a model such that it may be evaluated against the NOE data? For nucleic acids, there are at least four ways to build a starting model.

Model Based on Known DNA Structure. Idealized DNA/RNA structures of A-, B- and Z-DNA can be easily constructed on the basis of standard parameters. (67,68).

Model Based on the Crystal Structure. If the crystal structure of the system of interest has been determined, it may be used as the starting model. For example, DNA dodecamer d(CGCGAATTCGCG) has been extensively studied by NMR (65) and the results compared with the crystal structure (44). Similarly, the model of echinomycin complexed to DNA oligonucleotides for NMR studies (69,70) has been built and it is consistent with the crystal structure of the triostine-d(CGTACG) complex (37). Recently, we compared the solution and crystal structure of the nogalamycin-d(CGTACG) complex and concluded that they are substantially similar (23,71,72). This method will continue to remain as a powerful source of a starting model for NMR study of drug-DNA interactions.

Model Constructed by Molecular Modeling and Minimization. If no example of a 3D structure exists previously for the molecule that is being studied, then a new model needs to be built de novo using a number of model-building computer programs, such as MIDAS (73) or QUANTA (74). For example, DNA/RNA hairpin structures play important roles in the function of nucleic acids. No crystal structure is available for many sequences that form hairpin structures. A number of those molecules have been studied by NMR spectroscopy (75–77), and their starting models were constructed via a computer with special attention paid in the loop structure of the hairpin.

Distance Geometry. An algorithm based on a mathematical approach called *distance geometry* (DG) (78) has become popular in the protein structure determination. This method converts geometric constraints (e.g., distance constraints derived from NOE data) into three dimensional molecular coordinates. Various computer programs have been developed and discussed extensively by Kuntz et al. (79). Usually one or more starting models may be generated by those programs for subsequent molecular dynamics refinement.

5. *Structure Refinement*

A critical step in solving the solution structure of biological macromolecules is to reliably measure the NOE volumes of all the cross-peaks associated with any given proton. However, many cross-peaks of the NOESY spectrum are severely overlapped. In order to calculate the volume of each cross-peak, those overlapped peaks have to be deconvoluted. In our approach, the deconvolution function of an overlapped region is estimated on the basis of the calculated NOE volumes derived from a molecular model. The theoretical NOEs are calculated NOE volumes derived from a molecular model. The theoretical NOEs are calculated using a full-matrix relaxation approach (MARASS) developed by Gorenstein and his colleagues (80). The advantage of this approach is that in principle only the NOE spectrum at a single mixing time (e.g., τ_m = 200 msec) needs to be measured, substantially reducing the amount of time for collecting the NOE data. In practice, more than one NOE spectra are collected at different mixing times.

At this point, a relaxation rate matrix is calculated on the basis of all observed NOE crosspeak intensities at a given experimental mixing time. These rates are then used, along with the molecular correlation time (τ_c), to define internuclei distances for constraints to be used by a refinement program such as the energy minimization/molecular dynamics program X-PLOR (40). The resulting minimized model from the X-PLOR refinement becomes the new starting model for the calculation of the deconvolution function for the next cycle of refinement (Fig. 7). The process is iterated several times. The progress of the refinement is monitored by an NMR discrepancy factor (*R*-factor), analogous to the crystallographic *R*-factor.

$$\text{R-factor (NMR)} = \frac{\Sigma \mid N_o - N_c \mid}{\Sigma \, |N_o|} \qquad (12)$$

Here the N_o and N_c are the respective observed and calculated off-diagonal NOE volumes. A reasonable model should produce an R-factor of lower than 20%. In addition, a simulated NOE spectrum may be produced to inspect the agreement between the NOEs from the model and the experimental data (Fig. 7).

Additional information such as the sugar pucker conformation derived from the *J*-coupling constants may be added as constraints in the refinement procedure. It is usually impossible to measure the coupling constants of large molecule from an 1D spectrum. A number of useful variations of COSY, including double quantum-filtered COSY (DQF-COSY) (81), phase-sensitive COSY (82), PE-COSY (83, 84), total correlated spectroscopy (TOCSY) (85) and HOHAHA (86), have been developed. Figure 16 illustrates an expanded diagram (corresponding to region (g) in Fig. 12) of the H1′-H2′/H2″ crosspeaks using phase sensitive PE-COSY45 (84). The coupling constants between those protons can be more reliably measured, as compared to other prototypes of COSY methods. This is shown in Figure 17, where the phase-sensitive COSY45 clearly resolves the cross-peaks of the

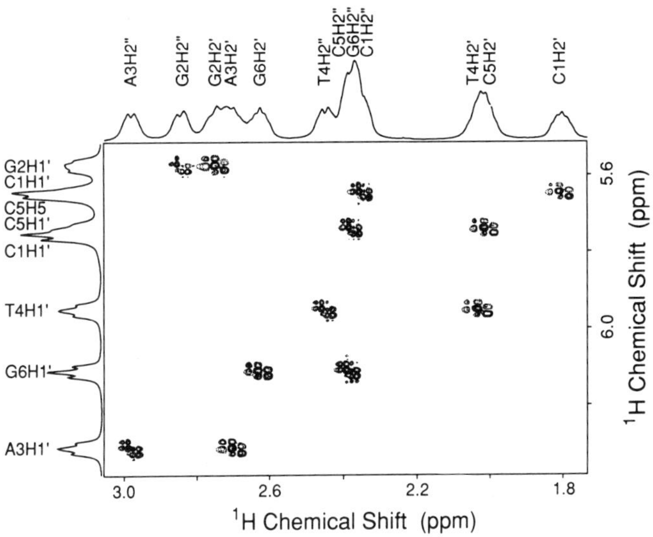

Figure 16 An expanded region (corresponding to region (g) in Fig. 12) of the COSY spectrum showing the cross-peaks between H1′ and H2′/H2″ protons related by *J*-coupling. This figure shows the phase-sensitive PE-COSY45 spectrum which is significantly more informative by providing the well-resolved coupling pattern. The detailed analysis of the cross-peaks provides information of the coupling constants between the coupled protons, in this case between the H1′ and the H2′/H2″ protons.

T4 residue. The coupling constants so measured may be used to calculate the torsion angles of the C1′-C2′ bond by the modified Karplus equation (62).

Similarly, the coupling constants between the proton (e.g., H5′/H5″) and ^{31}P may be measured by heteronuclear COSY method (87). They may be used to define the backbone conformation of the phosphodiester linkage.

Last, but not least, is the recent development of the three- or four-dimensional NMR spectroscopy which permit additional dimensions (e.g., ^{15}N) be added to resolve highly overlapped regions. Experimentally, the time required for the data collection of these new 3D or 4D spectra increases almost exponentially. Methods are being developed to overcome the time burden imposed by the new methods. For example, Bax et al. suggested a new way to record 2D NMR spectra with limited phase cycling (88). These same techniques can be applied to higher dimensionality (3D/4D) NMR spectra.

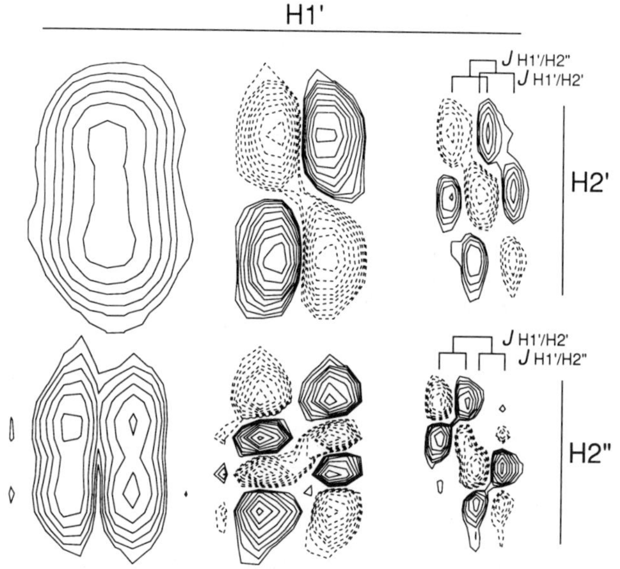

Figure 17 Different types of COSY spectra exemplified by the coupling of H1′ and the H2′/H2″ protons of the T4 residue. The magnitude COSY experiment incorporates both absorptive and dispersive signals and is thus very broad. The phase-sensitive COSY allows the absorptive signals to be seen alone, however with a complex sum of 16 in-phase (dash lines) and out-of-phase (solid lines) signals whose interpretation requires simulation. The PE-COSY45 experiment deemphasizes 8 of the 16 peaks to reveal a pattern from which *J*-coupling constants can be directly extracted.

B. Dynamic Properties in Solution

An important and powerful application of NMR spectroscopy on nucleic acids and their complexes with antitumor drugs is to study the dynamic aspects of their binding interactions. In solution, the molecules under study are in constant motion (translation, rotation, vibrations, etc.). The time scale for the vibrational motions is too short to be detected by NMR technique, instead they are studied by infrared (IR) or Raman spectroscopies. The molecular picture presented by the NMR technique is an averaged view of a large ensemble of conformations in rapid equilibrium. Some of those motions affect the line shape of the spectrum. For example, if the molecule is sufficiently large, its tumbling motion (which dominates the correlation time) becomes slow which would result in a broad spectrum. By raising the temperature of the solution, molecules will tumble faster, resulting in sharper resonances.

The melting of nucleic acid double helix into a single stranded molecule is a well-known phenomenon and the melting behavior can be followed by a temperature-dependent study using NMR. Again the helix-coil transition is a rapid process for NMR. One should keep in mind that the NMR spectrum at any given temperature reflects an averaged image of the multiple species, e.g., duplex and single strand, at certain ratios. For structural studies, it is desirable to record the NOESY spectrum at low temperature, as the tertiary structure of the molecule is better preserved.

For some DNA sequences, alternative helical structures are possible. As mentioned earlier, a DNA molecule with alternating dG-dC sequence adopts left-handed Z-DNA conformation under certain conditions (e.g., 4M NaCl or 40% ethanol solution). The conversion of the right-handed B-DNA structure to left-handed Z-DNA structure requires substantial rearrangement of the nucleotide backbone conformation. For example, dG nucleotide changes from the C2′-endo/anti conformation in B-DNA to the C3′-endo/syn in Z-DNA. This complicated conformational rearrangement suggests that the interconversion between the two structures would be a slow process. This is substantiated by a number of NMR studies of DNA oligonucleotides with alternating dG-dC sequences. The hexamer d(m^5C-G)$_3$ exists as B-DNA in low salt aqueous solution, but as Z-DNA in 40% methanol solution (89). At 20% methanol solution, the NMR spectrum of the hexamer exhibits two sets of resonances simultaneously of roughly equal populations, one from those of B-DNA and the other from those of Z-DNA. Therefore the two conformations are shown to be in slow exchange on the NMR time scale ($\sim 10^{-8}$–10^{-9} sec). (A slow exchange implies that the chemical shift difference between the corresponding resonances of the two species is much larger than the exchange rates.) Interestingly the NOESY spectrum (at long mixing time, e.g., 1000 msec) of this sample (in 20% methanol) shows cross-peaks between the *same* protons from the two molecular species (B-DNA and Z-DNA). This kind of observation has been pointed out previously and called chemical exchange (90). This was the first time that an NOE is observed between two slowly exchanged DNA conformers.

The slow chemical exchange between different molecular species is frequently observed in the systems of drug-DNA complexes. This is easily detected when the ratio of the drug to its DNA binding sites is less than stoichiometric. We will use the binding of nogalamycin to d(CGTACG) duplex as an example (72). The structure of the 2 : 1 complex has been determined at 1.3 Å resolution and is shown in Figure 18. We have shown that nogalamycin has a sequence specificity toward CpN (or NpG) (73). For the 2 (drugs): 1 (duplex) complex, the nogalamycins bind at the two availa-

Figure 18 (see also Color Plate following Chapter 1) Molecular structure of the 2 : 1 complex between nogalamycin and a modified d(CGTACG) hexamer DNA at 1.3 Å resolution. Two nogalamycins are intercalated between the CpG steps at both ends of the distorted right-handed B-DNA hexamer duplex. The aglycone spans the base pairs such that the sugars are in both grooves (nogalose in the minor groove and the aminoglucose in the major groove) of the duplex. Specific hydrogen bonds were found between the drug and DNA.

ble CpG sites and forms a very stable complex. When the solution contains one drug per DNA duplex, three species (free DNA, 1 : 1 complex and 2 : 1 complex) coexist simultaneously, with the population ratio of 1 : 2 : 1 at low temperature. As the binding process of nogalamycin to DNA is slow, all three species are visible in the same spectrum. Interestingly, this equilibrium is temperature dependent as shown in Figure 19. At high temperature (50°C), only the 2 : 1 complex and free single-stranded DNA remain. Thus, the melting of the free DNA duplex has shifted the equilibrium such that the 1 : 1 complex disappears. Many antitumor drugs, including but not limiting to actinomycin D (91,92), chromomycin A_3 (93,94), echinomycin (70,71), netropsin (95), and distamycin (96), have a similar slow exchange behavior.

NMR has also been used to determine the exchange rate of protons in

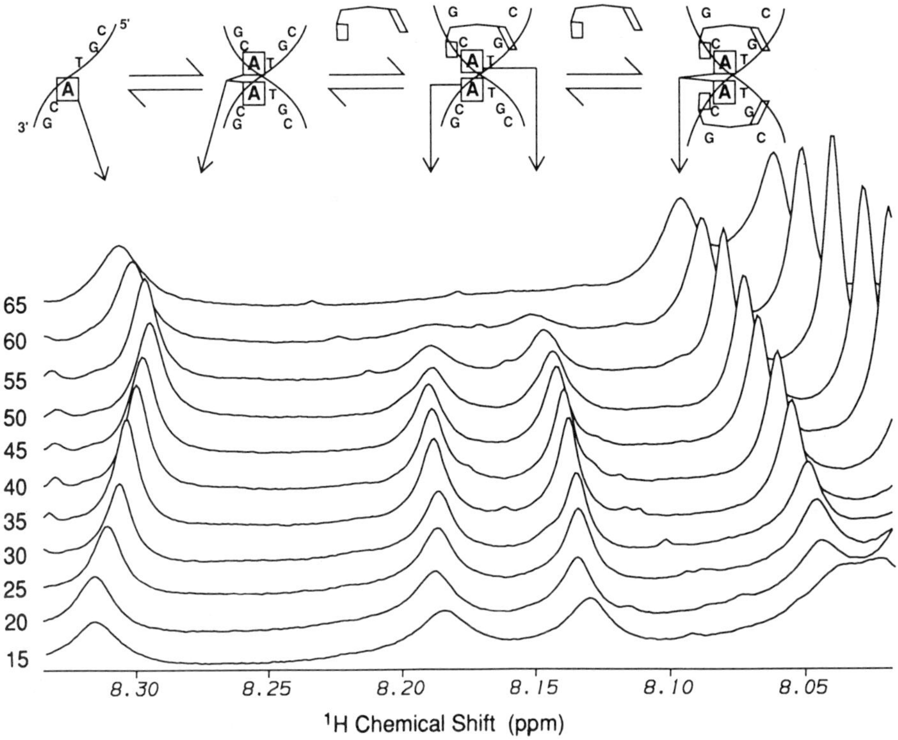

Figure 19 A temperature dependent study of the nogalamycin-d(CGTACG) complex. The solution is a 1 : 1 mixture of nogalamycin and d(CGTACG), which contains free DNA, 1 : 1 complex, and 2 : 1 complex. Peaks from each species are resolved indicating a slow exchange rate of the drug molecule between different binding sites (CpG). Only the region associated with the H8 proton of adenine is shown here. At low temperature the AH8 proton becomes four individual resonances resulting from the different protons of the molecular species shown at the top of the figure. The middle two resonances associated with the 1 : 1 complex disappear at high temperature due to the shifting of equilibrium toward the single-stranded DNA species.

nucleic acids. The exchange of the imino protons with the solvent water molecules requires that the base pair opens up transiently due to the structural fluctuation of the double helix. The rate of the imino proton exchange in B-DNA has been measured by NMR, and its lifetime has been determined to be in the range of milliseconds (97). Surprisingly, the exchange rate of protons in Z-DNA is significantly slower. It has been shown that the exchange rate of protons of the N2 amino group of guanine nucleotide in Z-

DNA is about 330 minutes (98)! This method is applicable to drug-DNA interactions which provides useful information regarding the binding kinetics of the interaction (99).

IV. DISCUSSION

A. Comparison of Solution and Crystal Structures

Both techniques (x-ray diffraction and NMR spectroscopy) are very powerful for elucidating 3D molecular structure, and they are clearly complementary to each other.

The undisputable advantage of x-ray diffraction analysis is that a carefully refined crystal structure provides an unprecedentedly detailed view of not only the three-dimensional structure of the molecules, but also of all the solvent molecules and ions in the surrounding regions. A high-resolution (1.0–2.0 Å) crystal structure can provide distance information between atoms of the order of 0.02 to 0.05 Å accuracy. Furthermore, there is no intrinsic size limit of the system that can be studied. Very large molecular assemblies, such as the nucleosome, whole virus particles and the 50S subunit of ribosome, with molecular weight over a million, have been crystallized (100) and their crystal structures either solved or in the process of being solved. Another interesting property of the crystallization process is that it can select a minor conformational species to form the crystal. For example, d(CGCGCG) hexamer crystallized in the rare Z-DNA conformation, instead of the more abundant B-DNA species. Similarly, many DNA sequences crystallized as A-DNA fragments. Those rare conformations may play important biological roles, and they might be overlooked in NMR studies.

However, there are limitations inherent in x-ray crystallography. Foremost is the requirement of a suitable crystal. No crystal, no structure. Another limitation is related to the fact that data collection of a complete set of diffraction data normally takes days. Therefore the molecular image in the crystal is a time-averaged, static one. But this situation may be changing by the synchroton Laue method discussed earlier.

Another issue concerns the lattice forces which may affect the conformation of the molecules. But this concern has diminished substantially in recent years. There have been many structures of the same molecule determined from different crystal lattices, and it is gratifying to see that most, if not all, of them have very similar conformation.

In contrast to the x-ray diffraction method, NMR spectroscopy does not require a crystal, which means that NMR may be applied to many biological systems. In addition to the ability of solving the 3D structure in solution, a powerful aspect of NMR study is its potential for revealing the dynamic behavior of the system.

On the other hand, several limitations of NMR studies are noteworthy. NMR inherently is a low-sensitivity experiment. The linewidth of the resonances is related to the tumbling rate of the molecules. The slower it tumbles (as in large molecules), the broader the linewidth is. This places an upper limit of the molecular weight of around 10–15K with the current technology. But several developments may push the limit up to ~30K in the next few years. Another related problem is that several conformational states may coexist in equilibrium, causing a serious broadening of the spectrum due to exchange between conformers. In addition, in many cases, the solubility of the molecules severely limits the sample concentration, making it difficult for NMR work.

A serious problem is associated with the paucity of the hydrogen atoms on nucleic acid molecule. For example, there are only 10–11 nonexchangeable protons per nucleotide, while there are about the same number of torsion angles around various single (C—C, C—O, C—N, and P—O) bonds to be defined. Furthermore, even without those problems and under optimum conditions, the accuracy of the NMR 3D structure is not very high, as it is inherently rooted in the uncertainty of the NOE measurements. Even though the number of NOEs per hydrogen atom, therefore per torsion angle, is relatively high (7 ~ 8), it is still far lower than that of a high resolution crystal structure (Table 1). Another problem is that nucleic acids are linear polymers (except in certain higher ordered structures like hairpin). NOE is a short-range phenomenon, providing distance information only to 5 ~ 7 Å at most. The errors in the NOE measurement may accumulate and propagate in going from one end to the other end of the molecule. Consequently, the NOE refinement may result in a significantly different model from the correct structure.

Thus it is apparent that the combination of the two methods would provide an extremely powerful approach to the structural and dynamic aspects of many biological problems, including the interactions of antitumor drugs with nucleic acids. Indeed this integrated approach is becoming more widely used. For example, the crystal structure analysis of the triostin A-d(CGTACG) complex revealed a highly unexpected finding of the existence of the Hoogsteen base pair. This was later confirmed by two independent NMR studies (70,71). Another example may be illustrated by the recent studies of the interactions between nogalamycin and several DNA oligomers. The solution NMR studies (72,101,102) have substantiated the 1.3 Å resolution crystal structure (23) and further augmented the understanding of sequence specificity of the drug and the dynamic property of the binding process. This congruence between the crystal structure and the solution structure of the 2 : 1 nogalamycin-d(CGTACG) complex is illustrated in Figure 20, which compares the experimental NOESY spectrum and the simulated NOESY spectrum based on the crystal structure. Numerous additional examples will be described in the subsequent chapters in this monograph.

In conclusion, it can be expected that this synergistic interactions between x-ray diffraction and NMR will become a standard way to study the biological macromolecules and their interactions with ligands.

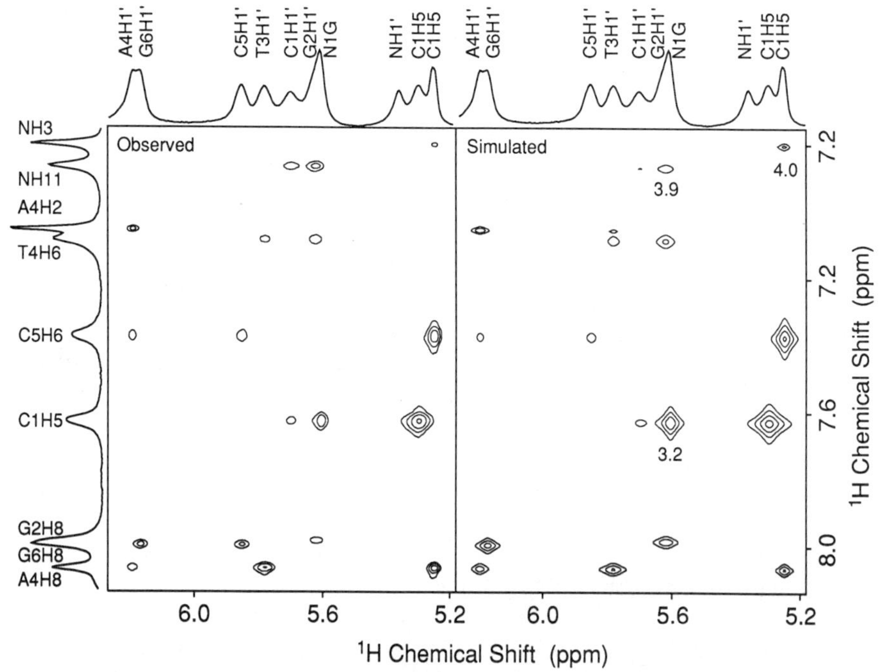

Figure 20 A comparison of the experimental (left) and simulated NOESY (right) spectra of the nogalamycin-d(CGTACG) 2 : 1 complex (23). The simulated spectrum was calculated based on a model initially derived from the 1.3 Å resolution crystal structure and subsequently refined using the NOE data by the refinement procedure described in the text. Some interproton distances between the bound nogalamycin (e.g., NH3, NH11, and N1G) and DNA, calculated on the basis of the crystal structure, are marked in the simulated spectrum. The similarity between the two NOESY patterns suggests that the solution structure is very close to the crystal structure.

V. FUTURE CHALLENGES

During the last decade, we have witnessed the explosion of our knowledge in DNA/RNA structures and their interactions with proteins and antitumor drugs, accumulated through the x-ray diffraction and NMR work of many investigators. There is little doubt that the same progress will be seen in the next decade. Many important structures associated with new and significant biological implications remain to be investigated. For example, it would be interesting to explore the higher-order nucleic acid structures such as triple-stranded, cruciform, or pseudoknot structures. How do these structures interact with antitumor drugs? Similarly, are there drugs that may interact with unusual DNA structures, such as Z-DNA? RNA molecules remain mostly an uncharted territory. Very few RNA structures are available so far (103–106), but this will change rapidly in the future. It would be highly significant to solve the structure of ribozymes, both in solid state and in solution.

Those difficult problems require further advances in both techniques. For x-ray crystallography, a reliable method for obtaining suitable crystals remains as the main bottleneck. Much work is being pursued in this area. For NMR, rapid progress in 3D- and 4D-NMR spectroscopy (107–108) coupled with selective isotope editing promises to push the frontier further.

ACKNOWLEDGMENTS

This work has been supported by grants from NSF and NIH to AHJW. We thank Y.-C. Gao, M. Sriram, and Z. Wu for technical assistance.

REFERENCES

1. For a recent review, see Pullman, A., and J. Jortner, eds., *Molecular Basis of Specificity in Nucleic Acid-Drug Interaction*, Kluwer, Dordrecht (1990).
2. Wang, A.H.-J., G.J. Quigley, F.J. Kolpak, J.L. Crawford, J.H. van Boom, G.A. van der Marel, and A. Rich, Molecular structure of a left-handed double helical DNA fragment at atomic resolution. *Nature 282*: 680–686 (1979).
3. Rich, A., A. Nordheim, and A.H.-J. Wang, The chemistry and biology of Z-DNA. *Ann. Rev. Biochem. 53*: 791–846 (1984).
4. Henderson, E., C.C. Hardin, S.K. Walk, I. Tinoco, Jr., and E.H. Blackburn, Telomeric DNA oligonucleotides from novel intramolecular structures containing guanine-guanine base pairs. *Cell 51*: 899–908 (1987).
5. Williamson, J.R., M.K. Raghuraman, and T.R. Cech, Monovalent cation-induced structure of telomeric DNA: The G-quartet model. *Cell 59*: 871–880 (1989).
6. Sen, D., and W. Gilbert, Formation of parallel four-stranded complexes by guanine-rich motifs in DNA and its implications for meiosis. *Nature (London) 334*: 364–366 (1988).

7. Sundquist, W.I., and A. Klug, Telomeric DNA dimerizes by formation of guanine tetrads between hairpin loops. *Nature (London) 342*: 825–829 (1989).
8. Crothers, D.M., T.E. Haran, and J.G. Nadeau, Intrinsically bent DNA. *J. Biol. Chem. 265*: 7093–7096 (1990).
9. Wang, A.H.-J., Interactions between antitumor drugs and DNA. In *Nucleic Acids and Molecular Biology*, Eckstein, F. and Lilley, D.M. eds., Vol. 1: 32–54, Springer, Berlin (1987).
10. Wang, A.H.-J., and M.-k. Teng, Molecular recognition of DNA minor groove binding drugs. In *Crystallographic and Modeling Methods in Molecular Design*, 123–150, Springer-Verlag, New York (1990).
11. Wang, A.H.-J., G. Ughetto, G.J. Quigley, and A. Rich, The interactions between an anthracycline antibiotic and DNA: Molecular structure of daunomycin complexed to d(CGTACG) at 1.2 Å resolution. *Biochemistry 26*: 1152–1163 (1987).
12. Quigley, G.J., A.H.-J. Wang, G. Ughetto, G.A. van der Marel, J.H. van Boom, and A. Rich, Non-Watson-Crick GC and AT base pairs in a DNA-antibiotic complex. *Science 232*: 1255–1258 (1986).
13. Coll, M., C.A. Frederick, A.H.-J. Wang, and A. Rich, A bifurcated hydrogen-bonded conformation in the d(A.T) base pairs of the DNA dodecamer d(CGCAAATTTGCG) and its complex with distamycin. *Proc. Natl. Acad. Sci. USA 84*: 8385–8389 (1987).
14. Blundell, T.L., and L.N. Johnson, *Protein Crystallography*. Academic Press, New York (1976).
15. Ladd, M.F.C., and R.A. Palmer, *Structure Determination by X-ray Crystallography*, 2nd ed. Plenum Press, New York (1985).
16. Wyckoff, H.W., C.H.W. Hirs, and S.N. Timasheff, eds., *Diffraction Methods for Biological Macromolecules. Methods in Enzymology*, Vols. 114–115, Academic Press, New York (1985).
17. Wuthrich, K., *NMR of Proteins and Nucleic Acids*. Wiley, New York (1986).
18. Ernst, R.R., G. Bodenhausen, and A. Wokaun, *Principles of Nuclear Magnetic Resonance in One and Two Dimensions*. Oxford University Press (Clarendon), London (1987).
19. Oppenheimer, N.J., and T.L. James, eds., *Nuclear Magnetic Resonance. Methods in Enzymology*, Vols. 176–177. Academic Press, New York (1989).
20. Eisenberg, D., and C.P. Hill, Protein crystallography: more surprises ahead. *Trends in Biochem. Sci. 14*: 260–264 (1989).
21. Hol, W., Protein crystallography and computer graphics-toward rational drug design. *Angew. Chem. Int. Ed. Engl. 25*: 767–778 (1986).
22. McPherson, A., *Preparation and Analysis of Protein Crystals*. Wiley, New York (1982).
23. Liaw, Y.-C., Y.-G. Gao, H. Robinson, G.A. van der Marel, J.H. van Boom, and A.H.-J. Wang, Antitumor drug nogalamycin binds DNA in both grooves simultaneously: molecular structure of nogalamycin-DNA complex. *Biochemistry 28*: 9913–9918 (1989).

24. Sherman, S.E., D. Gibson, A.H.-J. Wang, and S.J. Lippard, X-ray structure of the major adduct of the anticancer drug cisplatin with DNA: cis-[Pt(NH_3)$_2${d(pGpG)}]. *Science 230*: 412–417 (1985).
25. Wang, A.H.-J., and Y.-G. Gao, Crystallization of oligonucleotides and their complexes with antitumor drugs. In *Methods: A companion to Methods in Enzymology*, Vol. 1: 91–99 (1990).
26. Gessner, R.G., C.A. Frederick, G.J. Quigley, A. Rich, and A.H.-J. Wang, The molecular structure of the left-handed Z-DNA double helix at 1.0 Å atomic resolution. *J. Biol. Chem. 264*: 7921–7935 (1989).
27. Margaritondo, G., *Introduction to Synchrotron Radiation*. Oxford University Press, New York (1988).
28. Shenoy, G.K., P.J. Viccaro, and D.M. Mills, Characteristics of the 7-GEV advanced photon source: A guide for users. *Argonne National Laboratory Technical Report ANL-88-9* (1988).
29. Hamlin, R., Multiwire area detector x-ray diffractometers. In *Methods in Enzymology*, Wyckoff, H.W., Hirs, C.H.W., and Timasheff, S.N. eds., Vol. 114: 416–452, Academic Press, New York (1985).
30. Arndt, U.W., and G.A. in't Velt, Further development of an x-ray television detector. *Adv. Electr. Electron Phys. 74*: 285–296 (1988).
31. Miyahara, J., K. Takahashi, Y. Amemiya, N. Kamiya, and Y. Satow, A new type of x-ray area detector utilizing laser stimulated luminescence. *Nucl. Instr. Meth. A246*: 572–578 (1990).
32. Strauss, M.G., E.M. Westbrook, I. Naday, T.A. Coleman, M.L. Westbrook, D.J. Travis, R.M. Sweet, J.W. Pfulgrath, and M. Stanton, CCD-based detector for protein crystallography with synchrotron x-rays. *Nucl. Instr. Meth. A297*: 275–295 (1990).
33. Moffat, K., D. Szebenyi, and D. Bilderback, X-ray Laue diffraction from protein crystals. *Science 223*: 1423–1425 (1984).
34. Hajdu, J., P.A. Machin, J.W. Campell, T.J. Greenhough, I.J. Clifton, S. Zurek, S. Gover, L.N. Johnson, and M. Elder, Millisecond X-ray diffraction and the first electron density map from Laue photograph of a protein crystal. *Nature 329*: 178–181 (1987).
35. Karle, J., Recovering phase information from intensity data. *Science 232*: 837-843 (1986).
36. Hauptman, H., The direct methods of x-ray crystallography. *Science 233*: 178-183 (1986).
37. Wang, A.H.-J., G. Ughetto, G.J. Quigley, T. Hakoshima, G.A. van der Marel, J.H. van Boom, and A. Rich, The molecular structure of a DNA-triostin A complex. *Science 225*: 1115–1121 (1984).
38. Rabinovich, D., and Z. Shakked, A new approach to structure determination of large molecules by multiple-dimensional search methods. *Acta Crystallogr. A40*: 195–200 (1984).
39. Fitzgerald, P.M.D., MERLOT, an integrated package of computer programs for the determination of crystal structures by molecular replacement. *J. Appl. Crystallogr. 21*: 273–278 (1988).

40. Brunger, A.T., *X-PLOR, Version 2.1*, The Howard Hughes Medical Institute and Yale University (1990).
41. Wang, A.H.-J., T. Hakoshima, J.H. van Boom, G.A. van der Marel, and A. Rich, AT base pairs are less stable than GC base pairs in Z-DNA: The crystal structure of d(m^{5}CGTAm^{5}CG). *Cell 37*: 321–331 (1982).
42. Wang, A.H.-J., R.G. Gessner, G.A. van der Marel, J.H. van Boom, and A. Rich, A crystal structure of Z-DNA without an alternating purine-pyrimidine sequence. *Proc. Natl. Acad. Sci. USA 82*: 3611-3615 (1985).
43. Ho, P.S., C.A. Frederick, G.J. Quigley, G.A. van der Marel, J.H. van Boom, A.H.-J. Wang, and A. Rich, G.T wobble base-pairing in Z-DNA at 1.0 Å atomic resolution: The crystal structure of d(CGCGTG). *EMBO J. 4*: 3617–3623 (1985).
44. Drew, H.R., and R.E.Dickerson, Structure of a B-DNA dodecamer. *J. Mol. Biol. 151*: 535–556 (1981).
45. Kopka, M.L., C. Yoon, D. Goodsell, P. Pjura, and R.E. Dickerson, The molecular origin of DNA-drug specificity in netropsin and distamycin. *Proc. Natl. Acad. Sci. USA 82*: 1376–1380 (1985).
46. Coll, M., J. Aymami, G.A. van der Marel, J.H. van Boom, A. Rich, and A.H.-J Wang, The molecular structure of netropsin and d(CGCGATATCGCG) complex: DNA conformation in an alternating AT segment. *Biochemistry 28*: 310–320 (1989).
47. Carrondo, M., M. Coll, J. Aymami, A.H.-J Wang, G.A. van der Marel, J.H. van Boom, and A. Rich, Binding of a Hoechst dye to d(CGCGATATCGCG) and its influence on the conformation of the DNA fragment. *Biochemistry 28*: 7894–7895 (1989).
48. Hendrickson, W.A., J. Smith, and S. Sheriff, Direct phase determination based on anomalous scattering. In *Methods in Enzymology*, Wyckoff, H.W., Hirs, C.H.W., and Timasheff, S.N. eds., Vol. 115: 41–55, Academic Press, New York (1985).
49. Guss, J.M., E.A. Merritt, R.P. Phizackerley, B. Hedman, M. Murata, K.O. Hodgson, and H.C. Freeman, Phase determination by multiple-wavelength x-ray diffraction: crystal structure of a basic "blue" copper protein from cucumbers. *Science 241*: 806–811 (1988).
50. Yang, W., W.A. Hendrickson, R.J. Crouch, and Y. Satow, Structure of ribonuclease H phased at 2 Å resolution by MAD analysis of the selenomethionine protein. *Science 249*: 1398–1405 (1990).
51. Ogata, C.M., W.A. Hendrickson, Y. Satow, and X. Gao, 1.8 Å x-ray crystal structure of a chromomycin-DNA complex. Abstract of XVth Congress of International Union of Crystallography, p. C146.
52. Bricogne, G., Methods and programs for direct-space exploitation of geometric redundancies. *Acta Crystallogr. A32*: 832–847 (1976).
53. Podjarny, A.D., T.N. Bhat, and M. Zwick, Improving crystallographic macromolecular images: The real-space approach. *Ann. Rev. Biophys. Biophys. Chem. 16*: 351–373 (1987).
54. Wang, B.C., Resolution of phase ambiguity in macromolecular crystallo-

graphy. In *Methods in Enzymology*, Wyckoff, H.W., Hirs, C.H.W., and Timasheff, S.N. eds., Vol. 115: 90–112, Academic Press, New York (1985).
55. Jones, T.A., Interactive computer graphics: FRODO. In *Methods of Enzymology*, Wyckoff, H.W., Hirs, C.H.W., and Timasheff, S.N. eds., Vol. 115: 157–171, Academic Press, New York (1985).
56. Wang, A.H.-J., Y.-G. Gao, Y.-C. Liaw, and Y.-K. Li, Formaldehyde crosslinks daunorubicin and DNA efficiently: HPLC and x-ray diffraction studies. *Biochemistry 30*: 3812–3815 (1991).
57. Hendrickson, W.A., Stereochemically restrained refinement of macromolecular structures. In *Methods in Enzymology*, Wyckoff, H.W., Hirs, C.H.W., and Timasheff, S.N. eds., Vol. 115: 252–270, Academic Press, New York (1985).
58. Westhof, E., P. Dumas, and D. Moras, Crystallographic refinement of yeast aspartic acid transfer RNA. *J. Mol. Biol. 184*: 119–145 (1985).
59. Branden, C.-I, and T.A. Jones, Between objectivity and subjectivity. *Nature 343*: 687–689 (1990).
60. Wuthrich, K., Protein structure determination in solution by NMR spectroscopy. *J. Biol. Chem. 265*: 22059–22062 (1990).
61. Karplus, M., Vicinal proton coupling in nuclear magnetic resonance. *J. Amer. Chem. Soc. 85*: 2870–2971 (1963).
62. Haasnoot, C.A.G., F.A.A.M. de Leeuw, and C. Altona, The relationship between proton-proton NMR coupling constants and substituent electronegativities. *I. Tetrahedron 36*: 2783–2792 (1980).
63. Pieters, J.M.L., E. de Vroom, G.A. van der Marel, J.H. van Boom, T.M.G. Koning, R. Kaptein, and C. Altona, Hairpin structure in DNA containing arabinofuranosylcytosine: a combination of nuclear magnetic resonance and molecular dynamics. *Biochemistry 29*: 788–799 (1990).
64. Jeener, J., Ampere International Summer School, Basko Polje, Yugoslavia (1971).
65. Hare, D., D.E. Wemmer, S.H. Chou, G. Drobny, and B.R. Reid, Assignment of the nonexchangeable proton resonances of d(C-G-C-G-A-A-T-T-C-G-C-G) using two-dimensional nuclear magnetic resonance methods. *J. Mol. Biol. 171*: 319–336 (1983).
66. LeMaster, D.M., Uniform and selective deuteration in two-dimensional NMR of proteins. *Ann. Rev. Biophys. Biophys. Chem. 19*: 243–266 (1990).
67. Saenger, W., *Principles of Nucleic Acid Structure*. Springer-Verlag, New York (1984).
68. Wang, A.H.-J., G.J. Quigley, F.J. Kolpak, G.A. van der Marel, J.H. van Boom, and A. Rich, Left-handed double helical DNA: variations in the backbone conformation. *Science 211*: 171–176 (1981).
69. Gilbert, D.E., G.A. van der Marel, J.H. van Boom, and J. Feigon, Unstable Hoogsteen base pairs adjacent to echinomycin binding sites within a DNA duplex. *Proc. Natl. Acad. Sci. 86*: 3006–3010 (1989).
70. Gao, X., and D. Patel, NMR studies of echinomycin bisintercalation complexes with d(A1-C2-G3-T4) and d(T1-C2-G3-A4) duplexes in aqueous solu-

tion: Sequence-dependent formation of Hoogsteein A1.T4 and Watson-Crick T1.A4 base pairs flanking the bisintercalation site. *Biochemistry 27*: 1744–1751 (1988).
71. Gao, Y.-G., Y.-C. Liaw, H. Robinson, and A.H.-J Wang, Binding of the antitumor drug nogalamycin and its derivatives to DNA: structural comparison. *Biochemistry 29*: 10307–10316 (1990).
72. Robinson, H., Y.-C. Liaw, G.A. van der Marel, J.H. van Boom, and A.H.-J Wang, NMR studies on the binding of antitumor drug nogalamycin to DNA hexamer d(CGTACG). *Nucl. Acids Res. 18*: 4851–4858 (1990).
73. Ferrin, T.E., C.C. Huang, L.E. Jarvis, and R. Langridge, The MIDAS display system, *J. Mol. Graphics 6*: 13–20 (1988).
74. QUANTA. Polygen Corporation, Waltham, MA.
75. Hare, D.R., and B.R. Reid, Three-dimensional structure of a DNA hairpin in solution: two dimensional NMR studies and distance geometry calculations on (CGCGTTTTCGCG). *Biochemistry 25*: 5341–5350 (1986).
76. Bloomers, M.J., J.A.L.I. Walters, C.A.G. Haasnoot, J.M.A. Aelen, G.A. van der Marel, J.H. van Boom, and C.W. Hilbers, Effects of base sequence on the loop folding in DNA hairpins. *Biochemistry 28*: 7491–7498 (1989).
77. Cheong, C., G. Varani, and I. Tinoco, Jr., Solution structure of an unusually stable RNA hairpin, 5′GGAC(UUCG)GUCC. *Nature 346*: 680–682 (1990).
78. Blumenthal, L.M., *Theory and Applications of Distance Geometry*. Chelsea, New York, 1970.
79. Kuntz, I.D., J.F. Thomason, and C.M. Oshiro, Distance geometry. In *Methods in Enzymology*, Oppenheimer, N.J. and James, T.L. eds. Vol. 177: 159–204, Academic Press, New York (1989).
80. Post, C.B., R.P. Meadows, and D.G. Gorenstein, On the evaluation of interproton distances for three-dimensional structure determination by NMR using a relaxation rate matrix analysis. *J. Amer. Chem. Soc. 112*: 6796–6803 (1990).
81. Range, M., W. Chazin, C. Dalvit, and P.E. Wright, Multiple-quantum nuclear magnetic resonance. In *Methods in Enzymology*, Oppenheimer, N.J. and James T.L. eds., Vol. 176: 114–134, Academic Press, New York (1989).
82. Marion, D., and K. Wuthrich, Application of phase sensitive two-dimensional correlated spectroscopy (COSY) for measurements of ^{1}H-^{1}H spin-spin coupling constants in proteins. *Biochem. Biophys. Res. Commun. 113*: 967–974 (1983).
83. Mueller, L., P.E.COSY, a simple alternative to E.COSY. *J. Mag. Reson. 72*: 191–196 (1987).
84. Bax, A., and L. Lerner, Measurement of ^{1}H-^{1}H coupling constants in DNA fragments by 2D NMR. *J. Mag. Reson. 79*: 429–4438 (1988).
85. Braunschweiler, L., and R.R. Ernst, Coherence transfer by isotropic mixing: application to proton correlation spectroscopy. *J. Mag. Reson. 53*: 521–528 (1983).
86. Bax, A., Homonuclear Hartmann-Hahn Experiments. In *Methods in Enzymology*, Oppenheimer, N.J. and James T.L. eds., Vol. 176: 151–168, Academic Press, New York (1989).

87. Powers, R., C.R. Jones, and D.G. Gorenstein, Two-dimensional ^{1}H and ^{31}P NMR spectra and restrained molecular dynamics structure of an oligodeoxyribonucleotide duplex refined via a hybrid relaxation matrix procedure. *J. Biomol. Struct. Dyn. 8*: 253–294 (1990).
88. Marion, D., M. Ikura, R. Tschudin, and A. Bax, Rapid recording of 2D NMR spectra without phase cycling. Application to the study of hydrogen exchange in proteins. *J. Mag. Reson. 85*: 393–399 (1989).
89. Feigon, J., A.H.-J. Wang, G.A. van der Marel, J.H. van Boom, and A. Rich, A one- and two-dimensional NMR study of the B to Z transition of (m^{5}dC-dG) in methanolic solution. *Nucl. Acids Res. 12*: 1243–1263 (1984).
90. Jeener, J., B.H. Meier, P. Bachmann, and R.R. Ernst, Investigation of exchange processes by two-dimensional NMR spectroscopy. *J. Chem. Phys. 71*, 4546–4553 (1979).
91. Zhou, N., T.L. James, and R.H. Shafer, Binding of actinomycin D to $[d(ATGCAT)]_2$: NMR evidence of multiple complexes. *Biochemistry 28*: 5231–5239 (1989).
92. Jones, R.L., E.V. Scott, G. Zon, L.G. Marzelli, and W.D. Wilson, An NMR investigation of the binding of the anticancer drug actinomycin D to oligodeoxyribonucleotides with isolated 5′d(GC)3′ binding sites. *Biochemistry 27*: 6021–6026 (1988).
93. Gao, X., and D.J. Patel, Solution structure of the chromomycin-DNA complex. *Biochemistry 28*: 751–762 (1989).
94. Banville, D.L., M.A. Keniry, M. Kam, and R.H. Shafer, NMR studies of the interaction of chromomycin A_3 with small DNA duplexes: binding to GC-containing sequences. *Biochemistry 29*: 6521–6534 (1990).
95. Patel, D.J., and L. Shapiro, Sequence-dependent recognition of DNA duplexes. *J. Biol. Chem. 261*: 1230–1239 (1986).
96. Pelton, J.G., and D.E. Wemmer, Structural characterization of a 2 : 1 distamycinA.d(CGCAAATTTGCG) complex by two-dimensional NMR. *Proc. Nat. Acad. Sci. USA 86*: 5723–5727 (1989).
97. Leroy, J.L., M. Kochoyan, T. Huynh-Dihn, and M. Gueron, Characterization of base-pair opening in deoxynucleotide duplexes using catalyzed exchange of the imino proton. *J. Mol. Biol. 200*: 223–238 (1988).
98. Kochoyan, M., J.L. Leroy, and M. Gueron, Process of base-pair opening and proton exchange in Z-DNA. *Biochemistry 29*: 4799–4805 (1990).
99. Leupin, W., Experimental proofs of a drug's DNA specificity. In *Molecular Basis of Specificity in Nucleic Acid-Drug Interaction*, Pullman, A., and Jortner, J. eds., 579–603, Kluwer, Dordrecht (1990).
100. Yonath, A., F. Frolow, M. Shoham, J. Mussig, I. Makowski, C. Glotz, W. Jahn, S. Weinstein, and H.G. Wittmann, Crystallography of ribosomal particles. *J. Crystal Growth 90*: 231–244 (1988).
101. Zhang, X., and D.J. Patel, Solution structure of the nogalamycin-DNA complex. *Biochemistry 29*: 9451–9466 (1990).
102. Searle, M.S., J.G. Hall, W.A. Denny, and L.P.G. Wakelin, NMR studies of the interactions of the antibiotics nogalamycin with the hexadeoxyribonucleotide duplex $d(5'GCATGC)_2$. *Biochemistry 27*: 4340–4349 (1988).

103. Kim, S.H., F.L. Suddath, G.J. Quigley, A. McPherson, J.L. Sussman, A.H.-J. Wang, N.C. Seeman, and A. Rich, Three-dimensional tertiary structure of yeast phenylalanine transfer RNA. *Science 185*: 435–440 (1974).
104. Wang, A.H.-J., S. Fujii, J.H. van Boom, G.A. van der Marel, S.A. van Boeckel, and A. Rich, Molecular structure of r(GCG)d(TATACGC): a DNA-RNA hybrid helix joined to double helical DNA. *Nature 299*: 601–604 (1982).
105. Dock-Bregeon, A.C., B. Chervier, A. Podjarny, J. Johnson, J.S. de Bear, G.R. Gough, P.T. Gilham, and D. Moras, Crystallographic studies of an RNA helix: $[U(UA)_6A]_2$. *J. Mol. Biol. 209*: 459–474 (1989).
106. Rould, M.A., J.J. Perona, D. Soll, and T.A. Steitz, Structure of *E. coli* glutaminyl-tRNA synthetase complexed with $tRNA^{phe}$ and ATP at 2.8 Å resolution. *Science 246*: 1135–1142 (1989).
107. Griesinger, C., O.W. Sorensen, and R.R. Ernst, A practical approach to three-dimensional NMR spectroscopy. *J. Mag. Reson. 73*: 574–579 (1987).
108. Clore, G.M., L.E. Kay, A. Bax, and A.M. Gronenborn, Four-dimensional $^{13}C/^{13}C$-edited nuclear Overhauser enhancement spectroscopy of a protein in solution: Application to interleukin 1β. *Biochemistry 30*: 12–18 (1991).
109. Teng, M.-K., C.A. Frederick, N. Usmann, and A.H.-J. Wang, The molecular structure of the complex of Hoechst 33258 and the DNA dodecamer d(CGCGAATTCGCG). *Nucleic Acids Res. 16*: 2671–2690 (1988).

3

Modeling Drug–Nucleic Acid Interactions

An Exercise in Computer Graphics and Computational Chemistry

Shashidhar N. Rao

Searle Research and Development, Division of G. D. Searle and Company, Skokie, Illinois

I. INTRODUCTION

The mode of drug action has received attention in numerous investigations during the last four decades. A number of drugs are believed to mediate biological effects through specific interactions at membrane-bound extracytosolic receptors or by disrupting membrane fluidity in a nonspecific manner (1). In contrast, other drugs are known to enter the cytosol and the nucleus to exert their action by way of interactions with specific macromolecular targets. Nucleic acids and proteins are generally understood to be important targets for such drugs (2–4). In addition to interactions with their macromolecular hosts, a number of factors are crucial for the overall action of drugs. These include transport to the site of action, transport across the cellular plasma membrane, absorption into cytosol, and metabolism (1). This chapter does not deal with these factors, and for further details the reader is referred to reviews and publications in the appropriate literature.

The main focus of this chapter is the molecular modeling aspects of theoretical investigations on the interactions of drugs which act by way of modulating the structures and, consequently, the functions of nucleic acids. A brief description of the background on the structures of nucleic acids and the problems associated with the structures of drug molecules is given. This is followed by a survey of various computer graphics and computational chemistry packages commonly used in modeling nucleic acids. The use of some of this software in our laboratory is exemplified through two cases of drug-DNA interactions. We show that stereochemically feasible models which are qualitatively consistent with experimental observations can be obtained. The chapter concludes by highlighting the usefulness of the molecular modeling applications to drug–nucleic acid interactions, particularly when very limited high-resolution structural information is available.

Drug–nucleic acid interactions are known to form basic mechanism of action of a number of antitumor and antiviral agents (2a). Comprehensive reviews on this subject have been published, and the reader is referred to them for details (2b,c). Only a brief survey of the chief modes of such interactions are presented here, to outline the background for theoretical studies highlighted in this chapter.

A. Classes of Drug–Nucleic Acid Interactions

Three general categories of drug–nucleic acid interactions have been documented: (a) intercalation, (b) covalent and noncovalent groove binding, and (c) nonspecific. We briefly outline these interactions through a few well-known examples.

Simple rigid planar molecules like 9-aminoacridine (**1**), proflavin, and ethidium are well-known DNA intercalators. The planar aromatic moieties in these drugs are sandwiched between neighboring base pairs which are separated to distances approximately twice that found in normal unintercalated DNA structures (5–7). Intercalation is believed to obey the neighbor exclusion principle, according to which intercalators occupy every other site along the length of a DNA double helix and are forbidden to occupy neighboring sites simultaneously (8).

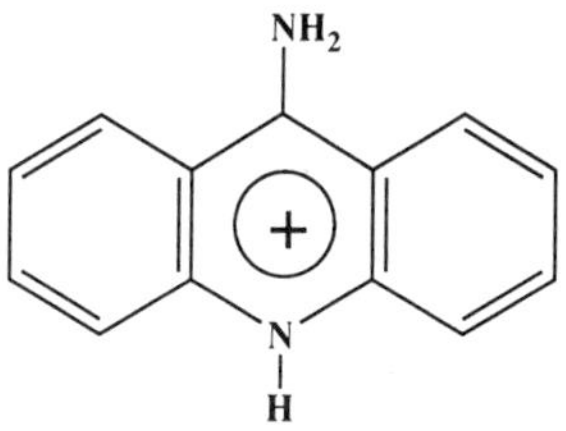

9-amino acridine : **1**

Antitumor agents such as mitomycin, anthramycin, CC-1065, and their analogs bind covalently to DNA in its minor groove and alkylate the exocyclic amino group of guanines (9–11). Mitomycin C (**2**), a very potent member of the family of mitomycins, is known to cause lethal biological effects through DNA cross-linking (12). Mitomycin A (**3**), closely related to mitomycin C, is not as efficacious as the latter in whole animal studies, although it is considerably more potent than **2** (13). In addition to covalent groove-binding drugs, DNA structure is also modulated by drugs like netropsin, which bind in the minor groove without the formation of a covalent linkage (14). Covalent and noncovalent groove binders exhibit DNA sequence specificity.

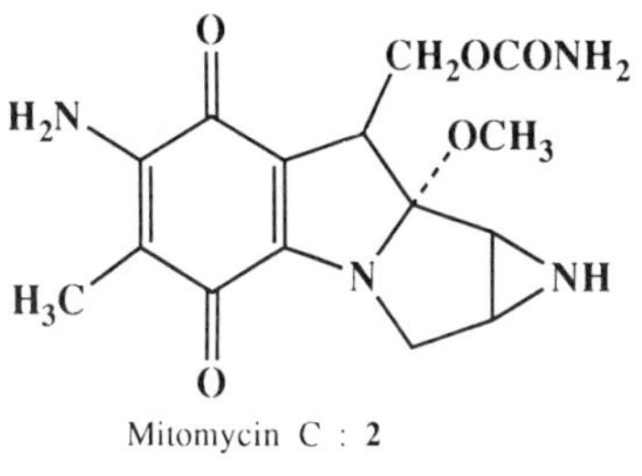

Mitomycin C : **2**

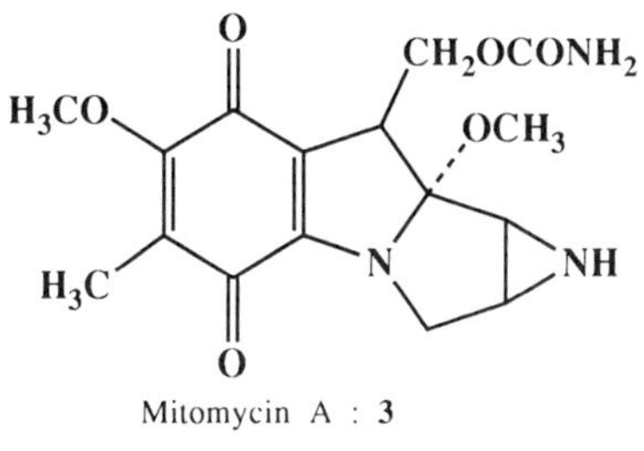

Mitomycin A : **3**

Nonspecific interactions of drugs with nucleic acids usually refer to electrostatic and hydrogen bonding interactions with the backbone phosphates and/or the counterion and solvation atmosphere surrounding them. They are not as thoroughly characterized as the interactions in groove-binding and intercalation complexes. Nonspecific interactions constitute a part of the total interactions between drugs and DNA.

A number of antitumor drugs such as actinomycin and daunomycin have been demonstrated through x-ray crystallographic studies to have intercalating and groove-binding components contributing *together* to their interactions with nucleic acids (15–17). In the former component, the aromatic part of the drugs intercalates between two neighboring base pairs, and in the latter component the so-called tail portion interacts with adjacent nucleotide residues, usually through hydrogen bonding interactions. In addition to the three classes of drug-DNA interactions, a novel mode of binding has been discovered recently. Bacterial agents belonging to the family of calicheamicins and esparamicins characterized by ene-diyne moieties are believed to cause DNA strand scission by abstracting one of its backbone protons (18).

In view of the diverse nature of drug–nucleic acid interactions, their understanding is a key factor in elucidating the mode of action of potentially therapeutic agents which act on DNA and RNA. As mentioned, this chapter deals *only* with aspects of molecular modeling which contribute structural insight into such an understanding. Before highlighting the salient features of such an exercise, we draw attention to three key issues in drug-DNA interactions, namely, (a) the structure of the drug, (b) the structures of the DNA, and (c) the changes in the structures of drug and DNA due to interactions between them.

B. Drug Structures

Determination of the bioactive structures and conformations of small to medium-sized drug molecules poses challenging problems to investigators in the experimental and theoretical fields. For drug molecules which are known to act at a membrane-bound receptor of unknown structure and composition, existence of structurally rigid active analogs is vital to the determination of the so-called bioactive conformations. On the other hand, for drugs which act at the "active site" of an enzyme or through binding to DNA, the challenge is considerably simplified. This is because the macromolecules usually have a well-defined structure of the active site, and the interacting drugs adopt suitable conformations to maximize their interactions with them.

C. DNA Structures

In contrast to the vast number of structural possibilities for small to medium-sized drug molecules due to their conformational flexibility, the structural diversity of a macromolecule like DNA is relatively restricted in the context of biologically important phenomena. Following the proposal of the well-known DNA double helix by Watson and Crick in 1953 (19), several x-ray fiber diffraction studies have been carried out on nucleic acids (20). A number of polymorphic forms of DNA and RNA were model built and refined to fit the somewhat restricted amount of fiber data (21). For example, the B polymorphic family (inclusive of A, B, C, and D forms of DNA and A and A′ forms of RNA) of double helixes was investigated through such studies and transitions between the members of this family were explained as a function of key physicochemical parameters such as humidity and salt concentration (21). In 1979, a novel form of DNA, called Z-DNA, was proposed for an alternating purine-pyrimidine hexamer based on high-resolution x-ray data obtained through single-crystal studies of $d(CGCGCG)_2$ and $d(CGCG)_2$ (22,23). In contrast to the predominantly right-handed forms of duplex DNA proposed earlier, this structure had a left-handed twist with a zigzag shape of the sugar-phosphate backbone (22,23). The limited number of structural classes accessible to DNA is complemented by the variations in the nucleotide (sugar-phosphate backbone linked to a purine or a pyrimidine base) sequence.

D. Structures with Drug-DNA Interactions

The above-described generalized structural features of drugs and nucleic acids lead to a multitude of possibilities for the structures of their complexes, only some of which would be biologically significant. A number of physicochemical and biochemical techniques have been employed to elucidate such structures (24). X-ray crystallographic studies have provided a limited amount of information on the structures of a few drug-oligonucleotide complexes (14–17,25). More recently, a number of theoretical investigations have characterized such complexes, leading to models qualitatively consistent with experimental data (26).

While x-ray crystallography provides high-resolution data on structures of drug-DNA complexes, the extent of conformational space explored is limited. Molecular modeling of such complexes is a useful tool in exploring the multidimensional conformational space accessible to them. This technique can provide a set of structures which are not only consistent with x-ray crystallographic data, but also help in the interpretation of one- and two-dimensional NMR experiments through correlations with observed coupling constants (J values) and NOEs (related to proton-proton distances). In what follows, the techniques employed in modeling two classes of drug-DNA complexes, namely, intercalation and groove binding (through a covalent link) will be outlined. Applications to the cases of DNA-mitomycin C and DNA-acridine bis-intercalation complexes will be specifically exemplified. It will be demonstrated that even in the absence of x-ray data on these complexes, it is possible to employ theoretical techniques to build stereochemically feasible models which are qualitatively consistent with experimental observations, thereby serving as a stimulus for further experimental studies.

II. MODEL BUILDING

A. Nomenclature

Standard nomenclature for the description of atoms of the DNA constituents (27) has been adopted in this chapter. In all the schematic representations of the double-stranded oligonucleotides, the strand of the left-hand side runs from the 5′ (top) to 3′ (bottom), while the complementary strand on the right-hand side runs from 5′ (bottom) to 3′ (top). The torsion angles are described following the notation of Seeman et al. (28). The notations for the bases, sugars, and phosphates are illustrated schematically in Figures 1 and 2. Residues are denoted by the names of the corresponding bases, and the numbering is continuous into the complementary strand. Thus, the bases in residues GUA1 and CYT20 are hydrogen bonded in the Watson-Crick configuration.

B. Computational Software

Most laboratories engaged in theoretical studies on drug-DNA interactions model the appropriate complexes through a combination of model building and energy optimization techniques, using computer-gaphics-based computational packages such as MIDAS (29), SYBYL (30), InSight/Discover (31), Quanta-CHARMm (32), MacroModel (33), and/or computational programs such as AMBER-UCSF (34), GROMOS (35), and BOSS (36). Some of these (e.g., AMBER, Discover, CHARMm, and GROMOS) provide force fields needed to reproduce well-documented experimental observations on nucleic acids, while packages such as MacroModel and BOSS are geared toward the force field to treat small to medium-sized drug molecules with a restricted amount of conformational flexibility. Packages such as MIDAS, Quanta, InSight, and SYBYL are used for graphical analyses of energy-refined structures, while AMBER, CHARMm, GROMOS, and BOSS possess modules that perform groupwise energy component analyses.

C. Model Building the DNA Component

A typical DNA oligomer structure used in the modeling of its complexes with drug molecules consists of two antiparallel strands in which the bases are held together in the Watson-Crick hydrogen bonding configuration. The 5′ and 3′ ends of both the strands have hydroxyl groups which are oriented so that they *do not* form intramolecular hydrogen bonds with other DNA constituents (*vide supra*). The starting models of the oligonucleotides are obtained from either the Brookhaven Protein Data Bank (PDB) (37) or computer-graphics-assisted model building using any of the previously listed packages or through the use of NUCGEN module in the AMBER-UCSF (34).

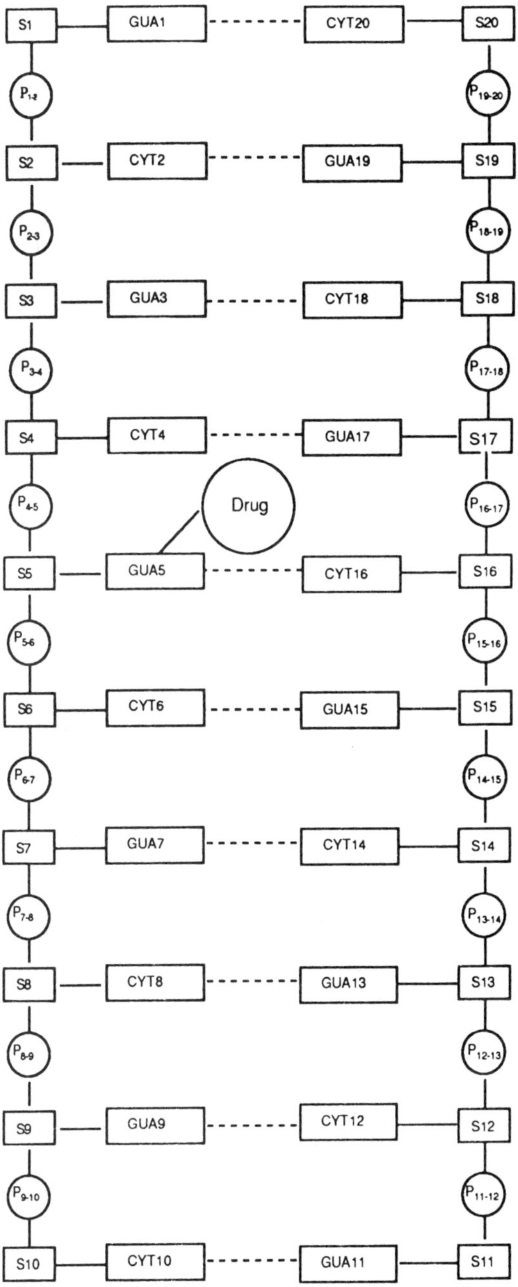

Figure 1 Schematic representation of d(GCGCGCGCGC)$_2$ (GC10) used in the studies on DNA-mitomycin complexes.

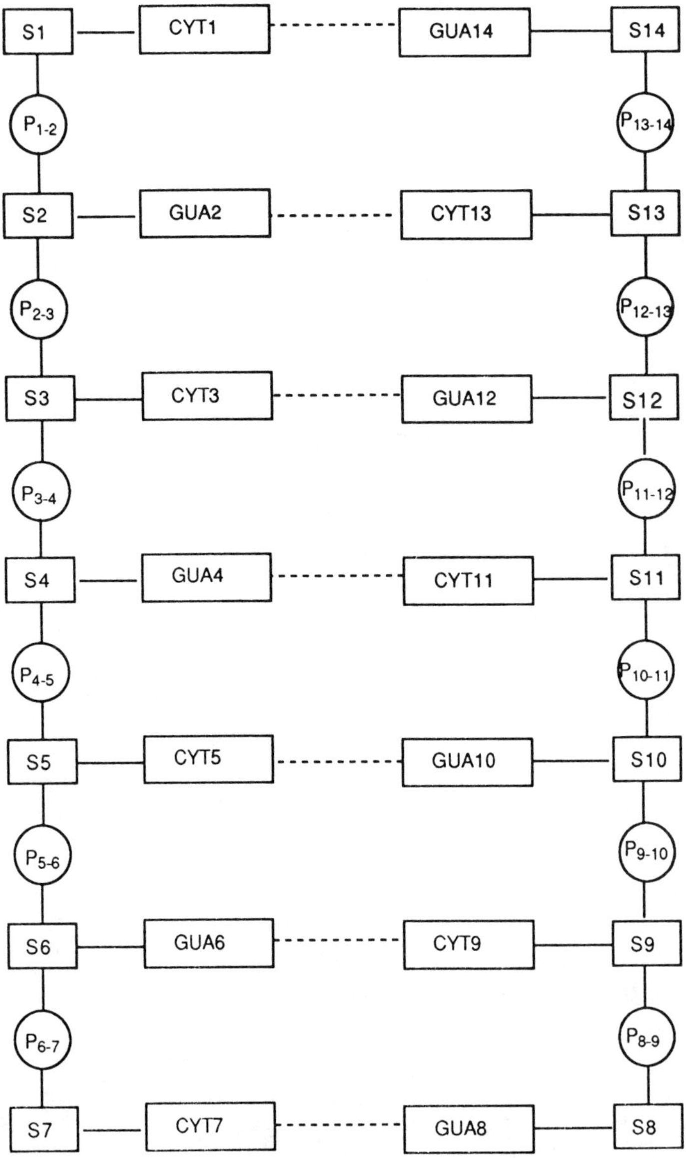

Figure 2 Schematic representation of d(CGCGCGC)·d(GCGCGCG) (GC7) used in the studies on DNA-9-aminoacridine intercalation complexes.

The PDB structures, obtained from high-resolution x-ray crystallography, are usually devoid of hydrogen positions, and these need to be generated and added before the model building of the complexes is undertaken. Most graphics packages have options to add hydrogens with appropriate geometries by properly recognizing the geometric environment around the heavy atoms (carbons and heteroatoms). The structures obtainable from NUCGEN in AMBER-UCSF and from the graphics packages are based on low-resolution x-ray fiber diffraction data. Typically, NUCGEN generates structures with all hydrogens on the heavy atoms in the polymorphic forms right-handed, A, B, A′, B′ and left-handed B. Models of left-handed Z oligonucleotides are generated based on the crystal structures from PDB.

In our laboratory, a combination of AMBER-UCSF, MIDAS, and SYBYL has been used for model building, energy refinement, and analyses of drug-DNA interactions. These procedures will be presented with reference to two specific cases. The first case is that of the DNA-mitomycin complex, whose model building is illustrated for mitomycin C (**2**) but is just as applicable for mitomycin A (**3**). The second case is that of the DNA-9-aminoacridine bis-intercalation complexes. An effort will be made to indicate parallel procedures obtainable and usable in other commercially available software. The DNA components chosen in the two cases are a decamer (Fig. 1) and a heptamer (Fig. 2) of alternating guanosines and cytosines, respectively, modeled in the B form of DNA (38). These will be referred to as GC10 and GC7 in the following text.

D. DNA-Mitomycin Complexes

The starting model of GC10 was obtained by using NUCGEN and displayed on an Evans and Sutherland PS2 that used MIDAS along with the crystal structure of a derivative of mitomycin C, which has an aziridine ring. The intermediate of mitomycin C (**4**) that is believed to covalently attach to the exocyclic nitrogen of guanine was modeled by removal of appropriate bonds and atoms from the x-ray crystal structure of the mitomycin C derivative.

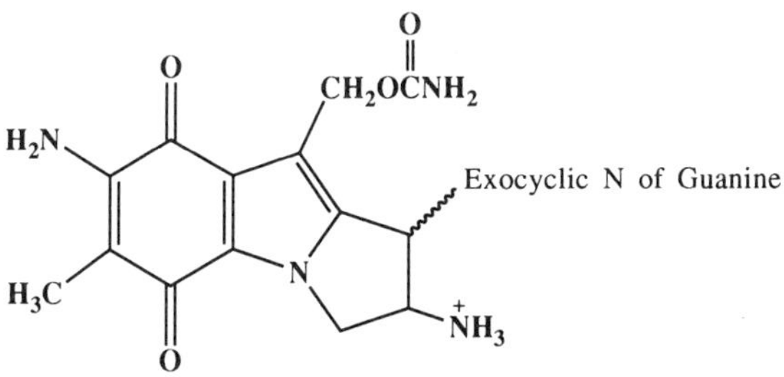

4

The drug was then docked into the minor groove of GC10, which was held fixed with its helix axis coincident with the *Y* axis of rotation. The orientation of the drug relative to DNA was altered through the movement of a joystick, as no automated docking algorithm was available in the framework of MIDAS. Whenever necessary, torsional rotations were also carried out about conformationally flexible bonds of the carbamate side chain. A satisfactory optimal position of the drug was found in which its hydrogen-bond-forming groups were in reasonable proximity and orientation relative to those on GC10. The complex thus modeled was saved as a file in the PDB format.

An important consideration in any “docking experiment” carried out as described is the validity of the complex modeled. Any experimental data available on the complex should be incorporated in such a docking procedure. While x-ray crystallography provides quantitative information at very high resolution, no such data are available in the case of complexes between DNA fragments and mitomycin C and its analogs. In such a situation, semiquantitative or even qualitative information obtainable from physicochemical and biochemical studies provide useful guidelines into model building.

In the case of DNA-mitomycin C complexes, no specific information was available on the key functionalities on the drug *vis-à-vis* their interactions with complementary DNA constituents. However, it is well established that the drug binds in the minor groove (9), not in the major groove as believed earlier (39). Hydrodynamic studies have shown that the overall structure of the complex is largely unchanged relative to the unbound DNA, while indicating that local distorions in the structures were not ruled out (40). In this light, it was important to model-build the minor groove complex in a manner that did not result in a large distortion in the double helical structure of the DNA decamer.

Recent versions of some of the commercial software packages [such as MacroModel (33) and Discover/Insight (31)] do provide automated docking algorithms which could be used to carry out the docking described above. Such algorithms usually have a component that computes intermolecular interaction energies using an approximate and simple force field (e.g., inclusive of only van der Waals and electrostatic terms) and a reasonable cutoff distance, in real time, relative to the docking speed.

E. DNA-9-Amino Acridine Complexes

As mentioned, intercalation complexes were studied with a DNA heptamer GC7 (Fig. 2) whose starting coordinates for model building of its complexes with 9-aminoacridine, were obtained by using the NUCGEN module of AMBER (34). Unlike in the case of groove binding complexes, the DNA structure in the intercalation complexes undergoes nontrivial changes in the region of intercalation. Very few high-resolution x-ray crystallographic studies on the complexes of well-known intercalating drugs with DNA oligomers (more than four nucleotides) have been reported (15–17,25). However, a number of complexes of dinucleotide monophosphates complexed with simple intercalators have been studied through single-crystal x-ray analyses by Sobell and co-workers (5–7,41). These provide useful data points in generating models of oligonucleotides complexed with intercalators, which can be done in one of the following ways. First, starting with the dinucleotide monophosphate complexes, one could extend the nucleotide chains on either side (3′ and 5′) of the intercalator to the desired oligomeric length. A drawback of this approach is that the resultant model could be subject to significant distortions confined to the intercalation site, with the rest of the structure relatively undistorted.

Alternatively, starting from the x-ray structure of DNA hexamer-intercalator complexes, one could replace the intercalator with the desired molecule, with similar functionalities in the two intercalators occupying similar positions relative to DNA. This approach has the disadvantage of introducing or deleting conformational features irrelevant to the replacing intercalator of interest.

In addition to the above approaches, one could also build intercalation sites by inserting the drug between the desired base pairs in an intact DNA. Naturally, such a complex would be characterized by numerous short contacts between the drug and the base pairs between which it is intercalated. An energetically reasonable model may then be obtained by subjecting the resultant complex to an energy refinement procedure that employs the steepest-decent minimizer (e.g., the MIN module in AMBER). In our experience, this approach has led invariably to at least one (*trans,trans*) phosphodiester conformation, which is only rarely observed in the x-ray crystal structures of di- and oligonucleotides and their complexes with intercalating drugs.

In our studies on the complexes between DNA and simple intercalators, we have adopted a fourth approach. A starting model of DNA oligomer (GC7) was subjected to a series of steps, which could be collectively termed modeling surgery, to insert the intercalator (9-aminoacridine as an example) and ensure reasonable distribution of conformational distortions. This is schematically illustrated in Figure 3.

A cut is inserted into the DNA backbone by breaking the sugar-phosphate link at the intercalation site, resulting in the creation of two segments in the oligonucleotide (Fig. 3a). The choice of inserting a cut at the 3′ end or the 5′ end relative to the intercalation site is an arbitrary one since intercalation requires two sets of base-paired nucleotides.

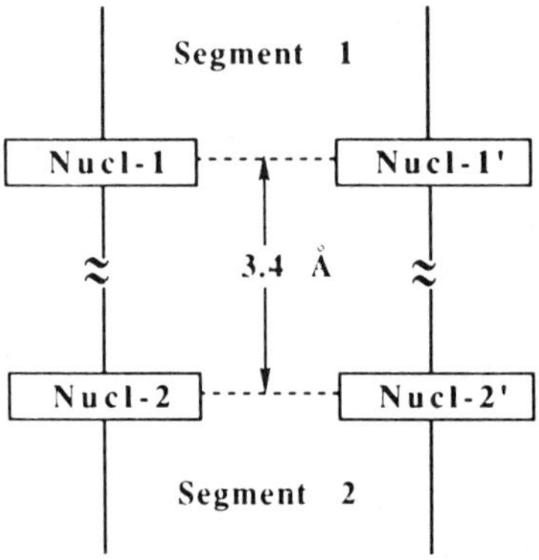

(a)

Figure 3 Creation of the intercalation site between two base-paired nucleotide pairs. (a) Insertion of a "cut" in the phosphodister bond of the intercalation site resulting in two segments (1 and 2) of the oligonucleotide. (b) Separation of the two segments to a distance of around 7 Å. (c) Placement of the intercalating drug and rotation of the two segments about the helix axis is to achieve maximal overlap, necessary rotation of the backbone torsions, followed by sealing of the "cut" bonds.

The upper and lower segments (Fig. 3a) are then separated about 6.5 to 7.5 Å (approximately twice the distance between pairs of base pairs in unintercalated DNA), as illustrated in Figure 3b. This is followed by the docking of the intercalator between the two base pairs on either side of the cut such that maximal visual overlap is obtained between the aromatic rings in the intercalator and the base pairs (Fig. 3c). In order to achieve such an overlap, typically the two segments are rotated about the helix axis relative to the intercalator.

The cut sites on the sugar-phosphate backbone are then "resealed," prior to which rotations are performed about the rotatable bonds to minimize the distances between bonding atoms. For example, the P—03′ and P—05′ bonds between the base pairs binding to the intercalator could be cut on the first and the second strand, respectively. Then rotations about C3′—03′, C5′—05′, and P—05′ are carried out on the first strand, and those about C3′—03′, C5′—05′, and P—03′ are carried out on the second strand to minimize the P . . . 03′ and P . . . 05′ distances. In the absence of any straightforward modeling manipulation of sugar geometry in almost all commercial software specializing in nucleic acid modeling, any manual computer graphic conformational changes are restricted to the freely rotating backbone and glycosidic torsions. The DNA-intercalator complex thus

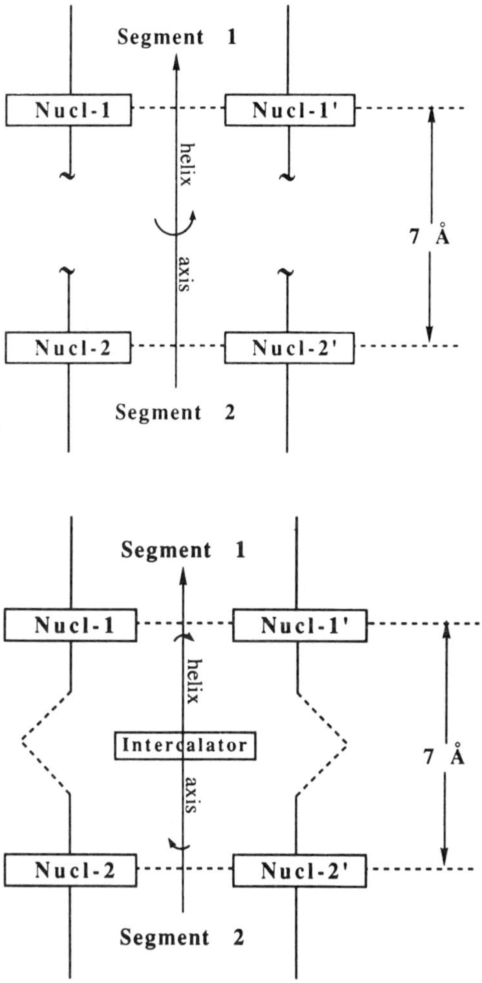

(b)

(c)

modeled is ready for further computational treatment, as will be described below.

In our studies on understanding the physical basis of the neighbor exclusion rule in bis-intercalation complexes, similar modeling strategies were employed (42). Neighbor-excluded bis-intercalation sites (Fig. 4a) and neighbor-exclusion-violated bis-intercalation sites (Fig. 4b) were model-built. In both complexes, the second intercalation site was created after the first one was fully modeled as described above.

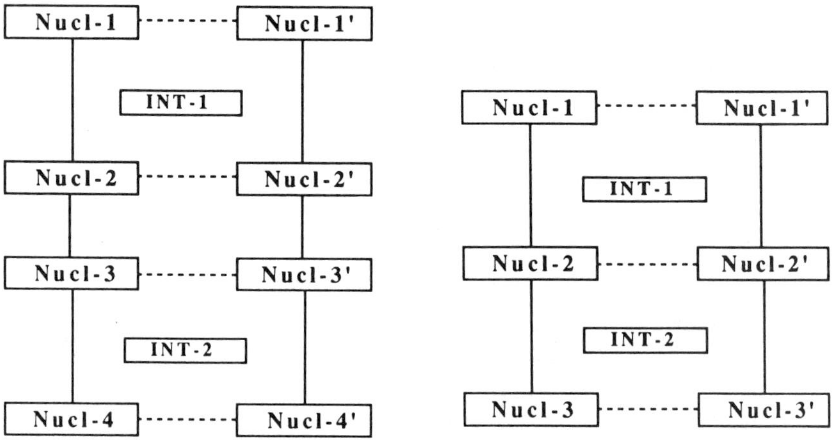

(a) **(b)**

Figure 4 Schematic illustration of two intercalators interacting with DNA while obeying (a) and violating (b) the neighbor exclusion rule.

F. Energy Refinement

In the next stage of the theoretical studies, the computer graphics model-built structures were subjected to energy refinement using the program AMBER-UCSF (34). A detailed description on the use of this program or any other computational package is beyond the scope of this chapter, and the reader is referred to appropriate documentation on such programs. It must be pointed out that software such as Discover/Insight (31), QUANTA/CHARMm (32), MacroModel (33) and SYBYL (30) do offer modules that perform energy refinement of model-built structures without exiting the program session.

Since the AMBER program contains force-field parameters (partial atomic charges, equilibrium bond lengths, angles and their corresponding force constants, barrier heights for torsions and their multiplicity, van der Waals parameters for atoms involved in one to five interactions, and hydrogen bonding parameters) for amino acids and nucleotides, additional parameters were defined for the atoms in mitomycin C, as deemed necessary. Wherever possible, similar force-field parameters were assigned from those of the chemically closely related atoms in the library of amino acids and nucleotides.

Consistent with the AMBER protocol, partial atomic charges were derived from quantum chemically derived electrostatic potentials calculated using a STO-3G basis set in GAUSSIAN80-UCSF (43). In light of the limitations on the then existing computational hardware facilities (processor speed and disk space availability) and limitations on the number of orbitals possible to be considered in GAUSSIAN80-UCSF, the drug molecule was broken up into "acceptable" fragments. Each fragment was masked by methyl groups at the points of fragmentation in the molecule. This was done because in our calculations the net charge on a typical methyl group was 0.0 ± 0.1. In arriving at the partial atomic charges for the atoms in entire molecules, suitable adjustments were made while combining the charges on various fragments to ensure neutrality without alteration in their quality. Typically the unbalanced charges were distributed across the functionalities which are not likely to be involved in any hydrogen bonding or stabilizing electrostatic interactions.

Energy refinements were carried out in several stages. In all the calculations without explicit solvent environment included, a distance-dependent dielectric (r_{ij}) was employed. A number of previous investigations on nucleic acids and their complexes with small molecules have employed such a dielectric value and demonstrated qualitative consistency of the models with the available experimental data. In the case of DNA-mitomycin C complexes, the first stage consists of minimization of the drug alone in the minor groove of the DNA oligomer. This permits the optimization of the manually model-built drug orientation to seek maximal interactions with DNA. The corresponding first stage in the DNA-intercalator complexes includes the optimization of the intercalator together with the base-paired nucleotides in the immediate vicinity of the intercalator. In the second, the entire complex is minimized with all the degrees of freedom allowed to move in the Cartesian minimization. Typically, all minimizations are done initially with a steepest-descent optimizer (500 to 1000 steps) and then with a conjugate gradient minimizer until an energy gradient of 0.1 kcalmole^{-1} Å^{-1} is attained. The structures thus obtained are analyzed for components of energies, hydrogen bonding interactions, and conformational profiles.

G. Analysis of Energy Components

The component analyses of energies of interactions are carried out to estimate the relative energetic stabilities of various complexes in each of the two classes considered here. Relative binding interactions of analogs of the drugs with the polynucleotides are compared through their net binding energies. The latter are calculated by subtracting the helix distortion and drug distortion energies from the total drug-DNA interaction energy. These distortion energies reflect induced fits that permit stronger intermolecular interactions. Drug and helix distortion energies are obtained by subtracting the energies of the minimized isolated drug or helix from their energies that occur in the complexes. It is easy to see that the choice of the conformation of the isolated drugs and DNA oligomers will quantitatively influence the net binding energies. Such an influence can be more in the case of the DNA which exhibits considerable conformational flexibility. On the other hand, it is computationally a challenging and perhaps a meaningless task to evaluate the "global minimum energy" for a given DNA oligomer. Therefore, we have used the strategy of pooling all the structures of uncomplexed DNA from the complexes with analogous drugs and selecting the one with lowest energy. In some of the earlier investigations on drug-DNA interactions, the energy-minimized B-DNA based on x-ray fiber diffraction data was used to evaluate distortion energies of oligonucleotides. In addition to the above sets of interactions, energies of interaction between the drugs and spatially close individual nucleotide residues (sugars, phosphates, and bases) schematically illustrated in Figures 1 and 2 are also evaluated. In addition to the component analyses, the refined structures were graphically analyzed using MIDAS and SYBYL, for the drug-DNA hydrogen bonding interactions.

The following section describes the results of the investigations on the DNA-mitomycin and DNA-9-aminoacridine complexes which were built and refined in the manner detailed above. In the case of DNA-mitomycin complexes, the structures of GC10-mitomycin C and GC10-mitomycin A are discussed. In the intercalation complexes, double intercalation by 9-aminoacridine in two modes corresponding to obeying and violating the neighbor exclusion rule (8) are considered.

It must be reemphasized that the goal of this chapter is to outline the usefulness of the above-described approaches of computer-assisted molecular model building and energy computations to derive structural features which are qualitatively consistent with the observed experimental data on drug-DNA interactions. For further details on the theoretical models of the above complexes and on the examples of modeling other related drug-DNA complexes, the reader is referred to the literature (26,44,45).

III. RESULTS AND DISCUSSION

A. DNA-Mitomycin Complexes

Table 1 describes the energy components of the refined models of GC10-mitomycin C and GC10-mitomycin A complexes. Stereo pictures of the two complexes are computer graphically illustrated in Figure 5, and color plate 1 shows a stereo pair of the GC10-mitomycin C complex. The two drugs have practically identical orientations in the minor groove of GC10, and the DNA distortions in the two complexes are very similar. Significant sugar-phosphate backbone conformational changes are confined to a few sugar rings and P—03′ torsions in the immediate vicinity of the site of covalent

Table 1 Energy Components in the Complexes between GC10 in Right-handed B Form of DNA and Mitomycin C (GC10-MC) and Mitomycin A (GC10-MA)

	Intermolecular					Intramolecular					
Complex	Total	Elec[a]	vdw	$E_{d\text{-}h}$[b]	He11[c]	He12[d]	E_{dis}[e]	E_{d1}[f]	E_{d2}[g]	E_{dd}[h]	E_{netb}[i]
GC10-MC	−895.2	−101.2	−28.3	−116.3	−792.8	−812.0	19.2	5.0	−4.3	9.3	−87.8
GC10-MA	−890.7	−94.6	−29.1	−110.5	−795.2	−812.0	16.8	6.1	−2.4	8.5	−85.2

[a]Total electrostatic energy.
[b]Total intermolecular energy.
[c]Energy of the helix in the complex.
[d]Energy of the helix minimized from the complex with the drug removed.
[e]Helix distortion energy.
[f]Energy of the drug in the complex.
[g]Energy of the drug alone.
[h]Drug distortion energy.
[i]Net binding energy ($E_{d\text{-}h} + E_{dis} + E_{dd}$).

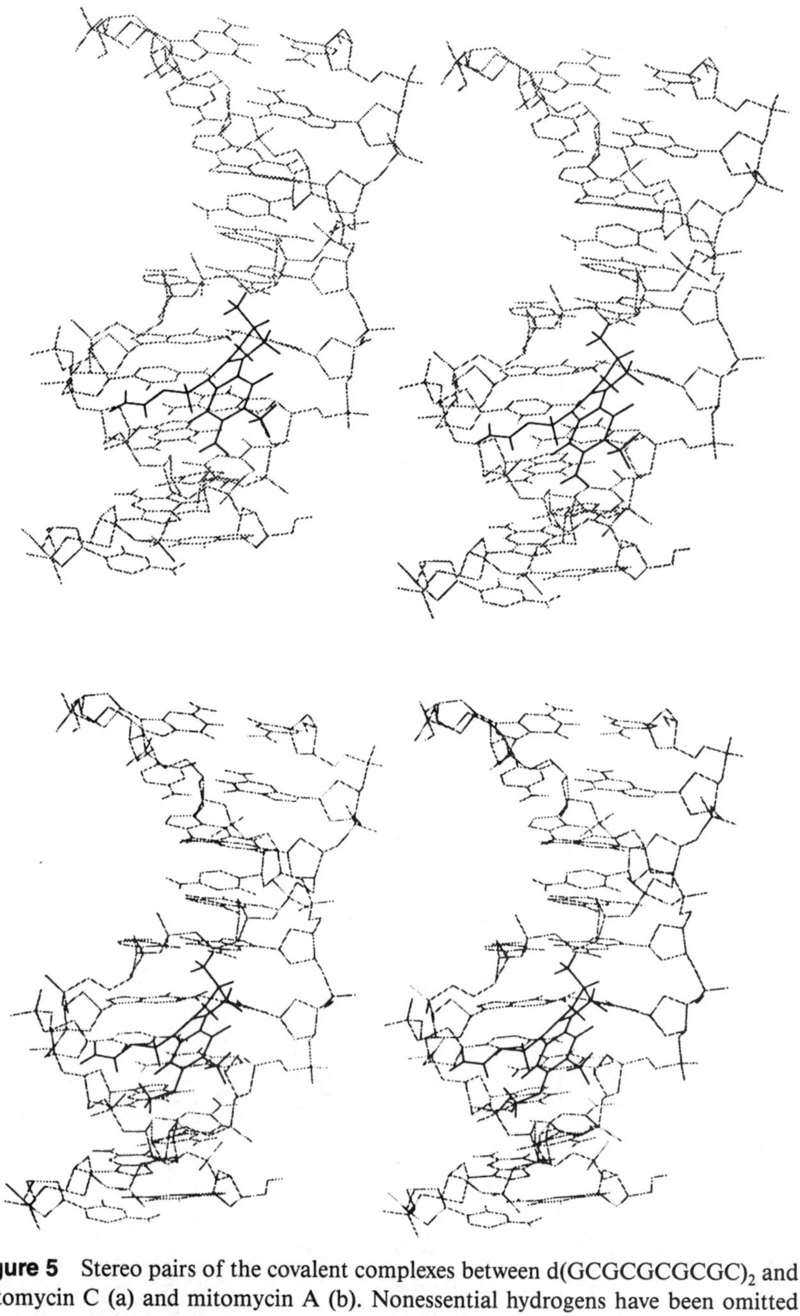

(a)

(b)

Figure 5 Stereo pairs of the covalent complexes between d(GCGCGCGCGC)$_2$ and mitomycin C (a) and mitomycin A (b). Nonessential hydrogens have been omitted for clarity of the structures.

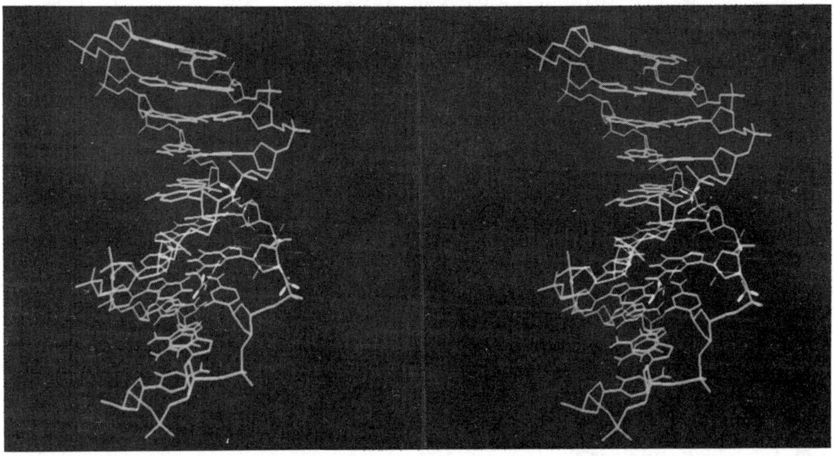

Plate 1 (see also Color Plate following Chapter 1) Stereo pair of the covalent complex between d(GCGCGCGCGC)$_2$ and mitomycin C. The DNA and the drug components are colored cyan and red, while the guanine nucleotide covalently linked to mitomycin C is shown in yellow.

attachment. The net binding energies favor the interactions of mitomycin C with GC10 over those of mitomycin A. This is rationalized by the important difference between the two drugs in the substituent at C1. The methoxy group in mitomycin A is incapable of forming hydrogen bonds with the nearby phosphate residue, while the amino group in the corresponding position in mitomycin C is "strongly" solvated by the neighboring phosphates. This distinction is further emphasized in Tables 2 and 3, which list hydrogen bonding parameters and individual component energies between the drugs and neighboring GC10 residues.

The results are in *qualitative* agreement with the experimentally observed weaker efficacy of mitomycin A compared to mitomycin C (13a–c). However, the calculated differences in drug-DNA interactions of mitomycins C and A do not correlate with the observed higher potency of mitomycin A compared to mitomycin C (13). This has been attributed to a more favorable quinone reduction potential of mitomycin A compared to mitomycin C, a factor not included in the computations of binding energies. This lack of correlationship is also seen for a number of analogs of mitomycin C (13a).

The lack of significant overall helix distortions in both the complexes is consistent with the results of hydrodynamic and physicochemical studies (40,46). Models similar to these have also been built and used to rationalize the observed biological activity of a few analogs of mitomycin C and antibi-

Table 2 Hydrogen Bonding Parameters[a] Involving Interactions of GC10 in the Right-handed B Form in the Complexes with MC and MA

Complex	Donor (X-H)	Acceptor atom (Z)	Length (Å)	Angle (deg)
GC10-MC	N3-HN3B(MC)	O2(CYT6)	1.72	167.6
	N3-HN3C(MC)	O1′(GUA7)	2.03	152.5
	N1-HN1B(MC)	OB($P_{17\text{-}18}$)	2.18	123.1
	N2-HN2B(GUA17)	O5(MC)	2.07	154.7
GC10-MA	N3-HN3B(MA)	O2(CYT6)	1.71	167.2
	N3-HN3C(MA)	O1′(GUA7)	2.02	155.8
	N2-HN2B(GUA17)	O5 (MA)	2.11	152.0

[a]In a hydrogen bond $X-H \cdots Z$, X and Z are, respectively, the donor and the acceptor atoms. The hydrogen bond length corresponds to the distance between H and Z, while the angle is $X-H \cdots Z$.

otics belonging to the pyrrolo[1,4]-benzodiazepine class (viz., anthramycin, tomaymycin, neothramycin).

B. DNA-9-Aminoacridine Complexes

Table 4 lists the energy components of the refined models of GC7-9-aminoacridine double-intercalation complexes, in one of which the neighbor exclusion principle is obeyed (OBEY), and the other where this principle is violated (VIOL). Stereo pairs of the two complexes are computer graphically

Table 3 Energies (kcal mole^{-1}) of Interactions between the Drug and the Components of GC10 in the Right-handed B Form in the Complexes with MC and MA

GC10 Group	Mitomycin A	Mitomycin C
$P_{4\text{-}5}$	−3.2	−3.0
$P_{5\text{-}6}$	−10.4	−9.8
$P_{6\text{-}7}$	−23.2	−23.7
$P_{7\text{-}8}$	−7.8	−7.7
$P_{8\text{-}9}$	−3.1	−3.1
$P_{15\text{-}16}$	−3.9	−3.9
$P_{16\text{-}17}$	−7.0	−7.1
GUA17	−6.8	−7.0
$P_{17\text{-}18}$	−11.7	−18.9
$P_{18\text{-}19}$	−3.0	−3.0

Table 4 Conformational Parameters (in degrees) in the Double Intercalation Complexes between d(CGCGCGC)·d(GCGCGCG) and 9-Aminoacridine Where the Neighbor Exclusion Rule Is Obeyed (OBEY) and Violated (VIOL)

	OBEY				VIOL			
Torsion	Chain 1		Chain 2		Chain 1		Chain 2	
$\zeta(5')$	160	124	147	164	148	159	145	158
α	180	182	186	179	180	190	186	191
β	275	264	262	224	311	270	185	274
γ	286	288	278	273	187	177	188	170
δ	173	175	178	166	190	173	196	171
ϵ	58	61	58	55	182	180	176	179
$\zeta(3')$	132	147	150	88	158	147	158	143
$\chi(5')$	75	61	68	107	67	68	87	72
$\chi(3')$	72	63	76	32	68	70	72	66

The notation of Seeman et al. (27) is followed in describing the backbone and glycosidic torsions in the oligonucleotide. Chain 1 and Chain 2 refer to the two strands of the double helical oligonucleotide. The two columns for each of the chains correspond to the two intercalation sites.

illustrated in Figure 6, while color plate 2 provides a closeup look at the intercalation sites. The drug molecules in the latter are displayed together with their van der Waals surfaces.

Unlike in the case of DNA-mitomycin complexes, the GC7 structures in OBEY and VIOL models undergo a few significant conformational changes compared to the corresponding uncomplexed B-DNA. Table 4 lists the backbone torsion angles of GC7 in the nucleotides around the two intercalation sites. Two unusual conformations characterize the intercalation models. One of them is a *trans,trans* phosphodiester conformation in VIOL at one of the two intercalation sites, leading to a somewhat stretched structure for this complex. The other is a *gauche*$^+$ conformation about C5′—05′ in the OBEY model, away from any of the intercalation sites. This conformation is predominantly *trans* in most crystal structures of oligonucleotides and in models of DNA derived on the basis of fiber diffraction data. These unusual conformations are also reflected in the distortions of the helical parameters characterizing the heptanucleotides in the two models. The helix in VIOL is longer than that in OBEY by about 1.5 Å, while the overall helix unwinding is larger in OBEY (20°) than in VIOL (10°).

The total energies of the two models (OBEY, -750.3 kcalmole^{-1} and VIOL, -750.6 kcalmole^{-1}) differ by less than 0.5 kcalmole^{-1}. However, the total drug-DNA interactions are favored in OBEY by about 5 kcalmole^{-1}

(a)

(b)

Figure 6 Stereo pairs of the double intercalation complexes between d(CGCGCGC)·d(GCGCGCG) and 9-aminoacridine (a) with the neighbor exclusion rule obeyed and (b) with the neighbor exclusion rule violated. Nonessential hydrogens have been omitted for clarity of the structures.

(OBEY, −145.9 kcalmole^{-1} and VIOL, −140.8 kcalmole^{-1}) while the helix distortions favor the VIOL model by about the same amount (OBEY, 56.1 kcalmole^{-1} and VIOL, 51.2 kcalmole^{-1}). The interactions between the two intercalators are not significantly different in the two models despite a distance-dependent dielectric.

The similarity in the total energies of the two classes of models is also emphasized through molecular dynamics simulations of these complexes (42). This implies that stereochemically feasible models of double intercalation complexes could be built incorporating the violation of the neighbor exclusion principle, without significant energy penalty.

Why then do the simple molecules like 9-aminoacridine and ethidium not violate the neighbor exclusion rule while intercalating into nucleic acids? A qualitative answer was sought through normal mode analysis and thermochemistry calculations on the two classes of models. The model OBEY has a higher vibrational entropy (by 8 calmole^{-1}K^{-1}) than VIOL, corresponding to a difference of 2.4 kcalmole^{-1} at 300K. The higher entropy in OBEY could possibly be attributed to a more stretched nature of the helix in VIOL, and has been argued to be one of the driving forces in the nonoccurrence of VIOL class of models for simple charged intercalators (42).

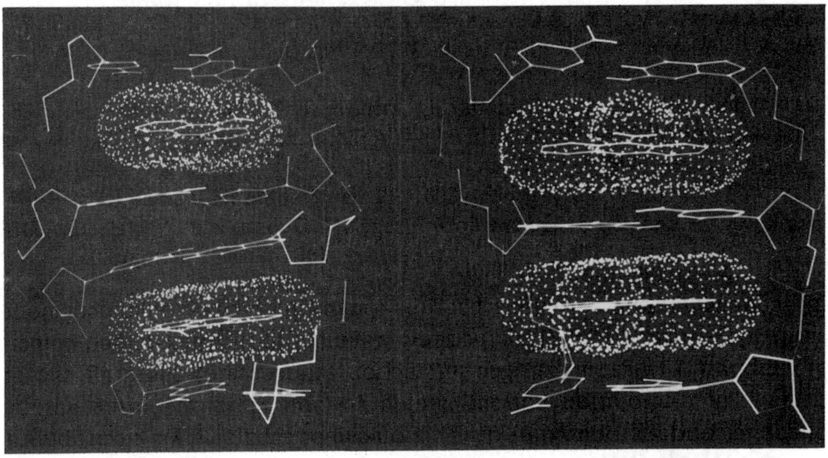

Plate 2 (see also Color Plate following Chapter 1) Double intercalation complexes between d(CGCGCGC)·d(GCGCGCG) and 9-aminoacridine with the neighbor exclusion rule obeyed (left) and with the neighbor exclusion rule violated (right). Base pairs not directly involved in intercalation with the 9-amino acridines are not shown.

IV. GENERAL REMARKS

From the applications described above, it is clear that theoretical models qualitatively consistent with physicochemical and biological studies can be built using model building in combination with simplistic approaches presented by molecular mechanics. It must be pointed out that in most molecular mechanics approaches, the electrostatic contributions tend to be overemphasized due to the fact that they are calculated using a simple Coulombic relationship, with either a constant or a distance-dependent dielectric. A few investigations have employed less simplistic approaches to treating electrostatics calculated based on partial atomic charges, by the use of nonlinear damping functionalities. Such investigations have in general yielded qualitatively similar results to the ones using the approach presented here and in several such studies. It is nevertheless gratifying to note that the present approach can be used to arrive at useful models capable of semipredictive power.

While the above two sets of models exemplify the utility of the molecular modeling approach in the field of nucleic acids, by no means do they cover or represent the entire spectrum of drug-DNA/RNA complexes as well as other modifications to nucleic acids. For example, the theoretical approaches described here have been applied to the studies of photodimerized DNA (47), formation of kinks and bends in nucleic acids (48), novel DNA structures such as parallel DNA (as opposed to antiparallel DNA most commonly found) (49), and complexes of DNA with tethered intercalators (50). It must also be stressed here that this chapter has concentrated only on the most basic and simple approaches in understanding the intermolecular interactions *vis-à-vis* drug-DNA complexes. Recent advances in computer technology which have led to significantly fast and superfast computers have permitted the application of the computationally more rigorous approaches such as molecular dynamics/free-energy perturbation coupled with high level ab initio calculations to make quantitative, in addition to qualitative predictions in such systems. The description of these methods and their applications are clearly beyond the scope of this chapter.

V. CONCLUSIONS

In light of the biological significance of drug–nucleic acid interactions, molecular modeling studies have been described. This chapter has demonstrated through two examples, one belonging to the class of intercalation and one to groove-binding complexes, the utility of theoretical methods in arriving at stereochemically feasible models which satisfy the constraints imposed by experimental data. Such methods combine a judicious use of computer-graphics-assisted molecular model building, manual docking of drugs into the "binding sites" on DNA, and computational chemistry packages consisting of molecular mechanics, molecular dynamics, and normal mode/thermochemistry calculations. In the groove-binding complexes, relative binding energies have been correlated with experimentally observed drug efficacies, implying the usefulness of the models in a predictive capacity to qualitatively estimate the binding of an analog which could be a useful target for synthesis. In the intercalation complexes, the physical basis behind one of the most significant principles has been explained. The computationally derived models could serve as useful starting points in the interpretation of future high-resolution solution and solid-state structural studies.

ACKNOWLEDGMENTS

I owe deep gratitude to Professors William Remers (University of Arizona, Tucson) and Peter Kollman (University of California, San Francisco) for their very fruitful collaboration with me in the area of modeling drug–nucleic acid complexes. I thank the members of the Drug Design Group at Searle R&D for the computational facilities used in part of the research presented in this chapter. I thank Dr. Chandra Singh and the members of the computer graphics laboratory, University of California, San Francisco, from whom much of my experience in using AMBER-UCSF and in computer-graphics-assisted molecular modeling was gained. Financial support for most of the research presented in this chapter, provided through grant NCI-CA-25644, is gratefully acknowledged.

REFERENCES

1. Fingl, E., and D.M. Woodbury, In *The Pharmacological Basis of Therapeutics*, 5th ed. Goodman, L.S., and Gilman, A. eds., 1–46, Macmillan, New York (1975).

2a. Remers, W.A. In *Antineoplastic Agents*, Remers, W.A. ed., 83–261, Wiley-Interscience, New York (1984).

2b. See *Chemistry and Physics of Drug-Ligand Interactions* (1990) N.R. Kallenbach, ed. New York, Adenine Press, for reviews.

2c. Craik, D.J. and Higgins, K.A. *Annu. Rep. NMR Spectrosc.* 22:61–138 (1990).

3. Calabresi, P., and R.E. Parks. In *The Pharmacological Basis of Therapeutics*, 5th ed. Goodman, L.S., and Gilman, A. eds., 1248–1307, Macmillan, New York (1975).

4. Buick, R.N., In *Antineoplastic Agents*. Remers, W.A. ed., 1–40, Wiley-Interscience, New York (1984).

5. Shieh, H.S., H.M. Berman, M. Dabrow, and S. Neidle, *Nucleic Acids Res. 8*: 85–97 (1980).

6. Tsai, C.-C., S.C. Jain, and H.M. Sobell. *J. Mol. Biol. 114*, 301–305 (1977).

7. Sakore, T. D., B.S. Reddy, and H.M. Sobell. *J. Mol. Biol. 135*, 763–785 (1979).

8. Crothers, D.M., *Biopolymers 6*, 575–584 (1968).

9. Tomasz, M., R. Lipman, D. Chowdary, J. Pawlak, G.L. Verdine, and K. Nakanishi. *Science 235*, 1204 (1987).

10. Kohn, K.W., and C.L. Spears. *J. Mol. Biol. 51*, 551 (1970).

11. Hurley, L.H., V.L. Reynolds, D.H. Swenson, G. L. Petzold, and T.A. Scahill. *Science 226*, 843–844 (1984).

12. Szybalski, W., and V.N. Iyer. *Antibiotics 1*, 211–245 (1967).

13a. Kunz, K.R., B.S. Iyengar, R.T. Dorr, D.S. Alberts, and W.A. Remers, *J. Med. Chem. 34*: 2281–2286 (1991).

13b. Moriguchi, I., and K. Komatsu. *Chem. Pharm. Bull. 25*: 2800–2802 (1977).

13c. Iyengar, B.S., H.-J. Lin, L. Cheng, and W.A. Remers *J. Med. Chem. 24*, 975–981 (1981).

14. Kopka, M. L., C. Yoon, D. Goodsell, P. Pjura, and R. E. Dickerson. *J. Mol. Biol. 183*, 553–563 (1985).

15. Takusagawa, F., M. Dabrow, S. Neidle, and H.M. Berman. *J. Biol. Chem. 259*, 4714–4715 (1984); Sobell, H. M., and S. C. Jain, *J. Mol. Biol.* **68**, 21–34 (1972).

16. Moore, M.H., W.N. Hunter, B. Langlois d'Estaintot, and O. Kennard. *J. Mol. Biol. 206*, 693–705 (1989).

17. Wang, A.H.-J., G. Ughetto, G.J. Quigley, and A. Rich. *Biochemistry 26*, 1152–1163 (1987).
18. Zein, N., M. Poncin, R. Nilakantan, and G.A. Ellestad. *Science 244*, 697–699 (1989).
19. Watson, J.D., and F.H.C. Crick. *Nature 171*, 737–738 (1953).
20. Saenger, W. In *Principles of Nucleic Acid Structure*, 44–45, Springer-Verlag, New York (1984).
21. Saenger, W. In *Principles of Nucleic Acid Structure*, 200–241, 411–422, Springer-Verlag, New York (1984).
22. Wang, A.H.-J., G.J. Quigley, F. J. Kolpak, J.L. Crawford, J.H. van Boom, G. van der Marel and A. Rich. *Nature 282*, 680–686 (1979).
23. Drew, H., T. Takano, S. Tanaka, K. Itakura, and R.E. Dickerson. *Nature 286*, 567–573 (1980).
24. D. Live, I.M. Armitage and D. Patel, eds., *Frontiers of NMR in Molecular Biology*, 215–274. Wiley-Liss, New York (1990).
25. Wang, A.H.-J., G. Ughetto, G.J. Quigley, and A. Rich. *J. Biomolec. Str. Dyn. 4*, 319–342 (1986).
26. Lybrand, T.P., S.C. Brown, S. Creighton, R.H. Shafer, and P.A. Kollman. *J. Mol. Biol. 191*, 495–507 (1986); Chen, K.-X., N. Gresh, and B. Pullman. *J. Biomolec. Str. Dyn. 3*, 445–466 (1985); Remers, W.A., S.N. Rao, T.P. Wunz, and P.A. Kollman. *J. Med. Chem. 31*, 1612 (1988); Rao, S.N., and W.A. Remers. *J. Med. Chem. 33*, 1701–1707 (1990).
27. IUPAC-IUB Joint Commission on Biochemical Nomenclature *Eur. J. Biochem. 131*, 9–15 (1983).
28. Seeman, N.C., J.M. Rosenberg, F.L. Suddath, J.J.P. Kim, and A. Rich. *J. Mol. Biol. 104*, 109–144 (1976).
29. Ferrin, T.E., C.C. Huang, L. Jarvis, and R. Langridge. *J. Mol. Graph. 6*, 2–12 (1988); Ferrin, T.E., C.C. Huang, L. Jarvis, and R. Langridge. *J. Mol. Graph. 6*, 13–27 (1988); Ferrin, T.E., G.S. Couch, C.C. Huang, E.F. Pettersen, and R. Langridge. *J. Mol. Graph. 9*, 27–32 (1991).
30. SYBYL version 5.32. Molecular modeling software developed and released by Tripos Associates, Inc., St. Louis MO 63144. Copyright 1986, 1989, 1990 and 1991.
31. Insight/Discover is a registered trademark of BIOSYM Technologies, Inc., 10065A, Barnes Canyon Road, San Diego, California.
32. Quanta/CHARMm is the computer graphics and computational chemistry package developed and released by Polygen Corporation, Waltham, Massachusetts. CHARMm is a registered trademark of Polygen Corporation and is a version of CHARM written and developed at Harvard University; Brooks, B.R., R.E. Bruccoleri, B.D. Olafson, D.J. Statos, S. Swaminathan, and M. Karplus. *J. Comp. Chem. 4*, 187 (1983).
33. Mohamadi, F., N.G.J. Richards, W.C. Guida, R. Liskamp, M. Lipton, C. Caufield, G. Chang, T. Hendrickson, and W.C. Still. *J. Comp. Chem. 11*, 440–467 (1990).
34. Singh, U.C., P.K. Weiner, J.W. Caldwell, and P.A. Kollman. *Computational Program AMBER(UCSF) Version 3.0*, Department of Pharmaceutical Chemistry, University of California, San Francisco CA (1986).

35. van Gunsteren, W.F., and H.J.C. Berendsen. *Groningen Molecular Simulations Library*, University of Groningen, The Netherlands (1987); van Gunsteren, W.F., and H.J.C. Berendsen, *Mol. Phys. 34*, 1311 (1977); Berendsen, H.J.C., J.P.M. Postma, W.F. van Gunsteren, A. DiNola, and J.R. Haak *J. Chem. Phys. 81*, 3684 (1984).
36. Jorgensen, W.L., *BOSS Version 2.8*; Yale University: New Haven, CT (1990).
37. Bernstein, F.C., T.F. Koetzle, G.J.B. Williams, E.F. Meyor, Jr., M.D. Brice, J.R. Rodgers, O. Kennard, T. Shimanouchi, and M. Tasumi. *J. Mol. Biol. 112*, 535–542; Abola, E.E., F.C. Bernstein, S.H. Bryant, Koetzle, and J. Weng. Protein Data Bank. In *Crystallographic Databases—Information Content, Software Systems, Scientific Applications*, F.H. Allen, G. Bergerhoff, and R. Sievers, eds., 107–132, Data Commission of the Int'l Union of Crystallography, Bonn, Cambridge, Chester (1987).
38. Arnott, S., P. Campbell-Smith, and R. Chandrasekaran. In *CRC Handbook of Biochemistry*, Fasman, G.D. ed., Vol. 2, 411–422, CRC, Cleveland, OH (1976).
39. Tomasz, M., R. Lipman, J.K. Snyder, and K. Nakanishi. *J. Amer. Chem. Soc. 105*, 2059–2063 (1983), Tomasz, M. *Ciochim. Biophys. Acta 213*, 288–295 (1970).
40. Kaplan, D.J., and M. Tomasz. *Biochemistry 21*, 3006–3013 (1982).
41. Jain, S.C., and H.M. Sobell. *J. Biomolec. Str. Dyn. 1*, 1161–1177; *J. Biomolec. Str. Dyn. 1*, 1179–1194 (1984).
42. Rao, S.N., and P.A. Kollman. *Proc. Natl. Acad. Sci. USA 84*, 5735–5739 (1987).
43. Singh, U.C., and P.A.J. Kollman. *Comp. Chem. 5*, 129–144 (1984).
44. Remers, W.A., M. Mabilia, and A.J. Hopfinger. *J. Med. Chem. 29*, 2492 (1986); Boyd, F.L., S. Cheatham, W.A. Remers, G.C. Hill, and L.H. Hurley. *J. Amer. Chem. Soc. 112*, 3279–3289 (1990); Boyd, F.L., D. Stewart, W.A. Remers, M.D. Barkley, and L.H. Hurley. *Biochemistry 29*, 2387–2403 (1990).
45. Lybrand, T.P., A. Dearing, P.K. Weiner, and P.A. Kollman. *Nucl. Acids Res. 9*, 6995–7011 (1982); Lybrand, T.P., and P.A. Kollman, *Biopolymers 24*, 1863–1879 (1985); Singh, U.C., N. Pattabiraman, R. Langridge, and P.A. Kollman. *Proc. Natl. Acad. Sci. USA 83*, 6402–6406 (1986).
46. Mercado, C.M., and M. Tomasz. *Biochemistry 16*, 2040–2046 (1977). Tomasz, M., J.K. Baron, C. Magliozzo, D. Tucker, E.M. Lafer, and B.D. Stollas. *Proc. Natl. Acad. Sci. USA 80*, 2874–2878 (1983); Wu, H.M., N. Dattagupta, and D.M. Crothers, *Proc. Natl. Acad. Sci. USA 78*, 6808–6811 (1981).
47. Rao, S.N., J.W. Keepers, and P.A. Kollman, *Nucleic Acids Res. 12*, 4789–4807 (1984); Rao, S.N., and P.A. Kollman, *Photochem. Photobiol. 42*, 465–475 (1985).
48. Haasnoot, C.A.G., C.W. Hilbers, G. van der Marel, J.H. van Boom, U.C. Singh, N. Pattabiraman, and P.A. Kollman, *J. Biomolec. Str. Dyn. 3*, 843–857 (1986).
49. Pattabiraman, N. *Biopolymers 25*, 1603–1606 (1986).
50. Cieplak, P., S.N. Rao, C. Helene, T. Montenay-Garestier, and P.A. Kollman, *J. Biomol. Str. Dyn. 5*, 361–382 (1987).

4
Sequence Specificity of Drug-DNA Interactions

James C. Dabrowiak, Allison A. Stankus, and Jerry Goodisman
Syracuse University, Syracuse, New York

I. INTRODUCTION

DNA sequencing methodology has revolutionized the study of drug-DNA interactions. In particular, the development of high-resolution polyacrylamide gels and efficient procedures for labeling and sequencing DNA have made it possible to identify the sites of drug action on naturally occurring

DNA substrates. Since the first review article by one of the authors in 1983 (1), the literature documenting the sequence specificity of anticancer drugs, carcinogens, and mutagens has greatly expanded. In this chapter we will outline the various methods used for studying drug-DNA interactions using sequencing methodology. No attempt will be made to exhaustively review the extensive literature on the subject, but rather the various techniques will be illustrated by discussing a number of published examples. In addition to the classic textbook on the subject by Gale *et al.* (2) and the more recent work by Niedel and Waring (3), additional information on specificity can be found in the excellent reviews by Dervan (4), Boyd and Hurley (5), and Pullman (6).

The first part of the chapter will focus on the qualitative aspects of specificity. It will be shown that the method for determining sequence specificity is dependent on the molecular mechanism by which the drug interacts with DNA. Specific examples include drugs and agents which nick, alkylate, or covalently bind to DNA as well as those which interact with the polymer in an equilibrium fashion. In addition to describing how to identify drug interaction sites using sequencing methodology, this chapter will outline the method used to obtain direct thermodynamic information through quantitative footprinting analysis. This technique allows for the measurement of individual site isotherms, which can be used to calculate drug-binding constants as a function of sequence on DNA. Binding kinetics are known to play a central role in the therapeutic effectiveness of many anticancer drugs. The use of DNA sequencing methodology in measuring sequence-dependent dissociation rates for drugs will also be presented and discussed.

II. METHODS FOR STUDYING SEQUENCE SPECIFICITY

A. General Qualitative Techniques

1. *Drugs that Produce DNA Strand Scission*

Determination of a drug's binding specificity generally begins by selecting a DNA restriction fragment and labeling it with a radioactive atom such as ^{32}P, followed by incubation of the fragment with the drug of interest. The segments of DNA used in these studies are typically 50–200 base pairs long, and they have the radioactive atom at a unique point in the sequence, generally at one end of one DNA strand. If the drug can cleave DNA, the oligonucleotides produced by drug action can be separated via electrophoresis in a high-resolution polyacrylamide gel and visualized using autoradiography (7) (Figs. 1, 2). The interaction sequences of the drug can be easily identified by subjecting the same fragment to the Maxam and Gilbert (8) sequencing protocols and separating the resulting DNA oligomers along

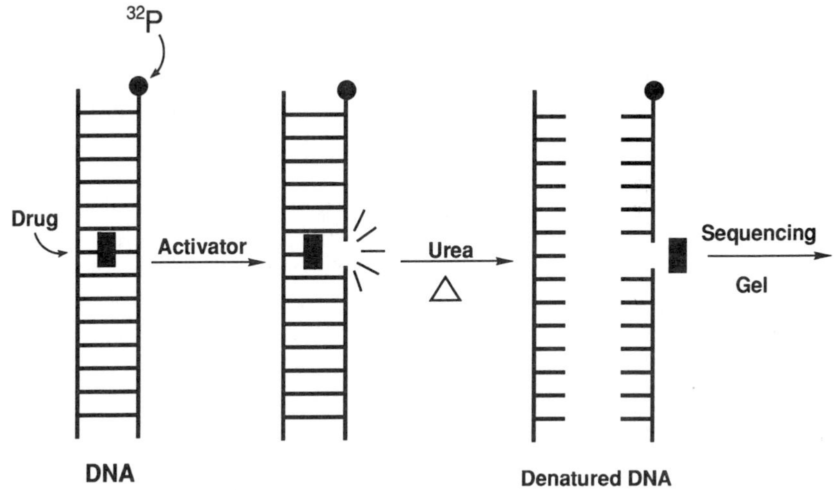

Figure 1 Drugs which produce DNA strand scission. A restriction fragment is selected, and it is single-end labeled with a radioisotope, e.g., ^{32}P (•), using standard techniques. Cleavage of the labeled fragment by the drug in the presence of an activating agent results in a break in the sugar-phosphate backbone of DNA. The cleavage products are separated using a DNA sequencing gel.

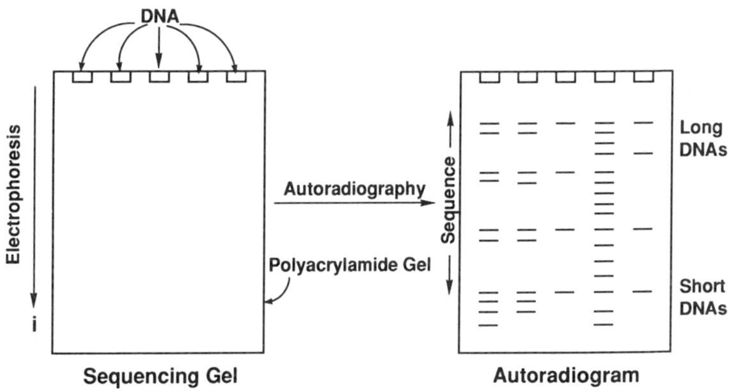

Figure 2 Shown is a cartoon of the electrophoresis and autoradiographic steps in the analysis of oligonucleotide cleavage products. The products from drug action are loaded into the wells of a polyacrylamide gel and electrophoresed in the direction shown. After electrophoresis, the gel is subjected to autoradiography (producing an autoradiogram) to reveal the locations of radioactive DNA in the gel. The location (sequence) of drug-produced breaks can be identified by coelectrophoresis of oligonucleotide fragments produced by cleavage of DNA using sequencing protocols.

with the drug-produced products via electrophoresis. From the intensity of the spots on the autoradiogram it is possible to obtain the relative affinities of the drug for the various sites on the restriction fragment. Although the basic experiment is relatively simple in concept, a number of factors should be considered in order to maximize the amount of useful information which can be obtained.

It is important to terminate the digest reaction early. If the total amount of cleavage on the fragment is low and within the so-called "one-hit" kinetic limit, each fragment which is separated in the sequencing gel will be the result of a single cleavage in the full-length DNA molecule. This ensures that specificity toward the uncleaved substrate DNA is being measured. Practically, this condition is best achieved by stopping the cleavage reaction early so that 70–80% of the original DNA remains uncleaved in the process. In this case the measured intensities are proportional to the degree of site loading and independent of the distance between the site and the radiolabel. This, of course, assumes that the cleavage rate constants for all of the sites are the same and that the intensity of a spot after a certain digest time is proportional to the amount of drug actually bound at the site. If cleavage is extensive, the amount of the shorter oligomers will be increased at the expense of the larger products. In this case using uncorrected intensities to assign relative affinities along the length of the polymer will not be valid.

In cases where the specificities of several agents are being compared, it is important to adjust the conditions so that all produce the same level of cleavage (ideally single-hit kinetics) on the labeled fragment. When this occurs, one can directly compare the intensities of bands derived from different cleavage agents and determine how effective different drugs are in cleaving a given DNA site. Their ability to cleave DNA in general can be obtained by noting the amount of drug which needs to be added to produce a common level of cleavage on the fragment.

In certain cases the nature of the cleavage chemistry remaining on the DNA strand after scission can be identified using electrophoresis techniques. This is done by comparing the electrophoretic mobilities of drug-produced fragments with the corresponding oligomers having known terminal chemistry. If the mobilities of the shortest fragments from both sources are the same, the fragments are likely to have the same terminal groups.

2. *Drugs that Require Posttreatment of DNA to Produce a Strand Break*

A large class of agents interact with DNA through alkylation. Unlike the previous case examples, treatment of DNA with alkylating drugs does not immediately cause a DNA strand break. In this case, it is necessary to treat the drug-modified DNA with some other agent to produce a break in the polymer so that the sites of interaction can be analyzed using a DNA sequencing gel. An excellent discussion of alkylating agents and how they chemically modify DNA can be found in the recent reviews by Warpehoski and Hurley (9), Neidle and Waring (3), and in the work of Church et al. (10).

Numerous alkylating agents interact with DNA by alkylating N(7) of guanine in a manner analogous to the Maxam and Gilbert sequencing reaction for guanine (8). Aklylation at N(7) of the purine base ultimately leads to depurination at the site and the creation of an alkali labile region of DNA. Treatment of DNA which has been previously exposed to the alkylating agent with hot piperidine results in a strand break at the site of alkylation. The DNA terminal chemistry in this case is identical to that produced by the Maxam and Gilbert sequencing protocols, i.e., 5′, 3′-phosphate.

In the case of adenine, alkylation, which occurs at N(3), located in the minor groove or N(7) in the major groove of double helical DNA, also results in an apurinic site (8). Treatment of the site with base produces a break having 3′- and 5′-phosphate terminal chemistry. Although the chemistry is less certain, alkylation at sites on cytosine and thymine can also occur (10). As with guanine and adenine this type of substitution weakens the N-glycosyl bond, ultimately causing a strand break in DNA.

3. *Drugs that Interact Covalently Without a Strand Break*

For the metal-containing anticancer drug cisplatin and certain alkylating agents such as mitomycin C, interaction with DNA does not lead to a strand break. Identifying the sites of binding using a sequencing gel is made difficult by the fact that normal workup and denaturing protocols do not remove the drug from DNA. Since drug-modified fragments have electrophoretic mobilities which are different than the corresponding oligomers without the attached drug, it is generally difficult to gain sequence information in these experiments.

An interesting technique which obviates the need to produce a strand break in DNA or to remove drug in order to examine specificity is the primer extension method developed by Gralla and co-workers (11) (Fig. 3). In this method an *unlabeled* double-stranded DNA restriction fragment can be treated with a drug such as cisplatin which covalently interacts at oligoguanine sequences of DNA. Following platination, the two strands of DNA can be separated by heat denaturation. In order to observe the sites of platinum binding on the now single-stranded DNA polymers, a single end-labeled primer (a small DNA oligomer 10–20 nucleotides in length) is hybridized to the proper portion of one of the single-stranded DNA molecules. The sequence of the primer is chosen such that it will hybridize to one of the strands in a region where specificity information is being sought. If the DNA is now treated with *E. coli* DNA polymerase in the presence of the

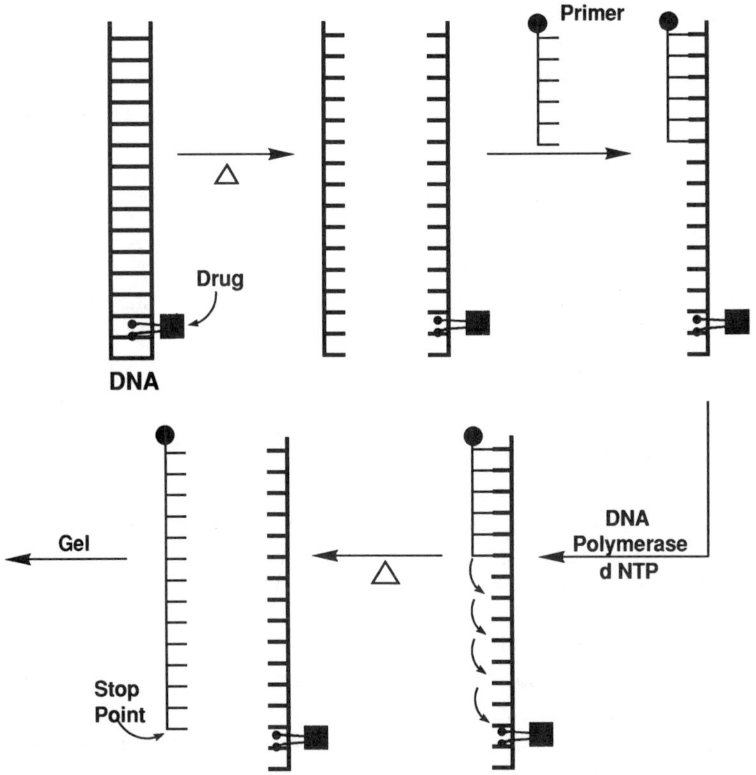

Figure 3 Shown is the primer extension method for identifying the binding site of a drug which interacts in a covalent manner on DNA. An *unlabeled* DNA restriction fragment is exposed to drug action, and the resulting adduct subjected to heat denaturation. A *radiolabeled* primer is annealed to the "cold" drug-bound strand and extended with DNA polymerase to produce an oligomer with a common labeled end and a "cold" end which terminates at the location of the bound drug on the template strand (stop point). The resulting oligonucleotides are analyzed using a standard DNA sequencing gel.

four mononucleotide triphosphates, the radioactive oligonucleotide primer will be extended in its 3′ direction. When DNA polymerase encounters a blocking lesion on the template strand, i.e., the attached platinum nucleus, it stops DNA synthesis, creating a radioactive DNA strand which still has the 5′ end of the original primer but now has a unique 3′ end corresponding to the blockage site. In this method no break in DNA is created, and it is only the termination points of the synthesized strand which mark the positions of drug binding on the opposite strand. The precise position of blockage may be identified by a separation of the extended primers on a high-resolution DNA sequencing gel followed by autoradiography. Only extended primers are observed upon autoradiography, since the ^{32}P end label on the primer is the only source of radioactivity.

Other methods have also been used to detect cisplatin binding sites on DNA (12,13). In one assay, DNA which has been platinated is subjected to digest using exonuclease III. Since this enzyme degrades DNA from its 5′ terminus, the sites of platinum binding appear as "stop points" in the degradation process. In order to identify the sites of binding using a DNA sequencing gel, it is first necessary to treat the digested DNA with cyanide ion or thiourea. These agents attack the DNA-bound platinum ion and remove it from the polymer. Electrophoresis of the resulting oligonucleotides along with the Maxam-Gilbert-produced products allows identification of the platinum binding sites on DNA. It was found that the exonuclease was inhibited from cutting one to three nucleotides preceding runs of two or more guanines on the same strand of DNA.

Certain drugs such as mitomycin C react by cross-linking opposing strands of DNA. In spite of this complication the specificity of the drug at single-nucleotide resolution has recently been reported by Millard *et al.* (14). After cross-linking a single end-labeled DNA restriction fragment with mitomycin C under reducing conditions, the modified fragment was subjected to cleavage by Fe(II) EDTA–hydrogen peroxide–ascorbic acid (15) (Fig. 4). The cleavage reaction by the latter reagent is generally believed to be mediated by hydroxyl radical and results in degradation of the sugar, breaking the original DNA chain to yield two new DNA fragments, one bearing the 5′ terminal phosphate and the other a mixture of 3′ terminal phosphates and phosphoglycolate. If the duplex is cross-linked by a mitomycin C molecule, cleavage between the site of cross-linking and the radiolabel results in short radioactive DNA oligomers. If cleavage is on the radiolabeled strand, but on the portion of DNA such that the cross-linked site is between the cleavage site and the label, the resulting DNA oligomer after denaturalization will be relatively long. This is because the cross-link will join the radiolabeled strand to the cold strand, thus producing a long DNA oligomer. By examining the ladder of oligomers produced in these experiments using electrophoresis, it is possible to identify the DNA sites which are adjacent to the drug-produced cross-link.

4. Drugs that Do Not Leave a Permanent Chemical Record on DNA

The largest class of DNA-binding drugs include those which interact with the polymer in an equilibrium manner. Unlike the previous examples, these agents do not leave a permanent record of their residence position on DNA, and mapping drug binding sites is done by cleavage of DNA in a process

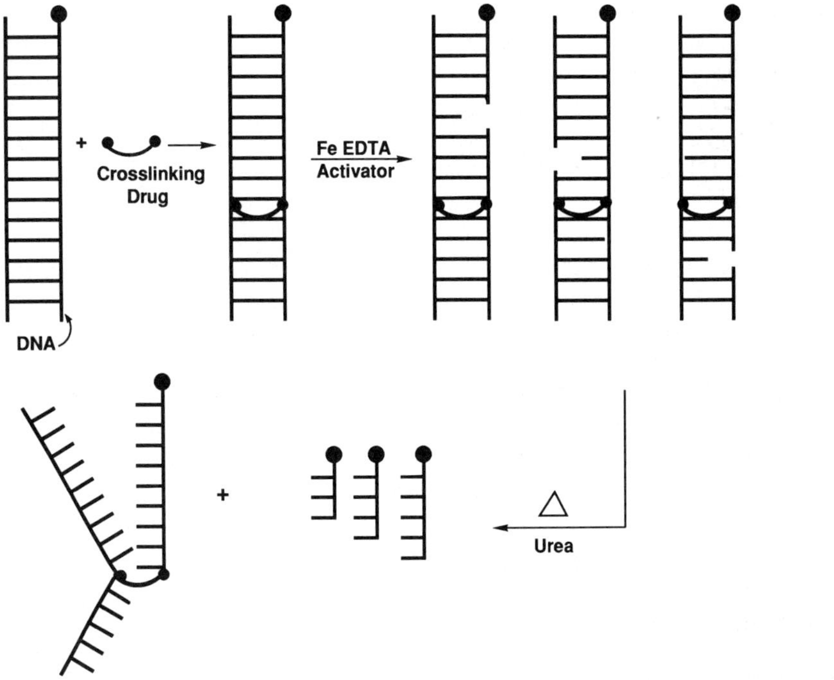

Figure 4 Cartoon showing how to identify the locations of drugs which cross-link both DNA strands. An end-labeled DNA restriction fragment containing drug-produced cross-links is subject to cleavage by a cleavage agent. The denatured products consist of a "nested set" of radiolabeled oligomers, the longest of which terminates at the site of the cross-link, and high molecular weight DNA having the "cold" strand linked by the drug to the labeled strand. These products are separated using a standard DNA sequencing gel. The longest oligonucleotide members of the "nested set" identify the location of the crosslinks.

termed the footprinting experiment (Fig. 5). In this experiment a drug is equilibrated with a single end-labeled DNA restriction fragment, and the resulting drug-DNA complex is treated with an agent capable of cleaving the fragment at many sites. The sites of drug binding are identified by measuring the amounts of the fragments produced in the cleavage reaction, after separating them using polyacrylamide gel electrophoresis. If the drug is able to prevent the cleavage agent from binding in a certain region of DNA, the extent of cutting at sites in that region will be less than in the absence of the drug. The autoradiogram is inspected for regions of omission (footprints) and in some cases scanned with a microdensitometer. The densities can be

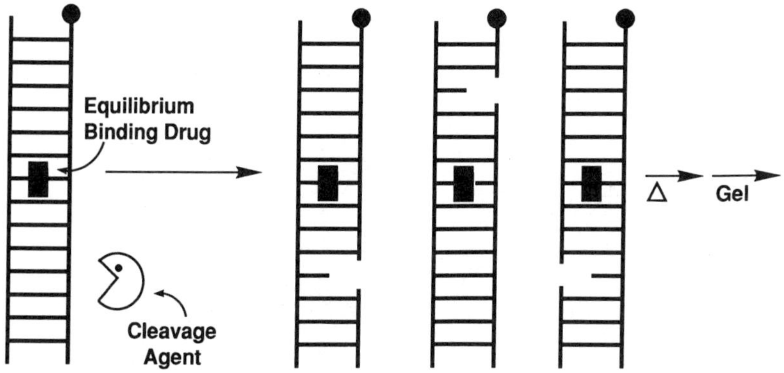

Figure 5 Identifying binding sites of equilibrium binding drugs with footprinting analysis. A drug-bound DNA restriction fragment which is single-end labeled is subjected to cleavage by a chemical or enzymatic cleaving agent. Following heat denaturing, the oligonucleotide products are analyzed using a standard DNA sequencing gel. The sites of drug binding are identified by a reduction in the amounts of oligomers which terminate within a drug-binding site as exhibited by a footprint on the autoradiogram.

used to calculate site-specific binding constants for the drug on the restriction fragment. A typical footprinting autoradiogram is shown in Figure 6.

A number of fundamental points concerning the footprinting experiment should be made. As has been shown by Goodisman and Dabrowiak (16), the amount of protection within a drug-binding site, a quantity which is central to measuring binding constants from footprinting data, is *independent* of the association and dissociation rates of a drug for a given DNA site. As long as the system is at equilibrium prior to beginning the digest and remains so during the course of the reaction, the autoradiographic spot intensities associated with a site, after a minor correction, reflect the occupancy of the site by the drug molecule. This is because the various kinetic processes leading to formation of the oligomer which appears in the gel and "scores" the site are only dependent on the *ratio* of the association and dissociation rates of the drug and thus the drug's equilibrium constant.

The reporting of site loading information by the probe or cleavage agent is *independent* of the mechanism by which the probe cleaves DNA. For example, the enzyme DNase I is an often-used footprinting agent which binds in the minor groove and cleaves by hydrolyzing the phosphodiester linkage of DNA (17). The reagent Fe-MPE, on the other hand, is an intercalating agent which facilitates hydroxyl radical attack on the sugar groups of DNA (18). Despite the dramatic difference in binding and cleavage mecha-

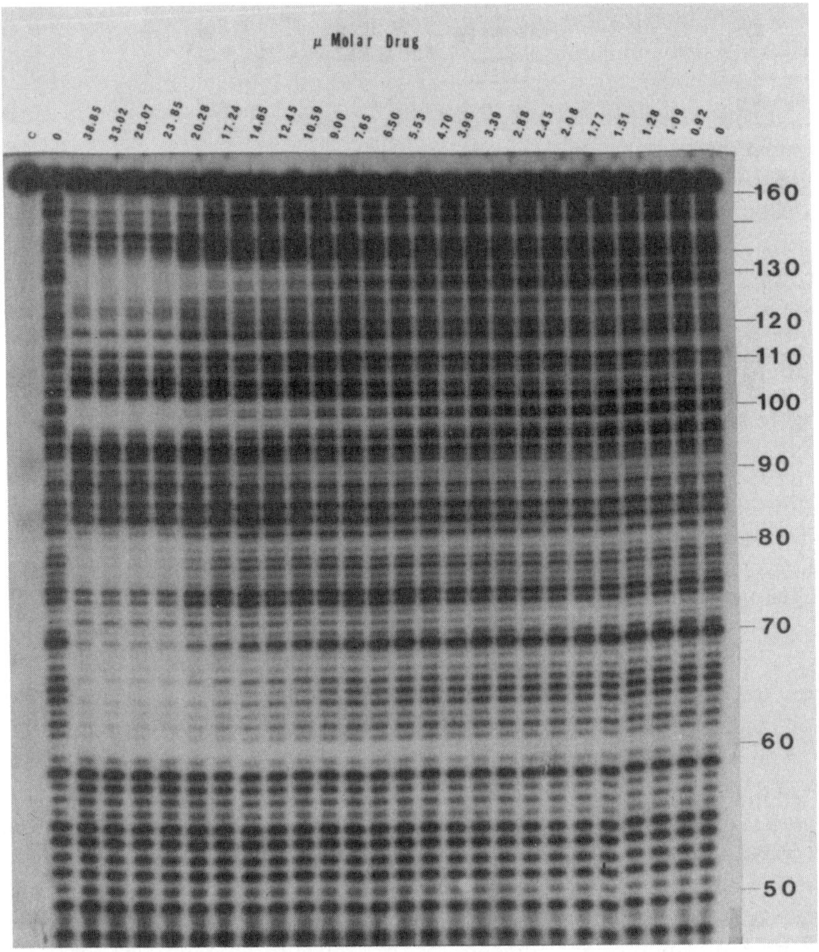

Figure 6 Footprinting autoradiogram involving the anticancer drug actinomycin D, a 139-base-pair fragment of pBR322 DNA, and the cleavage agent DNase I. The sequence of the fragment is indicated by the index on the right of the autoradiogram while the drug concentrations are given at the top. Sites of drug binding are revealed by inhibitions in cleavage as the concentration of actinomycin is increased. Additional experimental details can be found in Ref. 120.

nisms of these two probes, footprinting experiments for both can be analyzed in the same manner to derive binding constants (19).

While the reporting of site loading information is independent of the binding/cleavage mechanism of the probe, it is dependent on the probe's specificity. Although this point will be discussed in greater detail in the section addressing quantitative footprinting methods, probes which have low specificity, as evidenced by the fact that they cleave at nearly every nucleotide position, require a different model from probes with high specificity to derive values of the binding constants. However, in the qualitative experiment, where only the ranking of affinities is sought, the specificity of the cleavage agent is not a major concern.

Cleavage specificity of the probe plays an important role in the cleavage rate enhancements often observed for sites between drug-binding sites in the footprinting experiment. These are commonly attributed to drug-induced structural changes in DNA. Although structural changes may exist and be detectable through rate processes involving the cleavage agent, the nature of the equilibrium between the probe and DNA needs to first be addressed in order to definitively identify those DNA sites having altered structure. This topic will also be presented in greater detail in a later section.

Finally, the detection of a drug-binding event in the footprinting experiment does not depend solely on the blocking of the cleavage agent at the site by the drug. While steric inhibition is the clearest way to explain the existence of a footprint caused by the drug, other mechanisms can be used to identify the sites of binding. For example, binding in the minor groove may alter DNA structure and/or the distribution of ions and solvent in the environs of the site. Thus, it should be possible to detect drug binding not only with probes which attack the minor groove but also with those which attack the major groove and/or the phosphate sites on DNA. It should be kept in mind that it is necessary only to observe a change in the cleavage rate of the probe, which depends on drug concentration, in order to monitor binding in the footprinting experiment.

B. Sequence Specificity and Binding of Drugs by Class

In the following section the various methods used for studying drug-DNA interaction using sequencing methods will be illustrated with specific literature examples. No attempt will be made to be comprehensive, but the reader will be provided with an overview of the drugs' binding/cleavage mechanism; more detailed information can be obtained from the references which are cited. In addition to information derived from sequencing gels, where relevant the results of x-ray structural analyses will also be presented and discussed.

1. Drugs that Produce DNA Strand Scission

Bleomycin. The antitumor antibiotic bleomycin (Fig. 7, BLM) is the prototypic DNA cleavage agent. The drug, a naturally occurring glycopeptide, is in wide clinical use for the treatment of squamous cell tumors, malignant lymphomas, and testicular cancer (20). The major shortcoming of bleomycin is that it produces pulmonary side effects which begin as pneumonitis and progress to pulmonary fibrosis, which can be fatal (20). The literature on bleomycin is extensive, addressing every aspect of the drug's molecular mechanism of action, pharmacology, and oncological effects. For additional information the interested reader should consult earlier reviews (20–22) as well as the more recent work by Stubbe and Kozarich (23).

Bleomycin possesses three structural features which are important to its biological mechanism of action. The two sugars of the antibiotic, gulose and mannose, are believed to be essential for biological transport of the drug *in vivo*. Once inside the cell, bleomycin binds to the minor groove of DNA, possibly through an intercalative mode, via its 2-4′-bithiazole moiety (23–26). The drug is a strong chelator of transition-metal ions (27). In the presence of iron ions, which bind to the pyrimidine-imidazole portion of the drug, and reducing agents, bleomycin is capable of causing DNA strand scission in an oxygen-dependent manner (22,28,29). Early studies on the sequence specificity of the cleavage process showed that the Fe complex of the antibiotic most often cleaves at GC and GT sequences of DNA (30–31). The cleavage process involves hydroxyl and/or superoxide radicals which break DNA by removal of the hydrogen on the C-4′ position of the deoxyri-

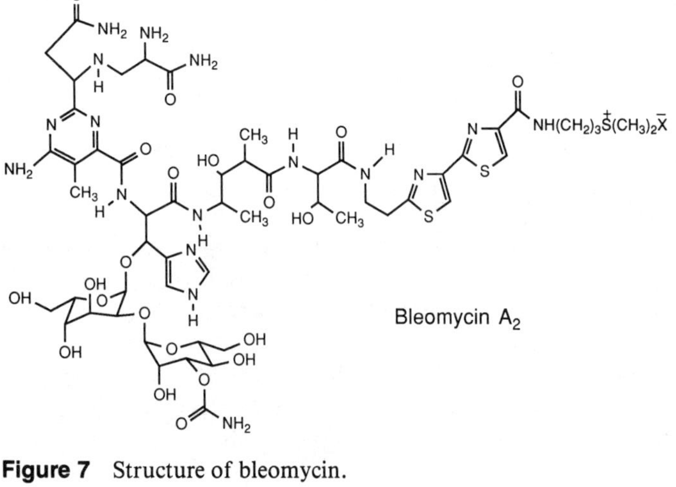

Figure 7 Structure of bleomycin.

bose moiety of a pyrimidine nucleotide adjacent to guanosine. Two types of products have been isolated and identified during Fe-BLM mediated degradation of DNA: free nucleic acid bases and base propenals. Recent studies by Stubbe and Kozarich (32) involving placement of radioactive labels on the deoxyribose moiety of DNA demonstrated that the cleavage process proceeds by two distinct routes, only one of which requires molecular oxygen. When the radioactive atom is located on the 5′ end of a DNA strand, the BLM-produced cleavage products exhibit a slightly greater electrophoretic mobility than their Maxam-Gilbert-produced (8) counterparts. This is because BLM damage to DNA leaves a phosphoglycolate moiety on the 3′ position of the sugar adjacent to the site of pyrimidine release rather than a phosphate group, as is the case with the Maxam and Gilbert cleavage product. In contrast, if the radioactive label is located on the 3′ end of a DNA strand, the BLM-produced product comigrates with the Maxam-Gilbert-produced oligomers, thus showing that BLM damage to DNA leaves a phosphate group on the 5′ side adjacent to the site of pyrimidine release.

In addition to Fe-BLM the sequence specificity of Co-BLM has also been investigated. Interest in Co-BLM stems primarily from the fact that ^{57}Co is a γ-emitting metal, and when bound to bleomycin, the metal complex may be a clinically useful diagnostic tool for detecting tumors (33). It has been observed that exposure of Co-BLM to light in the absence of oxygen induces cleavage of DNA (34). This process occurs with the release of oligonucleotides possessing 3′-phosphoglycolate termini without the formation of base propenals. Photodamage of DNA by Co-BLM also produces sites having a missing DNA base. Posttreatment with strong base and high temperature produces a break in the DNA strand at the modification site. Initial studies by Chang and Meares (34,35) showed that the sequence specificity of light-activated Co-BLM is the same as that of Fe-BLM, i.e., cleavage at GT and GC sequences of DNA. However, a reinvestigation of this system using bleomycin A5 indicated that light-activated Co-BLM showed a preference for cutting at the dinucleotide GA in addition to cleavage at every GT dinucleotide (36).

In addition to bleomycin, the cleavage specificity of a number of related antibiotics has been studied. Phleomycin, which has a reduced double bond in the bithiazole portion of the antibiotic, exhibits cleavage specificity which is the same as that of bleomycin (37). However, talisomycin which possesses an L-talose group, exhibits cleavage specificity which is different from that of bleomycin and phleomycin (38). Since talisomycin shows a preference for GT and GA sequences of DNA, it appears that the L-talose group of the antibiotic may in some fashion be bound to DNA, thus influencing cleavage specificity.

Esperamicins and Calicheamicins. The esperamicins and calicheamins (Fig. 8) are a potent new class of anticancer agent (39,40). The compounds are natural products which possess an unusual 1,5-diyn-3-ene ring system, referred to as the warhead, an aromatic moiety, and four sugar groups. As a class these compounds show extraordinary potency against various murine tumors and are a 1000-fold more active than adriamycin (41,42). In the presence of thiols, the trisulfide moiety attached to the warhead portion of the drug undergoes conversion to a thiolate anion. The anion in turn enters into an internal Michael addition reaction ultimately converting the 1,5-diyn-3-ene moiety into a phenylene diradical which is capable of cleaving DNA. Since two radicals are formed from a single activation event, an

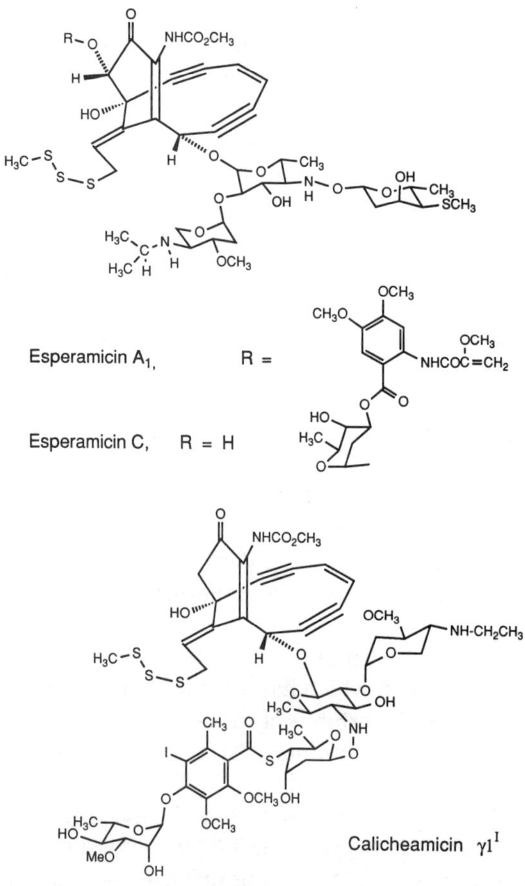

Figure 8 Structures of esperamicin and calicheamicin.

activated drug molecule can simultaneously react with opposing DNA strands to produce a double-strand break. Recent studies with closed circular PM2 DNA (43) have shown that the probability of double-versus single-strand cleavage is dependent on the number and arrangement of sugars attached to the warhead portion of the drug. While the arrangement of sugars in the calicheamin series of compounds is optimal for double-strand cleavage of DNA (41,42), the sugar arrangement for esperamicin A_1 leads to substantial single-strand breakage events. However, if the deoxyfucose sugar and its attached aromatic group are absent, e.g., esperamicin C, cleavage occurs mainly through a double-strand process.

Recent studies using DNA sequencing methods show that the calicheamin series of compounds are more specific than the esperamicins (41,42). Cleavage of a 275-base-pair fragment of pBR322 DNA with calicheamin γ_1^I showed that the compound most often cleaves at a tetranucleotide sequence TCCT. This specificity sharply contrasts with that of the esperamicins, which cleave at many sequences on DNA (41,44). Cleavage involving radiolabeling opposing strands of DNA and competition with other DNA binding drugs (44) indicate that the esperamicins and calicheamins bind via the minor groove of DNA. Although the details of the cleavage process have not yet been elucidated, the compounds probably attack the sugar group of DNA. NMR studies indicate that the phenylene diradical of calicheamin γ_1^I abstracts hydrogen atoms from DNA and not from solvent which surrounds the polymer (45). Greater number of oligonucleotide cleavage products are observed in 3′-end-labeled experiments than are observed for 5′-end labeled experiments. This suggests that the drugs produce more than one terminus on the 5′ side adjacent to the site of strand scission. Based on the electrophoretic mobilities of the Maxam and Gilbert produced cleavage products relative to those resulting from esperamicin-calicheamin cleavage of DNA, a phosphate group most likely occupies the 3′ side of the scission site (41,42).

Neocarzinostatin. Neocarzinostatin (NCS) (Fig. 9) is the first antitumor antibiotic found to contain a diyne-ene chromophore (46,47). The drug, isolated from culture filtrates of *Streptomyces*, consists of a biologically active chromophore (M_r = 659) and an apoprotein (M_r = 11,000), which serves as a carrier and protection for the chromophore from damage *in vivo*. The NCS chromophore binds to the minor groove of B-form DNA in a two-step process (48). External binding of the drug to the polymer involves electrostatic interactions between the positively charged amino sugar of the chromophore and the negatively charged phosphate backbone of DNA and is followed by intercalation of the warhead between adjacent DNA base pairs. Addition of a thiol cofactor such as dithiothreitol results in

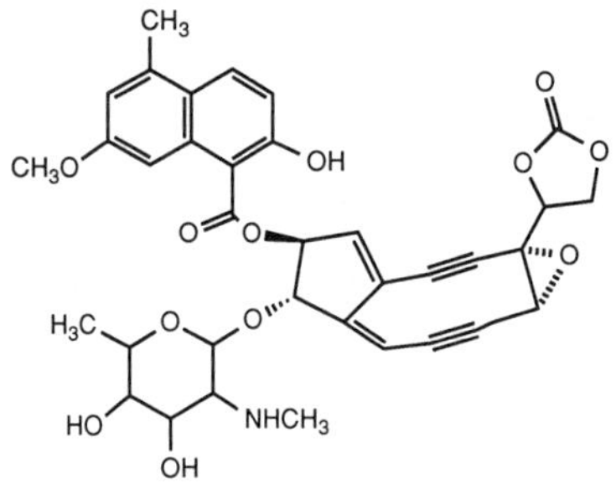

Neocarzinostatin (Chromophore)

Figure 9 Structure of the chromophore of neocarzinostatin.

opening of the epoxide group of the chromophore. Subsequent rearrangement creates an intermediate bearing radical centers at positions C-2 and C-6. This diradical species is implicated as the active form of the drug (49,50).

Strand breakage by NCS is initiated by hydrogen atom abstraction by the activated form of the drug at the C-5′ position of the deoxyribose. This O_2-dependent process exhibits sequence selectivity (46,47). When double-stranded restriction fragments were 5′-end-labeled with ^{32}P and cleaved by NCS, integration of the spot intensities of the products separated by polyacrylamide gel electrophoresis revealed a reactivity order of base attack of $T > A \gg C > G$. Comparison of band migration rates to those of products of Maxam-Gilbert cleavage (8) of the same substrates indicated that the 3′-end chemistry involves production of phosphate groups. However, when labeling was carried out on the 3′-ends of the fragments, the bands produced by drug cutting migrated approximately two nucleotides more slowly than the corresponding Maxam-Gilbert-produced bands. Damage to the 5′-end of DNA involves oxidation of the C-5′ carbon of the nucleotide, producing 5′-aldehyde groups. Studies using synthetic oligonucleotides six base pairs in length indicate a preferred binding site of GN_1N_2, where N_2 = T or A (51). The chromophore, which is intercalated at the GN_1 site, is orientated for attack at C-5′ of the deoxyribose of the N_2 nucleotide.

Another process of strand breakage by an alternative pathway involves release of free bases (52). Polymer decomposition products include base ($T > A \gg C > G$), formate, and a 4-carbon sugar fragment. The gaps left at the scission site contain phosphate groups at both ends. Less than 20% of strand breakage by NCS is accounted for by this mechanism.

Sequence selectivity is also observed in the ability of NCS to create double-stranded breaks in the DNA polymer (46). For example, in certain sequences such as AGC : TCG radical attack at C-1 of the cytosine residue of the top strand leads to base release and formation of a 2-deoxyribonolactone with intact phosphodiester linkages in the DNA strand (53,54). Strand cleavage then can be effected by treatment with alkali, leaving phosphate groups at each end of the strand break. This type of reactivity is found to be accompanied by direct strand breakage at the thymidine residue on the bottom strand two nucleotides to the 3′ side of the basic site (52).

Two types of NCS-DNA adducts have been identified (46). An as yet uncharacterized labile adduct is postulated to involve a modified portion of the deoxyribose moiety linked to the drug. A covalent adduct resistant to decomposition even by treatment with strong acid is believed to originate from coupling of a radical on the C_{12} subunit of NCS and a radical at the C-5′ position of the deoxyribose. The same base selectivity is observed in covalent adduct formation as is found in the strand breakage pathway, indicating that both types of DNA lesion may originate from similar DNA damage (46).

2. *Drugs that Require Posttreatment of DNA To Produce a Strand Break*

As noted previously, a number of drugs interact with DNA via alkylation of the heterocyclic bases. In certain cases this leads to a weakening of the bond between the sugar and the base, causing the loss of the base and the creation of a site which is sensitive to alkali. Treatment of the site with hot base results in a break in the DNA strand. The resulting DNA oligomers can then be separated using a standard sequencing gel.

Nitrosoureas. Methylnitrosourea (MNU) and 1-(2-chloroethyl)-1-nitrosourea (CNU, see Fig. 10) are members of a class of agents active against cancer (55,56). These and the related compounds nitrogen and sulfur mustards, aziridines, and oxiranes exert their cytotoxic effects by alkylating sites on DNA (3). The most important location of alkylation is N(7) of guanine, which lies within the major groove of double helical DNA. However, drugs such as CNU, which contain two alkylating functions, are able to alkylate N(7) of guanine as well as N(3) of cytosine on the opposing DNA strand. This creates a cross-link between both strands which is believed to be responsible for the anticancer activity of the bifunctional nitrosoureas (3).

Figure 10 Structures of methylnitrosourea (MNU) and 1-(2-chloroethyl)-1-nitrosourea (CNU).

An interesting group of alkylating agents possess an *N*-(2-chloroethyl)-*N*-nitrosourea function covalently linked to a dipeptide (10). These compounds bind within the minor groove of DNA, and they induce DNA cleavage at cytosine, adenine, and thymine but not guanine residues which are adjacent to the sites of binding. Thus, linking the CNU moiety to minor groove binders is a way to control the bases and groove attacked by the alkylating agent.

Aflatoxins. The aflatoxins (Fig. 11) are members of an important class of potent carcinogens and mutagens which alkylate DNA (57). Similar to the nitrosoureas, the compounds interact through the major groove of DNA and alkylate N(7) of guanine. This produces an alkali labile site which upon treatment with hot piperidine results in a DNA strand break. The alkylation reaction requires activation of aflatoxin to its reactive form, the 8,9-epoxide, which then alkylates N(7) of guanine. However, the reactivity of all guanines on random sequence DNA is not the same, and a recent study involving aflatoxin B_1 using DNA sequencing methodology showed that reactivity occurs in the pentanucleotide sequence, 5′-WX$\underline{G}$YZ-3′ (58). In this sequence the central $\underline{G}$ represents the active guanine, and W, X, Y, and Z can be any of the nucleotide bases. Using microdensitometry to quantitate the sequencing autoradiogram, it was determined that the X (5′ side) base and the Y z(3′ side) base are most influential in determining guanine reactivity. The influence of the bases on the 5′-position (X) is 5′-G (1.0) > C(0.8) > A(0.3) > T(0.2), while the influence of the bases in the 3′-position (Y) is 3′-G(1.0) > T(0.8) > C(0.4) > A(0.3).

3. *Drugs that Interact Covalently Without a Strand Break*

Mitomycin C. Mitomycin C (Fig. 12) is a well-studied antitumor antibiotic (59,60). The compound, a natural product, contains a quinone moiety, aziridine, and carbamate groups. Reductive activation of mitomycin results in the expulsion of the elements of methanol from certain carbons in the structure. In the presence of DNA, the activated form of the drug alkylates the 2-NH_2 group of guanine, which is located in the minor groove. Reduction produces two active sites on mitomycin C, causing the compound to

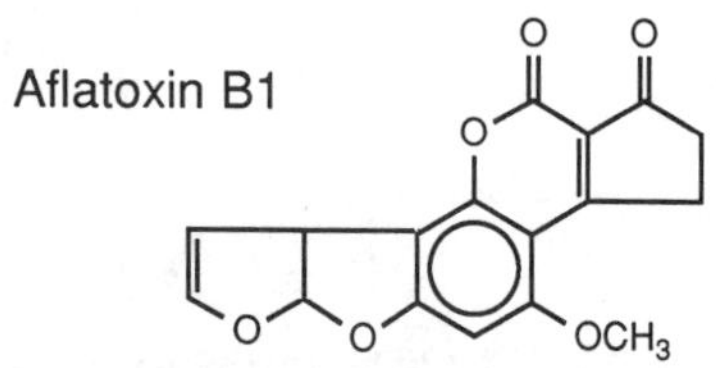

Figure 11 Structure of aflatoxin B_1.

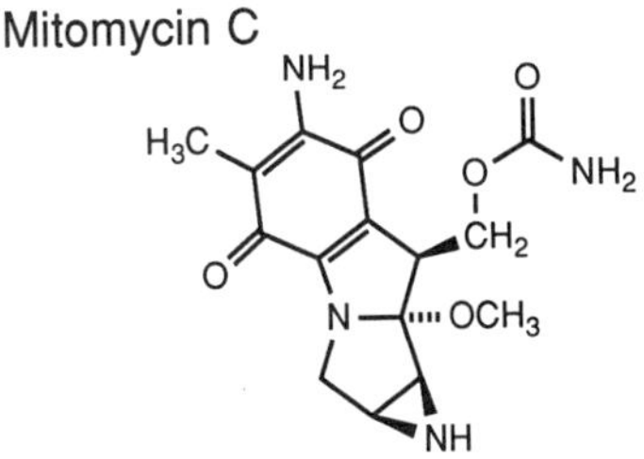

Figure 12 Structure of mitomycin C.

form cross-links with guanine bases located on adjacent strands of DNA. Since modification of the exocyclic amino group of guanine does not lead to depurination and the formation of alkli labile sites, mitomycin damage to DNA does not result in a strand break. This feature of the drug-DNA interaction makes detection of binding sites somewhat difficult, but techniques have been developed to determine mitomycin C specificity on random sequence DNA.

Using an assay involving cleavage of DNA with Fe(II) EDTA–hydrogen peroxide–ascorbic acid (Fig. 4) Millard *et al.* (14) showed that mitomycin cross-links deoxyguanosine residues at duplex 5′-CG-3′ sites in strong preference to 5′-GC-3′ sites and numerous sites rich in A and T. Sequences flanking the CG site influence cross-linking; for example, 5′-GCGC-3′ is more extensively cross-linked than 5′-TCGA-3′. Molecular mechanics calculations on the monoadduct of mitomycin C at N(2) of deoxyguanosine shows that the cross-links is more favorable with the deoxyguanosine on the opposing strand for the sequence 5′-CG-3′ than for 5′-GC-3′.

A recent study using a gel electrophoresis assay and several small oligonucleotide duplexes showed that mitomycin prefers to bind to the sequence CG over GC (61). Molecular modeling studies further show that the distance between the exocyclic amino groups in the sequence CG is different than in the sequence GC, and that the distance for the former is nearly the same as that of the cross-linked adduct. This suggests that the favorable placement of the second reaction center in CG greatly accelerates the second step in the cross-linking reaction. Interestingly, mitomycin C binds at AT and GC sequences equally well in a noncovalent manner, even though the only sequence with appreciable cross-linking is CG.

Anthramycin. Anthramycin (Fig. 13) is a carbonol amine-containing antitumor antibiotic which binds covalently to N-2 of guanine located in the minor groove of DNA (62). The DNA binding specificity of anthramycin has been studied using footprinting techniques with the probe MPE-Fe(II)

Anthramycin

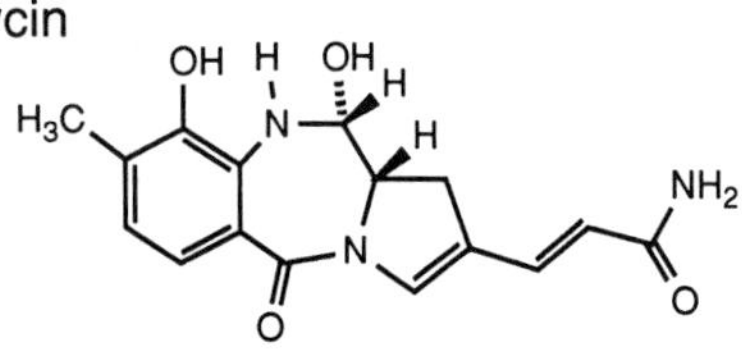

Figure 13 Structure of anthramycin.

(63). In these studies anthramycin was found to primarily protect sequences of the type PuGPu with bonding to the middle guanine in the sequence. Theoretical calculations indicated that anthramycin might interact with the A-form of DNA, which has a minor groove wider and shallower than that of B-DNA (64).

Cisplatin. The transition-metal complex *cis*-diamminedichloroplatinum(II), cisplatin (Fig. 14) is clinically used as an anticancer drug. The antitumor effects of the drug are in part related to its ability to platinate DNA at N-7 of guanine, which is located in the major groove. Since platination does not involve the release of the base and the subsequent formation of an alkali labile site, heating platinated DNA in the presence of piperidine will not result in a strand break.

As described earlier, the binding site of cisplatin on DNA has been determined by two methods. In one, the stop points of exonuclease III on platinated DNA were analyzed to reveal that the enzyme stops at oligoguanine sequences (12,13). This is consistent with cisplatin forming a cross-link with adjacent guanines on the same DNA strand.

The binding site of cisplatin has also been determined using the primer extension method of Gralla and co-workers (11) (Fig. 3). However, in this case it is not necessary to treat the platinated DNA with reagents to remove the platinum prior to electrophoresis. This is because the platinated DNA is used as a template in the primer extension involving DNA polymerase, and the newly synthesized DNA strand does not contain platinum. Using this assay Gralla *et al.* (11) found that the polymerase stopped at points corresponding to platination on the opposite strand or template strand at the

Cisplatin

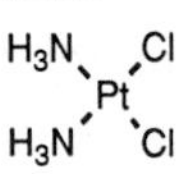

Figure 14 Structure of cisplatin.

sequence GG. This agrees with the binding model that cisplatin creates an intrastrand cross-link at the dinucleotide sequence GG.

CC-1065. The compound CC-1065 (Fig. 15) is an extremely potent antitumor antibiotic produced by *Streptomyces zelensis*. The drug, extensively studied by Hurley and co-workers (9,65,66), binds covalently in the minor groove of B-form DNA. Since it attacks N-3 of adenine, depyrination occurs resulting in an alkali labile site. Analysis using sequencing gels show that CC-1065 is selective for AT-rich regions of DNA, attacking with greatest affinity the 5′-terminal adenine of the sequences 5′-PuNTTA and AAAAA. When a DNA restriction fragment modified by reaction with CC-1065 is heated to 100°C for 30 min and analyzed on a sequencing gel, the drug-induced breakage appears to be adjacent to an adenine on the 5′ side of the site. If, on the other hand, the fragment is treated with piperidine prior to loading on the DNA sequencing gel, the drug-induced break appears to take place at the last adenine in the interaction sequence. Thus, the piperidine-dependent shift of one nucleotide, presumably due to β-elimination, results from loss of the CC-1065 N-3 adenine adduct. A more detailed discussion of this interesting antitumor agent will be found in Chapter 8.

4. *Drugs that Do Not Leave a Permanent Chemical Record*

A large number of anticancer drugs exert their effects by binding to DNA in an equilibrium manner. Since these agents do not chemically modify DNA,

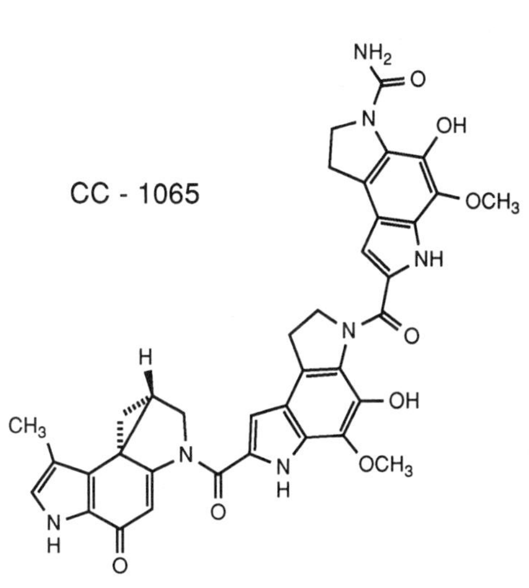

Figure 15 Structure of CC-1065.

their residence positions on the polymer are usually determined using footprint methods (Fig. 5).

Netropsin. The antiviral agent netropsin (Fig. 16) (67) has been the subject of numerous footprinting investigations. The earliest footprinting experiments involving netropsin, which were nonquantitative in nature, confirmed the earlier observations that netropsin prefers to bind to AT-rich regions of DNA (68,69). Single-crystal x-ray analyses have shown that the specificity of the antiviral agent is related to its ability to hydrogen bond to O-2 of thymine and N-3 of adenine located in the minor groove of DNA (70,71). An additional element controlling specificity appears to be related to the fact that the floor of the minor groove at AT-rich regions of DNA is relatively flat and featureless. This allows deep penetration of the drug into the minor groove, producing close van der Waals contacts and hydrogen bonds between the agent and DNA. An additional factor which may influence specificity in the case of netropsin is that in AT-rich regions of DNA the minor groove is charged negatively (72,73). Since the drug is a dication, electrostatic interactions between it and DNA may enhance AT specificity.

In quantitative footprinting experiments using DNase I and Mn-T4 as probes, binding constants for netropsin on a restriction fragment from pBR322 DNA were determined (74–76). The sites on the fragment fell into two groups: strong sites of type $(T \cdot A)_4$ and weaker sites of type $(A \cdot T)_3(G \cdot C)$. The binding constants for the strong sites spanned about two orders of magnitude, 10^6–10^8 M^{-1}, and the weaker sites had affinities of $\sim 10^5$ M^{-1}. From quantitative footprinting experiments conducted in the temperature range 0–45°C, it was possible to obtain a full thermodynamic profile of netropsin, $\Delta G°$, $\Delta H°$, and $\Delta S°$, as a function of sequence on the 139-base-pair restriction fragment (77). The binding enthalpies for all sites but one on the 139-mer were found to be exothermic, roughly correlating with the free energy of binding. The values of the entropy, $\Delta S°$, were positive as expected for association of two solvated species, the drug and DNA.

Netropsin

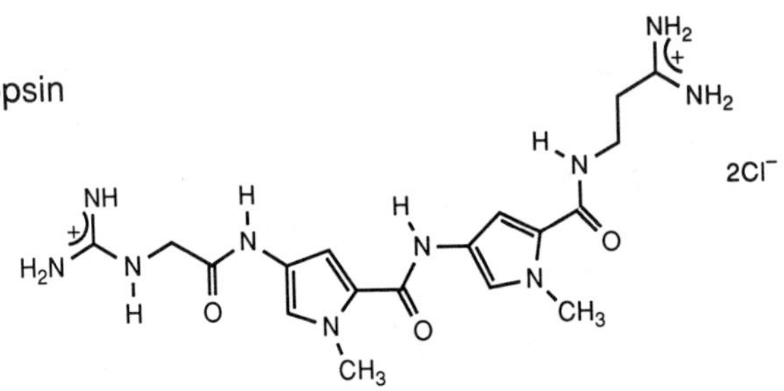

Figure 16 Structure of netropsin.

Although the specific origins of these effects are difficult to determine, the positive entropy may be due to solvent being released from DNA to the ordered structure of bulk water.

Certain footprinting studies involving netropsin have suggested that the drug is able to distort DNA upon binding, as evidenced by increased cleavage by DNase I between drug-binding sites (78). Although increased cleavage is in fact observed when this drug binds to DNA, the origin of the effect may be due to factors other than a drug-induced structural change in DNA. This point will be discussed later.

Distamycin. The antiviral agent distamycin (Fig. 17) is a tripeptide structurally related to netropsin (67). Footprinting studies using DNase I and MPE confirmed that the drug binds in the minor groove at AT-rich regions of DNA (69,79). A single-crystal x-ray structural analysis of the drug bound to the DNA duplex $[d(CGCAAATTTGCG)]_2$ revealed that, like netropsin, distamycin hydrogen bonds in a bifuncated manner to O-2 of thymine and N-3 of adenine located in the minor groove (80). Recently,

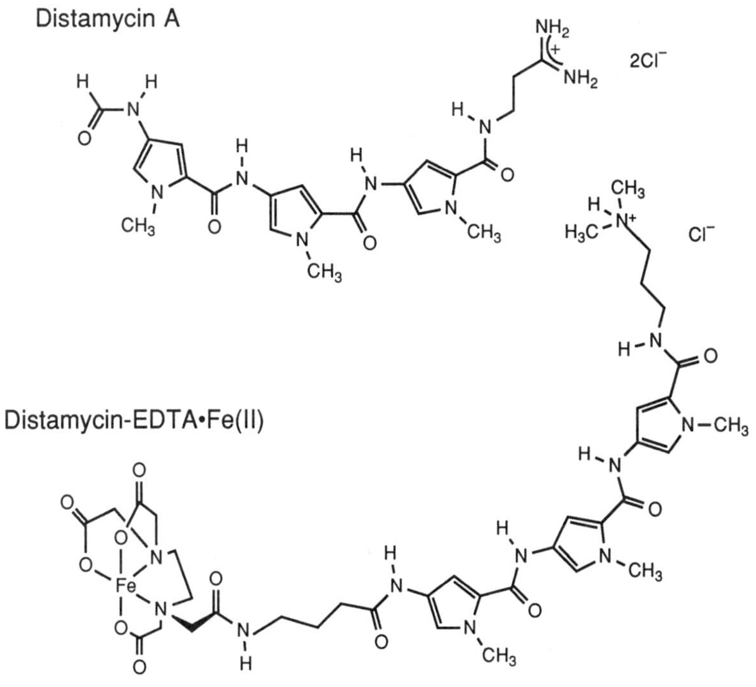

Figure 17 Structures of distamycin A and the related affinity cleavage agent, distamycin-EDTA•Fe(II), ED-Fe(II).

Pelton and Wemmer (81) reported the unusual observation that distamycin can form a 2 : 1 complex with the duplex d(CGCAAATTGGC)·d(GCCAATTTGCG). Remarkably, both drugs bound simultaneously in the minor groove of the central AAATT segment and were in close contact with both DNA and one another. Molecular modeling further indicated that the positive propylamidinum groups of distamycin are directed toward opposite ends of the helix and that the minor groove must expand relative to the 1 : 1 complex to accommodate both drugs. The fact that the 1 : 1 complex can be observed at low drug-to-DNA ratios indicated that the drug does not form an association dimer outside of DNA prior to forming the 2 : 1 complex on DNA. The second drug binds with negative cooperativity, having a binding constant which is 20% of the binding constant associated with the 1 : 1 complex. Interestingly, the duplex having the core sequence 5′-AAATTT-3′ binds two drugs but with positive cooperativity (82). To what extent this occurs through a drug monomer or through a preformed drug dimer is unclear, but the binding constant of the second drug is twice that of the first. For distamycin the length of binding sequence clearly influences the stoichiometry of the drug-DNA complex which forms. Only one distamycin molecule is capable of interacting with the sequence 5′-AATT-3′ (83).

Dervan and co-workers synthesized a large number of DNA binding ligands termed affinity cleaving agents (4). All of the molecules are equipped with an Fe(II)-EDTA moiety which is covalently linked to some part of the ligands' framework. Thus, instead of identifying the ligands' binding sites using conventional footprinting methods, it is only necessary to bind the ligand to DNA and activate the Fe(II)-EDTA portion so that strand scission occurs. The oligonucleotide products which result are analyzed using a standard DNA sequencing gel (Figs. 1, 2).

Studies with two distamycin affinity cleavage agents, one having the Fe(II)-EDTA group on the carboxy terminus of the *N*-methylpyrrole tripeptide ED·Fe(II) (Fig. 17) and a second having the cleavage group on the amino terminus of the drug DE·Fe(II), showed that the high-affinity sites are of type $(T \cdot A)_5$ (84). The unequal intensity of cleavage on each side of the site permits assignment of the orientation of the ligand at the site. A model consistent with the cleavage pattern indicates one preferred binding site for ED·Fe(II) and DE·Fe(II) is AAATT, with the amino end of the tripeptide oriented to the 3′ end of the thymine-rich strand. However, since this study was done prior to that of Pelton and Wemmer (81), it is not known if and to what extent a complex with a 2 : 1 stoichiometry affected the cleavage pattern of the affinity cleavage agents.

Dihydrokikumycin B and Anthelvencin A. Dihydrokikumycin B and anthelvencin A (Fig. 18) are natural products structurally related to netrop-

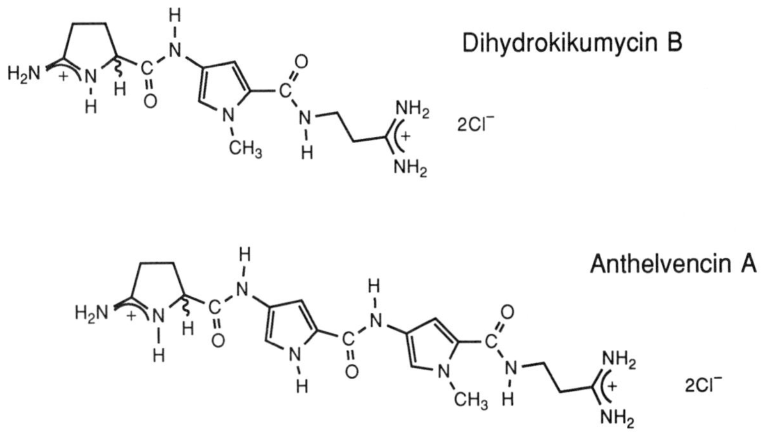

Figure 18 Structures of dihydrokikumycin B and anthelvencin A.

sin. In each case the compounds are pyrrole peptides appended to a chiral 2-amino-1-pyrrolinium group. DNase I footprinting studies show that the compounds utilize the same binding sites on a 139-base-pair segment of pBR322 DNA as does netropsin (85,86). However, due to size, DNase I footprinting studies show that the binding region for dihydrokikumycin B is smaller than that of netropsin or anthelvencin A. The presence of the chiral center in the charge terminus of the drugs suggests that the different enantiomers may have different affinities for specific DNA sites. In the case of dihydrokikumycin B the natural 4S-(+)-isomer exhibits significantly higher stereoselectivity than its enantiomer. Anthelvencin A is a tripeptide, whereas dihydrokikumycin B is a dipeptide. The fact that the enantiomeric anthelvencins exhibit weak stereoselectivity suggests that ligand size may influence stereoselectivity. The stereoselectivity exhibited by dihydrokikumycin B is similar to that reported for a hydroxy-bis(netropsin)succinamide (87). These compounds possess two chiral centers positioned between two strong nonchiral DNA binding netropsin groups. Apparently the high binding affinity of the netropsin groups forces the chiral recognition element to contact DNA sites, thereby enhancing stereoselectivity.

Hoescht 33258. The compound Hoescht 33258 (Fig. 19) is a well-known stain for DNA. There are three x-ray crystallographic studies of Hoescht 33258 bound to small oligonucleotide duplexes. Two of these studies involve the duplex $[d(CGCGAATTCGCG)]_2$ (88,89), and in a third study the dye is bound to a duplex having the sequence $[d(CGCGATATCGCG)]_2$ (90). In the

Hoescht 33258

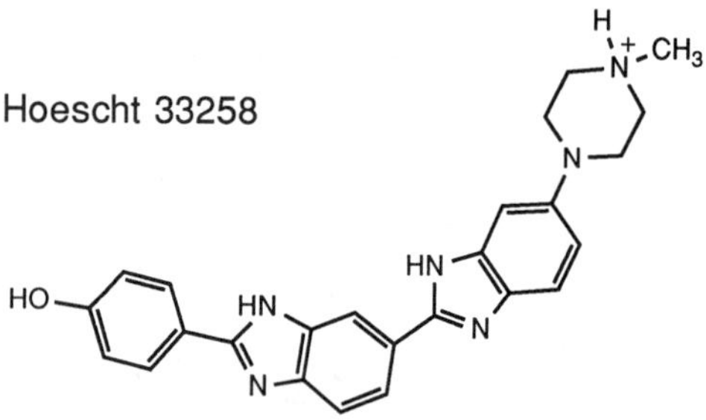

Figure 19 Structure of Hoescht 33258.

case of one of the duplexes, the dye molecule either straddles the AATT central sequence symmetrically or is displaced by one base pair so that the piperazine ring is located over a wider region of DNA having a GC base pair. In the case of the DNA duplex having the ATAT central core, the drug covers three of the four AT base pairs and extends its piperazine ring to the first GC base pair adjacent to the alternating AT segment. In this case the binding of the dye to the duplex modifies the structure of the central part of the duplex, suggesting that DNA exhibits considerable plasticity when bound by the dye.

Footprinting studies show that the dye spans 5 ± 1 base pairs and predominantly binds to AT-rich regions of DNA (91). An interesting analog of Hoescht 33258 contains an atom ^{125}I as part of its structure (92). Iodine 125 is a β-emitting isotope and in the presence of DNA produces strand scission. Thus, in this case, instead of carrying out a footprinting experiment, it is only necessary to bind the drug to DNA, thereby producing strand cleavage. A recent study showed that the sequence specificity of the labeled compound on DNA present in intact cells was found to be identical to that of purified high molecular weight DNA (92).

In addition to footprinting and x-ray crystallographic studies, 2D ^{1}H NMR has been used to study Hoescht 33258 bound to DNA (93). The duplex $[d(CTTTTGCAAAAG)]_2$ binds two dye molecules cooperatively in symmetry-related orientations at the two AT-rich sites. The dyes bind with their positively charged *N*-methylpiperazine moieties pointed toward the center of the duplex while the phenol groups are disposed toward the 3′-ends of the sequence.

Berenil. Berenil (Fig. 20) is an antitrypanosomal agent which binds in the minor groove of DNA. DNase I footprinting studies showed that, like netropsin, the agent prefers to bind to AT-rich regions which are at least four base pairs in length (94). Berenil shares with netropsin and distamycin the

Berenil

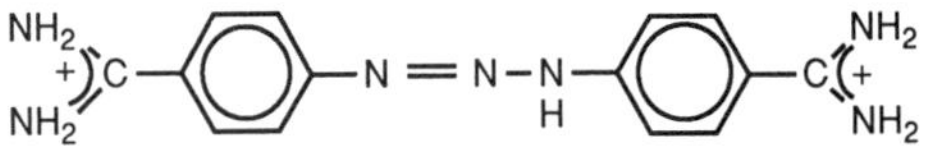

Figure 20 Structure of berenil.

property of enhanced rates of cleavage for DNase I at certain sequences flanking berenil binding sites. Competitive footprinting studies with the chromosomal protein HMG-I, which binds to AT-rich regions of DNA, show that berenil is a weaker binding ligand than is netropsin and distamycin (95). Yoshida *et al.* (96) characterized the 1 : 1 complex with the duplex $[d(GCAATTGC)]_2$ using NMR. Berenil binds via the minor groove to central AT portion of the duplex. The NMR study also showed that the duplex is in the B conformation with the berenil amidine protons either bound to O-2 of the external thymines or to N-3 of the internal adenines.

The recent x-ray structure of the berenil complex with $d(CGCGAATTCGCG)_2$ shows the antitrypanosomal agent bound to the 5′-AAT-3′ region of the oligonucleotides (97). At one end of the drug, the amidinium group is in hydrogen-bonded contact with N-3 of the adenine base complementary to thymine of the AAT. The other amidinium group of the agent does not make direct contact with DNA but instead is hydrogen-bond-bridged to a water molecule which is bonded to an adenine base and a deoxyribose-ring oxygen atom.

Daunomycin and Adriamycin. Adriamycin (Fig. 21) is the most widely used anticancer drug. Although daunomycin, a close structural relative of

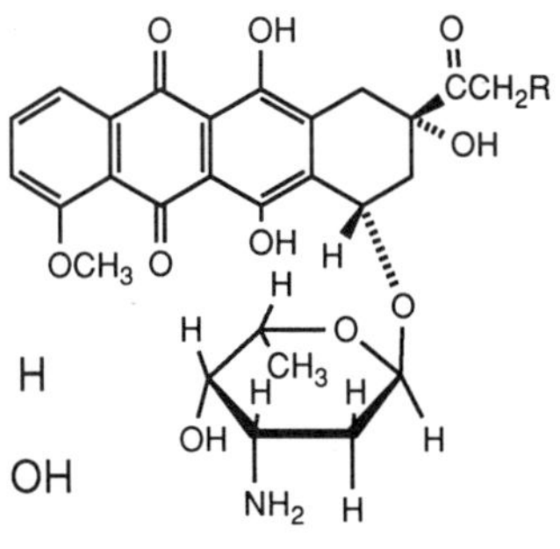

Daunomycin R = H

Adriamycin R = OH

Figure 21 Structures of daunomycin and adriamycin.

adriamycin, has not enjoyed widespread chemotherapeutic use, it is one of the best-characterized DNA intercalating agents (2). X-ray structural analyses of daunomycin-oligonucleotide complexes show that the drug is able to intercalate via the minor groove of DNA (98,99). The sugar group, daunosamine, in one structure (99) lies in the minor groove with its protonated amino function not directed at any particular phosphate group of DNA. In the other structure (98) this group forms a hydrogen bond with O-2 of thymine. These features have also been emphasized in modeling studies involving the drug (100). The drug strongly prefers B-DNA and will selectively discriminate against alternative DNA conformations such as left-handed or *Z*-DNA (101).

The sequence specificity of daunomycin, adriamycin, and the β-anomer of adriamycin have been investigated using DNase I footprinting techniques (102). The fact that the clearest inhibition is observed at lower temperatures (e.g., 4°C) indicates that the resonance time of the drug on DNA and/or the temperature dependence of the binding constant may be important in obtaining a footprint. These anthracycline antibiotics were found to recognize a triplet binding site containing adjacent GC base pairs of variable sequence flanked by an AT base pair.

Recent footprinting studies on a 160-base-pair fragment with daunomycin in the concentration range 0–3.0 μM have revealed the susceptibility of various sites to attack by DNase I (103). After normalizing the footprinting intensity data to a single site, four types of behavior were observed: (i) protection from DNase I cleavage; (ii) protection, but only after reaching a critical total daunomycin concentration; (iii) enhanced cleavage; (iv) no effect of added drug. Seven of the 10 sites on the fragment possessed the sequences (A or T)GC and (A or T)CG, while the remaining 3 had the sequence (A or T) C (A or T). This result is in general agreement with other footprinting studies showing that the high-affinity sequence for daunomycin and adriamycin is CA, while modest-affinity sites are GC, CG, CT, TC, CC and AC (104,105).

Chromomycin, Mithramycin, and Olivomycin. The aureolic acids chromomycin, mithramycin, and olivomycin (Fig. 22) are members of an important class of antibiotic and antiviral agents which bind to double helical DNA (2). Chromomycin, the most well studied member of the class, consists of an aglycone having five appended hexopyranoses. An interesting feature of the aureolic acids is that they require Mg^{2+} for binding to DNA.

Two footprinting studies involving these anticancer agents have been reported (106,107). The first, employing two DNA restriction fragments 117 and 168 base pairs in length (106), showed that the recognition sequence is approximately three base pairs in length. Many binding sites are similar for

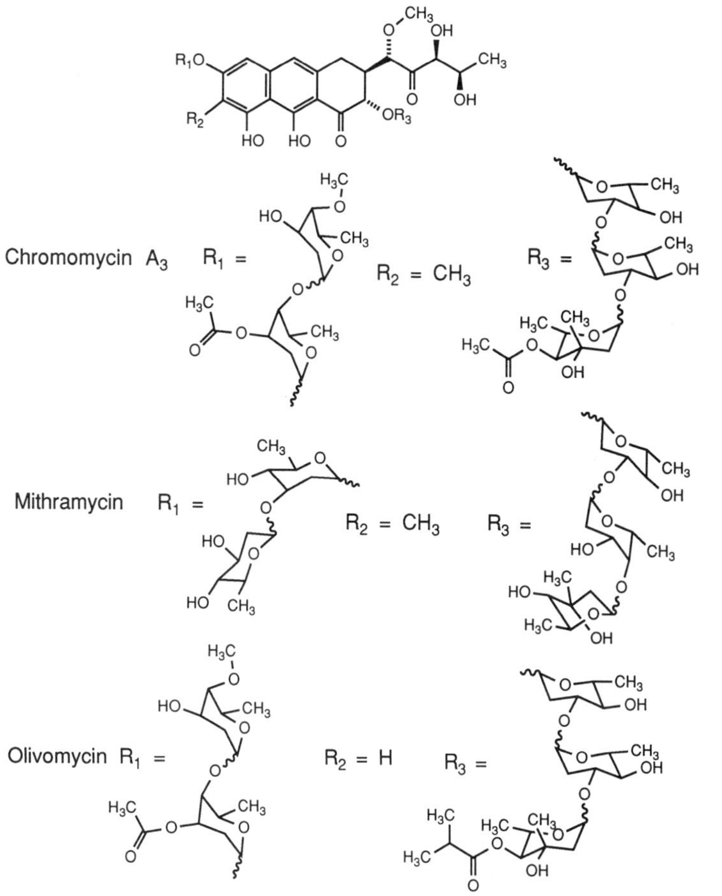

Figure 22 Structures of chromomycin, mithramycin, and olivomycin.

these three antibiotics; chromomycin and olivomycin binding sites are nearly identical. For chromomycin the preferred sites on a 70-base-pair segment of DNA, examined using Fe-MPE as a cleavage agent, were, in order of decreasing affinity, GGG, AGC > GCC, CCG > AGC, TCC, GTC. The sequence AGC appears to have different affinities, indicating the importance of either flanking sequences or the presence of a nearby bound drug molecule.

Based on the results of a second footprinting analysis of the three aureolic acids (107), it was concluded that binding requires two contiguous

GC base pairs with the preferred site having the sequence GG. The reaction of dimethylsulfate with N-7 of guanine was modified by the presence of mithramycin, indicating either that the drug may bind via the major groove of DNA, or that binding is in the minor groove but that rate of alkylation in the region opposite the bound drug is modified.

Although a single-crystal x-ray analysis of one of the aureolic acids bound to a small segment of DNA has not been reported, the structure of the chromomycin-DNA complex in solution has been studied using NMR (108,109). The NMR results demonstrate that chromomycin binds as a symmetrical dimer to the self-complementary duplex having the sequence [d (TTGGCCAA)]$_2$ with the retention of two-fold symmetry in the complex. The two chromomycin molecules in the dimer share the minor groove at the central portion of the duplex, i.e., GGCC, in such a way that each hydrophilic edge of the chromophore is located next to a GG site. The drug-bound duplex appears to have a shallow minor groove and a deep major groove, suggesting that DNA is in the A conformation.

The binding of chromomycin A_3 and mithramycin to the duplex d(ATGCAT)$_2$ has also been studied by NMR (110). NOE contacts from the sugars of mithramycin to several deoxyribase protons suggests that two of the sugars are oriented along the sugar-phosphate backbone of G_3-C_4 while two other sugars are located along the sugar-phosphate backbone of A_5-T_6. These drug-DNA contacts are very similar to those found for chromomycin binding to d(ATGCAT)$_2$.

Nogalamycin. Nogalamycin (Fig. 23) is an anthracycline antibiotic which binds to DNA. Attached to the aglycone portion of the drug is a sugar, nogalose, and a bulky bicycloamino sugar giving the molecule a dumbbell shape. Although this shape appears to preclude intercalation, a recent x-ray structural analysis of nogalamycin bound to the self-complementary DNA hexamer [d(^{m5}CGTsA^{m5}CG)]$_2$ shows that the drug is threaded between the phosphodiester backbone of the duplex (111). While the three aromatic rings of the aglycone are intercalated within DNA, the nogalose lies in the minor groove of DNA, and the bicycloamino sugar is in the major groove.

DNase I footprinting studies with four different restriction fragments revealed that nogalamycin most often binds to regions of alternating purine-pyrimidine sequence, e.g., TG (CA) and GT (AC) (112). This suggests that the preferred binding sites may contain all four nucleotides, and/or the peculiarities of the dynamics of DNA may be important to binding. In addition to enhancing DNase I cleavage outside of drug-binding sites, nogalamycin selectively inhibits cleavage of DNA at certain guanine-containing sequences by the G-specific photosensitized reaction with methylene blue. The drug also inhibits base alkylation using dimethylsulfate. These results suggest that intercalation of nogalamycin into DNA may be of two

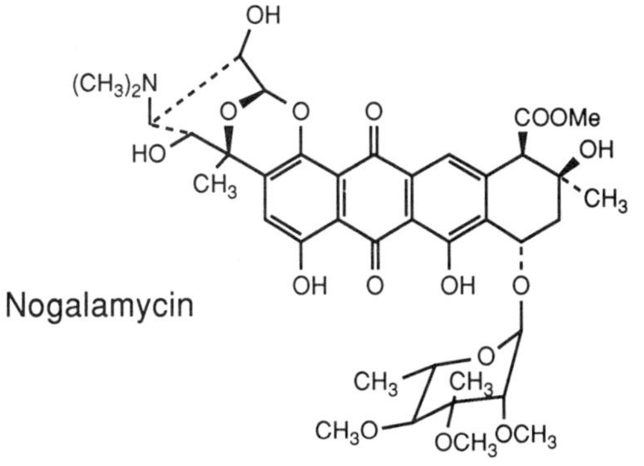

Figure 23 Structure of nogalamycin.

types, one with the nogalose moiety occupying the minor groove and one with this sugar positioned in the major groove of DNA.

Actinomycin D. Actinomycin D (Fig. 24) is a drug which binds to DNA by intercalation (2). The binding mechanism involves intercalation of the phenoxazone ring system between the base pairs at GC sequences of DNA. Based on a single-crystal x-ray analysis as well as on model building, the GC specificity of the agent is believed to be related to its ability to hydrogen bond to the 2-amino group of guanine, which is located in the minor groove of DNA (2). Actinomycin D has been the subject of numerous footprinting

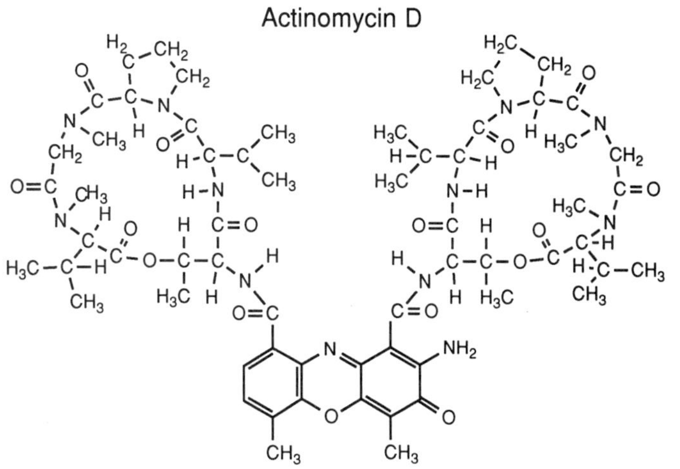

Figure 24 Structure of actinomycin D.

investigations. While early nonquantitative footprinting studies with the compound confirmed that it binds with high affinity to sites having the dinucleotide sequence GC (68,69,113), later studies have more extensively documented the specificity of the agent. Using 7-azidoactinomycin D, a compound which can be photoactivated to cleave DNA, Rill and co-workers (114) showed that, while GC doublets are important, flanking sequences influence the amount of binding which takes place. By quantitively analyzing the cleavage sites on several DNA restriction fragments induced by 7-azidoactinomycin D, it was found that GC doublets were strongly preferred only if the 5′ flanking base was a pyrimidine and the 3′ flanking base was not cytosine. In addition, the central GG doublet of the sequence TGGG, which does not contain the sequence GC, is also a strong actinomycin D binding site. The fact that non-GC sequences can strongly bind actinomycin D was also noted by Snyder *et al.* (115). Using batch calorimetry and other techniques, these workers showed that the duplex $[d(CGTCGACG)]_2$ is able to bind two actinomycin D molecules in a strongly cooperative manner. Although the mode of binding of the drug to this duplex was not determined, the aberrant spectroscopic and calorimetric behavior of the interaction suggests that classical intercalation via the minor groove of DNA may not be occurring with this duplex.

Using quantitative footprinting methods, Rehfuss *et al.* (116) measured the binding constant of actinomycin D toward the duplex $[d(TAGCGCTA)]_2$. The analysis required considering competitive equilibria between the drug and the cleavage agent, DNase I, and DNA. Although the measured binding constant of $4 \times 10^6 M^{-1}$ was in agreement with those measured by other methods on small duplexes containing GC sites, it was higher than that found for the same sequence embedded in polymeric DNA (117–119). Quantitative footprinting studies on a 139-base-pair fragment of pBR322 DNA also showed that the sequence CGTC, studied by Snyder *et al.* (115), is a strong actinomycin D binding site. All the other high-affinity sites on a 139-mer possessed the classical binding sequence GC, flanked on the 5′ side by either T or C but not G. These sites have binding constants in the range 1–6 $\times 10^6 M^{-1}$.

Since actinomycin D is a intercalating drug, it may distort DNA upon binding. In the various footprinting studies involving this anticancer drug, cleavage rate enhancements, often observed outside of binding sites, are commonly attributed to drug-induced structural changes in DNA (79). This assumption seems reasonable, since a structural change is expected to lead to a change in the cleavage rate constant at nearby sites, thus producing altered (increased or decreased) cleavage rates. Recent quantitative analysis of an actinomycin D DNA footprinting experiment revealed that the bulk of the rate increases outside of the highest affinity actinomycin binding sites on DNA can be explained by simple mass-action effects associated with the enzyme DNase I and are not structural in origin (119,120). This mass-action effect occurs because, as DNA binds drug, the number of sites to which DNase I can bind and cleave is reduced. Since there is an equilibrium between bound and unbound enzyme to DNA, this causes the effective concentration of DNase I to increase at all non-drug-binding sites, thereby leading to enhanced cleavage at these sites. More recent attempts to detect actinomycin-induced structural changes in DNA using other cleavage probes, Mn-T4 and Fe-MPE, met with limited success (118). The analysis of the initial parts of footprinting plots, which are nearly linear, revealed that actinomycin D can cause anomalous rate enhancements at a few sites adjacent to drug sites when binding is probed with DNase I or the metalloporphyrin Mn-T4 but not when Fe-MPE is used. In light of the DNA structural features recognized by DNase I (17) and Mn-T4 (121), it appears that actinomycin D may alter minor groove width and/or partially melt sequences adjacent to the site of binding. Huang *et al.* (122) have recently shown that actinomycin D causes a change in DNA conformation at a distance from its binding site on the DNA duplex $[d(ATATATAGCTATATAT)]_2$ that does not occur in the duplex $[d(AAAAAAAGCTTTTTTT)]_2$. The study suggested that DNase I rate enhancements associated with actinomycin D binding are caused by ligand-induced structural changes in DNA.

Echinomycin, Triostin A, and TANDEM. Echinomycin, triostin A, and TANDEM (Fig. 25) are members of a family of antibiotics which are characterized by cross-linked heterodetic octadepsipeptide rings bearing two quinoxaline-2-carboxamide chromophores (2,123). An interesting feature of the molecules is that the two chromophores are arranged on the same side of the peptide ring, and their planes are approximately parallel. This geometry allows the chromophores to simultaneously bind to DNA by intercalation. Since the spacing between the chromophores is 3.5 Å, the intercalation process sandwiches two base pairs of DNA between the two intercalating chromophores. This phenomenon of *bis*-intercalation for echinomycin was first documented by Waring and Wakelin (124), who showed that the drug caused twice the unwinding seen with classical intercalating drugs like ethidium, as well as twice the DNA helix extension.

The molecular basis of specific recognition between echinomycin and DNA, recently reviewed by Waring (123), appears to be due to hydrogen bonding between the drug's alanine carbonyl groups and the 2-amino group of guanine. Footprinting studies using DNase I and MPE as probes indicate that the preferred binding site is the dinucleotide CG, although the actual binding site size is at least four base pairs, and there is some evidence that AT base pairs may be preferred alongside the central CG (125). A circular dichroism study of echinomycin binding to 11 different DNAs concluded

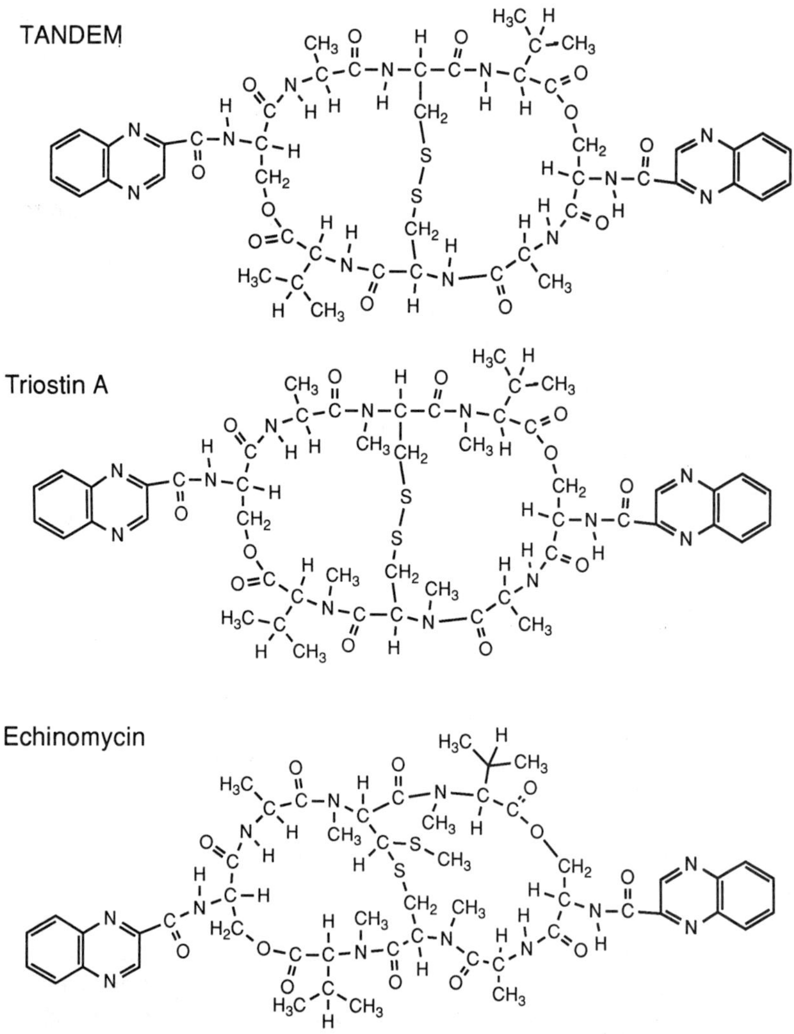

Figure 25 Structures of echinomycin, triostin A, and TANDEM.

that the preferred sites for intercalation contain at least one guanine or cytosine, CC, GC, CG, AC, with GC preferred to CG (126).

In addition to *bis*-intercalation, one of the striking features of the echinomycin-DNA complex is that the antibiotic causes base pairs other than those sandwiched between the quinoxoline chromophores to change from Watson-

Crick base pairing to Hoogsteen base pairing. This can be seen in the structure of the hexamer having the sequence $[d(CGTACG)]_2$, which possesses two bound echinomycin molecules at the CG sites at the end of the sequence (127). The central AT dinucleotide in this complex exists with Hoogsteen base pairing.

The existence of Hoogsteen base pairing has been confirmed in solution using NMR (109), and it has been studied with footprinting methods (128–130). Although footprinting investigations noted rate increases for DNA modification outside of echinomycin binding sites, it is presently not clear if these changes are due to an alteration in the cleavage rate constants at the sites as would be caused by a structural change in DNA or if they are due to mass-action effects associated with redistribution of the probe molecules on DNA as drug is added to the system.

While the sequence specificity of triostin A appears to be the same as that of echinomycin, i.e., the preferred sequence is CG, the synthetic compound TANDEM has altered specificity (123). This compound lacks peptide *N*-methyl groups which causes a dimunition in the binding constant for natural DNA and a complete change in specificity when compared to echinomycin.

III. QUANTITATIVE FOOTPRINTING ANALYSIS

A. General Considerations

The objective of the quantitative footprinting study is to obtain drug-binding constants by using the autoradiographic spot intensities which appear on the sequencing autoradiogram (131). In a quantitative footprinting experiment, it is important to carry out a number of digest reactions with the same DNA and probe concentrations but with different amounts of drug added to the reaction mixtures. Initially, these reactions are carried out over a wide range of drug concentrations to determine the concentration range over which binding events on the restriction fragment take place. These studies may be done in the presence or absence of unlabeled carrier DNA. Because it is experimentally difficult to determine the concentration of a radiolabeled fragment present in the mixture, it is common to add a known amount of carrier DNA. Since the carrier is generally present in large excess relative to radiolabeled fragment, it serves as a drug buffer and controls the pool of free and bound drug present in the system.

After these initial experiments, a second set of footprinting titrations is carried out over a narrow drug concentration range. The range of concentrations is such that only the strongest binding sites on the polymer are accepting drug. A common mistake is to choose drug concentrations which are too few in number or are too high, corresponding to concentrations which produce significant site loading on the polymer. Although high concentrations give a dramatic "footprint" on the sequencing autoradiogram, they are not particularly useful in defining the shape of the binding isotherm associated with the high-affinity sites. It must be kept in mind that it is very important to measure the early part of the isotherm, which is best done by using a number of concentration points at relatively low drug concentration. Also, as was earlier stated, it is important to terminate the digest reaction early, so that the products which are separated in the electrophoresis experiment are statistically the result of single cuts in the full-length fragment. In this case, the measured autoradiographic spot intensities associated with the binding site decrease linearly with the occupancy of the site by a drug molecule. After electrophoresis, the footprinting autoradiogram is scanned with a microdensitometer to yield intensities which are proportional to the oligonucleotide concentrations. These intensities form the basis of quantitative footprinting analysis.

1. *Total-Cut Plot*

The quantitative footprinting experiment is an inherently "noisy" experiment. The method utilizes a set of digests of a DNA fragment in the presence of various amounts of drug, all supposedly carried out under identical reaction conditions. However, it is difficult to load an identical amount of radiation into each of the wells of the gel, so, for this reason as well as other reasons, the total density in a lane on the gel, corresponding to a particular drug concentration, may vary irregularly from lane to lane. In order to correct for this and reduce noise in the experiment, all the individual spot intensities in a given lane can be multiplied by a common factor. The factor is determined by constructing what is referred to as the *total-cut plot*, which is a plot of the total integrated density on the autoradiogram for each lane on the gel versus the drug concentration of that lane. The total-cut plot serves two purposes. First, the trend of the total cut (usually, constancy with drug concentration) confirms the theoretical model, as explained below. Second, it is used to correct for incorrect gel loading and other errors. The correction factor necessary to adjust the intensity in a given lane is the factor necessary to adjust the value of the corresponding point on the total-cut plot to a smooth curve derived by considering all of the points. Once this correction to the experimental data is made, a series of plots showing the autoradiographic spot intensity as a function of drug concentration for each oligomer is made. These plots, referred to as *footprinting plots*, are analyzed to calculate binding constants from the footprinting data.

In the following section we will analyze the interaction of actinomycin D with a 139-base-pair segment of pBR322 DNA as probed with the enzyme DNase I (117–119). The footprinting autoradiogram of this experiment is shown in Figure 6. The analysis, involving 26 different drug concentrations, leads to the calculation of site specific binding constants for the drug on the restriction fragment. The sequence of the 139-mer and the location of the strong drug binding sites are shown in Figure 26. Figure 27 shows the total-cut plot associated with the act D–DNA interaction. This plot, the sum of the intensities of cut fragments as a function of total drug concentration, was fitted to a straight line to derive the correction factor mentioned above. Although the fitted curve has a nonzero slope, the value of the mean-square deviation permits a curve with zero slope (constant total cut) to also fit the experimental data with acceptable accuracy.

2. *Footprinting Plots*

Using correction factors derived from the total-cut plot, we produce footprinting plots, some of which are shown in Figure 28. The basic equation determining the footprinting plots is

$$(\text{rate})_i = k_i[\text{probe}]_i(1 - \nu_i) \qquad (1)$$

In this expression, the cleavage rate at cleavage site *i*, $(\text{rate})_i$, is equal to the product of the cleavage rate constant at the site (k_i), the concentration of cleavage agent or probe at the site ($[\text{probe}]_i$), and the fraction of the site *i not blocked* by drug ($1 - \nu_i$), where ν_i is the fraction of cleavage sites *i* blocked by drug binding at a site near cleavage site *i*. Note that the drug-binding site may not be identical to the cleavage site. In the single-hit regime, the autoradiographic spot intensity is proportional to $(\text{rate})_i$. The proportionality constant, multiplied by k_i, gives a new constant.

For a nucleotide position or cutting site within a drug-binding site, cleavage *decreases* as drug is added to the system. These footprinting plots show the classic footprinting phenomenon, the result of drug-blocking cleavage by the probe. Examples of these plots can be found in Figure 28, sites 66, 67,

```
5'-AGCTTTAATGCGGTAGTTTATCACAGTTAAATTGCTAACGCAGTCAGGCACCGTGTATGAAATCTAACAA
    30        40        50        60        70        80        90
3'-   AAATTACGCCATCAAATAGTGTCAATTTAACGATTGCGTCAGTCCGTGGCACATACTTTAGATTGTT

 TGCGCTCATCGTCATCCTCGGCACCGTCACCCTGGATGCTGTAGGCATAGGCTTGGTTATGCCGGTACTGCCG-3'
100       110       120       130       140       150       160       170
 ACGCGAGTAGCAGTAGGAGCCGTGGCAGTGGGACCTACGACATCCGTATCCGAACCAATACGGCCATGACGGC-5'
```

Figure 26 The 139-base-pair fragment from pBR322 DNA, with strong actinomycin D binding sites indicated with rectangles.

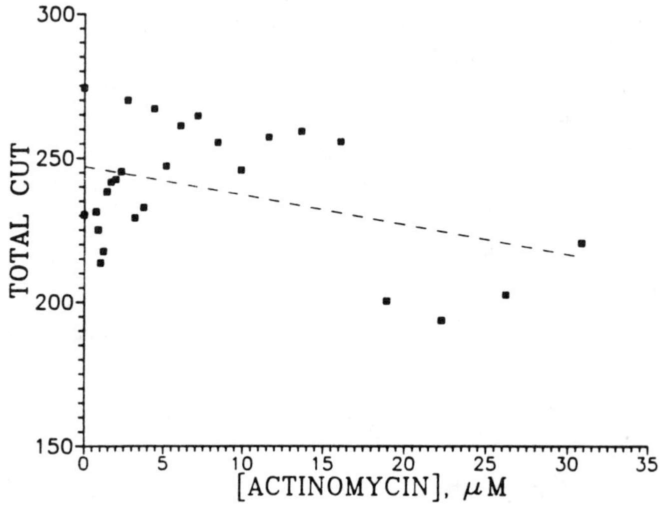

Figure 27 Plot of total cleavage on the 139-mer as a function of actinomycin D concentration. Dashed line is linear fit to the total cleavage on the restriction fragment.

68, for actinomycin D binding to the 139-base-pair segment of pBR322 DNA (118,119). With relatively long, multisite DNA molecules, intensities for cleavage sites between drug-binding sites are observed to *increase* as drug is added to the system (19). For DNase I, Fe-MPE and other equilibrium binding agents, the increase in cleavage is caused by the drug blocking a certain fraction of the DNA present in the system, thereby making the effective concentration of the cleaving agent increase at all sites not blocked by drug; i.e., there is an increase in $[\text{probe}]_i$ in Eq. (1) (118–120,131). The DNA fragment behaves as if the total number of probe molecules per fragment is constant with drug concentration, as indicated by the constancy of the total amount of cleavage on the fragment. Examples of this enhancement are seen in Figure 28, sites 54, 55, and 58. The enhancement occurs only if the DNA is not saturated with cleavage agent; if it is, the concentration of bound probe outside of binding sites does not change as DNA accepts drug.

In some cases, drug binding to DNA may induce a structural change in the polymer, causing the cleavage rate constant at certain sites, k_i, to be altered. These effects, should they occur, may *increase* or *decrease* the rate from that expected from the just-mentioned mass action or redistribution effect. Because actinomycin D is an intercalating drug, it is more likely to cause a distortion in DNA upon binding than is a groove-binding agent such as netropsin (2). That this may be happening when act D binds to the 139-

mer is indicated in the footprinting plots of sites 54 and 59 of Figure 28. The low-drug portions of these footprinting plots, near 2 μM drug, exhibit changes in the relative amount of cleavage as a function of added drug which are inconsistent with simple mass action or binding effects, and they may be the result of changes in k_i in Eq. (1).

Another type of footprinting plot typically found in the footprinting experiment is a composite of the first two, namely an enhancement followed by a decrease in cleavage rate due to binding. Such curves are associated with drug sites having binding constants lower than those of the primary sites, i.e., secondary binding sites. Examples of this type of behavior can be found (see Fig. 28, sites 150, 154, and 158) for the act D binding sites associated with the sequence GGC, which are secondary drug-binding sites.

Although footprinting plots resemble titration curves and in general are nonlinear in drug concentration, the low-drug parts of these curves are relatively linear. Thus, in addition to the footprinting plots themselves, it is

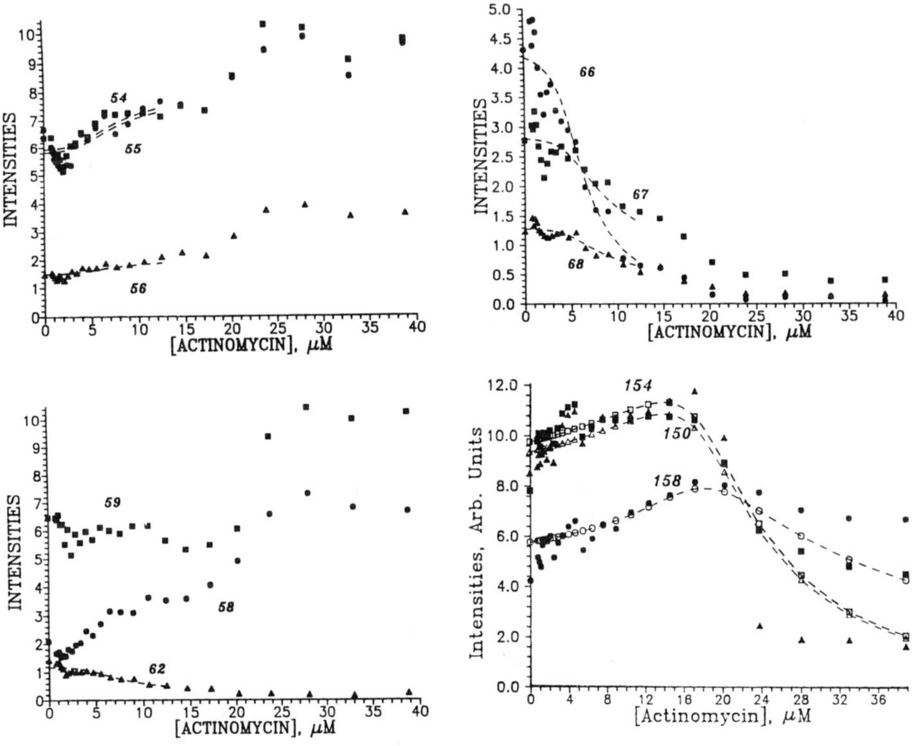

Figure 28 Representative footprinting plots at various sites on the 139-mer for actinomycin D are given. Dashed lines are theoretical fits using various models.

sometimes useful to calculate their initial slopes and divide by the zero-drug intercepts to get "initial relative slopes." When portrayed on the sequence of the fragment being studied, the initial relative slopes clearly show the positions of drug binding, potential sites of drug-induced structural changes, and enhancements due to mass-action effects on the polymer (118). Such a plot for the actinomycin D–139-mer interaction is illustrated in Figure 29.

B. Constructing the Model

The analysis of footprinting plots now requires a quantitative theoretical picture of how the presence of bound drug affects cleavage by the probe at the various sites on the polymer being studied. Usable footprinting data will not be available for all sites. This is especially true for DNase I, cutting by which is sensitive to groove width and flexibility of DNA (17). Thus, for some sites, intensities will be low, and reliable information on how cleavage is affected by addition of drug cannot be obtained. An additional complicating feature of the footprinting experiment is that single-site resolution cannot be obtained over the entire sequencing autoradiogram. As the length of the oligomer increases, resolution along the DNA helix decreases, making it impossible to definitively identify sites which are far away from the radiolabel.

1. Drug Size and the Cleavage Characteristics of the Probe

One can readily distinguish sites where drug binding is occurring by inspecting the footprinting plots or their initial slopes. In order to construct a

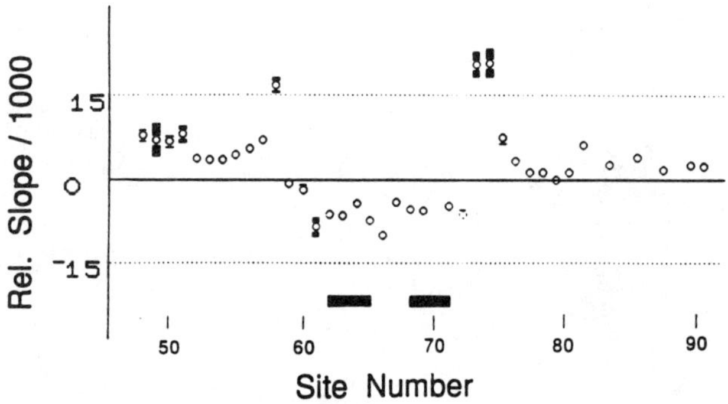

Figure 29 Initial relative slopes with errors for the DNase I–act D–139-mer experiment. Horizontal rectangles indicate the location of strong drug-binding sites on the 139-mer.

useful model, it is important to have some estimate of the size of the inhibition region when the drug is bound to DNA. In the case of actinomycin D, the preferred intercalation site for the phenoxazone ring is GC with the two cyclic pentapeptides displaced an additional base pair on either side of the intercalation site (2). Thus, a single bound act D molecule would "cover" about four base pairs of DNA. The enzyme, on the other hand, possesses a small loop which fits into the minor groove of DNA (17). This loop, which is important for the binding of DNase I to double-stranded DNA, spans three to four base pairs of DNA (131). Since the catalytic site is located at one "end" of the loop, an inhibition pattern of about three to four base pairs to the 3′ side of a single bound act D molecule is expected. The enzyme-DNA contacts on the "other side" of the catalytic site are relatively weak compared to the loop, and it is probable that the enzyme can cleave near the 5′ edge of a drug site, thus making the inhibition region 7-8 base pairs long. This size inhibition pattern is, in fact, realized for netropsin, a groove-binding agent with molecular dimensions similar to act D (75). However, despite the drugs' similarities in size, the footprinting data reveal that act D produces a smaller inhibition region (six to seven base pairs) than netropsin (seven to eight base pairs). The reason for this difference is not clear, but it may be that the cationic residues on the enzyme more effectively block approach to netropsin, which is a cation, than to act D, which is a neutral ligand. An inhibition size of six to seven base pairs for act D is consistent with the strong inhibition observed at site 66 of the 139-mer (Fig. 28). This inhibition can be explained by act D binding at 63-64 or 69-70, GC, either of which would block cleavage at 66. The molecular implications of this observation are that act D blocks DNase I cleavage about four base pairs to the 3′ side of the intercalation site and about three base pairs to the 5′ side of the site, producing an inhibition region six to seven base pairs in length.

2. *Drug- and Probe-Binding Equilibria*

We now turn to the factor $1 - \nu_i$ in Eq. (1), which reflects the drug-binding equilibria. In the case of an isolated drug-binding site, the binding constant K_j at the jth drug site is given by

$$K_j = \frac{c_j}{(c - c_j)D_0} \tag{2}$$

In this expression c_j is the concentration of sites j at which drug is bound, c is the total concentration of sites j, and D_0 is the free-drug concentration. Since ν_i is $(c - c_j)/c$ in this case, one has

$$1 - \nu_i = (1 + K_jD_0)^{-1} \tag{3}$$

If two drug sites j and k are near one another, an enzyme cleavage site i may be blocked by drug binding at either site j or site k, each of which is governed by an equation like (2). Then

$$1 - \nu_i = \left(1 - \frac{c_j}{c}\right)\left(1 - \frac{c_k}{c}\right) = \frac{1}{(1 + K_jD_0)(1 + K_kD_0)} \tag{4}$$

The two drug-binding equilibria are considered mutually independent. On heterogeneous sequence DNA, drug sites are often adjacent to one another so that the binding of a drug molecule at one site may prevent (exclude) the binding of a second drug at an adjacent site on the same DNA fragment. If the binding of drugs at sites j and k are mutually exclusive events, the probability of binding at j is $K_jD_0/(1 + K_jD_0 + K_kD_0)$, with a similar expression for binding at k. Since $1 - \nu_i$ is the probability that there is no drug bound at either j or k,

$$1 - \nu_i = (1 + K_jD_0 + K_kD_0)^{-1} \tag{5}$$

in this case. Because the two drug-binding events are not independent, one cannot simply multiply the probabilities for no drug binding together as in Eq. (4).

The concentration of probe, which is the factor $[\text{probe}]_i$ in Eq. (1), depends on the theoretical model for drug-probe competition. If the total amount of cleavage is constant as a function of drug concentration, the system is behaving as if the amount of probe bound to DNA is constant. This implies that, as drug is added to the system, blocking some DNA, the effective concentration of probe at unblocked sites increases. The concentration of probe available to cut at site i should be inversely proportional to the number of non-drug-blocked sites present in the system, since the same number of probe molecules is distributed over these sites. Thus, we write $[\text{probe}]_i$ as a constant T divided by $1 - K_ec_b$, where K_e is a parameter whose value is to be determined and c_b is the total concentration of drug bound to the fragment. The constant T is absorbed into k_i in Eq. (1), and $[\text{probe}]_i$ is written as just $(1 - K_ec_b)^{-1}$, the enhancement factor. Of course, this factor cannot be valid when c_b becomes comparable to K_e^{-1}; increasing amounts of bound drug will eventually displace probe from the fragment, decreasing the total amount of cleavage on the fragment.

In cases in which the probe, like the ligand being examined, is specific for certain sites, or if it is not specific, like DNase I, but the DNA molecule being studied is small so the drug and the probe compete for a common site, it is convenient to consider the competitive equilibria between the drug and the cleavage agent. The factor $[\text{probe}]_i$ is determined by the simultaneous equilibrium expressions

$$K_i = \frac{c_i}{(c - c_i - c_{pi})D_0}, \qquad K_{pi} = \frac{c_{pi}}{(c - c_i - c_{pi})P_0} \tag{6}$$

Here c_i is the concentration of site i at which drug is bound and c_{pi} is equal to the concentration of site i at which probe is bound, so c-c_i-c_{pi} is the concentration of site i at which neither is bound. One expects $K_i \gg K_{pi}$. Also, D_0 and P_0 are free-drug and free-probe concentration for isolated sites

$$D_0 = D_t - \Sigma c_{pi}, \qquad P_0 = P_t - \Sigma c_{pi} \tag{7}$$

with D_t and P_t the known total drug and probe concentrations. When the binding regions of the drug and the probe are not identical and there are overlapping sites, the situation can be much more complicated (76). Equations (6) and (7) are to solved simultaneously. The value of c_{pi} for given values of D_0 and P_0 is used for $[\text{probe}]_i$ in Eq. (1).

3. *Carrier DNA: The Drug Buffer*

Often unlabeled carrier DNA is present in addition to the DNA fragment whose cleavage products are directly monitored in the electrophoresis experiments. If the carrier DNA and the fragment being studied possess approximately the same base composition, the number of drug-binding sites on the carrier can be estimated by considering the density of drug sites on the labeled fragments. It is usually adequate to represent drug binding to the carrier as if the carrier possessed an effective concentration of sites c_c with binding constant K_c, an average of the binding constants of the different sites actually present on the carrier. Then the amount of drug bound to the carrier, c_a is given by

$$K_c = \frac{c_a}{(c_c - c_b - c_a)D_0} \tag{8}$$

Usually, the concentration of drug bound to sites on the labeled fragment c_b may be neglected compared to c_a, and the determination of D_0 is simplified. The value of K_c may be determined along with $\{K_i\}$ in the minimization (following section). Sometimes c_c can also be determined.

When the carrier has a base composition which is very different from the fragment being studied, its effects appear in the total-cut curve. For example, suppose that the carrier possesses more drug-binding sites than the radiolabeled fragment. This is the case with netropsin when footprinting studies using DNase I and the 139-mer are carried out in the presence of polydAdT·polydAdT as carrier (131). Considerably more drug binds to the carrier than the fragment during the early part of the titration, thereby releasing probe molecules to the solution. Since the fragment is also in equilibrium with the probe, an increase in probe concentration in solution will cause increased probe binding to the fragment and increased cutting.

In another case, total cleavage on the fragment may not be constant as drug is added to the system even though the base composition of the fragment and carrier are essentially the same. For example, netropsin and the DNA-cleaving metalloporphyrin MnT4 both bind to sequences which are rich in adenine and thymine (70,74,76). However, due to size difference, the porphyrin is able to bind to certain sites which are too small to accommodate the antiviral agent. Thus, whether total cut increases, decreases, or remains constant depends on the number of porphyrin-or-drug and porphyrin-"only" sites present on the carrier and fragment (76). In the case of the 139-mer and calf thymus DNA, there exists an imbalance of such sites between the carrier and radiolabeled fragment, and total cleavage on the fragment is observed to *increase* as drug is added to the system.

C. Deriving Binding Constants

1. Parameter Search

Once the model is established, the individual site binding constants K_I and other parameters are determined to give the best fit between theoretical and experimental footprinting plots. The fit is defined by

$$D = \Sigma(I_{ij} - \tilde{I}_{ij})^2 \tag{9}$$

where I_{ij} is the jth intensity (jth drug concentration) for site i, and $\tilde{I}_{ij}$ is the corresponding intensity calculated according to Eq. (1) with some set of values for the parameters. These are varied to minimize D.

The factor $[\text{probe}]_i$ in Eq. (1) depends on the parameter K_e and, through the concentration of bound drug, $\{K_i\}$, whereas $1 - \nu_i$ is calculable in terms of the $\{K_i\}$. However, all of the expressions for the fraction of site bound by drug at the various sites involve the free drug concentration. This must be calculated in terms of D_t, the known total drug concentration, by combining $D_0 = D_t - \Sigma c_i$ with equilibrium expressions. If a carrier is present, binding to the carrier must be included. The amount of drug bound to carrier is modeled by assuming a single kind of site of concentration c_c with an unknown equilibrium constant (another parameter). In addition to the equilibrium constants and K_e, $\tilde{I}_{ij}$ also involves the linear parameters k_i, which are essentially normalizing constants. The best values for these can be found analytically. To minimize D with respect to the nonlinear parameters for a sizable fragment having a number of drug sites, a systematic search algorithm in multiparameter space is required. The searching procedure which we have used, the simplex method (132), will be briefly described.

A set of values for the parameters to be determined in quantitative footprinting analysis, the equilibrium constants $K_1, K_2, \ldots, K_n$, plus the enhancement constant K_e, may collectively be thought of as a point in a space of $n + 1$ dimensions. A simplex is a collection of $n + 2$ such points, which are neither collinear nor coplanar. Suppose that D has been calculated for each of the $n + 2$ points and that point m has the largest value of D. The simplex search procedure involves moving from m to a new point which is not one of the original $n + 2$ points. The new point is obtained by calculating the center of gravity of all the points of the simplex except m, and moving, in a line from m to the center of gravity, twice the distance from m to the center of gravity. The value of D at the new point will almost always be lower than at least some of the values of D for the other points of the simplex. After replacing m by the new point, one looks for the point of the new simplex for which D is largest and repeats the process.

If D at the new point is higher than at any other point, one can move backward along the direction to the center of gravity. There also exists a procedure for accelerating the search when D at the new point is lower than at all other points of the simplex. Both procedures alter the size of the simplex: the former decreases it, and the latter increases it. Eventually, when convergence to the minimum is obtained, one finds a simplex such that every new step increases D. It is still conceivable that there are several relative minima, as well as the absolute minimum, in D as a function of the parameters, and one has no guarantee that the minimum found is the true absolute minimum. To check this, one normally carries out the simplex search several times with different starting points.

The value of D reflects the deviations of I_{ij} from $\tilde{I}_{ij}$. If the size of these deviations is no larger than the fluctuations of the experimental I_{ij} from one drug concentration to the next, the model is said to fit the data. Naturally, use of a more complicated model with more parameters, e.g., incorporation of cooperativity and/or anticooperativity effects associated with drug binding, can always give a smaller value of D, but it makes no sense to increase the number of parameters when the deviations $I_{ij} - \tilde{I}_{ij}$ are already smaller than experimental errors. Also, as the number of parameters is increased, it becomes more likely that linear dependence between parameters will arise.

2. *Errors*

The error in an equilibrium constant is estimated by examining the effect of a change in that equilibrium constant on D. It is usually found that a change of a few percent in any one equilibrium constant changes D by 10% or more so that equilibrium constants are reported to two significant figures. Several things must be noted in making this statement, however: (a) It is usually possible to make a change of more than a few percent in any one constant without changing D by 10% if the other constants are allowed to adjust. (b)

When there are several binding constants of different sizes, D is much less sensitive to the values of the smaller constants than to the values of the larger ones, especially if these describe overlapping sites. (c) How well a constant is determined depends on how many data points have been used to find the constant, which usually means how many footprinting plots measure that particular binding event.

If the site is not overlapped and in a well resolved region of the autoradiogram, a number of footprinting plots, all monitoring the same event, can be used to measure the binding constants for the site. In this case, even if the errors in individual $I_i(D_{tj})$ are relatively large, the binding constant can be precisely determined. This situation contrasts with a single site in a poorly resolved region of the gel where only one or two intensities measure the loading event. Such a condition produces large errors in the determined value of K_i.

IV. KINETIC DATA FOR DRUG-DNA INTERACTIONS USING A SEQUENCING GEL

In addition to thermodynamic factors, kinetics play an important role in the molecular mechanism of action of anticancer drugs. Prior to the development of DNA sequencing methodology, it was not possible to obtain association-disassociation information of a drug for a specific sites on natural DNA molecules. In 1986 Phillips and Crothers (133) described an assay for obtaining site-specific kinetic data for DNA interacting drugs (Fig. 30).

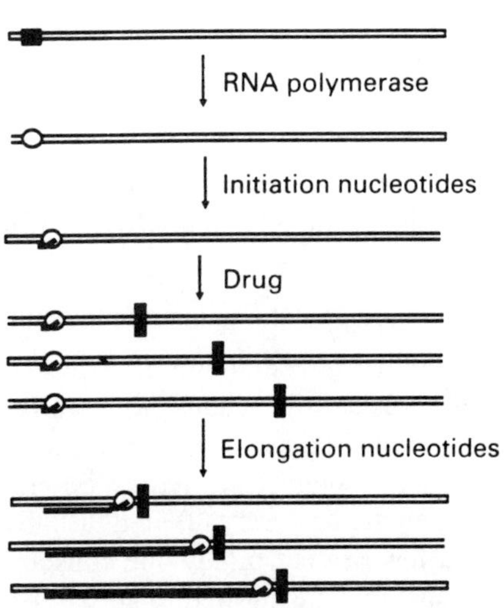

Figure 30 Diagrammatic representation of major steps in the transcription assay.

When a restriction fragment containing a promoter for RNA polymerase is treated with the polymerase and nucleotide triphosphates, transcription takes place. However, additions of DNA binding drugs to the fragment causes transcriptional blockage, ultimately yielding RNA molecules which have a common origin but which terminate at drug-binding sites. If the transcript is radiolabeled, the resulting ribo-oligonucleotide products can be separated by electrophoresis and visualized using autoradiography. After measuring natural pausing sites for RNA polymerase, the spot intensities derived from sequencing autoradiograms can be used to calculate site-specific dissociation rates for the drug. Using this transcription assay, White and Phillips (134) measured the sequence specificity and dissociation kinetics at six actinomycin D binding sites on DNA during conditions of active transcription of the DNA from the lac UV5 promoter. The dissociation kinetics for these sites, spanning two orders of magnitude, showed that sequences flanking the GC intercalation site are important in the kinetics of drug binding. An interesting feature of these studies is that termination of transcription for three of the actinomycin D sites occurred 5 to 10 nucleotides downstream of the actual drug-binding site. Since these termination sites occur shortly after regions on the RNA which have the potential to form stable hairpin loops, it has been proposed that the presence of the drug delays RNA polymerase sufficiently to enable the formation of an RNA helix close to the transcription complex.

Aside from obtaining site-specific kinetic data, transcription footprinting can be used to measure the sequence specificity of drugs. In the case of actinomycin D, the blocked transcripts are invariably observed up to the base pair comprising an intercalation site but are sometimes also seen corresponding to one base pair prior to the site. Using this approach, Phillips *et al.* (135) measured the sequence specificity of actinomycin D, distamycin, echinomycin, mithramycin, adriamycin, and nogalamycin. A recent extension of this technique has yielded the new technique of bidirectional transcriptional footprinting (136). By using two counterdirected promoters on the same DNA restriction fragment, it is possible to carry out transcription footprinting on both strands of the promoter in order to locate drug-binding sites. This can be done by selectively deactivating each promoter using a restriction enzyme. By comparing transcripts from adjacent strands of the fragment, it is possible to locate individual drug-binding sites with a high degree of accuracy, ± 1 base pair. This resolution exceeds that of conventional DNase I footprinting.

V. CONCLUSIONS

Sequencing methodology was first used to study drug-DNA interactions 12 years ago (30,31). As is evident from this review, there has been intense activity since the initial reports, and the sequence specificities of many DNA binding drugs are presently known. The reoccurring theme in all of these studies is that the sequence specificity of these agents is not high, and binding occurs at many sites on DNA. Since low DNA binding specificity may be the basis for harmful side effects, it is important to identify new agents capable of attacking only certain sites on DNA. This concept holds the promise of treating specific diseases at the DNA sequence level with a minimum of adverse side effects. It is hoped that this chapter will be of value to those using the classical techniques of identifying new drugs through screening procedures of natural products as well as those building on the exciting advances in molecular biology to create new types of totally synthetic drugs possessing unprecedented degrees of selectivity.

ACKNOWLEDGMENTS

We wish to thank Ms. Juli Neri, Ms. Joyce Lagoe, and Ms. Julie Edenfield for typing the manuscript. Special thanks also to Ms. Lorraine G. Hoogs for drawing all of the structures found in the manuscript. This work was supported by a grant from the American Cancer Society, NP-681.

REFERENCES

1. Dabrowiak, J.C., Sequence specificity of drug-DNA interactions. *Life Sci. 32*: 2915–2931 (1983).
2. Gale, E.F., E. Cundliffe, P.E. Reynolds, M.H. Richmond, and M.J. Waring, *The Molecular Basis of Antibiotic Action*. Wiley, London (1981).
3. Neidle, S., and M.J. Waring, *Molecular Aspects of Anticancer Drug Action*. Verlag-Chemie, Weinheim, Germany (1983).
4. Dervan, P.B., Design of sequence specific DNA molecules. *Science 232*: 464–471 (1986).
5. Hurley, S.H., and F.L. Boyd, Approaches toward the design of sequence-specific drugs for DNA. In *Annual Reports in Medicinal Chemistry*, Bailey, D.M. ed., Vol. 22: 259–268 (1987).
6. Pullman, B., Molecular mechanisms of specificity in DNA-antitumor drug interactions. *Adv. Drug Res. 18*: 1–113 (1989).
7. Sambrook, J., E.F. Fritsch, and T. Maniatis, *Molecular Cloning, A Laboratory Manual. 2nd ed.*, Cold Spring Harbor Laboratory Press, Cold Spring Harbor, NY (1989).
8. Maxam, A., and W. Gilbert, Sequencing end-labeled DNA with base-specific chemical cleavages. *Meth. Enzymology 65*: 499–560 (1980).
9. Warpehoski, M.A., and L.H. Hurley, Sequence selectivity of DNA covalent modification. *Chem. Res. Toxicol. 1*: 315–333 (1988).
10. Church, K.M., R.L. Wurdeman, Y. Zhang, F.-X. Chen, and B. Gold, *N*-(2-

Chloroethyl)-*N*-nitrosoureas covalently bound to nonionic and monocationic lexitropsin dipeptides. Synthesis, DNA affinity binding characteristics and reactions with ^{32}P end-labeled DNA. *Biochemistry 29*: 6827–6838 (1990).

11. Gralla, J.D., S. Sasse-Dwight, and L.G. Poljak, Formation of blocking lesions at identical DNA sequences by the nitrosourea and platinum classes of anticancer drugs. *Cancer Res. 47*: 5092–5096 (1987).
12. Royer-Pokora, M., L.K. Gordon, and W.A. Haseltine, Use of exonuclease III to determine the site of stable lesions in defined sequences of DNA: The cyclobutane pyrimidine dimer and *cis* and *trans* dichlorodiammine platinum II examples. *Nucleic Acids Res. 9*: 4595–4609 (1981).
13. Tullius, T.D., and S.J. Lippard, *cis*-Diaminedichloro-platinum(II) binds in a unique manner to oligo(dG)·oligo(dC) sequences in DNA—A new assay using exonuclease III. *J. Am. Chem. Soc. 103*: 4620–4622 (1981).
14. Millard, J.T., M.F. Weidner, S. Raucher, and P.B. Hopkins, Determination of the DNA cross-linking sequence specificity of reductively activated mitomycin C at single-nucleotide resolution: Deoxyguanosine residues at CpG are cross-linked preferentially. *J. Am. Chem. Soc. 112*: 3637–3641 (1990).
15. Tullius, T.D., B.A. Dombroski, M.E.A. Churchill, and L. Kam, Hydroxyl radical footprinting: A high resolution method for mapping protein-DNA contacts. *Meth. Enzymology 155*: 537–558 (1987).
16. Goodisman, J., and J.C. Dabrowiak, Theoretical analysis of the footprinting experiment. *J. Biomol. Struct. Dynamics 2*: 967–979 (1985).
17. Suck, D., A. Lahm, and C. Oefner, Structure refined to 2 Å of a nicked DNA octanucleotide complex with DNase I. *Nature 332*: 465–468 (1988).
18. Hertzberg, R.P., and P.B. Dervan, Cleavage of DNA with methidiumpropyl-EDTA-Iron (II): Reaction conditions and product analyses. *Biochemistry 23*: 3934–3945 (1984).
19. Dabrowiak, J.C., K. Kissinger, and J. Goodisman, Quantitative footprinting analysis of drug-DNA interactions: Fe(III) methidium-propyl-EDTA as a probe. *Electrophoresis 10*: 404–412 (1989).
20. Carter, S.K., S.T. Crooke, and H. Umezawa, Bleomycin. In *Current Status and New Developments*, Academic Press, New York (1987).
21. Hecht, S.M., *Bleomycin, Chemical, Biochemical and Biological Aspects*. Springer-Verlag, New York (1979).
22. Dabrowiak, J.C., Bleomycin. *Adv. Inorg. Biochem. 4*: 69–113 (1982).
23. Stubbe, J., and J.W. Kozarich, Mechanisms of bleomycin-induced DNA degradation. *Chem. Rev. 87*: 1107–1136 (1987).
24. Chien, M., A.P. Grollman, and S.B. Hortwitz, Bleomycin-DNA interactions: Fluorescence and proton magnetic resonance studies. *Biochemistry 16*: 3641–3647 (1977).
25. Povirk, L.F., M. Hogan, and N. Dattagupta, Binding of bleomycin to DNA: Intercalation of the bithiazole rings. *Biochemistry 18*: 96–101 (1979).
26. Booth, T.E., T.T. Sakai, and J.D. Glickson, *Biochemistry 22*: 4211–4217 (1983).
27. Dabrowiak, J.C., F.T. Greenway, W.E. Longo, M. Van Husen, and S.T.

Crooke, A spectroscopic investigation of the metal binding site of bleomycin A_2. *Biochem. Biophys. Acta 517*: 517–526 (1978).

28. Sausville, E.A., R.W. Stein, J. Peisach, and S.B. Horwitz, Properties and products of the degradation of DNA by bleomycin and iron (II). *Biochemistry 17*: 2746–2754 (1978).
29. Sausville, E.A., J. Peisach, and S.B. Horwitz, A role for ferrous iron and oxygen in the degradation of DNA by bleomycin. *Biochem. Biophys. Res. Commun. 73*: 814–822 (1976).
30. Takeshita, M., A.P. Grollman, E. Ohtsubo, and H. Ohtshubo, Interaction of bleomycin with DNA. *Proc. Natl. Acad. Sci. USA. 75*: 5983–5987 (1978).
31. D'Andrea, A.D., and W.A. Haseltine, Sequence specific cleavage of DNA by the antitumor antibiotics neocarzinostatin and bleomycin. *Proc. Natl. Acad. Sci. USA 75*: 3608–3612 (1978).
32. Kozarich, J.W., L. Worth, Jr., B.L. Frank, D.F. Christner, D.E. Vanderwall, and J. Stubbe, Sequence-specific isotope effects on the cleavage of DNA by bleomycin. *Science, 245*: 1396–1399 (1989).
33. Nunn, A.D., The relative stability of bleomycin metal complexes to metal exchange. *Eur. J. Nucl. Med. 2*: 53–57 (1977).
34. Chang, C.-H., and C.F. Meares, Light-induced nicking of deoxyribonucleic acid by cobalt (III) bleomycins. *Biochemistry 21*: 6332–6334 (1982).
35. Chang, C.-H., and C.F. Meares, Cobalt-bleomycins and deoxyribonucleic acid: Sequence-dependent interactions, action spectrum for nicking, and indifference to oxygen. *Biochemistry 23*: 2268–2274 (1984).
36. McLean, M.J., A. Dar, and M.J. Waring, Differences between sites of binding to DNA and strand cleavage for complexes of bleomycin with iron or cobalt. *J. Mol. Recogn. 1*: 184–192 (1989).
37. Fox, K.R., G.W. Grigg, and M.J. Waring, Sequence-selective binding of phleomycin to DNA. *Biochem. J. 243*: 847–851 (1987).
38. Takeshita, M., L.S. Kappen, A.P. Grollman, M. Eisenberg, and I.H. Goldberg, Strand scission of deoxyribonucleic acid by neocarzinostatin, auromomycin, and bleomycin: Studies on base release and nucleotide sequence specificity. *Biochemistry 20*: 7599–7606 (1981).
39. Golik, J., G. Dubay, G. Groenewold, H. Kawaguchi, M. Konishi, B. Krishnan, H. Ohkuma, K. Saitoh, and T.W. Doyle, Esperamicins, a novel class of potent antitumor antibiotics. 3. Structure of esperamicins A_1, A_2 and A_{1b}. *J. Am. Chem. Soc. 109*: 3462–3464 (1987).
40. Lee, M.D., T.S. Dunne, M.M. Siegel, C.C. Chang, G.O. Morton, and B.D. Borders, Calichemicins, a novel family of antitumor antibiotics. 1. Chemistry and partial structure of calichemicin γ_1^I. *J. Am. Chem. Soc. 109*: 3464–3466 (1987).
41. Long, B.H., J. Golik, K. Forenza, B. Ward, R. Rehfuss, J.C. Dabrowiak, J.J. Catino, S.T. Musial, K.W. Brooshire, and T.W. Doyle, Esperamicins, a class of potent antitumor antibiotics: Mechanism of action. *Proc. Natl. Acad. Sci. USA 86*: 2–6 (1989).
42. Zein, N., A.M. Sinha, W.J. McGahren, and G.A. Ellestad, Calichemicin

γ_1^I: An antitumor antibiotic that cleaves double-stranded DNA site specifically. *Science 240*: 1198–1201 (1988).

43. Kishikawa, H., Y.-P. Jiang, J. Goodismann, and J.C. Dabrowiak, Coupled kinetic analysis of cleavage of DNA by esperamicin and calicheamicin. *J. Am. Chem. Soc. 113*: 5434–5440 (1991).
44. Suguira, Y., Y. Uesawa, Y. Takahashi, J. Kuwahara, J. Golik, and T.W. Doyle, Nucleotide-specific cleavage and minor-groove interaction of DNA with esperamicin antitumor antibiotics. *Proc. Natl. Acad. Sci. USA 86*: 7672–7676 (1989).
45. Zein, N., W.J. McGahren, G.O. Morton, J. Ashcroft, and G.A. Ellestad, Exclusive abstraction of nonexchangeable hydrogens from DNA by calichemicin γ_1^I. *J. Am. Chem. Soc. 111*: 6888–6890 (1989).
46. Goldberg, I.H., *Free Radical Biol Med. 3*: 41 (1987).
47. Goldberg, I.H., Mechanism of neocarzinostatin action: Role of DNA microstructure in determination of chemistry of bistranded oxidative damage. *Acc. Chem. Res. 24*: 193–202 (1991).
48. Dasgupta, D., D.S. Auld, and I.H. Goldberg, Cyrospectrakinetic evidence for the mode of reversible binding of neocarzinostatin chromophore to poly-(deoxyadenylic-thymidylic acid). *Biochemistry 24*: 7049–7054 (1985).
49. Myers, A.G., Proposed structure of the neocarzinostatin chromophore-methyl thioglycolate adduct; A mechanism for the nucleophilic activation of neocarzinostatin. *Tetrahedron Lett. 28*: 4493–4496 (1987).
50. Myers, A.G., and P.J. Proteau, Evidence for spontaneous, low temperature biradical formation from a highly reactive neocarzinostatin chromophore-thiol conjugate. *J. Am. Chem. Soc., 111*: 1146–1147 (1989).
51. Lee, S.H., and I.H. Goldberg, Sequence-specific, strand selective, and directional binding of neocarzinostatin chromophore to oligodeoxyribonucleotides. *Biochemistry 28*: 1019–1026 (1989).
52. Povirk, L.F., C.W. Houlgrave, and Y.-H. Han, Neocarzinostatin-induced DNA base release accompanied by staggered oxidative cleavage of the complementary strand. *J. Biol. Chem. 263*: 19263–19266 (1988).
53. Kappen, L.S., and I.H. Goldberg, Identification of 2-deoxyribonolactone at the site of neocarzinostatin-induced cytosine release in the sequence d(AGC). *Biochemistry 28*: 1027–1032 (1989).
54. Kappen, L.S., C.-Q. Chen, and I.H. Goldberg, Atypical abasic sites generated by neocarzinostatin at sequence-specific cytidilate residues in oligodeoxynucleotides. *Biochemistry 27*: 4331–4340 (1988).
55. Montgomery, J.A., Chemistry and structure—Activity studies of the nitrosoureas. *Cancer Treat Rep. 60*: 651–664 (1976).
56. Hartley, J.A., N.W. Gibson, K.W. Kohn, and W.B. Mattes, DNA sequence selectivity of guanine-N7 alkylation by three antitumor chloroethylating agents. *Cancer Res. 46*: 1943–1947 (1986).
57. Busby, W.F., Jr., and G.N. Wogan, In *Chemical Carcinogenesis*, Searle, C.E. ed., Vol. 2: 945–1136, American Chemical Society, Washington, DC (1984).
58. Benasutti, M., S. Ejadi, M.D. Whitlow, and E.L. Loechler, Mapping the

binding site of aflatoxin B_1 in DNA: Systematic analysis of the reactivity of aflatoxin B_1 with guanines in different DNA sequences. *Biochemistry 27*: 472–481 (1988).

59. Iyer, V.N., and W. Szybalski, Mitomycins and porfiromycin: Chemical mechanism of activation and cross-linking of DNA. Isolation and structure of a covalent cross-link adduct between mitomycin C and DNA. *Science 145*: 55–58 (1964).
60. Tomasz, M., R. Lipman, D. Chowdary, J. Pawlak, G.L. Verdine, and K. Nakaniski, *Science 235*: 1204–1208 (1987).
61. Teng, S.P., S.A. Woodson, and D.M. Crothers, DNA sequence specificity of mitomycin cross-linking. *Biochemistry 28*: 3901–3907 (1989).
62. Thurston, D.E., and L.H. Hurley, A rational basis for development of antitumor agents in the pyrrolo[1,4]benzodiazepine group. *Drug Future CIPS 8*: 947–971 (1983).
63. Hertzberg, R.P., S.M. Hecht, V.L. Reynolds, I.J. Molineux, and L.H. Hurley, DNA sequence specificity of the pyrrolo[1,4]benzodiazepine antitumor antibiotics. Methidiumpropyl-EDTA-iron(II) footprinting analysis of DNA binding sites for anthramycin and related drugs. *Biochemistry 25*: 1249–1258 (1986).
64. Zakrzewska, K., and B. Pullman, A theoretical investigation of the sequence specificity in the binding of the antitumor drug anthramycin to DNA. *J. Biomol. Struct. Dyn. 4*: 127–136 (1986).
65. Hurley, L.H., V.L. Reynolds, D.H. Swenson, G.L. Petzold, and T.A. Scahill, Reaction of the antitumor antibiotic CC-1065 with DNA: Structure of a DNA adduct with DNA sequence specificity. *Science 226*: 843–844 (1984).
66. Hurley, L.H., and D.R. Needham-VanDevanter, Covalent binding of antitumor antibiotics in the minor groove of DNA. Mechanism of action of CC-1065 and the pyrrolo (1,4) benzodiazepines. *Acc. Chem. Res. 19*: 230–237 (1986).
67. Zimmer, C., and U. Wähnert, Nonintercalating DNA-binding ligands: Specificity of the interaction and their use as tools in biophysical, biochemical and biological investigations of the genetic material. *Prog. Biophys. Molec. Biol. 47*: 31–112 (1986).
68. Lane, M.J., J.C. Dabrowiak, and J.N. Vournakis, Sequence specificity of actinomycin D and netropsin binding to pBR322 DNA analyzed by protection from DNase I. *Proc. Natl. Acad. Sci. USA 80*: 3260–3264 (1983).
69. Van Dyke, M.W., R.P. Hertzberg, and P.B. Dervan, Map of distamycin, netropsin, and actinomycin binding sites on heterogeneous DNA: DNA cleavage inhibition patterns with methidiumpropyl-EDTA Fe(II). *Proc. Natl. Acad. Sci. USA 79*: 5470–5474 (1982).
70. Kopka, M.L., C. Yoon, D. Goodsell, P. Pjura, and R.E. Dickerson, The molecular origin of DNA-drug specificity in netropsin and distamycin. *Proc. Natl. Acad. Sci. USA 82*: 1376–1380 (1985).
71. Coll, M., J. Aymami, G.A. van der Marel, J.H. van Boom, A. Rich, and A.H.-J. Wang, Molecular structure of the netropsin-$[d(CGCGATATCGCG)]_2$

complex: DNA conformation in an alternating AT segment. *Biochemistry 28*: 310–320 (1989).

72. Pullman, B., Sequence selectivity in DNA-antitumor drug interactions. In *Structure and Expression*, Sarma, R.H. and Sarma, M.H. eds., Vol. 2: 237–249, Adenine Press, Albany, NY (1988).
73. Zakrzewska, K., R. Lavery, and B. Pullman, Theoretical studies of the selective binding to DNA of two non-intercalating ligands: Netropsin and SN 18071. *Nucleic Acids Res. 11*: 8825–8839 (1983).
74. Ward, B., R. Rehfuss, and J.C. Dabrowiak, Quantitative footprinting analysis of the netropsin-DNA interaction. *J. Biomol. Struct. Dyn. 4*: 685–695 (1987).
75. Ward, B., R. Rehfuss, J. Goodisman, and J.C. Dabrowiak, Determination of netropsin-DNA binding constants from footprinting data. *Biochemistry 27*: 1198–1205 (1988).
76. Dabrowiak, J.C., B. Ward, and J. Goodismann, Quantitative footprinting analysis using a DNA-cleaving metalloporphyrim complex. *Biochemistry 28*: 3314–3322 (1989).
77. Dabrowiak, J.C., J. Goodisman, and K. Kissinger, Thermodynamic data from drug-DNA footprinting experiments. *Biochemistry 29*: 6139–6145 (1990).
78. Portugal, J., and M.J. Waring, Comparison of binding sites in DNA for berenil, netropsin and distamycin. *Eur. J. Biochem, 167*: 281–289 (1987).
79. Fox, K.R., and M.J. Waring, DNA structural variations produced by actinomycin and distamycin as revealed by DNase I footprinting. *Nucleic Acids Res. 12*: 9271–9285 (1984).
80. Coll, M., C.A. Frederick, A.H.-J. Wang, and A. Rich, A bifurcated hydrogen bonded confirmation in the d(A·T) base pairs of the DNA dodecamer d(CGCAAATTTGCC) and its complex with distamycin. *Proc. Natl. Acad. Sci. USA 84*: 8385–8389 (1987).
81. Pelton, J.G., and D.E. Wemmer, Structural characterization of a 2 : 1 distamycin A d(CGCAAATTGGC) complex by two-dimensional NMR. *Proc. Natl. Acad. Sci. USA 86*: 5723–5727 (1989).
82. Pelton, J.G., and D.E. Wemmer, Binding modes of distamycin A with $d(CGCAAATTTGCG)_2$ determined by two-dimensional NMR. *J. Am. Chem. Soc. 112*: 1393–1399 (1990).
83. Kelevit, R.E., D.E. Wemmer, and B.R. Reid, ^{1}H-NMR studies on the interaction between distamycin A and a symmetrical DNA dodecamer. *Biochemistry 25*: 3296–3303 (1986).
84. Taylor, J.S., P.G. Schultz, and P.B. Dervan, DNA affinity cleaving. Sequence specific cleavage of DNA by distamycin-EDTA Fe(II) and EDTA-distamycin Fe(II). *Tetrahedron 40*: 457–465 (1984).
85. Lee, M., R.G. Shea, J.A. Hartley, J.W. Lown, K. Kissinger, J.C. Dabrowiak, G. Vesnaver, K. Breslauer, and R.T. Pon, Molecular recognition between oligopeptides and nucleic acids. Sequence specific binding of (4S)-(+)- and (4R)-(−)-dihydrokikumycin B to DNA deduced from ^{1}H NMR, footprinting studies and thermodynamic data. *J. Mol. Recogn. 2*: 6–17 (1989).

86. Lee, M., R.G. Shea, J.A. Hartley, K. Kissinger, R.T. Pon, G. Vesnaver, K.J. Breslauer, J.C. Dabrowiak, and J.W. Lown, Molecular recognition between oligopeptides and nucleic acids: Sequence-specific binding of the naturally occurring antibiotic (4*S*)-(+)-anthelvencin A and its (4*R*)-(−) enantiomer to deoxyribonucleic acids deduced from ^{1}H NMR, footprinting, and thermodynamic data. *J. Am. Chem. Soc. 111*: 345–354 (1989).
87. Griffin, J.H., and P.B. Dervan, Sequence-specific chiral recognition of right-handed double-helical DNA by (2*S*,3*S*) and (2*R*,3*R*)-dihydroxybis (netropsin) succinamide. *J. Am. Chem. Soc. 108*: 5008–5009 (1986).
88. Pjura, P.E., K. Grzeskowiak, and R.E. Dickerson, Binding of Hoechst 33258 to the minor groove of *B*-DNA. Helix parameters for the Hoechst 33258-DNA complex. *J. Mol. Biol. 197*: 257–271 (1987).
89. Teng, M., N. Usman, C.A. Frederick, and A.H.-J. Wang, The molecular structure of the complex of Hoechst 33258 and the DNA dodecamer d(CGCGAATTCGCG). *Nucleic Acids Res. 16*: 2671–2690 (1988).
90. Carrondo, M., M. Coll, J. Aymami, A.H.-J. Wang, G.A. van der Marel, J.H. van Boom, and A. Rich, Binding of a Hoechst dye to d(CGCGATATCGCG) and its influence on the conformation of the DNA fragment. *Biochemistry 28*: 7849–7859 (1989).
91. Harshman, K.D., and P.B. Dervan, Molecular recognition of B-DNA by Hoechst 33258. *Nucleic Acids Res. 13*: 4825–4835 (1985).
92. Murray, V., and R.F. Martin, Sequence specificity of ^{125}I-labelled Hoechst 33258 in intact human cells. *J. Mol. Biol. 201*: 437–442 (1988).
93. Searle, M.S., and K.J. Embrey, Sequence-specific interaction of Hoescht 33258 with the minor groove of an adenine-tract DNA duplex studied in solution by ^{1}H NMR spectroscopy. *Nucleic Acids Res. 18*: 3753–3762 (1990).
94. Laughton, C.A., T.C. Jenkins, K.R. Fox, and S. Neidle, Interaction of berenil with the *try* T DNA sequence studied by footprinting and molecular modelling. Implications for the design of sequence specific DNA recognition agents. *Nucleic Acids Res. 18*: 4479–4488 (1990).
95. Wegner, M., and F. Grummt, Netropsin, distamycin and berenil interact differentially with a high-affinity binding site for the high mobility group protein HMG-I. *Biochem. Biophys. Res. Comm. 166*: 1110–1117 (1990).
96. Yoshida, M., D.L. Banville, and R.H. Shafer, Structural analysis of $d(GCAATTGC)_2$ and its complex with berenil by nuclear magnetic resonance spectroscopy. *Biochemistry 29*: 6585–6592 (1990).
97. Brown, D.G., M.R. Sanderson, J.V. Skelly, T.C. Jenkins, T. Brown, E. Gasman, D.I. Stuart, and S. Neidle, Crystal structure of a berenil-dodecanucleotide complex: the role of water in sequence specific ligand binding. *EMBO J. 9*: 1329–1334 (1990).
98. Moore, M.H., W.N. Hunter, B. Langlois d'Estaintot, and O. Kennard, DNA-drug interactions. The crystal structure of d(CGATCG) complexed with daunomycin. *J. Mol. Biol. 206*: 693–705 (1989).
99. Wang, A.H.-J., G. Ughetto, G.J. Quigley, and A. Rich, Interactions between an anthracycline antibiotic and DNA: Molecular structure of daunomycin

complexed to d(CpGpTpApCpG) at 1.2 Å resolution. *Biochemistry 26*: 1152–1163 (1987).

100. Chen, K.-X., N. Gresh, and B. Pullman, A theoretical investigation on the sequence selective binding of daunomycin to double-stranded polynucleotides. *J. Biomol. Struct. Dyn. 3*: 445–466 (1985).

101. Chaires, J.B., Long range allosteric effects on the B to Z equilibrium by daunomycin. *Biochemistry 24*: 7479–7486 (1985).

102. Chaires, J.B., K.R. Fox, J.E. Herrera, M. Britt, and M.J. Waring, Site and sequence specificity of the daunomycin-DNA interaction. *Biochemistry 26*: 8227–8236 (1987).

103. Chaires, J.B., and J.E. Herrera, Preferential binding of daunomycin to $5'{}^{A}_{T}CG$ and $5'{}^{A}_{T}GC$ sequences revealed by footprinting titration experiments. *Biochemistry 29*: 6145–6153 (1990).

104. Skorobogaty, A., R.J. White, D.R. Phillips, and J.A. Reiss, Elucidation of the DNA sequence preferences of daunomycin. *Drug Design and Delivery 3*: 125–152 (1988).

105. Skorobogaty, A., R.J. White, D.R. Phillips, and J.A. Reiss, The 5′-CA DNA-sequence preference of daunomycin. *FEBS Lett. 227*: 103–106 (1988).

106. Van Dyke, M.W., and P.B. Dervan, Chromomycin, mithramycin, and olivomycin binding sites on heterogeneous deoxyribonucleic acid. Footprinting with (methidiumpropyl-EDTA)Iron(II). *Biochemistry 22*: 2373–2377 (1983).

107. Fox, K.R., and N.R. Howarth, Investigations into the sequence-selective binding of mithramycin and related ligands to DNA. *Nucleic Acids Res. 13*: 8695–8714 (1985).

108. Gao, X., and D.J. Patel, Solution structure of the chromomycin-DNA complex. *Biochemistry 28*: 751–762 (1989).

109. Gao, X., and D.J. Petal, Antitumor drug-DNA interactions: NMR studies of echinomycin and chromomycin complexes. *Quart. Rev. Biophys. 22*: 93–138 (1989).

110. Banville, D.L., M.A. Keniry, and R.H. Shafer, NMR investigation of mithramycin A binding to $d(ATGCAT)_2$: A comparative study with chromomycin A_3. *Biochemistry 29*: 9294–9304 (1990).

111. Williams, L.D., M. Egli, Q. Gao, P. Bash, G.A. van der Marel, J.H. van Boom, A. Rich, and C.A. Frederick, Structure of nogalamycin bound to a DNA hexamer. *Proc. Natl. Acad. Sci. USA 87*: 2225–2229 (1990).

112. Fox, K.R., and M.J. Waring, Nucleotide sequence binding preferences of nogalamycin investigated by DNase I footprinting. *Biochemistry 25*: 4349–4356 (1986).

113. Scamrov, A.V., and R.Sh. Beabealashvilli, Binding of actinomycin D to DNA revealed by DNase I footprinting. *FEBS Lett. 164*: 97–101 (1983).

114. Rill, R.L., G.A. Marsh, and D.E. Graves, 7-Azido-actinomycin D: A photoaffinity probe of the sequence specificity of DNA binding by actinomycin D. *J. Biomol. Struct. Dyn. 7*: 591–605 (1989).

115. Snyder, J.G., N.G. Hartman, B. Langlois D'Estantoit, O. Kennard, D.P. Re-

meta, and K.J. Breslauer, Binding of actinomycin D to DNA: Evidence for a nonclassical high-affinity binding mode that does not require GpC sites. *Proc. Natl. Acad. Sci. USA 86*: 3968–3972 (1989).

116. Rehfuss, R., J. Goodisman, and J.C. Dabrowiak, Quantitative footprinting analysis. Binding to a single site. *Biochemistry 29*: 777–781 (1990).

117. Goodisman, J., and J.C. Dabrowiak, Quantitative aspects of DNase I footprinting. In *Advances in DNA Sequence Specific Agents*, Hurley, L.H. ed. Jai Press, in press.

118. Rehfuss, R., J. Goodisman, and J.C. Dabrowiak, Quantitative footprinting analysis of the actinomycin D-DNA interaction. In *Molecular Basis of Specificity in Nucleic Acid-Drug Interactions*, Pullman, B. ed., 157–177, Kluwer, Norwood, MA (1990).

119. Goodisman, J., R. Rehfuss, B. Ward, and J.C. Dabrowiak, Site specific binding constants for actinomycin D on DNA determined from footprinting studies, *Biochemistry*, in press.

120. Ward, B., R. Rehfuss, J. Goodisman, and J.C. Dabrowiak, Rate enhancements in the DNase I footprinting experiment. *Nucleic Acids Res. 16*: 1359–1369 (1988).

121. Raner, G., J. Goodisman, and J.C. Dabrowiak, Porphyrins as probes of DNA structure and drug-DNA interactions. In *Metal-DNA Chemistry*, Tullius, T. ed., *ACS Symp. Ser. Vol. 402*: 74–89 (1989).

122. Huang, Y.-Q., R.P. Rehfuss, S.R. LaPlante, E. Boudreau, P.N. Borer, and M.J. Lane, Actinomycin D induced DNase I cleavage enhancement caused by sequence specific propagation of an altered DNA structure. *Nucleic Acids Res. 16*: 11125–11139 (1988).

123. Waring, M.J., The molecular basis of specific recognition between echinomycin and DNA. In *Molecular Basis of Specificity in Nucleic Acid-Drug Interaction*, Pullman, B. ed., Kluwer, Norwood, MA (1990), pp. 225–245.

124. Waring, M.J., and L.P.G. Wakelin, Echinomycin: A bifunctional intercalating antibiotic. *Nature 252*: 653–657 (1974).

125. Loretta Low, C.M., H.R. Drew, and M.J. Waring, Sequence-specific binding of echinomycin to DNA: Evidence for conformational changes affecting flanking sequences. *Nucleic Acids Res. 12*: 4865–4879 (1984).

126. Jones, M.B., U. Hollstein, and F.S. Allen, Site specificity of binding of antitumor antibiotics to DNA. *Biopolymers 26*: 121–135 (1987).

127. Ughetto, G., A.H. Wang, G.J. Quigley, G.A. van der Marel, J.H. van Boom, and A. Rich, A comparison of the structure of echinomycin and triostin A complexed to a DNA fragment. *Nucleic Acids Res. 13*: 2305–2323 (1985).

128. Portugal, J., K.R. Fox, M.J. McLean, J.L. Richemberg, and M.J. Waring, Diethyl pyrocarbonate can detect a modified DNA structure induced by the binding of quinoxaline antibiotics. *Nucleic Acids Res. 16*: 3655–3670 (1988).

129. Jeppesen, C., and P.E. Nielsen, Detection of intercalation-induced changes in DNA structure by reaction with diethyl pyrocarbonate or potassium permanganate. Evidence against the induction of Hoogsteen base pairing by echinomycin. *FEBS Lett. 231*: 172–176 (1988).

130. Mendel, D., and P.B. Dervan, Hoogsteen base pairs proximal and distal to echinomycin binding sites on DNA. *Proc. Natl. Acad. Sci. USA 84*: 910–914 (1987).
131. Dabrowiak, J.C., and J. Goodisman, Quantitative footprinting analysis of drug-DNA interactions. In *Chemistry and Physics of DNA-Ligand Interactions*, Kallenbach, N.R. ed., 143–174, Adenine Press, Guilderland, NY (1989).
132. Fletcher, R., *Practical Methods of Optimization, Vol. 1*, Chap. 2, Wiley, New York (1980).
133. Phillips, D.R., and D.M. Crothers, Kinetics and sequence specificity of drug-DNA interactions: An in vitro transcription assay. *Biochemistry 25*: 7355–7362 (1986).
134. White, R.J., and D.R. Phillips, Transcriptional analysis of multisite drug-DNA dissociation kinetics; Delayed termination of transcription by actinomycin D. *Biochemistry 27*: 9122–9132 (1988).
135. Phillips, D.R., R.J. White, H. Trist, C. Cullinane, D. Dean, and D.M. Crothers, New insight into drug-DNA interactions at individual drug binding sites probed by RNA polymerase during active transcription of the DNA. *Anti-Cancer Drug Design 5*: 21–29 (1990).
136. White, R.J., and D.R. Phillips, Bidirectional transcription footprinting of DNA binding ligands. *Biochemistry 28*: 6259–6269 (1989).

5

Construction of Quantitative Structure-Activity Relationships (QSARs) from Ligand-DNA Molecular Modeling Studies

A. J. Hopfinger and Mario Gustavo Cardozo
University of Illinois at Chicago, Chicago, Illinois

Yoshiyuki Kawakami
Tsukuba Research Laboratories, Eisai Company, Ltd., Tsukuba, Ibaraki, Japan

I. INTRODUCTION

The philosophy of this chapter is based upon the belief that the purpose of doing ligand-DNA molecular modeling studies is to *understand* the molecular mechanisms of binding in order to *predict* biological actions. We will focus upon intercalating ligands by way of example, but the methods, strategies, and problems are, largely, transferable to modeling any type of DNA-ligand interaction. DNA-intercalating agents are, as a group of compounds, diverse in chemical structure, but do possess the common structural feature of a flat aromatic ring system. The planar aromatic chromophore inserts (intercalates) between adjacent DNA base-pairs. This intercalation process results in a tight, but reversible, binding of the ligand to DNA.

DNA intercalating agents exhibit a wide range of biological actions. Indeed, it remains a major unanswered question as to the physicochemical requirements of the ligand needed to express a specific type of biological activity. In some cases it appears that DNA intercalation is a precursor catalyst to alkylation and/or cross-linking reactions of a ligand to DNA. The alkylation of the exocyclic amino group of guanidine by benzo(α)pyrene diol epoxides appears to involve intercalation of the polycyclic aromatic hydrocarbon ring system prior to alkylation (1). The ring-opening, DNA cross-linking reaction of dynemecin also appears to follow the intercalation of its aromatic chromophore (2).

The primary medicinal interest in DNA intercalating ligands is their potential as anticancer drugs (3). The intercalating agent doxorubicin(I), a

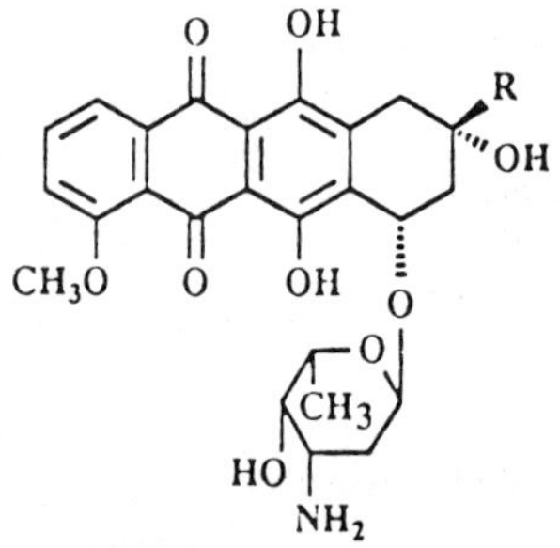

I

R = $COCH_2OH$: doxorubicin

natural product discovered by random screening, has proved to be the first broad-spectrum anticancer drug even though it also exhibits severe cardiotoxicity (4). The success of doxorubicin has provided an impetus to develop other DNA intercalation ligands as anticancer agents. A reasonable amount of information on structure-activity relationships for intercalating anticancer agents has been generated and is summarized in the next section.

II. OBSERVED STRUCTURE-ACTIVITY RELATIONSHIPS

Studies of the structure-activity relationship of DNA-intercalating ligands has focused largely on establishing relationships between DNA binding and anticancer activity with the following general findings:

1. DNA intercalation is a necessary, but not a sufficient condition for anticancer activity. For example, a study of alkyl-substituted derivatives of amacrine(II) demonstrated that a *t*-butyl group at the 3- or 4-positions prevented intercalation binding of these analogs to DNA. Correspondingly, the in vivo antitumor activity was also lost (5).

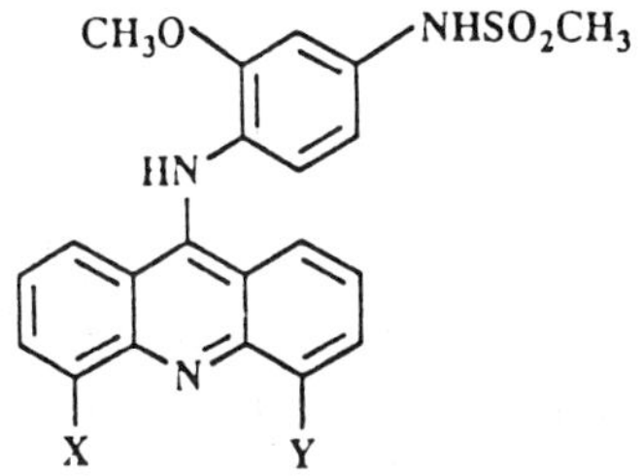

II

X = Y = H : amsacrine

2. Intercalation binding strength of a ligand shows a positive correlation with cytotoxic potency (anticancer activity) in several, but not all, intercalating ligand analog series. A set of forty-eight 9-anilinoacridines exhibit a significant correlation between the association constants, *K*, for binding to poly-d(A-T), and measured potencies of the analogs against L1210 leukemia in vivo (6). However, bis-intercalating agents, such as the diacridines, III, bind to DNA nearly as tightly as regulatory proteins, $K \sim 10^9\ M^{-1}$, but do not exhibit corresponding anticancer potency in any of the screens in which they were tested. Overall, a high level of intercalative binding seems to be a necessary, but again, not a sufficient condition for anticancer activity (7,8).

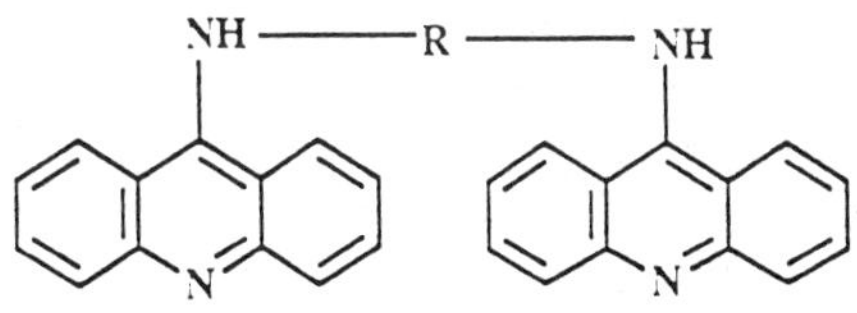

III

3. The limited available data suggests that there is a correlation between anticancer activity and ligand-DNA residence time. The dissociation

kinetics of a set of 9-aminoacridine-4-carboxamides, IV, indicated a direct correlation between anticancer (antitumor) potency and the kinetic stability of the ligand-DNA intercalation complex (9).

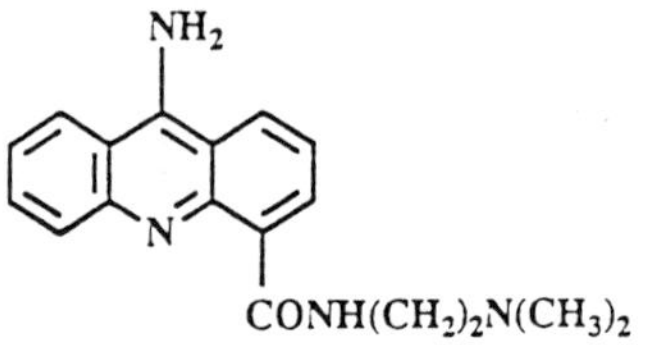

IV

4. The majority of intercalating ligands show no base-pair sequence specificity, or exhibit a preference for guanine-cytosine base pairs, particularly 5′-CpG sites (10). This limited selectivity has been attributed to (a) net ligand polarizability (11), (b) ease of locally unwinding the DNA helix (12), and (c) additional ligand-specific H-bonding to GC sites (13). Mono-intercalators, regardless of their structural complexity, have relatively small binding sites, smaller than the 10–15 base pairs needed for drugs to act as specific gene expression inhibitors. There is no trend between intercalation sequence selectivity of ligands and their corresponding biological activity, including anticancer activity.

III. MODES OF ACTION

Realistic molecular modeling of anticancer intercalation ligand receptor interactions depends upon having some understanding of the modes of action of these types of compounds. Three principle modes of action are known. The primary mode of action is believed to be inhibition of DNA synthesis. Amacrine(II), for example, is a selective inhibitor of DNA synthesis (14) and exhibits a linear dose-response relationship for inhibition of DNA polymerase I in a cell-free system (15). There are correlations, for some intercalating ligands, between cytotoxicity and DNA damage (16,17). However, a variety of other compounds do not exhibit any correlation between intercalation propensity and extent of DNA damage (18).

More recently, it has been found that some intercalating agents, such as amacrine, stimulate the production of DNA breaks by topoisomerase II (19). Production of such DNA breaks appear to occur as a result of the intercalator binding preferentially to the "cleavable" DNA-enzyme complex, and thereby altering the equilibrium of the enzyme to bond breaking over bond making.

IV. CURRENT STATUS OF MOLECULAR MODELING OF LIGAND-DNA INTERACTIONS

The number of molecular modeling studies of ligand-DNA interactions, whether the intermolecular process is intercalation, or otherwise, is small in comparison to the number of ligand-protein interaction modeling investigations (3). Just like most protein-ligand modeling studies, the focus of DNA-ligand computational investigations has been on the structural aspects of the interaction. The very limited number of computational studies of intercalation processes have determined the minimum energy free-space DNA-ligand geometries as a function of groove insertion, and, to a lesser extent, base-pair sequence. Such studies can provide structural templates as hypotheses for ligand analog design. Generally, the impetus for developing structural templates is to provide a rationale for selecting substituents to a parent ligand which will yield more tightly intercalated analogs via local substituent-DNA interactions.

The idea of using calculated intercalation properties along with other molecular descriptors, usually derived from the ligand, to develop a quantitative structure-activity relationship (QSAR) as part of a modeling study has been rare. A deficit to not developing a QSAR, in terms of drug design, is to only have an assumed, and qualitative, relationship between biological activity and the structural features of the computed intercalation complex. An underlying implicit assumption in non-QSAR modeling studies is that biological activity is solely a function of the intercalation geometry. This is clearly an oversimplification as can be gleaned from Section II. For example, the lipophilicity of an anticancer intercalating ligand can be critical for its in vivo delivery to a tumor site.

The case studies of DNA-intercalating modeling discussed below all involve deriving QSARs as part of the study. QSARs are needed to understand the intercalation processes in terms of making meaningful assessments of the set of molecular components contributing to biological action.

V. DNA-LIGAND INTERCALATION MODELING PROBLEMS

Before discussing some intercalation modeling studies designed to generate QSARs, it is useful to discuss some problems explicit to the computational exploration of the intercalation process. We have encountered five types of problems which require further consideration in order to increase the reliability and significance of modeling systems containing DNA (and probably RNA).

A. Exploration of the Degrees of Freedom

This problem is analogous to that encountered in protein-ligand systems and, owing to the symmetry of DNA structures, may be less severe than for protein-ligand systems. One approach we have found effective is to first perform intra- and/or intermolecular energy searches over the major geometric degrees of freedom of the DNA-ligand system. The major degrees of geometric freedom are usually selected subjectively by the investigator. However, there are some tricks to meaningfully identifying major geometric variables. For example, the ligand contains a planar structural moiety that inserts parallel between adjacent base pairs. Therefore, the intermolecular degrees of freedom corresponding to translation along the DNA fiber axis (Z, see Fig. 1 (20)) can be restricted to small excursions from being midway between the adjacent base pairs for all intercalation assemblies. Moreover, the planar-ring intercalated portion of the ligand must be nearly planar to the two adjacent base pairs between which it is inserted. Therefore, two of the intermolecular degrees of orientational freedom (α and B of Fig. 1) must be restricted to values that correspond to such a planar alignment. Of course the ligand must be allowed to flip over with respect to a direction along the DNA helix axis. The planar restriction on two of the orientation degrees of freedom induces constraints on the remaining orientational variable (γ of Fig. 1) which corresponds to in-plane rotations of the ligand. Only a very limited range of in-plane ligand rotations can be sterically tolerated for intercalation geometries. Finally, insertion of the ligand into the DNA structure must occur from specific entry points (major and/or minor grooves for B-form DNA), and only a limited range of distances of the ligand from the DNA structure correspond to intercalation

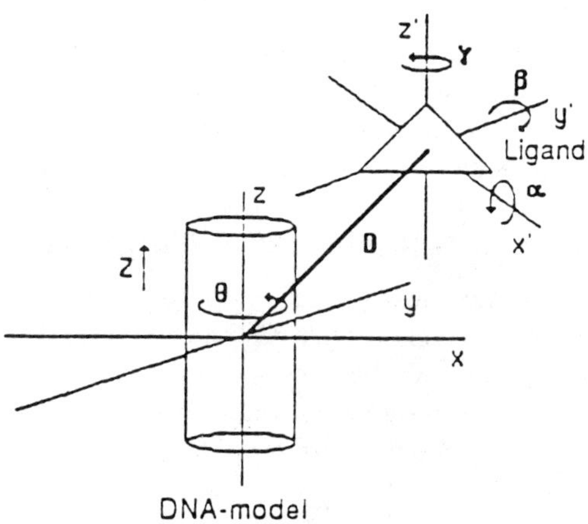

Figure 1 The intermolecular degrees of freedom used to define the interaction of the intercalating ligand (triangle) with a DNA model (cylinder).

assemblies. Thus, a set of search constraints can be imposed on θ and D as defined in Figure 1.

The intramolecular degrees of freedom can be pruned down in two ways. First, the intercalation geometry of the DNA structure, normally based upon x-ray structures (21,22), is held rigid in the initial geometric scan/search process. Secondly, the ligand is restricted to being close in conformation to minimum-energy intramolecular states. Moreover, if the ligand contains large, highly flexible substituents, these substituents are temporarily stripped off the ligand, and preliminary intercalation scans/searches carried out for the structurally simplified ligand molecule. In the case of studying an analog series of intercalators, relatively rigid analogs devoid of large, flexible substituents are often available. If these analogs are known to intercalate, then they can be used, in a rigorous modeling fashion, to perform the initial geometric intercalation scan/searches.

Overall, the technique of searching described above permits an investigator to often consider all major geometric degrees of intercalation freedom in a search for the set of proximate stable intercalation complex structures.

B. Modeling a Polymer Using Oligomer-End Effects

This problem of how large a DNA model polymer to select in a molecular dynamics (MD) simulation involves determining the balance between computational practicality and structural reliability in the MD simulation of an intercalation complex. In the absence of any computer resource constraints, a direct solution to this problem is to use a large DNA-duplex segment, often 30 or more base pairs (23,24).

An alternate approach to represent the dynamic behavior of a chain segment of a high polymer is to use an oligomer in which the motions of the oligomer ends are damped. This damping to emulate the motions of a high polymer can be achieved to some degree by lowering temperature, but in a more controlled fashion by the addition of fictitious masses on the ends of the oligomer. Selection of the specific masses and their locations is largely a trial-and-error procedure to reproduce experiment and/or computational data on conformational and/or motion behavior in large (high-polymer) chains. One specific attempt to take into account end-effect geometries and motions in an MD simulation is given below as part of the case study report on the QSAR analysis of a set of anticancer benzothiopyranoindazole (BTPI) analogs.

C. Inclusion of Counterion Effects

The DNA molecules are very likely ionized, with various types of counterions nearby, under many relevant physiological drug action conditions.

Many proteins operate in similar chemical states for relevant biological conditions. Still, relatively limited modeling and computational exploration of ionization/counterion effects on ligand binding behavior have been reported. This is probably due, in the main, to limitations in force-field representation, both in terms of mathematical form and in the corresponding parameterization, of interactions involving charged species. Perhaps the preferred way to currently include ion charge interactions in a modeling study is to determine how sensitive molecular geometry and/or thermodynamics are to the representation of the ionic interactions in the force field. That is, the force-field contributions of ionic species are treated as scaling variables in the modeling study in order to compensate for the uncertainty in their behavior.

D. Role of Base-Pair Sequence

Experimental evidence cited above suggests that base-pair sequence does not play a large role on the specific nature of most intercalation complexes. The only general observation appears to be that G-C-rich regions are preferred intercalation sites. However, there are some cases where base-pair sequence specificity is part of the intercalation mechanism. Moreover, it is desirable to be able to explore the role of base-pair specificity upon DNA-ligand interactions as part of the overall modeling strategy. The good news is that the computation of DNA-ligand binding as a function of base-pair sequence is straightforward once a sequence is selected. The bad news is that the relatively time-consuming and complex sequence of searching, minimizing, and simulating a DNA-ligand complex, as a function of its geometry, must be repeated for each base-pair sequence considered in the representation of the DNA.

Two strategies to economize the modeling process as a function of base-pair sequence can be suggested. The first is to bracket the modeling study by first considering only $(G\text{-}C)_n$ and $(A\text{-}T)_n$ models in the study. To some extent, these two base-pair sequences will provide information on the largest geometric and thermodynamic variations on an intercalation process that can be realized as a function of base-pair sequence. If the stable intercalation complexes for these two base-pair sequence models are very similar, then it is reasonable to expect that the corresponding properties of the DNA-ligand intercalation complexes will also be similar. The other modeling strategy is to use the stable intercalation geometries found for one particular base-pair sequence ($(G\text{-}C)_n$ is a good choice) as starting points in energy minimization studies for other base-pair sequence models. This strategy seems reasonable because it appears the overall geometries of intercalation are not too sensitive to base-pair sequence. Rather, the relative stability of different classes of intercalation geometries varies with base-pair sequence. It is possible to eliminate the time-consuming process of searching the geometric degrees of freedom for intercalation when this strategy can be applied.

E. Role of Solvent on Ligand-Receptor Binding

There is no report of the successful prediction of the absolute values of ligand receptor binding constants, K_i, using simulation, mainly MD methods. At best, differences in binding free energies, ΔG's, of members of highly congeneric ligands series to a common receptor macromolecule (an enzyme in most cases), in the presence of (aqueous) solvent, have been computed using free-energy perturbation theory (FEPT) (25,26). Even in these cases, however, the FEPT calculations are found to be metastable in that relatively small changes in the perturbation degrees of freedom can lead to large changes in the estimated ΔG's.

There are three general classes of interactions that must be considered in solvated ligand-receptor binding: (a) ligand-solvent, (b) receptor-solvent, and (c) (ligand-receptor) complex-solvent. Our studies suggest that limitations in computing the (ligand-receptor) complex-solvent term is the primary source of not being able to accurately estimate ligand-receptor thermodynamics. The usual way of treating the (ligand-receptor) complex-solvent system is to first construct a trial ligand-receptor system in the absence of solvent. This system may be subjected to energy minimization and/or limited MD simulation to alleviate any local high-energy interactions. The resulting geometry is then solvated using one of the proposed schemes (25,27). The resulting system is then the initial point in an MD simulation from which thermodynamic properties are estimated using the trajectory data. This simulation procedure can, although seldom done, be repeated using different geometric starting points for the solvated ligand-receptor complex.

The "achilles heel" in this simulation procedure seems to reside in the addition of the solvent (water) molecules to the ligand-receptor complex. This approach generates a "blanket" of solvent molecules over the surface of a tightly joined ligand-receptor geometry. There is little opportunity for solvent molecules to "squeeze" between the ligand and the receptor during the MD simulation. We have found that the resulting thermodynamics of this type of system can be significantly different from the case where the solvated ligand and the solvated receptor are forced to coalesce into one another, and, ultimately, bind to one another (28). In this case, the ligand and receptor are partially desolvated, and solvent molecules can be bound "between" the ligand and the receptor in the resulting binding state.

The simulated coalescence of the solvated ligand and solvated receptor is probably a more realistic modeling of the binding process than is the blanket of solvent molecules over the naked ligand-receptor complex. In principle, the two models should converge to a common equilibrium state at long MD simulation times. The coalescence model approximates a starting point on the MD simulation trajectory farther along the time march to equilibrium. However, it seems this simulation time advantage must come at the expense of forcing the ligand and receptor over a joint partial desolvation energy barrier which precedes their actual binding. If the forced trajectory over the partial desolvation energy barrier is not correct (the lowest barrier pathway), then the resulting binding trajectory, and corresponding thermodynamics, can be in error. At this time there are some untested ideas on how to realistically traverse the desolvation zone in short simulation times to realize the binding process. One simple thought is to artificially enhance the temperature (kinetic energy) of the system during desolvation. Overall, the coalescence and blanket solvation models for ligand binding are plagued with long simulation times to either traverse the desolvation zone (coalescence), or for solvent molecules to diffuse into the naked ligand-receptor complex (blanket).

VI. QSAR INTERCALATION STUDIES

A. Intercalation of Doxorubicin and Daunomycin Anthracycline Analogs

By the end of the 1970s, detailed molecular models had been proposed for the intercalation of daunomycin(V) and doxorubicin(I) with DNA (29–32). In these models the drug intercalates into the major groove of the DNA such that the long axis of the anthraquinone ring is nearly parallel to the long axes of the adjacent DNA base pairs. The proposed intercalation models differed from one another with respect to modes of intermolecular hydrogen bonding, and formation of an electrostatic bond between the charged sugar amine of the drug and a phosphate of the DNA backbone.

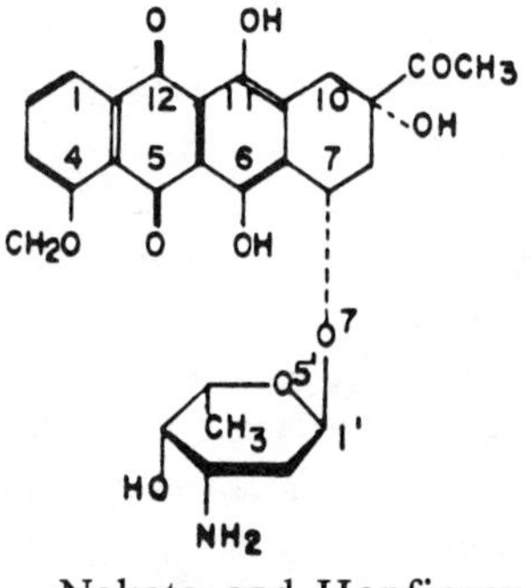

V

Nakata and Hopfinger (33) modeled the intercalation of doxorubicin with dinucleotide dimer sequences using the search and minimization techniques described above. Dimer sequence, type of sugar ring puckering, entry into the major and minor grooves of the dinucleotide dimer minihelix, ligand conformation, and charged state of the primary sugar amine group were also considered in the study.

The nonbonded potentials developed by Hopfinger (34) were used with a monopole-monopole electrostatic potential in which all charge densities were computed using the CNDO/2 method (35). A total of 144 systematically selected intermolecular starting positions, involving both the major and minor grooves, were chosen in each drug-dinucleotide energy minimization. The dinucleotide sequences

$$\Big\downarrow\begin{pmatrix}\text{C-G}\\ \text{P P}\\ \text{G-C}\end{pmatrix}\Big\uparrow \quad \text{and} \quad \Big\downarrow\begin{pmatrix}\text{T-A}\\ \text{P P}\\ \text{C-G}\end{pmatrix}\Big\uparrow$$

were considered in our study.

The dinucleotide dimer structure of the ethidium ion CpG crystal complex (22) and the most stable, unwound B-form-like dimer structure, found in DNA unwinding calculations (36), were each used for both sequences in the study. In the case of the calculated dimer structure, both C-3′ *endo*-C-2′ *endo* and C-2′ *endo*-C-2′ *endo* sugar ring puckerings were independently considered. The dinucleotide geometry was held fixed in each intermolecular energy minimization.

Three initial conformations of doxorubicin were used: the bromoacetyl crystal structure (37), and the two conformations found to be intramolecular energy minima (38).

The minimum energy complexes, with respect to major and minor groove intercalation, for each dinucleotide dimer sequence and conformation, are reported in Table 1. A consistent finding is that minor groove intercalation is preferred except for the (i^5UpA · Et) dinucleotide dimer geometry. The energy preference for minor groove intercalation decreases for the ionized form of doxorubicin. Minor groove intercalation is characterized by the anthraquinone ring inserting between base pairs with its long axis nearly perpendicular to the long axes of the base pairs. This mode of intercalation had not been observed in any intercalation crystal structure, but was suggested by Patel (39), from NMR studies of solution complexes of nucleic acid oligomeric dimers with daunomycin. Shortly after these NMR studies were reported, this new mode of intercalation was found in the crystal structure of a d(CpGpTpApCpG) dimer-daunomycin intercalation complex by Rich and co-workers (22b). The only significant structural difference between the doxorubicin-(CpG) dimer structure and the d(CpGpTpApCpG)-daunomycin complex is the conformation of the sugar rings. They are flipped 180° with respect to one another when the anthraquinone ring is intercalated. The difference in intercalation energies for

Table 1 Intercalation Properties of Doxorubicin with Two Dinucleotide Dimer Sequences

Dinucleotide		Doxorubicin		Intermolecular energy[e]	
Sequence[a]	Position[b]	Conformation[c]	Form[d]	Complex	Isolated dinucleotide
D1	Minor	AM	NH_2	−124.08	−16.06
		AX		−118.38	−17.11
		AD		−124.16	−16.62
	Major	AM	NH_2	−112.43	−16.06
		AX		−116.63	−17.11
		AD		−118.61	−16.62
D2	Minor	AM	NH_2	−117.74	−15.80
		AX		−117.48	−16.96
		AD		−121.89	−16.46
	Major	AM	NH_2	−114.73	−15.89
		AX		−103.78	−16.96
		AD		−104.27	−16.46
D3	Minor	AM	NH_2	−123.87	−40.77
		AX		−127.99	−41.33
		AD		−132.60	−40.84
	Major	AM		−123.05	−40.77
		AX		−104.73	−41.33
		AD		−105.99	−40.84
D3	Minor	AM	NH_3^+	−109.20	−38.72
		AX		−107.91	−39.80
		AD		−113.98	−39.42
	Major	AM		−111.66	−38.72
		AX		−105.31	−39.80
		AD		−96.43	−39.42
r1	Minor	AM	NH_2	−73.62	−44.42
		AX		−90.73	−45.46
		AD		−61.21	−44.42
	Major	AM		−101.02	−44.42
		AX		−94.38	−45.46
		AD		−94.49	−44.95

[a]D1-(TpC-ApG), C2′ *endo*-C2′ *endo*; D2-(TpC-ApG), C3′ *endo*-C2′ *endo*; D3-(CpG-GpC), C3′ *endo*-C2′ *endo*, r1-(i^5UpA·ET) from Ref. 22.
[b]Refers to entry into the major or minor groove of the minihelix of the dinucleotide dimer.
[c]AM - global intramolecular minimum energy conformation as reported in Ref. 38. AX - secondary intramolecular minimum energy conformation as reported in Ref. 38. AD - *N*-bromoacetyldaunomycin crystal conformation from Ref. 37.
[d]Refers to sugar amine group being neutral or charged.
[e]Complex refers to the minimum energy intercalation structure, the energy units are kcal/mole.

the two flipped sugar ring conformations are $\Delta E = 0.5$ kcal mole^{-1} for doxorubicin and $\Delta E = 0.7$ kcal mole^{-1} for daunomycin. These small energy differences suggest that crystal packing forces may be ultimately responsible for the crystal structure sugar ring conformation of the intercalated daunomycin molecule.

The predicted minor groove intercalation structure is stabilized by complementary electrostatic interactions between the adjacent base pairs with the anthraquinone ring. A weak hydrogen bond–electrostatic interaction is realized between the sugar amine and a backbone phosphate group. The 9-position hydroxyls also have favorable electrostatic interactions with either phosphates or base pair groups, depending upon drug conformation and dinucleotide sequence.

The preferred models of major and minor groove intercalation are shown in stereo in Fig. 2 for (CpG). The mode of major groove intercalation is similar to the early proposed models in that the anthraquinone ring is nearly "parallel" to the base pairs. Proposed major groove intercalation models (29–32) also postulate a strong electrostatic bond between the NH_3^+ group of the drug and a second-nearest-neighbor nucleotide residue phosphate group. Thus, these models involve a trinucleotide dimer as the intercalation "substrate." Nevertheless, tetranucleotide dimer-doxorubicin calculations still predict the minor groove intercalation complex to be more stable than major groove complexes.

The most significant aspect of the intercalation part of this computational study is the *correct prediction of a new mode of intercalation* that is supported by both crystal structure and NMR data.

Fink et al. (40) carried out a QSAR analysis of 29 anthracyclines of the general form of VI which are reported in Table 2. The antitumor activity of these compounds against B-16 melanoma have been measured in terms of $\log(1/C)$. Here, C is the molar concentration (moles kg^{-1}) producing a t/c of 125 against B-16 melanoma introduced intraperitoneally into mice, where t is the increased life-span of the population given the intercalating agent compared to that of the control (c) population not given the intercalator.

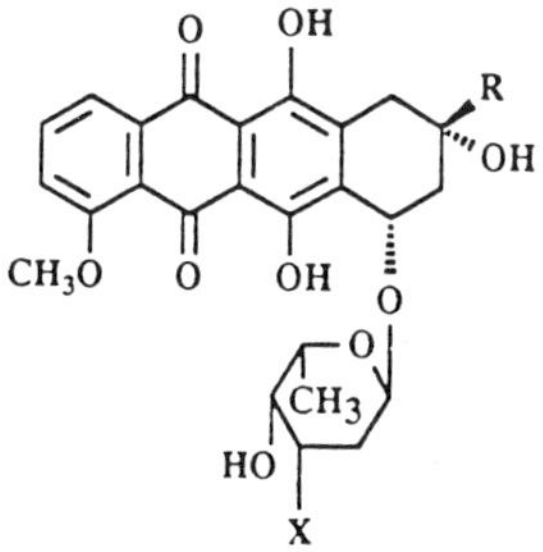

VI

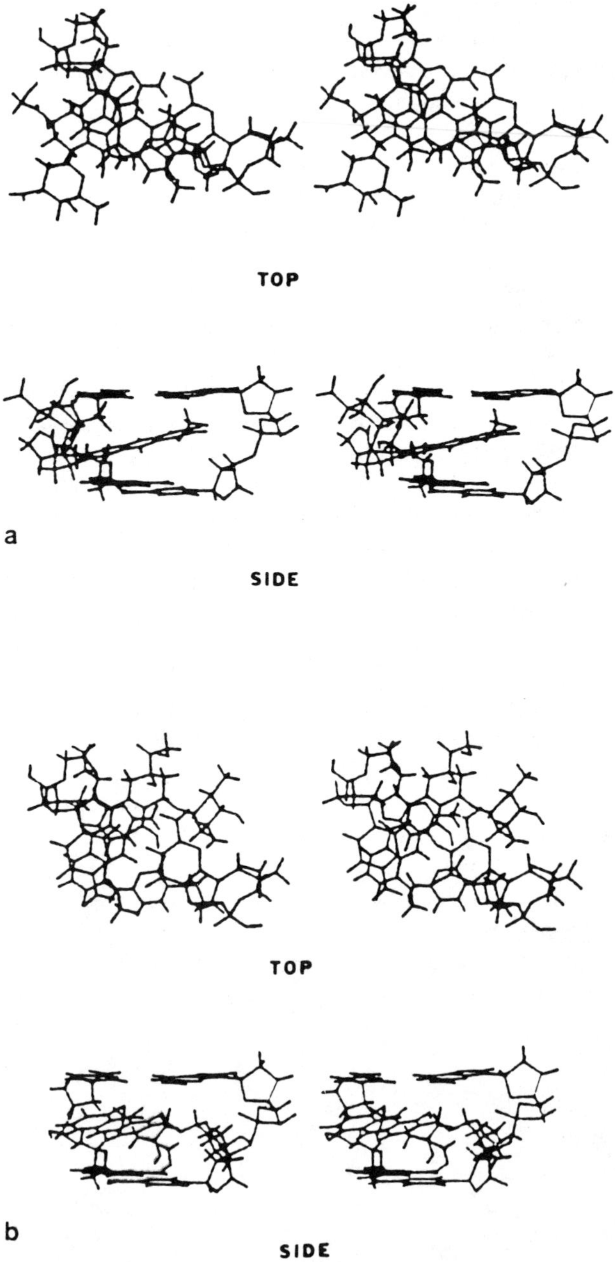

Figure 2 Preferred mode of the intercalation of doxorubicin with (CpG) having mixed sugar ring puckering: (a) major groove, (b) minor groove.

Fink et al. (40) were able to construct a QSAR for the antitumor activity of 23 of the 29 anthracyclines in Table 2.

$$\log(1/C) = 0.41[\log P] + 0.48[I_0] + 0.8[I_1] + 6.57$$
$$N = 23 \quad R = 0.874 \quad S = 0.288 \quad AE = \pm 12.0\% \qquad (1)$$

where N is the number of compounds considered, R is the correlation coefficient, S is the standard deviation, and AE is the average error of prediction. Also, P is the water-octanol partition coefficient (log P values are reported in Table 2), and I_0 and I_1 are indicator variables having values of 0 or 1: I_0 is set equal to 1 only if the 4-OCH_3 of (VI) is converted to OH or H. I_1 is set equal to 1 for the five hydrazones having "large lipophilic moieties." Six compounds in the data base (Table 2) could not be included in the construction of Eq. (1).

The minor groove intercalation geometry found in our structural modeling calculations for doxorubicin (see Fig. 2b) was used as the starting point for intercalation energy minimization for each of the 29 analogs in Table 2. The resulting intercalation energies (I.E.), were then used as QSAR correlation descriptors to explain the observed variance in log(1/C) values reported in Table 2.

The resulting correlation equation for antitumor activity is

$$\log(1/C) = 0.246[\log P] - 0.030[\log P]^2 - 0.423[\text{I.E.} \times 10] - 0.27$$
$$N = 29 \quad R = 0.901 \quad S = 0.289 \quad AE = \pm 11.9\% \quad \log P_{opt} = 4.17 \qquad (2)$$

Log P_{opt} corresponds to that value of log P which maximizes its contribution to increasing log(1/C). Equation (2) involves only two descriptors, log P and I.E., and can be used to explain the activity of all 29 anthracyclines in Table 2 with about the same accuracy as Eq. (1). Thus the S and AE of Eqs. (1) and (2) are, respectively, about the same, but the value of R in eq. (2) is significantly larger than that of Eq. (1).

The most significant aspect of the QSAR part of this study is that the *computation of I.E. of a hypothetical (R,X)-VI ligand with a (CpG) dimer, along with estimation of its log P, permits the prediction of antitumor activity in advance of synthesis.*

B. Prediction of Initial Reduction Potentials of Compounds Related to Anthracyclines

The major drawback of doxorubicin, daunomycin, and related anthracyclines in anticancer therapy is their very significant cardiotoxicity. The biochemical mechanism which controls the cardiotoxic behavior of anthracyclines is not fully understood. However, there is increasing evidence that the redox behavior of these molecules is involved in their cardiotoxicity

Table 2 Anthracycline Data Base, Structure-Activity Descriptors, and Observed and Predicted Antitumor Activities

No.	R	X	log P	I.E. × 10 (kcal mole^{-1} atom^{-1})	log(1/*C*) antitumor activity Obs.	log(1/*C*) antitumor activity Pred.
1	$-CH(CH_3-NH_2$	NH_2	1.84	−9.91	4.34	4.27
2	$COCH_3$	$N(CH_2\text{-}\phi)_2$	5.99	−10.96	4.37	4.78
3	$COCH_2OH$	$N(CH_3)_3$	−.95	−12.44	4.54	4.73
4	$COCH_2OCO(CH_2)CH_3$	$NHCOCF_3$	1.83	−11.37	4.55	4.88
5	$COCH_3$	OH	1.76	−10.98	4.64	4.71
6	H	NH_2	1.92	−11.59	5.05	4.99
7	$COCH_2OH$	$NCO(CH_2)_{10}CH$	6.57	−11.30	5.15	4.84
8	$COCH_3$	$N(CH_3)_2$	2.74	−12.38	5.17	5.41
9	$COCH_2OH$	$NCH_2\phi$	3.48	−11.80	5.18	5.22
10	$-C(CH_3)=NOH$	NH_2	2.80	−12.73	5.20	5.57
11	$-C(CH_3)=NNHCOCH_2OH$	NH_2	1.82	−13.05	5.58	5.60
12	$-C(CH_2OH)=NNHCO$ Cl Cl	NH_2	3.97	−12.74	5.66	5.63

13	$-C(CH_2OH){=}NNHCO-(CH_2)_6CH_3$	NH_2	4.26	−13.00	5.68	5.74
14	(5-Iminos)$-COCH_3$	NH_2	1.74	−13.62	5.75	5.82
15	$COCH_2OH$	$N(CH_3)_2$	2.18	−12.62	5.78	5.46
16	$C(CH_3){=}NNHCO$ Cl	NH_2	4.22	−12.75	5.78	5.63
17	$C(CH_3){=}NNHCO$	NH_2	3.51	−13.09	5.80	5.76
18	$COCH_2OCO$	NH_2	1.27	−14.51	5.90	6.13
19	(2,3-diCH_3;4-des-OCH_3)$-$ $COCH_2OH$	NH_2	2.28	−14.51	5.94	6.27
20	$C(CH_3){=}NNHCO$ NO_2	NH_2	3.69	−12.69	5.96	5.60
21	$COCH_3$	NH_2	1.83	−15.23	5.97	6.51
22	(4-des-OCH_3)$-COCH_3$	NH_2	1.85	−14.71	6.11	6.30
23	$COCH_2OCOCH_2OCO(CH_2)_{16}CH_3$	NH_2	1.27	−15.92	6.25	6.72
24	(4-des-CH_3)$-CH(CH_3)OH$	NHCOH[a]	1.44	−14.05	6.32	5.96
25	$COCH_2OH$	NH_2	1.27	−14.43	6.33	6.09
26	$COCH_2OCO(CH_2)_6CH_3$	NH_2	1.27	−14.09	6.34	5.94
27	(4-des-OCH_3)$-COCH_2OH$	NH_2	1.29	−14.23	6.46	6.01
28	(4′-epi)$-COCH_2OH$	NH_2	1.27	−14.31	6.50	6.04
29	(4-des-CH_3)$-COCH_3$	NH_2	1.73	−15.09	6.78	6.44

[a]NHCOH → NH_2

(41). In particular, the redox behavior of the quinone moiety appears to involve the generation of semiquinone and reactive oxygen species (42). Some chromophore-modified anthracyclines exhibit a correlation between the suppression of redox activity and corresponding cardiotoxic effects. Examples to date include 5-imino analogs (41,43,44,45) and anthrapyrazole analogs (46–49). The final QSAR study of intercalation anticancer agents to be discussed in this chapter are a set of anthrapyrazoles which exhibit low cardiotoxicity. However, before turning to a discussion of this anticancer QSAR study, we present one possible approach to predicting cardiotoxicity as a function of the ring structure of anthracycline analogs.

Two groups of workers (49,50) report the redox potentials of various anthracyclines in aqueous media with respect to a standard aqueous saturated calomel electrode at 25°C. Table 3 contains the half-wave potentials (HWPs) that have been assigned to the initial reduction of 18 anthracyclines, which were considered in this study (51) and whose structures are given in Figure 3. The structures in Table 3 are abbreviated by maintaining the chromophore, but simplifying the side chains.

The molecular geometries of each of the 18 compounds in Table 3 were built by using standard bond lengths and angles with the CHEMLAB-II molecular modeling package (52). These structures were initially optimized by using the molecular mechanics MMFF option in CHEMLAB-II. The resulting structures were then processed through the CNDO/2 (35) program to obtain partial atomic charges. The initially optimized structures with atomic partial charges were then reoptimized in MMFF, taking into account electrostatic interactions in the optimization. The reoptimized structures were then rerun through CNDO/2 and also optimized, yet a third time, by using MNDO (53). Sets of electronic descriptors including highest occupied molecular orbital energy (HOMO), lowest unoccupied molecular orbital energy (LUMO), (HOMO-LUMO) energy, and partial atomic charges were generated from both the CNDO/2 and MNDO calculations.

All combinations of the electronic descriptors were correlated against the observed HWPs by using multidimensional linear regression analysis.

The LUMO energy levels from both the CNDO/2 and MNDO calculations strongly correlate with the observed HWP. The correlation equations are

$$\begin{gathered} \text{HWP} = 0.020(\text{LUMO}) + 0.538 \\ \text{CNDO/2: } N = 18 \quad R = 0.957 \quad SD = 0.066 \quad F = 175.5 \end{gathered} \tag{3}$$

and

$$\begin{gathered} \text{HWP} = 0.021(\text{LUMO}) + 1.414 \\ \text{MNDO: } N = 18 \quad R = 0.963 \quad SD = 0.057 \quad F = 206.3 \end{gathered} \tag{4}$$

Table 3 Half-Wave Potentials (HWPs) and LUMO Energy Levels of Antitumor Anthracyclines

	-HWP obsd (V)		-HWP					
	Peters et al. (1986)[b]	Showalter et al. (1986)[c]	Predicted (V)		Residuals		LUMO (kcal mole^{-1})	
Compounds			Eq. (1)	Eq. (2)	Eq. (1)	Eq. (2)	CNDO/2	MNDO
anthraquinone[a] (AQS)		0.456	0.3437	0.3749	0.1123	0.0811	−9.7006	−50.2253
doxorubicin (DXR)	0.64		0.6364	0.5956	0.0035	0.0444	4.9364	−39.5602
4-de-*O*-methyl-11-deoxy-DXR	0.53		0.6297	0.6034	−0.0997	−0.0734	4.6013	−39.1837
11-deoxy-DXR	0.54		0.6186	0.6474	−0.0786	−0.1074	4.0456	−37.0601
daunorubicin (DAU)	0.64	0.625	0.6364	0.5956	0.0035	0.0444	4.9364	−39.5602
4-de-*O*-methyl-11-deoxy-DAU	0.53		0.6297	0.6034	−0.0997	−0.0734	4.6013	−39.1837
11-deoxy-DAU	0.54		0.6186	0.6474	−0.0786	−0.1074	4.0456	−37.0601
5-imino-DAU		0.675	0.6964	0.6777	−0.0214	−0.0026	7.9321	−35.5963
N,N-dibenzyl-5-imino-DAU	0.745		0.6964	0.6777	0.0486	0.0673	7.9321	−35.5963
5-imino-DXR	0.72		0.6964	0.6777	0.0236	0.0423	7.9321	−35.5963
5-imino-DAU-ol	0.70		0.6964	0.6777	0.0036	0.0223	7.9321	−35.5963
mitoxantrone (MTX)		0.775	0.6514	0.7464	0.136	0.0287	5.6834	−32.2810
anthrapyrazole (APZ)		1.001	1.0089	1.0147	−0.0079	−0.0137	23.5597	−19.3068
7-OH-APZ		1.040	1.0230	1.0359	0.0170	0.0041	24.2617	−18.2854
10-OH-APZ		0.983	1.0036	1.0235	−0.0206	−0.0405	23.2938	−18.8845
7,10,-$(OH)_2$-APZ		1.045	1.0162	1.0364	0.0288	0.0085	23.9223	−18.2577
7,9,10-$(OH)_3$-APZ		1.085	1.0722	1.0359	0.0128	0.0491	26.7212	−18.2849
7,8,10-$(OH)_3$-APZ		1.047	1.0182	1.0210	0.0288	0.0260	24.0226	−19.0018

[a] 2-Anthraquinonesulfonic acid sodium salt.
[b] See Ref. 41.
[c] See Ref. 49.

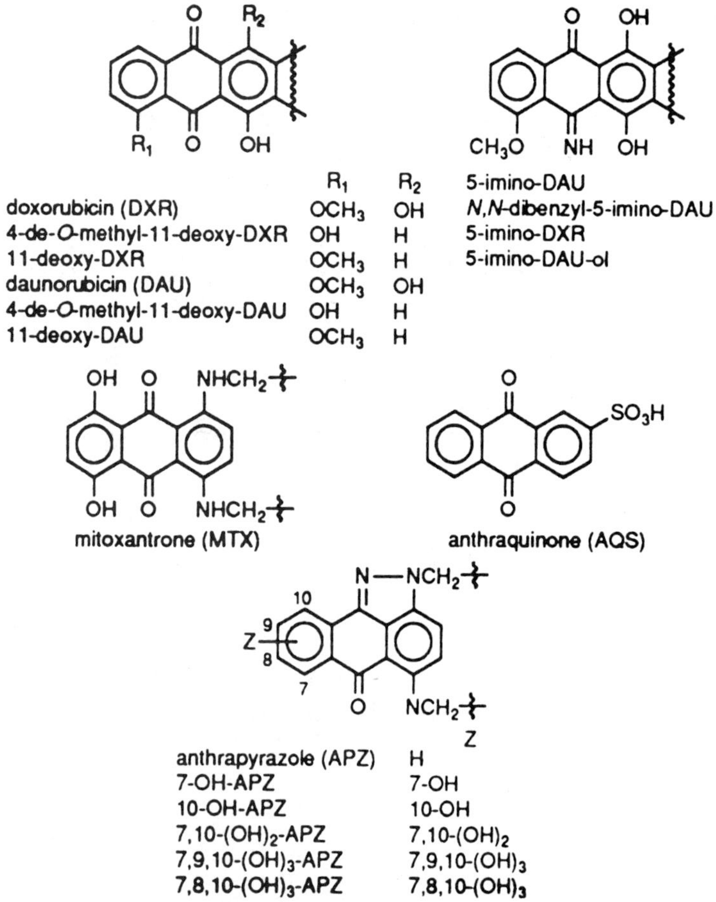

Figure 3 Structures of the 18 anthracycline-related analogs used in the (HWP, LUMO) QSAR analyses.

The CNDO/2 and MNDO LUMO energies are expressed in kcal $mole^{-1}$ and the HWPs are given in volts; N is the number of compounds, R is the correlation coefficient, SD is the standard deviation of fit, and F is the statistical significance of fit. The observed and predicted HWPs and differences for HWP are given as part of Table 3 along with the computed values of the LUMO energies. Figure 4 contains plots of (a) Eq. (3) and (b) Eq. (4) as well as the observed HWP values in HWP versus LUMO space. It can be argued that the 18 compounds studied separate into two major clusters, namely, the anthrapyrazoles and the rest of the anthracyclines.

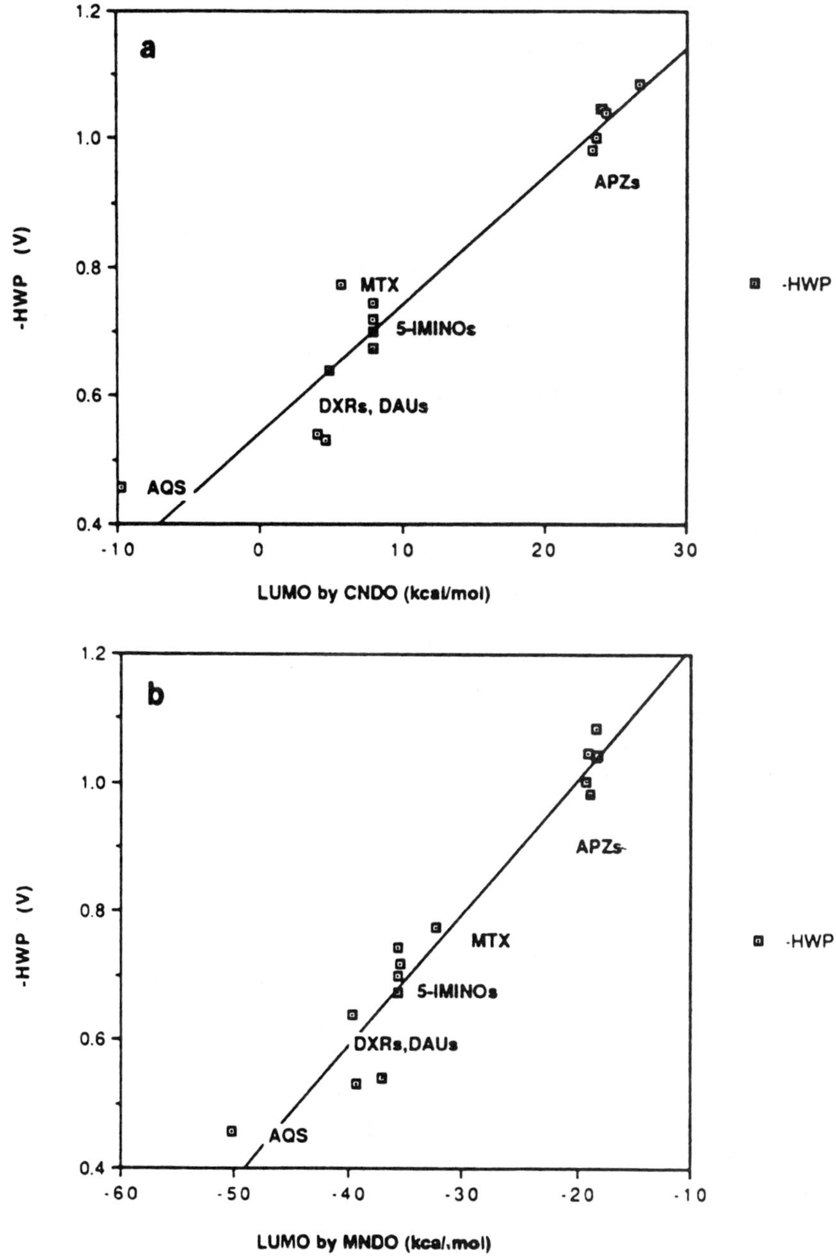

Figure 4 LUMO versus HWP for (a) the CNDO/2 study and (b) the MNDO analysis.

Closer analysis indicates that the nonanthrapyrazoles separate into 5-imino analogues and the remainder of the compounds. This clustering behavior is unfortunate in terms of applicability of regression analysis. It would be better if the HWP and LUMO values were uniformly distributed over their respective ranges. Nevertheless, the regression equations are quite significant. Moreover, the clustering behavior indicates that HWP is, as one expects, largely characteristic of the central ring structure and only depends in a minor way on ring substituents. The corresponding clustering of LUMO values indicates that the same is true for LUMO. The findings of this study are consistent with the combined experimental and theoretical studies of Tempczyk et al. (54), where superoxide anion radical formation of anthraquinone antibiotics may occur by single-electron transfer.

Quinones form an important group of substrates for several enzymes, including NADPH-cytochrome P-450 reductase, NADH dehydrogenase, xanthine oxidase, and NADH-cytochrome b_5 reductase and undergo one-electron reduction (55–59). Half-wave potentials, considered an approximate value of the free energy of reduction, were found to be negatively correlated with both the results of the microsomal oxygen consumption test and the production of superoxide anion.

It is possible that the LUMO value indicates the ease of initial one-electron reduction based upon the correlation between HWP and LUMO, since one-electron MO methods have been used. Moreover, the finding that a single calculated molecular property, LUMO, can be used to explain the variance in HWP for a structurally diverse set of anthracyclines may have useful drug design applications. It should be possible to predict the HWP and thereby infer the relative cardiotoxicity of a new anthracycline related compound, relative to the 18 investigated in this work, in advance of its synthesis.

C. Intercalation of Benzothiopyranoindazole Anticancer Analogs

The benzothiopyranoindazoles, see VII, are a novel class of chromophore-modified anthracenediones related to mitoxantione which exhibit a broad spectrum of in vivo anticancer activity (60). These analogs represent a class of third-generation DNA-complexing agents which have applications against a number of malignant diseases. While the BTPIs, as well as other anthrapyrazoles, do not appear to be particularly more potent anticancer agents than clinical anthracyclines, they do appear to be considerably less cardiotoxic (60,61). The modification of the central quinone to a quasi-iminoquinone, as compared to doxorubicin for example, results in reduced superoxide dismutase oxygen consumption in rat liver microsomal screens, and, as we have already seen, greater resistance to electrochemical reduction (61).

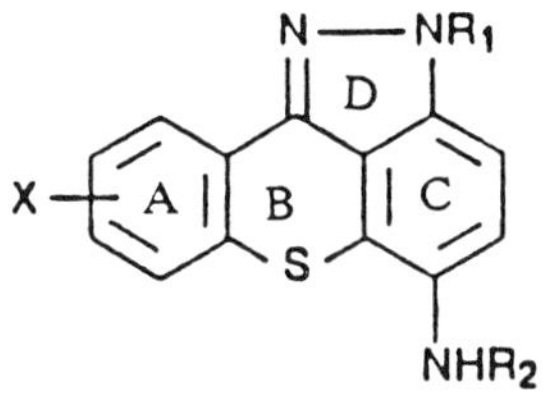

VII

One of the objectives of the study described here was to explore how the introduction of the D-ring in the BTPIs, modifies the intercalation properties of these compounds with DNA relative to the anthracyclines. There is no direct evidence, but only indirect data (62), that the BTPIs intercalate with DNA. The assumption has been made that these compounds can intercalate, and this hypothesis was tested by seeing if the calculated intercalation binding energy correlates with observed antitumor potency in a set of analogs of VII. In order to effectively explore the full molecular geometry–molecular energy relationship for intercalation, full Cartesian molecular dynamics has been employed. The exploration of the geometric degrees of freedom was a concern in the QSAR study of the doxorubicin analogs reported above. Molecular dynamics was not available to relax and search about the apparent minimum energy intercalation complexes found in the search and minimization procedures.

The geometries of the BTPIs used in this study were obtained by energy minimization of sterically allowable structures using the quantum mechanical method MOPAC (63) and the MM method MMFF (52). The intramolecular minimum energies for the flexible substituents were obtained from fixed valence conformational energy scans at 30° increments about the principal torsion angles.

G-C tetramer and hexamer sequence duplexes were used to model the DNA polymer. The central base-pair dimer of each oligomer was initially assigned the crystal geometry of the GpCp-3/CpGp-ethidium ion complex (22a). The remaining base pairs were added above and below the central dimer so as to mimic the B-form DNA conformation (21). The resulting initial geometries were then optimized in MMFF with the constraint that the center two base pairs remained separated at the observed intercalation separation distance. Electrostatic interactions in MMFF were calculated using a columbic representation with the molecular dielectric constant equal to 3.5, and atomic charges calculated by the CNDO/2 method (35). The ionic phosphate charges were defined by placing half the unit charge on the P atom, and the other half of the unit charge equally distributed over the oxygen atoms.

Eight analogs of VII which have only one, common flexible side chain, $R_1 = CH_2CH_2NEt_2$, see Table 4, were considered in terms of their interac-

Table 4 The Fixed Valence Geometry, Static Minimum Energy Intercalation Complexes of Semirigid Substituent Analogs of BTPI with a $(G\text{-}C)_4$ Duplex. ΔE corresponds to the intermolecular ligand $(G\text{-}C)_4$ interaction energy.

Compound no.	X	R_1	R_2	Anticancer activity[a] $\log(1/IC_{50})$	Intercalation no.	Amber Z^b	θ^b	γ^b	ΔE^c
1	9-OH	$CH_2CH_2NEt_2$	H	7.96	R1	0.3 Å	12°	186	0.0
					M1	0.2	7	4	0.3
					R2	−0.2	18	177	0.5
					R3	−0.1	−10	171	0.5
					M2	−0.2	−12	−5	1.6
					M3	−0.1	−19	−8	5.2
2	10-OH	$CH_2CH_2NEt_2$	H	7.21	M1	0.2 Å	16°	3°	0.0
					R1	0.3	−9	184	0.5
					M2	−0.1	18	7	1.5
					R2	−0.2	4	169	4.3
					M3	−0.1	12	−9	5.3
3	H	$CH_2CH_2NEt_2$	H	7.00	M1	0.0 Å	8°	6°	0.0
					R1	−0.1	−4	172	0.4
					M2	0.2	12	−9	0.5
					M3	0.0	−16	187	2.0

4	8-OH	$CH_2CH_2NEt_2$	H	6.72	M1	0.2 Å	22°	11°	0.0
					R1	−0.1	−9	188	0.2
					R2	−0.1	19	165	0.6
					M2	0.3	18	−13	4.3
5	9-OMe	$CH_2CH_2NEt_2$	H	5.82	M1	0.1 Å	13°	20°	0.0
					R1	−0.1	−9	169	0.2
					R2	0.1	17	193	0.6
					M2	−0.1	−7	−26	4.3
6	10-OMe	$CH_2CH_2NEt_2$	H	5.70	M1	0.2 Å	−9°	−8°	0.0
					R1	0.1	1	169	0.1
7	7-OMe	$CH_2CH_2NEt_2$	H	Inactive ≤ 5.00	M1	0.3 Å	16°	21°	0.0
					M2	−0.3	−11	−13	1.6
					M3	0.1	29	−16	2.3
					R1	−0.2	3	161	4.8
8	8-OMe	$CH_2CH_2NEt_2$	H	Inactive ≤ 5.00	M1	0.0 Å	7°	−4°	0.0
					M2	0.1	13	2	0.4
					R1	0.0	−9	186	0.4
					M3	−0.2	−8	12	2.8
					R2	−0.1	−5	173	3.2

[a]As defined in Ref. 67.
[b]As defined in Figure 1.
[c]kcal $mole^{-1}$ per $(G\text{-}C)_4$ complex. Global minimum $\Delta E = 0$.

tions with the DNA models. The interaction geometries of these pseudo rigid intercalating ligands with the GpCpGpCp-3/CpGpCpG tetramer were searched in the manner described earlier. A nonbonded MM force field composed of dispersion/steric, electrostatic, and hydrogen bond contributions was used to determine the intermolecular energy at each grid point. In all cases both the ligand and (G-C) tetramer were fully charged and a molecular dielectric, ϵ, of 3.5 was assumed. However, the dispersion/steric force field was represented by the AMBER (64) potentials. The calculated relative binding energies, ΔE, include only the interactions between ligand and the DNA model. The apparent intermolecular energy minima identified in the fixed valence geometry scans were subsequently used as starting points for fixed valence geometry energy minimizations of the six intermolecular degrees of freedom for each analog and its torsion angles for substituents X and R_1. This minimization strategy is, again, that described in Section VA. Part A of Table 4 contains the results of these minimization calculations. The minimum energy intercalation geometries are described in terms of the intermolecular degrees of freedom defined in Figure 1. The intercalation number, *ai*, in Table 4 refers to major groove ligand entry such that the D-ring is on the G-side of the base pair above the ligand (a = M), or on the C-side (a = R). The i refers to the *ith* deepest M or R minimum energy complex. Only complexes within 5 kcal mole^{-1} per complex of the global minimum are given in Table 4.

The most stable classes of intercalation geometries, see Table 4 and Figures 5–10, correspond to insertion of the ligand into the $(G\text{-}C)_4$ duplex along the major groove. This is in contrast to the anthraquinones, where the long axis of the ring system inserts "perpendicular" to the DNA base pair hydrogen bond axes along the minor groove of the DNA model, as discussed for doxorubicin analogs. The computed source of intermolecular stabilization energy for major groove intercalation of the BTPIs, and, by analogy, the anthrapyrazoles in general, is the strong hydrogen bond/ionic interaction between the charged nitrogen at the 5-position of the ligand and the nearby phosphate oxygens of one of the sugar phosphate backbones.

The stabilizing intercalation interactions for major groove entry can be seen in Figure 5 where the active analog #1 of Table 4 is shown intercalated in its global minimum energy complex. These static energy–fixed valence geometry calculations were repeated for a $(G\text{-}C)_6$ duplex model of DNA using some of the same analogs of VII as given in Table 4. The resulting minimum energy complex geometries are given in Table 5. The precise intercalation geometries, as defined by the intermolecular degrees of freedom in Figure 1, are quite similar using both the $(G\text{-}C)_4$ and $(G\text{-}C)_6$ duplexes. This suggests that the longer-range ligand-receptor interactions present in the $(G\text{-}C)_6$ duplex do not seriously modify the intercalation behavior

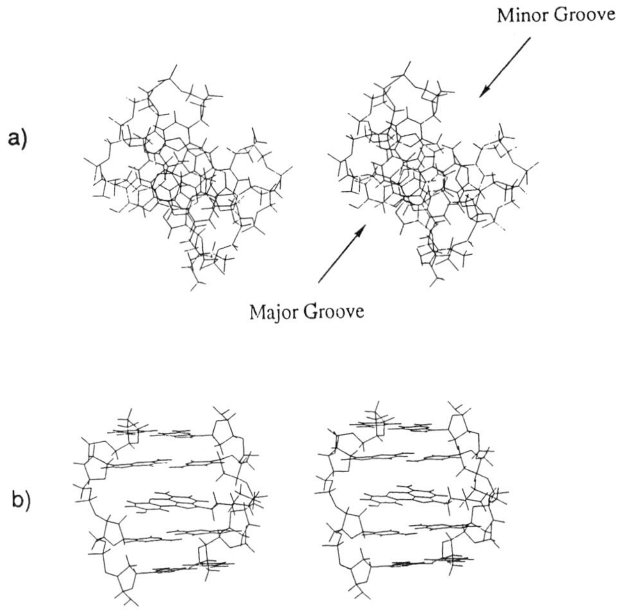

Figure 5 The intercalation geometry of compound no. 1 (X = 9-OH, R_1 = $CH_2CH_2NEt_2$, R_2 = H, -log(IC_{50}) = 7.96) in the R_1 complex as defined in Table 4: (a) top view; (b) side view.

of the rigid analogs of VII, as compared to the limited range interactions inherent to the $(G\text{-}C)_4$ duplex model.

Molecular dynamics intercalation simulations and relaxations were next performed using the eight analogs given in Table 4 using the MOLSIM molecular dynamics software package (65). A set of six additional analogs of VII having a flexible substituent, R_2 = $-CH_2CH_2NH_2$, whose corresponding anticancer activities span the observed range for the entire set of analogs, were also considered in the MD analysis.

The C(1)pG(2)pC(3)pG(4)pC(5)pG(6)-3/G(1)pC(2)pG(3)pC(4)pG(5)-pC(6)-5 hexamer duplex geometry used in the static scan/minimize fixed valence geometry studies was selected as the initial structure for the DNA model. The static minimum energy intercalation complexes reported in Table 4 were selected as starting points in the MD simulations/relaxations.

As discussed earlier, MD can be useful in simulating a relaxed representation of molecular structure, including the valence geometry, and in exploring both the intra- and intermolecular degrees of freedom. In particular, MD simulations allow sampling of the conformations of the flexible substituents of analogs of VII as they intercalate with the DNA models.

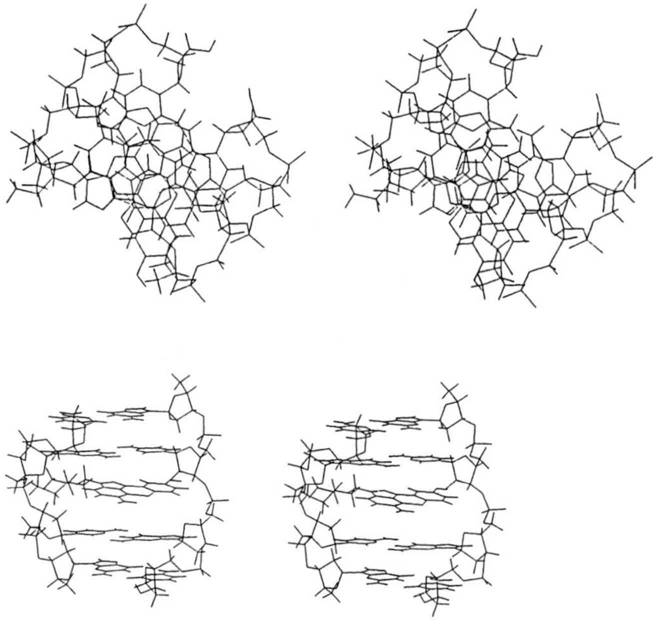

Figure 6 Same as Figure 5, but for complex M1 of Table 4.

The starting conformations assigned to the analogs having flexible substituents were the intramolecular minimum energy conformations identified in fixed valence geometry conformational scans. Overall, the MD simulations/relaxation studies appear to provide a more realistic, and complete, "picture" of the intercalation process involving analogs of VII with a $(G\text{-}C)_6$ duplex model for DNA, than the static calculations by themselves.

The molecular dielectric constant ϵ was varied from 3.5 to 50 in the analyses of some analogs in order to roughly take into account aqueous solvent interactions on the intercalation process. Variation of ϵ does not seriously alter the relative ordering of the intercalation binding energies of the analogs analyzed; see Table 6. However, the relative *differences* in binding energies, for equivalent intercalation geometries, as well as the absolute energies of intercalation change significantly with ϵ.

The role of formal charge on both the ligand and on the DNA model was also investigated. In all cases when either the ligand, DNA, or both are neutral in charge, intercalation from the minor groove is preferred. The stabilizing charge-charge interaction between the ligand nitrogen and a local (to the intercalation site) phosphate group realized for major groove intercalation geometries is, at least, partially lost.

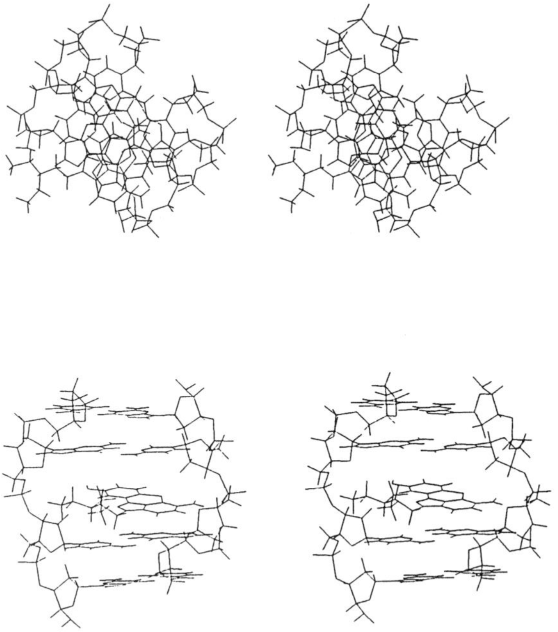

Figure 7 Same as Figure 5, but for complex R2 of Table 4.

This study also included devising a means of treating "end effects" of the DNA model in the MD simulations. That is, as discussed in the section on modeling problems, a method is needed to model the high-polymer chain in MD simulations using a relatively short chain-segment model. The chains on both sides of an arbitrarily selected chain segment within a high molecular weight linear polymer act as momentum sinks upon the chain segment. The chain segment, by itself, would be expected to undergo more motions than when embedded in a long linear chain. We have modeled this momentum sink effect, in part, by lowering the simulation temperatures to 100K to freeze out some end-effect motions. However, the major reason for lowering the simulation temperature was to minimize motion processes that would yield highly destabilizing steric interactions between the intercalated ligand and the DNA model, causing a loss in intercalation or, at least, leading to an intercalation geometry far from equilibrium.

The principal means of accounting for the high-polymer momentum sink

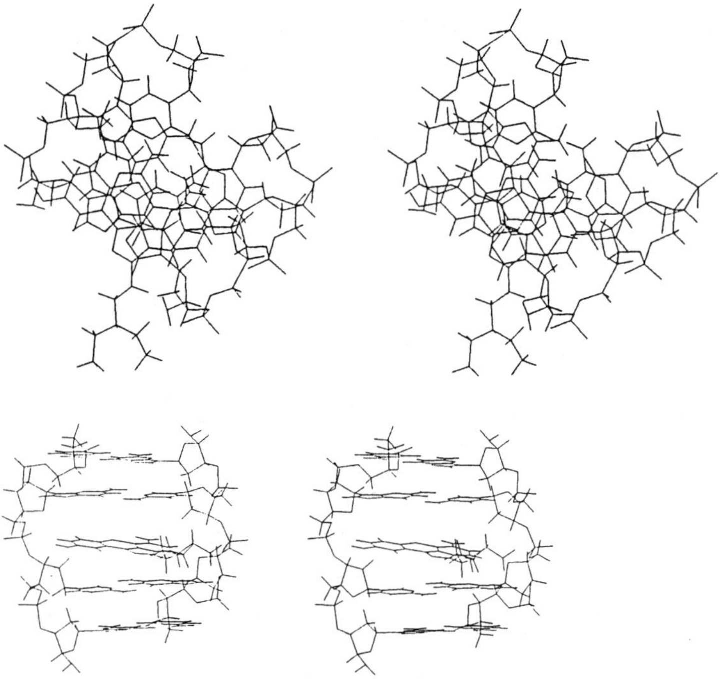

Figure 8 Same as Figure 5, but for complex R3 of Table 4.

was to increase the masses at the ends of each of the sugar phosphate chain segments. Such heavy masses reduce motions in the backbone chains, but the reduction in motion is not highly coupled to the base pairs, which continue to undergo unrealistically large motions in spite of heavy masses (5000 amu) on the ends of the backbone chains. Thus, the masses of each of the nonhydrogen atoms of the noncentral two base pairs, C(1)pG(2)p/G(1)pC(2) and C(5)pG(6)/G(5)pC(6), were also increased. Different enhanced mass treatments were considered. Increasing the atomic masses in G(2)/C(2) and C(5)/G(5) by a factor of 10 and in C(1)/G(6) and G(1)/C(6) by a factor of 100, yields average RMS atomic displacements in the $(G\text{-}C)_6$ duplex that are almost the same as the central interior six-base-pair structure of a $(G\text{-}C)_{10}$ duplex. Consequently, this particular distribution of mass enhancements was selected throughout the analysis.

Equilibration of the MD simulation was performed prior to generating the intercalation trajectories for each analog. The equilibrium process started with placing the ligand in the static minimum energy intercalation geometry to be explored. The temperature was set to 20K and the MD run

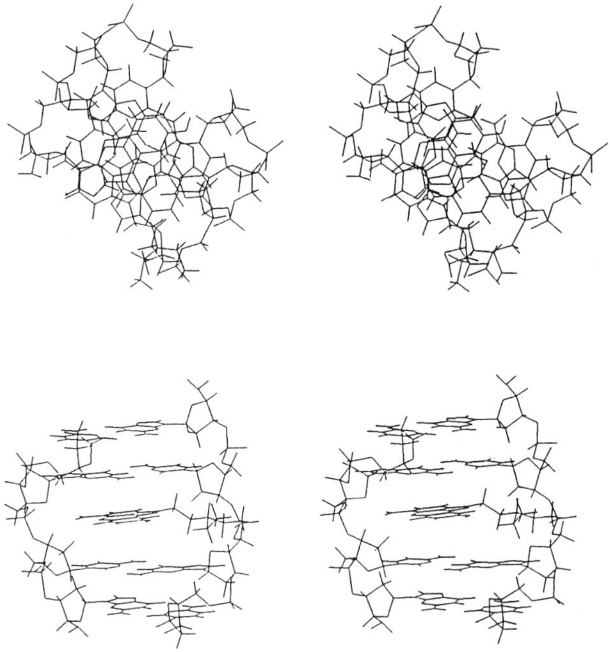

Figure 9 Same as Figure 5, but for complex M2 of Table 4.

for 0.2 psec. The temperature was then increased to 30K, 40K, 60K, and 100K at 0.2-psec intervals. The purpose of slowly increasing temperature over time was to prevent the occurrence of "instant" erratic motions at higher temperatures (100K) and short simulation times due to repulsive steric interactions. Such erratic motions could cause the MD trajectory to be vectored to sampling high-energy, unrealistic intercalation states. However, it appears the ligand -$(G\text{-}C)_6$ intercalation complexes are well behaved and quickly go to equilibrium. This can be seen from the equilibration times versus total complex energy plot at 100K for the most active analog (X = 9-OH, $R_1 = -(CH_2)_2NEt_2$, R_2 = H) #1 in Tables 4–6, which is shown in Figure 11. The MD simulation system was loosely coupled (J = 0.1 psec) to a constant temperature bath. The intercalation geometry quickly (~0.5 psec) relaxes to what appears to be a steady-state intercalation structure. The same steady-state MD structure was realized from alternate MD starting geometries that were significantly perturbed from the apparent minimum found in the static energy calculations. Still, there is the possibility that, overall, the intercalation geometries are quite flexible, and it takes considerable simulation time to find lower energy intercalation

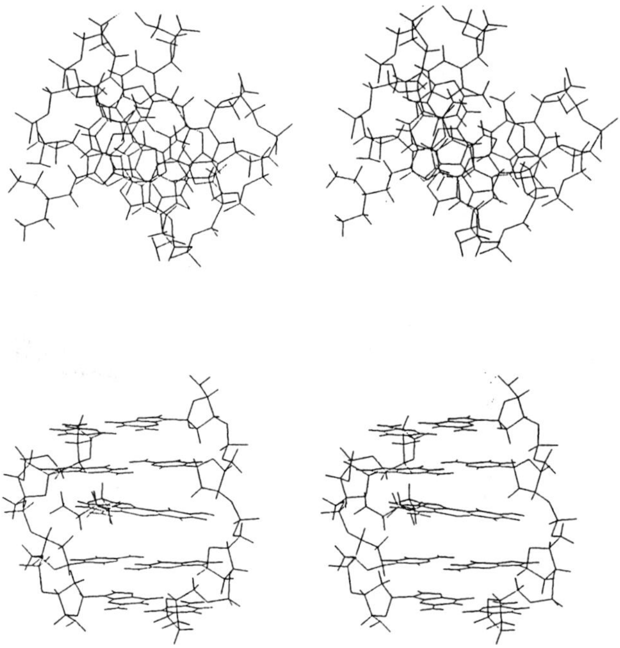

Figure 10 Same as Figure 5, but for complex M3 of Table 4.

states. This potential entropic complication was explored by simply running trajectories for longer periods of time. The energy remains, essentially, constant, as does the intercalation geometry over 30-psec sampling simulations. This finding led us to assume that 10-psec MD simulations, following equilibration, were sufficient to identify relaxed, low-energy intercalation complexes. Table 6 contains the results of the MD simulations of each of 14 analogs of VII intercalating with the $(G\text{-}C)_6$ duplex. The E corresponds to the lowest ligand-$(G\text{-}C)_6$ duplex interaction energy of the MD trajectory for each analog.

The conformational behavior of the substituents of BTPI analogs having flexible side chains in the presence of the DNA model remain ambiguous. That is, the DNA model does not serve as a conformational entropy reservoir and lock the substituents into preferred binding states. Rather the side chains interconvert among conformer classes characteristic of the intramolecular energy minima even when the ligand is intercalated. Correspondingly, the different side-chain conformer states yield about the same binding energy to $(G\text{-}C)_6$, suggesting a thermodynamic basis for the conformational flexibility in the side chains of bound ligands.

Table 5 The Fixed Valence Geometry, Static Minimum Energy Intercalation Complexes of Rigid Substituent Analogs of BTPI with a $(G\text{-}C)_4$ Duplex. ΔE corresponds to the intermolecular ligand $(G\text{-}C)_6$ interaction energy.

Compound no.	X	R_1	R_2	Anticancer activity[a] log($1/IC_{50}$)	Intercalation no.	Amber Z^b	Amber θ^b	Amber γ^b	Amber ΔE^c
1	9-OH	$CH_2CH_2NEt_2$	H	7.96	R1	0.3 Å	15°	187	0.0
					M1	0.1	6	7	0.2
					R2	−0.2	16	178	0.6
					R3	−0.2	−8	173	0.7
					M2	−0.1	−11	−8	2.1
					M3	−0.1	−15	−14	5.0
3	H	$CH_2CH_2NEt_2$	H	7.00	M1	0.0 Å	7°	7°	0.0
					R1	−0.1	−7	170	0.3
					M2	0.2	15	−13	0.5
					M3	0.0	−14	184	2.1
4	8-OH	$CH_2CH_2NEt_2$	H	5.82	M1	0.1 Å	26°	8°	0.0
					R1	−0.1	−11	187	0.2
					R2	−0.1	13	163	0.5
					M2	0.2	12	−11	4.0
8	8-OMe	$CH_2CH_2NEt_2$	H	Inactive	M1	0.0 Å	5°	−9°	0.0
				≤ 5.00	M2	0.1	15	2	0.3
					R1	0.0	−7	185	0.4
					M3	−0.1	−9	13	3.0
					R2	−0.1	−6	171	3.1

[a]As defined in Ref. 67.
[b]As defined in Figure 1.
[c]kcal $mole^{-1}$ per $(G\text{-}C)_6$ complex. Global minimum $\Delta E = 0$.

Table 6 The MD Global Low-Energy Intercalation Complexes of 14 Substituent Analogs of VII with a $(G\text{-}C)_6$ Duplex. *E* corresponds to the actual intermolecular ligand $(G\text{-}C)_6$ energy. ϵ is the assigned molecular dielectric constant.

Compound[a] no.	X	R_1	R_2	Anticancer[b] activity $-\log(IC_{50})$	Intercalation no.
1	9-OH	$CH_2CH_2NEt_2$	$CH_2CH_2NEt_2$	8.59	R1
2(1)	9-OH	$CH_2CH_2NEt_2$	$CH_2CH_2NEt_2$	7.96	M1
3	H	$CH_2CH_2NEt_2$	$CH_2CH_2NEt_2$	7.52	M1
4	10-OH	$CH_2CH_2NEt_2$	$CH_2CH_2NEt_2$	7.32	M2
5(2)	10-OH	$CH_2CH_2NEt_2$	$CH_2CH_2NEt_2$	7.21	M1
6	8-OH	$CH_2CH_2NEt_2$	$CH_2CH_2NEt_2$	7.15	M1
7(3)	H	$CH_2CH_2NEt_2$	$CH_2CH_2NEt_2$	7.00	M1
8(4)	8-OH	$CH_2CH_2NEt_2$	$CH_2CH_2NEt_2$	6.72	M1
9	9-OMe	$CH_2CH_2NEt_2$	$CH_2CH_2NEt_2$	6.68	R1
10(5)	9-OMe	$CH_2CH_2NEt_2$	$CH_2CH_2NEt_2$	5.82	R1
11(6)	10-OMe	$CH_2CH_2NEt_2$	$CH_2CH_2NEt_2$	5.70	M1
12(7)	7-OMe	$CH_2CH_2NEt_2$	$CH_2CH_2NEt_2$	Inactive ≤ 5.00	M1
13(8)	8-OMe	$CH_2CH_2NEt_2$	$CH_2CH_2NEt_2$	Inactive ≤ 5.00	M2
14	8-OMe	$CH_2CH_2NEt_2$	$CH_2CH_2NEt_2$	Inactive ≤ 5.00	M2

[a]Same numbers as Table 4 in ( ).
[b]As described in Ref. 67.
[c]As defined in Figure 1.
[d]Absolute energy in kcal $mole^{-1}$ per $(G\text{-}C)_6$ complex.

Like the static MM scan/energy minimization calculations, the MD simulations suggest major groove intercalation is preferred when the ligand and DNA are charged. If the ligand has no formal charge, minor groove intercalation is energetically preferred. Major groove entry is preferred for charged ligands because of the favorable hydrogen bond/electrostatic interaction between the highly positive charged nitrogen of the ligand and a nearby DNA phosphate group. This stabilizing interaction cannot be realized to such an extent for minor groove intercalation.

Table 7 contains the computed free-space "binding energies" ($\epsilon = 40$) for the 14 analogs of VII considered in the MD-based QSAR analysis. The binding energies, ΔE, of Table 7 are the lowest ligand-$(G\text{-}C)_6$ interaction energies determined over the course of each intercalation simulation. The selection of $\epsilon = 40$ is ultimately based upon optimizing the statistical significance of the QSAR as a function of ΔE, which, in turn, is a function of ϵ. However, a previous study of charged dye-polymer interactions in solution

$\epsilon = 3.5$				$\epsilon = 20$				$\epsilon = 40$			
Z^c	θ^c	γ^c	E^d	Z^c	θ^c	γ^c	E^d	Z^c	Θ^c	γ^c	E^d
0.1	−6	12	−159.1	0.1	−6	9	−76.4	0.1	−4	8	−67.3
0.3	8	188	−113.6	0.3	9	185	−61.1	0.3	10	184	−58.0
0.1	−3	7	−157.6	0.0	−2	5	−74.1	0.0	−2	5	−63.1
0.1	−9	172	−151.5	0.1	−7	177	−78.9	0.1	−6	178	−64.8
0.2	10	−6	−138.1	0.2	12	−3	−62.2	0.2	12	−2	−57.9
0.2	−9	22	−149.0	0.2	−7	19	−76.8	0.2	−6	18	−63.7
0.0	9	13	−130.3	0.0	7	9	−60.8	0.0	7	9	−54.3
0.1	17	8	−133.1	0.0	18	12	−68.2	0.2	19	13	−56.1
0.1	−9	16	−154.9	0.1	−5	19	−70.4	0.1	−6	18	−64.9
0.1	17	17	−131.7	0.1	17	14	−69.2	0.1	16	14	−57.8
0.2	−12	−8	−139.3	0.2	−9	−10	−62.0	0.2	−7	−10	−58.6
0.3	10	18	−130.7	0.2	11	21	−61.7	0.3	14	23	−56.0
0.1	13	−7	−141.2	0.1	10	−8	−64.8	0.0	9	−3	−58.2
0.2	−12	15	−158.0	0.2	−9	13	−71.9	0.2	−9	13	−63.5

indicated that $\epsilon = 40$ was needed in order to quantitatively estimate the ΔH's of complexing (66).

The anticancer activity is also reported in Table 7 and expressed as log(1/C), where C is the IC_{50} molar value in a murine L1210 leukemia in vitro screen as described by Baguley (67). Even a cursory analysis of the log(1/C) and ΔE values in Table 7 indicates a lack of any common trend, let alone a correlation. Certainly it is possible that no correlation exists because anticancer potency is not dependent upon binding energy in this class of ligands. Less severe, anticancer potency may depend on ΔE, but also be dependent upon some other physicochemical property so that a ΔE-log(1/C) relationship is masked by the influence of an additional molecular feature. This additional physicochemical property was found to be the lipophilicity of the intercalator in the QSAR analysis of the doxorubicin analogs. Ligand lipophilicity is not found to be a QSAR descriptor, in conjunction with ΔE, for the BTPIs. Alternately, the model used for the estimation

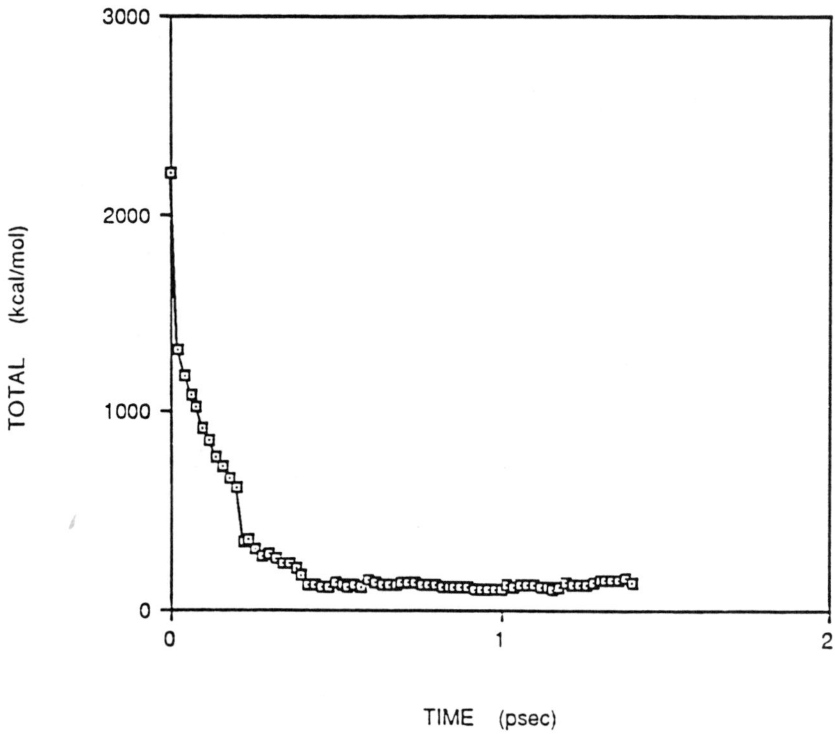

Figure 11 The equilibration curve, time (psec) vs. energy (kcal $mole^{-1}$ $complex^{-1}$) using complex M1 of compound no. 1 of Table 5. The ends of the chains have been mass-weighted as described in the text.

of the binding energy may also be unrealistic. The neglect of explicitly including water molecules in the MD simulations may lead to calculated binding energies that are not relevant to the actual biological state of action—BTPIs binding to DNA in a largely aqueous environment.

A limited number of MD intercalation simulations of the most and least active analogs of the BTPIs, #1 and #14 of Table 7, respectively, were carried out using both the blanket and desolvation models of aqueous solvation to estimate binding energies (25,27,67). Succinctly, each of these solvent simulation models gave significantly different ($\Delta E + \Delta E_D$) values from one another for a common ligand. The reason for this variation in estimated aqueous solvation binding energies seems to be that the simulations are far from equilibrium. The practical simulation period, 30 psec, is not nearly enough time to realize total equilibration of the water-ligand-DNA system. As mentioned earlier, this lack of equilibration, due to the

Table 7 The Free-Space Intercalation Energies, ΔE, and the Corresponding Aqueous Desolvation Energies, ΔE_D, for the 14 BTPI Analogs Studied

Compound[a] No.	X	R_1	R_2	Anticancer[b] activity log(1/C)	ΔE kcal mole^{-1}	ΔE_D kcal mole^{-1}	$\Delta E + \Delta E_D$ kcal mole^{-1}
1	9-OH	$CH_2CH_2NEt_2$	$CH_2CH_2NH_2$	8.59	−67.3	17.9	−49.4
2(1)	9-OH	$CH_2CH_2NEt_2$	H	7.96	−58.0	9.6	−48.4
3	H	$CH_2CH_2NEt_2$	$CH_2CH_2NH_2$	7.52	−63.1	14.9	−48.2
4	10-OH	$CH_2CH_2NEt_2$	$CH_2CH_2NH_2$	7.32	−64.8	18.0	−46.8
5(2)	10-OH	$CH_2CH_2NEt_2$	H	7.21	−57.9	10.8	−47.1
6	8-OH	$CH_2CH_2NEt_2$	$CH_2CH_2NH_2$	7.15	063.7	17.3	−46.4
7(3)	H	$CH_2CH_2NEt_2$	H	7.00	−54.3	8.0	−46.3
8(4)	8-OH	$CH_2CH_2NEt_2$	H	6.72	−56.1	10.5	−45.6
9	9-OMe	$CH_2CH_2NEt_2$	$CH_2CH_2NH_2$	6.68	−64.9	19.8	−45.1
10(5)	9-OMe	$CH_2CH_2NEt_2$	H	5.82	−57.8	13.6	−44.2
11(6)	10-OMe	$CH_2CH_2NEt_2$	H	5.70	−58.6	14.9	−43.7
12(7)	7-OMe	$CH_2CH_2NEt_2$	H	Inactive ≤ 5.00	−56.0	14.6	−41.4
13(8)	8-OMe	$CH_2CH_2NEt_2$	H	Inactive ≤ 5.00	−58.2	15.8	−42.4
14	8-OMe	$CH_2CH_2NEt_2$	$CH_2CH_2NH_2$	Inactive ≤ 5.00	−63.5	21.2	−42.3

[a]Same number as Table 4 in ( ).
[b]As described in Ref. 67.

need for long simulation times, is likely responsible for the shortcomings in successfully predicting ligand binding to enzymes using MD and/or Monte Carlo methods.

Overall, MD simulation modeling does not yet provide a practical means to reliably estimate a set of relative aqueous binding energies, $(\Delta E + \Delta E_D)$, for use in a QSAR study. However, the use of a relatively simple aqueous hydration shell model (68,69) to estimate solute hydration energies turns out to provide meaningful data in the construction of a QSAR. In this modeling the aqueous desolvation energy, ΔE_D, is defined as

$$\Delta E_D = E_C - (E_L + E_R) \tag{5}$$

where E_C is the aqueous solvation energy of the lowest energy intercalation complex determined in the free-space MD simulation with $\epsilon = 40$, E_L is the aqueous solvation energy of the isolated BTPI ligand in the conformation it adopts in the lowest energy intercalation complex, and E_R is the aqueous solvation energy of the isolated $(G\text{-}C)_6$ duplex in the conformation it adopts in the lowest energy intercalation complex.

The values of ΔE_D, using the aqueous hydration shell model are given in Table 7. The additive use of ΔE_D with ΔE, the free-space binding energy, permits the construction of a simple, but highly significant, QSAR,

$$\log(1/C) = -0.46(\pm 0.02)[\Delta E + \Delta E_D] - 0.46$$
$$N = 14 \quad R = 0.97 \quad S = 0.13 \tag{6}$$

The QSAR expressed by Eq. (6) suggests that the anticancer potency of a BTPI analog can be increased in two interrelated ways; first, increase the intercalation binding energy, ΔE, of the BTPI analog to DNA, and, secondly, minimize the loss in favorable aqueous solvation energy for both the BTPI congener and the DNA due to intercalation. The predicted log(1/C) values using the QSAR (Eq. (6)), as well as the residual anticancer activity fits, $\Delta\log(1/C)$, are given in Table 8. It should be noted that none of the 14 BTPI analogs are significant outliers with respect to their respective predicted log(1/C) values.

The key contribution of ΔE_D to the QSAR is not, by way of example, in differentiating the aqueous solvation energetics among analogs having the same X substituents, but at different ring locations. Rather, ΔE_D, using a hydration shell model, meaningfully permits inclusion of the intrinsic differences in aqueous solvation energies of different substituents, *due to their chemical structures*, in the QSAR. Molecular dynamics simulation modeling using explicit water molecules apparently cannot meaningfully account for changes in solvation energetics as a function of substituent chemistry.

The results of the free-space intercalation MD simulations, combined

Table 8 Summary of the QSAR, Eq. (6), for the 14 BTPI Analogs Studied

				Anticancer activity			
Compound[a] no.	X	R_1	R_2	log $(^1/c)^b$ obs.	log$(^1/c)$ cal.	Δ[log$(^1/c)$]	$\Delta E + \Delta E_D$ kcal mole^{-1}
1	9-OH	$CH_2CH_2NEt_2$	$CH_2CH_2NH_2$	8.59	8.41	0.18	−49.4
2(1)	9-OH	$CH_2CH_2NEt_2$	H	7.96	7.95	0.11	−48.4
3	H	$CH_2CH_2NEt_2$	$CH_2CH_2NH_2$	7.52	7.86	−0.34	−48.2
4	10-OH	$CH_2CH_2NEt_2$	$CH_2CH_2NH_2$	7.32	7.21	0.11	−46.8
5(2)	10-OH	$CH_2CH_2NEt_2$	H	7.21	7.35	−0.14	−47.1
6	8-OH	$CH_2CH_2NEt_2$	$CH_2CH_2NH_2$	7.15	7.03	0.12	−46.4
7(3)	H	$CH_2CH_2NEt_2$	H	7.00	6.98	0.02	−46.3
8(4)	8-OH	$CH_2CH_2NEt_2$	H	6.72	6.66	0.06	−45.6
9	9-OMe	$CH_2CH_2NEt_2$	$CH_2CH_2NH_2$	6.68	6.43	0.26	−45.1
10(5)	9-OMe	$CH_2CH_2NEt_2$	H	5.82	6.01	−0.19	−44.2
11(6)	10-OMe	$CH_2CH_2NEt_2$	H	5.70	5.78	−0.08	−43.7
12(7)	7-OMe	$CH_2CH_2NEt_2$	H	Inactive ≤ 5.00	4.72	0.28	−41.4
13(8)	8-OMe	$CH_2CH_2NEt_2$	H	Inactive ≤ 5.00	5.18	−0.18	−42.4
14	8-OMe	$CH_2CH_2NEt_2$	$CH_2CH_2NH_2$	Inactive ≤ 5.00	5.13	−0.13	−42.3

[a]Same number as Table 4 in ( ).
[b]As described in Ref. 67.

with previous calculations, suggests a relatively simple model to predict the thermodynamic preference for either major or minor groove intercalation complexes. If the BTPI ligand analog possesses a positive charge *and* the DNA phosphate groups are negatively charged, then major groove intercalation is preferred, owing to favorable ionic/hydrogen bonding between the N^+H of the ligand side chain and the oxygens of a nearby phosphate backbone group. This stabilizing interaction is more fully realized on steric grounds for major groove insertion than for minor groove intercalation. If either the ligand and/or DNA does not possess formal charge(s), then minor groove intercalation is preferred on the basis of enhanced intermolecular attractive dispersion interactions. This charge model to predict intercalation behavior can be compromised depending upon the actual distance-dependent behavior of the local molecular dielectric. It is generally thought that ligand intercalation occurs in a largely aqueous medium having a pH corresponding to formal charges on both the DNA phosphate groups and on the BTPIs. Obviously, intercalation in a more lipid-like environment could alter the thermodynamic preference for intercalation.

We have made an implicit assumption in these modeling and simulation studies that G-C environments are preferable receptor sites to those involving A-T base pairs. This assumption follows from many experimental studies indicating a general selectivity for intercalation at G-C-rich duplex domains as mentioned and referenced earlier. A further consideration in future modeling is to evaluate base-pair sequence on intercalation preference (binding energy) for the BTPIs. Still, the QSAR using G-C duplex sequences to compute the intermolecular intercalation energy, as a correlation descriptor, indirectly supports a preference for G-C-rich sequences as preferred receptor sites.

VII. SUMMARY

It is to be emphasized that the examples detailed in this chapter of the generation of QSARs for intercalation agents would not be possible using only intermolecular modeling, or only Hansch-type (70) methods. A marriage of descriptors derived from geometric searching, energy minimization, molecular dynamics simulation, linear free-energy, and group additivity methods is needed to supply the physicochemical information necessary to construct meaningful relationships to observed anticancer activities. These QSARs can be interpreted in terms of biochemical action models that are consistent with known mechanisms of action.

REFERENCES

1. Nakata, Y., D. Malhotra, A.J. Hopfinger, and D.R. Bickers, *J. Pharm. Sci. 72*, 809–811 (1983).

2. (a) Long, B.H., J. Golik, S. Forenza, B. Ward, R. Rehfuss, J. Dabrowiak, J. Catino, S.T. Musial, K.W. Brookshire, and T.W. Doyle, *Proc. Natl. Acad. Sci. USA 86*, 2-6 (1986).
   (b) Zein, M., A.M. Sinha, W.J. McGahren, and G.A. Ellestad, *Science, 240*, 1198-1201 (1988).
3. (a) Lown, J.W., ed., *Anthracycline and Anthracenedione-Based Anticancer Agents*, Elsevier, Amsterdam (1988).
   (b) Denny, W.A., *Anti-Cancer Drug design, 4*, 241-264 (1989).
   (c) Wakelin, L.P.G., *Med. Res. Rev. 6*, 275 (1986).
4. Young, R.C., R.F. Ozols, and C.E. Myers, *New Engl. J. Med. 305*, 139–153 (1981).
5. Denny, W.A., S.J. Twigden, and B.C. Baguley, *Anti-Cancer Drug Design 1*, 215–222 (1986).
6. Baguley, B.C. and B.F. Cain, *Molecular Pharmacology, 22*, 486–493 (1982).
7. Hartley, J.A., K. Reszko, E.T. Zuo, W.D. Wilson, A.R. Morgan, and J.W. Lown, *Molecular Pharmacology 33*, 265–271 (1988).
8. Valentini, L., V. Nicolella, E. Vannini, M. Menuzzi, S. Penco, and F.M. Arcamone, *Il Farmaco Edizione Scientifica 40*, 376–382 (1985).
9. Wakelin, L.P.G., G.J. Atwell, G.W. Rewcastle, and W.A. Denny, *J. Med. Chem. 30*, 855–861 (1988).
10. Wilson, W.D., Y.-H. Wang, S. Kusuma, S. Chandrasekaran, N. C. Yang, and D.W. Boykin, *J. Am. Chem. Soc. 107*, 4989–4993 (1985).
11. Muller, W., and D.M. Crothers, *European J. Biochem. 54*, 267–278 (1975).
12. Neidle, S., *Biochem. J. 243*, 1–12 (1987).
13. Sobel, H.M., and S.C. Jain, *J. Molec. Biology, 68*, 21–29 (1972).
14. Burr-Furlong, N., J. Sato, T. Brown, F. Chavez, and R.B. Hurlbert, *Cancer Res. 38*, 1329–1334 (1978).
15. Gormley, P.E., V.S. Sethi, and R.L. Cysyk, *Cancer Res. 38*, 1300–1308 (1978).
16. Raj, A.S., and J.A. Heddle, *Mutation Res. 78*, 253–261 (1980).
17. Wilson, W.R., N.M. Harris, and L.R. Ferguson, *Cancer Res. 44*, 4420–4427 (1984).
18. Crossen, P.E., *Mutation Res. 68*, 295–299 (1979).
19. Drlica, K., and R.J. Franco, *Biochem. 27*, 2253–2259 (1988).
20. Malhotra, D., R.A. Pearlstein, O. Kikuchi, Y. Nakata, and A.J. Hopfinger, *Ann. N.Y. Acad. Sci. 367*, 295. (1981).
21. Arnot, S., P.J. Campbell-Smith, and J. Chandrasekaran, In *Handbook of Biochemistry and Molecular Biology*, Fasman, O.D. ed., Vol. 2: 411, CRC Press, Cleveland, OH (1976).
22. (a) Jain, S.C., C.-C. Tsai, and H.M. Sobell, *J. Mol. Biol. 114*, 317–331 (1977).
   (b) Quigley, G.J., A.H.-J. Wang, G. Ughetto, G. van der Marel, J.H. van Boom, and A. Rich, *Proc. Natl. Acad. Sci. USA 77*, 7204–7208 (1980).
   (c) Liaw, Y.-C., Y.-G. Gao, H. Robinson, G.A. van der Marel, J.H. van Boom, and A.H.-J. Wang, *Biochemistry 28*, 9913-9917 (1989).
23. Grootenhuis, P.D.J., P.A. Kollman, G.L. Seibel, R.L. Des Jarlais, and I.D. Kuntz, *Anti-Cancer Drug Design 5*, 237–242 (1990).
24. Jones, G.B., C.L. Davey, T.C. Jenkins, A. Kamal, G.G. Kneale, S. Neidle,

G.D. Webster, and D.E. Thurston, *Anti-Cancer Drug Design 5*, 249-264 (1990).
25. McCammon, J.A., and S.C. Harvey, *Dynamics of Proteins and Nucleic Acids,* Cambridge University Press, New York (1987).
26. (a) First described by Zwanzig, R.W., *J. Chem. Phys. 22*, 1420–1426 (1954). (b) Kollman, P.A., *J. Am. Chem. Soc. 111*, 5649-5655 (1989).
27. (a) Jorgensen, W.L., and K.J. Buckner, *J. Phys. Chem. 91*, 6083-6089 (1987). (b) Karplus, M., *Isr. J. Chem. 27*, 121-129 (1986).
28. Unpublished results presented at the Drug Information Association meeting on *Structural Chemistry of Biological Systems*, Hilton Head Island, SC, Jan. 1991.
29. Henry, D.W., In *Cancer Chemotherapy*, Sartorelli, A.C. ed., 15-57, American Chemical Society, Washington (1976).
30. Neidle, S., In *Topics in Antibiotic Chemistry*, Sammes, P.G. ed., Vol. 3: 240-278, Ellis Horwood, Chichester (1978).
31. Pigram, W.J., W. Fuller, and L.D. Hamilton, *Nature New Biol. 235*, 17-19 (1972).
32. Neigle, S., *Cancer Treat. Rep. 61(5)*, 923-929 (1977).
33. Nakata, Y., and A.J. Hopfinger, *Biochem. Biophys. Res. Common. 95*, 583 (1980).
34. Hopfinger, A.J., *Conformational Properties of Macromolecules*, Academic Press, New York (1973).
35. Pople, J.A., and D.C. Beveridge, *Approximate Molecular Orbital Theory*, McGraw-Hill, New York (1970).
36. Malhotra, D., and A.J. Hopfinger, *Nucleic Acid Res. 8*, 5289-5296 (1980).
37. (a) Neidle, S., and G. Taylor, *Biochim. Biphys. Acta 479*, 450-459 (1977). (b) Courselle, C., B. Busetta, S. Geofire, and M. Hospital, *Acta Crystallog. B35*, 764-767 (1979). (c) Anguili, R., E. Foresti, L. Riva DiSanserverino, N.W. Isaacs, O. Kennard, W.D.S. Motherwell, D.L. Wamplter, and F. Arcamone, *Nature New Biol. 234*, 78-80 (1971).
38. Nakata, Y., and A.J. Hopfinger, *FEBS Lett. 117*, 259-266 (1980).
39. (a) Patel, D.J., *Biopolymers 18*, 553-569 (1979). (b) Patel, D.J., S.A. Kozlowski, and J.A. Rice, *Proc. Natl. Acad. Sci. USA 78*, 3333-3337 (1981).
40. Fink, S.I., A. Leo, M. Yamakawa, C. Hansch, and F.R. Quinn, *Il Farmaco 35*, 965 (1980).
41. Peters, J.H., G.R. Gordon, D. Kashiwase, J.W. Lown, S-F. Yen, and J.A. Plambeck, *Biochem. Pharmacol. 35*, 1309-1323 (1986).
42. Lown, J.W., *Mol. Cell Biochem. 55*, 17-26 (1983).
43. Lown, J.W., H.-H. Chen, J.A. Plambeck, and E.M. Acton, *Biochem. Pharmacol. 28*, 2563-2568 (1979).
44. Acton, E.M., and G.L. Tong, *J. Med. Chem. 24*, 669-673 (1981).
45. Tong, G.L., D.W. Henry, and E.M. Acton, *J. Med. Chem. 22*, 36-39 (1979).
46. Showalter, H.D.H., J.L. Johnson, L.M. Werbel, W.R. Leopold, R.C. Jackson, and E.F. Elsalger, *J. Med. Chem. 27*, 253-255 (1984).

47. Showalter, H.D.H., J.L. Johnson, J.M. Hoftiezer, L.M. Werbel, J.L. Schilis, and J. Plowman, *Am. Assoc. Clin. Res. Abstr.* 352-352 (1984).
48. Fry, D.W., T.J. Boritzki, J.A. Besserer, and R.C. Jackson, *Biochem. Pharmacol. 34*, 3499-3508 (1985).
49. Showalter, H.D. H., D.W. Fry, W.R. Leopold, J.W. Lown, J.A. Plambeck, and K. Reszka, *Anticancer Drug Des. 1*, 73-85 (1986).
50. Rao, G.M., J.W. Lown, and J.A. Plambeck, *J. Electrochem. Soc. 124*, 195-204 (1977).
51. Kawakami, Y., and A.J. Hopfinger, *Chem. Res. in Tox. 3*, 244-248 (1990).
52. *CHEMLAB-II User's Guide-V 11.0*, Chemlab Inc., Lake Forest, IL (1991).
53. Dewar, M.J.S. and W. Thiel, *J. Am. Chem. Soc. 99*, 4899-4908 (1977).
54. Tempczyk, A., J. Tarasiuk, T. Ossowicki, and E. Borowski, *Anti-Cancer Drug Des. 2*, 371-385 (1988).
55. Bachur, N.R., S.L. Gordon, M.V. Gee, and H. Kon, *Proc. Natl. Acad. Sci. USA 76*, 954-957 (1979).
56. Lown, J.W., and H.-H. Chen, *Can. J. Chem. 59*, 3212-3217 (1981).
57. Karasch, E.D., and R.F. Novak, *Biochem. Pharmacol. 30*, 2881-2884 (1981).
58. Karasch, E.D., and R.F. Novak, *Arch. Biochem. Biophys. 224*, 682-694 (1983).
59. Doroshow, J.H., *Cancer Res. 43*, 460-471 (1983).
60. Werbel, L.M., E.F. Elslager, D.F. Ortwine, J.L. Shillis, H.D.H. Showalter, D.F. Worth, and J. Plowman, *J. Proc. Am. Assoc. Cancer Res. 26*, 254 (1985).
61. (a) Showalter, H.D. H., J.L. Johnson, J.M. Hoftiezer, W.R. Turner, L.M. Werbel, W.R. Leopold, J.L. Shillis, R.C. Jackson, and E.F. Elslager, *J. Med. Chem. 30*, 121-131 (1987).
    (b) Showalter, H.D. H., J.L. Johnson, and J.M. Hoftiezer, *J. Heterocycl. Chem. 23*, 1491-1501 (1986).
62. (a) Showalter, H.D. H., L.M. Werbel, W.R. Leopold, D.W. Fry, W.D. Klohs, and R.C. Jackson, In *Anthracycline and Anthracenedione-Based Anticancer Agents*, Lown J.W. ed., 201-244 Elsevier, Amsterdam. (1988).
    (b) Fry, D.W., T.J. Boritzki, J.A. Besserer, and R.C. Jackson, *Biochem. Pharmacol. 34*, 3499-3508 (1985).
63. Stewart, J.J.P., and F.K. Seiler, *QCPE 455* (Ver. 4.0), (1987).
64. Weiner, S.J., et al., *J. Am. Chem. Soc. 106*, 765 (1984).
65. *MOLSIM Users Guide (Ver. 2.0)*, The Chem21 Group, Lake Forest, IL (1990).
66. Orchard, B.J., J.S. Tan, and A.J. Hopfinger, *Macromolecules 17*, 169-176 (1984).
67. Baguley, B.C., and R. Nash, *Eur. J. Cancer 17*, 671-679 (1981).
68. Hopfinger, A.J., *Intermolecular Interactions and Biomolecular Organization*, Wiley-Interscience, New York (1977).
69. (a) Hopfinger, A.J., *Macromolecules, 4*, 731-738 (1971).
    (b) Forsythe, K.H., and A.J. Hopfinger, *Macromolecules 6*, 423-431 (1973).
70. Hansch, C., and T. Fujita, *J. Am. Chem. Soc. 86*, 1616-1621 (1964).

6
The Molecular Basis of Sequence-Specific DNA-Protein Interactions

Thomas Kodadek
The University of Texas at Austin, Austin, Texas

I. INTRODUCTION

Double-stranded DNA is the storehouse of genetic information in free-living cells. As might be expected, the biochemical pathways involved in the maintenance, modification, utilization, and perpetuation of this information are complex and are only now beginning to be understood in molecular detail. Almost all of these processes involve proteins that recognize specific DNA sequences. These factors exhibit selectivities that range from extremely stringent to somewhat permissive, depending on their biological function. In this chapter, I will review what we know about the chemical basis of these exquisite examples of molecular recognition.

Why should this topic be of concern to the medicinal chemist? A central goal of DNA targeted drug design is to be able to manipulate the activity of individual genes by creating compounds that recognize and/or modify specific DNA sequences. While DNA-DNA recognition strategies, such as triple helix formation, are currently the most highly developed approaches, mimicking the action of natural DNA binding proteins is at least as appealing. For example, consider the following case. It is now known that certain transcription factors can stimulate the expression of target genes by 1000-fold via a mechanism that requires, among other things, sequence-specific binding of the transcription factor to distant elements known as enhancers. It has recently been found that the protein products of some oncogenes are members of this family of transcription factors and many cancers are now thought to be due to the transcriptional regulatory process careening out of control. A drug able to bind to specific enhancers and complete with the oncogene products for the control elements would be of considerable therapeutic utility. Indeed, this strategy for depressing gene expression is complementary to the antisense oligonucleotide scheme for inactivating mRNA which has received so much attention. However, the DNA-targeted approach is more appealing, since there are always far fewer copies of a gene present in a cell than molecules of the corresponding mRNA. Therefore, smaller doses of the drug would be required.

This chapter will first provide a broad outline of DNA metabolism, in order that the chemistry to be discussed later be viewed in its proper biological context. A discussion of the various mechanisms theoretically available to proteins for sequence recognition will then be presented, followed by a presentation of the commonly found structural motifs in DNA-binding proteins. These include the "leucine zipper," "helix-turn-helix," "homeobox," and "zinc finger" domains, as well as others. I will focus on examples where some structural information is available. Proteins that recognize specific structures rather than sequences (such as DNA repair proteins) or bind to DNA nonspecifically (such as histones) will not be considered in detail.

II. DNA METABOLISM: A BRIEF OVERVIEW

The genetic information stored in the DNA double helix is (a) replicated with a very high degree of fidelity upon cell division, (b) transcribed into messenger RNA as the first step of the pathway that ultimately translates nucleotide sequence to amino acid sequence, (c) maintained by DNA repair in the face of the inherent chemical instability of nucleic acids and the deleterious effects of the environment, and (d) rearranged by a recombination process that can be general or sequence-specific (Fig. 1). All of these processes are complex, and appear to involve "protein machines" (1), spe-

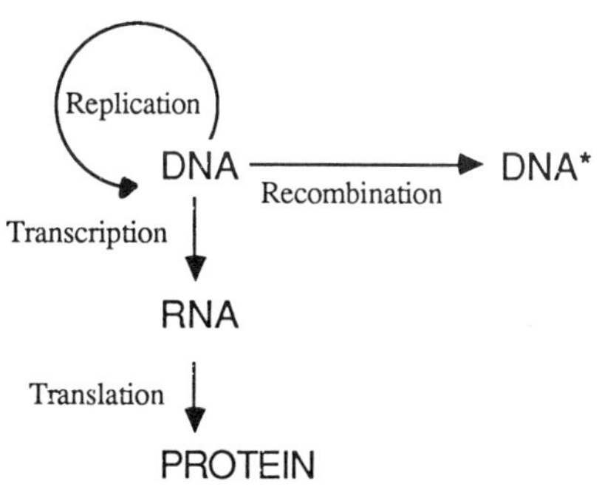

Figure 1 The "central dogma" of molecular biology, representing the flow of genetic information. DNA* = rearranged DNA.

cific complexes of polypeptides that modify one another's properties to produce a sum that is greater than the parts.

A. DNA Replication

Replication of duplex DNA genomes proceeds in three phases: initiation, fork progression, and termination (2). Initiation occurs at specific DNA sequences known as origins and involves loading one or two DNA polymerase holoenzymes onto the substrate. In most organisms of concern to medicinal chemists, including *E. coli* and mammals, two replication forks proceed bidirectionally from the origin (Fig. 2). Some simple organisms such as *E. coli* have a single chromosomal origin, called *oriC* in *E. coli*. More complex species possess many origins, which divide their much larger genomes into units known as replicons.

Reconstitution of in vitro systems for the faithful initiation of DNA replications has proven extremely difficult and biochemical details are available only for *E. coli* and a few viral systems. In general, the picture that has emerged (particularly from studies of *E. coli* and phage lambda) is that the first step of initiation involves sequence-specific origin binding by several copies of an initiation factor, for example the dnaA protein of *E. coli* (3) which binds to *oriC*, or the *O* protein of lambda which binds to *ori*λ (4). The initiation complex, in which the DNA appears to be wrapped around the dnaA or O protein (5), attracts an "adaptor" protein, the dnaC

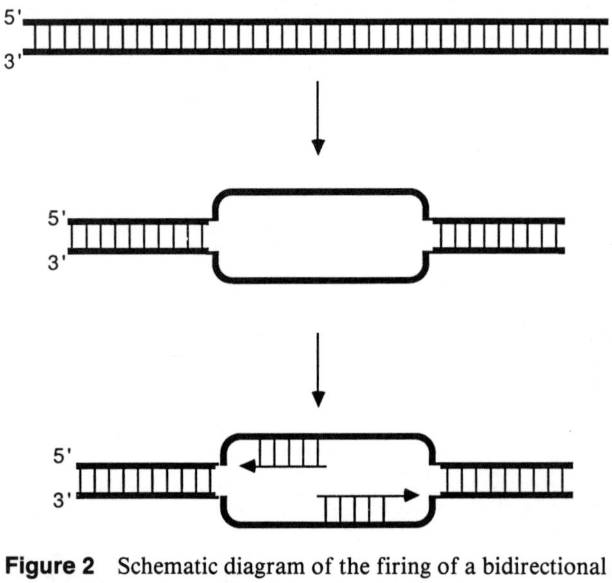

Figure 2 Schematic diagram of the firing of a bidirectional replication origin.

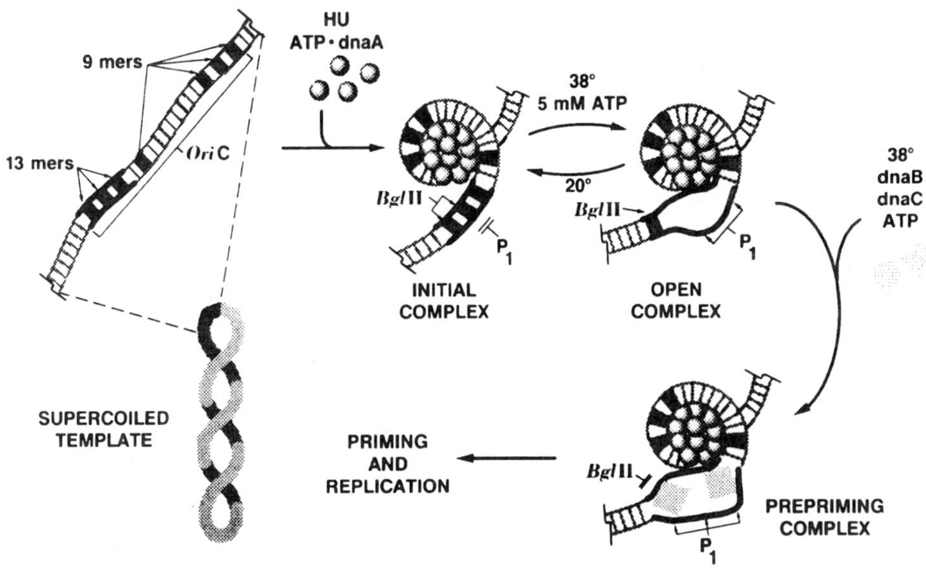

Figure 3 A scheme for oriC initiation as proposed by Bramhill and Kornberg (Ref. 3).

protein of *E. coli* (6) or the P protein of λ (7), which is complexed with a DNA helicase. This is the dnaB protein in both cases (8,9). The helicase is thereby loaded onto the DNA at the origin and proceeds to unwind the duplex; the A-T-rich sequences found at the origins of both organisms may be important in this step (10). The function of the origin unwinding process is to provide enough room for the very large DNA polymerase III holoenzyme (11) to insert itself between the strands (Fig. 3). Once this is accomplished a RNA primer is synthesized from ribonucleotide triposphates (rNTPs) complementary to a sequence near the origin (this reaction is probably catalyzed by the dnaG protein, which is the primase component of the holoenzyme (12)) and employed by the polymerase to begin leading strand synthesis which proceeds continuously in a 5′ to 3′ direction. The fundamental reaction catalyzed by DNA polymerases is the formation of a phosphodiester bond between the 3′-OH of a primer-template complex and a deoxynucleotide triphosphate (dNTP) whose base is complementary in a Watson-Crick sense to the template. Soon thereafter, lagging strand synthesis begins. Again, the dnaG protein lays down small RNA primers which are extended by a second molecule of DNA polymerase III present in the holoenzyme. Because of the intrinsic 5′ to 3′ polarity of polymerase-catalyzed DNA synthesis, the lagging strand must be synthesized discontinuously in pieces of one to a few thousand nucleotides in length, known as Okazaki fragments (13). Once the lagging strand has reached the 5′-end of the previous Okazaki fragment, the lagging strand polymerase lets go of its template and searches for a new primer-template complex (Fig. 4). The DNA recognized by the primases of various organisms as a primer site vary (14–16), but in all cases they are short, specific sequences that occur frequently. Thus, the primase must be a sequence-specific binding protein, but one that recognizes a specific *single-stranded* region of DNA. Only a small fraction of potential Okazaki primer site are actually used. The mechanism that regulates the average length of Okazaki fragments is not well understood. Throughout the replication process, the single-stranded DNA of the lagging strand is stabilized by a single-stranded DNA binding protein (the *ssb* gene product).

The dnaB helicase, which exhibits a 5′- to 3′-polarity of "movement" along a DNA strand (9), continues along in advance of the replication fork, opening the duplex and serving as the replicative helicase. The extreme processivity of this enzyme makes it well suited to proceed all the way around the large, circular *E. coli* genome once it is loaded onto the DNA. A number of accessory proteins present in the polymerase holoenzyme imbue polIII with similarly high processivity (11). Maintenance of this complex requires periodic hydrolysis of ATP. The leading and lagging strand machinery are thought to be physically coupled, though there is little direct evidence for this model (1,17), shown in Figure 4.

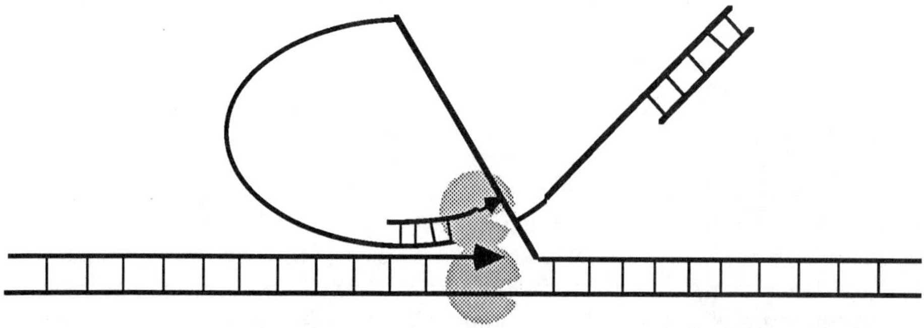

Figure 4 The "trombone model" for fork progression. The polymerase holoenzymes (stipled circles) that catalyze leading and lagging strand synthesis are thought to be physically coupled and both travel in the same direction. When the lagging strand polymerase encounters the 5′-end of the previous Okazaki fragment it releases the lagging strand template and finds another primer site. Synthesis on the leading strand is thought to be continuous.

During fork progression, incorrectly inserted nucleotides are removed by virtue of polIII's 3′- to 5′-exonuclease activity. It is estimated that an incorrect base is stably incorporated only about once in every 10^9 polymerization events (18). *E. coli* and presumably all other organisms have specific DNA repair systems to correct these rare errors (19).

Chromosomal replication in *E. coli* is terminated at a specific location on the chromosome called *terC*, which is approximately halfway around the circular chromosome from *oriC* (20). Recently, a *terC*-binding protein has been purified and characterized that is indispensable for replication termination (21). Interestingly, this factor has been shown to be a "contra-helicase," acting to terminate the helix separation activity of the dnaB protein which is required for continued fork progression.

The final steps of the replication process involve "cleaning up" the lagging strand by removal of the RNA primers of the individual Okazaki fragments by RNase H and gap filing and ligation by DNA polymerase I and *E. coli* DNA ligase (2). The product of the complete replication process is a topologically linked dimer of chromosomes which is decatenated by DNA Gyrase, a type II topoisomerase (22,23).

Much less is known about the biochemistry of eukaryotic DNA replication. However, studies with SV40, an animal virus, have shown that its origin binding protein, the large T antigen is a DNA helicase (24), suggesting that origin unwinding to facilitate holoenzyme loading is a fundamental step in DNA replication.

From the point of view of sequence-specific DNA-binding, the most interesting DNA replication factors are the dnaA and O proteins, the dnaG primase and its relatives in other organisms that recognize Okazaki start sites, and the *terC*-binding factor. Unfortunately, very little is currently known about the structural basis of sequence recognition for these proteins. An understanding of environmental factors influencing the DNA-binding activity of the dnaA protein would be particularly fascinating, since this factor constitutes at least part of the switch which triggers the onset of replication. Recent work from Kornberg's laboratory suggests that the active state of the dnaA protein is a monomer and that membrane material and the dnaK heat shock protein influence the distribution between this form and an inactive aggregated state (25).

B. Transcription

The biochemistry of RNA synthesis is perhaps the most intensively studied process in all of molecular biology due to the obvious importance of gene expression in development, oncogenesis, adaptation of an organism to the environment, and a host of other biological phenomena.

Like DNA replication, RNA synthesis involves the joining of nucleotide triphosphate precursors in a pattern defined by a DNA template by nucleophilic attack of a 3′-OH on the α-phosphate of, in this case, a rNTP. Unlike DNA synthesis, a primer-template complex is not required and the copy does not remain stably associated with template (Fig. 5). The process is catalyzed by a complex enzyme known as RNA polymerase, which is composed of several polypeptide subunits. (For a review of these subunits and their individual functions in prokaryotic and eukaryotic RNA polymerases, see Refs. 26–30.)

Polymerase recognizes specific DNA sequences known as promoters that act as assembly sites for the transcriptional machinery. Every gene does not have a unique promoter. Especially in prokaryotic cells, many genes with related functions are grouped into operons having a single promoter (31). Once bound to a promoter in what is called the "closed complex," the polymerase must open the double helix to form an active "open complex." It then begins RNA synthesis, a process that is sometimes aborted after a few nucleotides have been incorporated. Polymerization is eventually terminated by the action of proteins such as the *E. coli* rho factor (32,33), or in response to specific termination signals in the DNA sequence that

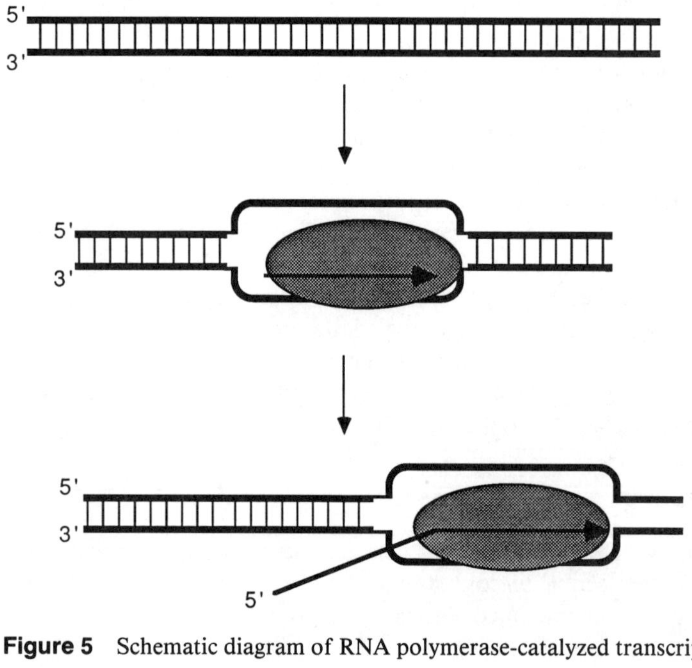

Figure 5 Schematic diagram of RNA polymerase-catalyzed transcription.

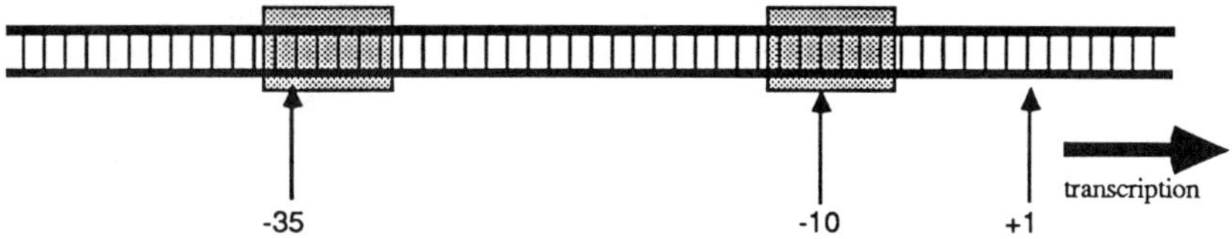

Consensus Promoter: TTGACA---17 bp---TATAAT

-35 -10

Figure 6 Schematic diagram of a consensus prokaryotic promoter element.

are thought to induce a particular secondary structure into the RNA product (34).

An analysis of many *E. coli* promoters has led a consensus sequence (35). While the region covered by RNA polymerase is about 60 bp, only two small regions within the promoter appear to be highly conserved. These are the "-10" and "-35" regions (Fig. 6) and are presumably the sites of direct DNA-protein contact. How RNA polymerase recognizes these particular sites is unknown. Natural promoters differ to a greater or lesser extent from this ideal consensus in the sequence of the -10 and -35 regions as well as the spacing between them. In general, the "strength" of a promoter appears to be correlated to how well it resembles the consensus, strength being correlated with the basal level of transcription from a particular promoter in the absence of other regulatory signals.

Most eukaryotic promoters contain a "TATA box" which has a consensus sequence: TATA(A/T)A(A/T), which is similar to the bacterial -10 sequence, except that it is located at -25. Three other sequences are often present. These are the "CAAT box" (consensus: GG(C/T)AATCT), located around -80 and two GC boxes (consensus: GGGCGG), located at about -50 and -95. The structure of the well-studied thymidylate kinase promoter is shown in Figure 7. Messenger RNA synthesis in eukaryotes is

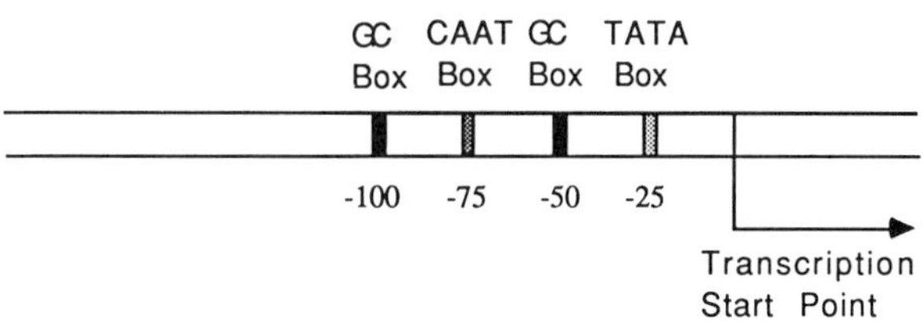

Figure 7 Schematic diagram of the thymidylate kinase promoter.

catalyzed by RNA polymerase II. Besides this complex, multi-subunit protein, several general transcription factors (TFs) are required, including TFIIA, TFIIB, TFIID, TFIIE, TFIIF, and TFIIG (36–40).

Binding of the polymerase (and in the case of eukaryotes, its attendant transcription factors) to a promoter in a manner competent to subsequently begin transcription is a major point of control. It appears that other proteins which recognize DNA sequences partially or completely distinct from the promoter act to regulate these events. As we shall see, this regulation may be of a positive or negative nature and, in the former case, appears to involve specific protein-protein interactions between the polymerase and the control factors. Since this topic is far too broad to review even briefly here, I will concentrate on two systems that serve to highlight the important general concepts and help us place the structural material to follow in its proper biological context. These systems are the phage lambda "switch" from a lysogenic to a lytic life-style (a good example of a typical prokaryotic regulon) and the expression of galactose metabolism genes in yeast, a useful prototype for eukaryotic transcriptional control. For the reader interested in learning more, I would recommend the exceptionally readable and clear presentation of the lambda switch in a recent book by Ptashne (41). The lambda system and that of the very closely related phage 434 are especially important to chemists and pharmacologists interested in correlating structure with function, since some of the protein-DNA complexes important in control of gene expression have been characterized crystallographically. While the yeast galactose system is not well understood at the atomic level, it will serve to highlight present problems and opportunities in this area.

1. The Switch from Lysogeny to Lysis in Lambda-Infected E. coli Cells

Lambda is a simple bacteriophage (42) that exhibits two distinct types of life cycles. Upon infection it can either integrate its DNA into the host chromosome at a specific site (known as attB) through a site-specific recombination reaction, or replicate itself many times over, synthesize new protein coats, and lyse the cell. The former state is known as lysogeny and a cell chronically infected with lambda is called a lysogen. The lysogenic state is quite stable, and the incorporated DNA can be replicated along with the host chromosome through many generations of cell division. However, when a lysogen is treated with UV light or some other DNA-damaging agent the lytic pathway is induced, thus allowing the phage to desert a sinking ship, so to speak. This switch is regulated at the level of transcription. While we will not be concerned with how a lysogenic state is achieved in the first place (see Ref. 41 for a discussion), we consider here how it is stably maintained and how environmental stress causes the induction of lytic growth.

The control mechanism involves three proteins (host RNA polymerase, repressor, and Cro), two promoters (P_R and P_{RM}), and three DNA sequences known as operators (O_R1, O_R2, and O_R3). The promoters, whose designations stand for *p*romoter *r*ightward and *p*romoter of *r*epressor *m*aintenance, point in opposite directions and are adjacent but do not overlap (Fig. 8). Immediately to the right of P_R is the *Cro* gene, and farther downstream are located the so-called very early genes required for lytic growth, including *O* and *P*. *If transcription from P_R is efficient, the phage proceeds into the lytic pathway.* Immediately to the left of P_{RM} is the gene *cI*, which codes for the repressor protein. As we shall see in a moment, repressor acts to silence P_R but not P_{RM}, thereby blocking transcription of the genes that would initiate lysis, but does not hinder expression of its own gene (indeed, quite the opposite is true). *So long as a high concentration of intact repressor is present in the cell, the lysogenic state is stably maintained.* An important point is that P_{RM} is a weak promoter, but P_R is a very strong one. Overlapping the promoters are three operator sites, O_R1–3. These are similar, but not identical, 17 bp sequences that have approximate two-fold symmetry. The organization of the DNA sites is shown in Figure 8.

In a stable lambda lysogen, the only active phage gene is *cI*. Its protein product, the Repressor, is composed of distinct N- and C-terminal domains linked by a 40-amino-acid bridge. Repressor is predominantly a dimer in solution. These contacts are primarily due to residues in the C-terminal domain (Fig. 9). In addition, repressor dimers have some affinity for one another when placed in juxtaposition (vide infra). The N-terminus of Repressor is a sequence-specific DNA binding domain. The target sites are the 17-bp operators. As will be discussed in great detail later in this chapter, each N-terminal domain contains a "helix-turn-helix" motif that allows it to bind to an operator "half-site." The dimer sits on one face of the helix cylinder (Fig. 9). When the intrinsic affinity of the repressor dimer for each operator site was measured in the absence of the other two, it was found that the protein binds very tightly to O_R1, while O_R2 and O_R3 are recognized about equally well, but more weakly than O_R1. However, when all three

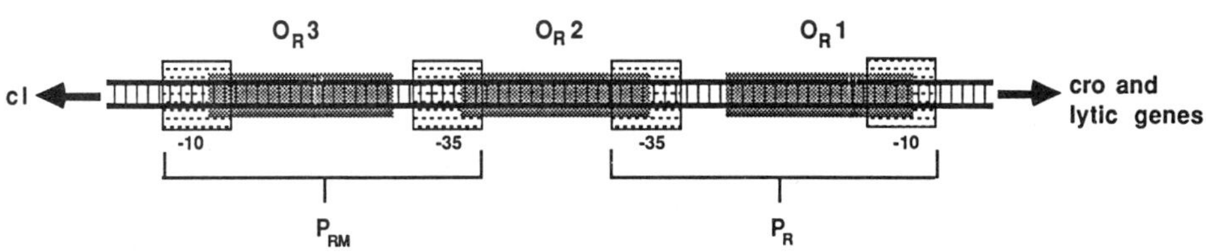

Figure 8 Organization of the DNA transcriptional control elements that function in the lambda lysogen/lysis switch.

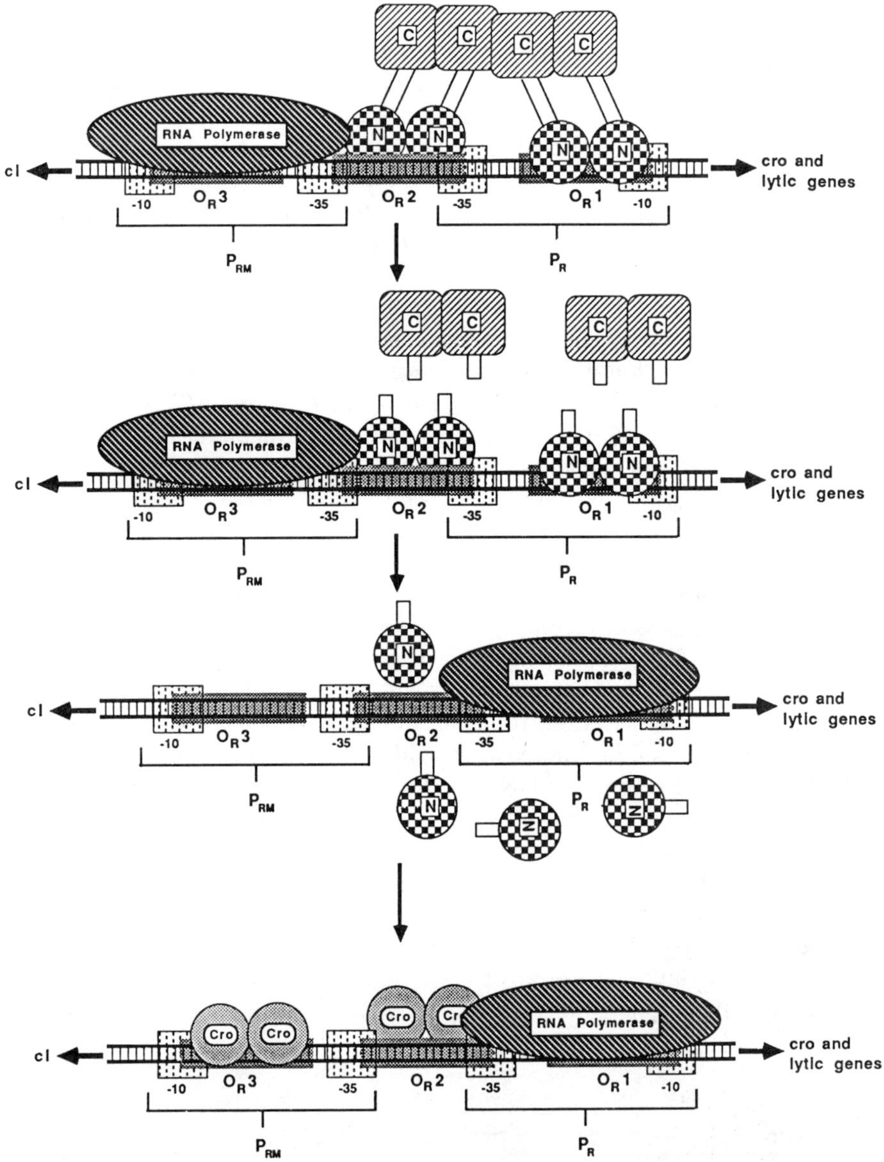

Figure 9 Summary of the biochemical events that lead to induction of lytic growth in λ lysogens. During lysogeny, P_R is blocked by binding of repressor to O_R1 and O_R2. Upon damage to the *E. coli* chromosome, RecA protein-mediated cleavage of the λ Repressor occurs, greatly destabilizing the binding of the N-terminal domains. Once these polypeptides fall off the DNA, RNA polymerase is free to utilize the powerful P_R promoter and high-level transcription of Cro and the lytic genes begins.

operators are present in their natural context, a significant preference for O_R2 over O_R3 is observed. This is due to cooperativity between repressor dimers bound to O_R1 and O_R2. Recall that repressor dimers have some affinity for one another (primarily through their C-terminal domains). Because of the intrinsically high affinity of O_R1 for the repressor, this site will be occupied to the largest extent at any particular nonsaturating repressor concentration. Binding of a second molecule of dimer to O_R2 is then facilitated by interactions with the O_R1-bound repressor dimer, i.e., the binding to O_R1 and O_R2 is cooperative (Fig. 9). This more than compensates for the lower intrinsic affinity of the repressor for O_R2. However, because of the geometry of the complex, a repressor dimer bound to O_R3 cannot touch the repressor-O_R1-O_R2 complex, so binding at O_R3 is not cooperative. Therefore, in vivo, O_R1 and O_R2 will normally be occupied, but O_R3 will only fill at high repressor concentrations. It should be noted that a situation can be contrived in which binding to O_R1 is not observed by mutating its sequence. In this case, cooperative binding between repressor dimers at O_R2 and O_R3 is observed as expected, since the distance between the centers of O_R2 and O_R3 is very similar to that between O_R1 and O_R2.

The biological implications of these physical properties are profound. O_R1 (completely) and O_R2 (partially) overlap P_R (Fig. 8). Therefore, when these sites are occupied by repressor, transcription from P_R is silenced; repressor recognizes some of the same nucleotides as RNA polymerase and is present at far higher concentrations than polymerase in lambda lysogens. This is the case because repressor binding to O_R2 *stimulates* rather than depresses the level of transcription of *cI* from P_{RM} (occupation of O_R1 has no effect on P_{RM} since the two sites do not overlap). O_R2 overlaps with P_{RM} to a slightly lesser degree than with P_R. This fact apparently not only allows polymerase to bind to P_{RM} when O_R2 is occupied, but in addition allows the repressor and the polymerase to form a specific complex that acts to stimulate binding (and perhaps opening) of this weak promoter. The *cI* gene is not expressed at ridiculously high levels however, because at a sufficiently higher repressor concentration O_R3 will be filled, thereby shutting off transcription from P_{RM}. So, through a combination of negative and positive regulation of initiation from the divergent promoters, the repressor protein acts to keep its own concentration high and maintains the lysogenic state (Fig. 9).

The switch to the lytic state involves proteolytic destruction of the repressor protein. Upon UV irradiation or some other environmental insult, a biochemical response known as the SOS pathway (43) is induced in *E. coli* cells. A key step in this process is the activation of the protease activity of the host RecA protein (44,45). Though normally involved in homologous recombination (46), this remarkable protein can also act as a protease when polymerized along single-stranded DNA. It cleaves only at a specific, accessible protein sequence which is found in the bacterial LexA repressor. This factor, whose global organization is not unlike the lambda repressor, controls the expression of a host of genes involved in DNA repair and various proteolytic pathways. Induction of these genes helps the cell to survive the otherwise lethal effects of irradiation. The requirement for extensive patches of single-stranded DNA to reveal the RecA protein's protease activity is an elegant alarm mechanism, since they would not be present under normal circumstances. The bridge between the N- and C-terminal domains in the lambda repressor contains a cleavage site. Therefore, the protein is snipped in two just as is the LexA repressor (Fig. 9). Loss of the C-terminal domain leaves the repressor N-termini bereft of the benefits of monomer-monomer and dimer-dimer interactions, thus drastically lowering the affinity of the DNA-binding domain for the operator sites. As soon as O_R1 and O_R2 are vacated, high-level expression of the genes downstream of the very strong $P_R$promoter begins, inducing the switch to the lytic growth cycle (Fig. 9).

The final facet of the switch we need to consider is how it is made irreversible. Induction of RecA's proteolytic activity is a transient phenomenon. Once the RecA filaments dissociate, why is not more repressor made from transcription of the *cI* gene from P_{RM}? Part of the answer is the greater strength of the P_R promoter. However, the difference in the basal level of transcription from the two promoters is not sufficient to explain the very clear-cut switch from lysogeny to lysis. The solution lies with the Cro protein, whose gene is expressed at high levels upon activation of P_R. Cro is also a dimeric, sequence-specific DNA-binding factor that recognizes the operator sequences (Fig. 9). However, *the relative affinities of Cro for the three sites is exactly the opposite of those exhibited by repressor* and the Cro dimers do not bind cooperatively. Thus, as the Cro concentration increases, O_R3 is occupied to the greatest extent, followed only later by O_R2 and O_R1. The Cro-O_R3 complex greatly depresses transcription from P_{RM} by denying access to RNA polymerase, thus eliminating any chance that repressor might make a "comeback" and reassert the lysogenic state. Only at high Cro concentrations is P_R silenced. By this time, there is a sufficient concentration of the very early gene products present for lysis to be far along.

From the above discussion, we can see that the control mechanism is extremely dependent on the ability of the repressor and Cro proteins to recognize very similar operator sequences with different affinities. The molecular basis of this selectivity is beginning to be understood through a great deal of biochemical work and the crystallographic analysis of protein-operator complexes. These studies are described in detail later in the chapter. The negative regulation exhibited by Cro and repressor through mutual exclusion is a common theme in gene expression. In addition, the concept of simulation of gene expression through protein-protein interactions, as in the case of the repressor-polymerase interaction at O_R2, is central to current models of the control of eukaryotic gene expression.

2. *The Yeast Galactose Regulon*

When the yeast *Sacharromyces cerevisea* is grown on galactose as the sole carbon source a number of genes involved in its utilization are induced at high levels. When the galactose concentration drops, expression of the *gal* genes drops again to much lower basal levels. In addition, when glucose is present in the medium, the gal genes are not induced even when high levels of galactose are also present (47).

The biochemistry underlying this eukaryotic transcriptional regulon is moderately well understood, though some important questions remain unanswered. In addition to the promoter, another DNA sequence about 250 bp away from the target genes is required for high-level transcription (48) (Fig. 10). This is called an upstream activation sequence (UAS) in yeast and is functionally equivalent to enhancers in higher eukarotes (49). These are defined as DNA sequences that are required to stimulate transcription of a gene and function in an orientation and (more or less) distance-independent manner. The UAS_G (G denotes the particular UAS important

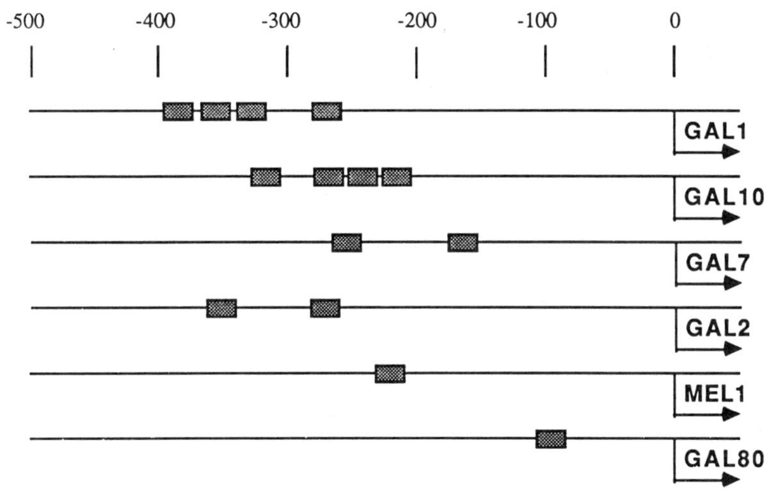

Figure 10 Organization of the related 17-base-pair GAL4 protein-binding sites (stipled boxes) upstream of genes that are positively regulated by this factor.

in stimulating the *gal* genes) includes four similar 17 bp sequences that are the binding sites of the GAL4 protein, a transcriptional activation factor. The observation is that when GAL4 is bound to the UAS_G, transcription of the target genes is stimulated greatly, up to 500-fold. Like the lambda operators, these 17mers are variations on a symmetric near consensus sequence and each particular sequence has a certain "strength" of activation (50). Although copperative binding of GAL4 to separate 17mers has been observed, this is not required for activation. A single copy of the near consensus 17mer will substitute for the natural UAS_G (50).

How a protein-DNA binding event several hundred base pairs from the promoter can affect the efficiency of transcriptional initiation is not absolutely clear, but there is rapidly mounting circumstantial evidence that the process involves direct protein-protein contacts between a specific domain of the GAL4 protein (51) and some as yet unidentified component of the transcriptional apparatus. The intervening DNA is proposed to "loop out" to facilitate this contact (52) (Fig. 11). This particular interaction is prototypical of a large number of observations in which distant DNA sequences interact in a functional sense and a great deal of effort has been expended to test various models for this fundamentally important phenomenon. These experiments have almost completely ruled out scenarios involving "tracking" of a factor from one site to another, or a protein-induced DNA conformational change that is propagated through the helix over a distance (53). DNA looping, the currently favored model, has been observed recently in an in vitro transcription system. Su et al. (54) looked at DNA fragments containing an enhancer-like sequence that is recognized by the *E. coli* NtrC protein separated by 390 bp from the *glnA* promoter. Purified NtrC protein and *E. coli* RNA polymerase were introduced and the samples examined by electron microscopy. In a significant percentage of the complexes, loops were observed of about the expected size.

How the putative GAL4-transcription complex interaction enhances the efficiency of transcriptional initiation is not clear. A number of models have been proposed, including a GAL4-induced conformation shift, GAL4-induced rejection of an inhibitory factor, etc. However, the only direct clue we have is the observation that the binding of transcription factor IID (a TATA-binding protein) is stimulated by contact with a GAL4 fragment bound nearby (55). The general model clearly suggested by this finding is that GAL4 promotes assembly of the transcription complex at the promoter, although the details of these events are very obscure. Until we have a better understanding of the components of the yeast RNA polymerase II enzyme and its many attendant transcription factors, this question will be difficult to address.

A simple mechanism for negatively regulating the activity of the *gal*

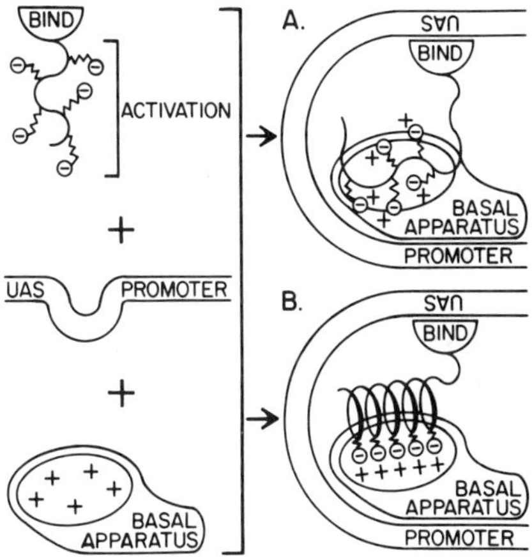

Figure 11 Cartoon of the looping model for transcriptional activation. A transcriptional activation factor (top left) recognizes an upstream activation sequence (UAS), while the basal transcriptional apparatus (RNA polymerase II and its attendant transcription factors) binds to the promoter. The activator is thought to physically contact some component of the basal machinery by virtue of a discreet activation domain. Since one prominent family of such domains is highly acidic, the activation domain is represented as a "negative noodle" in this figure. Scenarios A and B (right side) reflect the two most currently popular theories for protein-protein contacts. The first (A) suggests that the activation domain is completely unstructured and that ionic contacts are of prime importance. The other (B) proposes that the activation domain is an amphipathic α-helix and that the alignment of charge along one face of the helix is important in protein-protein contacts. There is no direct evidence for either model. In any case, the contact somehow allows the basal transcriptional machinery to initiate transcription more efficiently, perhaps by aiding its assembly at the promoter.

genes would be for the affinity of the GAL4 protein for the 17mers be dependent on the galactose concentration through some sort of allosteric process. However, this is not the case. Basal level transcription is instead maintained by a second protein, the GAL80 gene product, which binds specifically to GAL4 (56–59). This interaction prevents the GAL4 protein from activating transcription even though GAL80 does not apparently influence the DNA binding activity of GAL4. The current model is that GAL80 simply covers part of the GAL4 domain that contacts the transcriptional machinery. As might be expected then, GAL80 is a galactose sensor. When high levels of galactose are present in the medium, the protein is thought to undergo a conformational change that structurally alters the GAL4-GAL80 complex, allowing high-level transcription to occur (Fig.

11). At present, it is not clear whether GAL80 binds galactose itself, or some metabolite of the sugar. It is also not clear how high glucose concentrations short-circuit the entire system.

Some of these ideas will probably be tested directly in the near future, since two laboratories have independently reported the purification of a small amount of a 1 : 1 GAL4 : GAL80 complex from yeast cells (57,58), and Fukasawa and co-workers have purified small quantities of GAL80 alone (59).

Our understanding of some of the details of the gal regulon have profited greatly from an extensive analysis of the domainal organization of the 881 residue GAL4 protein (Fig. 12). The DNA-binding domain, which contains an unusual "zinc finger" structure, is located in the first 74 amino terminal residues (51). However, the intact GAL4 protein binds to DNA as a dimer and the residues required for dimer stability are located between residues 74 and 147 (60). The details of GAL4-DNA complexation will be considered in detail in the section on zinc finger proteins. The domain responsible for interaction with GAL80 is surprisingly small; it has been shown that a fragment containing only the C-terminal 30 amino acids of GAL4 will suffice (61,62). The residues on GAL80 that are involved in complex formation have not been characterized. The primary transcriptional activation domain is located at the C-terminal end of the protein (63,64). A second, cryptic, activation domain that can function when the C-terminus of GAL4 is deleted has been mapped to the middle of the gene (64). However, the biological relevance of this observation is unclear. The fact that the GAL80-binding region and the major transcriptional activation domain of GAL4 overlap is consistent with the current model for GAL80 function.

We can now see that some of the mechanisms employed by phage lambda are also used in the yeast *gal* regulon, albeit with a few twists. In particular,

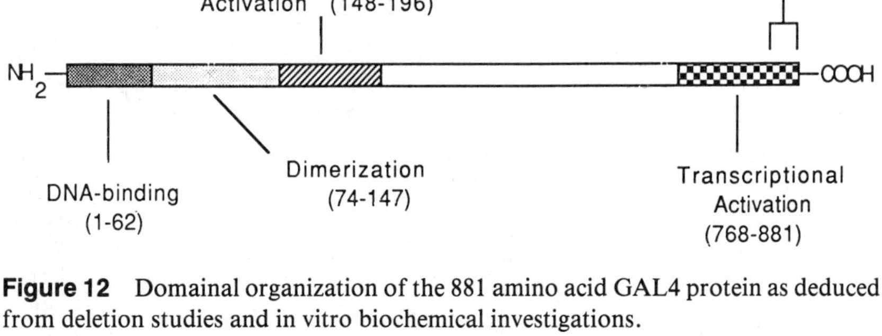

Figure 12 Domainal organization of the 881 amino acid GAL4 protein as deduced from deletion studies and in vitro biochemical investigations.

the strategy of positively regulating transcription by using protein-protein interactions to attract polymerase to a promoter appears to be common to both, although in the yeast system the operator equivalent is not adjacent to the promoter. It should be noted that transcriptional activation from a distance is not a uniquely eukaryotic phenomenon. Upstream activation sequences have also been discovered in some *E. coli* operons (65,66).

C. Recombination, Repair, and Restriction

The chromosome of any particular organism is a stable, but not static entity. It is true that the integrity of the information contained in the nucleotide sequence must be maintained. Thus, it is not surprising that elaborate DNA repair systems have evolved to remove and correct damage due to environmental factors, the inherent instability of DNA and "errors" in DNA metabolic processes, especially replication (67). The proteins which bind to damaged DNA and attract the repair machinery to the appropriate site do so by recognizing unique structural perturbations in the nucleic acid. While some factors, such as the uvrABC complex of *E. coli*, will recognize a broad spectrum of lesions, others are quite specific for certain types of damage. While the molecular details of the mechanism by which these factors "know" that DNA is damaged is certainly a fascinating topic, it is beyond the scope of this chapter; therefore, we will not consider DNA repair mechanisms in detail.

Likewise, the elaborate enzymatic machinery involved in homologous recombination (68) is blind to sequence and will therefore also not concern us. However, there are several types of site-specific recombination and transposition processes which require little or no homology between DNA substrates, but do require specific DNA sequences. These reactions are widespread in nature and several such systems are of keen medicinal interest. For example, the DNA-form of the AIDS virus, HIV-1, integrates into the host chromosome via a site-specific recombination system, the diversity of immunoglobulin proteins is generated by site-specific recombination of gene cassettes (69) and the spread of antibiotic resistance among many hospital bacterial strains involves transposition.

In all cases, site-specific recombination involves the staggered breakage and reunion of specific sites (70). This protein-catalyzed process, which does not require the presence of a high-energy cofactor, is accomplished by a topoisomerase-like mechanism in which a hydroxyl side chain of the recombinase cleaves a phosphodiester bond, then rejoins it with the appropriate end of the other recombination site. Depending on the individual system and the orientation of the recombination sites, a number of outcomes are possible (Fig. 13a), including integration, resolution, and inver-

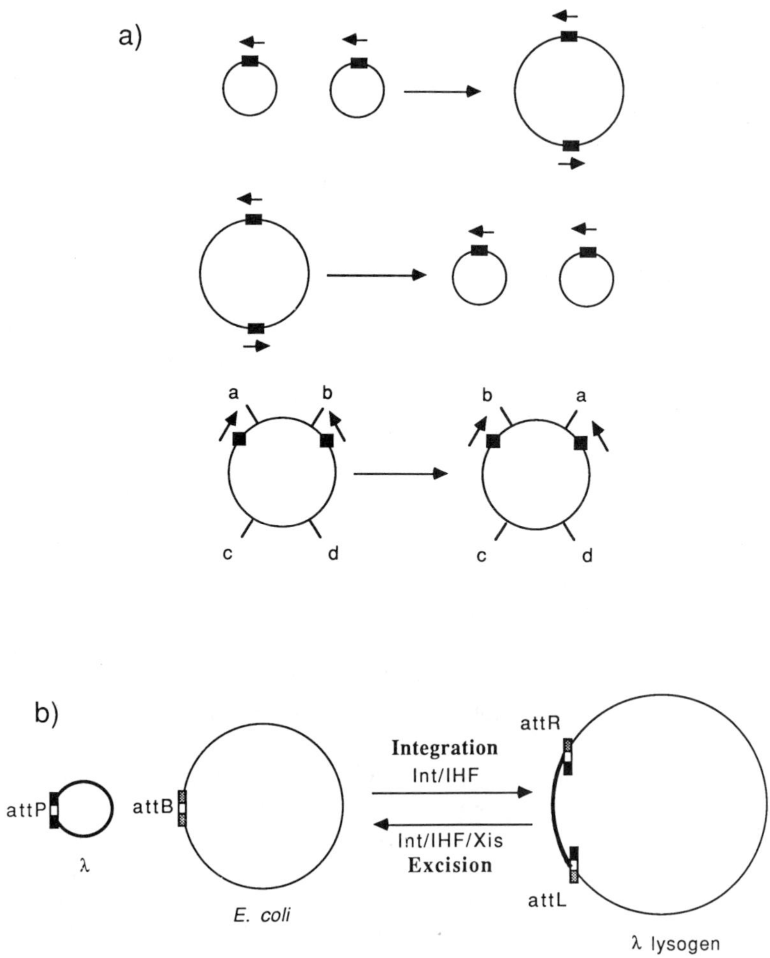

Figure 13 Site-specific recombination reactions. (a) Possible outcomes of a recombination reaction, including dimerization, resolution, and inversion. Boxes represent the recombination sites. (b) Schematic representation of lambda integration and excision.

sion. The paradigm for site-specific recombination is the integration/excision system of phage λ (Fig. 13b). As mentioned above, integration occurs by breakage/reunion of sites on the phage and bacterial chromosomes known as *attP* and *attB*, respectively (71). This process requires the phage Integrase protein (the product of the *int* gene) and is greatly stimulated by the bacterial integration host factor (IHF), which is composed of two polypeptides (the products of the *himA* and *hip* genes). The attP site is rather large, about 240 base pairs, and contains several specific sites for both Integrase and IHF binding (72). AttB is much smaller, 25 base pairs, and only binds Integrase, not IHF. The two attachment sites share 15 base pairs of homology, known as the *core*, the actual site of recombination, within which the Int protein makes and reseals seven base pair staggered breaks. Integrase recognizes two distinct DNA sequences (73), one of which is at the core; the other is at the flanks (Fig. 13). IHF recognizes only a single sequence, which it bends severely upon binding. The structure of the "intosome," including the two att sites and several molecules of Integrase and IHF, is not known but has been thought of as being nucleosome-like. The elaborate structure is probably required to bring about proper alignment of the DNA sites.

Excision of the λ chromosome from *E. coli* also involves Integrase and IHF, but also requires another phage factor, Xis, which is also a sequence-specific DNA-binding protein and binds to *attL* and *attR*, the sites involved in excision.

Not much is known about the structural basis of DNA-protein recognition in this or other related site-recombination systems, with the notable exceptions of the Hin recombinase of *Salmonella typhimurium*, which is thought to have a helix-turn-helix domain and the catalytic domain of the γ,δ-resolvase protein, which has been crystallized by Steitz and co-workers (74). Thus, this area remains a fruitful venue for detailed studies of DNA-protein interactions.

Transposition differs from site-specific recombination in a number of ways, most particularly the requirement for at least some DNA synthesis and the lack of a requirement for any sequence homology. Some transposons exhibit significant target sequence preferences, but others can insert themselves into any DNA sequence. The breakage/joining events thought to be involved in bacteriophage Mu transposition (one of the best-understood systems) are illustrated in Figure 14. The Mu A and B proteins, along with the host HU protein, make staggered cuts in the target DNA with 5′ overhangs. These ends are then joined to 3′ Mu ends to generate a structure that has two primer/template junctions. At this point, the pathway can bifurcate. If replication initiation is efficient, leading strand synthesis will generate a cointegrate that, after ligation, leads to one possible product. Alternatively, if replication initiation is not efficient, the junction between the Mu and target DNA can be cut, producing a simple insert product (75). While there is a considerable body of biochemical knowledge concerning the activities of transposases, we again do not have any structural information regarding any of these proteins.

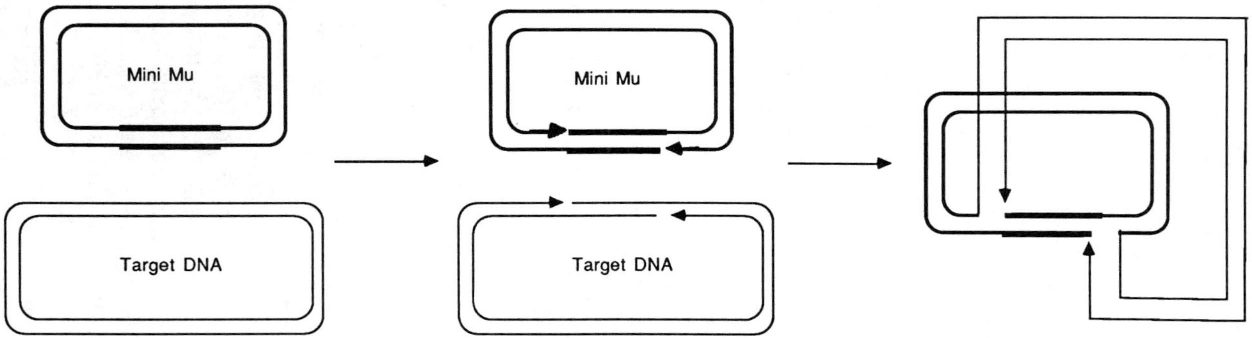

Figure 14 Proposed pathway phage Mu transposition, as postulated by Mizuuchi and co-workers. The substrates represent a "mini-Mu" plasmid (smaller oval) containing the recombination site (thick lines) and target DNA (large oval). A staggered cut in the target molecule and a nick on opposite strands at each end of the Mu DNA leads to generation of the intermediate shown. DNA replication, utilizing the free 3′-ends can lead to a cointegrate. Alternatively, degradation of the DNA that flanks the Mu sequences followed by gap repair can lead to a simple insert.

D. Summary

It is my hope that the above review of DNA metabolism, though necessarily superficial, will provide the nonexpert reader with a basic idea of the general mechanisms employed by nature to efficiently employ, maintain, and replicate the genetic information contained in the chromosome. While it is not absolutely necessary to digest the above material to understand the information on the chemistry of sequence-specific DNA-protein interactions presented in the next several sections, that discussion will be much more meaningful when considered in its proper biological context. For example, in the review of the phage lambda repressor-operator crystal structure in Section III, considerable time is spent on what the structure tells us about the ability of the repressor to distinguish between similar natural operator sequences. Although an interesting question in itself, this aspect of the DNA binding chemistry of repressor acquires a deeper meaning when one realizes that the lysogenic life-style of this organism is completely dependent upon the observed selectivity. In addition, as stated at the beginning of the chapter, one of the aims of this review is to provide interested scientists with an intellectual framework for understanding how natural macromolecules recognize specific DNA sequences, in the hope that these principles will help guide the design of synthetic binding compounds. If one is thinking about eventually using such compounds as drugs, an appreciation of the role of the protein which inspired the design will obviously help to predict the biological consequences of drug administration.

III. SEQUENCE-SPECIFIC PROTEIN-DNA INTERACTIONS: THEORETICAL CONSIDERATIONS

The huge degree of interest in the biochemistry of DNA metabolism has resulted in the characterization of hundreds of sequence-specific DNA binding proteins. Comparison of the amino acid sequences of these factors has suggested that *there are a limited number of general structural motifs for DNA binding that have been conserved through evolution.* In a few cases, this hypothesis has been confirmed by comparison of the three-dimensional structures of proteins thought to be members of the same "family." While individual proteins in any particular family might recognize fairly divergent DNA sequences, this is thought to be the result of the variability of a few to several key residues in the binding domain that make important DNA contacts; the overall structural motif is conserved.

We are currently familiar with several structural motifs employed in sequence-specific DNA binding, including several types of "zinc fingers," the helix-turn-helix, and the homeodomain, among others. There are undoubtedly many more to be discovered. Before plunging into a discussion of the structurally characterized DNA-protein complexes from these families, it is useful to reflect on the various ways that, in theory, a polypeptide might discriminate between different DNA sequences.

A. Direct Readout of the DNA Bases

The most obvious mechanism for sequence discrimination involves base specific hydrogen bonding of amino acid side chains in the major and minor grooves. As pointed out by Seeman et al. (76), the edges of the four possible DNA base pairs, A-T, T-A, G-C, and C-G (Fig. 15), present very different opportunities for hydrogen bonding and hydrophobic contacts. This is most easily seen by overlaying two of the base pairs (Fig. 16), looking down the helix axis. From this comparison it can be seen that there are six possible sites for interaction with a protein in the major groove (W1, W2, W3, W3′, W2′, and W1″; W stands for *w*ide groove) and three sites in the minor groove (S1, S2, and S1′; S stands for *s*mall groove). Any

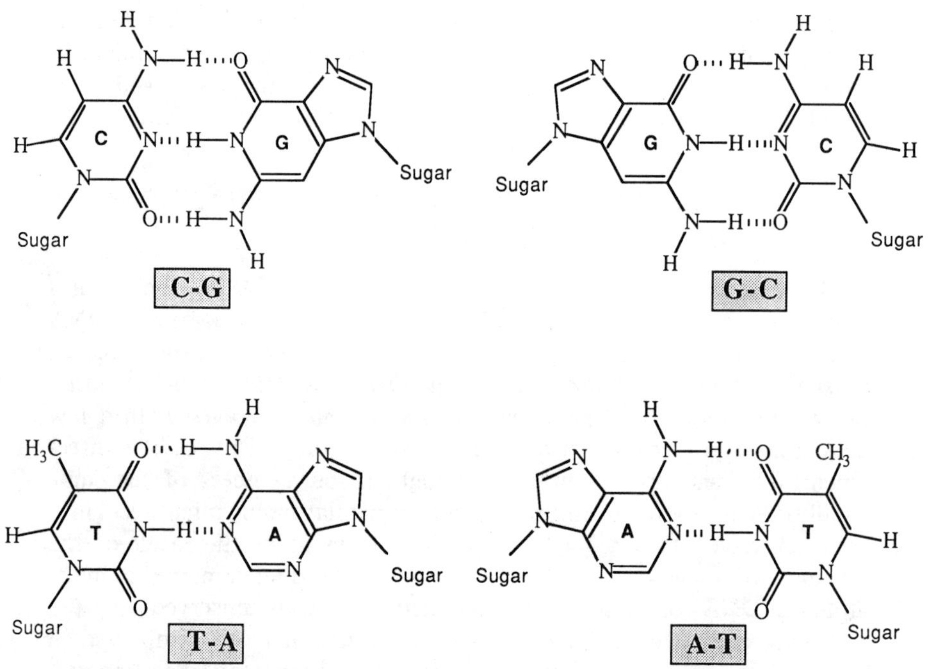

Figure 15 Structures of the four Watson-Crick DNA base pairs.

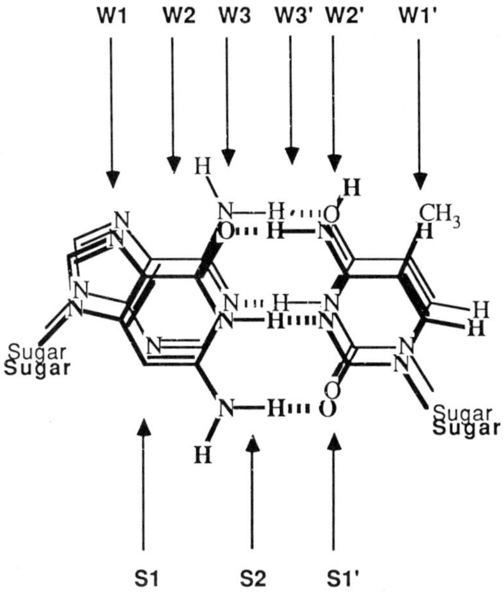

Figure 16 Overlay of G-C (bold) and A-T (plain) base pairs, illustrating the positions of various sites in the major (W sites) and minor (S sites) grooves available for interaction with a protein.

particular base pair will only have four W sites; W1, W2, W3′, and W1′ for a pyrimidine-purine base pair, or W1, W3, W2′, and W1′ for a purine-pyrimidine pair. The atoms in these sites have different potentials to interact with amino acids side chains depending on the identity of the base. For example, moving right to left across the major groove of a G-C base pair, we have: W1 = H-bond acceptor, W2 = blank, W3 = H-bond acceptor, W3′ = blank, W2′ = H-bond donor, and W1′ = C—H bond, which cannot form a hydrogen bond but could make a weak hydrophobic contact. On the other hand, an A-T pair would have the array: W1 = H-bond acceptor, W2 = blank, W3 = H-bond donor, W3′ = blank, W2′ = H-bond acceptor, and W1′ = methyl group, which can make a strong hydrophobic contact. Clearly, these arrays are chemically distinct. As shown in Table 1, this is true for all of the possible base pairs. Therefore, the major groove edges of each base pair will "look" different to a potential protein partner. One can go through a similar exercise for the minor-groove residues, the results of which are shown in Table 2. Note that interactions in the minor groove cannot distinguish between A-T and T-A base pairs nor G-C and C-G pairs. As we shall see, distinction between the chemical nature

Table 1 Description of the Interaction Sites in the Major Groove for Each of the Four DNA Base Pairs

Base Pair	W1	W2	W3	W3′	W2′	W1′
C-G	C-H	D	B	A	B	A
G-C	A	B	A	B	D	C-H
A-T	A	B	D	B	A	Me
T-A	Me	A	B	B	B	A

A = H-bond acceptor, B = blank, D = H-bond donor, C-H = H5 of cytidine, Me = CH_3 of thymine.

of the base edges is indeed a major mechanism for sequence-specific recognition. As might be expected, the more versatile major groove is the dominant site of recognition.

An important point that comes out of this analysis is that a single hydrogen bond with a peptide partner is almost certainly insufficient to uniquely identify a base pair. This can be appreciated from the following argument. While an interaction at W1 could easily distinguish a purine from a pyrimidine (imidazole nitrogen versus C—H or C—CH_3), it could not distinguish between A and G. Therefore, W1 (and W1′) cannot be used to uniquely identify a base. However, W2 (or W2′) will be occupied only if the left base is a pyrimidine and furthermore will be a hydrogen bond donor (NH_2) for C and a hydrogen bond acceptor (carbonyl oxygen) in the case of T. Similarly, W3 will be occupied only if the base is a purine and will be either an H-bond donor (NH_2) or acceptor (carbonyl oxygen) for A and G, respectively. Therefore, in theory, a single hydrogen bonding interaction

Table 2 Description of the Interaction Sites in the Minor Groove for Each of the Four DNA Base Pairs

Base pair	S1	S2	S3′
C-G	A	D	A
G-C	A	D	A
A-T	A	C-H	A
T-A	A	C-H	A

A = H-bond acceptor, D = H-bond donor, C-H = H6 of adenine.

with an amino acid side chain could discriminate completely between the four possible base pairs through interaction with W2 or W3 (or W2′ and W3′). However, this is not the case in practice. W2 and W3 are very close to one another in space, about 1.1 Å distant (76). It would be almost impossible for a protein side chain to be held rigidly enough to "see" a W2 site but not W3. Therefore, adenine could not be distinguished from cytosine, for example, since W2 and W3 are both occupied by an amino group in this case. Clearly, then, at least two hydrogen bonds are required to uniquely recognize a base pair.

In theory, any conceivable mode of noncovalent bonding could be used in DNA-protein complexation and we shall see that there are indeed a large variety of types found even in the few complexes that have been structurally characterized to date. There are examples of interactions between single bases and almost all of the polar protein side chains, branched interactions where one protein side chain hydrogen bonds to another, thereby holding the latter in the correct position to contact a base. Bidentate H-bonds in which a bifunctional residue such as an amide employs both heteroatoms in hydrogen bonds with a single base or even adjacent bases have been identified, as well as many other types of hydrogen bonding. In addition, hydrophobic and van der Waals contacts are very common and in some cases seem to contribute to sequence selectivity, particularly in the case of packing against the C-6 methyl group of T. Indeed, about the only type of potential interaction that has not been observed in sequence-specific protein-DNA complexes is intercalation of aromatic side chains between the bases. However, this is a strong component of binding in complexes of single-stranded DNA and helix-destabilizing proteins (77).

B. Indirect Readout by Sequence-Specific Conformational Preferences

It is now well accepted that the detailed structure of DNA is dependent on its sequence, even within the general B-DNA family. This realization has come from analysis of high resolution crystal structures of oligonucleotides that exhibit significant local variation in helical twist, base roll, propeller twist, etc. (78,79). Therefore, it stands to reason that certain sequences will be better able to adopt unusual conformations than others and that this property can contribute to sequence-specific DNA-protein recognition. For example, suppose that strong compression of the minor groove is required for formation of a hypothetical DNA-protein complex. A DNA sequence that contains a G at the site of compression will not readily adopt this conformation because of the presence of a bulky amino group in the minor groove and will therefore be disfavored. A number of other scenarios can be imagined, many of which will be discussed below. The important point is that these conformational preferences allow the sequence of DNA to be "read" by a protein *even if it does not contact the base directly.*

A recent crystallographic study by Otwinowski et al. of the trp repressor-operator complex (80) appears to provide a stunning example of this principle, since the structure revealed no direct hydrogen bonds between the protein and the nucleotide bases. However, as will be discussed in more detail below, it has been suggested that these workers used an oligonucleotide that does not correspond to the natural operator and that the structure they derived reflects nonspecific DNA biding. Nonetheless, other studies do suggest that conformational effects can play an important role in sequence recognition when used in concert with direct contacts. Very often, these deformations from ideal B-DNA are induced by strong backbone contacts, including hydrogen and ionic bonds to the phosphates. Backbone interactions are also important for two other reasons. They serve as a scaffolding to correctly position the side chains that do directly contact the bases and contribute to the overall binding energy of complexation.

IV. MAJOR STRUCTURAL MOTIFS FOR SEQUENCE-SPECIFIC DNA BINDING: THE HELIX-TURN-HELIX

In the early days of sequence-specific protein-DNA investigations, a very significant question was whether there exists a code of some sort that would allow one to deduce the sequence recognized by a particular DNA binding protein simply by knowing the amino acid sequence of the DNA binding domain (81). It is now clear that there is no simple code. Rather, specificity is achieved through the convergence in three dimensions of several groups that contact many sites on the DNA. However, there are conserved structural motifs employed over and over again by Nature for this task. In this section, we shall review those motifs that have been structurally characterized or at least investigated, beginning with the best known, the helix-turn-helix domain.

A number of years ago, it was postulated that a number of prokaryotic transcriptional regulatory factors share a conserved 20-residue structural motif now called the helix-turn-helix (82–85). This proposal was based on empirical considerations concerning the propensity of individual residues to form α-helical structures. The x-ray crystal structures of four of these proteins, or a DNA binding fragment derived proteolytically from the intact protein, have been solved, substantiating the helix-turn-helix hypothesis. In addition, the high resolution structure of the corresponding DNA-protein complex is available in three of these cases, providing very detailed information concerning the mechanism of sequence recognition. These studies, combined with a considerable body of biochemical and genetic data, make the helix-turn-helix family the best understood motif for sequence-specific DNA recognition.

A. Phage 434 Repressor-Operator Complex

In 1987, Anderson, Ptashne, and Harrison (86) reported the high-resolution structure of a complex between the 69-amino-acid DNA binding domain of the phage repressor (R(1-69)) and a synthetic operator. 434 is a close relative of phage lambda, and its repressor fulfills a function similar to its lambda analogue, recognizing six similar, but not identical, 14-bp operators of approximate twofold symmetry in the phage genome.

The crystals originally employed in this study diffracted to 3.2 A in some directions and 4.5 Å in others and employed a 14-bp oligonucleotide. More recently, the same group has improved the resolution of their data to 2.5 Å by using a 20-bp operator-containing DNA that forms better crystals (87). The structure of the free protein has also been solved (88). The sequence of this oligonucleotide is

```
        1L 2L 3L 4L 5L 6L 7L 7R 6R 5R 4R 3R 2R 1R
5' T A T A C A A G A A A G T T T G T A C T -3'
3'   T A T G T T C T T T C A A A C A T G A A
```

The operator, which corresponds to the natural $O_R 1$, is underlined and overlined (the nucleotides on the bottom strand are referred to as 1L′, etc.).

1. Protein Conformation

R(1-69) is a cluster of four α-helices plus a short C-terminal helical extension which represents the cleaved bridge to the C-terminal domain. Two of the helices, α2 and α3, constitute the helix-turn-helix motif (Fig. 17). As will be discussed below, the helix axes of the 434 and lambda repressor DNA-binding domains are nearly superimposable, substantiating the notion (89) that this motif for DNA binding is structurally very well conserved. The overall structure is stabilized by packing of several hydrophobic side chains in the interior of the helix bundle. Several polar interactions participate as well, notably Glu35-Arg5 and Gln17-Glu32.

R(1-69) binds to the operator as a dimer. Each "half-site" is contacted by a single protein monomer, two of which interact across the dyad. Anderson et al. suggest that these contacts may be important in correctly orienting the DNA-binding surfaces of the two monomers, thereby allowing cooperative binding to occur (86). Both half-site-monomer contacts are, for all practical purposes, identical and this discussion will focus on one half-site.

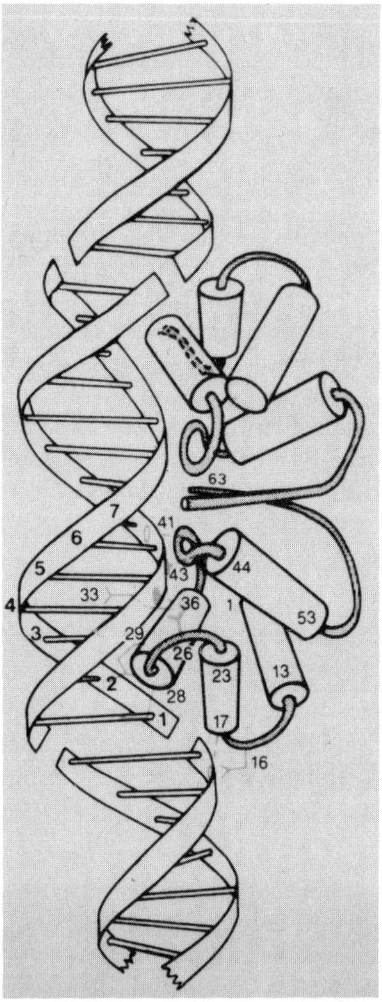

Figure 17 Overview of the structure of the 434 R(1-69)-operator complex. The α-helices are represented as cylinders. (Adapted from Ref. 86.)

2. *DNA Conformation*

The DNA in the protein-DNA complex is B-form throughout its length. Interestingly, the oligonucleotides stack upon one another in the crystal, forming a long, pseudocontinuous helix. The local helical twist from one base to another varies from about 9.2 to 12.3 base pairs per turn, a range also observed in crystal structures of uncomplexed B-DNA (78,79). With respect to the 10.5-base-pair periodicity of idealized B-DNA, the operator is overwound at its center and underwound at the ends (Fig. 18). There is some bend to the helix axis, with the sharpest departure from linearity occurring between base pairs 4 and 5 and 10 and 11. The sum effect of the bending and overwinding near the center of the operator is to compress the minor groove in this region. In addition, these structural deformations lead to a novel three-centered two-bond interaction between the A7L and the highly propeller twisted T6L and T7R (Fig. 18). However, the major groove is not drastically enlarged and the width of the minor groove gradually returns to "normal" toward the end of the oligonucleotide. This unusual arrangement partially explains the preference for certain sequences in the center of the operator that do not make sequence-specific contacts with R(1-69). A comparison of the operator conformation in the complex with an idealized B-helix is provided in Figure 18a.

3. *Protein-DNA Contacts*

The fit of the helix-turn-helix unit for each R(1-69) monomer is shown in Figure 19. The recognition helix $\alpha 3$ lies in the major groove, while the loop connecting it to $\alpha 4$ runs along the DNA backbone. Important backbone contacts (vide infra) occur on both sides of the major groove. As mentioned previously, these nonspecific contacts serve to dock the protein against the DNA in the correct orientation for sequence-specific interactions to occur.

The specific hydrogen bonds formed between the recognition helix and base pairs 1R-4R (and by symmetry 1L-4L), which are strongly conserved in all 16 natural half-sites, are made by three glutamine residues (Gln28, Gln29, and Gln33) located on one face of the recognition helix. Each participates in a completely different type of interaction (Fig. 20). Gln28 employs the nitrogen and oxygen of its amide side chain to hydrogen bond to N7 and N6, respectively, of adenine in the first T-A base pair. Interestingly, this type of interaction had been predicted by Seeman et al. many years before as a plausible mechanism for adenine recognition (in this case, binding to W1 and W3) (76). As an aside, the side chain of Gln 28 also makes a van der Waals contact with T − 1 (minus one). This interaction, which is outside of the canonical 14-bp operator, may help to explain the slight preference for operators with T in this position.

The ϵ-NH2 of Gln29 forms a bifurcated pair of hydrogen bonds with N-7 and O-4 of guanine in the C-G pair that occupies position 2. The ϵ-NH_2 of Gln33 forms a single hydrogen bond with O-4 of the T at position 4, while the amide oxygen interacts with a tightly bound water molecule.

Hydrophobic contacts seem to be as important as hydrogen bonds for sequence recognition. For example, there are no specific hydrogen bonds between the R(1-69) and the strictly conserved A-T base pair at position 3

(a)

(b)

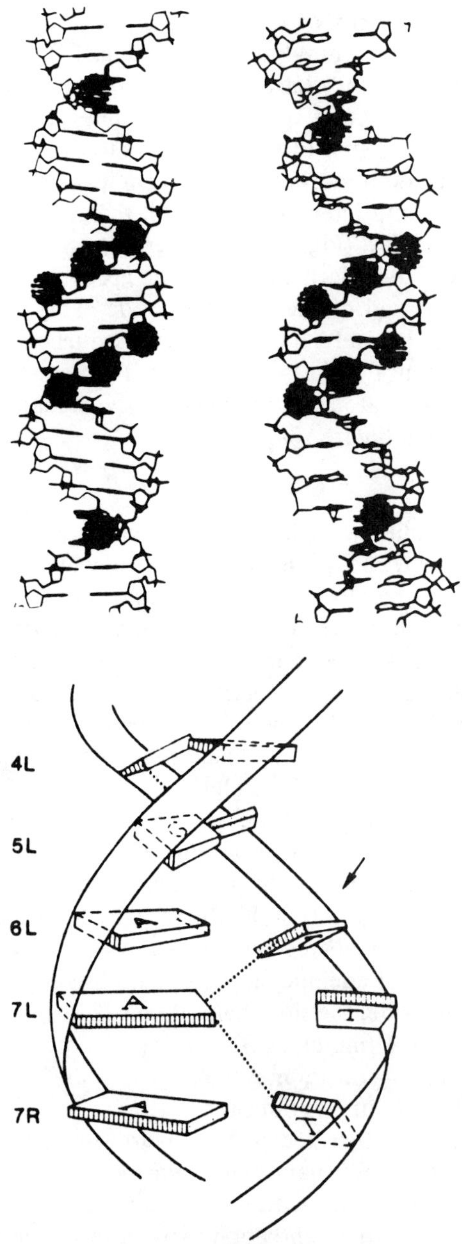

Figure 18 (a) Conformation of the 434 operator in the co-crystal (right) and idealized B-DNA (left). The contacted phosphates are represented by black circles. (b) View highlighting the unusual non-Watson-Crick bonds (dotted lines) in the center of the operator. The protein is not shown for clarity. (Adapted from Refs. 86 and 87.)

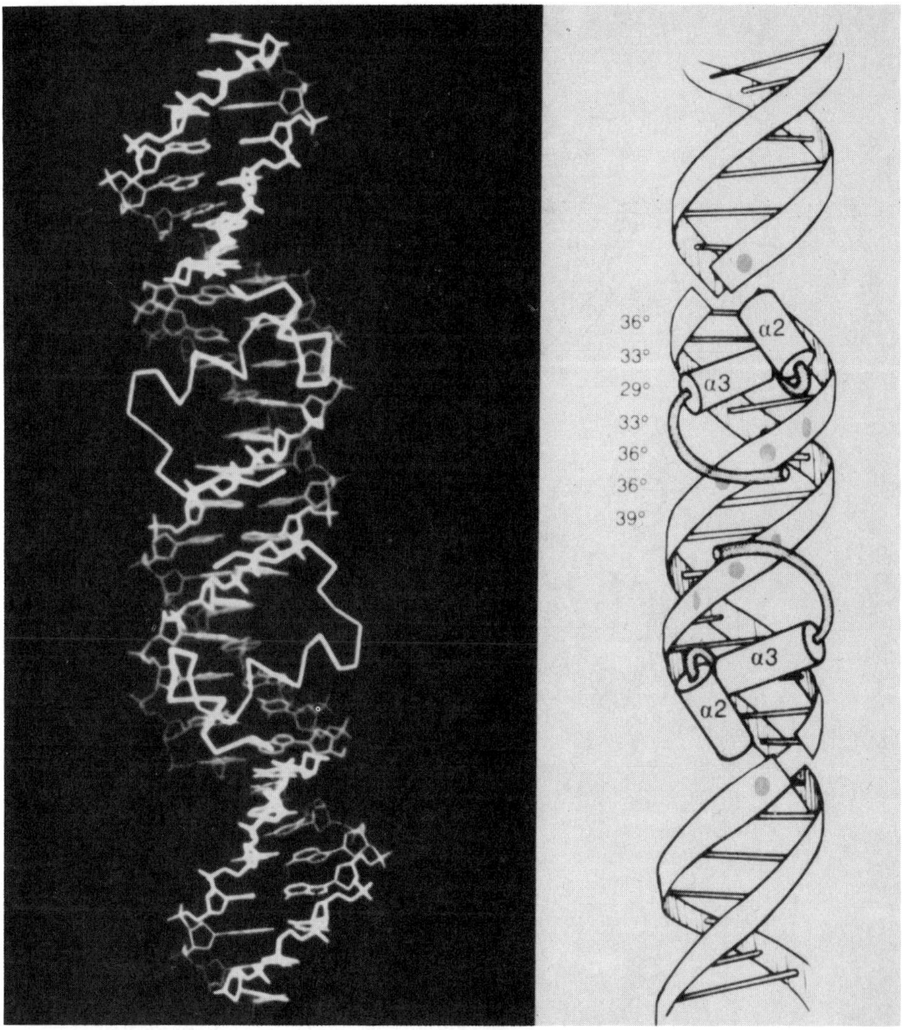

Figure 19 Schematic of the way the helix-turn-helix unit sits in the major groove in the 434 R(1-69)-operator complex. (Adapted from Ref. 86.)

of the operator. However, the side chains of Gln29 and Thr27 form a hydrophobic pocket to receive the methyl group of thymine. This is an apparent exception to the "rule" discussed earlier regarding the necessity of two interactions to uniquely specify a base pair. The thymine methyl group is unique at position W1′. The fact that the hydrophobic pocket that ac-

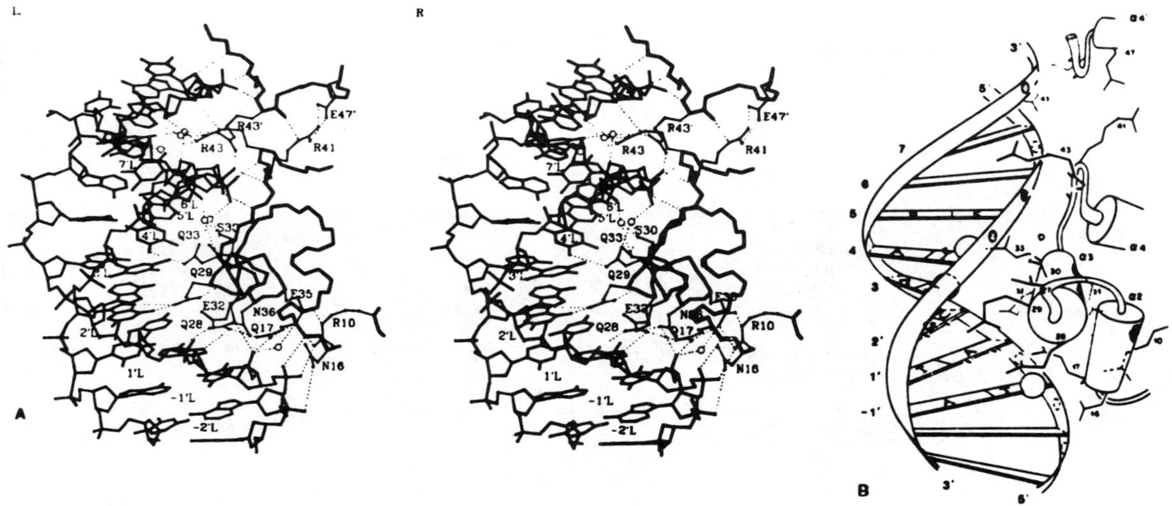

Figure 20 Stereoview and schematic of the details of the protein-DNA interactions in the R(1-69)-operator complex. Only one of the monomers is shown with the exception of a few residues from the other monomer at the top of the drawing. (Adapted from Ref. 87.)

cepts it is rigid enough to give rise to such a high specificity however, is due to an extensive net of interactions over the surface of the protein-DNA complex (see below). Another important van der Waals contact is observed between the methyl group of thymine in the fourth base pair and Gln29 and Ser30. Together with the direct hydrogen bond between Gln 33 and O-4 of T4, these interactions serve to uniquely recognize the last of the strongly conserved base pairs. It should be noted that of the sixteen natural half sites, only the right half of O_R3 does not have an A-T base pair at position 4, but rather contains a G-C pair. It is probably not a coincidence that G also has a carbonyl oxygen suitably positioned to accept a hydrogen bond from Gln33. Thus we are reminded of the inability of a single hydrogen bond to distinguish between like atoms in positions W2 and W3.

The precise positioning of the groups described above is due an extensive set of contacts along the sugar-phosphate backbone (Fig. 20) supported by a number of interconnecting hydrogen bonds, some of which involve tightly bound solvent molecules. The five phosphates contacted most strongly in the crystal structure correspond precisely to those that were identified as being important for protein-DNA complexation by phosphate ethylation interference experiments (90). The phosphate 5′ of nucleotide −1 accepts hydrogen bonds from Arg10, and the main chain N—H of Gln17. Both of these residues are located in $\alpha 2$. Gln17 and Asn36, which are located at the C-termini of $\alpha 2$ and $\alpha 3$ respectively, converge to hydrogen bond to the phosphate 5′ of A1. In addition, the sugar of A1 is in van der Waals contact with the side chain of Gln32. Formation of all of these contacts require the ends of the operator to bend toward the repressor.

The many other backbone contacts important in complex formation can be seen in Figure 20. Some of the most important for the following discussion are the result of the fact that the main chain of the loop between $\alpha 3$ and $\alpha 4$ runs along the backbone across the minor groove in the center of the operator. This geometry results in hydrogen bonds between three main-chain N-H groups and two phosphates: Lys40 and Arg41 to the phosphate 5′ of nucleotide 6′ and Arg43 to the phosphate 5′ of nucleotide 5′. The side chains of Arg43 in each monomer project into the minor groove to participate in further contacts that differ somewhat in the two half-sites. These contacts are the only examples of significant crossover interactions between the left repressor and the R half-site and vice versa.

A final point of interest in this region is that the Arg43 side chains appear to be located at favorable distances from the surrounding phosphates for Coulombic interactions to occur (Arg43L with phosphates 5′R and 6′L and Arg43R with 5′R and 6′L). The authors suggest that the presence of these positively charged side chains in the minor groove probably stabilizes its compression (86), since this deformation brings the phosphate groups much closer together than is usually the case in B-DNA. These salt bridges appear to be the only close range Coulombic interactions in the current model. This is surprising, given the high percentage of Arg and Lys residues in the repressor. However, the vast majority of phosphate contacts involve hydrogen bonds.

4. *Correlation of the Structure with Genetic and Biochemical Studies*

It is very important to remember that the conclusions derived from crystallographic studies such as the one reviewed above are models based on a rather complex set of manipulations of diffraction patterns. Indeed, as shall be discussed below, better data refinement sometimes leads to new models that require serious revision of the proposed structure. An important check on the structural model therefore, is whether it readily explains the biochemical and genetic data that is usually available for the protein in question. This is especially true for DNA binding proteins. Almost invariably in these cases, the crystal structure is preceded by biochemical analyses of the DNA-protein interactions that include random or site-directed mutagenesis studies which implicate particular amino acids or nucleotides as being directly or indirectly involved in sequence-specific recognition.

One of the most satisfying aspects of the 434 repressor (1-69)-operator structure is that the correlation between the crystallography and the biochemistry is excellent. For example, all of the phosphates implicated as being important by phosphate ethylation interference studies are involved in hydrogen bonds in the cocrystal structure (90). On the protein side, the crystal structure readily rationalizes the activity of all the 434 repressor mutants described to date. For example, replacing Arg43, whose important role was discussed above, with Ala reduces the binding constant by a factor of 500 (91). In another study, random mutagenesis was employed to generate all possible amino acid changes at positions 28 and 29, the glutamines that directly contacts the base pairs. No viable mutants were detected, as expected from the crystal structure. However, one somewhat puzzling result is that mutation of Gln33 to Leu does not alter sequence discrimination at base pair 4. This point will be discussed further when we consider binding of the Cro protein to 434 operators.

5. *Specificity Mutants Define Important Residues for DNA Recognition*

The structure also rationalizes the results of mutagenesis experiments more subtle than these "knockout" studies. In order to define (independent of the crystal structure) amino acid residues that directly contact the bases, Ptashne and co-workers searched for "new specificity" mutants. This elegant method allows one to address the classic quandary of whether the result of the original mutagenesis experiment (knockout of sequence specific DNA binding in this case) is simply due to the production of a protein that folds incorrectly, or really is due to the loss of a crucial side-chain functional group. The basic idea is to take a mutant or mutants that lacks the ability to recognize the natural operators (in this case various substitutions for Gln28 made by site-directed methods) and search for unnatural operator sequences that are recognized sequence specifically by the mutant repressor. A positive result argues very strongly that the mutant protein is properly folded and that the residue in question is in close proximity to the base pairs.

In order to rapidly screen a number of mutants, Wharton and Ptashne set up an in vivo assay (Fig. 21) in which the *E. coli lacZ* gene was modified such that the wild-type lac operator could be replaced with various synthetic 434 operators mutants (92). These "tester strains" were then transformed with a plasmid that expressed a mutant 434 repressor. Controls with the wild-type repressor and operator showed that formation of a repressor-operator complex depressed the expression of β-galactosidase in the tester strains about twofold, a level sufficient to cause these colonies to appear light red on MacConkey lactose agar plates. Normal levels of β-galactosidase expression resulted in bright red colonies. Twelve mutant operators were synthesized that contained each of the possible substitutions in the first base pairs and the symmetric change in the other half-site. The 12 tester strains derived from inserting these oligonucleotides into the *lac* operator and one containing the wild-type operator were then transformed with plasmids expressing mutants of the 434 repressor in which Gln28 had been changed to either Ala, Leu, Ser, Met, Arg, His, Lys, or Pro. This matrix of experiments thus generated 104 plates of transformants, only one of which had light red colonies. This was the one carrying a mutant operator with T rather than A at position 1L (and 1′R) and a plasmid that expressed a Ala28 repressor mutant. A quantitative measurement of the β-galactosidase level in these transformants indicated that 434 repressor [Gln28Ala] repressed β-galactosidase expression about threefold. None of the mutant repressors recognized the wild-type operator.

Having thus identified a potential positive, the mutant repressor was purified and its interaction with the wild-type and mutant operators studied biochemically. DNaseI protection experiments showed that the repressor [Gln28Ala] half-occupied the mutant operator at about the same concentration ($3 \times 10^{-8} M$) as required for the wild-type repressor to half-occupy the natural operator ($1 \times 10^{-8} M$). No interaction between the mutant repressor and the wild-type operator or the wild-type repressor and the mutant operator could be detected.

The most likely explanation of this result is that Ala28 in the mutant operator forms a van der Waals interaction with the 5-methyl group of thymine in the unnatural operator. Indeed, model building using the

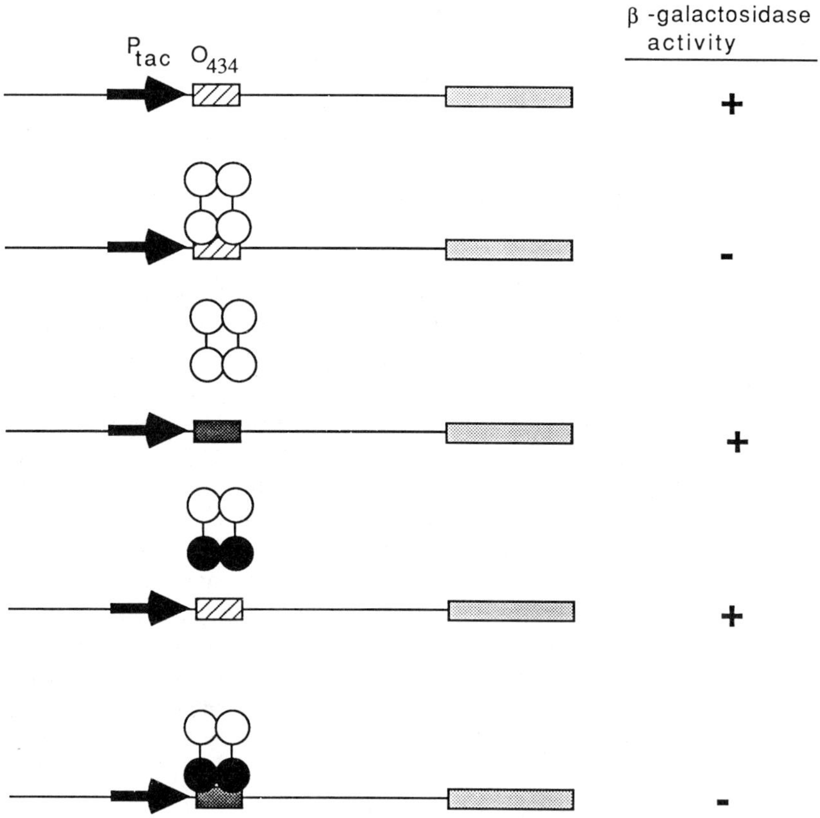

Figure 21 Design of the "specificity mutant" experiment. The 434 operator was placed adjacent to the potent P_{tac} promoter (arrow) and transformed into an *E. coli* cell that expresses wild-type or mutant 434 repressor. When the repressor is bound, transcription of the lacZ gene (which encodes β-galactosidase) is not observed. Therefore, the combination of a wild-type repressor (open circles) and operator (striped box) results in no β-galactosidase activity, but mutating only the repressor (black circle) or the operator (black box) does, since the protein-DNA complex is strongly destabilized. However, lacZ will be shut off in the interesting case of a mutant repressor recognizing a mutant operator, thus allowing cells containing these combinations to be identified by color on MacConkey plates, which contain an indicator for β-galactosidase.

crystal-structure-derived coordinates shows that no backbone deformation is required to accommodate the change from a Gln-A interaction to a Ala-T interaction (Fig. 22). This interpretation is strongly supported by an examination of the affinity of repressor [Gln28Ala] for synthetic operators bearing unconventional bases in position 1L (and 1R′), uracil, or 5-methylcytosine. The uracil-containing operator did not bind to the mutant repressor, but the 5-methylcytosine-containing oligonucleotide did, albeit with an affinity 15-fold lower than the operator in which T occupies these positions.

While the specificity mutant method is somewhat labor intensive, it is a very powerful tool for identifying residues involved in sequence recognition that can be applied to numerous other systems.

6. The Helix Swap and Redesign Experiments

Another type of directed mutagenesis experiment that has been used to great effect in the 434 system is the "domain swap" method. This elegant study was designed to demonstrate that the recognition helix does contact the DNA and is the determinant of sequence specificity. This was accomplished by replacing the recognition helix of the 434 repressor and replacing it with that of the phage P22 repressor by genetic engineering (92). P22 is yet another lambdoid phage that also has a repressor to help govern the switch from lysogeny to lytic growth. However, the sequences of the P22 operators are distinct from those of phage 434. It was found that the hybrid repressor recognized the P22 operator, but not the 434 operator.

In order to further demonstrate that the key residues involved in se-

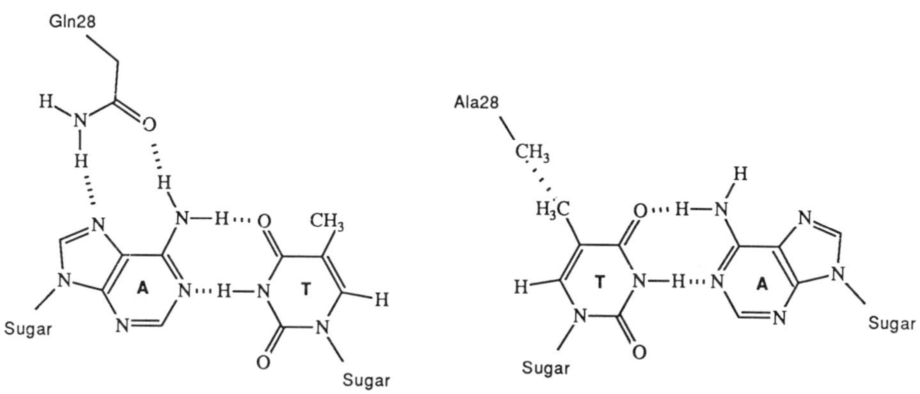

Figure 22 Structural model for binding of the mutant 434 repressor-mutant operator complex. It is proposed that a van der Waals contact between Ala28 and the thymine methyl group replaces the two hydrogen bonds that normally occur between Gln28 and the adenine of the wild-type base pair.

quence recognition are located on one face of the putative amphipathic recognition helix, the DNA binding specificity of an artificial repressor was examined. The protein created by replacing the residues of the 434 repressor thought to be exposed in the major groove with the analogous amino acids of the bacteriophage P22 repressor, had the complete sequence of neither recognition helix (Fig. 23). Except for these five substitutions, the rest of the protein sequence was identical to that of wild-type 434 repressor. DNA binding studies of the hybrid protein showed that it binds tightly to the P22 operator, but not the 434 operator, in vivo and in vitro. While we now appreciate that sequence specific DNA-protein complex formation involves numerous residues and not only the ones implicated by this study, the two repressors are both members of the same family of DNA binding factors and share considerable homology. Presumably, there is sufficient structural conservation to more or less preserve the overall contact surface presented to the operator. Like the specificity mutant analysis, the domain swap can provide a great deal of information about the residues important in DNA sequence recognition in closely related proteins. As we shall see, a similar

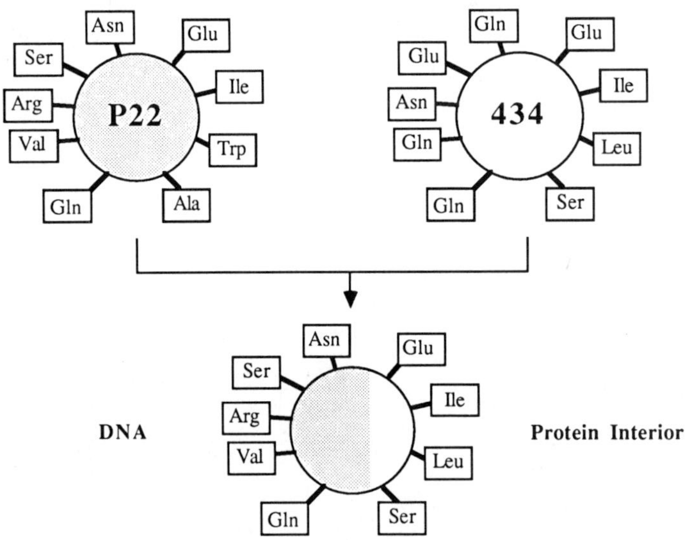

Figure 23 Redesign of a recognition helix alters DNA binding specificity. The amino acids of the 434 repressor's recognition helix that are postulated to contact the DNA were replaced by the analogous residues from the phage P22 recognition helix. The new repressor which contained only those five substitutions, recognized the P22 operator but not the 434 operator.

approach has been used to great affect in the study of the GAL4 class of zinc finger proteins.

B. The 434 Cro Protein-Operator Complex

As mentioned in the review of prokaryotic transcription, the other important sequence specific DNA binding factor used by the lambdoid phages to control transcription is the Cro protein. Cro recognizes the same operator sequences as repressor, but the differential affinities of the two proteins for the various operator sites is opposite.

Harrison and co-workers have solved the crystal structure of the 434 Cro protein alone and complexed to a 14-base-pair oligonucleotide corresponding to O_R2 (95,96), the same DNA that was employed in the original, lower resolution 434 repressor-operator work. Thus, we have the unique opportunity to deduce the structural basis of the differing relative affinities of these two proteins for the various naturally occurring operators. Repressor binds most poorly to O_R3, which is the only operator with a half site in which the fourth base pair is not A-T (it is G-C in the right half of O_R3). However, Cro exhibits a higher affinity for O_R3 than O_R1 or O_R2. Thus, we will be particularly interested in comparing the interactions of the two proteins with the operator half-sites in the vicinity of base pair 4.

The phage 434 repressor DNA-binding domain (residues 1–69) and Cro protein are quite similar in structure and sequence. Thus, it was a surprise when the Cro and R(1-69)-operator complexes crystallized in different forms (96). A detailed analysis of the structure showed a number of interesting differences, some of which help to rationalize the hierarchy of affinities of the two proteins for the three operators. One of the major differences between the two complexes is that the DNA in the Cro operator complex is not significantly bent and is overwound throughout its length, not only at the center (Fig. 24). The different DNA conformations are appently due to the fact that the Cro and repressor monomers have significantly different orientations in the dimer, even though the backbones of the four helix bundle of a particular monomer are almost superimposable (Fig. 25). This difference in quarternary structure results from the fact that the two proteins differ in sequence along the dimer interface. Cro contains two bulky residues (Phe46 in $\alpha4$ and Tyr61 in $\alpha5$) that repressor does not. Therefore, although the protein-DNA contacts at the ends of the operator are similar in both complexes, the orientation of the Cro dimer is such that the DNA does not need to bend to contact the appropriate residues. This result is consistent with gel mobility shift experiments which suggest that binding of R(1-69) to the operator bends it, but binding of Cro does not (97).

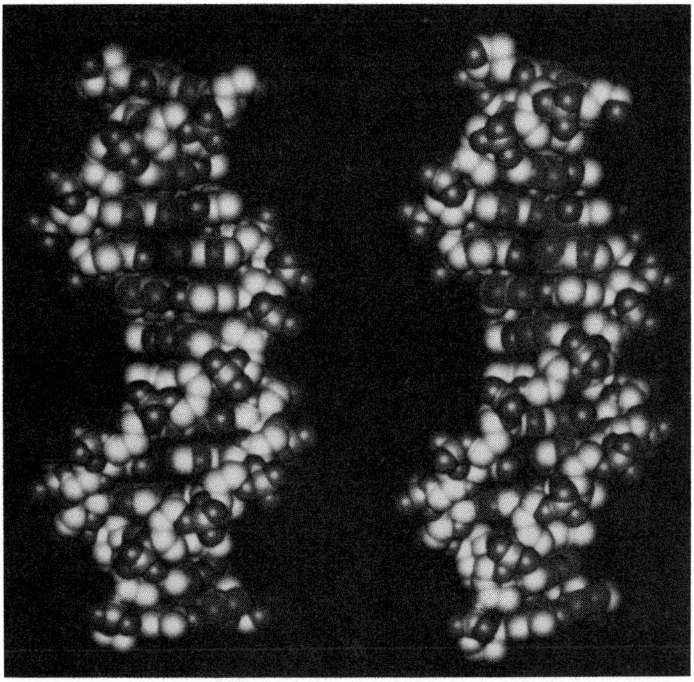

Figure 24 Space-filling model comparing the DNA structure of the 434 operator in the 434 Cro (left) and R(1-69) (right) cocrystals. The operator is essentially straight in the Cro complex but is bent significantly at the ends in the repressor complex. (Adapted from Ref. 96.)

The differing dimer orientations also lead to the most significant difference between the way the two proteins interact with the DNA backbone farther towards the center of the operator. In the Cro-DNA complex, the loop between $\alpha 2$ and $\alpha 3$ lies very close to the DNA backbone, allowing contacts between Lys27 and phosphates 3′ and 4′ (a main-chain N—H hydrogen bond to P4′ and a salt bridge via the side chain to P3′; see Figure 26). These contacts are absent in the R(1-69) complex because the analogous residue is Thr, not Lys, and because the DNA backbone is about 2.5 Å farther from the connecting loop. Thus, P4′ is out of the hydrogen bonding range of the main-chain N—H of Thr27. In addition, the branched side chain would probably interfere with this contact even if the groups could approach more closely.

Like repressor, Cro has Gln residues at positions 28 and 29 (in the recognition helix) that appear to directly contact the base pairs, although the resolution of this structure (3.2 Å is one direction and 5.5 Å perpendicu-

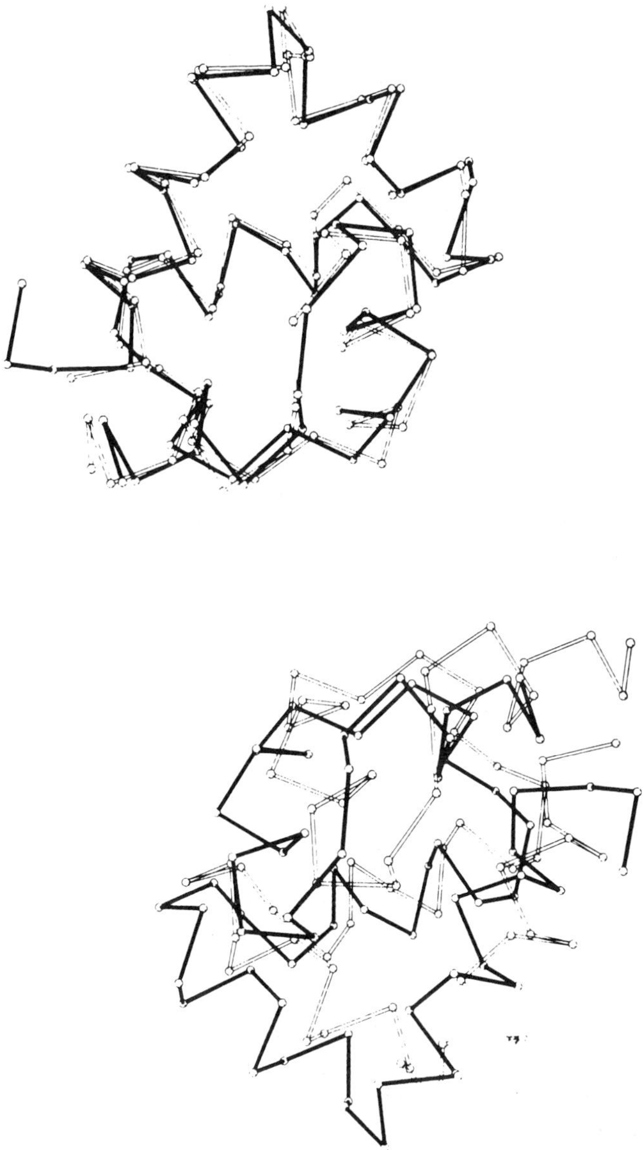

Figure 25 Superposition of the 434 Cro and R(1-69) dimers. The protein monomers (top) superimpose quite well but the other monomers of the dimers (bottom) differ in position due to the different dimer interfaces. Cro = black lines, R(1-69) = open lines. (Adapted from Ref. 96.)

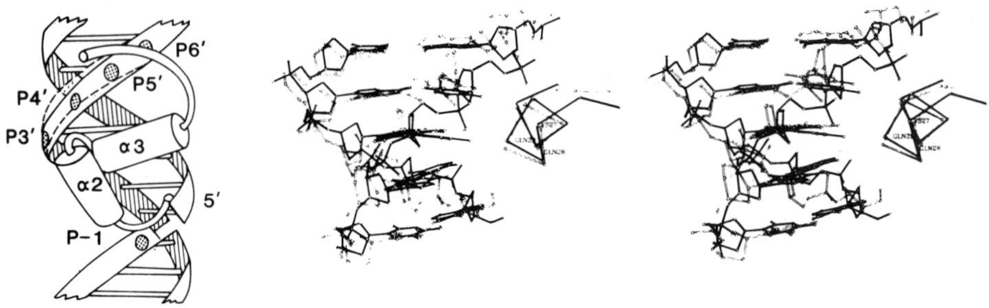

Figure 26 Relative positions of the Cro and R(1-69) proteins and the DNA backbone in their respective complexes. Cro = black lines, R(1-69) = open lines. (Adapted from Ref. 96.)

lar to it) precluded visualization of most side chain conformations. In addition, the nonpolar segments of the Gln29 and Lys27 side chains can be positioned to form van der Waals contacts with the methyl group of T3. These contacts are almost identical to those found in the R(1-69) operator complex.

The interactions with Cro and repressor with the thymine at base pair 4 differ, as expected. Recall that Gln33 of R(1-69) forms a hydrogen bond with O-4 of T4 and the methyl group of this base is in van der Waals contact with the side chains of Gln29 and Ser30. Cro has Leu at position 33, but a Gln at position 32. However, it is not clear that Gln32 forms a hydrogen bond to T4. The thymine methyl group lies very close to Cβ of Gln29 and Cδ of Leu33; thus a van der Waals interaction at this site is probably preserved. However, this contact appears to be much tighter than the analogous one in the repressor complex and Wolberger et al. (96) suggest that its removal by substituting cytosine for thymine at this position could result in a relief of steric strain that might compensate for the loss of the tight hydrophobic contact. Another way to put this is that the strength of a van der Waals interaction is extremely sensitive to interatomic distance, weakening rapidly if the participating atoms are too far away or too close. It is possible that the relevant distances in the Cro complex are such that ΔG is not as negative for the Cro-T4 methyl contact. This may help explain why Cro is not as sensitive as repressor to substitution of C for T at position 4. It also rationalizes why Cro binds poorly to operators with purines in this position, since there is no nearby group to hydrogen bond with N-7, the group that takes the place of the thymine methyl group. Analysis of better crystals that diffract to higher resolution will be required to address this definitively.

C. The Bacteriophage Lambda Repressor-Operator Complex

Pabo and co-workers have succeeded in crystallizing the lambda *cI* repressor DNA binding domain (residues 1–92) alone (98) and more recently as a complex with a 20 bp oligonucleotide (see below) containing the natural O_L1 sequence (overlined and underlined) (99). The protein does not undergo major conformational shifts upon complex formation

1 2 3 4 5 6 7 8 9 10 11 12 13 14 15 16 17
T A T A T C A C C G C C A G T G G T A T
T A T A G T G G C G G T C A C C A T A A

(although some side chains are reoriented) so we will focus on the DNA-protein complex. The overall structure of the lambda repressor-operator complex is very similar to that observed in the 434 studies described above. The DNA binding domain is composed of five α-helices of which $\alpha2$ and $\alpha3$ make up the helix-turn-helix motif (Fig. 27). The N-terminus of $\alpha2$ is

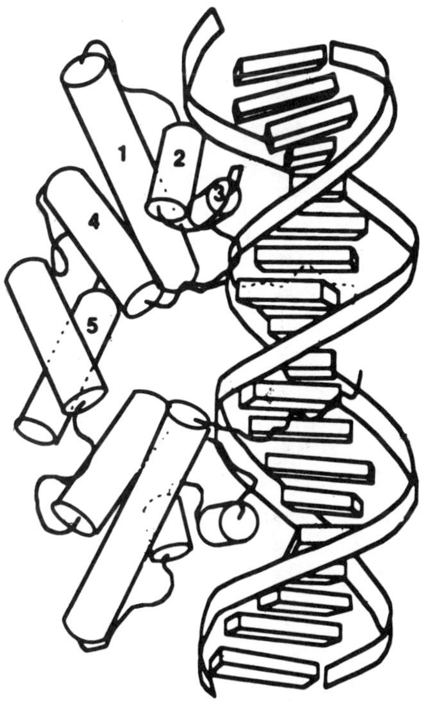

Figure 27 Overview of the λ Repressor fragment-operator complex. (Adapted from Ref. 99.)

pointed towards the duplex and $\alpha 3$ sits in the major groove. Helix 5 protrudes from the globular domain defined by the first four helices to form the dimer interface. Unlike the 434 repressor, the lambda repressor does not strongly perturb the structure of the operator; only a slight bend of the DNA ends toward the protein is observed. In addition, as we shall discuss below, the N-terminal arm of lambda repressor wraps around the helix and makes contacts in the major groove on the opposite face of the operator, a mode of interaction not observed in the 434 studies.

D. Protein-Base Contacts in the Major Groove

λ repressor contacts atoms along the edges of the base pairs and phosphate oxygens. Gln44, the first residue of $\alpha 3$, forms two hydrogen bonds with the adenine at base pair two (amide NH_2 to N-7 and N-6—H to amide oxygen) (Fig. 28). This contact is stabilized by a hydrogen bond between

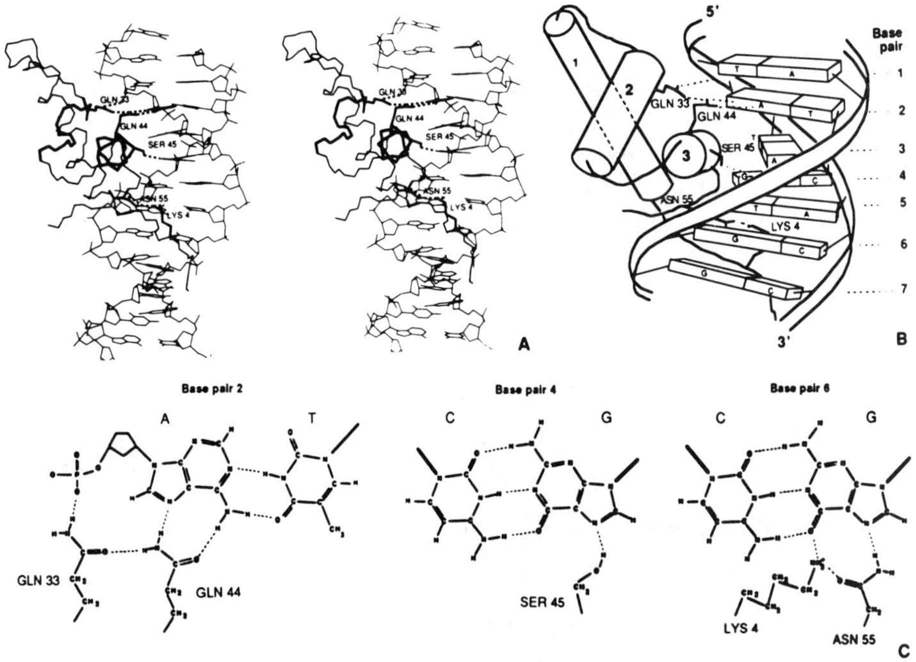

Figure 28 Stereoview (A) and schematic (B) of the details of the interactions between the helix-turn-helix unit of the λ repressor fragment and the operator. The chemical contacts thought to be important in sequence recognition are sketched in (C). (Adapted from Ref. 99.)

Gln44 and Gln33 (the first residue of $\alpha 2$). Gln33 also hydrogen bonds to phosphate (P) 2, providing another example of how backbone contacts orient and stabilize sequence-specific base contacts. Recall that a very similar arrangement is seen in the 434 repressor-operator structure.

The guanine in base pair 4 forms a hydrogen bond with Ser45, the second residue of the recognition helix. There is clearly an interaction between the side-chain hydroxyl and N-7 and perhaps also a weaker one with O-6. The side chain of Asn55, located in the loop between $\alpha 3$ and $\alpha 4$, interacts with the guanine of base pair 6. A direct hydrogen bond is observed between the amide N—H and N-7 of the base as well as a novel bridged interaction in which the terminal amino group of Lys4 is sandwiched between the amide oxygen of Asn55 and O-6 of guanine (Fig. 28). Although the N-terminal arm on which Lys4 is located projects toward the center of the operator, the side chain is able to wrap back around towards base pair 6.

Wrapping of the amino terminal arm of the protein along the path of the major groove toward the center of the operator was predicted from biochemical data which showed protection of bases on the back side of the operator (100) and the structure of the free protein (98), which showed a disordered arm able to adopt this stretched conformation. Besides the Lys4 contact with base pair 6, the current model suggests that Thr2 contacts base pair 8. However, the exact nature of this interaction was not described since the electron density map for the first five residues was not as clear as for other regions of the complex.

Very tight hydrophobic contacts are observed between protein residues and thymines at position 3 and 5. The methyl group of the former base occupies a pocket defined by the β-carbon of Ala49 and the terminal carbon of Ile54. At base pair 5 another hydrophobic pocket composed of Gly46, Ser45, and Ala49 accepts the methyl group.

As expected, numerous important protein-phosphate contacts are observed. Lys19 and Tyr22, which are on $\alpha 1$, hydrogen bond to the phosphate oxygens of P1 and P2 pointing away from the major groove. These interactions are the result of the fact that $\alpha 1$ runs almost parallel to the outer edge of the sugar-phosphate backbone (Fig. 29). P1 is also contacted on the other oxygen by the main-chain N—H of Gln33. The side chain of this residue and that of Asn52 (these are the first and last residues in helix 3, respectively) hydrogen bond to P2. The contacts on the other edge of the major groove involve Gly43 (main chain N—H) Asn58 and Asn61 and are detailed in Figure 32. As in the 434 repressor operator structure, it appears that an intricate network of hydrogen bonds is employed to orient the protein surface correctly.

As was the case in the 434 system, a great deal of biochemical and mutagenesis work has been done with the lambda repressor. While a de-

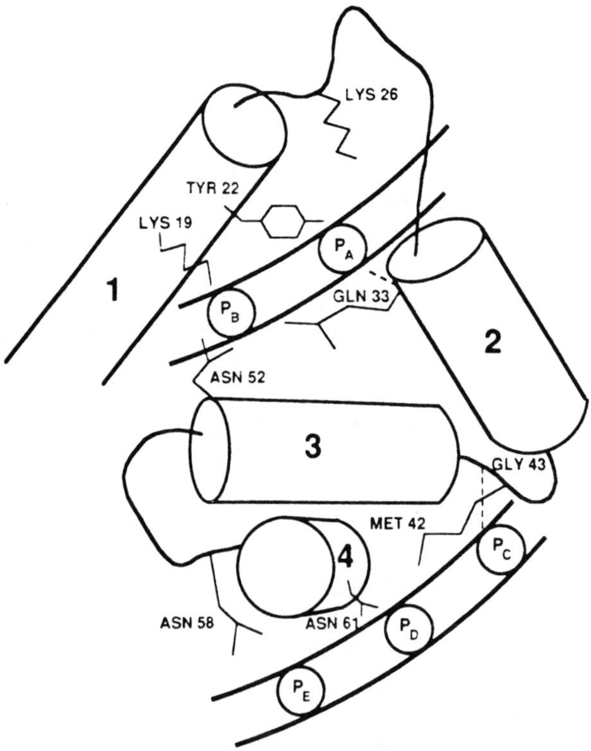

Figure 29 Sketch showing the interactions between the phosphates of the operator and the relevant side chains of the λ repressor fragment. (Adapted from Ref. 99.)

tailed discussion will not be presented here, the crystal structure appears to be fully compatible with all of these data.

E. Comparisons Between the 434 and Lambda Protein-DNA Complexes

A comparison between the structures of related proteins is always more enlightening than data from a single species. Since we have in hand two high-resolution structures of repressor-operator complexes and a lower resolution structure of the 434 Cro operator complex, what can we say about the mechanism of sequence recognition by helix-turn helix proteins?

While we know that at least some of the amino acid side chains that interact directly with the bases are located on one face of the recognition helix $\alpha 3$, this knowledge is not sufficient to predict the sequence specificity of the protein a priori. That is, there is no simple code relating the primary sequence of a DNA binding domain and the sequence it recognizes. Indeed, even with a knowledge of the separate protein and DNA structures, it would have been impossible to even predict the alignment in the protein-DNA complex. For example, the first two residues of both repressor recognition helices make site specific contacts with the edge of the bases in the major groove, yet Gln28 and Gln29 in the 434 protein contact base pairs one and two, while there is an intervening base pair between the sites contacted by Gln44 and Ser45 of the λ protein. Indeed, one of the major take home lessons of these studies is that specific binding is a complicated phenomenon that involves numerous contacts over the entire contact surface of the complex. While many of these are termed nonspecific because they do not involve the bases directly, this is really a misnomer. The structures make clear that the extensive network of backbone contacts are important in orienting and anchoring the residues that do make contacts with the bases.

Nevertheless, a number of generalizations can be made. The overall organization of the Cro and repressor complexes are very similar, although the 434 repressor dimer is oriented such that the operator must distort significantly to interact with the protein near its ends. This is not the case for 434 Cro or the lambda repressor. The helix-turn-helix unit is oriented very similarly in all three proteins and in several cases corresponding residues in the two repressors within this region make similar contacts (101). Recall, for example, that a Gln in each recognition helix contacts an adenine and that this interaction is stabilized by a side-chain–side-chain hydrogen bond with another Gln which is itself hydrogen bonded to a phosphate oxygen, thus explaining why Gln occurs so frequently as the first amino acid in helix-turn-helix motifs. Outside the helix-turn-helix unit, contacts are less well conserved.

F. Other Helix-Turn-Helix Proteins—The Trp and Lac Repressors

Two other repressors which contain helix-turn-helix motifs merit discussion since they teach us important, though very different, lessons. These are the *E. coli* trp and lac repressors. The former has been the subject of extensive structural work by Sigler and co-workers (102–104), who have been able to obtain crystals of the apo-repressor (tryptophan not bound), the repressor (tryptophan bound) and, purportedly, the trp repressor-operator complex. In simple terms, the trp repressor functions as a tryptophan sensor and shuts off transcription of genes involved in amino acid metabolism when tryptophan builds up (105). It does so by changing the conformation of the

DNA binding domain, adopting a high-affinity structure when the amino acid is bound. This mechanism of transcriptional repression is extremely common, but the trp repressor system is the only one for which we have detailed structural information.

Trp repressor is a highly helical protein whose basic structural organization is shown in Figure 30. Dimers of the trp repressor are formed by extensive contacts between helices A, B, C, and F, which provide an extremely stable core. From this core helices D and E, the helix-turn-helix unit, protrude from each monomer. This unit has been called a "flexible reading head" for reasons that will become obvious. A comparison of the structures of the repressor and apo-repressor has shown that by far the major effect of tryptophan binding is to reorient the helix-turn-helix with respect to the rest of the protein. There is very little change in quarternary structure. The nature of this movement (Fig. 30) is a tilting of the amino terminal end of the E helix (the putative recognition helix) towards the dyad axis. The D helix and the loop between them are dragged along. The molecular basis for this transition is shown in Figure 31. The tryptophan binding pockets are wedged between the central core and the flexible "reading head" and include a hydrophobic slot defined by Cα of Gly85, the proximal three methylenes of Arg84 (top) and the proximal three methylenes of Arg54 (bottom), which accepts the indole ring. Arg84 and Gly85 are part of the E helix. The carboxyl and amino groups are hydrogen

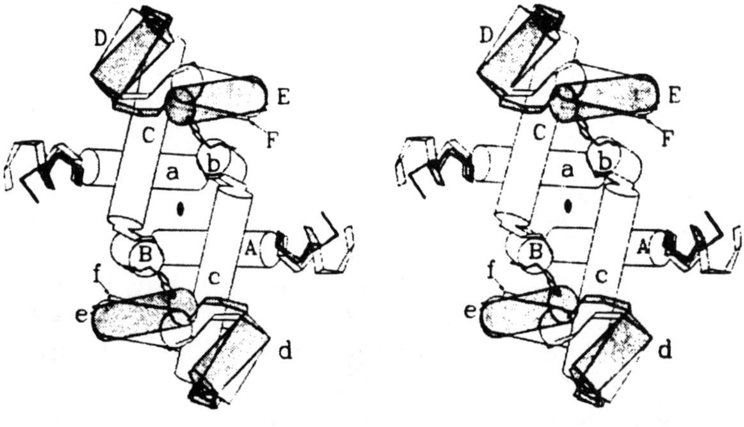

Figure 30 Stereoview showing a comparison of the trp aporepressor and repressor. Regions showing the greatest difference are shaded. The shift in the position of the helix-turn-helix unit (helices d and e and D and E) helps to rationalize the different DNA binding affinities of the two forms. (Adapted from Ref. 104.)

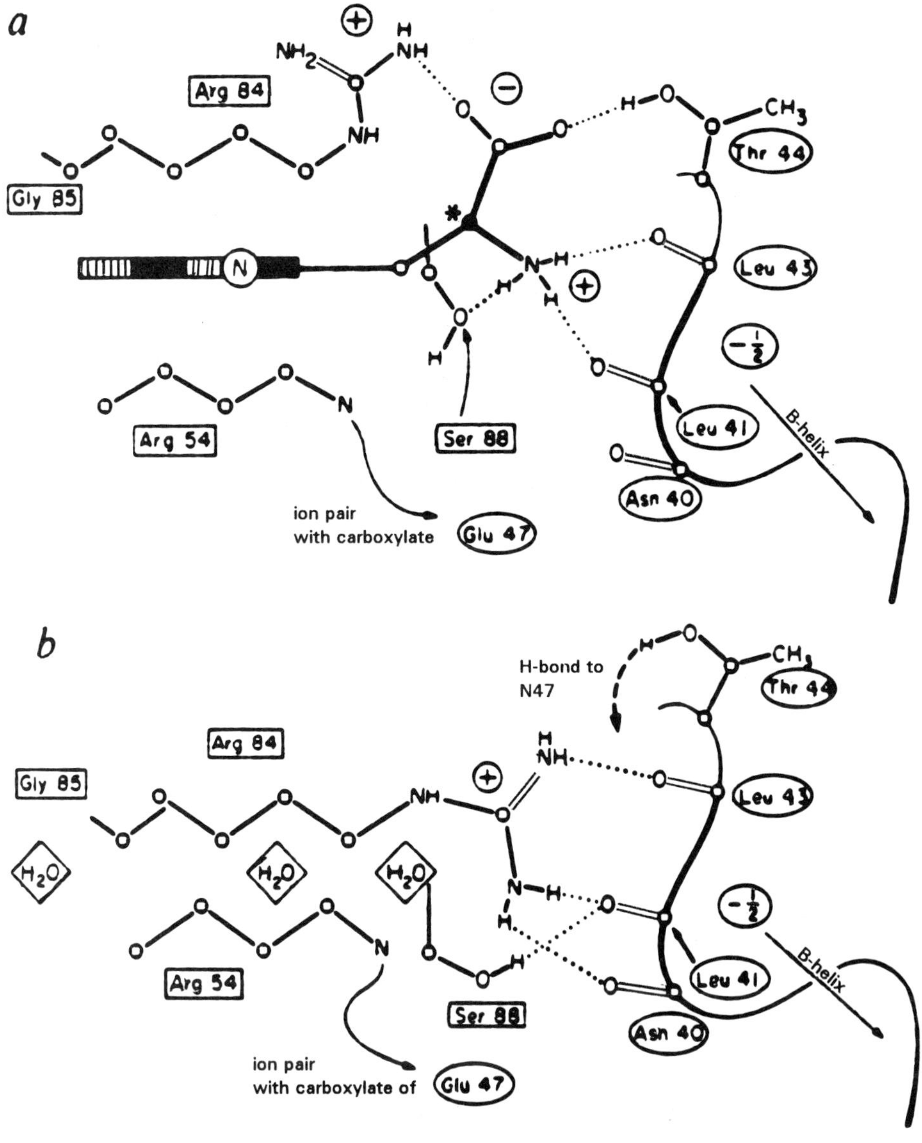

Figure 31 Atomic level representation of the changes that occur in the trp repressor (a) to aporepressor (b) conversion. (Adapted from Ref. 104.)

bonded to various protein residues as shown in the figure. In the aporepressor structure, both terminal nitrogens of the guanidinium group of Arg84 are involved in hydrogen bonds with residues from the other monomer. When tryptophan is bound, the carboxyl group is inserted into this position, resulting in rotation of the Arg84 side chain away from the B helix of the other subunit. This movement, along with the displacement required to accommodate the aromatic ring, results in the movement of the helix-turn-helix unit. Model building suggests that the recognition helix will not fit into the major groove until moved away from the central core by tryptophan binding, providing a satisfying model to explain the large change in DNA affinity induced by amino acid binding.

Sigler and co-workers have also reported the structure obtained from a cocrystal of the trp repressor and a synthetic oligonucleotide which they believe includes the complete operator sequence (80). However, as we shall see below, it has been argued that this is not the case. The startling conclusion of this study was that the protein did not make *any* specific contacts with the edges of the operator bases in the major groove! Considerable deformation of the operator was observed, leading Otwinowski et al. to propose that the trp repressor-operator complex is an extreme case of a protein "reading" a sequence based on the DNAs ability to tolerate deviations from the normal B-form structure (80).

The consensus sequence for Trp repressor binding, obtained by comparison of the natural binding sites, is shown in Figure 32. The classical model postulates that the trp repressor binds as a dimer about the central axis of symmetry. Based on this model, Otwinowski et al. employed an overlapping 19mer duplex oligonucleotide that includes the bases about the central symmetry element (80).

It has been noted that there are two additional axes of approximate symmetry in the consensus Trp repressor binding site (106) (Fig. 32), but

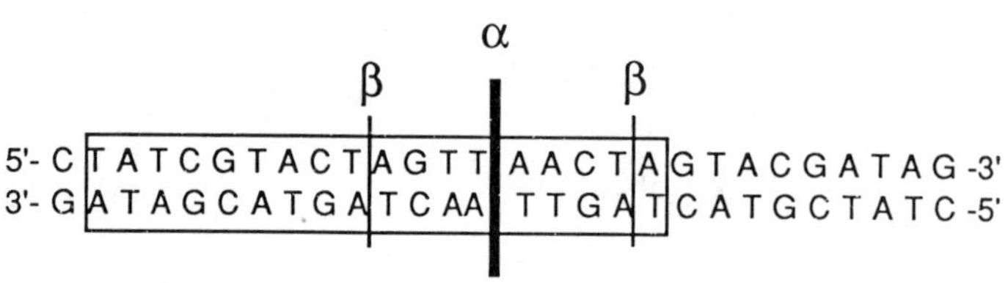

Figure 32 Consensus sequence of the trp repressor DNA binding site. The overall center of symmetry is represented by α while the flanking symmetry elements are designated β. The oligonucleotide shown by Müller-Hill and colleagues to bind the trp repressor tightly is boxed. A 19-base-pair oligonucleotide symmetric about α did not bind the repressor nearly as well.

not much attention has been paid to this observation until recently. Because of the extremely surprising nature of the structural results, Staacke et al. decided to explore an alternative model of Trp repressor binding in which repressor monomers bind symmetrically across the outer axes (107). In this view, the Trp repressor binds as a tetramer, or more properly a dimer of dimers, with the *overall* center of symmetry being in the middle, but with the dimer symmetry axes located farther out (Fig. 32). If correct, this would mean that the structure solved by Sigler's laboratory is that of a nonspecific complex, since the 19mer is simply too short to accommodate the trp repressor tetramer.

To test this idea, gel retardation and methylation protection experiments were conducted with the oligonucleotide employed in the crystallographic study (with the ends made flush), and a shorter construct including the bases about the outer axis of symmetry (boxed in Fig. 32). The result was that the 18mer was bound by the trp repressor very tightly at low concentrations of protein, but the oligonucleotide used by Otwinowski et al. was not. Retardation of this oligonucleotide does occur at much higher protein concentrations, suggesting that the classical model is incorrect and that the crystal structure is indeed that of a nonspecific complex. However, this conclusion remains controversial (P. Sigler, personal communication). Given the uncertainty surrounding the cocrystal structure, it will not be discussed in detail.

The lac repressor has been extremely well characterized genetically and biochemically (108) but unfortunately, it has so far been impossible to grow diffraction quality crystals of this factor. Nonetheless, a number of elegant in vitro biochemical studies and an NMR solution structure investigation of the DNA-binding domain (the "headpiece") demonstrate that the helix-turn-turn unit in this protein is of a different class than the ones discussed above.

The lac repressor normally exists as a tetramer, but proteolytic fragments that retain DNA-binding activity are dimeric. The operator sites are related pseudosymmetric sequences. The NMR study (109) employed the oligonucleotide shown below which includes an operator half-site (overlined and underlined, the diamonds represent the center of symmetry for the full operator) and two G-C base pairs at each end to prevent "fraying,"

1 2 3 4 5 6 7 8 9 10
5′-GGAATTGTGAGG◊GG-3′
3′-CCTTAACACTCG◊CC-5′

and a 56-amino-acid proteolytic fragment which Boelens et al. call HP56 (109). This fragment includes three major helical regions, including the proposed helix-turn-helix unit (residues 6–13, 14–16, and 17–25). Despite the complexity of the spectrum, a number interesting intermolecular cross peaks could be identified in the 2D-NOE spectrum. They clearly demonstrate that the helix-turn-helix unit does sit in the major groove. Due to the somewhat long mixing times used in the experiment, the NOEs cannot be used to measure distances quantitatively because of spin diffusion effects. However, they show qualitatively that the orientation of the helix-turn-helix unit is opposite that observed in the 434 and lambda complexes (Fig. 33, see also Figs. 19 and 27). That is, the N-terminus of the lac repressor's recognition helix enters the major groove close to the operator's dyad axis with the C-terminus situated farther towards the end. For example, a model

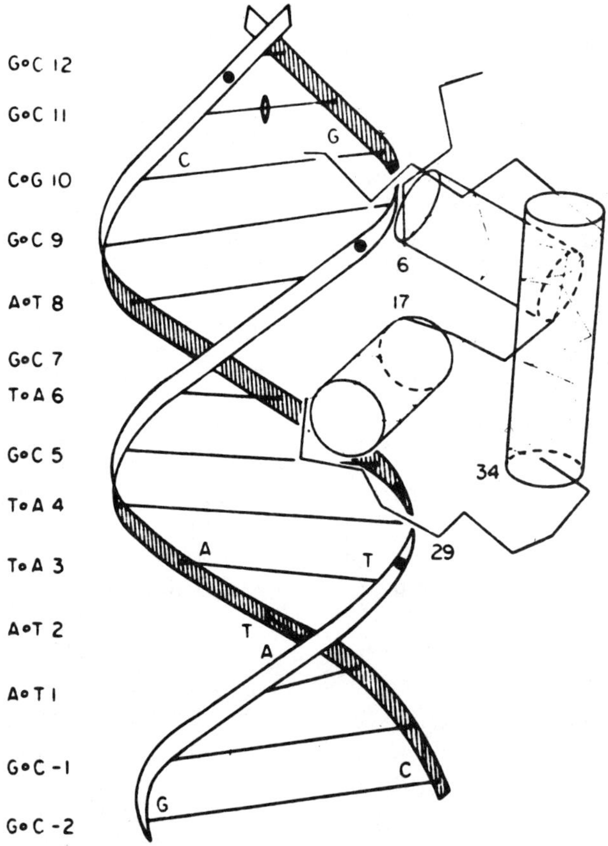

Figure 33 Model of the lac headpiece-operator complex deduced from NMR data. The orientation of the recognition helix is opposite that seen in the λ and 434 repressor structures. (Adapted from Ref. 109.)

in which the lac repressor's helix-turn-helix unit is docked in the "normal" orientation seen in the phage protein structures would predict close contacts between Tyr4 with base pairs 3 and/or 4, and His29 with base pairs 9 and 10. However, there are strong NOEs between aromatic protons of Tyr7 and base protons on C-9 and G-10 of the half-site.

The opposite orientation of the recognition helix is also strongly supported by the biochemical studies of Muller-Hill and colleagues, who carried out "specificity mutant" studies of the type described above for the 434 repressor (110,111). These workers first set about optimizing the DNA sequence to come up with an operator that would provide the maximum possible repression of the lacI gene in the presence of the wild-type repressor. This DNA sequence is shown below. Replacements in the first two residues of the

```
1 2 3 4 5 6 7 8 9 10
A A T T G T G A G C ◊ G C T C A C A A T T
T T A A C A C T C G ◊ C G A G T G T T A A
                      10' 9' 8' 7' 6' 5' 4' 3' 2' 1'
```

recognition helix were found that yielded mutant repressors able to recognize operator variants at base pair 7 and 7′. Later, the same group found that a mutant repressor with Asn at position six of the recognition helix bound to an operator variant containing a G-C to A-T substitution at position 5 and 5′. Thus, with two helix-operator contacts defined, an orientation could be assigned, which was the same one suggested by the NMR study (Fig. 33).

The finding that the lac repressor's helix-turn-helix unit is oriented opposite that of the phage helices in the repressor-operator complexes serves as a warning sign not to make more out of sequence or even structural homologies than is warranted by the hard data.

V. THE HOMEODOMAIN: A EUKARYOTIC VERSION OF THE HELIX-TURN-HELIX MOTIF

The homeobox was first noticed as a 180-bp DNA sequence common to the *Antennapedia* (*Antp*) (112,113) and *fushi tarazu* (*ftz*) (114) genes involved in controlling *Drosophila* segment identity and number. Subsequently, homeoboxes were found in several other *Drosophila* and vertebrate developmental regulatory genes (115). It is now becoming clear that the proteins encoded by these genes are transcriptional factors and that the homeodomain (the 60-residue protein fragment encoded by the homeobox) is involved in sequence specific DNA recognition. An analysis of the amino acid sequence of the various homeodomains suggested that they might contain a helix-turn-helix motif (116, 117). This hypothesis was confirmed by Gehring, Wuthrich, and co-workers who solved the solution structure of a polypeptide containing the *Antp* homeodomain by NMR spectroscopy (118,119). This protein fragment was purified from an *E. coli* clone containing an overexpression vector with the homeobox inserted downstream of a powerful T7 promoter. Because of the way in which the plasmid was constructed, the homeodomain is flanked by an N-terminal methionine and seven non-homeotic *Antp* residues at the C-terminus. The family of backbone structures shown in Figure 34 was obtained through an energy minimization procedure whose input contained about 700 distance constraints determined by two-dimensional NOESY and 118 scalar coupling constants derived from 2D-COSY. The model clearly reveals the presence of three major helical segments, including residues 10–21, 28–38, and 43–59. Qian et al. (119) propose that the third helix is actually two directly adjacent helices distinguished by a 30° kink between residues 52 and 53, but we shall discuss the *Antp* fragment as a three helix bundle since, as we shall see, this kink may not be a general feature of homeodomains but the three helix motif certainly is. Helices II and III correspond to the helix-turn-helix unit. Helices I and II run almost exactly antiparallel to one another, with helix III in a roughly perpendicular orientation. Many of the amino acids strongly conserved in many homeodomains are located in this region. For example, Trp48 and Phe49 are found in all known homeodomains (117) and can be seen to be crucial for structural stability. These hydrophobic residues pack against helix I in the interior of the globular structure and anchor helix III, the putative recognition helix, with respect to rest of the structure. Gln51 and Arg53, on the other face of the amphi-

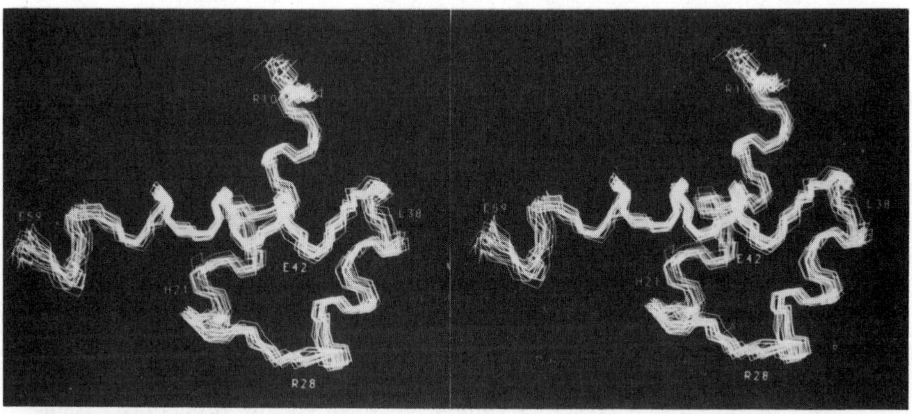

Figure 34 Family of structures for the *Antp* homeodomain obtained from distance geometry calculations based on NOE constraints. (Adapted from Ref. 119.)

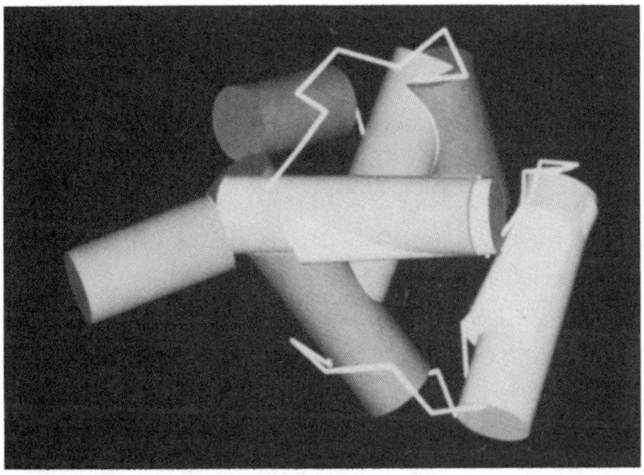

Figure 35 Comparison of the *Antp* homeodomain with the phage 434 repressor DNA binding domain. α-helices are represented as cylinders. The helix-turn-helix motifs are shown in a lighter shade. The helix-turn-helix is the only structural element conserved in the three-dimensional structure of the proteins. (Adapted from Ref. 119.)

pathic helix, seem to be well situated for interaction with DNA assuming that helix III sits in the major groove.

The analogy to the prokaryotic helix-turn-helix proteins was also substantiated by this structure. Figure 35 shows a superposition of the phage 434 R(1-69) structure with that of the *Antp* homeodomain. The alignment was carried out to provide maximal overlap between the helix-turn-helix units of the two polypeptides. It can be seen that besides a small difference in the angle between the helical axes, helices II and III of the *Antp* homeodomain are spatially similar to the helix-turn-helix unit of the 434 repressor. However, the eukaryotic helix-turn-helix unit consists of longer helices and there is almost no overlap outside of this unit.

Gel retardation and DNase footprinting assays allowed the characterization of the minimum DNA sequence recognized by *Antp* (120), which is shown below. The half-life

1 2 3 4 5 6 7 8 9 10 11 12 13 14
5′-C T C T A A T G G C T T T C-3′
3′-G A G A T T A C C G A A A G-5′

of the complex formed between this 14mer and the *Antp* homeodomain was approximately 90 min ($K_D = 10^{-9}/M$) allowing Otting et al. (120) to carry out an NMR study of the complex. Titration experiments clearly showed a 1 : 1 protein-DNA stoichiometry. Thus, the homeodomain binds as a monomer, unlike the prokaryotic repressor proteins. The methods employed to obtain sequence-specific resonance assignments will not be described here, but it should be noted that the chemical shifts of the protein and the DNA in the complex were sufficiently different from those observed in the unbound state, that a complete reassignment of the spectrum was necessary. This required the use of uniformly ^{15}N-labeled polypeptide which allowed Otting et al. to perform a three-dimensional ^{15}N-correlated [^{1}H,^{1}H]-NOESY experiment. Spreading the data out in three dimensions aided the assignment procedure greatly.

Very shortly after the NMR study of the *Antp* homeodomain-DNA complex appeared, Pabo and colleagues reported the crystal structure of a complex between the *engrailed* homeodomain (another *Drosophila* homeotic protein) and a 21mer oligonucleotide (Fig. 36B) that binds to the 61 residue polypeptide with a K_D of about $10^{-9}/M$ (121). Since the major features of both structures are similar, we will focus on the *engrailed* complex (determined at 2.8 Å resolution) and comment on significant differences where appropriate. The amino acid numbering scheme has been standardized to align conserved residues in each homeodomain. The structure of the engrailed homeodomain alone is undergoing refinement and has not been published at the time of this writing.

The engrailed homeodomain (Fig. 36A), like that of *Antp*, is composed

A

Met Asp Glu Lys Arg Pro Arg Thr Ala Phe Ser
1
[Ser Glu Gln Leu Ala Arg Leu Lys Arg Glu Phe Asn Glu] Asn Arg
10 20
Tyr Leu Thr [Glu Arg Arg Arg Gln Gln Leu Ser Ser Glu] Leu Gly
30
Leu Asn [Glu Ala Gln Ile Lys Ile Trp Phe Gln Asn Lys Arg Ala
40 50
Lys Ile Lys Lys] Ser

B

1 2 3 4 5 6 7 8 9 10 11 12 13 14 15 16 17 18 19 20 21
5'- T T T T G C C A T G T A A T T A C C T A A -3'
3'- A A A C G G T A C A T T A A T G G A T T A -5'

Figure 36 (A) Amino acid sequence of the engrailed homeodomain used in the crystallographic study of Kissinger et al. (121). The α-helical regions of the polypeptide are boxed. (B) Nucleotide sequence of the DNA used in the crystallographic study. Note that this may not be identical to the in vivo binding site of the engrailed protein.

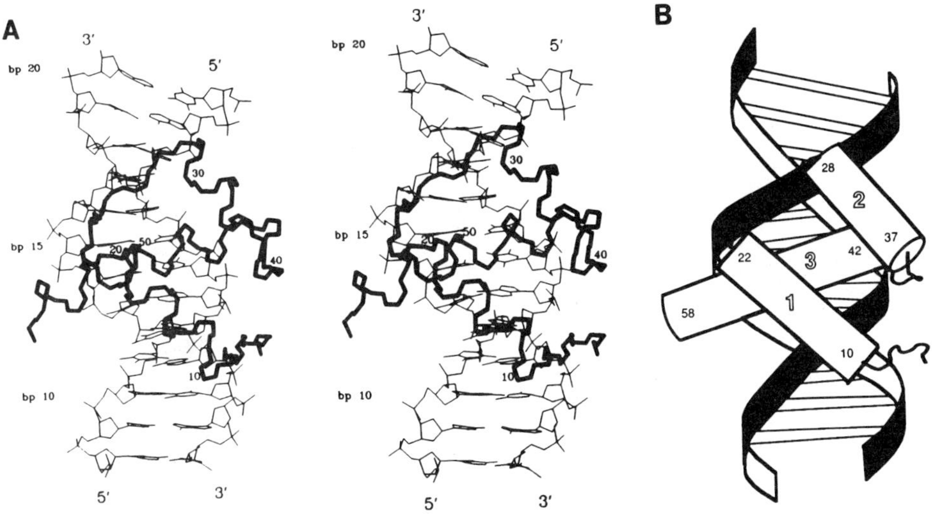

Figure 37 Stereoview (A) and schematic (B) of the engrailed homeodomain-DNA complex. Only the backbone structure is shown. (Adapted from Ref. 121.)

primarily of three α-helices, however the kink observed in the *Antp* structure is not present in the engrailed homeodomain. Both structures confirm that helix III is indeed the recognition helix and sits in the major groove of the DNA. The monomeric protein recognizes primarily a TAAT site in the middle of the oligonucleotide (base pairs 11–14) (Fig. 37). Sequence specific contacts are made by Asn51 and Ile47 in the engrailed structure. Ile47 forms a van der Waal's interaction with the thymine methyl group of base pair 14 and the amide side chain of Asn51 forms a bidentate hydrogen bond with the adenine in base pair 13 (Fig. 38). Gln50 has been implicated as being a key residue for DNA recognition by mutagenesis studies (122,123) but, surprisingly, does not form a base-specific hydrogen in the engrailed structure, although a van der Waals contact between its side chain and the methyl group of T16 is observed. However, Kissinger et al. (121) note that a small rotation of the side chain would bring the amide into position to hydrogen bond to O-4 of T14. In the *Antp* solution structure, a strong NOE is observed between the γ-methylene protons of Gln50 and H-5 of the cytosine in base pair 8 (Fig. 39), which is consistent with the amide group being in position to hydrogen bond, but does not prove that it does. Therefore, while it seems very likely that Gln50 is involved directly in sequence recognition, this has not been proven by the structural data. It should be noted that the in vivo DNA binding site for engrailed has not yet

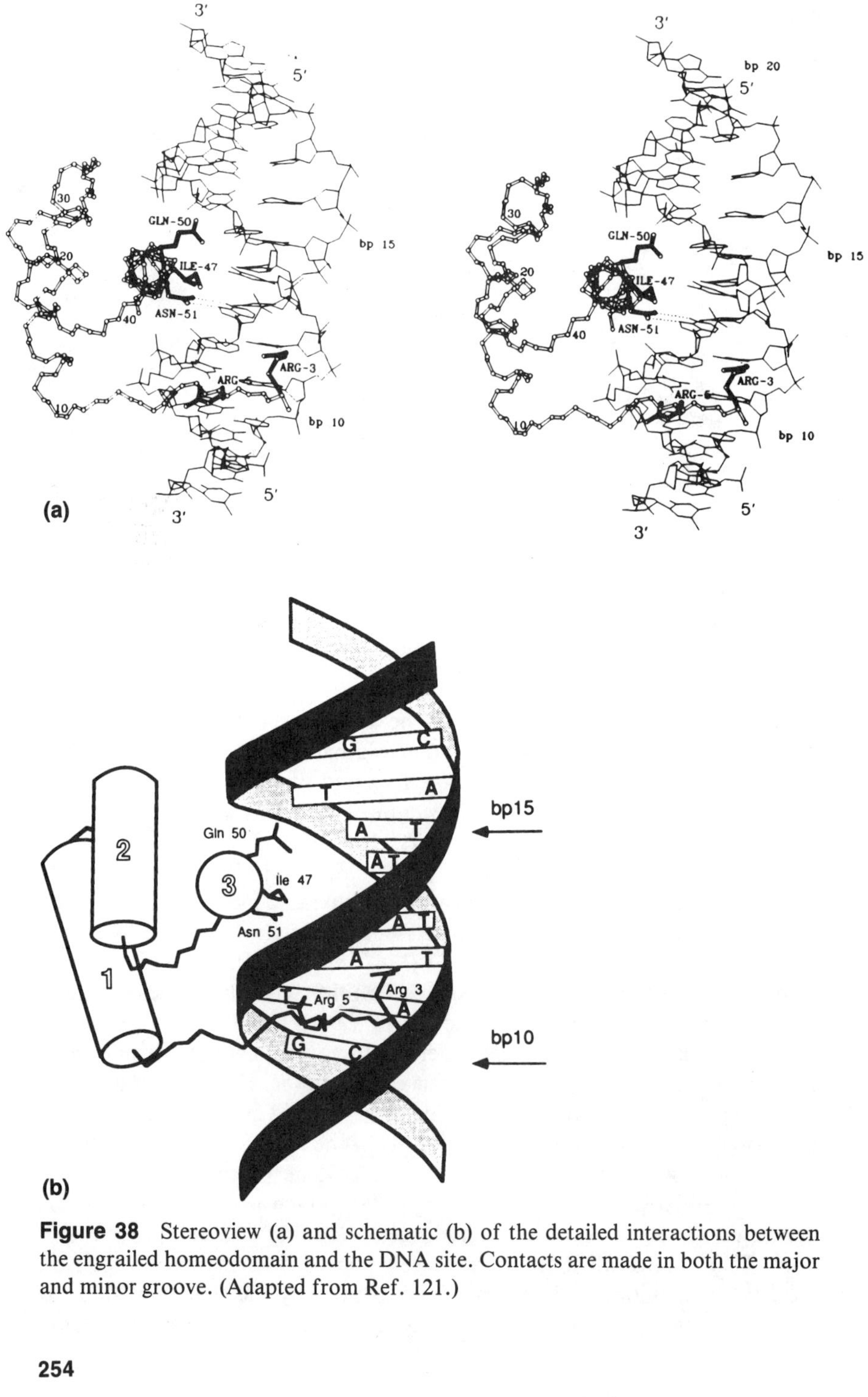

Figure 38 Stereoview (a) and schematic (b) of the detailed interactions between the engrailed homeodomain and the DNA site. Contacts are made in both the major and minor groove. (Adapted from Ref. 121.)

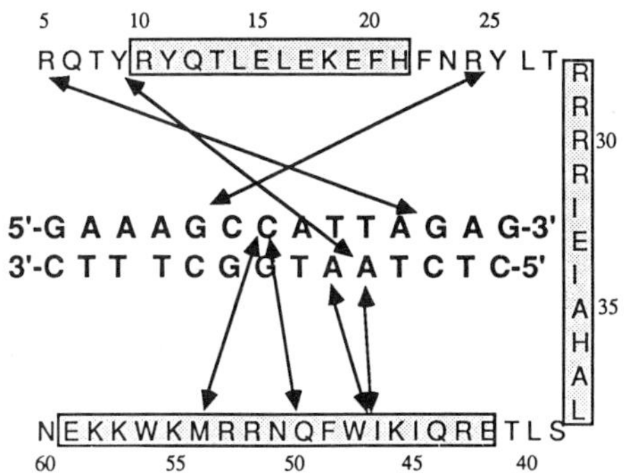

Figure 39 Summary of intermolecular NOEs (arrows) detected in an NMR study of the *Antp* homeodomain-DNA complex by Otting et al. (120). Helical regions of the protein are boxed.

been determined precisely and that the oligonucleotide used in this study may not be the optimal partner for the homeodomain. Indeed, Gln50 appears to be important in distinguishing between different homeodomain binding sites, suggesting that it contacts a specific base pair outside the core TAAT recognition site (122,123). It is conceivable that with a different DNA sequence, a direct interaction between a base edge and Gln50 would be observed.

One of the most interesting features of both structures is that there are sequence-specific contacts in the minor groove. These are made by the N-terminal arms of the polypeptides (Figs. 37 and 40). In the engrailed structure, Arg5 forms a hydrogen bond with O2 of the thymine in base pair 11 and Arg3 appears to form an identical interaction with T12, although the electron density for Arg3 is not as well defined as for Arg5 in the engrailed structure. These contacts probably contribute greatly to the preference for the TAAT site which the engrailed homeodomain recognizes, since a G-C or C-G base pair would have an amino group in the minor groove which is electronically and sterically incompatible with the two Arg side chains. However, as mentioned in the section on theoretical considerations, these hydrogen bonds would probably not discriminate between A-T and T-A base pairs since O-2 of thymine and N-3 of adenine occupy similar positions in each.

As seen in previous structures, extensive contacts along the sugar-phosphate backbone are observed. These are summarized in Figure 40 and shown in stereo in Figure 41. Most of these contacts flank the region that helix III occupies in the major groove. These contacts do not greatly perturb the DNA structure, which is similar to an idealized B-form helix

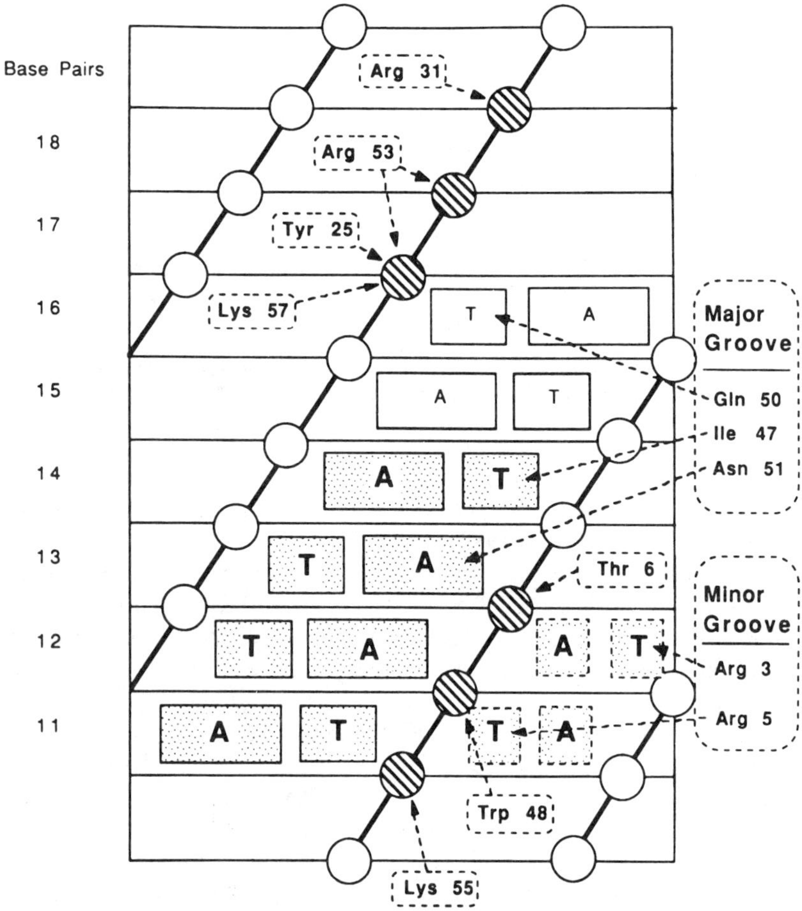

Figure 40 Sketch summarizing the DNA-protein contacts in the engrailed homeodomain-containing complex. (Adapted from Ref. 121.)

throughout its length with the exception of a slight widening of the major groove where the recognition helix is located.

Although the helix-turn-helix units of the prokaryotic and eukaryotic proteins are quite similar, different positions along the helices are used to contact the DNA. In essence, it is as if the DNA in the homeodomain complexes has been shifted to the C-terminal end of the recognition helix relative to the lambda and 434 structures.

With regard to the biological role of the homeotic proteins, which is to regulate gene expression and development, it seems rather odd that the

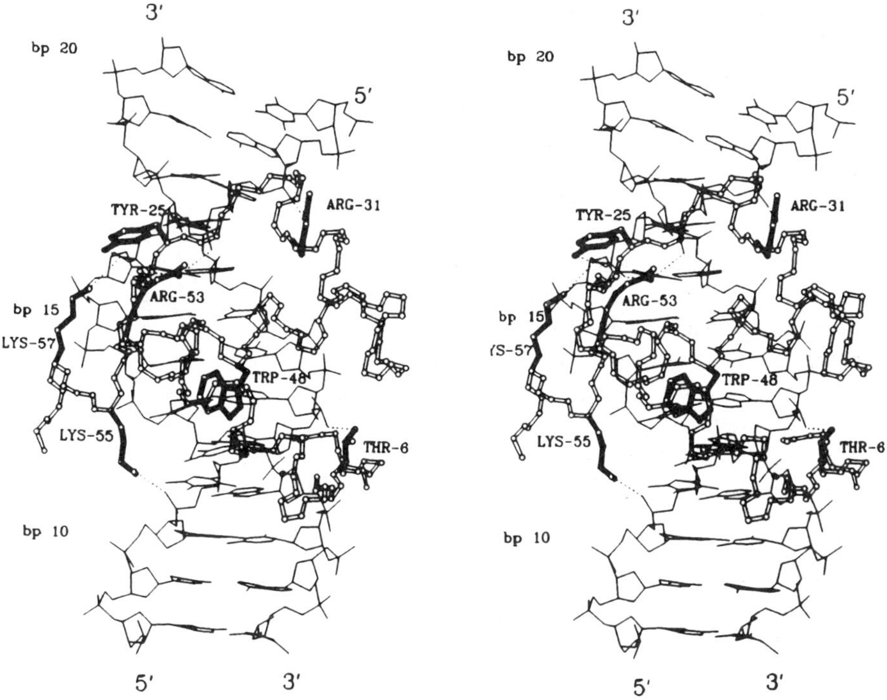

Figure 41 Stereoview highlighting the important contacts between the engrailed homeodomain and the phosphates of the DNA. (Adapted from Ref. 121.)

protein recognizes essentially a 4-bp sequence: 5″-TAAT-3′. This would appear to be too small to serve as a unique site for binding of a transcription factor. It is possible that the intact protein might contain other DNA-binding domains that further contribute to sequence specificity. Alternatively, these proteins might associate with other DNA binding factors inside of the cell. This combinatorial strategy would be an attractive way to amplify the binding selectivity of any particular set of proteins (124). Regardless of the actual in vivo situation, the *Antp* and engrailed-DNA structures provide our first glimpse at how homeodomains interact with DNA and provide a basic framework for thinking about other, more complicated, scenarios.

VI. ZINC FINGER PROTEINS

The *Xenopus laevis* transcription factor IIIA (TFIIIA) has a dual role. It binds sequence specifically to the 5S gene, thereby stimulating its transcription, and also complexes the RNA product, forming a 7S storage particle that accumulates in large quantities in the oocyte. Therefore, TFIIIA is a sequence-specific RNA and DNA-binding protein. In 1985, based on an analysis of the DNA-derived sequence of the protein (125), Klug and co-workers proposed that TFIIIA contains repetitive zinc-binding domains of about 30 amino acids that include two cysteines and two histidines each which are involved in sequence-specific DNA recognition (126). This seminal hypothesis is supported by a number of observations. The nine repeats suggested by inspection of the amino acid sequence of TFIIIA (Fig. 42), corresponds to the number of bound zinc atoms found by atomic absorption spectroscopy using highly purified protein. The protein is inactive when zinc in removed. Limited proteolysis of a 7S particle containing TFIIIA revealed the presence of fragments with the expected molecular weight of one or more fingers. Extended x-ray absorption fine structure (EXAFS) analysis of TFIIIA shows the presence of zinc atoms coordinated by two nitrogens (2.0 Å) and two sulfurs (2.3 Å) (127). Also, a peptide corresponding to a single TFIIIA finger has been shown to fold only in the presence of added metal (128). Finally, as will be discussed below, NMR experiments with chemically or biologically produced polypeptides have confirmed the finger hypothesis and revealed their solution structure.

```
MGEKALPVVYKR

        YICSFADCGAAYNKNWKLQ-AHLC-KH 37
TGEK-PFPCKEEGCEKGFTSLHHLT-RHSL-TH 67
TGEK-NFTCDSDGCDLRFTTKANMK-KHFNRFH 98
NIKICVYVCHFENCGKAFKKHNQLK-VHQF-SH 129
TQQL-PYECPHEGCDKRFSLPSRLK-RHEK-VH 159
AG----YPCKKDDS CSFVGKTWTLYLKHVAECH 188
QD---LAVC--DVCNRKFRHKDYLR-DHQK-TH 214
EKERTVYLCPRDGCDRSYTTAFNLR-SHIQSFH 246
EEQR-PFVCEHAGCGKCFAMKKSLE-RHSV-VH 276

DPEKRKLKEKCPRPKRSLASRLTGYIPPKSKEKNAS
VSGTEKTDSLVKNKPSGTETNGSLVLDKLTIQ 344
```

Figure 42 Amino acid sequence of Xenopus TFIIIA aligned to highlight the repeated motif. Conserved hydrophobic residues are shown in outline. Coordinating residues are in boldface and underlined.

Since 1985, hundreds of other proteins have been implicated by sequence analysis as possessing zinc finger domains (129). Many of these are very similar to those of TFIIIA and some diverge considerably. Indeed, an important point is that we now know that there are several distinct subfamilies of these proteins. While we will retain the term *zinc finger* for all DNA binding factors that require one or more bound metal atoms to function properly, it is important to remember that this is a loose definition and does not necessarily imply structural similarity. Although the importance of the zinc finger domains has been appreciated for only a few years, a tremendous amount of work has been done and a great deal is now known about their structure and biochemical properties (for reviews, see Refs. 129–131). Each subfamily, defined by the nature of the metal coordination sphere, will be reviewed below.

A. The "His_2Cys_2" Fingers: TFIIIA and Its Relatives

The class of zinc fingers that includes TFIIIA itself has over 100 members and have been found in a variety of eukaryotic organisms including man, yeast, frogs, and *Drosophila*. Many of these proteins have been shown to be transcription factors (130), suggesting that the TFIIIA-like zinc finger is indeed a conserved unit for sequence-specific DNA binding. The consensus sequence derived from a comparison of these proteins is shown below (X = any amino acid).

$$[(Tyr,Phe)\text{-}X\text{-}Cys\text{-}X_{2,4}\text{-}Cys\text{-}X_3\text{-}Phe\text{-}X_5\text{-}Leu\text{-}X_2\text{-}His\text{-}X_3\text{-}His\text{-}X_{2\text{-}6}]$$

In addition to the ligating residues, there are a number of conserved hydrophobic residues. It was originally speculated that these might form a hydrophobic core in the three-dimensional structure of the finger since oily side chains are often used for this purpose. In addition, the spacing between individual fingers in multifinger proteins is often roughly conserved, which is probably important in the phasing of the protein-DNA contacts.

1. *Structure of the Zinc Finger*

While a specific zinc-finger-DNA complex has not yet been structurally characterized, we know a great deal about the structure of the fingers themselves. This is principally because the compact modular nature of the zinc finger domains is ideal for high-resolution 2D-NMR studies. Klevit and co-workers have reported the solution structure of one of the two zinc fingers of the yeast ADR1 protein (ADR1a), a positive activator of the alcohol dehydrogenase gene (132). The model derived from these studies is shown in Figure 43. The two histidine residues are contained in an α-helix almost three turns in length. The presence of the hydrophobic core was also confirmed. It is formed by the conserved leucine and phenylalanine and also involves another leucine. This orients the α-helix with respect to a loop containing the two cysteines. The experimentally derived model is similar to one proposed by Berg on the strength of sequence comparisons to structurally characterized metal-binding proteins (133), except that the Berg model predicted that the cysteine-containing region would be an antiparallel β-strand.

Both this NMR study and an analysis by Frankel et al. of a 30-residue peptide corresponding to one of the TFIIIA fingers (128) clearly show that the metal atom is required for correct folding of the finger. For example, the TFIIIA apopeptide is much more sensitive to cleavage by trypsin than the Zn-peptide and the two exhibit very different CD spectra (128,132). Likewise, the aromatic and methyl ^{1}H resonances of the Zn-ADR1a complex undergo substantial shifts when an excess of EDTA is introduced to remove the metal atom. Therefore, the metal is intimately involved in the formation of the proper secondary structure in the His_2Cys_2 fingers.

A second TFIIIA-like finger solution structure has been reported by Lee et al. (134). They studied a 25-residue peptide corresponding to the 31st finger of the *Xenopus* Xfin protein (135), which they call Xfin-31. This particular finger was chosen because it so closely resembles the consensus sequence for the Cys_2His_2 fingers. The derived backbone structure plus the side chains whose positions were well defined by NOESY constraints is shown in Figure 44. Like ADR1a, the Xfin-31 peptide has a helix that extends from Glu12 to Lys24 and includes the two ligating histidines. Because of the higher resolution of this structure, it could be seen that the helix form changes from α at the beginning to 3_{10} at the end. As expected, the conserved hydrophobic residues were found to pack in the core of the structure. In the Xfin-31 structure, the antiparallel β-strand predicted in

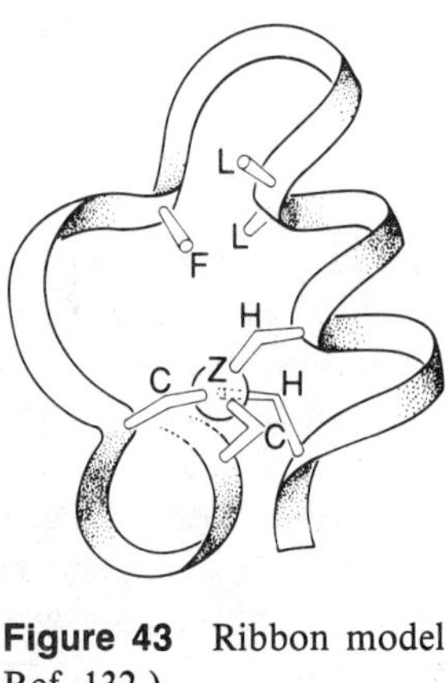

Figure 43 Ribbon model for the *ADR1*, a zinc finger peptide. (Adapted from Ref. 132.)

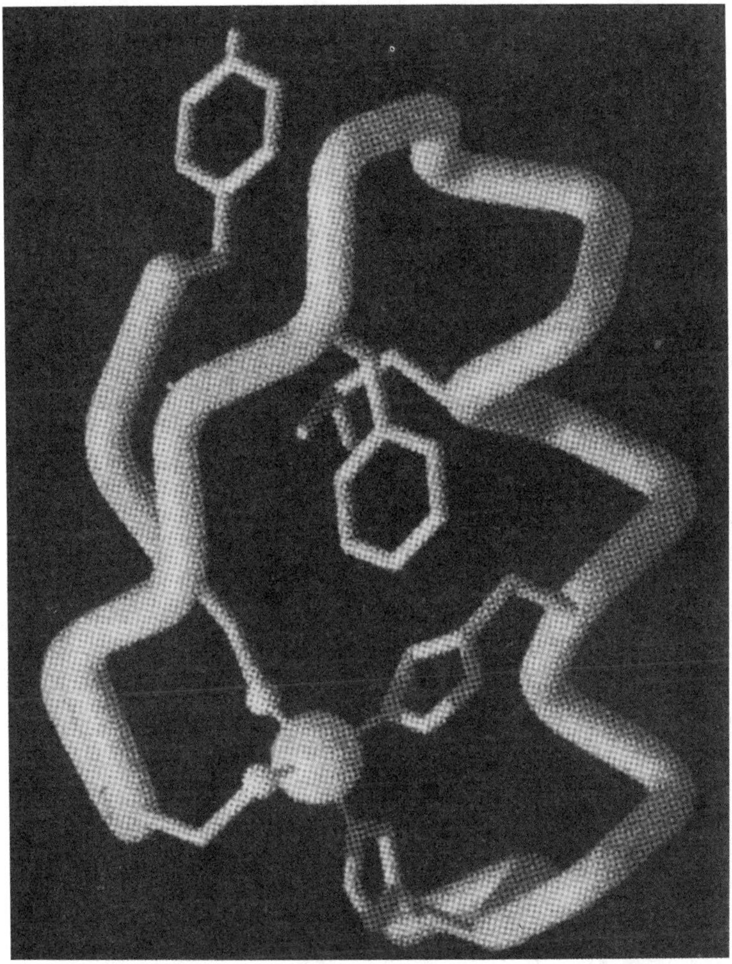

Figure 44 Model of the *Xfin*-31 zinc finger peptide. (Adapted from Ref. 134.)

the Berg model is observed, with characteristic hydrogen bonds between the N—H of Tyr1 and CO of Phe10 and the N—H of Phe10 and the CO of Tyr1. An important aspect of this structure, like that of the ADR1a peptide, is that the N- and C-termini are at the opposite ends of the molecule and point in different directions. As discussed below, this arrangement has important implications for models of protein-DNA interactions.

2. *Mode of DNA Binding*

Almost all TFIIIA-like zinc finger proteins contain at least two and often many more zinc fingers (the only exception I am aware of is the terminus protein of *Drosophila* (136)). For example, the *Xenopus* Xfin protein has 37 zinc fingers. Further, the DNA-binding sites of these proteins can be quite large; TFIIIA recognizes a sequence of about 50 nucleotides within the 5S gene, called the internal control sequence (ICR) (137). These facts suggest that each zinc finger interacts with a short stretch of DNA. The results of various chemical protection studies show that these interactions are in the major groove (138). Therefore, in addition to trying to explain the mode of interaction of a single zinc finger with DNA, an additional challenge to model builders is to correctly predict how the fingers are linked together over such a long stretch of DNA. The protein must either lie along one face of the helix and project its fingers into the major groove from the central spine, or the protein backbone may trace the path of the major groove. Of course, the latter model has some severe topological constraints given that the TFIIIA DNA binding site encompasses about five turns of the helix and therefore implies a discontinuity over the length of the protein-DNA complex.

To distinguish between these possibilities, the TFIIIA-DNA complex has been examined with nuclease and chemical probes (vide infra). From these experiments, as well as simple model building, have come two models for the structure of the TFIIIA-DNA complex (Fig. 45). One, proposed by Berg (133) [a very similar model was proposed independently by Gibson et al. (139)], is sometimes called the "zinc links" model. It supposes that the fingers run along the major groove, with the linkers running along the

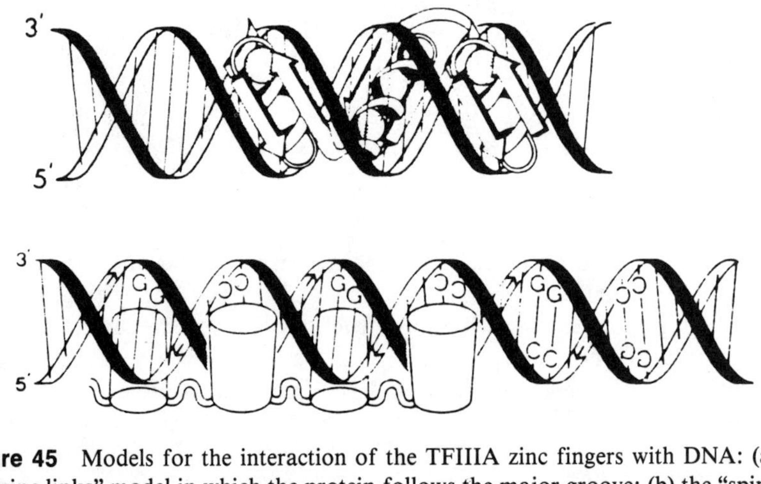

(a)

(b)

Figure 45 Models for the interaction of the TFIIIA zinc fingers with DNA: (a) the "zinc links" model in which the protein follows the major groove; (b) the "spine with fingers" model in which the protein sits on one face of the DNA cylinder. (Adapted from Refs. 133 and 138.)

phosphodiester backbone. This helps to define a "sausage link" arrangement of the fingers, hence the name. Model-building based on this overall architecture and using B-form DNA suggests that each finger recognizes between three and five base pairs.

An alternative model is that the spine of the protein sits on one face of the DNA helix and the fingers project away from the spine and into the major groove (138). This model, proposed by Klug and co-workers, supposes that each finger of TFIIIA contacts a 5.5-bp stretch. This fits a sequence and structural periodicity noticed by these workers in the TFIIIA binding site and that of other zinc finger proteins such as SP1 (140). Based partly on these data, these workers also proposed that the binding sites of TFIIIA and similar proteins may exist in an underwound conformation such as A-form. This suggestion was supported by a crystal structure of 9-bp oligonucleotide corresponding to part of the TFIIIA binding site (141). However, a solution NMR study of the same oligonucleotide showed it to be B-form (142) and the CD spectrum of the entire TFIIIA binding site seems to more closely resemble B-form than A-form DNA, both alone and in the protein complex (143). However, this point is still being argued in the literature (144).

While not definitive, the currently available data seem to favor the models of Berg (133) and Gibson et al. (139). Perhaps the most important fact to be considered in distinguishing between the two schemes is the position of the N- and C-termini of the individual fingers deduced by NMR. The fact that they are on opposite sides of the finger and point in different directions fits naturally into the Berg model, but it is difficult to see how this geometry can be incorporated into the "spine and finger" model. Various footprinting studies have also been done on the complex, including some that employed fragments of the TFIIIA (145). While some of these have been interpreted as suggesting the protein lies along one face of the helix for the length of the ICR and that adjacent fingers interact in a regular manner with each 5.5-bp repeat (138), the fragment studies, which included very high resolution hydroxyl radical footprinting (146) suggest otherwise. These studies (145) employed N- and C-terminal truncated TFIIIA fragments produced by in vitro transcription/translation to probe local contacts by the various fingers. This was feasible because many of the fragments retained sufficient avidity for the ICR to be footprinted. The clear conclusion was that the interaction is not regular over the entire length of the DNA site (Fig. 46), but that the finger-DNA contacts are clustered. There is an intimate contact between fingers 1 and 2 with 17 bp near the 3′-end of the binding site, including six Gs on the noncoding strand that are critical for binding (147). The protein emerges from the major groove at about bp 80 to lie on one face of the helix for about 20 bp. A number of contacts in

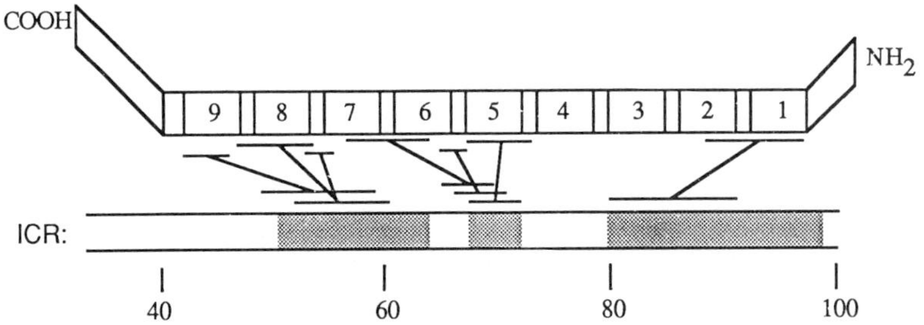

Figure 46 Summary of contacts between TFIIIA and the ICR as defined by a high-resolution footprinting analysis of complexes containing truncated proteins (145). Regions of the protein (top) that are not involved in DNA binding are tilted up. The zinc fingers are represented as numbered boxes. Areas of contact are indicated with lines.

the minor groove may be made in this region. The protein again appears to follow the major groove for another 20 bp in the 5′-end of the DNA site. Given the topological problem mentioned previously, it seems unlikely that the protein backbone trace exactly follows the major groove for two turns of the helix. The exact arrangement will probably not be known until the complex can be characterized crystallographically. The important point is that the fragment mapping study does strongly support the notion that the protein-DNA complex is not regular along its length.

B. The Cys_4 Zinc Finger Proteins

A search for other proteins that have sequences suggestive of metal-binding domains has resulted in the realization that other DNA binding proteins have motifs similar to, but clearly distinct from, the zinc fingers of TFIIIA (148). One of these families, the steroid receptors (149), all contain the sequence shown below which includes nine cysteines.

$$\text{Cys-X}_2\text{-Cys-X}_{13}\text{-Cys-X}_2\text{-Cys-X}_{15\text{-}17}\text{-Cys-X}_5\text{-Cys-X}_9\text{-Cys-X}_2\text{-Cys-X}_4\text{-Cys}$$

These proteins are the sequence-specific binding factors involved in hormone-dependent regulation of transcription of specific genes. Among the best characterized members are the glucocorticoid, estradiol, and progesterone receptors, which have been shown to bind as dimers to their target sites (150). A glucocorticoid receptor fragment able to recognize the glucocorticoid response element (GRE, the site of binding of the receptor) was expressed to high levels in *E. coli* and shown to bind two atoms of zinc (151). Indeed, specific GRE binding was shown to be dependent upon the presence of zinc or cadmium. The XAFS is consistent with a tetrahedral geometry in which the metal is coordinated by four sulfur atoms. As expected from these results, mutagenesis and chemical modification experiments have shown that one of the nine conserved cysteines is dispensible for metal and DNA binding (152). Therefore, it appears that the DNA binding domain of these proteins, and presumably those of the other members of the family, contains two zinc fingers of the Cys_4 type.

Very recently a solution NMR structure of the zinc-binding domain of the glucocorticoid receptor (GR) was reported (153,154) and, soon thereafter, the solution structure of an estradiol receptor (ER) fragment (155). The sequence of the 71-residue GR fragment employed by Kaptien and co-workers and the backbone structural model obtained are shown in Figure 47. The ER structure is shown in Figure 48. Since the structures are so similar, we will describe only the GR in detail. The most outstanding elements of secondary structure are two α-helices spanning Ser459 to Glu469 and Pro493 to Gly504 respectively. The residues between the second and third cysteines in each zinc coordination sphere form loops of different shapes, one of which (from the second finger) protrudes from the oblate core. A type II β-turn is seen in residues 475 to 478 and a type I reverse turn from Arg479 to Cys482. These two turns connect the first two cysteines of the second finger. Mutation of these residues has been found to abolish the cooperativity of binding normally observed (154), so they are probably involved in protein-protein contacts.

The α-helices are oriented almost perpendicular to one another with their hydrophilic sides facing the solvent. Several hydrophobic residues from these helices and the extended peptide regions form a hydrophobic core that presumably stabilizes the overall structure. Many of these residues are widely conserved in the steroid receptor family. Structurally, the role of the two zinc atoms appears to be to hold the α-helices and the extended regions in a precise spatial orientation (Figs. 47 and 48).

What does this tell us about the mode of DNA-binding by these proteins? The boxed residues in Figures 47 and 48 are those that are essential for discrimination between the GRE and the estrogen receptor element (ERE) (156–158). The symmetrized consensus sequences of the two sites are

GRE: 5′-A G **A A** C A X X X T G **T T** C T-3′
ERE: 5′-A G **G T** C A X X X T G **A C** C T-3′

where X = any nucleotide and the divergent residues are in boldface. Thus, any model will logically involve close contact between the boxed residues and the divergent bases in the ERE and GRE. In addition, the DNA binding site of the thyroid receptor, a related protein, is identical to the ERE, except that the three-nucleotide spacer is absent. Therefore, the ER, at least, must

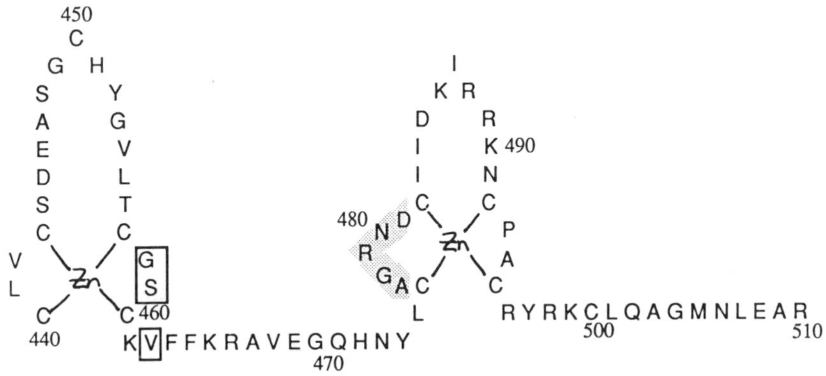

(a)

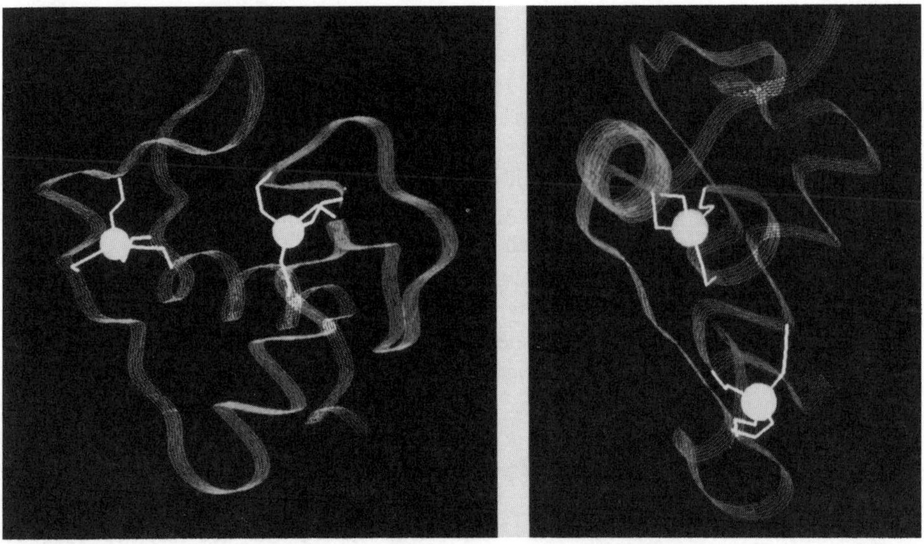

(b)

Figure 47 Amino acid sequence (a) and structure (b) of the DNA binding domain of the glucocorticoid receptor. Residues important in distinguishing between response elements are boxed, those thought to be involved in dimerization are shaded (a). Zinc atoms in (b) are shown as balls. (Adapted from Ref. 154.)

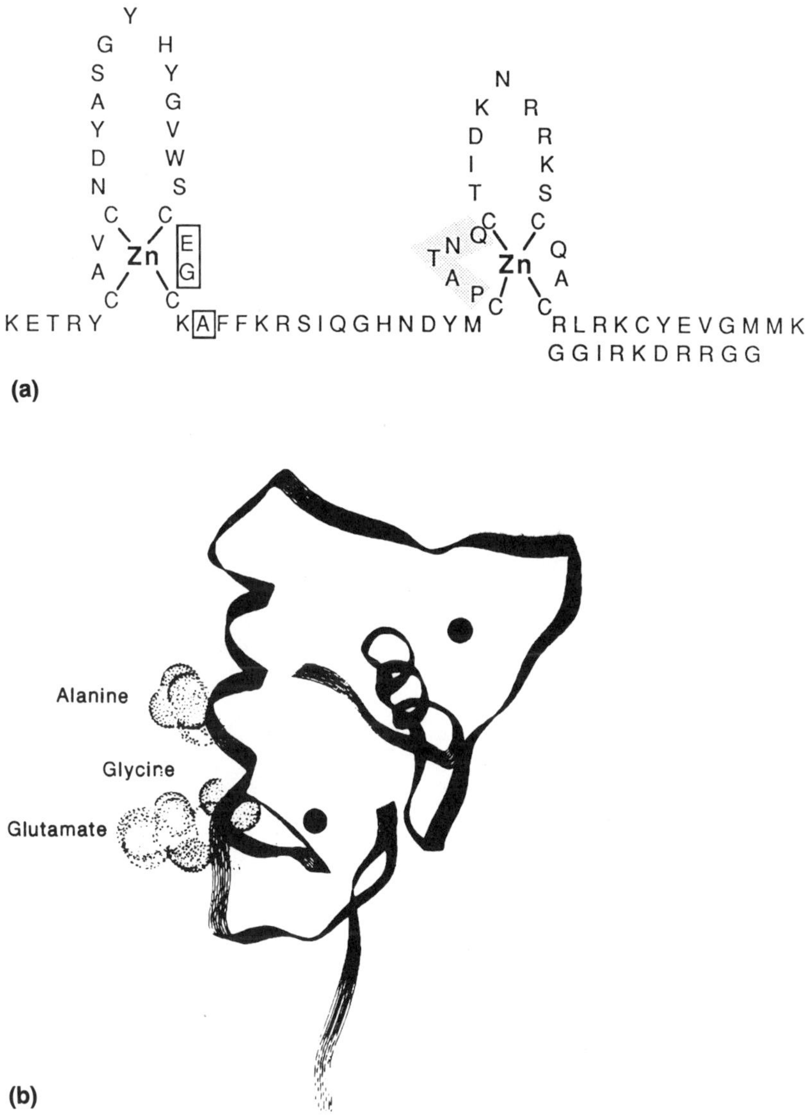

Figure 48 Amino acid sequence (a) and structure (b) of the DNA binding domain of the estrogen receptor receptor. Residues important in distinguishing between response elements are boxed, those thought to be involved in dimerization are shaded (a). Zinc atoms in (b) are shown as balls. The residues important in response element recognition are shown as shaded balls in (b). (Adapted from Ref. 155.)

be able to sense precisely the spacing between the half-sites. The boxed amino acids, Gly458, Ser459, and valine462 for the GR and Glu25, Gly26, and Ala29 for the ER, are at the base of, or in, the first helix. Thus, it seems likely that this unit is a recognition helix which sits in the major groove of the half-site (154,155). Interestingly, if two GR or ER monomers are independently positioned in a symmetric fashion with the recognition helix in the major groove and the boxed amino acids near the divergent bases, the residues in the turn between the first and second cysteines in the second finger (circled in Figs. 47 and 48) are brought into close contact (Fig. 49). Recall that mutagenesis of these amino acids abolishes cooperativity of binding. Therefore, the independently proposed models for DNA binding of the glucocorticoid and estrogen receptors to their cognate response elements (154,155) (Figure 50) are in accord with the current biochemical and genetic evidence. However, a deeper understanding of the DNA-protein contact surfaces will require a structural characterization of a protein-DNA complex.

C. The "Zn_2Cys_6" Fingers: GAL4 and Its Relatives

As mentioned in the first section, GAL4 is a transcriptional activation protein that stimulates expression of several galactose responsive genes. The DNA binding sites of GAL4 in vivo are several related 17-bp sequences that all deviate to a small extent from a symmetrical consensus sequence

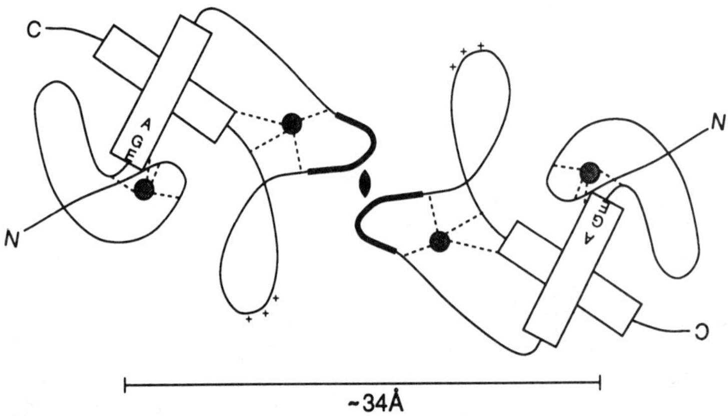

Figure 49 Cartoon of the proposed dimeric arrangement of the ER DNA-binding domain, showing the required 34 Å spacing between the recognition elements that would allow them to fit into adjacent major grooves. (Adapted from Ref. 155.)

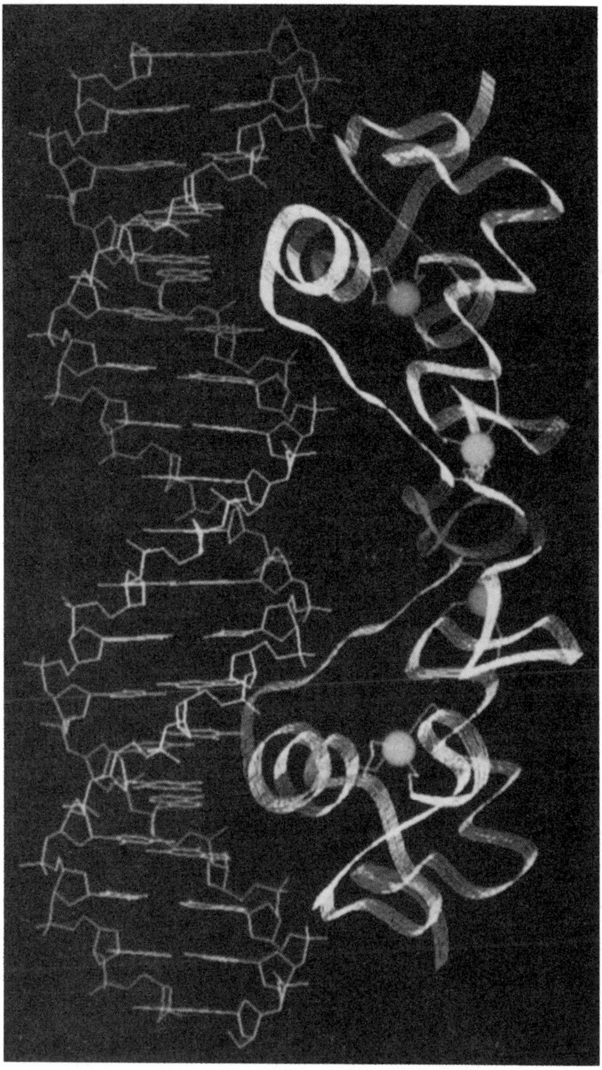

Figure 50 A model for the GR-GRE complex proposed by Hard et al. (154).

(Fig. 51). For convenience, this sequence is called "the 17mer" for the duration of this discussion.

Ptashne and co-workers found that a fusion construct containing the first 74 residues of GAL4 and β-galactosidase retained the sequence-specific binding properties of the native protein (51), localizing the DNA binding domain to this region. Mutagenesis experiments by M. Johnston and co-

Rotationally symmetric site

5'-C G G A G G A C T G T C C T C C G-3'
3'-G C C T C C T GA C A G G A G G C-5'

Natural UAS$_{GS}$

5'-C G G A T T A G A A G C C G C C G-3'
3'-G C C T A AT C T T C G G C G G C-5'

5'-C G G G T G A C A G C C C T C C G-3'
3'-G C C C A C T G T C G G G A G G C-5'

5'-A G G A A GA C T C T C C T C C G-3'
3'-T C C T T C T G A G AG G A G G C-5'

5'-C G C G C C G C A C T G C T C C G-3'
3'-G C G C G G C G T G A C G A G G C-5'

5'-C G G A C A A C T G T T G A C C G-3'
3'-G C C T G T T G A C AA C T G G C-5'

5'-C G G a X X a c **A**
T c a X X a G G C-5'
Consensus

Figure 51 Sequence of the DNA sites recognized by the GAL4 protein.

workers are in agreement with this finding (159–161). Almost all of the point mutants generated in a saturation mutagenesis study that strongly affect sequence-specific DNA-binding were located in the first 54 residues of GAL4 (Fig. 52). It should be noted that some of the residues implicated in DNA binding by this analysis lie outside the zinc-binding domain as defined by the six cysteine residues. There is also a considerable degree of sequence homology in this region between GAL4 and several other fungal transcription factors (159), such as the LAC9 and PPR1 proteins, suggesting that the structural motif found in the GAL4 protein is widespread.

Biochemical analyses of purified GAL4 fragments have shown that the native GAL4 protein binds to its site as a dimer and requires a dimerization domain located somewhere between residues 74 and 147 for extremely tight

Ala-**Cys**-Asp-Ile-**Cys**-Arg-Leu-Lys-Lys-Leu-Lys-**Cys**-Ser-Lys-Glu-Lys-Pro-Lys-**Cys**-
10 20
Ala-Lys-**Cys**-Leu-Lys-Asn-Asn-Trp-Glu-**Cys**-Arg-Tyr-Ser-Pro-Lys-Thr-Lys-Arg-Ser-
30 40
Pro-Leu-Thr-Arg-Ala-His-Leu
50

Figure 52 Primary sequence of the zinc-binding domain ofthe GAL4 protein. Residues that are involved in ligation are in boldface, while residues that, when mutated, abolish sequence-specific DNA binding are underlined.

binding (60). The tight binding of the β-gal fusion construct was a fortuitous result of the tetrameric nature of β-galactosidase. Phosphate ethylation interference and hydroxyl radical footprinting experiments with a GAL4(1-147)-DNA complex containing the 17mer sequence showed that the GAL4 monomers contact chiefly the strands 3′ of the central T-A base pair (60). The spacing of these contacts with respect to the helical periodicity of the DNA (Fig. 53) suggests that if the GAL4 dimer sits entirely on

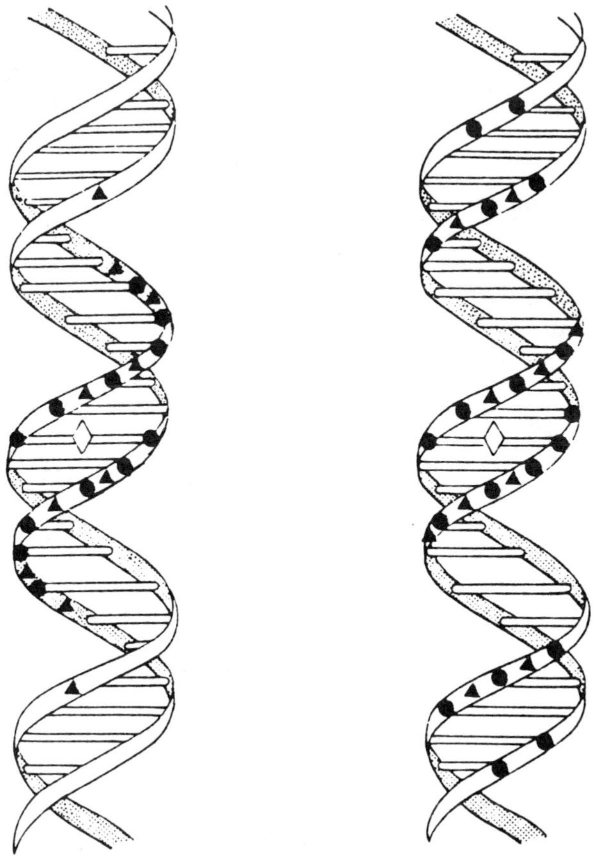

Figure 53 Schematic representation of the protein-DNA contacts made in the GAL4 protein-consensus 17mer complex and, for comparison, those made in the λ repressor-operator complex. Sugars protected from hydroxyl radicals are represented as circles, while phosphates that interfere with binding when ethylated are represented as triangles. (Adapted from Ref. 60.)

one face of the helical cylinder, as in the case of the 434 repressor-operator complex, the 17mer must exist in an underwound conformation such as A-DNA. Alternatively, the 17mer could retain a more classical B-form structure if the GAL4 DNA-binding domain wraps itself to some extent around the double helix. A recent spectroscopic study of the GAL4 binding site shows that the DNA is clearly B-form both alone and in the complex with the GAL4 fragment (162). Indeed, the CD spectrum of a complex containing a GAL4(1-140) fragment and an oligonucleotide including the 17mer sequence is only slightly different than that of the free oligonucleotide, suggesting that the protein may effect changes in base rolls, etc., but does not affect the global helix geometry. The conclusion that the GAL4 protein must wrap itself around the helix is consistent with the observation that cooperative binding between GAL4 dimers on separated 17mers is insensitive to the spacing between the sites (i.e., half or whole turns of the helix) (50). The pattern of protein-DNA contacts suggests that the dimerization domains of each monomer contact one another over the center of the binding site, as is observed with a number of dimeric DNA binding proteins. This dimer interface is probably a relatively stiff structure, since deletion of the central T-A base pair or the presence of an addition base pair between the half-sites lowers the affinity for GAL4(1-140) about 50-fold relative to the 17mer (163). This result was somewhat surprising in light of the tight binding exhibited by the β-gal-GAL4(1-74) fusion protein which is unlikely to have the GAL4 binding domains spaced in precisely the same manner they are in the native protein. Interestingly, the 50-fold decreased binding constant observed for the "spacing mutants" in the GAL4 system is about the same as that found for an oligonucleotide containing the correct spacing, but with a mutation in one of the half-sites (163). Taken together, these results strongly suggest that the binding energy (K_D of about $2 \times 10^{-11}\ M^{-1}$) is dominated by a monomer-half-site contact and that reasonably (about 50-fold less) stable complexes can be formed when the other monomer of the dimer contacts DNA nonspecifically, which is apparently the case for the spacing mutants of the 17mer. This may also have been the case for the β-gal fusion protein.

What is the structure of the DNA binding domain of GAL4? Pan and Coleman have found that a 149-amino-acid fragment containing the first 147 residues of GAL4 (GAL4(149*)) contains about 1.5 atoms of bound zinc per monomer (164). The apoprotein does not bind the 17mer sequence specifically. Cadmium can replace zinc, but cobalt does so only poorly. These workers further characterized the Cd derivative of this polypeptide and observed two distinct signals in the ^{113}Cd-NMR spectrum (164). The chemical shift of each is consistent with ligation of three or four cysteines. Since there are only six cysteines in the DNA binding domain, this observation suggested the presence of a binuclear, bridged complex (Fig. 54). Elegant studies using heteronuclear 2D-COSY established that this is indeed the case (165,166). Pan and Coleman have reported 2D-NMR data that seems likely to reveal the backbone structure of the GAL4 zinc finger in the near future (165), but this structure was not available at the time of this writing.

The clues we have concerning the particular residues important in sequence recognition come from an interesting experiment performed by Corton and S. Johnston (167), which suggests that 14 amino acids C-terminal to the sixth cysteine of the binuclear complex are primarily responsible for sequence specific DNA recognition. This conclusion was reached by comparing the DNA-binding properties of hybrid proteins (Fig. 55) consisting of GAL4 and PPR1 residues. The PPR1 protein is a yeast transcription activator of the *URA1* and *URA3* genes which also binds to a near symmetric UAS (UAS_U), but one of a different sequence than the UAS_G. It shares a high level of sequence homology with GAL4 in the zinc-binding region. It was found that GAP3, which includes the metal-binding region of the PPR1 protein inserted in place of the analogous GAL4 sequences recognized neither UAS_U or UAS_G. However, *a point mutant of GAP3 in which* Lys_{23} *is substituted for* Gln_{23} *recognizes* UAS_G *but not* UAS_U (Lys is the amino acid found at position 23 in GAL4). Furthermore, GAP4N, a hybrid in which residues 2–52 of GAL4 were replaced with the analogous PPR1 residues, recognizes UAS_U but not UAS_G. This shows that most of the

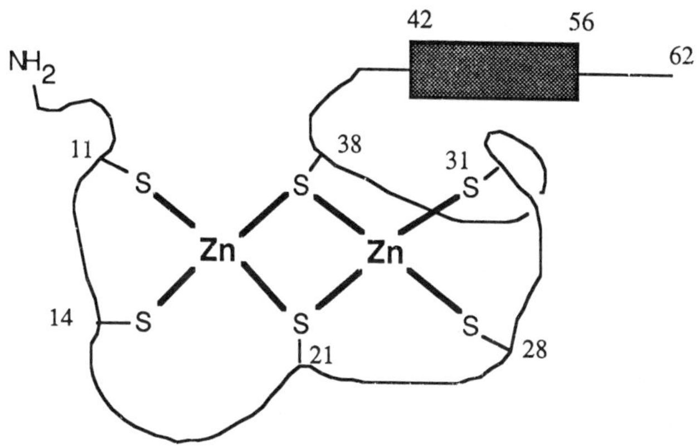

Figure 54 Sketch of the cloverleaf skeleton of the binuclear, bridging GAL4 zinc-binding domain. Residues that have been implicated in discrimination between different UASs are represented by the stippled box. The detailed structure of this domain is currently under investigation.

	1	11	39	53	DNA-binding UASG	DNA-binding UASU
		23				
GAL4	MKLLSSIEQA	**CDICRLKKLKCSKEKPKCAKCLKNNWEC**	RYSPKTKRSPLTRA	HLTEVE	+	-
GAP3	MKLLSSIEQA	**CKRCRLKKIKCDQEFPSCKRCAKLEVPC**	RYSPKTKRSPLTRA	HLTEVE	-	-
GAP3 (LYS23)	MKLLSSIEQA	**CKRCRLKKIKCDKEFPSCKRCAKLEVPC**	RYSPKTKRSPLTRA	HLTEVE	+	-
GAP4N	MIGISKSRTA	**CKRCRLKKIKCDQEFPSCKRCAKLEVPC**	VSLDPATGKDVPRS	HLTEVE	-	+
PPR1	NIGISKSRTA	**CKRCRLKKIKCDQEFPSCKRCAKLEVPC**	VSLDPATGKDVPRS	YVFFLE	-	+

METAL-BINDING REGION SPECIFICTY DOMAIN

Figure 55 Amino acid sequences of GAL4, PPR1 and several hybrid proteins. The ability of each protein to recognize the PPR1 (UAS_U) and GAL4 (UAS_G) protein binding sites is denoted by + or −. These data (167) implicate Lys23 and amino acids 39–53 of GAL4 as important in recognition of the UAS_G.

GAL4 zinc finger is not required for discrimination between DNA binding sites (except for Lys_{23}) but that the 14 amino acids downstream of the finger are crucial. It should be noted that, at least in the case of GAL4, it seems highly unlikely that the entire 14 amino acid specificity-determining region is α-helical since it contains three proline residues. The corresponding PPR1 domain has two prolines. Of course, a smaller helix might exist. Nonetheless, it seems likely that these proteins employ a motif that differs greatly from the recognition helix motif which, although oriented and positioned in many different ways, is found in so many sequence-specific DNA binding protein.

At present, it is not clear which residues in the 17mer are contacted by what residues of the protein, since no specificity mutants have been generated of the type discussed in the 434 repressor-operator system. However, an inspection of the in vivo GAL4 binding sites reveals that by far the most strongly conserved nucleotides are the three on the end: 5′-GCC (Fig. 51). Therefore, it seems likely that some of the residues in the 14-amino-acid C-terminal of the zinc cluster contact these bases. The highly basic zinc-binding region has at least one residue, Lys23, important in sequence-specific recognition and probably also contributes heavily to the overall stability of the DNA-protein complex through many non-specific contacts. Clearly, the study of this class of zinc fingers is still in its early stages.

D. Other Zinc Finger Proteins

There are other types of zinc and nucleic acid binding proteins, most prominently those that contain a "Cys_3His"core binding site (168). However, these factors are generally found to be RNA or nonspecific single-stranded DNA binding factors and, while quite interesting, are beyond the scope of this review.

Before closing, however, one other putative zinc finger region deserves comment, that of the T7 gene 4 protein. Gene product 4 is the primase/helicase of the T7 replication complex (169) and deletion analyses have suggested that the single Cys_4-type motif found in the protein is part of the primase (170). Thus, it might be that this finger is involved in the sequence-specific recognition of *single-stranded* DNA, since the protein must find the short sequence on the lagging strand that tells it where to begin the RNA primer for a new Okazaki fragment. There have been no biophysical studies of this zinc finger; thus, it remains an interesting target for further analysis.

VII. THE LEUCINE ZIPPER

In 1988, McKnight and co-workers noticed a regular repeat of leucine (or sometimes methionine) residues every seventh amino acid in several proteins known or thought to be transcription factors (Fig. 56) (171). These include the nuclear transforming proteins (i.e., the products of oncogenes) Myc, Fos, and Jun. The leucine repeat occurs in regions of these proteins that have been implicated in DNA binding for cases in which the domain organization of the protein has been studied. Landschulz et al. proposed that the leucine repeats form extended α-helices with the hydrophobic side chains on one face of the helix. Since leucine occurs every seventh residue (about twice the periodicity of a helix), they proposed that these helices could dimerize by interdigitation of the leucine side chains (Fig. 57), much like the teeth in a zipper. In their model, the helices are arranged in an antiparallel manner.

Deletion analysis of the GCN4 protein, a yeast transcription factor that contains a leucine repeat showed that this motif alone is insufficient to support sequence-specific DNA binding. A C-terminal 60-amino fragment of GCN4, containing a highly basic run of 30 residues followed by the leucine repeat, is able to recognize the GCN4 binding site sequence specifically (172–174). Deletion of the basic region *or* the leucine zipper abolishes DNA binding. Since basic residues are usually correlated with regions that directly contact DNA, this result suggested that the leucine zipper–leucine

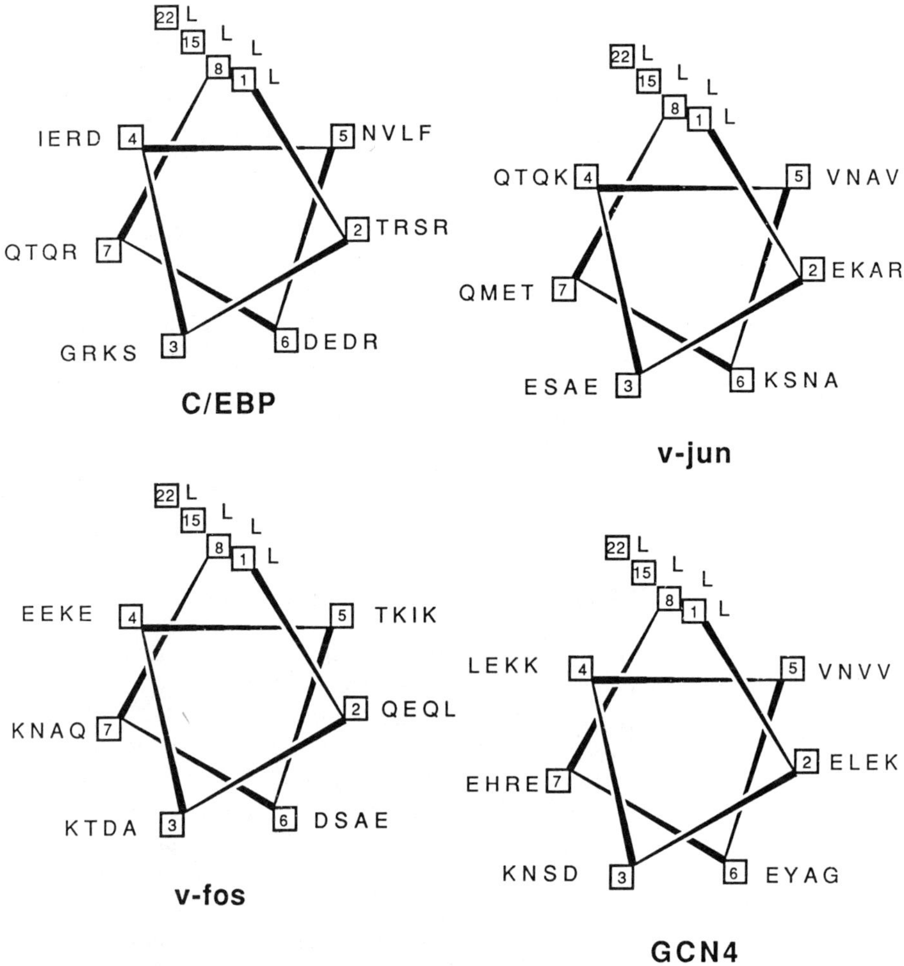

Figure 56 Helical wheel analysis of several protein sequences reveals a heptad repeat of leucines that creates a hydrophobic spine along one face of the putative helix.

zipper interaction may act to orient the basic DNA binding surface. This model is supported by the observation that GCN4 binds to DNA as a dimer.

A. Structure of the Leucine Zipper

While there are presently no reports of a crystal structure of a leucine-zipper-containing protein, the modest size of this domain and the adjacent basic region have allowed most of the important structural questions to be addressed by analysis of peptide models. The first of these studies, conducted by Kim and co-workers (175), focused on a peptide corresponding to the C-terminal 33 amino acids of GCN4 (called GCN4-p1, See Fig. 58). The CD spectrum of this peptide showed it to be essentially 100% α-helical, substantiating this aspect of the McKnight model. Under the conditions employed initially, the T_m of the helix (as measured by variable temperature CD spectroscopy) was 57°C, but this value was found to be temperature dependent, suggesting that the peptide self-associates. This was confirmed by equilibrium sedimentation experiments which showed that GCN4-p1 exists as a dimer over a broad concentration range. The observation of a dimer explains the very high solubility of the peptide in water, since this would allow the hydrophobic surface to be shielded from solvent.

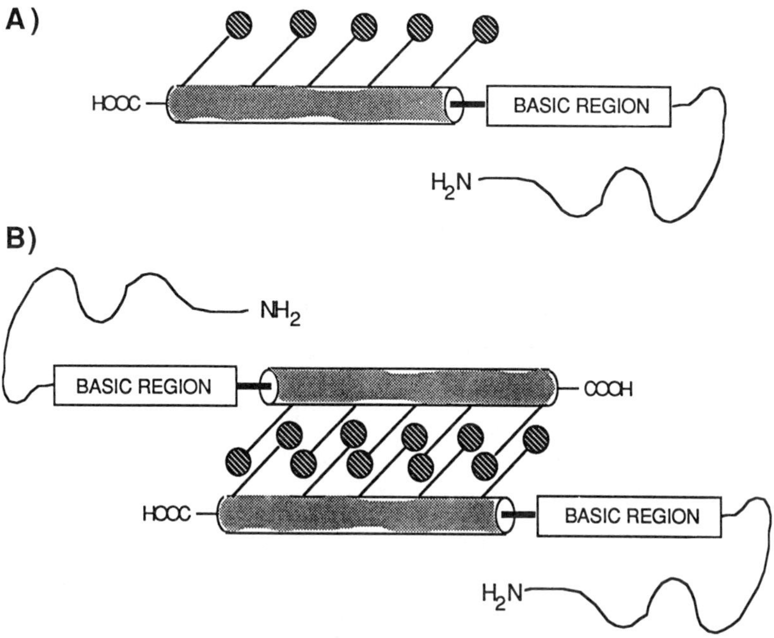

Figure 57 The original leucine zipper proposal for the C/EBP protein, which contains a basic region (required for DNA binding) adjacent to the leucine zipper (A). The leucine side chains are proposed to interdigitate in the manner of the teeth in a zipper, locking the two antiparallel helices together (B). Because the leucine side chains project from the helix at an angle, the antiparallel arrangement of the helices is a necessary facet of this model. This was later shown to be incorrect.

Having established that the GCN4 leucine zipper is indeed a dimer of very stable helices, O'Shea et al. (175) proceeded to determine the arrangement of the helices in the dimer; antiparallel, as suggested by Landschulz et al., or parallel. This was addressed by synthesizing two similar peptides, GCN4-p1N and GCN4-p1C, that included three extra amino acids at the N- or C-termini (Fig. 58). Two glycines were attached as a flexible spacer arm to connect a cysteine to the leucine zipper domain. GCN4-p1N and GCN4-p1C were then mixed together at equivalent concentrations and oxidized, resulting in disulfide bond formation. The striking result was that only homodimers were observed, heterodimers were not formed (the various dimeric species could be separated by HPLC). However, when the oxidation was carried out in a denaturing buffer, a statistical mixture of both homodimers and the heterodimer was obtained (Fig. 58). Furthermore, the parallel dimers, once refolded by removal of the guanidinium hydrochloride, exhibited a concentration-independent T_m, while the T_m of the antiparallel dimers remained concentration-dependent. These data clearly show that the leucine zipper consists of *parallel* α-helices, since the production of homodimers only under native conditions suggests that the cysteinyl sulfurs are in close contact only in the parallel orientation. Thus the Kim study supports some aspects of the original leucine zipper model (174), but not others. In particular, the parallel arrangement of the α-helices suggests that the leucine side chains cannot interdigitate in the manner suggested by Landschulz et al. (174) since the angle of projection from the helix backbone is incorrect. Therefore, Kim and co-workers suggested a coiled coil arrangement (175) (Fig. 59) akin to that found in fibrous proteins such as keratin. In this motif, the two right-handed helices coli about one another in a lazy left-handed superhelix. In addition to the regularly spaced leucines, this model requires another hydrophobic residue adjacent to the

Ac-RMKQ**L**EDKVEE**L**LSKNYH**L**ENEVAR**L**KKLVGER	GCN4-p1
Ac-CGGRMKQ**L**EDKVEE**L**LSKNYH**L**ENEVAR**L**KKLVGER	GCN4-p1N
Ac-RMKQ**L**EDKVEE**L**LSKNYH**L**ENEVAR**L**KKLVGERGGC	GCN4-p1C

Figure 58 Amino acid sequences of the peptides employed by Kim and co-workers to test the orientation of the two helices in the GCN4 leucine zipper. The first peptide, GCN4-p1, is identical to the natural GCN4 zipper sequence. The other two peptides contain the GCN4-p1 sequences but also include an N or C-terminal cysteine connected to the GCN4 residues by a flexible linker composed of two glycines. The cysteines allow the peptides to dimerize. Ac represents an acetylated N-terminus. The leucines of the repeat are underlined and in boldface.

leucine in a "4-3 repeat" arrangement, which is, in fact, observed in the members of this class.

A two-dimensional NMR study of GCN4-p1 was consistent with the coiled coil model (176). In particular, the helical nature of the peptide was confirmed and the symmetry observed in the spectrum was consistent only with a parallel dimer. However, details of the backbone and side-chain configurations remain to be determined.

Therefore, the leucine zipper appears to be a conserved motif for protein dimerization, but does not participate directly in contacting the DNA. Elegant experiments confirming this idea will be discussed below. Although it is now clear that the term *zipper* is not really correct, it appears to be simply too catchy for the biochemical community to discard and is still used routinely to designate this family of proteins.

While we will return to the point of how heterodimers may be formed selectively via leucine zipper interactions, let us first turn to the question of how adjacent regions might take advantage of the spatial orientation provided by the leucine zipper to bind to DNA sequence-specifically.

B. DNA Recognition via a "Scissor Grip" Motif

Like GCN4, many leucine zipper proteins include highly basic regions immediately N-terminal to the zipper. This subclass has been given the name bZIP, to emphasize the presence of two important domains. In addition to the deletion analysis of GCN4 mentioned above, a number of experiments strongly suggest that the basic region, not the leucine zipper, contacts the DNA. For example, a hybrid protein containing the leucine zipper of GCN4 and the basic region of C/EBP (a mammalian transcription factor of the leucine zipper family) binds DNA with the specificity of C/EBP (177). In addition, a synthetic peptide corresponding to the basic region of GCN4 (minus the leucine zipper) does not bind to DNA specifically, but a dimer of this species formed by a disulfide bond at its C-terminus does (178). Thus we can safely conclude that the role of the leucine zipper is limited to facilitating dimerization.

What is the mode of interaction of the basic region with DNA? Again, there is no direct structural information available, but a number of interesting biochemical and biophysical experiments using synthetic peptides have led to the "scissor grip" and "induced helical fork" models (Fig. 60). These very similar models were independently proposed by Vinson et al. (179) and DeGrado et al. (180), respectively. They propose that the dyad axis of the leucine zipper is oriented perpendicular to the DNA helix axis. At the N-terminus of the zipper, the DNA binding arms of the basic region branch out in opposite directions to form a Y-shaped molecule. In both proposals,

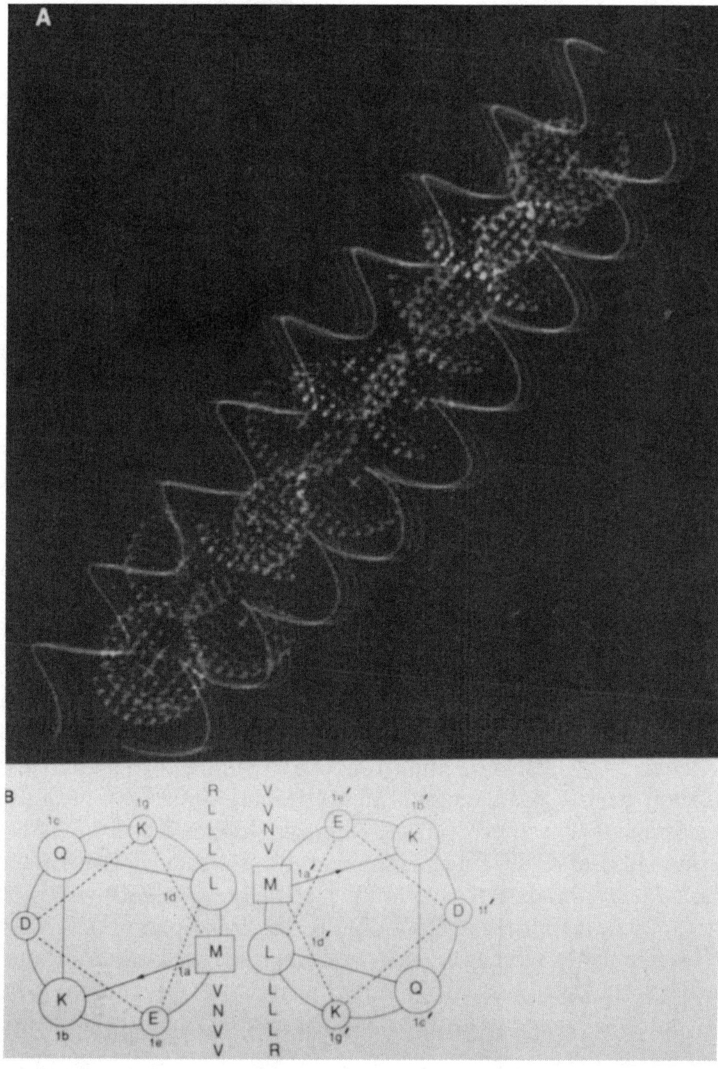

Figure 59 The coiled-coil model for leucine zippers. In this scheme, the leucine side chains are not interdigitated, but rather pack against each other and the adjacent residue in a 4-3 arrangement. This is the currently accepted structural model.

the arms are thought to be α-helical and in the scissor grip model a specific prediction is made that there is a sharp bend in the arm caused by an interruption in the helix at the highly conserved Asn (Asn-18 in C/EBP, using the numbering scheme described in Fig. 61).

Both proposals resulted from an inspection of the sequence of the basic region of bZIP proteins that bind to similar DNA sites, which include the sequence 5′-TGAC-3′. O'Neil et al. noticed that the position of the conserved residues in these proteins exhibit a periodic distribution with a repeat of 3.6 residues. In particular, Asn-18, Ala-15, Ala-14, and Cys-11 are absolutely conserved, with the minor exception of a Ser substitution at position -11 in GCN4, and at least one of these residues is different in bZIP proteins that recognize different DNA sequences. Furthermore, this quartet is surrounded by positively charged lysine and arginine residues. This led De-

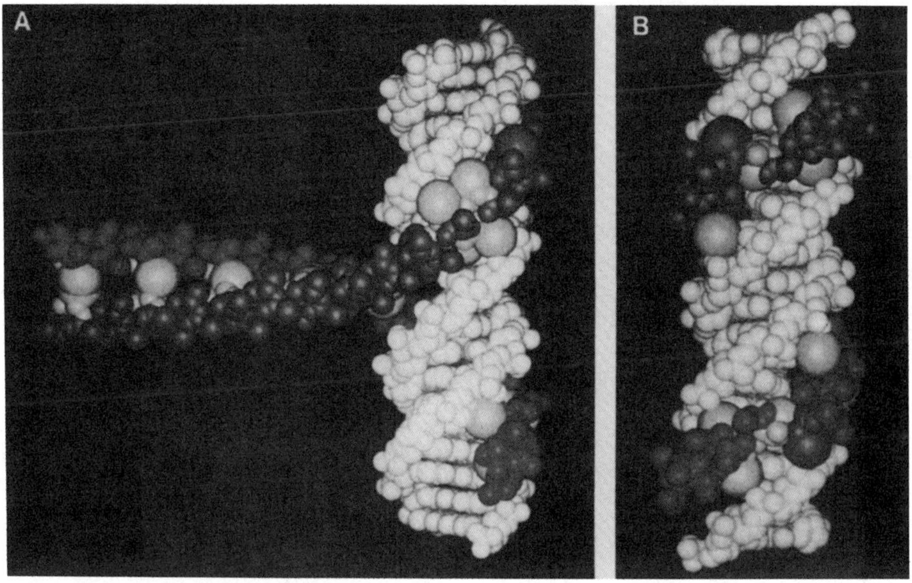

Figure 60 The "scissors grip" model of bZIP-DNA interactions proposed by Vinson et al. (179). The basic regions of the dimer are proposed to form bifurcating α-helices that enter the major groove of the symmetric DNA site. The model is named after the hold used in wrestling. DeGrado and co-workers have proposed a very similar model called the "induced helical fork" (180). See text for the minor differences between these schemes.

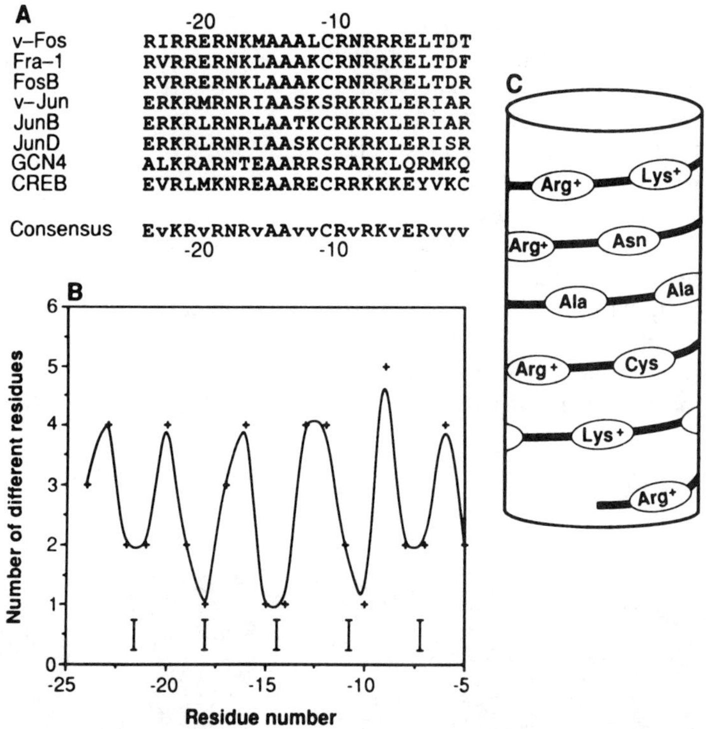

Figure 61 (A) Amino acid sequences from eight bZIP proteins that bind to similar DNA sequences. The negative numbering is relative to the first leucine of the repeat. The bottom row illustrates a consensus sequence of amino acids found in at least half of the proteins (capital letters). "v" denotes a variable position that is not well conserved. (B) The periodicity of the conserved amino acids is about 3.6 residues, the same as that of an α-helix, suggesting that the basic region is helical (C) and that the conserved residues sit on the face of the recognition helix that contacts the base pairs.

Grado and co-workers to propose that the four conserved residues actually contact the bases of the DNA site (180). The periodicity of the conserved residues further suggested that the basic region was is an α-helix, which is proposed to sit in the major groove of the binding site, consistent with methylation protection experiments (179,181,182). Therefore, if these models are correct (and we shall see that they probably are), the bZIP motif is yet another way to dock a recognition helix with residues in the major groove of a double helix.

The lack of conserved residues on the back side of the putative helix and the proposed extended nature of the entire bZIP structure led to the suggestion that the basic region might be disordered unless complexed with DNA. This facet of the model has been confirmed by CD and proteolysis experiments using various bZIP fragments. For example, Kim and co-workers showed that the molar ellipticity at 222 nm, indicative of an α-helix, increases dramatically when a disulfide-linked dimer of peptides corresponding to the basic region of GCN4 is mixed with an oligonucleotide containing the DNA binding site (178). O'Neil et al. performed a similar experiment with a larger synthetic dimer containing, in addition to the basic region of GCN4, an artificial leucine zipper designed to form an extremely stable coiled coil. Again, a large increase in the helicity of the basic region was observed upon DNA complexation (180). Finally, the DNA-binding domain of C/EBP becomes highly resistant to trypsin digestion when complexed to the C/EBP binding site. The protein fragment is readily degraded in the absence of DNA or in the presence of nonspecific DNA (183). When large amounts of trypsin were employed to cut the specific DNA-protein complex, cleavage was observed at the link between the zipper and the basic region.

These experiments clearly demonstrate that several aspects of the structural models for the bZIP domains are correct. The leucine zippers are coiled coils that act simply to mediate dimerization and the basic regions directly contact the DNA; these regions acquire an α-helical secondary structure in the complex. However, some of the more specific postulates of each model, in particular the nature of the protein-nucleotide contacts and the degree of bend in the basic region, require substantiation and probably await crystallographic characterization of a bZIP protein.

C. Homodimers Versus Heterodimers

Before we leave the leucine zipper motif, it is worth considering how leucine zippers tell one another apart. We know that this happens. For example, Fos and Jun form a heterodimer preferentially over homodimers. However, the current model does not suggest a mechanism for this discrimination. This may involve residues other than those of the 4-3 repeat or could involve the details of how the hydrophobic residues pack together. While some of these questions have begun to be addressed in natural proteins via mutagenesis, a complete answer will probably depend on careful equilibrium studies of synthetic leucine zipper peptides as well as a detailed crystallographic investigation.

VIII. THE EcoRI ENDONUCLEASE-DNA COMPLEX

Restriction endonucleases constitute another major class of sequence-specific DNA binding proteins. These useful proteins recognize palindromic sites of four, six, and sometimes eight base pairs, to which they bind as dimers. By far the best understood complex is that between the EcoRI endonuclease and its cognate site, 5′-GAATTC-3′. Rosenberg and co-workers have reported the x-ray crystal structure of a complex between EcoRI and an oligonucleotide containing the recognition site (184). Cleavage was prevented by excluding Mg^{2+}, a required cofactor, from the crystallization medium. However, it was found that soaking Mg^{2+} into the crystals did result in DNA cleavage, showing that the crystalline complex was catalytically competent. The original structural model could not be refined beyond an R factor of 0.25, and recent examination of isomorphorphous derivatives containing various halogenated oligonucleotides has resulted in a revised model (185). Some of the details of the protein-DNA contacts are not completely clear, but a number of important conclusions can be drawn from the presently available information.

The current backbone model is shown in Figure 62, along with the original model for comparison. The overall arrangement consists of a largely parallel, five-stranded β-sheet surrounded by α helices. The DNA recognition motif (Fig. 63A) is very different from that seen in the helix-turn-helix proteins and the structurally characterized zinc fingers. The crucial residues lie in the inner and outer "recognition helices," labeled $\alpha 4$ and $\alpha 5$ in Figure 62. The N-termini of these helices point into the major groove of each half-site. Each is part of a β-loop-α motif in which the loops have slightly different structures. Connecting $\beta 3$ and $\alpha 4$ is an extended chain spanning Met137 to Ala142. This chain runs through the major groove of each half-site and contains many of the residues important in DNA recognition (Fig. 63B), as evidenced by the fact that the region of the gene encoding these residues is a mutational hotspot. Details of the side-chain–base interactions derived from the old model are discussed in Ref. 185, but it is not clear that all of these will survive further analysis, so they will not be described here. However, mutagenesis studies have clearly revealed that Glu144 is crucial in sequence-specific recognition (186), as predicted by the crystal structure. Glu111, which is probably directly involved in strand cleavage (187), is near the C-terminus of $\beta 3$ and is situated very near the scissile bond, which engenders added confidence in the revised model.

A striking feature of this protein-DNA complex is the highly perturbed nature of the DNA (Fig. 64) (184). The protein drastically unwinds the center of the site by approximately 25°, an arrangement called a type I neokink. The principal effect of this perturbation is to widen the major groove, which seems to be required for protein binding. A pair of symmetrically related, somewhat less severe kinks have also been noted. Thus, the EcorRI-DNA complex represents the best example to date of the impor-

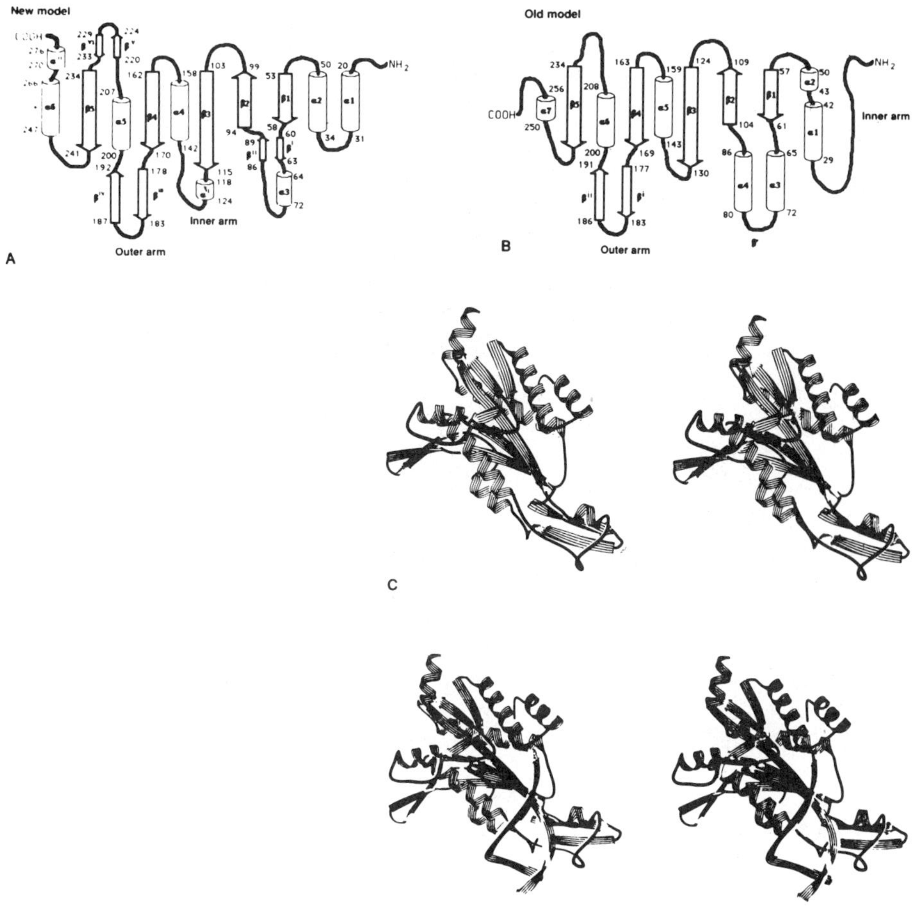

Figure 62 Original (A) and revised (B) structural models for the EcoR1 endonuclease in the DNA-protein cocrystal. (C) stereoview of the protein in the revised model. (D) Stereoview of the protein-DNA complex. The DNA strands are drawn as arrows. (Adapted from Ref. 185.)

tance of DNA deformation in protein-nucleic acid recognition and provides a strikingly different motif for DNA binding.

IX. THE Arc/Mnt/Met REPRESSOR FAMILY

The most recent addition to the ranks of structurally characterized sequence-specific DNA binding proteins is the Arc repressor of the *Salmo-*

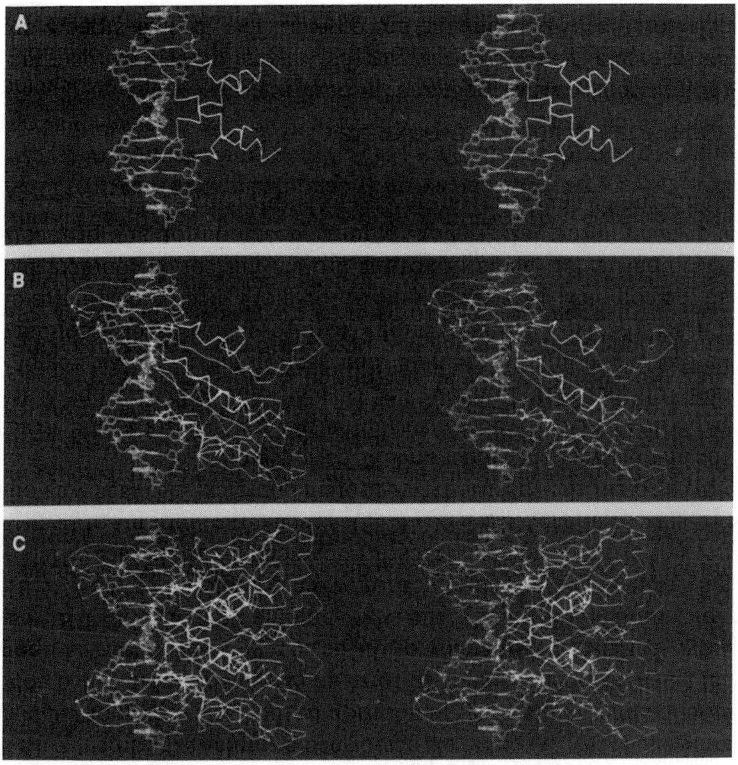

Figure 63 Various partial structures highlighting important parts of the EcoR1-DNA complex. (A) View perpendicular to the double helix showing the recognition helices and the arms that wrap around the DNA. (B) Cα atoms of one monomer and the DNA. (C) Both subunits of the dimer and the DNA.

nella bacteriophage P22 (which like the lambda repressor, is also involved in controlling the switch from lysogeny to lysis) and its relatives, the Mnt and Met repressors. Sauer and co-workers have conducted extensive biochemical studies of the binding of the Arc and Mnt repressors to their cognate operators (188–193); the Mnt operator is shown below. The site is nearly symmetrical about the central G-C base pair (underlined). The Arc protein binding site (not shown) is

1 2 3 4 5 6 7 8 9 10 11 12 13 14 15 16 17 18 19 20 21
5′-**A T A G G T C C A C G G T G G A C C T G T**-3′
3′-**T A T C C A G G T G C C A C C T G G A C A**-5′

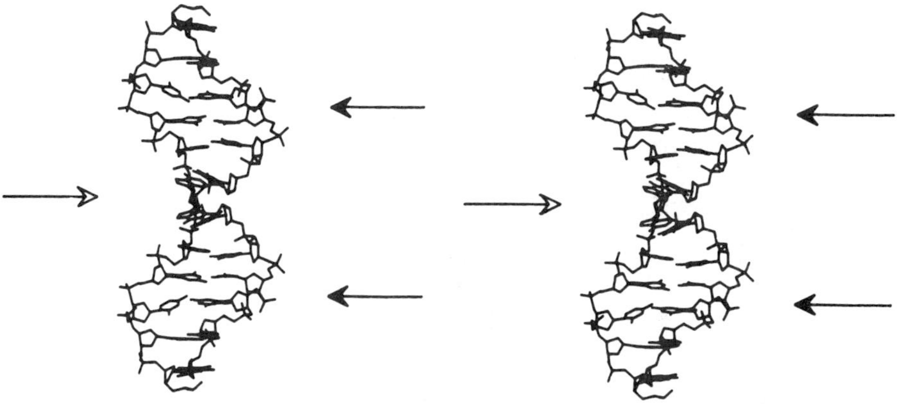

Figure 64 Stereoview of the DNA structure in the cocrystal. The arrows indicate the sites of severe kinks in the DNA.

also a symmetrical sequence. The proteins bind to their sites as tetramers, with a dimer binding to the major groove of each half-site. A summary of the biochemically detected contacts for the Mnt-operator complex is shown in Figure 65. From the many experiments designed to localize the DNA binding domains of the Arc and Mnt proteins, one of the most striking results was that a hybrid protein in which the N-terminal six amino acids of the Mnt repressor were replaced with nine N-terminal residues of the Arc protein recognized the Arc operator but not the Mnt binding site. Furthermore, extensive mutagenesis studies have shown that Arg2, His6, Asn8, and Arg10 of Mnt repressor are critical for high-affinity binding. Together, these results demonstrate that a small region at the N-termini of these related proteins define the sequence-specific DNA binding domain. In addition, a "specificity mutant" was discovered; a Mnt His6Pro mutant recognizes a mutant operator with symmetric substitutions at base pairs 5 and 17, presumably defining His6-bp5 and, by symmetry, His6-bp17 as one sequence-specific contact.

The first successful structural analysis of this class of proteins was reported by Rafferty et al., who solved the crystal structure of the *E. coli* Met repressor (194), which controls genes encoding enzymes involved in methionine and SAM biosynthesis. The Arc, Mnt, and Met repressors share extensive homology (Fig. 66). The structure shows highly intertwined monomers composed of three α-helices (residues 30–45, 52–66, and 86–94) and a β-strand (residues 20–29) (Fig. 67). Most of the monomer-monomer contacts occur along the symmetrically related α2s and the β-strands, which

A

1 3 5 7 9 11 13 15 17 19 21

A T A G G T C C A C G G T G G A C C T G T

T A T C C A G G T G C C A C C T G G A C A

B

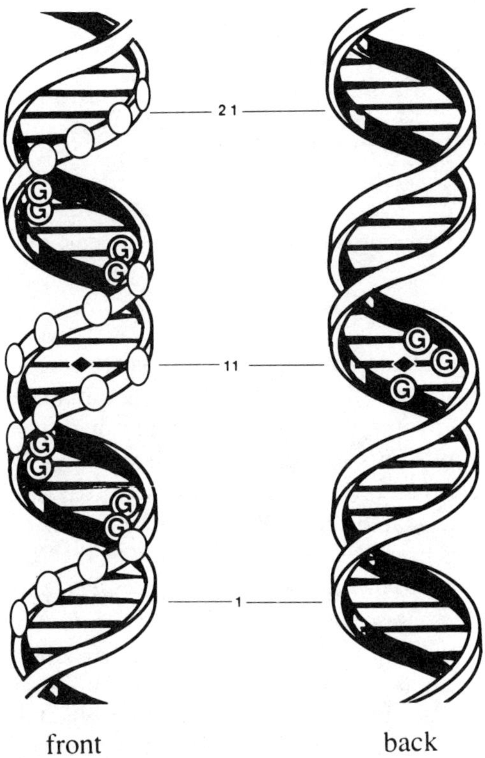

Figure 65 Summary of the protein-DNA contacts between the Mnt repressor and its cognate operator. Open circles indicate sites of protection from hydroxyl radicals and circled Gs represent guanines protected from methylation. (Adapted from Ref. 188.)

loop over one another in an antiparallel fashion. The Met repressor does not contain a helix-turn-helix, leucine zipper or zinc finger unit.

More recently, the NMR-derived solution structure of the Arc protein has been reported (195–197). In the area of sequence homology with the Met repressor, the structures are very similar. The N-terminus of the protein includes the β-strand (residues 9–13) and a disordered region (residues

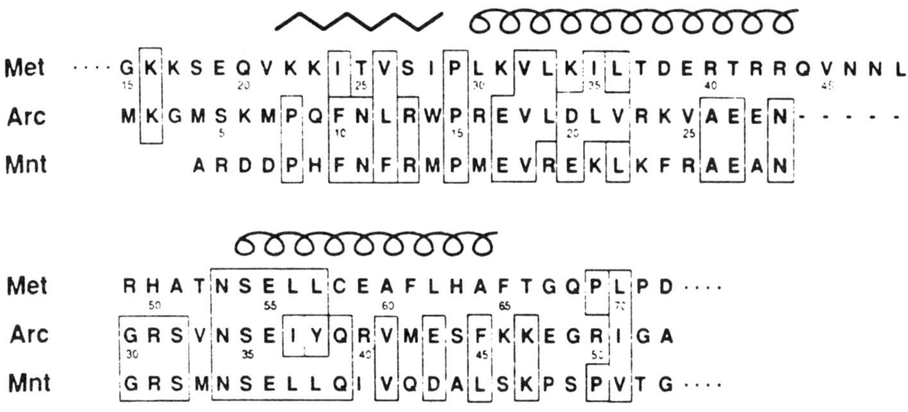

Figure 66 Amino acid sequences of the Met, Arc, and Mnt repressors.

1–8). Given the extensive biochemical data available for the Mnt repressor-operator interaction, a model for the DNA protein complex can be proposed (197) (Fig. 68). In it, the antiparallel β-strands of the dimer sit in the major groove of the operator half-site and the N-terminal residues, which are known to be crucial for determining binding specificity, extend away from the strands in different directions. This model accounts for all of the footprinting, phosphate ethylation interference, and mutagenesis data. If the model is correct, which seems likely, the Mnt repressor family of DNA-binding proteins (198) represents a completely new motif for DNA recognition which employs β-strands to dock with the major groove.

X. SUMMARY AND CONCLUSIONS

The preceding discussion has touched on the biochemical and structural properties of some of the better understood examples of sequence-specific protein-DNA recognition. Surely there are many other sequence-specific binding motifs yet to be discovered. Indeed, we already know of the existence of DNA binding domains that are crucial for sequence-specific recognition, but whose amino acid sequences do not appear to match those of members of the known structural families (199). Thus, a great deal of work remains to be done and it is a safe bet that new and surprising findings will be the rule rather than the exception in this young and growing field for some time to come.

Nevertheless, even with the limited data in hand, it is possible to come to a few tentative conclusions. The first, unfortunately, is that the details of protein-DNA recognition are imposingly complex. While a limited num-

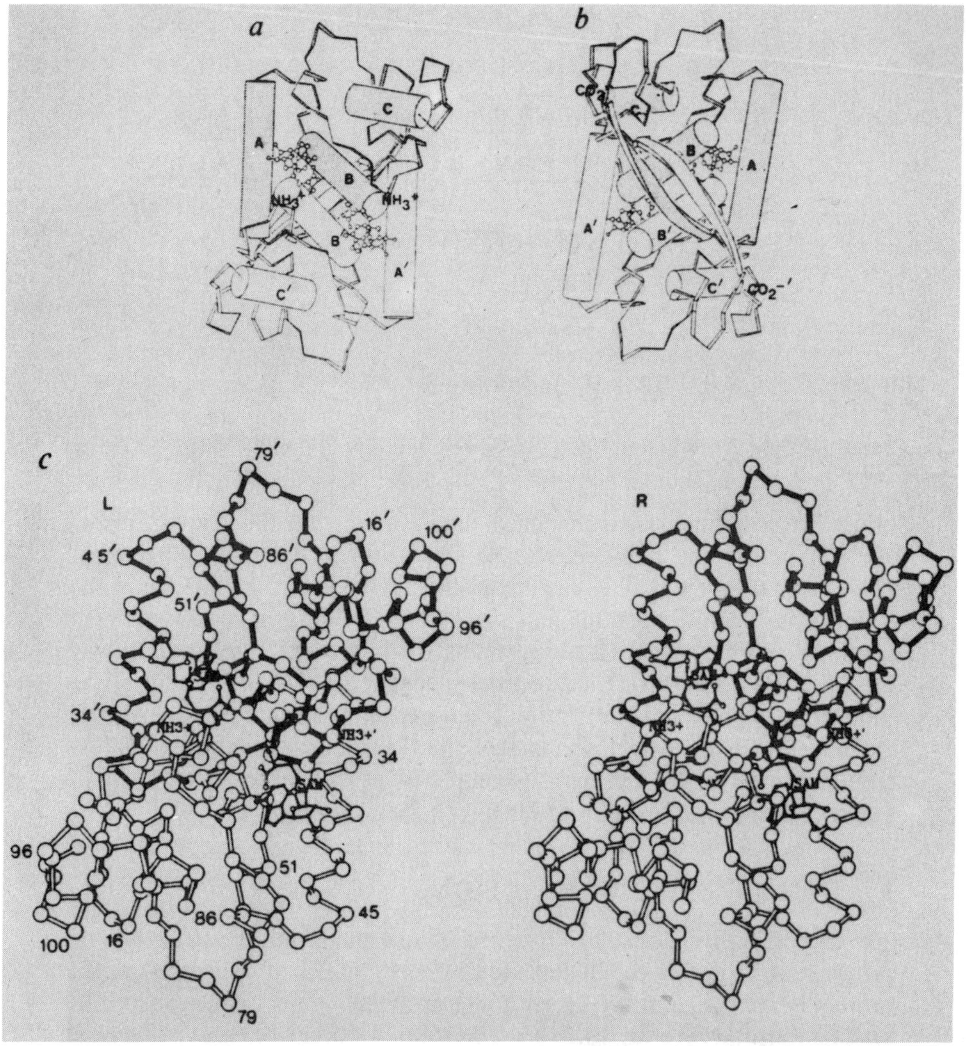

Figure 67 X-ray crystal structure model of SAM-bound Met repressor. (Adapted from Ref. 194.)

ber of amino acid residues are often important in direct readout of the nucleotide sequence, the three-dimensional positioning of these residues is crucially dependent on a vast array of interacting "nonspecific" interactions, many of which involve the DNA backbone. Often, in order for all of the potential contacts to be realized, the DNA helix must deform, a phenomenon known as indirect readout. The degree and type of deviation

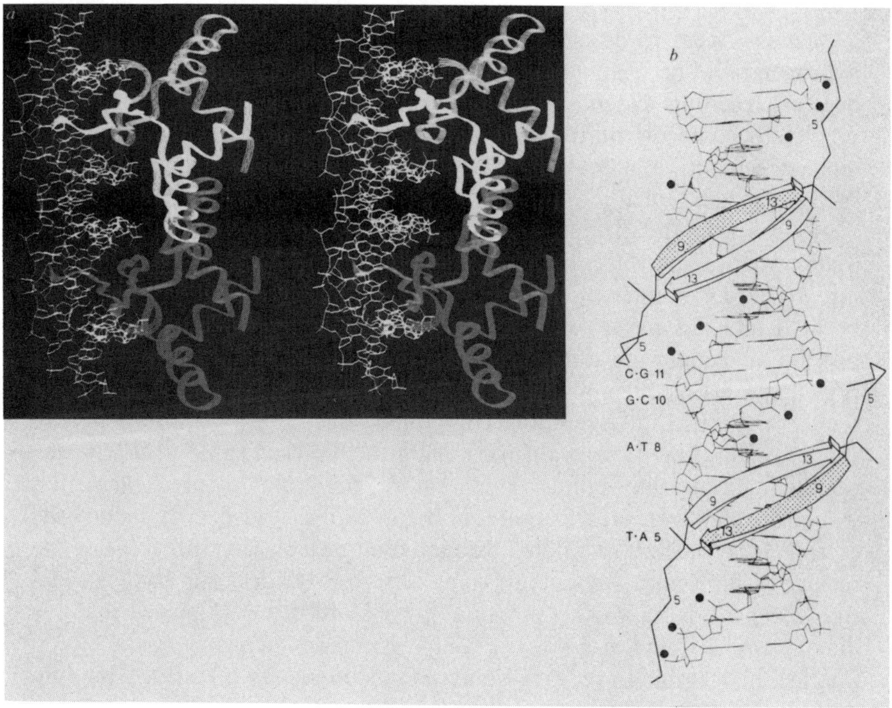

(a) (b)

Figure 68 Stereodiagram (a) and model (b) of the Arc repressor-operator complex derived from NMR studies. (Adapted from Ref. 197.)

from idealized B-DNA varies widely from complex to complex, even in the case of the 434 R(1-69) and Cro operator complexes, which include the same DNA sequences.

A second conclusion is that the most important interactions occur in the major groove of DNA, for the simple reason that it is large enough to accommodate various elements of protein secondary structure, most importantly the α-helix. While the homeodomain-DNA structures have provided the first direct evidence for sequence readout from minor groove contacts, these appear to primarily supplement major groove contacts. Invariably, groups on the edges of the bases in the major groove are recognized by protein side chains though either hydrogen bonds or van der Waals contacts. Though our current knowledge of the quantitative contributions of individual contacts in various complexes is quite thin, these points will undoubtedly begin to be assessed through the detailed study of appropriate mutants. While an impressive number of different types of base-specific hydrogen bonding schemes have been observed in various complexes, at least a few seem to be employed fairly often, for example the bidentate hydrogen bond between a Gln side chain and an adenine base. As our data base of structures expands in the coming years, it may become possible to predict to some extent the hydrogen bonding patterns important in sequence recognition, at least for new members of a well-characterized motif such as the helix-turn-helix.

Returning to the theme mentioned in the introduction, it is appropriate to ask what the above information tells us about the feasibility of mimicking protein-DNA interactions with synthetic compounds. Clearly, this will be difficult. The major hurdle lies in the apparent necessity to precisely position a recognition domain with respect to the binding site by means of a large array of supplementary contacts. For a completely synthetic sequence-specific DNA binding compound, this would require a precise spatial orientation of several functional groups poised to interact with the double helix at many points. Furthermore, the "rule" of preorganization would suggest that such a molecule be as stiff as possible to reduce the entropy cost inherent in binding through multiple contact points. While the science of molecular recognition is making great strides, our ability to design compounds in which functional groups are precisely placed in three dimensions is crude at best. Certainly, synthetic chemistry is not up to the task of matching the three-dimensional complexity of a DNA binding domain. Of course, this analysis can be turned on its head and be viewed as an extremely exciting challenge for the next generation of synthetic organic chemists. In the near future, the best hope appears to lie in "seminatural" compounds which include peptides corresponding to pieces of known DNA binding domains. Indeed, some progress has been made in designing sequence-specific DNA binding peptides, particularly of the bZIP variety (178,180). A reasonable next step might be to develop a set of synthetic building blocks able to orient the DNA binding peptides in specific ways. Once these are available, work can begin on replacing peptides with potential nonbiological recognition elements, eventually, perhaps, leading to the development of sequence-specific binding compounds composed completely of nonbiological components. Such species would constitute a completely new family of potential drugs able to influence gene expression and other important aspects of DNA metabolism.

REFERENCES

1. Alberts, B.M., *Cold Spring Harbor Symp. Quant. Biol.* 49:1–12 (1984).
2. Kornberg, A., *DNA Replication*, W.H. Freeman, San Francisco (1979).
3. Bramhill, D., and A. Kornberg, *Cell 52*:743–755 (1988).

4. Tsurimoto, T., and K. Matsubara, *Mol. Gen. Genet. 181*, 325–331 (1981).
5. Funnell, B.E., T.A. Baker, and A. Kornberg, *J. Biol. Chem. 262*:10327–10334 (1987).
6. Kornberg, A., T.A. Baker, L.L.Bertsch, D. Bramhill, B.E. Funnell, A.S. Lasken, H. Maki, S. Maki, K. Sekinizu, and E. Wahle. In *DNA Replication and Recombination*, Kelly, T.J. and McMacken, R. eds., Alan R. Liss, New York (1987).
7. Wickner, S., *Cold Spring Harbor Symp. Quant. Biol. 43*:303–310 (1979).
8. McMacken, R., L. Silver, and C. Georgopoulos. In *Eschericia coli and Salmonella Typhimurium: Cellular and Molecular Biology*, Ingraham, J.L., Low, K.B., Magasanik, B., Schaechter, M., and Umbarger, H.E. eds., American Society for Microbiology, Washington, DC, 564–612 (1987).
9. LeBowitz, J. H., and R. McMacken, *J. Biol. Chem. 261*:4738–4748 (1986).
10. Baker, T.A., K. Sekimizu, B.E. Funnell, and A. Kornberg, *Cell 45*:53–64 (1986).
11. O'Donnell, M., and P.S. Studwell, *J. Biol. Chem. 265*:1179–1187 (1990) and references therein.
12. Bouché, J.-P., L. Rowen, and A. Kornberg, *J. Biol. Chem. 235*:765–769 (1979).
13. Okazaki, T., Y. Kurosawa, T. Ogawa, T. Seki, S. Shinozaki, S. Hirose, A. Fujiyama, Y. Kohara, Y. Machida, F. Tamanoi, and T. Hozumi, *Cold Spring Harbor Symp. Quant. Biol. 43*:203–219 (1978).
14. Richardson, C.C., B.B. Beauchamp, H.E. Huber, R.A. Ikeda, J.A. Myers, H. Nakai, S.D. Rabkin, S. Tabor, and J. White, *UCLA Symp. Mol. Cell. Biol. 47*:151–171 (1987).
15. Liu, C.C., and B.M. Alberts, *J. Biol. Chem. 256*:2821–2829 (1981).
16. Rowen, L., and A. Kornberg, *J. Biol. Chem. 253*:770–774 (1978).
17. McHenry, C.S., *J. Biol. Chem. 257*:2657–2663 (1982).
18. Loeb, L.A., and T.A. Kunkel, *Ann. Rev. Biochem. 52*:429–457 (1982).
19. Lu, A.-L., S. Clark, and P. Modrich, *Proc. Nat. Acad. Sci. USA 80*:4639–4643 (1983).
20. Kuempel, P., S. Duerr, and N.R. Seely, *Proc. Nat. Acad. Sci. USA 74*:3927–3931 (1977).
21. Khatri, G.S., T. McAllister, P.R. Sista, and D. Bastia, *Cell 59*:667–674 (1989).
22. Sudin, O., and A. Varshafsky, *Cell 25*:659–669 (1981).
23. Wang, J.C., *Ann. Rev. Biochem. 54*:665–697 (1985).
24. Stahl, H., P. Dröge, and R. Knippers, *EMBO J. 5*:1939–1944 (1986).
25. Hwang, D.S., E. Crooke, and A. Kornberg, *J. Biol. Chem. 265*:19244–19248 (1990).
26. Burgess, R.R., A.A. Travers, J.J. Dunn, and E.K. Bautz, *Nature 221*:43–46 (1969).
27. Chamberlin, M. In *The Enzymes* Boyer, P.D. ed., Vol. 15:61–108, Academic Press, New York (1982).
28. Lewin, B., *Genes III*, 3rd ed., Wiley, New York (1987).
29. Dahmus, M.E., *J. Biol. Chem. 258*:3956–3960 (1983).

30. Conaway, J.W., and R.C. Conaway, *Science 248*:1550–1553 (1990).
31. Dickson, R.S., J. Abelson, W.M. Barnes, and W.S. Reznikoff, *Science 187*: 27–35 (1975).
32. Morgan, W.D., D.G. Bear, and P.H. von Hippel, *J. Biol. Chem. 258*:9553-9564 (1983).
33. Platt, T., *Ann. Rev. Biochem. 55*:339-372 (1986).
34. Farnham, P.J., and T. Platt, *Cell 20*:739-748 (1980).
35. Hawley, D.K., and W.R. McClure, *Nuc. Acids Res. 11*:2237–2255 (1983).
36. Nakajima, N., M. Horikoshi, and R.G. Roeder, *Mol. Cell. Biol. 8*:4028-4040 (1988).
37. Hahn, S., S. Buratowski, P.A. Sharp, and L. Guarante, *Cell 58*:1173–1181 (1989).
38. Sumimoto, H., Y. Ohkuma, T. Yamamoto, M. Horikoshi, and R.G. Roeder, *Proc. Nat. Acad. Sci. USA 87*:9158-9162 (1990).
39. Ohkuma, Y., H. Sumimoto, M. Horikoshi, and R.G. Roeder, *Proc. Nat. Acad. Sci. USA 87*:9163-9167 (1990).
40. Burton, Z., M. Killeen, M. Sopta, L.G. Ortolan, and J. Greenblatt, *Mol. Cell. Biol. 8*:1602–1613 (1988).
41. Ptashne, M., *A Genetic Switch: Gene Control and Phage λ*, Cell Press, Cambridge, MA (1986).
42. For a comprehensive review of lambda biology, see *Lambda II*, Hendrix, R.W., Robert, J.W., Stahl, F.W., and Weisberg, R. eds. Cold Spring Harbor, New York (1983).
43. Walker, G.C., *The SOS response of* Escherichia coli. In *Eschericia coli and Salmonella Typhimurium: Cellular and Molecular Biology*, Ingraham, J.L., Low, K.B., Magasanik, B., Schaechter, M., and Umbarger, H.E. eds., 1346-1357, American Society for MIcrobiology, Washington DC (1987).
44. Roberts, J.W., C.W. Roberts, and D.W. Mount, *Proc. Nat. Acad. Sci. USA 74*:2283–2287 (1977).
45. Sauer, R.T., M.J. Moss, and M. Ptashne, *J. Biol. Chem. 257*:4458–4462 (1982).
46. Cox, M.M., and I.R. Lehman, *Ann. Rev. Bochem. 56*:226–262 (1988).
47. Johnston, M., *Microb. Rev. 51*:458–476 (1987).
48. Bram, R.J., and R. Kornberg, *Proc. Nat. Acad. Sci. USA 82*:43–47 (1985).
49. Wang, J.C., and G.N. Giaever, *Science 240*:300–304 (1988).
50. Giniger, E., and M. Ptashne, *Proc. Nat. Acad. Sci. USA 85*:382–386 (1988).
51. Keegan, L., G. Gill, and M. Ptashne, *Science 231*:699–704 (1986).
52. Ptashne, M., *Nature 335*:683–689 (1988).
53. Craigie, R., and K. Mizuuchi, *Cell 45*:795–800 (1986).
54. Su, W., S. Porter, S. Kutsu, and H. Echols, *Proc. Nat. Acad. Sci. USA 87*: 5504–5508 (1990).
55. Horikoshi, M., M.F. Carey, H. Kakidani, and R.G. Roeder, *Cell 54*:665-669 (1988).
56. Lue, N.F., D.I. Chasman, A.R. Buchman, and R.D. Kornberg, *Mol. Cell. Biol. 7*:3446–3451 (1987).
57. Parthun, M.R., and J.A. Jaehning, *J. Biol. Chem. 265*:209–213 (1989).

58. Chasman, D.I., and R.D. Kornberg, *Mol. Cell. Biol. 10*:2916–2923 (1990).
59. Yun, S.-J., Y. Hiraoka, M. Nishizawa, K. Takio, K. Titani, Y. Nogi, and T. Fukasawa, *J. Biol. Chem. 266*:693–697 (1991).
60. Carey, M., H. Kakidani, J. Leatherwood, F. Mostashari, and M. Ptashne, *J. Mol. Biol. 209*:423–432 (1989).
61. Ma, J., and M. Ptashne, *Cell 50*:137–142 (1987).
62. Johnston, S.A., J.M. Salmeron, and S.S. Dincher, *Cell 50*:143–146 (1987).
63. Johnston, S.A., M.J. Zavortink, C. Debouck, and J.E. Hopper, *Proc. Nat. Acad. Sci. USA 83*:6553–6557 (1986).
64. Ma, J., and M. Ptashne, *Cell 48*:847–853 (1987).
65. Dunn, T., S. Hahn, S. Ogden, and R. Schlief, *Proc. Nat. Acad. Sci. USA 81*:5017–5020 (1984).
66. Dandanell, G., and K. Herman, *EMBO J. 4*:3333–3338 (1985).
67. Friedberg, E.C., *DNA Repair*, W.H. Freeman, New York (1984).
68. Weinstock, G.M. In *Eschericia Coli and Salmonella Typhimurium: Cellular and Molecular Biology*, Ingraham, J.L., Low, K.B., Magasanik, B., Schaechter, M., and Umbarger, H.E. eds., 1034–1043, American Society for Microbiology, Washington, DC (1987).
69. Oettinger, M.A., D.G. Schatz, C. Gorka, and D. Baltimore, *Science 248*: 1517–1523 (1990).
70. Craig, N.L., and N. Kleckner. In *Eschericia Coli and Salmonella Typhimurium: Cellular and Molecular Biology*, Ingraham, J.L., Low, K.B., Magasanik, B., Schaechter, M., and Umbarger, H.E. eds., 1054–1070, American Society for Microbiology, Washington, DC (1987).
71. Craig, N.L., and H.A. Nash, *Cell 35*:795–803 (1983).
72. Craig, N.L., and H.A. Nash, *Cell 39*:707–716 (1984).
73. Ross, W., and A. Landy, *Cell 33*:261–272 (1983).
74. Sanderson, M.R., P.S. Freemeont, P.A. Rice, A. Goldman, G.F. Hatful, N.D.F. Grindley, and T.A. Steitz, *Cell 63*:1323–1329 (1990).
75. Craigie, R., and K. Mizuuchi, *Cell 41*:867–876 (1985).
76. Seeman, N.C., J.M. Rosenberg, and A. Rich, *Proc. Nat. Acad. Sci. USA 73*:804–808 (1976).
77. Prigodich, R.V., Y. Shamoo, K.R. Williams, J.W. Chase, W.H. Konigsberg, and J.E. Coleman, *Biochemistry 25*:3666–3672 (1986).
78. Drew, H.R., R.M. Wing, T. Takano, C. Broka, S. Tanaka, K. Itakura, and R.E. Dickerson, *Proc. Nat. Acad. Sci. USA 78*:2179–2183 (1981).
79. Dickerson, R.E., and H.R. Drew, *J. Mol. Biol. 149*:761–786 (1981).
80. Otinowski, Z., R.W. Schevitz, R.-G. Zhang, C.L. Lawson, A. Joachimiak, R.Q. Marmorstein, B.F. Luisi, and P.B. Sigler, *Nature 335*:321–329 (1988).
81. However, for a hypothesis that a code may exist for the helix-turn-helix proteins, see K. Adler, K. Beyreuther, E. Fanning, N. Geisler, B. Gronenborn, A. Klemm, B. Muller-Hill, M. Pfahl, and A. Schmitz, *Nature 237*: 322–327 (1972).
82. Anderson, W.F., D.H. Ohlendorf, Y. Takeda, and B.W. Matthews, *Nature 290*:754–758 (1981).
83. McKay, D.B., and T.A. Steitz, *Nature 290*:744–749 (1981).

84. Dodd, I.B., and J.B. Egan, *Nuc. Acids Res. 18*:5019–5026 (1990).
85. Drummond, M., P. Whitty, and J. Wootton, *EMBO J. 5*:441–447 (1986).
86. Anderson, J.E., M. Ptashne, and S.C. Harrison, *Nature 326*:846–852 (1987).
87. Aggarwal, A.K., D.W. Rodgers, M. Drottar, M. Ptashne, and S.C. Harrison, *Science 242*:899–907 (1988).
88. Mondragon, A., S. Subbiah, S.S.C. Almo, M. Drottar, and S.C. Harrison, *J. Mol. Biol. 205*:189–200 (1989).
89. Pabo, C.O., and R.T. Sauer, *Ann. Rev. Biochem. 53*:293–321 (1984).
90. Bushman, F.D., J.E. Anderson, S.C. Harrison, and M. Ptashne, *Nature 316*:651–653 (1985).
91. Koudelka, G.B., S.C. Harrison, and M. Ptashne, *Nature 326*:886–888 (1987).
92. Wharton, R.P., and M. Ptashne *Nature 326*:888–891 (1987).
93. Wharton, R.P., and M. Ptashne *Nature 316*:601–605 (1985).
94. Wharton, R.P., E.L. Brown, and M. Ptashne, *Cell 38*:361–369 (1984).
95. Sauer, R.T., R.R. Yocum, R.F. Doolittle, M. Lewis, and C.O. Pabo, *Nature 298*:447–451 (1982).
96. Wolberger, C., Y. Dong, M. Ptashne, and S.C. Harrison, *Nature 335*:789–795 (1988).
97. Koudelka, G.B., P. Harbory, S.C. Harrison, and M. Ptashne, *Proc. Nat. Acad. Sci. USA 85*:4633–4637 (1988).
98. Pabo, C.O., and M. Lewis, *Nature 298*:443–447 (1982).
99. Jordan, S.R., and C.O. Pabo, *Science 242*:893–899 (1988).
100. Pabo, C.O., W. Krovatin, A. Jeffrey, and R.T. Sauer, *Nature 298*:441–443 (1982).
101. Pabo, C.O., A.K. Aggarwal, S.R. Jordan, L.J. Beamer, U.R. Obeysekare, and S.C. Harrison, *Science 247*:1210–1213 (1990).
102. Schewitz, R.W., Z. Otwinowski, A. Joachimiak, C.L. Lawson, and P.B. Sigler, *Nature 317*:782–786 (1985).
103. Joachimiak, A., R.W. Schevitz, R.L. Kelly, C. Yanofsky, and P.B. Sigler, *J. Biol. Chem. 258*:12641–12643 (1983).
104. Zhang, R.-g., A. Joachimiak, C.L. Lawson, R.W. Schevitz, Z. Otwinowski, and P.B. Sigler, *Nature 327*:591–597 (1987).
105. Yanofsky, C., and I.P. Crawford, In *Eschericia Coli and Salmonella Typhimurium: Cellular and Molecular Biology*, Ingraham, J.L., Low, K.B., Magasanik, B., Schaechter, M., and Umbarger, H.E. eds., 1453–1473, American Society for Microbiology, Washington, DC (1987).
106. Bennett, G.N., M.E. Schweingruber, K.D. Brown, C. Squires, and C. Yanofsky, *Proc. Nat. Acad. Sci. USA 73*:2351–2355 (1976).
107. Staacke, D., B. Walter, B. Kisters-Woike, B. v. Wilcken-Bergmann, and B. Müller-Hill, *EMBO J. 9*:1963–1967 (1990).
108. Beckwith, J. The lactose operon. In *Eschericia Coli and Salmonella Typhimurium: Cellular and Molecular Biology*, Ingraham, J.L., Low, K.B., Magasanik, B., Schaechter, M., and Umbarger, H.E. eds., 1444–1452, American Society for Microbiology, Washington, DC (1987).
109. Boelens, R., R.M. Scheek, J.H. vanBoom, and R. Kaptein, *J. Mol. Biol. 193*:213–216 (1987).

110. Lehming, N., J. Sartorius, S. Oehler, B. von Wilcken-Bergmann, and B. Müller-Hill, *Proc. Nat. Acad. Sci. USA 85*:7947–7951 (1988).
111. Lehming, N., J. Sartorius, B. Kirsters-Woicke, B. von Wilcken-Bergmann, and B. Müller-Hill, *EMBO J. 9*:615–621 (1990).
112. McGinnis, W., M. Levine, E. Hafen, A. Kuroiwa, and W.J. Gehring, *Nature 308*:428–433 (1984).
113. McGinnis, W., R.L. Garber, J. Wirz, A. Kuroiwa, and W.J. Gehring, *Cell 37*:403–408 (1984).
114. Scott, M.P., and A.J. Weiner, *Proc. Nat. Acad. Sci. USA 81*:4115–4119 (1984).
115. Gehring, W.J., *Science 236*:1245–1252 (1987).
116. Laughon, A., and M.P. Scott, *Nature 310*:25–31 (1984).
117. Scott, M.P., J.W. Tamkun, and G.W. Hartzell, *Biochem. Biophys. Acta 989*:25–48 (1989).
118. Otting, G., Y.-q. Qian, M. Müller, M. Affolter, W. Gehring, and K. Wüthrich, *EMBO J. 7*:4305–4309 (1988).
119. Qian, Y.-q., M. Billeter, G. Otting, M. Müller, W.J. Gehring, and K. Wüthrich, *Cell 59*:573–580 (1989).
120. Otting, G., Y.-q. Qian, M. Billeter, M. Müller, M. Affolter, W.J. Gehring, and K. Wüthrich, *EMBO J. 9*:3085–3092 (1990).
121. Kissinger, C.R., B. Liu, E. Martin-Blanco, T.B. Kornberg, and C.O. Pabo, *Cell 63*:579–590 (1990).
122. Triesman, J., P. Gönczy, M. Vashishtha, E. Harris, and C. Desplan, *Cell 59*:553–562 (1989).
123. Hanes, S.D., and R. Brent, *Cell 57*:1275–1283 (1989).
124. Hayashi, S., and M.P. Scott, *Cell 63*:883–894 (1990).
125. Ginsberg, A.M., B.O. King, and R.G. Roeder, *Cell 39*:479–489 (1984).
126. Miller, J., A.D. McLachlan, and A. Klug, *EMBO J. 4*:1609–1614 (1985).
127. Diakun, G.P., L. Fairall, and A. Klug, *Nature 324*:698–699 (1986).
128. Frankel, A.D., J.M. Berg, and C.O. Pabo, *Proc. Nat. Acad. Sci. USA 84*: 4841–4845 (1987).
129. Berg, J.M., *Ann. Rev. Biophys. Biophys. Chem. 19*:405–421 (1990).
130. Berg, J.M., *Prog. Inorg. Chem. 37*:143–185 (1989).
131. Berg, J.M., *J. Biol. Chem. 265*:6513–6516 (1990).
132. Párraga, G., S.J. Horvath, A. Eisen, W.E. Taylor, L. Hood, E.T. Young, and R.E. Klevit, *Science 241*:1489–1492 (1988).
133. Berg, J.M., *Proc. Nat. Acad. Sci. USA 85*:99–102 (1988).
134. Lee, M.S., G.P. Gippert, V. Soman, D.A. Case, and P.E. Wright, *Science 245*:635–637 (1989).
135. Altaba, A.R., H. Perry-O'Keefe, and D.A. Melton, *EMBO J. 6*:3065–3070 (1987).
136. Baldarelli, R.M., P.A. Mahoney, F. Salas, E. Gustavson, P.D. Boyer, M.-F. Chang, M. Roark, and J.A. Lengyel, *Devel. Biol. 125*:85–95 (1988).
137. Boegenhagen, D.F., W.M. Wormington, and D.D. Brown, *Cell 28*:413–421 (1982).
138. Fairall, L., D. Rhodes, and A. Klug, *J. Mol. Biol. 192*:577–591 (1986).

139. Gibson, T.J., J.P.M. Postma, R.S. Brown, and P. Argos, *Prot. Eng. 2*:209–218 (1988).
140. Rhodes, D., and A. Klug, *Cell 46*:123–132 (1986).
141. McCall, M., T. Brown, W.N. Hunter, and O. Kennard, *Nature 322*:661–664 (1986).
142. Aboul-ela, F., G. Varani, G.T. Walker, and I. Tinoco, *Nuc. Acids Res. 16*: 3559–3572 (1988).
143. Gottesfeld, J.M., J. Blanco, and L.L. Tennant, *Nature 329*:460–462 (1987).
144. Fairall, L., S. Martin, and D. Rhodes, *EMBO J. 8*:1809–1817 (1989).
145. Varna, K.E., M.E. Churchill, T.D. Tullius, and D.D. Brown, *Mol. Cell. Biol. 8*:1684–1696 (1988).
146. Tullius, T.D., and B.A. Dombrowski, *Proc. Nat. Acad. Sci. USA 83*:5469–5473 (1986).
147. Sakonju, S., and D.D. Brown, *Cell 31*:395–405 (1982).
148. Berg, J.M., *Science 232*:485–486 (1986).
149. Yamamoto, K.R., *Ann. Rev. Genet. 19*:209–252 (1985).
150. Kumar, V., and P. Chambon, *Cell 55*:145–146 (1988).
151. Freedman, L.P., B.F. Luisi, Z.R. Korszun, R. Basavappa, P.B. Sigler, and K.R. Yamamoto, *Nature 334*:543–546 (1988).
152. Zilliacus, J., K. Dahlman-Wright, J. Carlstedt-Duke, and J.A. Gustaffson, Unpublished results, cited in ref. 154 (1990).
153. Härd, T., E. Kellenbach, R. Boehlens, R. Kaptein, K. Dahlman, J. Carlstedt-Duke, L.P. Freedman, B. Maler, E.I. Hyde, J.-A. Gustafsson, and K.R. Yamamoto, *Biochemistry 29*:9015–9023 (1990).
154. Härd, T., E. Kellenbach, R. Boehlens, B.A. Maler, K. Dahlman, L.P. Freedman, J. Carlstedt-Duke, K.R. Yamamoto, J.-A. Gustafsson, and R. Kaptien, *Science 249*:157–160 (1990).
155. Schwabe, J.W.R., D. Neuhaus, and D. Rhodes, *Nature 348*:458–461 (1990).
156. Mader, S., V. Kumar, H. de Verneuil, and P. Chambon, *Nature 338*:271–274 (1989).
157. Umesono, K., and R.M. Evans, *Cell 57*:1139–1146 (1989).
158. Danielson, M., L. Hinck, and G.M. Ringold, *Cell 57*:1131–1138 (1989).
159. Johnston, M., *Nature 328*:353–355 (1987).
160. Johnston, M., and J. Dover, *Proc. Nat. Acad. Sci. USA 84*:2401–2405 (1987).
161. Johnston, M., and J. Dover, *Genetics 120*:63–74 (1988).
162. Hansen, A., M. VanHoy, and T. Kodadek, *Biochem. Biophys. Res. Comm. 175*:492–499 (1991).
163. VanHoy, M., and T. Kodadek, *Biochemistry*, in preparation (1991).
164. Pan, T., and J.E. Coleman, *Proc. Nat. Acad. Sci. USA 86*:3145–3149 (1989).
165. Pan, T., and J.E. Coleman, *Proc. Nat. Acad. Sci. USA 87*:2077–2081 (1990).
166. Pan, T., and J.E. Coleman, *Biochemistry 29*:3023–3029 (1990).
167. Corton, J.C., and S.A. Johnston, *Nature 340*:724–727 (1989).
168. Summers, M.F., T.L. South, B. Kim, and D.R. Hare, *Biochemistry 29*:329–340 (1990).
169. Tabor, S., and C.C. Richardson, *Proc. Nat. Acad. Sci. USA 78*:205–209 (1981).

170. Bernstein, J.A., and C.C. Richardson, *Proc. Nat. Acad. Sci. USA 85*:396-400 (1988).
171. Landshulz, W.H., P.F. Johnson, and S.L. McKnight, *Science 240*:1759-1764 (1988).
172. Hope, I.A., and K. Struhl, *Cell 46*:885-894 (1986).
173. Hope, I.A., and K. Struhl, *EMBO J. 6*:2781-2784 (1987).
174. Struhl, K., *Cell 50*:841-846 (1987).
175. O'Shea, E.K., R. Rutkowski, and P.S. Kim, *Science 243*:538-542 (1989).
176. Oas, T.G., L.P. McIntosh, E.K. O'Shea, F.W. Dahlquist, and P.S. Kim, *Biochemistry 29*:2891-2894 (1990).
177. Agre, P., P.F. Johnson, and S.L. McKnight, *Science 246*:922-926 (1989).
178. Talanian, R.V., C.J. McKnight, and P.S. Kim, *Science 249*:769-771 (1990).
179. Vinson, C.R., P.B. Sigler, and S.L. McKnight, *246*:911-916 (1989).
180. O'Neil, K.T., R.H. Hoess, and W.F. DeGrado, *Science 249*:744-777 (1990).
181. Nakabeppu, Y., and D. Nathans, *EMBO J. 12*:3833-3841 (1989).
182. Tabata, T., H. Takase, S. Takayama, K. Mikami, A. Nakatsuka, T. Kawata, T. Nakayama, and M. Iwabuchi, *Science 245*:965-967 (1989).
183. Shuman, J.D., C.R. Vinson, and S.L. McKnight, *Science 249*:771-774 (1990).
184. McClarin, J.A., C.A. Frederick, B.-C. Wang, P. Greene, H.W. Boyer, J. Grable, and J.M. Rosenberg, *Science 234*:1526-1541 (1986).
185. Kim, Y., J.C. Grable, R. Love, P. Greene, and J.M. Rosenberg, *Science 249*:1307-1309 (1990).
186. Hager, P.W., N.O.Reich, J.P. Day, T.G. Coche, H.W. Boyer, J.M. Rosenberg, and P.J. Greene, *J. Biol. Chem. 265*: 21520-21526 (1990).
187. Kim, K., S.J. Benkovic, and P. Modrich, *J. Biol. Chem. 264*:11807-11815 (1989).
188. Vershon, A.K., R.D. Kelley, and R.T. Sauer, *J. Biol. Chem. 264*:3267-3273 (1989).
189. Knight, K.L., and R.T. Sauer, *Proc. Nat. Acad. Sci. USA 86*:797-801 (1989).
190. Youderian, P., A.K. Vershon, S. Bouvier, R.T. Sauer, and M.M. Susskind, *Cell 35*:777-783 (1983).
191. Knight, K.L., and R.T. Sauer, *J. Biol. Chem. 264*:13706-13710 (1989).
192. Vershon, A.K., S.-M. Liao, W.R. McClure, and R.T. Sauer, *J. Mol. Biol. 195*:311-322 (1987).
193. Vershon, A.K., S.-M. Liao, W.R. McClure, and R.T. Sauer, *J. Mol. Biol. 195*:323-331 (1987).
194. Rafferty, J.B., W.S. Somers, I. Saint-Girons, and S.E.V. Phillips, *Nature 341*:705-710 (1989).
195. Zagorski, M.G., J.U. Bowie, A.K. Vershon, R.T. Sauer, and D.J. Patel, *Biochemistry 28*:9813-9825 (1989).
196. Breg, J.N., R. Boelens, A.V.E. George, and R. Kaptien, *Biochemistry 28*: 9826-9833 (1989).
197. Breg, J.N., J.H.J. van Opheusden, M.J.M. Burgering, R. Boelens, and R. Kaptien, *Science 346*:586-589 (1990).
198. Bowie, J.U., and R.T. Sauer, *J. Mol. Biol. 211*:5-6 (1990).
199. Murre, C., P.S. McCaw, and D. Baltimore, *Cell 56*:777-783 (1989).

APPLICATIONS

7

Netropsin and the Lexitropsins
The Search for Sequence-Specific Minor-Groove-Binding Ligands

Mary L. Kopka and Teresa A. Larsen
University of California at Los Angeles, Los Angeles, California

I. INTRODUCTION

The nucleic acids are prime targets of pharmaceuticals in the war on disease. These drugs interact with DNA or RNA molecules in proliferating cells, interfering with steps in the cell cycle and in some cases inhibiting further cell growth altogether. For example, the first clinically used AIDS drug, azidothymidine, or AZT (1), is inserted into the growing DNA chain, thereby interrupting chain elongation and impeding further replication of the genome. AZT functions as an antiviral agent; other drug molecules possess antibiotic, anticancer or antiparasitic activity.

It has become imperative to understand how these molecules bind and interact on the molecular level, so that newer, more effective therapies can be designed. In this age of molecular biology not only do we have designer genes, but designer drugs as well. After gathering all the relevant structural and chemical information needed, it is the chemist with the aid of a computer who has the advantage; drugs of the future will most certainly be found in this way.

II. HOW DRUGS BIND TO DNA

In the past 10 years, x-ray crystallographers as well as structural chemists have learned an enormous amount about the DNA molecule in its elementary forms: the A, B and Z-helices (2–6). Recently, more complicated DNA molecules—hairpins (7–10), triple helices (11,12), and quadruple helices (13)—have been found or proposed. Structures of drugs complexed with DNA have been reported and are reviewed and categorized in Refs. 14 and 15. An RNA–drug structure analysis also has appeared (16). The structural information in these papers has helped in unraveling the modus operandi of drug binding to the nucleic acids.

From this structural information perspective, along with all the physical and biochemical knowledge available, it is known that drugs bind to DNA in one of three ways:

1. Intercalation between base-pair planes
2. Formation of a covalent bond between drug and DNA
3. Insertion into the grooves via hydrophobic contacts, hydrogen bonds, and favorable electrostatic attraction

A. Intercalation

When a drug intercalates, the DNA unwinds in order to allow the incoming compound, consisting of an aromatic chromophore with or without side chains, to slip between the base pairs of the double helix. A number of crystallographic and NMR studies have illustrated the binding of these molecules. In the crystal structure of an actinomycin-deoxyguanosine complex, the drug molecule is sandwiched between two guanine rings in a manner suggestive of intercalation between two GC base pairs in a regular helix. This structure has become the model for intercalative binding of this antibiotic to DNA (17). Actinomycin is sequence-specific and prefers binding to GpC base pairs. The intercalator ethidium bromide (18), on the other hand, intercalates between base planes of any sequence.

Wang and co-workers have reported the structures of members of the anthracycline family of antibiotic intercalators, daunomycin (19,20), adriamycin (21), and nogalamycin (22,23). Another interesting series of papers (24–28) describes the structure analyses of the bis-intercalators triostin A and echinomycin, quinoxaline depsipeptides that opt for CpG sequences. Since it is not the purpose of this chapter to cover intercalative binding, the reader is referred to other chapters in this book dealing with this subject.

B. Covalent Binding

No covalently bound drug-DNA crystal structure analyses have appeared, except for the papers of the early 1980s that present the binding of cisplatin, *cis*-[diamminedichloroplatinum(II)], with a dinucleotide, pGpG (29,30), and with the DNA dodecamer CGCGAATTCGCG (31). However, NMR results are available on the binding of a number of other alkylating agents that form adducts with DNA. Two illustrative examples of this type of interaction are the antibiotic anthramycin (32,33), a pyrrolodiazepine antibiotic, and the carcinogen benzo[*a*]pyrene or BP (34–36). Both compounds bind covalently to the guanine N-2 amino group. These compounds are optically active, and difficulties, ensuing from the multiplicity of products formed, are encountered in their separation and purification. The preference of these compounds for guanine can be attributed to the ready participation of the 2-amino group in nucleophilic substitution and addition reactions. However, under certain conditions BP also can react with adenine. Other covalent binding ligands have been studied, and the reader is again referred to other chapters in this volume for details. The remainder of this chapter will focus on groove binding.

C. Groove Binding

How molecules interact with the major and minor grooves of the nucleic acids has intrigued chemists and biologists since Watson and Crick elucidated the structure of DNA in 1953 (37). How does a drug or protein molecule recognize the four base pairs AT, TA, GC or CG? Where is the information stored? Structural information distinguishing the base pairs is carried on the base edges of the major and minor grooves (38–40). A protein or drug must complement the particular base sequence it recognizes. The actual information readout occurs via (a) hydrogen bonds between the drug or amino acid and hydrogen bond acceptors and donors on the major and minor groove edges of the base pairs (illustrated in Fig. 1), (b) close van der Waals contacts, and (c) electrostatic attraction between drug and DNA. Both Table 1 and Figure 1 show clearly that the major groove has a four-symbol recognition code and the minor groove a two-symbol code. The pattern of hydrogen bond acceptors and donors in the major groove is unique for AT versus GC base pairs; and since the four symbol code is asymmetric, inverted base pairs can be distinguished. However, in the minor groove there is no distinction between an AT or TA base pair with

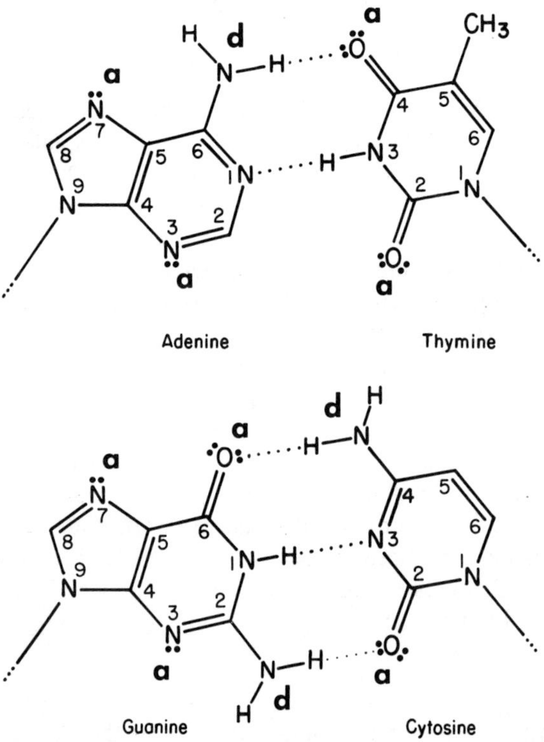

Figure 1 AT and GC base pairs, showing the pattern of hydrogen bond donors, **d**, and acceptors, **a**, on the major (upper) groove edge and minor (lower) groove edge of each base pair. Proteins and drugs "read" the base sequence by sensing the pattern of donors and acceptors on the groove floors.

Table 1 Hydrogen Bonding Patterns in DNA

Base pair	Major groove	Minor groove
A T	a d a (CH_3)	a - a
T A	(CH_3) a d a	a - a
G C	a a d	a d a
C G	d a a	a d a

a = acceptor, d = donor.

respect to the position of hydrogen bond acceptors and donors; the same is true for GC versus CG.

Research to date indicates that drugs bind primarily in the narrow minor groove of B-DNA, whereas larger molecules such as proteins utilize the wide major groove. Proteins use the major groove not only because it is wider and easily accepts the bulky protein, but primarily for fidelity of readout since the major groove has twice the information content. Very specific targeting between protein and nucleic acid occurs here; this specific recognition is important for accurate cellular control. Drugs and other small molecules, on the other hand, are easy to fit and position in the narrow minor groove of B-DNA and tend to bind in this groove.

III. MINOR GROOVE BINDERS

A. Netropsin and Distamycin A

Netropsin is a naturally occurring antibiotic isolated from *Streptomyces netropsis* (41,42) that has antitumor and antiviral activity (15). The basic chemical unit of netropsin and the closely related compound, distamycin A (Fig. 2) from *Streptomyces distallicus* (43), is a carboxamide followed by a pyrrole ring. Netropsin has two of these units and distamycin three. The ends of these molecules can be charged as netropsin is with its diamidino ends, or uncharged like the aldehyde (—CHO) of distamycin.

Both netropsin and distamycin bind only to double-stranded B-DNA (44). In fact, binding of netropsin to A or Z-DNA induces a transition to the B-form (45,46). The preference of netropsin for AT-rich sites has been noted from the earliest studies of the molecule (47). Two modes have been established for netropsin binding to poly d(AT)·poly d(AT), but only one mode to the homopolymer poly(dA)·poly(dT) (48). The first high-affinity binding mode localizes netropsin in the minor groove. The second mode may represent netropsin binding along the phosphates or, a less likely possibility, binding in the major groove.

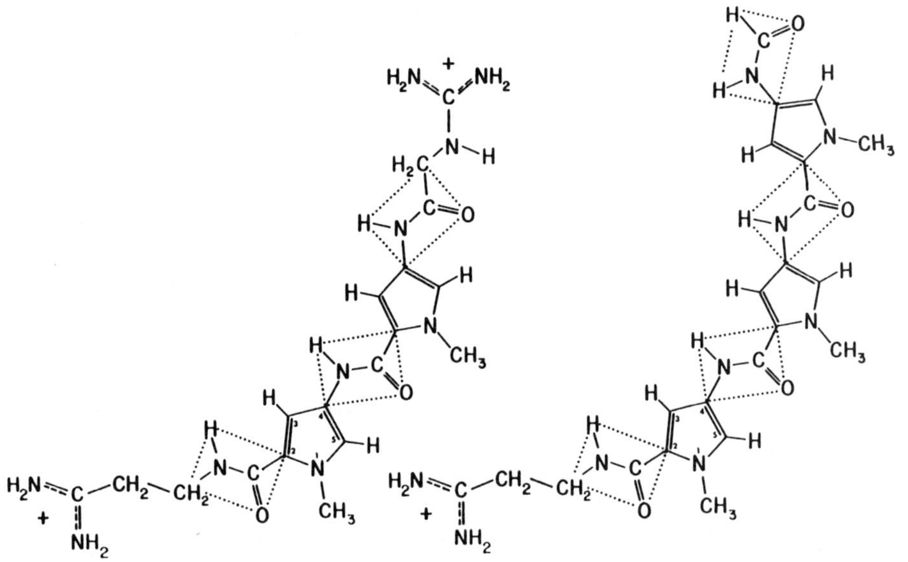

NETROPSIN DISTAMYCIN A

Figure 2 Molecular structures of netropsin with the guanidinium group at top of molecule followed by an amide linkage (in dotted box), *N*-methylpyrrole ring 1, amide, *N*-methylpyrrole 2, amide and ending with propylamidinium tail. Distamycin A with formamide at top, followed by *N*-methylpyrrole 1, amide, *N*-methylpyrrole 2, amide, *N*-methylpyrrole 3, amide and ending with the propylamidinium tail.

Longer analogs of netropsin with three or more pyrrole-carboxamide units do sometimes admit a GC base pair in the binding site (49,50), but the preferred choice of these antibiotics remains AT-rich sequences in the narrow minor groove of the DNA duplex (51,52). Krylov et al. (53) measured the free energy of binding of each carboxamide unit of distamycin to AT polymers as 2.0 kcal mol^{-1}, which is comparable to the formation of one hydrogen bond. The binding of a carboxamide to a GC unit was only 0.95 kcal mol^{-1}. The difference of 1.05 kcal mol^{-1} between the free energies of interaction of the distamycin amide with an AT or GC base pair results in a difference in the binding constant of three orders of magnitude. Dervan's footprinting and affinity cleavage experiments (54–56) on restriction DNA fragments have shown conclusively that synthetic analogs of netropsin and distamycin with n amides prefer a binding site of $n + 1$ base pairs; i.e., for netropsin with three amide linkages, four base pairs are covered by the drug (56). Prior to Dervan's work, binding sites of three, four, and five base pairs had been attributed to netropsin.

These drugs are known to inhibit the DNA and RNA polymerase reaction in vitro by interacting with the DNA template and preventing synthesis of DNA (50,57). Recently, experiments have shown that at low concentrations, netropsin and distamycin, by binding to specific promoter sites, can activate transcription as well (58,59). These sites can be either wild type or nonalternating AT-rich inserts. As the drug concentration is increased, the rate of transcription drops considerably. Footprinting experiments (59) have located the activating site in the promoter region. Eventually, at higher drug concentrations binding occurs at other loci and inhibition results. These small sequence specific drugs have the potential to alter the DNA structure locally, and this may be the reason for activation of transcription. Depending on the sequences of the DNA promoter, and/or the DNA template, as well as the drug concentration, either inhibition or activation of transcription takes place.

At present neither netropsin nor distamycin are in general clinical use except as topical applications for chicken pox, herpes zoster, and smallpox vaccination eruptions (60,61). They have become, nevertheless, a paradigm of minor groove binding for four reasons: (a) the abundant biochemical and physicochemical information, as reviewed by Zimmer (62); (b) the NMR structures of netropsin and distamycin complexed with DNA oligomers ranging from 8 to 12 base pairs (63–70); (c) the x-ray crystal structure analysis of the drug netropsin in 1979 (71); and (d) the ensuing single-crystal x-ray analyses of netropsin and distamycin bound to a variety of AT-rich sequences (72–75).

1. Netropsin + CGCGAATT^{Br}CGCG

The first x-ray structure of netropsin cocrystallized with the 9-bromo derivative of the B-DNA dodecamer, CGCGAATTCGCG (72,73), confirmed and also expanded on the wealth of experimental data in the literature. Patel (64) had previously investigated this complex by NMR using NOEs to establish the nonbonded contacts between the C2 hydrogens on adenine and the drug pyrrole and methylene groups. In the x-ray analysis netropsin binds as expected in the narrow AATT minor groove of the DNA helix by replacing the spine of hydration (Fig. 3, 4 top and Plate 1a), which fills the minor groove in the parent dodecamer (76,77). These water molecules are held in place by cross-strand hydrogen bonds spanning the groove in a network that runs down the length of the AT region from the N-3 and O-2 atoms on successive adenine and thymine base edges.

The crescent shaped drug slips into the groove, quasi-intercalating between the groove walls, with its three amides facing the floor of the groove (Fig. 4 bottom and Plate 1b). Specificity for AT base pairs is achieved by means of nonbonded contacts between the netropsin methylene or pyrrole CH groups and adenine C-2 hydrogen atoms on the floor of the minor

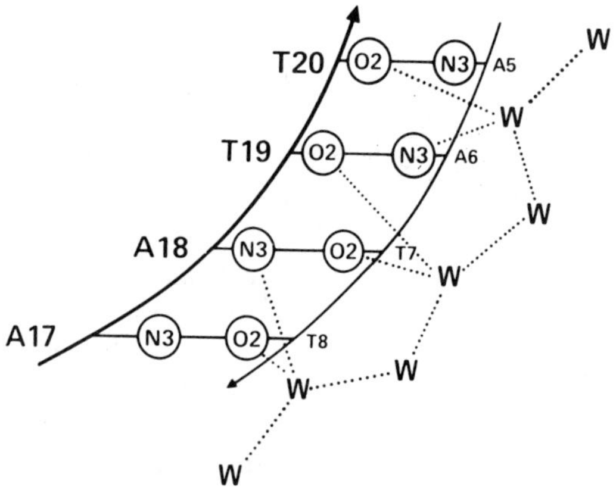

Figure 3 Spine of hydration in -AATT- region of CGCGAATTCGCG. Waters (W) make bifurcated hydrogen bonds (dotted lines) to O-2 atoms on thymine and N-3 atoms on adenine.

groove. The 2-amino group of guanine is disruptive to this close fit of the drug to the DNA. The absence of this NH_2 on adenine permits the close approach of the drug to AT base pairs. After "reading" the groove floor, replacing the waters there and positioning itself into the groove, the drug forms bifurcated or more correctly, according to Jeffrey, three-centered hydrogen bonds (78,79) between its amide NH groups and the base edge nitrogens and oxygens of the DNA. This pattern is reminiscent of the hydrogen bonding bridges of the water molecules of the spine, bridging the same N-3 atoms of adenine and O-2 atoms of thymine.

Four factors contribute to the binding process: van der Waals steric contacts, hydrogen bonds, electrostatic attraction and hydrophobic interactions. The sequence-specific "reading" of the DNA occurs via close van der Waals contacts, whereas hydrogen bonds help position the drug and the electrostatic and hydrophobic interactions contribute stability to the complex. Marky and Breslauer (80) conclude that the binding of netropsin to GCGAATTCGC is strong and enthalpy (ΔH) driven (Table 2). Deep penetration into the minor groove is necessary to form the molecular interactions that yield a strong enthalpic binding affinity: hydrogen bonds and van der Waals contacts. They reach this conclusion by comparing thermodynamic profiles of netropsin bound to DNA duplexes that permit deep minor groove penetration to those that do not. (See Table 2).

Pullman and Pullman (81,82) have proposed from theoretical calculations that the AT preference by netropsin for the deep minor groove arises

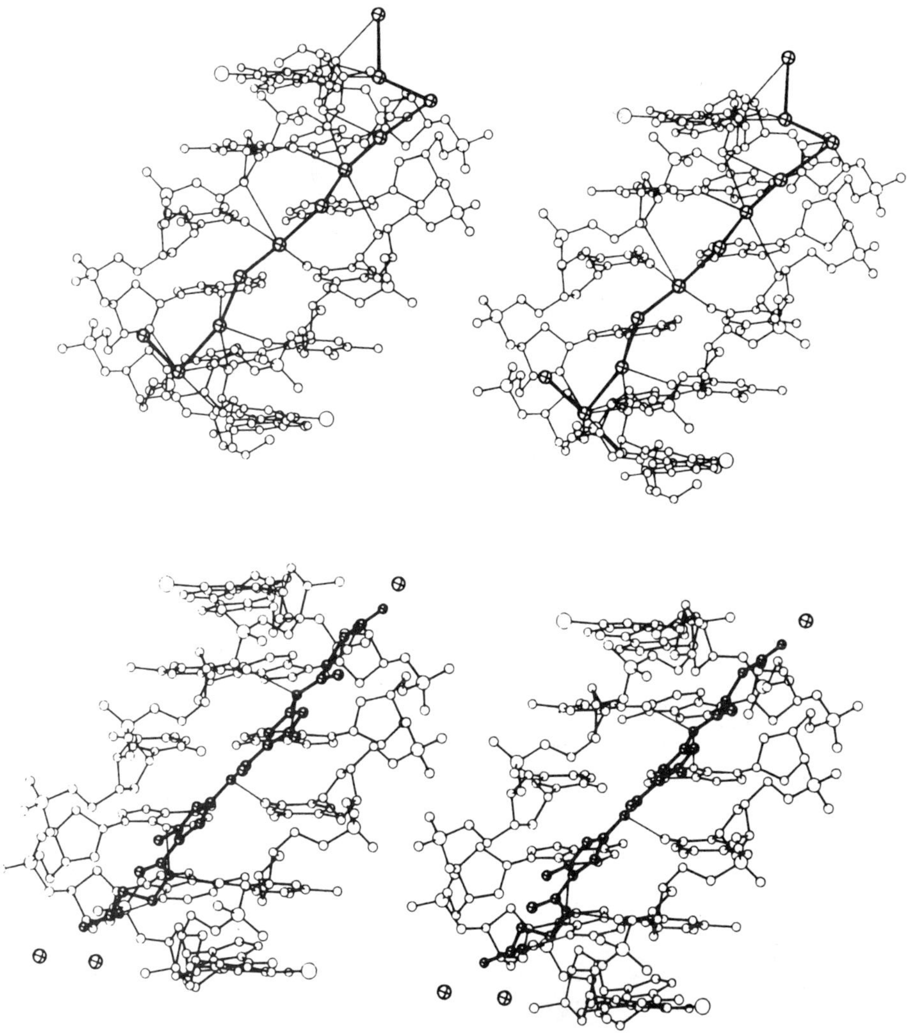

Figure 4 Spine of hydration (top) showing central six base pairs, -GAATT^{Br}C-, of CGCGAATT^{Br}CGCG with base pair G4-C21 at top. Crossed circles are water or netropsin molecules. Note the bifurcated hydrogen bonds from N-3 of adenines to O-2 of thymines "locking" the H_2O molecules down. Netropsin (bottom) binding to the same six base pairs. Guanidinium end of netropsin molecule is at top and propylamidinium is at the bottom. Note bifurcated hydrogen bonds from nitrogen amides on netropsin to base edge nitrogens and oxygens on adenine and thymine. Central nitrogen amide of drug makes a single bond because netropsin is slightly bowed in the center in order to fit in the groove with proper alignment of hydrogen bonding atoms to base edges. Notice water molecules within hydrogen bonding distance to tails of drug. Adenine N-3 and thymine O-2 distances to H_2O oxygens or drug NH atoms are 3.55Å or less.

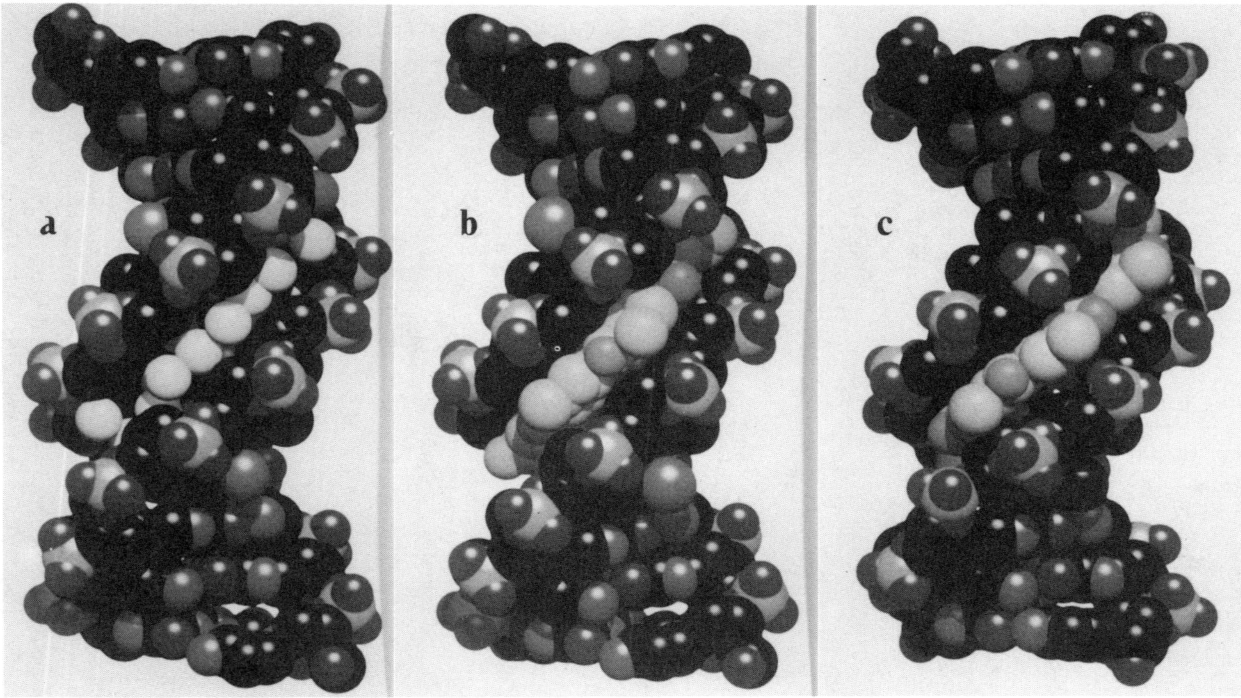

Plate 1 (see also Color Plate following Chapter 1) Minor groove view of CGCGAATT^{Br}CGCG with (a) "spine of hydration" filling the minor groove. Water molecules are pale blue. DNA carbon atoms are black, nitrogens – blue, oxygens – red, phosphorus – yellow, and bromine – violet. (b) With netropsin bound in shades of aqua for carbon and nitrogen atoms and oxygens in green. Note blue atom at bottom left of groove is a H_2O molecule. (c) CGCAAATTTGCG with distamycin bound in shades of yellow. Light yellow indicates carbon and nitrogen atoms with dark yellow for oxygen atoms. Note partially hidden H_2O molecule in groove at bottom left.

Table 2 Thermodynamics of Netropsin Binding to DNA

DNA Duplex	$\Delta G°$ (kcal mol^{-1})	$\Delta H°$ (kcal mol^{-1})	$\Delta S°$ (kcal mol^{-1})
polyd(AT) · polyd(AT)	−12.7	−11.2	+5.0
polyd(A) · polyd(T)	−12.2	−2.2	+33.0
polyd(GC) · polyd(GC)	−7.1	−4.3	+9.3
d(GCGAATTCGC)$_2$	−11.5	−9.3	+7.5

Data are from Ref. 80.

mainly from electrostatic forces. Their calculations suggest that a positively charged ligand the size of a water molecule or greater would not be attracted to the phosphate backbones, but to the negative potential in the bottom of the major and minor grooves. In the minor groove this negative potential is deeper for AT regions than for GC. Hence any cationic drug such as netropsin should move preferentially toward an AT-rich region of the minor groove. However, Marky and Breslauer (80) cast doubt on the electrostatic contribution to sequence specificity by measuring identical salt dependencies for the binding constants of netropsin to both alternating AT and GC sequences, reasoning that if the preference were electrostatic, salt would interfere with binding. They also find that the binding profile of netropsin to poly d(AT)·poly d(AT) is unchanged by removal of one of the charged ends from netropsin, with the same ΔH as the doubly charged compound (83). Their interpretation of these results is that it is not the charged ends, but rather the pyrrole-amide moiety interacting with the minor groove that contributes to the enthalpy-driven high binding affinity (84).

Notice, however that in contrast to the behavior observed with the alternating AT polymer, the binding of netropsin to the homopolymer poly(dA)·poly(dT) is dominated by the entropy (ΔS) term (Table 2). This entropic binding occurs with a variety of DNA binding drugs, both groove binders and intercalators, despite their different binding modes. Breslauer and co-workers conclude that this high entropy results from properties of the homopolymer DNA, rather than of the drugs themselves. They believe that the homopolymer is more heavily hydrated, and that disruption of this hydration accounts for the entropic binding.

2. *Netropsin + CGCGATATCGCG*

In 1989 Coll et al. solved the second crystal structure analysis of a netropsin-DNA complex (74). The DNA oligomer in this complex contains an alternating AT binding site in the minor groove in place of the AATT center. Netropsin binds to CGCGATATCGCG in two orientations (Fig. 5)

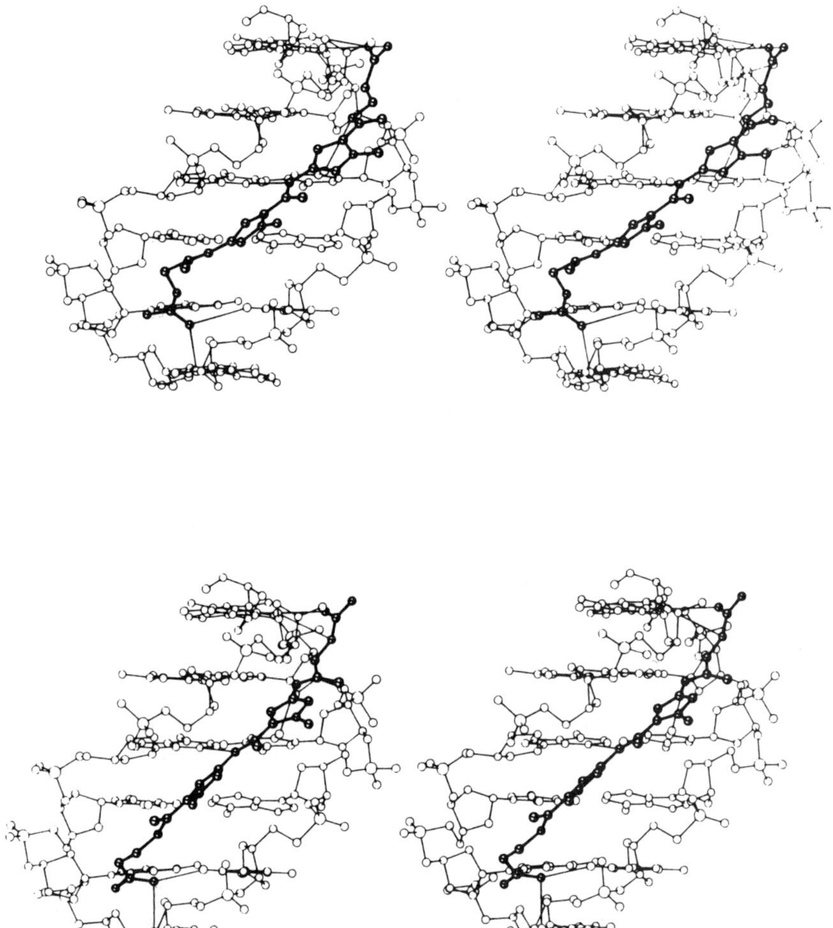

Figure 5 (top) Netropsin (crossed circles) bound to six central base pairs of CGC-GATATCGCG with the orientation of the propylamidinium tail at the top of the helix, making bonds from each propylamidinium nitrogen NH_2 to base edge atoms. Netropsin does not make bifurcated (three-centered) hydrogen bonds to the N-3 and O-2 atoms on the base edges of this alternating AT sequence, but instead makes single hydrogen bonds. The final amide NH at bottom of helix near the guadinium tail makes no hydrogen bonds. One guanidinium tail NH_2 makes hydrogen bonds with the base edge N and O atoms. (bottom) Netropsin bound to same helix, but in opposite orientation with propylamidinium tail down at the bottom of the helix. Again the final amide NH, which is now near the propylamidinium tail, does not make a bond with the base edges.

and, in neither case were bifurcated hydrogen bonds formed to the DNA. Instead, single hydrogen bonds are formed between the individual amide NH's of the drug and the O–2 or N–3 of thymine and adenine. (See Fig. 6A,B.) Coll attributes these single hydrogen bonds within the alternating AT site to the uneven spacing between either the two hydrogen-bonding O-2 or two hydrogen-bonding N-3 atoms across the groove, making the formation of bifurcated or three-centered bonds difficult. In this structure the O-2–O-2 separation across the groove is 4.8 Å (T6–T20) and 4.7 Å (T8–T18), whereas the N-3–N-3 separation is 3.3 Å (A7–A19).

Klug et al. (85) in a prior report discuss the alternating helical twist between ApT and TpA steps. Their model for the alternating B helix, based on the crystal structure of pATAT (86,87), has an alternation of helical twist between 33° for the efficiently stacked ApT step and 39° for the destacked TpA step. The x-ray structure of CGCATATATGCG (88), which confirms this model, has alternating twist angles in the AT region of 33°, 39°, 31°, 42°, and 33°. [See Fig. 4 of Yoon et al. (88).] This fluctuation in twist angle impedes the formation of three-centered hydrogen bonds from drug to DNA.

To understand why two orientations of a drug are possible in crystal structure analysis, a knowledge of what is occurring in solution as the drug and DNA are cocrystallizing is necessary. In solution, drug and free DNA are in equilibrium with bound complex, and the rate of association can be measured by NMR techniques, which yield on, off, and “flip” rates of the asymmetric drug binding to a DNA oligomer (65,89). If the on and off rate are sufficiently slow on the NMR time scale, then NMR can distinguish between the two orientations of the asymmetric drug on the DNA. Since any complex containing an asymmetric ligand can also crystallize in two possible orientations, both species may be deposited in the crystalline lattice as the crystal grows. This would generate a structure with twofold disorder, and both orientations would fit equally well into the electron density. Forces such as crystal packing can, however, favor the deposition of one orientation of the drug onto the lattice, and in this case, only this orientation would satisfactorily fit the electron density.

When netropsin cocrystallized with CGCGAATTCGCG (72,73), only one conformer with the propylamidinium tail pointing “down” toward the G12-C13 end of the DNA was found (Fig. 6C). However, an NMR study of netropsin bound to the octamer, GGAATTCC, found two orientations (65). An interesting feature of the Coll (74) alternating AT-netropsin crystal structure is that the netropsin molecule could be fitted to the electron density in the minor groove in two possible orientations, i.e., with the propylamidinium tail either “up” or “down” (Fig. 6A,B). There was no preference indicated by the electron density.

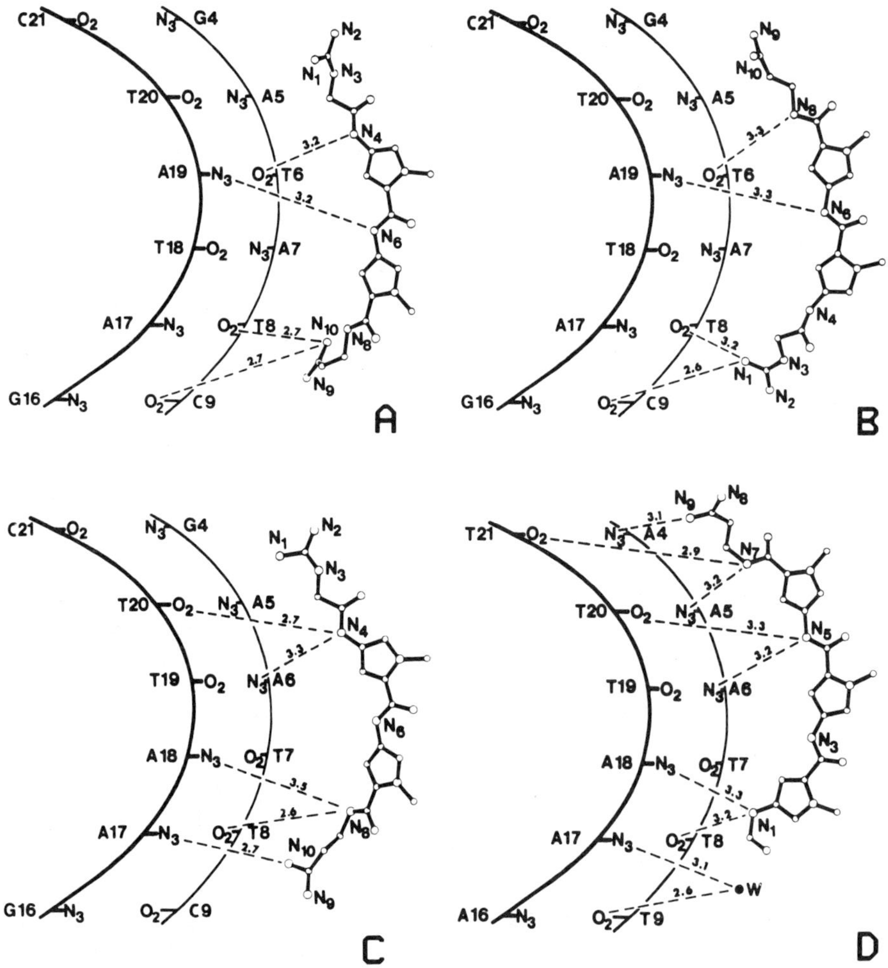

Figure 6 Comparison of different DNA-netropsin/distamycin crystal complexes. (A) and (B) Netropsin-CGCGATATCGCG with netropsin in two possible orientations with four single hydrogen bonds (74). (C) Netropsin-CGCGAATT^{Br}CGCG in one orientation only with two sets of bifurcated hydrogen bonds (72,73). (D) Distamycin-CGCAAATTTGCG in one orientation only with three sets of bifurcated hydrogen bonds (75). The propylamidinium tail is "down" in (A) and (C) and "up" in (B) and (D). (Reproduced with permission from Ref. 74.)

The affinity cleavage footprinting work of Dervan on long noncomplementary sequences (55,56,90) confirms that distamycin-type drugs containing three to five pyrrole rings can bind in two orientations. In some cases, however, there is definitely a preferred direction on the footprint, and Dervan reasons that flanking sequence may play some role in determining directionality (56). Additionally, these footprinting experiments (55,91,92) reveal that netropsin and distamycin bind most strongly to the homopolymer poly(dA)·poly(dT), and secondly that distamycin is more tolerant to GC base pairs near the end of the binding site (55). Dabrowiak and coinvestigators (91) found that the presence of the dinucleotide step TpA within any stretch of four AT bases to some extent discourages netropsin binding to DNA. They reach this conclusion from calculating binding constants of different AT sequences, and find that sequences containing TpA steps have binding constants as much as two orders of magnitude lower than those AT sequences that do not.

3. *Distamycin A + CGCAAATTTGCG*

The final crystal structure to consider from the netropsin family is that of distamycin complexed with the dodecamer, CGCAAATTTGCG, in which the drug is anchored to five base pairs in the AAATTT center (Fig. 7) by making three-centered (bifurcated) hydrogen bonds once again (75). This structure has three sets of three-centered hydrogen bonds instead of two because distamycin contains one more amide linkage than netropsin. In this crystal analysis, distamycin binds to the central AT region in the opposite direction to that of netropsin, with the propylamidinium tail positioned at the start of the AAATTT sequence at the first adenine, covering the five base -AAATT-sequence. In contrast the netropsin propylamidinium tail position is at the last thymine in the -AATT-segment of the Kopka et al. structure (Fig. 6C,D and Plate 1c).

The NMR work of Wemmer (68,69) and co-workers, studying the drug bound to CGCGAATTCGCG, positions the drug in the AATT region in one orientation, with the propylamidinium tail in the vicinity of the C9G16 base pair in the same orientation as the netropsin crystal structure analysis of this sequence. The authors also find that the O-4′ sugar atoms stack over the center of the three pyrrole rings, giving stability to the complex. other NMR investigation (70) of distamycin at a 2 : 1 drug to DNA ratio bound to (CGCAAATTGGC)·(GCCAATTTGCG) indicates that two drug molecules are bound simultaneously in the minor groove in close contact with the DNA and each other, oriented with each propylamidinium tail pointing towards opposite ends of the helix. The drug molecules cover the AATT base sequence in a 5′→3′ direction with the propylamidinium tail positioned below T8 for one molecule and above T19 for the other. Molecu-

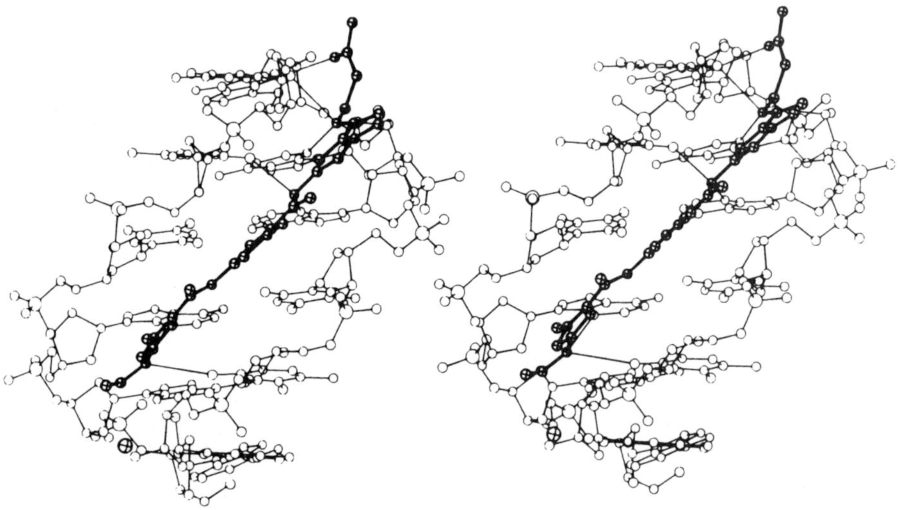

Figure 7 Stereo view of minor groove of CGCAAATTTGCG with distamycin A bound via bifurcated hydrogen bonds from the N-3 atoms on adenine and O-2 atoms on thymine to the drug NH amide atoms and to one hydrogen on the propylamidinium tail. Note that one amide NH does not make any bonds with the DNA because the drug is bowed outward in this region to allow a better fit with the DNA helix.

lar mechanics calculations indicate that for such a 2 : 1 complex the minor groove must expand significantly to accommodate two drug molecules side by side.

Major groove cross-strand hydrogen bonds were first noted by Nelson et al. (93) in the oligo dA·oligo dT tract found in the crystal structure of CGCAAAAAAGCG paired with its complement GCGTTTTTTCGC. In the crystal analysis of distamycin bound to CGCAAATTTGCG, Coll confirms the presence of these major groove bonds between the carbonyl O-4 of two adjacent thymines on one strand and the N-6 of an adenine on the other strand. A high degree of propeller twist occurs in the central A tract of both of these structures (see Table 2 of Ref. 74), bringing the N-6 and O-4 atoms into the proper position for the formation of these three-centered hydrogen bonds (75,93) and also improving purine-purine stacking.

After presenting the way these drugs orient and bind in these three crystal structures, how the drugs affect the local helical environment can be considered. First, the drugs themselves have a natural twist between their pyrrole planes, which can adjust during binding to make the drug "isohelical", i.e., twisted in a screw sense to match the pitch of the minor groove of DNA. For example, unbound netropsin has a natural twist of 20° and in the DNA-crystal complex increases to 33° (73). Second, in the Kopka (72,73) structure of netropsin bound to the -AATT- center, the drug bends the DNA back into the major groove by 8°. This bending that the drug initiates can be calculated from a plot of the normal vectors to the base planes as given in Figure 3 of Ref. 94 and Figure 9 of Ref. 73. (The NEWHEL90 program available from R.E. Dickerson at UCLA can be used for these calculations.) However, no drug-induced bending occurs in either the netropsin -ATAT- structure (74) or the distamycin -AAATTT- complex (75). Third, netropsin widens the minor groove by approximately 2 Å at the point of attachment of the propylamidinium tail in the Kopka structure. No statement can be made about netropsin or distamycin widening the grooves in the two Coll structures, because only the netropsin -AATT- structure has a native data set with which to compare groove width.

Three other minor groove binding drugs that are related to, but differ in some ways from, the netropsin-distamycin series have been complexed with DNA and their crystal structures will be introduced now before covering the design of the lexitropsins.

B. Hoechst 33258

In 1968 Loewe and Urbanietz (95), while synthesizing benzimidazole derivatives, prepared Hoechst 33258, a bis-benzimidazole with a methylpiperazine ring attached to one benzimidazole and a phenol ring attached to the other (Fig. 8). Clinical screening of this compound showed it to have anthelmintic properties, and initially it was used against intestinal worms (96,97). At present this highly fluorescent compound is employed mainly as a chromosomal stain both in vitro and in vivo (98–100). In addition, analytic chemists have begun to utilize this dye for DNA quantitation in tissue samples (101,102).

Hoechst 33258 exhibits two binding modes to DNA: (a) a high affinity, fluorescent, groove-binding mode, and (b) a low affinity, fluorescence-quenching, salt-dependent mode in which the drug molecule lies more parallel to the DNA bases (103–106). Equilibrium dialysis experiments (107) revealed the strong binding preference of Hoechst to sites containing at least three AT base pairs. Mikhailov and co-workers (108) measured a 45° angle of inclination of the dye to the helix axis, proving that the dye in its high-affinity mode was not an intercalator but bound in one of the grooves of the B-helix, specifically the narrow minor groove. Independent footprinting experiments (109–111) indicate that Hoechst covers at least four base pairs, but usually five, and probably makes three hydrogen bonds in

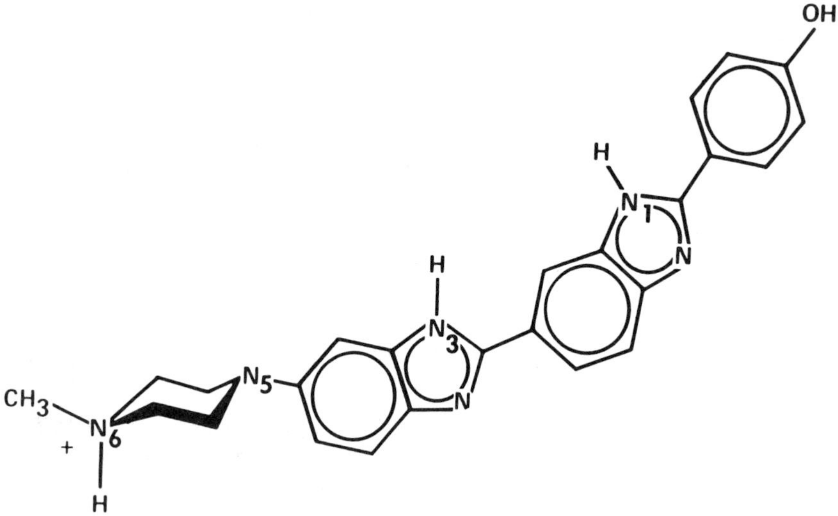

HOECHST 33258

Figure 8 Molecular structure of Hoechst 33258 with the phenol at top followed by two benzimidazole rings and ending with a piperazine ring. At pH ≈ 2, the piperazine rings would each have a positive charge. At neutral pH the structure would carry one positive charge as shown.

the recognition process, accounting for earlier reports of a three base pair binding site. These footprints reveal (a) a site of at least four base pairs is needed for binding (111,112); (b) guanine can be readily accommodated at the end of the binding site (109,111); (c) the presence of a TpA step can discourage binding (111,112).

Fluorescence of the dye is quenched by the introduction of bromodeoxyuridine, BrdU, into the DNA (113,114). Enhancement of fluorescence by late incorporation of thymidine into chromosomes substituted with BrdU is employed as a tool to determine regions of late DNA replication by Latt (113). Breslauer (115) makes use of the fluorescent properties of the dye to characterize the minor groove environment of Hoechst bound to alternating poly(AT). He finds that the minor groove of the complex is nonpolar with a dielectric constant ≈ 20 Debye (vs. bulk water ≈ 80 D). In such a low dielectric environment, hydrogen bonds and hydrophobic contacts are expected to contribute a strong enthalpic driving force.

To date, there has been one NMR structure of Hoechst bound to a dodecamer of sequence CTTTTGCAAAAG (116). In this study, Searle and

Embrey report that two drug molecules bind cooperatively in symmetry related orientations at the center of the 5′TTTT3′ and the 5′AAAA3′ regions, with the piperazine pointing towards the center of the duplex. The adenine N-3 and thymine O-2 groups are involved in three-centered hydrogen bonds with the NH groups on the benzimidazole rings. The bases participating in these hydrogen bonds are the outer three base pairs of each AT stretch, T2–T4 and A9–A11.

1. *Hoechst 33258 + CGCGAATTCGCG*

Like netropsin, several x-ray structures of Hoechst bound to different oligomeric DNA sequences have appeared in the current literature. These include two independent x-ray analyses of Hoechst bound to CGCGAATT-CGCG (117,118), with a third study being readied for publication (119). The first two analyses found the drug bound as expected in the minor groove, but differed in binding site by one base pair (Fig. 9C,D). Pjura et al. (117) report the binding sequence to be -ATTC-, while Teng and coworkers (118) find the molecule covering the -AATT- base pairs. The third structural study (119) again is an analysis of the drug bound to the AATT site, but with data collection at low temperatures: 0°, −25°, and −100°C, in the hope of improving the reliability of the analysis.

In the Pjura structure (117), the Hoechst molecule is shifted down one base pair from the expected AATT binding region to bind to ATTC. In this case, the nonplanar piperazine ring is positioned perpendicular to the walls of the wider GC region, where it is easily accommodated. Water molecules of the spine are still seen in the electron density near the first AT base pair, A4-T21, with one water hydrogen bonded to the terminal OH of the phenol. Pjura interpreted the density in the vicinity of the piperazine ring as containing two conformers. One conformer is not coplanar with the second benzimidazole (Bz2), but roughly 60° out of the plane of Bz2 (Fig. 10a). This moves the ring down into the GC region of the helix. In the other conformer position, the piperazine ring extends out of the groove. The other four crystal analysis found only one position for the piperazine ring.

The Teng Hoechst structure (118) has the same orientation of the drug as the Pjura structure, but with the drug situated entirely in the AATT region, binding as netropsin did to these base pairs (Fig. 10b). The electron density map is consistent with only one conformation for the piperazine ring. This ring is more nearly coplanar with Bz2 than in the Pjura structure, fitting into the groove parallel to the walls. Two solvent molecules sit close to the phenolic hydroxyl as in the -ATTC- drug complex. This orientation of the drug molecule, with the phenol positioned in the region of the G4-C21 base pair, is given further support by the presence of

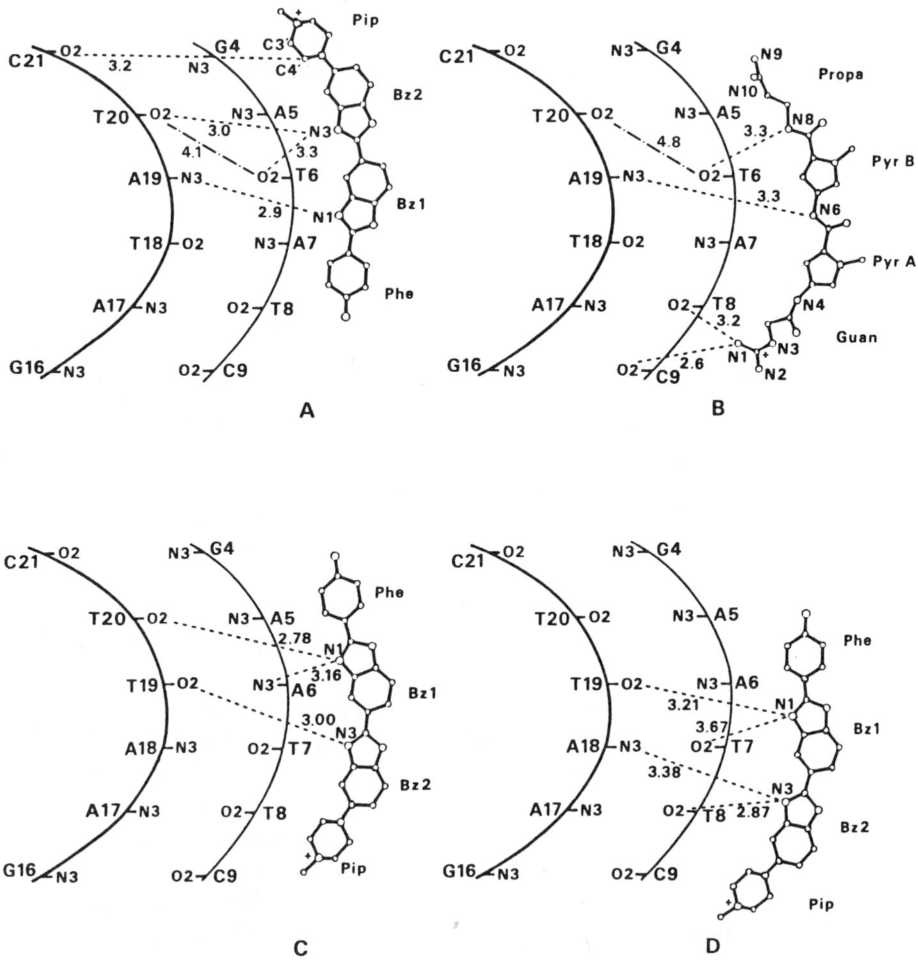

Figure 9 Comparison of different DNA-Hoechst and netropsin complexes. (A) and (B) show Hoechst and netropsin bound to CGCGATATCGCG. In (A) the piperazine ring is "up" and the Hoechst N-3 makes a three-centered bond with O-2 of T6, covering the GATA sequence. The distance between the C-4′ atom of the piperazine and the O-2 of cytosine is also shown. (B) Netropsin makes only a single hydrogen bond with O-2 of T6, covering ATAT. Note the different distances between O-2 of T20 and O-2 of T6 in both complexes. In the Hoechst interaction the T6 has twisted to make the shorter 4.1Å distance between T20 O-2 and T6 O-2. (C) and (D) show Hoechst bound to CGCGAATTCGCG (117,118). The drug orientation is opposite (A) with the piperazine "down". (C) covers the sequence AATT and (D) ATTC. (Reproduced with permission from Ref. 120).

these solvent molecules. A water molecule is not likely to hydrogen bond to the piperazine methyl group if it were there, but certainly can to the phenol hydroxyl.

The third structural analysis (119), in which the data was collected at low temperatures, also shows the drug residing in the AATT region of the minor groove (Fig. 10c and Plate 2d) as in the Teng structure. Both orientations were refined for all three temperatures: 0°, −25°, and −100°C. In the latter two structures the drug is positioned with the piperazine ring "down" at the bottom of the AATT region. The drug in the 0°C analysis can fit the density in either of two orientations, the piperazine "up" at the beginning of the AT sequence or piperazine "down" at the end of the AT sequence. When the two orientations were refined separately, they converged to a similar residual error or R-factor. If one orientation were favored: (a) A lower R-factor would result for that choice. (b) The hydrogen bonds would not be distorted, with DNA contacts optimal. (c) The map would have no missing or unaccounted electron density (Fig. 11). In contrast to the −25° and −100°C analyses, the 0°C analysis revealed no obvious preference for one orientation based on these criteria.

One may ask why all these Hoechst-DNA structures, while binding to the same sequence in the minor groove, differ so in detail. A number of reasons can be given. Each of the crystals was obtained from slightly different salt and spermine concentrations, types of anion, pH and DNA-drug ratios. Not only did crystals of different quality result, but data quality varied considerably as well. In a crystallographic structure analysis, the resolution of the data, given in ångstroms, is a very important factor. Higher resolution (a smaller number) means a better resolved electron density map. Table 3 lists the various crystals studied and compares the data used to determine each structure. Each of the data sets extends to a reasonable resolution of approximately 2.25 Å. However, nominal resolution is not sufficient to judge the quality of the analysis. The number of independent x-ray data (F_0) per atom of the structure is a better measure of the amount of information and thus the reliability of the analysis. The larger this data-to-parameter ratio, the better determined is the structure. It is apparent from Table 3 that the −25°C study is the best of the six. Weakness of the data collected at room temperature may have lead to noisy difference maps from which the drug position was determined. The other four structures have either stronger data or slightly higher resolution leading to a more believable drug position in the central -AATT- region of the dodecamer.

2. Hoechst 33258 + CGCGATATCGCG

Hoechst binds to the alternating AT minor groove of this sequence (120) in a manner similar to that of the Pjura structure, with the piperazine ring

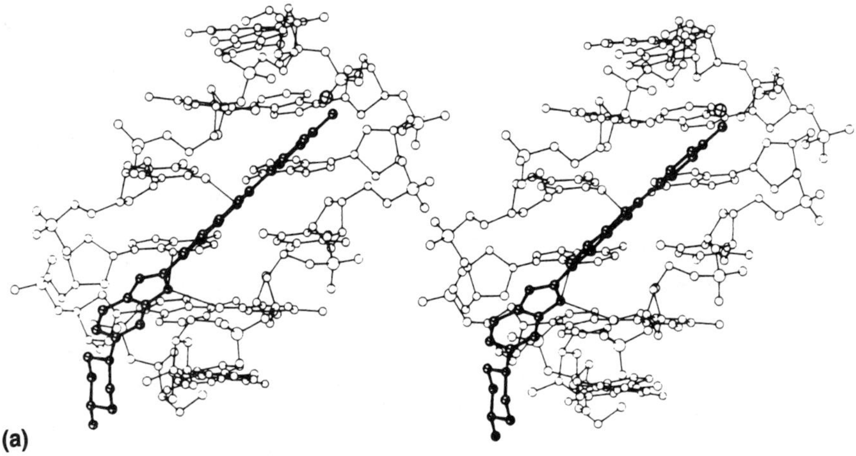

(a)

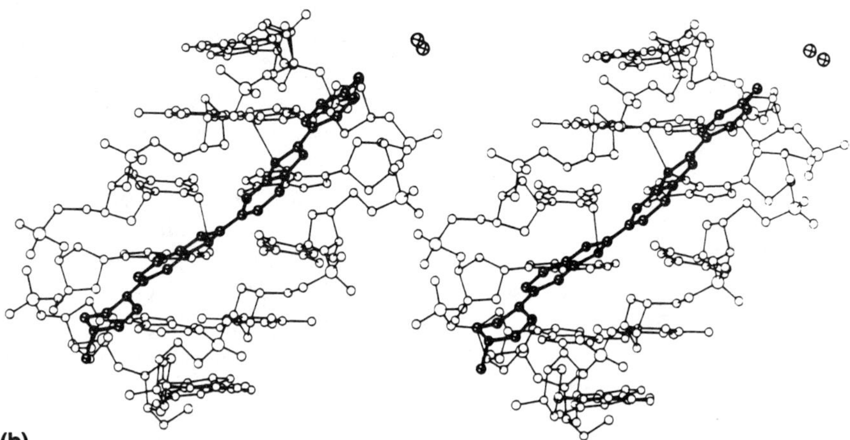

(b)

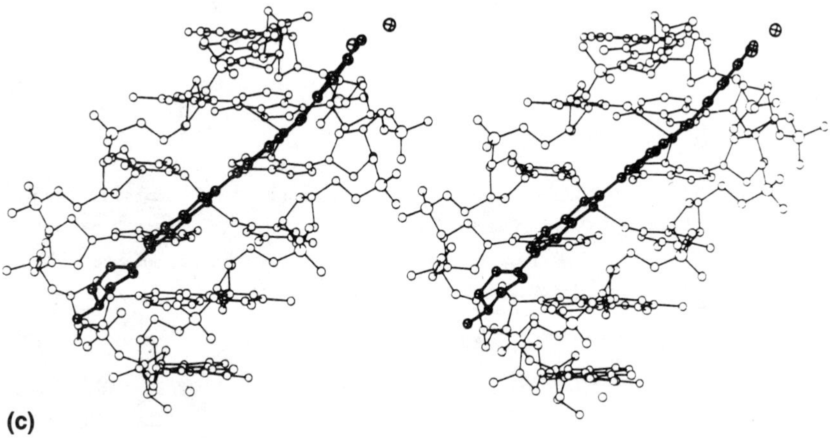

(c)

found in the GC region. However, in this crystal structure the drug is in the opposite orientation to that found in the AATT dodecamer; the piperazine is "up." The binding sequence is -GATA- instead of the expected -ATAT- (Fig. 9A,B). The binding of Hoechst induced a change in some of the helical parameters, including helical twist. As mentioned earlier, Netropsin + CGCCATATCGCG, an alternation of helical twist occurs between an ApT step (33°) and a TpA step (39°), but here the alternation is reversed, so that the ApT step has a larger twist (39°, 36°) than the TpA step (31°). In addition, Thymine 6 is rotated out of the plane of its Watson-Crick base pair in order to make a hydrogen bond to the N-3 nitrogen of Bz2. No stereo drawing of the binding of Hoechst to the alternating four-base AT sequence in this dodecamer was possible because the coordinates were not available.

In all of the Hoechst structures, regardless of minor groove sequence, one bifurcated hydrogen bond from the NH on one of the benzimidazoles is always present, and the second benzimidazole has either a bifurcated or a single hydrogen bond to the base atoms. The better resolved structures have two of these three-centered (bifurcated) bonds to the bases. As was just mentioned, the binding of Hoechst does affect the helical environment considerably in the ATAT structure. However, Hoechst binding to AATT sequences acts in a similar fashion to netropsin bound to these base pairs; Hoechst bends the DNA back toward the major groove $\approx 5°$. In the low temperature study at $-25°$ and $-100°$C the bend into the major groove increases slightly to $\approx 7°$. The drug also widens the groove between 2 and 3 Å

Figure 10 Stereo view of Hoechst 33258 (crossed circles) bound to central six base pairs of CGCGAATTCGCG. (a) Data taken at 25°C. Drug molecule has slipped down one base pair from netropsin binding site to bind to -ATTC-, positioning the piperazine in GC region of the helix. A single hydrogen bond is made from the nitrogen on benzimidazole 1 to O-2 on T19. The benzimidazole 2 nitrogen makes bifurcated hydrogen bonds to N-3 on A18 and O-2 on T8. (b) Data taken at 16°C. Drug molecule is positioned in central -AATT- region. A bifurcated hydrogen bond is made from the nitrogen on benzimidazole 1 to T20 O-2 and A6 N-3. Benzimidazole 2 makes a single hydrogen bond to T19 O-2. Note two water molecules (large crossed circles) at top. These waters at a 4.0-Å distance are further away than in structure below. (c) Data taken at $-25°$C. Drug molecule is positioned in central -AATT- region, making a bifurcated hydrogen bond to T20 O-2 and A6 N-3 from benzimidazole 1 and a bifurcated hydrogen bond from the nitrogen on benzimidazole 2 to T19 O-2 and T7 O-2. Note two water molecules (large crossed circles) at top within hydrogen bonding distance to phenol OH. Hydrogen bonds are drawn for distances less than 3.55Å. No hydrogen bonds are drawn to the waters.

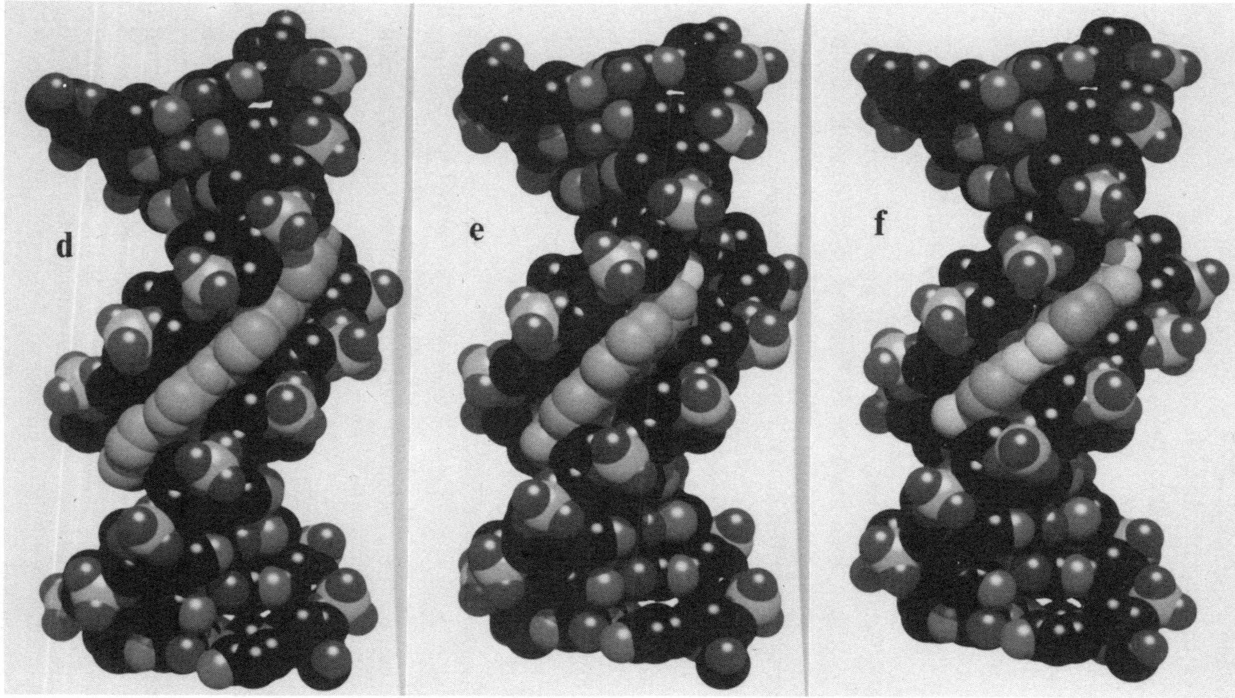

Plate 2 (see also Color Plate following Chapter 1) Minor groove view of CGCGAATTCGCG with (d) Hoechst 33258 bound in minor groove. DNA carbon atoms are black, nitrogens – blue, oxygens – red, and phosphorus – yellow. Hoechst molecule is shown in shades of violet with carbons in light violet and nitrogens in dark violet and phenol oxygen at right top of Hoechst in pink violet. (e) DAPI shown in shades of deep pink for carbons and lavender for nitrogens. Note bound H_2O molecule at top of minor groove. (f) Berenil in shades of salmon with carbon atoms dark salmon and nitrogens light salmon. Again H_2O molecule is at top of groove, but in a different position than in DAPI structure. Raster images calculated using RMS by D.S. Goodsell (200).

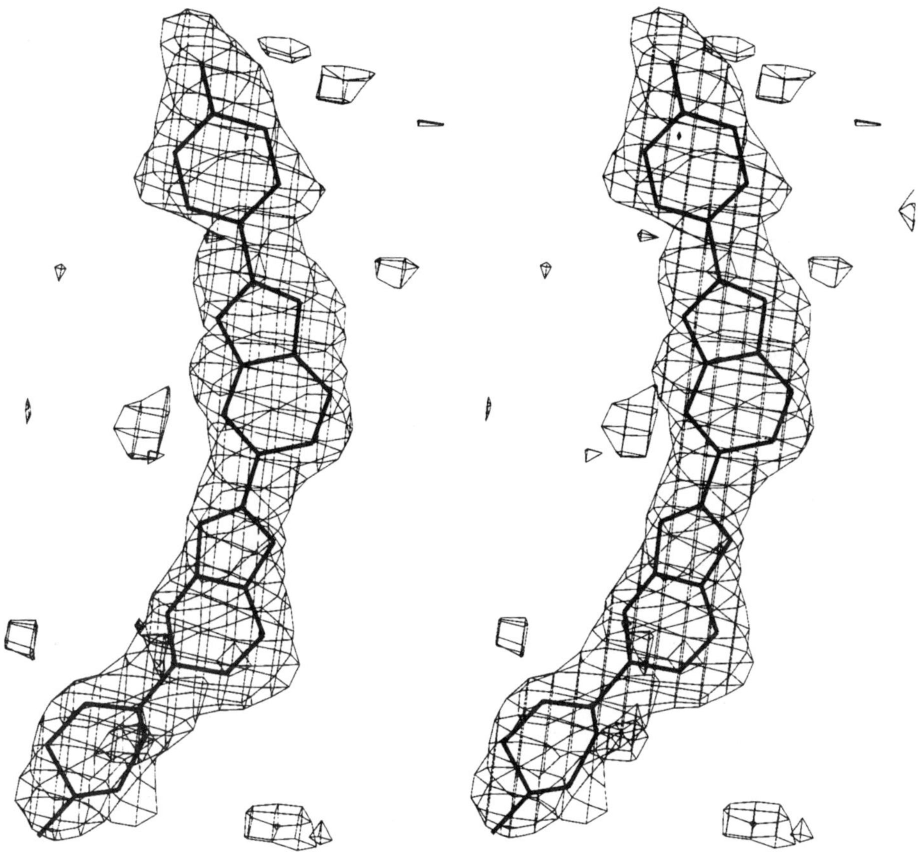

Figure 11 Final $\Delta(F_0 - F_c)$ map of the −25°C Hoechst molecule fitted in the density as it appears in the central region of the CGCGAATTCGCG-drug complex, R_{factor} = 15.2%. F_c's have been calculated with the drug deleted from the model. Notice the goodness of fit of the rings to the density.

Table 3 Comparison of Data for Hoechst Bound to:

	Temp. (°C)	Resol. (Å)	R-factor (%)	A No. of $F_0 > 2\sigma$	B No. of atoms	A/B ratio
A. CGCGAATTCGCG						
Pjura et al.	RT	2.2	14.0	1561	518	3.0 : 1
Teng et al.	15	2.25	15.7	2000	518	4.0 : 1
Quintana et al.	0	1.97	15.7	2540	518	5.2 : 1
	−25	1.9	15.2	2873	518	5.9 : 1
	−100	1.97	14.9	2072	518	4.2 : 1
B. CGCGATATCGCG						
Carrondo et al.	15	2.3	13.7	1573	518	3.2 : 1

in the vicinity of the GC base pair where the piperazine is positioned. Once again, since there is no native data set for the alternating AT sequence, no statement can be made about groove widening or bending induced by Hoechst.

C. DAPI

In the early 1970s after the successful use of diarylamidines against trypanosomes, a synthetic program to find even more effective trypanocides was undertaken. As a result, the compound DAPI, an acronym for 4′-6-diamidine-2-phenylindole (Fig. 12), was synthesized and proved to be a very useful agent against these protozoa (121). In some patients there can be undesirable side effects from this drug, therefore another diarylamidine, pentamidine, is more commonly used today. DAPI has also been shown to inhibit DNA and RNA polymerase but only with poly(dA·dT) as template (122,123). For this reason, De Clerq, in a study of the inhibitory effect of various compounds against oncornaviral DNA polymerase, found DAPI to have a slightly better than average inhibitory effect (124).

DAPI is a fluorochrome like Hoechst 33258, and the early investigation of this property by Kapuscinski (125) revealed its strong AT, AU, and IC preference. No fluorescence resulted when DAPI bound to GC-containing sequences. The fluorescent qualities of this ligand are used to advantage in tissue culture for cytological staining of AT-rich regions of DNA (126). The initial use of DAPI as a chromosomal stain (127–129) has expanded to include clinical tests for chromosomal abnormalities in neoplasias (130). Recently, automation of these methods using flow cytometry has resulted in a rapid method for screening cancer cells (131).

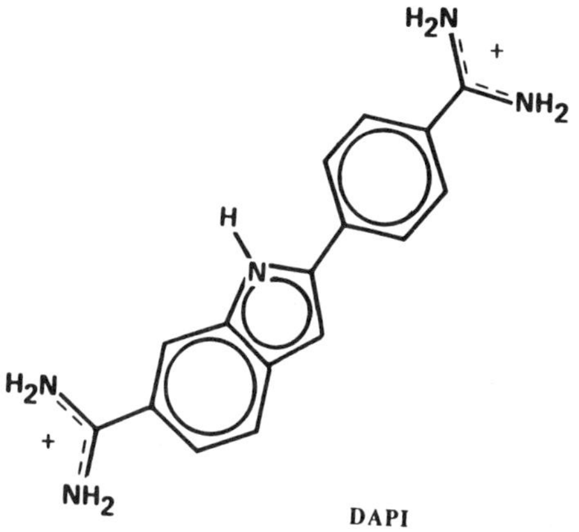

Figure 12 Molecular formula of DAPI, 4′-6′-diamidino-2-phenylindole, with a diamidino tail followed by the phenyl ring at the right and the indole ring at the left followed by the other diamidino tail.

DAPI binds to three base pairs (125,132). Spectroscopic data (125, 132,134) indicate two modes of binding to AT-rich nucleotides and a third nonspecific mode (133). By employing a combination of liner and circular dichroism as well as fluorescent techniques, Kubista and co-workers (133) demonstrated that at low binding ratios DAPI binds in its high-affinity mode with its long axis parallel to the grooves, preferentially binding to AT base pairs. The formation of this complex accounts for its increased fluorescent properties (125). At higher binding ratios, it can be inferred from the spectra that the second high-affinity, AT-binding mode is actually the mutual interaction in the groove of two close-lying DAPI molecules.

Cavatorta et al. (134) suggest that DAPI is not AT-specific. However, most agree that any binding to GC quenches the fluorescence (125,134), and a definite AT preference exists for the high-affinity binding mode which is salt-insentitive (132,133). The weaker, salt-sensitive mode is uninfluenced by base composition. Speculation about the nature of the nonspecific mode includes proposals that the major groove is involved, or the interaction is with the phosphates along the backbone with even a suggestion by some of intercalation (125,132,133).

As with Hoechst, partial quenching of the DAPI fluorescence occurs with the introduction of BrdU (127). However, Buys and co-investigators (135) propose that a different mechanism of quenching exists for each dye: molecular quenching by BrdU for Hoechst versus photolysis of BrdU substituted DNA by DAPI. In a recent study of DAPI binding to double-stranded RNA, Manzini et al. (136) measured the thermodymamic properties of DAPI binding to poly(dA)·poly(dU) as well as to calf thymus DNA. In both cases the enthalpy change (ΔH) is negative: -4.5 kcal mol^{-1} for calf thymus DNA and -8 kcal mol^{-1} for poly(dA)·poly(dU). This data demonstrates that the binding of DAPI to poly(dA)·poly(dU) with its more negative enthalpy is favored over binding to calf thymus DNA, which is a mixture of GC and AT bases. Furthermore, Manzini proposes that the three hydrogen donor groups on DAPI contribute hydrogen bonds to the drug-DNA complex (132).

1. DAPI + CGCGAATTCGCG

The appearance in the literature of the crystal structure of DAPI bound to the dodecamer, CGCGAATTCGCG, addresses the method of attachment of DAPI in its high affinity, AT-specific binding mode within the minor groove (137). To position the drug accurately, both DAPI orientations were refined in parallel. This refinement resulted in an unambiguous placement of the drug in one orientation with the its phenyl ring at the end of the AATT region (Fig. 13 and Plate 2e). The binding pattern is shown in the

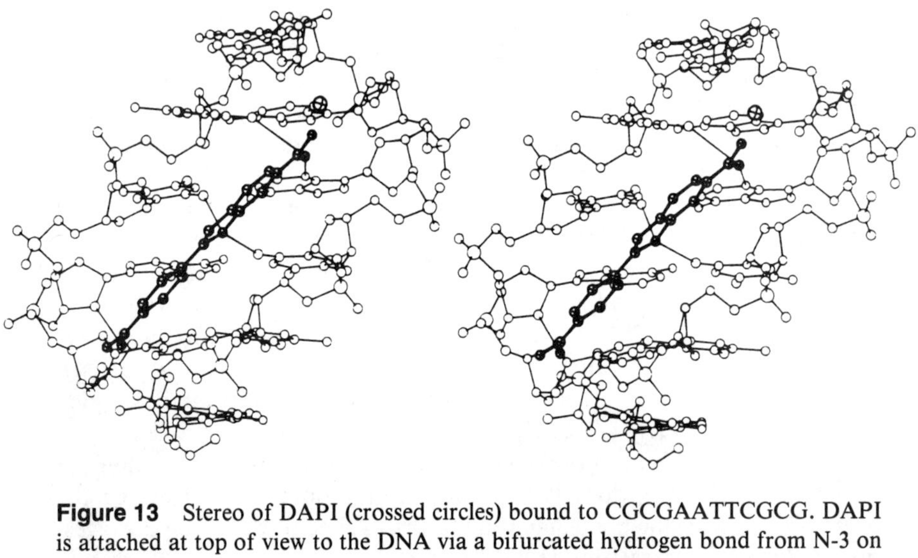

Figure 13 Stereo of DAPI (crossed circles) bound to CGCGAATTCGCG. DAPI is attached at top of view to the DNA via a bifurcated hydrogen bond from N-3 on the amidinium tail to thymine 20 O-2 and adenine 6 N-3 as well as to the water molecule (large crossed circles) at top of helix. The water molecule is in hydrogen bonding distance to both nitrogens, N-2 and N-3, on the amidinium tail. The central nitrogen in the phenylindole ring makes a bifurcated hydrogen bond across the groove to the O-2 atoms on thymines 19 and 7. At the bottom of the picture the DAPI amidine tail is in close contact with the O-4′ sugar on adenine 18.

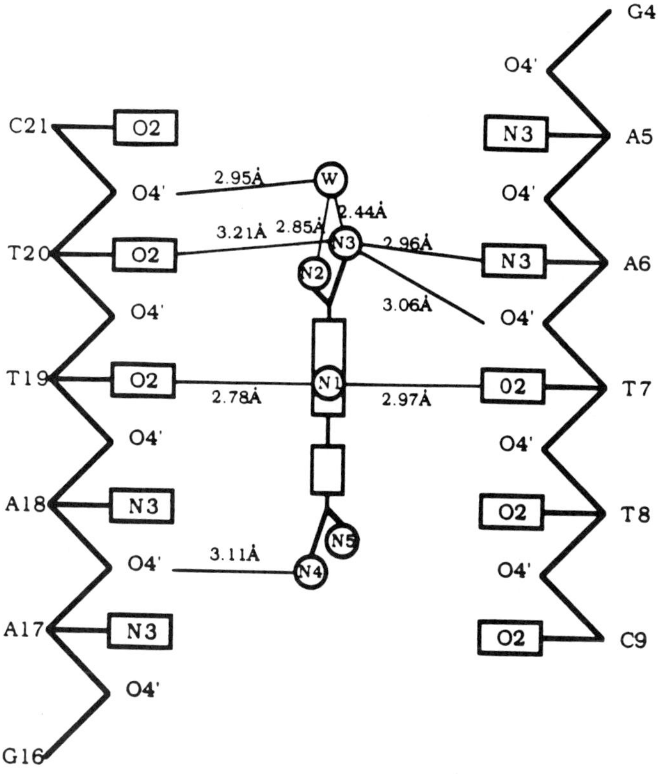

Figure 14 Distances between DAPI nitrogens (N-1–N-5), the bound water molecule (W) and potential hydrogen bond acceptors along the bottom and sides of the minor groove (thymine O-2, adenine N-3 and sugar O-4′). Note three-centered bond between water and C-21 O-4′, N-2, and N-3 of amidine tail. N-4 on the other amidine tail is hydrogen bonded to sugar O-4′ atom on A18.

schematic in Figure 14. This schematic shows the N-3 amidino nitrogen forming hydrogen bonds to the N-3 atom on A6 and O-2 on T20. One spine water molecule, not replaced by the drug, is within hydrogen bonding distance to these same N-2 and N-3 amidino nitrogens as well as to the O-4′ sugar on C-21. The indole ring N-1 nitrogen forms a three-centered hydrogen bond across the groove from O-2 of T7 to O-2 of T19 at the center of the sequence. The amidino nitrogen N-4 at the other end of DAPI makes a single hydrogen bond to the O-4′ sugar atom on A18. An attempt was made to fit the opposite orientation of the DAPI molecule into the density and a comparison of the two orientations is shown in Figure 15.

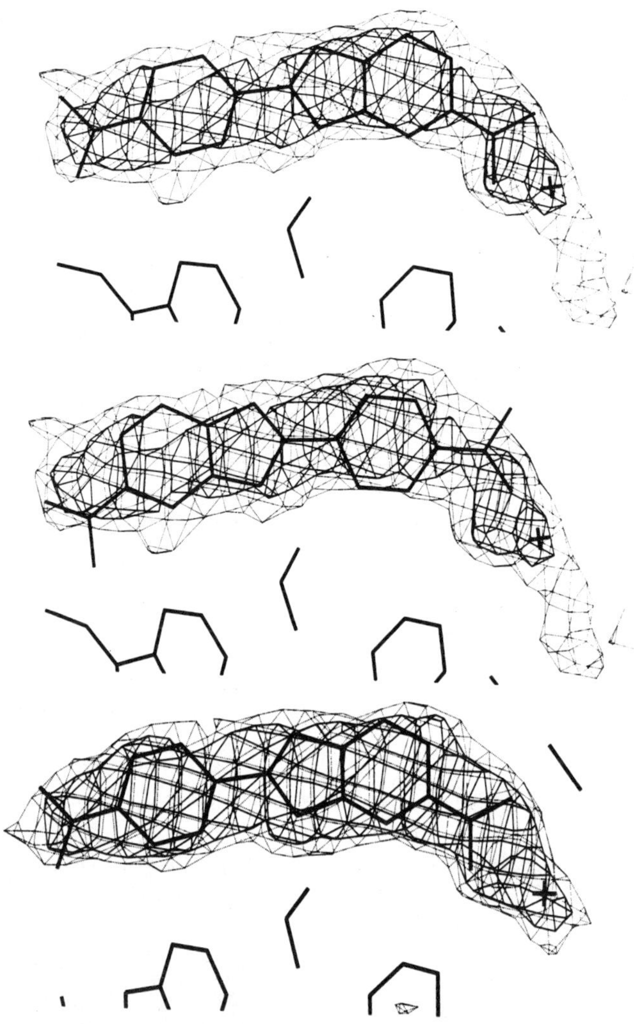

Figure 15 Difference maps in the plane of the DAPI molecule, that establish the position and orientation of the drug. (top) Difference map at $R_{\text{factor}} = 24.4\%$, contoured at the 3σ and 2σ levels as solid and dotted lines, respectively. The DAPI skeleton is fit in orientation A, and a water molecule is added at the cross. (middle) Same difference map with DAPI built in orientation B, again with a water at X. Now the shapes of the difference density and DAPI skeleton do not match. Sliding the DAPI molecule within the framework does not rectify the fit. (bottom) Final deletion difference map, ΔF, after refinement of DNA plus DAPI in orientation A plus 25 waters, $R_{\text{factor}} = 21.5\%$.

One can see that when this alternate orientation was fit to the density, the amidino tails were not positioned in the density properly and, therefore this orientation was eliminated.

In a similar manner to the other drugs presented, DAPI binds with its indole and phenyl rings parallel to the walls of the minor groove, covering the three-base-pair sequence of -ATT-. The local specificity is conferred by the tight van der Waals contacts of the phenyl and indole with the C-2 of adenine A6 and A18, respectively, while the amidine phenyl group senses the C-2 of adenine A17. The drug displaces all but one of the water molecules of the spine of hydration present in the AATT center of this dodecamer. Maximum widening of the minor groove of ≈ 2 Å occurs at the site of the phenyl ring, and the binding of the drug bends the helix axis backward by less than 4°.

D. Berenil

Berenil, synthesized in the early 1950s, is another of the diarylamidine compounds like DAPI, possessing anti-parasitic properties (138,139). It consists of two phenylamidine rings connected by a triazene bridge (Fig. 16). As an effective trypanocide in animals (140,141), it is still used in Africa today, although some resistance to its effectiveness has developed over the years (142–144).

The drug functions as a trypanocide by binding to kinetoplasts, specialized DNA containing regions of the mitochondrion (141). Appearance of a

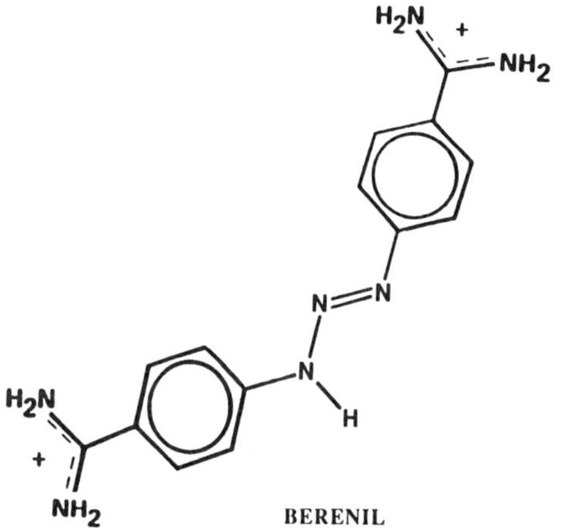

BERENIL

Figure 16 Molecular structure of berenil, 4,4′-diamidinodiazoaminobenzol, with two diamidine benzyl rings joined by a triazene bridge.

strong blue fluorescence is evidence of a drug-DNA complex. The preference of berenil for AT-rich DNA has been demonstrated by a number of researchers (145,146), with the prediction of the size of the binding site to be four nucleotides long (92,145). Intercalation as a binding mode has been ruled out by both Newton (147) and Waring (148). In a review of nonintercalative binding, Krey (149) proposed that berenil binds, as suggested for a related diarylamidine, hydroxystilbamidine, by lying across the minor groove with the amidine tails reaching from phosphates on one strand to phosphates on the opposite strand (150). A theoretical study on the binding of these two compounds postulated that, since berenil is a curved molecule, its amidinium tails would interact from its concave side primarily with oxygens on thymine, burying the berenil molecule deep in the minor groove (151). This same study used molecular modeling to show berenil spanning only two base pairs in the binding site.

The crystal structure of berenil alone (152) revealed that the molecule was indeed slightly curved, with a twist of 9.6° between the phenyl rings. Molecular modeling techniques were then used to investigate the binding of berenil to various AT oligomers. In these simulations, hydrogen bonds were formed from the O-2 atoms on thymine or the N-3 atoms on adenine to the amidine tails. Since formation of hydrogen bonds to the adenine N-3 atoms was somewhat hindered by the adenine H-2 atoms coming in close contact with the phenyl rings of berenil, the authors postulate that this might result in a binding preference for ApT over TpA steps. This supports the footprinting studies of Portugal and Waring, in which they imply that alternating AT sequences are favored, particularly ApT over TpA (92).

1. Berenil + CGCGAATTCGCG

The most recent structure analysis of a minor groove binding drug and DNA (153), that of berenil complexed with the -AATT- dodecamer, was published in 1990. The drug spans the three base pair -AAT- site (Fig. 17 and Plate 2f). The same researchers, using molecular modeling and energy minimization to measure conformational energy changes, found three sites that differed in energy by ≈ 1 kcal, aATT, aaTT, and AATt. The site with the lowest calculated energy was AAT, the binding site found in the x-ray analysis. (See Table 1 of Ref. 152).

Berenil, with its phenyl rings parallel to the walls of the groove, is held in place by hydrogen bonds from each amidinium tail to opposite strands on the DNA (Fig. 18). One amidino tail bonds directly through one NH_2 to the base edge N-3 atom of adenine A18 at a distance of 3.11 Å and with a somewhat longer bond of 3.4 Å to O-2 on thymine 8. The other amidino tail, positioned near A5, does not participate in any hydrogen bonds directly to this adenine base, but bonds instead to a water molecule. This

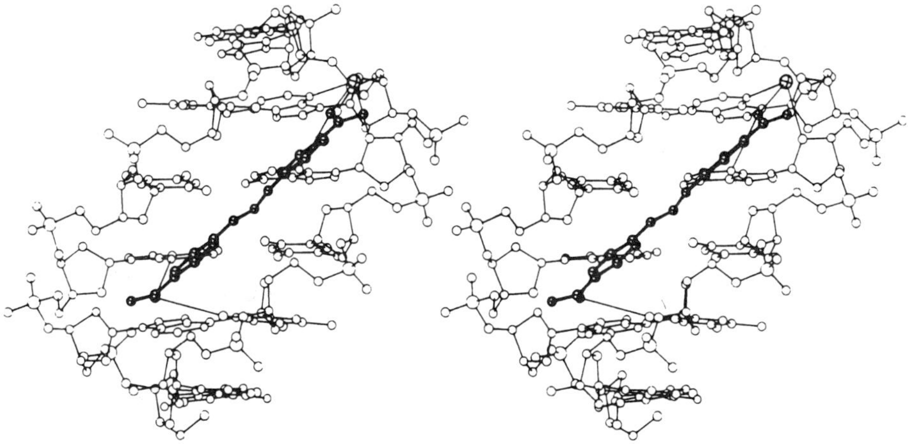

Figure 17 Minor groove stereo drawing of berenil (crossed spheres) bound to central six base pairs of CGCGAATTCGCG. At top the amidino tail is hydrogen bonded to the DNA through a water molecule (large crossed sphere) to adenine 5 N-3 and the O-4′ on adenine 6 and through a somewhat longer bond (3.4 Å) directly to adenine 6 N-3. At the bottom of the picture the drug is attached via a bifurcated bond from its amidino tail to N-3 on adenine 18 and O-2 on thymine 8. Hydrogen bonds are drawn for all distances less than 3.55Å.

water participates in hydrogen bonds with the N-3 of adenine 5, one NH_2 of the amidinium tail, and the O-4′ sugar atom of adenine 6 (Fig. 18). Neidle and co-workers state that this water is shifted relative to the position in the native structure (153). The triazene bridge faces away from the groove floor and therefore does not participate in base recognition. However, it is within hydrogen bonding distance to a water molecule. As seen with the other minor groove drugs, binding of berenil to the helix widens the minor groove by 1.5 Å.

Concurrent with the publication of the x-ray crystal analysis of the berenil-dodecamer complex, there appeared an NMR study of berenil binding to an octamer, GCAATTGC, having the same AATT center as the dodecamer (154). The authors propose two models for the binding interaction based on their NMR data. In one model the drug is anchored via two hydrogen bonds, one from each amidinium tail, to thymine oxygens located across the groove, i.e., from nitrogens on the tails to thymine 6 and thymine 14. In this structure the drug molecule spans four base pairs. Their second model, which they favor, has hydrogen bonds to adenines located on opposite strands, with bonds from the nitrogens on the tails to N-3 of adenine 4 and adenine 12. In this NMR structure, the drug spans only two base pairs.

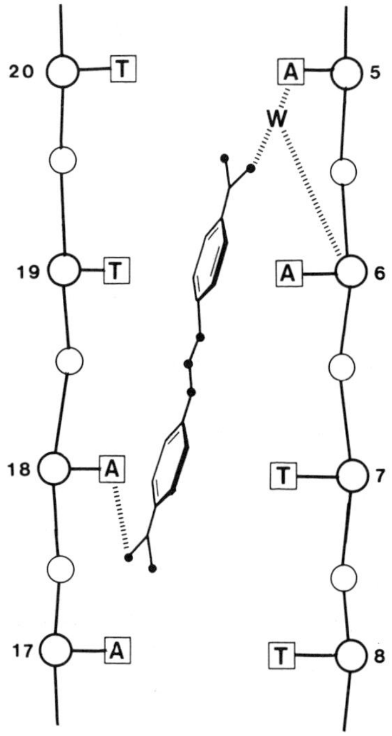

Figure 18 Schematic representation of the berenil molecule bound to CGC-GAATTCGCG. Dashed lines indicate hydrogen bonds from one amino group of each amidine to N-3 on adenine 18 and through a water bridge to N-3 on adenine 5. This water also bridges to the O-4′ sugar on adenine 6. Hydrogen bond distances of 3.2 Å or less are drawn. (Reproduced with permission from Ref. 153.).

Both models misjudge the length of the binding site. Energy calculations have found the longest site to be at most three base pairs (152), and this was corroborated by the crystal structure analysis (153). In the case of the proposed two-base-pair site (154), where the drug binds across the groove from adenine to adenine, the drug would have to be bowed or have one tail twisted back on itself to span such a short distance in the narrow minor groove. The long axis of the berenil molecule is 11–13 Å (147), and the width of the groove in the -AATT- center of the netropsin dodecamer is 5–5.5 Å. On the other hand, the four-base-pair site is too long a distance for the short berenil molecule to span.

This concludes coverage of all the known crystallographic structures of minor-groove-binding drugs and includes some of the NMR results as well. For those interested in molecular modeling, coordinates to model the crystal structures mentioned here can be obtained from the Protein Data Bank at Brookhaven National Laboratory. However, before any attempt is made to model a DNA-drug interaction, it is important to determine the proper binding sequence, length, and groove width of the DNA. When choosing a sequence, remember that solution studies reveal a strong homopolymer preference of netropsin and distamycin. However, the binding constants to mixed AT sequences are substantial enough to use these sequences as well for modeling. It is known that fiber diffraction coordinates yield a much wider minor groove in AT-rich regions than the crystal structures indicate. Single-crystal diffraction demonstrated that the AT minor groove regions can be on the average 2–3 Å narrower than the GC regions, and this should be taken into account.

IV. DRUG RECOGNITION PATTERNS

Although x-ray diffraction does not locate the hydrogens on the DNA-drug complex, it is possible to generate the hydrogen positions in their correct geometry for the various groups on the drugs and the DNA oligomers. A close look at the pattern of NH bonding groups and CH van der Waals contact groups of the four drugs: netropsin, Hoechst 33258, DAPI, and berenil are shown in Figure 19. One striking feature that is evident from this figure is the strong repeat motif of approximately 4.6 Å (range is 4.2–4.9 Å) between the NH hydrogen bonding groups along the netropsin molecule's concave edge. This 4.6 Å repeat occurs four times in netropsin and could be a factor contributing to its strong binding character, because it matches the spacing of the hydrogen bond acceptors on the floor of the groove.

The other three drugs do not have extended repeat patterns, but do have the 4.6 Å average repeat distance somewhere along their concave edges. In some instances this 4.6 Å repeat is between NH bonding groups, but also this same distance can be found between a hydrogen bonding NH and a CH van der Waals contact group. Hoechst shows the netropsin repeat pattern only once in the center of the molecule and this may contribute to its ability to accept GC base pairs at the end of the binding sequence where this pattern is absent. In the case of DAPI, there is one strong 4.46 Å repeat between NH hydrogen bonding atoms and then a long repeat of 6.20 Å between NH hydrogen bond donors of the phenylindole ring and the N-4 amidine tail amino group. In the DAPI-DNA crystal structure this 6.2 Å distance is not quite long enough to allow the N-4 amino group to make hydrogen bonds with the base edges of adenine 17 and thymine 9, but instead this amino group is in close contact with the O-4′ sugar atoms that

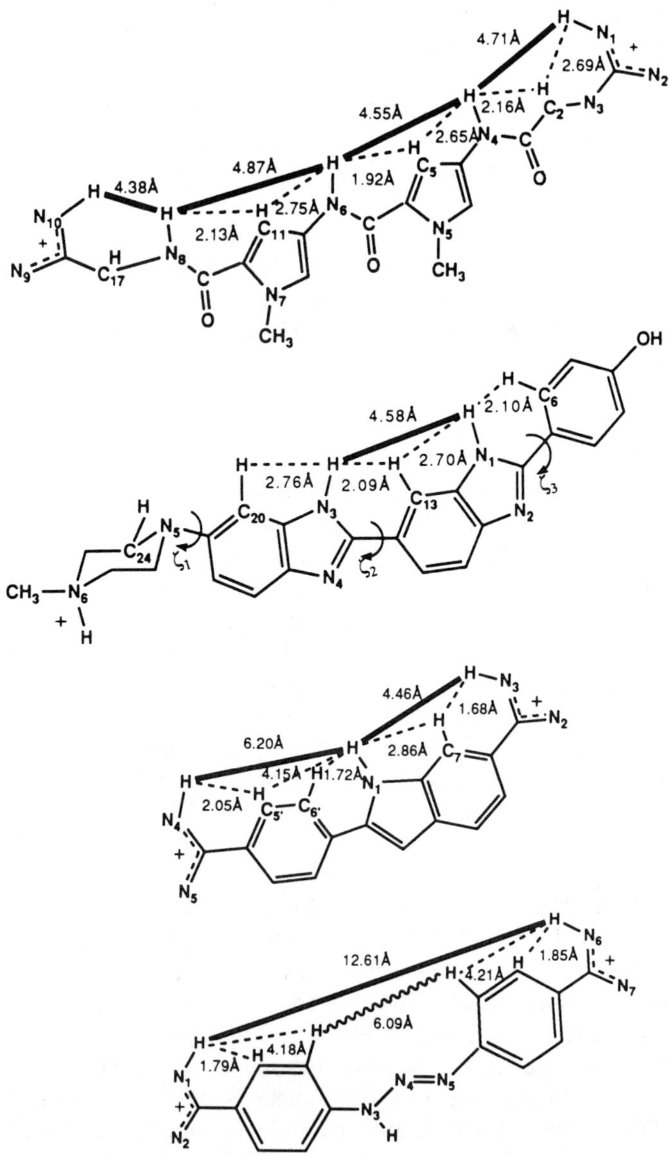

A

B

C

D

Figure 19 Hydrogen atom separations along concave edges of (A) netropsin, (B) Hoechst 33258, (C) DAPI, and (D) berenil. Solid lines connect NH hydrogens. Dashed lines connect CH hydrogens with other OH hydrogens or NH hydrogens. Wavy line indicates long distance between CH atoms in central portion of berenil molecule. Note repeat distance range of 4.2–4.9 Å between hydrogen bonding atoms.

precede these bases (Fig. 14). In the berenil molecule the repeat between NH binding atoms is 12.6 Å which is quite long. However, there is a 4.2 Å repeat between NH hydrogen bonding amino groups on the tail and CH van der Waals sensing groups which help to position the molecule.

V. GC RECOGNITION: THE LEXITROPSIN RATIONALE

The netropsin CGCGAATTCGCG dodecamer x-ray analysis (72,73), which recognized that the specificity of AT preference resided in the non-bonded contacts between netropsin methylene or pyrrole CH groups and adenine C-2 hydrogen atoms on the floor of the minor groove, opened the door to constructive design of drugs that would "read" the floor of the groove (Fig. 20). Two groups, Dickerson at UCLA (72,73) and Lown at the University of Alberta (155–157), realized that by substituting an imidazole (Im) or furan (Fu) for a pyrrole (Py), a potential GC reading element could be created. With these substitutions, a hydrogen bond can be made from the ring nitrogen of imidazole or the oxygen of furan to the guanine NH_2 (Fig. 20). Lown and co-investigators (155–157) developed the synthesis methods for preparing these compounds, and the subsequent chemical, physical and biological investigation of these new sequence-reading ligands is the focus of the remainder of this chapter. Although there have been no single crystal x-ray analyses of lexitropsins bound to DNA oligomers, nuclear magnetic resonance techniques have been employed to investigate lexitropsin binding in solution and help establish the sequence reading potential of these ligands.

A. Imidazole Versus Pyrrole Reading Elements

1. *Dicationic Lexitropsins*

In the first series of "lexitropsins" or groove-reading drugs synthesized by Lown (155), an *N*-methylimidazole ring replaced one or both *N*-methylpyrrole rings of netropsin, keeping the same two cationic end groups, the guanidinium (Gua^+) and propylamidinium ($CH_2CH_2Am^+$) (Fig. 21). Footprinting experiments (155,158) with these lexitropsins, which will be called series 1 lexitropsins, demonstrated that all still favor the same AT sites (Table 4). In some instances, the bis-imidazole, Gua^+-Im-Im-$CH_2CH_2Am^+$, will bind GC base pairs, but not exclusively. Pullman's electrostatic calculations help to explain the relatively strong proclivity of the Im-Im for GC regions (159,160). His complexation energy calculations (160) show that the energy of ligand binding decreases to a small negative value as the number of imidazoles replacing pyrroles on the drug increases, demonstrating that AT preference is diminished, but not replaced by a

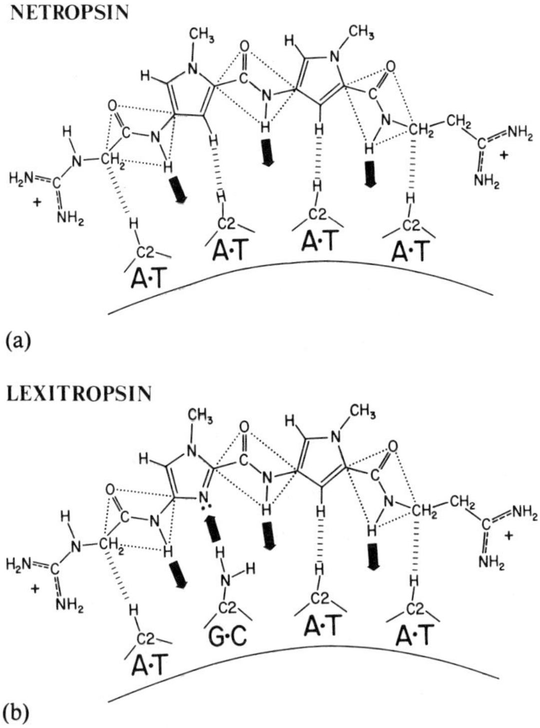

Figure 20 Binding of netropsin (top) and lexitropsin (bottom) to the minor groove of B-DNA. Filled arrows indicate hydrogen bonds, in the donor-to-acceptor direction. Barred lines mark close van der Waals nonbonded contacts between drug and DNA: more specifically, to the C-2 hydrogen atoms of adenine residues. (a) Netropsin binding; (b) modification of netropsin by replacing one pyrrole with imidazole to generate a lexitropsin capable of reading base sequence and accepting a GC base pair.

preference for GC. In series 1 lexitropsins, AT base pairs are admitted in the recognition sequence as well as GC. Another indicator of the diminution in AT preference is that the change in complexation energy in going from AT to GC, $\Delta_{complex}$(AT) − (GC), never becomes a positive quantity, which would imply a GC predilection. Instead of having a reading element capable of specific binding to GC, the end result is one that is indiscriminate, i.e. either accepting GC or AT (Fig. 22). Burckhardt et al. (161), reaching the same conclusion using circular dichroism (CD) methods, state that two or more imidazole residues are required to promote significant GC binding, but note that such a complex becomes unstable as the ionic strength increases, which is not the case for netropsin.

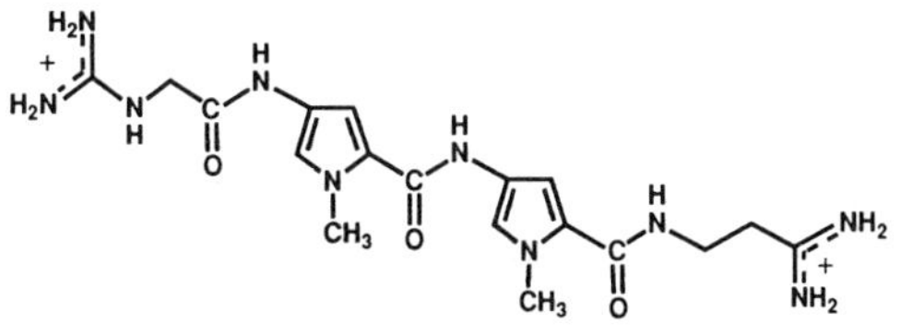

1 NETROPSIN

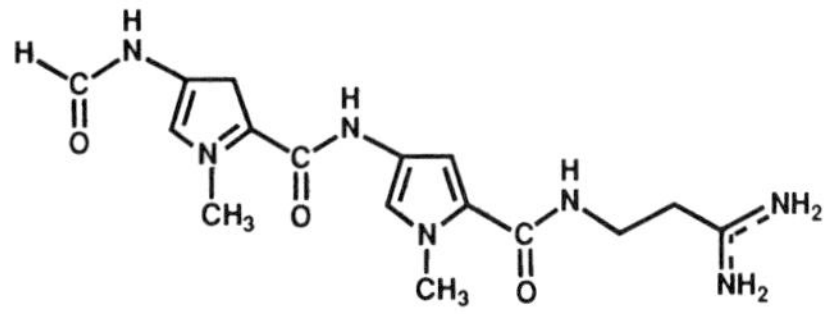

1 N-FORMYLNETROPSIN

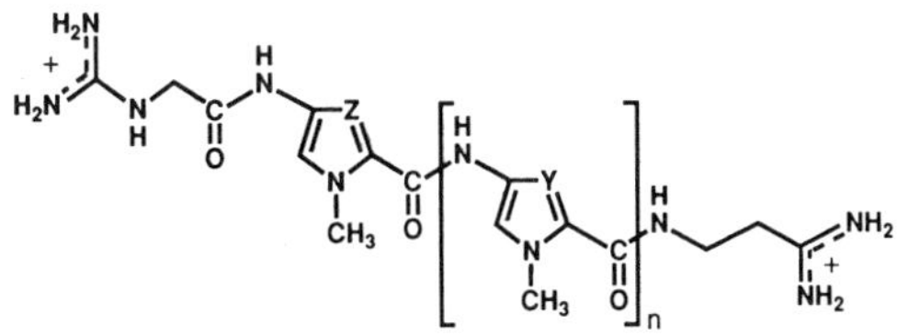

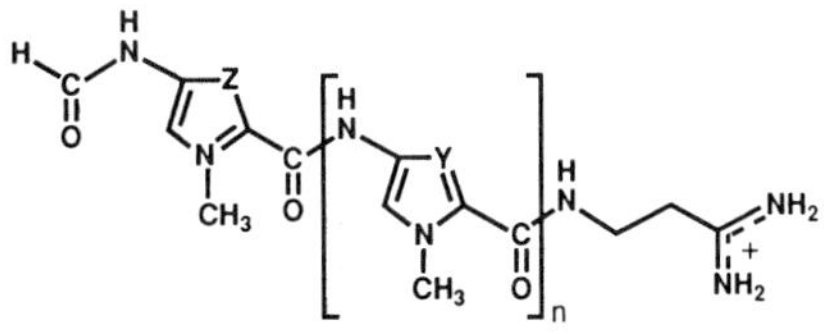

2 Z = N, Y = CH, n = 1

3 Z = CH, Y = N, n = 1

4 Z = N, Y = N, n = 1

5 Z = N, Y = N, n = 2

Figure 21 Series 1—dicationic lexitropsins, including the naturally occurring netropsin molecule and series 2—monocationic lexitropsins, including *N*-formylnetropsin. The substitution at position Y or Z of a nitrogen for a carbon changes the *N*-methylpyrrole ring to a *N*-methylimidazole.

Table 4 Lexitropsin Reading Elements

	Series 1 Dicationic Lexitropsins			Series 2 Monocationic Lexitropsins	
	AT	GC		AT	GC
Py-Py[a]	+		CHO-Py-Py[b]	+	(+)
Py-Im	+		CHO-Im-Py	+	(+)
Im-Py	+	(+)	CHO-Py-Im	(+)	+
Im-Im	(+)	+	CHO-Im-Im		+
Im-Im-Im		+			

Parentheses indicate weak acceptance.
[a]netropsin
[b]*N*-formylnetropsin
From Refs. 155 and 164.

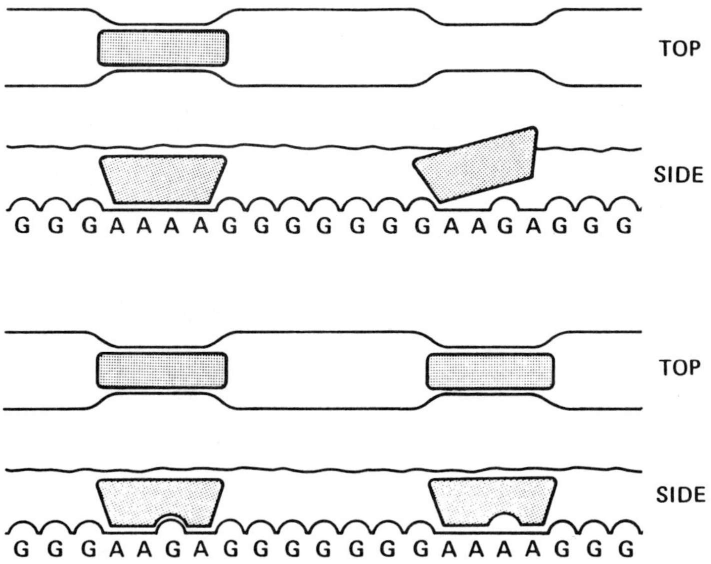

Figure 22 Schematic drawing to illustrate (1) the effect of minor groove width on AT base pair specificity of a drug molecule such as netropsin, and (2) a "lexitropsin" (shaded trapezoids), introducing potential GC specificity. A = site of an AT or TA base pair. G = site of a GC or CG base pair, with a semicircular bump symbolizing the C2 amine group of guanine. (upper left) The drug molecule binds well to the narrow and deep minor groove in a region of AT base pairs, and not to the wide and shallow groove characteristic of GC base pair regions. (upper right) Incorporation of an isolated GC base pair into an AT region introduces a protruding $-NH_2$ group that would push the drug molecule away. (lower left) A lexitropsin with a "notch" that can fit over the guanine amine (e.g., substitution of imidazole or furan for pyrrole, and hence —N— or —O— for $-CH_2-$) might bind to the GC-containing site. (lower right) A complication to this simple lexitropsin model is that the "notched" drug might still accept an adenine at its modified site, meaning that site on the drug molecule would have been made, not GC-specific, but merely nonspecific.

These compounds were tested for their antitumor activity (156) and several were found to be active against vaccinia virus, but none of the new ligands was more effective than netropsin or distamycin. Recently, the binding of two of these lexitropsin ligands was studied by NMR (162,163). The reader should be aware that in all of the NMR structures that will be reported, it was first established that the DNA was in the right-handed B-conformation before determining drug contacts.

NMR Results. 1a. Gua$^+$-Py-Im-CH_2CH_2Am$^+$ + CGCAATTGCG. The evidence from NOE, COSY, and NOESY experiments by Lee et al. (162) demonstrates that the dicationic Py-Im lexitropsin binds to the AATT minor groove of this B-DNA decamer by a series of hydrogen bonds to N-3 and O-2 atoms on the base edges on the floor of the minor groove.

1b. Gua$^+$-Im-Im-CH_2CH_2Am$^+$ + CGCAATTGCG. A second set of NMR experiments (163) investigating the binding of the dicationic bis-imidazole analog of netropsin, Gua$^+$-Im-Im-CH_2CH_2Am$^+$, to the same decamer located the drug primarily in the -AATT- site and very transiently on the NMR time scale at -CAAT-. The secondary site is reflected in the chemical shifts observed for the cytosine 3 and guanine 8 bases. It appears that this dicationic lexitropsin ligand has a greater affinity for -AATT- than for any GC-containing site.

Both of these NMR results corroborate the conclusion derived from footprinting experiments that no strong preference for GC exists with these series 1 lexitropsins. According to Lown, the electrostatic attraction of the dicationic ends to AT pairs in the minor groove is a major contributor to the lexitropsin ambiguity in binding (155), and this is affirmed by Pullman's aforementioned electrostatic analysis (159).

2. *Monocationic Lexitropsins*

Lown proposed to overcome the charged end attraction to AT regions by replacing one cationic end, the guanidinium group, with the uncharged *N*-formyl group found on distamycin (155,164). This next series of lexitropsins, series 2 (Fig. 21), is also evaluated in Table 4. These monocationic lexitropsins have a greater propensity for binding to GC base pairs than the dicationics as evidenced from the footprinting data (158,164). In series 2, *N*-formylnetropsin (CHO-Net) sometimes admits GC base pairs at the beginning of the binding site, and its sequence preference can be characterized as ranging from a four-base-pair site containing all AT bases $(AT)_4$ to the admittance of a single GC base pair anywhere within a $(GC)(AT)_3$ site. The strongest GC binder is the monocationic bis-imidazole, CHO-Im-Im, with the ability to admit as many as three GC pairs. Its binding can be characterized as $(GC)_3(AT)$ with an apparent preference for the sequence -CCGT-, based on footprinting results (162). The GC binding affinity of the monocationic lexitropsins in order of decreasing GC preference is CHO-Im-Im > CHO-Py-Im ≫ CHO-Im-Py > CHO-Net > Net. CHO-Im-Im and CHO-Py-Im have a definite disposition for GC while the others in the series do not.

The binding properties exhibited by these monocationic series 2 lexitropsins occur for a twofold reason according to Lown: (a) Not only is there a reduction in charge on the drug molecule, which reduces the attraction to AT regions, but (b) removal of the guanidinium methylene allows the NH_2 of guanine to be admitted at the *N*-formyl end of the lexitropsin molecule by eliminating a van der Waals contact (72,73) between the methylene group and the DNA. Lown attributes this replacement of the guanidinium tail by the *N*-formyl group as the reason for a GC base pair being accommodated there. Pullman's electrostatic energy computations on the bis-imidazole corroborates its GC-reading capability (165). Zimmer and co-workers (166) used circular dichroism to investigate the differences between the mono- and dicationic lexitropsins. They conclude that the monocationic imidazole-containing lexitropsins have a lower affinity to AT sequences than netropsin and distamycin and that this attraction for AT sequences decreases as the number of imidazoles increases in either series. These imidazole oligopeptides do not bind to pure GC sequences, but to GC flanked by at least one AT, because the methylenes on the propylamidinium tail necessitate reading an AT base pair. Zimmer also postulates that perhaps another AT pair may be necessary to stabilize the complex.

NMR Results. Once again NOE, COSY, and NOESY methods were used in the NMR analyses of the binding of the monocationic lexitropsin CHO-Im-Im to two DNA oligomers: (a) the self-complementary CATGGCCATG and (b) the noncomplementary (GATCCGTATG)·(CATACGGATC) (167,168), and (c) the binding of a "truncated" *N*-formylnetropsin to CG-CAATTGCG (169).

2a. CHO-Im-Im-$CH_2CH_2Am^+$ + CATGGCCATG. COSY results suggest that the bis-imidazole monocationic lexitropsin location is at -CCAT-. These results, coupled with NOE signals, indicate that methylene protons are in close proximity to the A8–T13 base pair (167). Confirmation of these assignments by chemical shift measurements place the drug with its propylamidinium tail in the region of adenine 8 and thymine 9 with the two imidazoles binding to cytosines 6 and 7, thus covering the four base sequence -CCAT-.

2b. CHO-Im-Im-$CH_2CH_2Am^+$ + (GATCCGTATG) · (CATACG-GATC). A combination of NOE, COSY and NOESY measurements established that CHO-Im-Im binds to the four-base-pair sequence -CCGT- with its propylamidinium tail positioned at the T7 end of these four bases. Here the lexitropsin is favoring a $(CG)_3(AT)$ binding sequence (168).

2c. A "truncated" N-formyl netropsin: CHO-Py-Py-CH_2Am^+ + CG-CAATTGCG. As has been shown by x-ray diffraction (72,73) the methylenes in the propylamidinium tail read AT base pairs by sensing the C2H on adenine, and this is one reason for the AT preference at the end of the sequence reading frame. Lown proposed (167) that removal of one of these methylenes would nullify the AT preference. In this NMR study of the binding of the "truncated" *N*-formyl netropsin, CHO-Py-Py-CH_2Am^+, Lee et al. (169) claim to have eliminated this AT bias by shortening the amidinium tail by one methylene group to CH_2Am^+. The attributed binding site on the DNA is -ATTG- and not the usual minor groove site of -AATT-. However, there is some disagreement whether shortening this tail by one methylene will produce a GC-reading end group. The possibility still exists for this remaining methylene, which is bonded to an amide nitrogen and cannot rotate freely, to clash sterically with a guanine NH_2. Also, the charge on the amidinium group will cause it to seek the favorable electrostatic potential found in AT regions.

This initial lexitropsin work established that the monocationic lexitropsins are stronger GC-reading ligands than the dicationic and confirmed the ability of the methylene group to sense AT bases. Lown and co-workers continued their synthetic work to include furans, compounds containing oxygen in place of nitrogen in the organic rings of the lexitropsin molecules, as proposed in the original lexitropsin design (72,73,155).

B. Furan Versus Pyrrole Reading Elements

The preparation of furanyl lexitropsins proved to be more challenging synthetically than the imidazoles, but eventually Lown and co-investigators were successful in preparing a monocationic bis-furan (Fu-Fu) drug (170). This lexitropsin with ring oxygens (Fig. 23) was subjected to the usual footprinting and NMR analysis. The footprinting data established that this lexitropsin prefers at most three GC base pairs $(GC)_3(AT)$ in a four-base-pair binding site, with a strict 3′AT requirement. The binding constant of the Fu-Fu molecule is one order of magnitude greater than a monocationic netropsin molecule measured under similar conditions, 10^6 versus 10^7, which indicates a somewhat weaker binding.

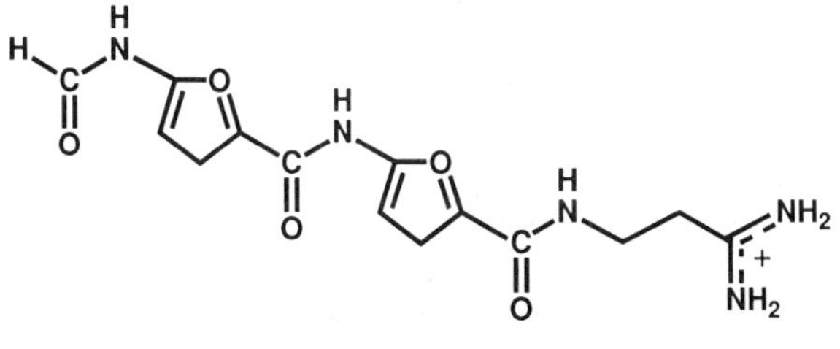

BIS-FURAN LEXITROPSIN

Figure 23 Molecular structure of bis-furan lexitropsin. Substitution of two furan rings for pyrrole rings on *N*-formylnetropsin creates the bis-furan lexitropsin.

1. NMR Results: CHO-Fu-Fu-$CH_2CH_2Am^+$ + CATGGCCATG

The suppression of the COSY CH5-CH6 cross-peaks of the C-6 and C-7 bases gives credibility to situating the bis-furan lexitropsin on -GCCA- (170). This interpretation is based on the proposal of Borah et al. (171) that the binding of a ligand on or near the cytosine in question would suppress this CH5-CH6 cross-peak. Coupled with NOE broadening of relevant proton signals (170), the COSY result and NOESY data substantiate the 5′-GCCA-3′ position. Analysis of the NMR and footprinting data indicates that the bis-furan lexitropsin recognizes GC base pairs by making hydrogen bonds between the furan oxygens and the N-2 amino groups of guanine as predicted.

C. Thiazole Versus Pyrrole Reading Elements

The latest members of the lexitropsin series of groove-reading drugs are the sulfur-containing thiazole (Th) ligands (172–175). Figure 24 shows the two types of thiolexitropsins that were synthesized, those with the ring sulfur atom on the concave side of the ligand (173) and those with the sulfur on the convex side, facing out with no contact with the floor of the groove (172,173).

Footprinting experiments by Rao et al. with these thiazole lexitropsins (174) demonstrate that compounds 6, 7 and 8 (Fig. 24) with the sulfur pointing out of the groove are more GC accepting than distamycin A or netropsin. This GC specificity is understandable, since 6, 7, and 8 mimic imidazole with the nitrogen atom, rather than the sulfur, facing the groove. The thiazole lexitropsins with their ring sulfur atoms facing toward the groove proved not to be GC accepting, but are closely related instead to distamycin in their high AT preference. The authors hypothesize that the bulky sulfur atom directed toward the floor of the groove is interfering with GC binding by sterically clashing with the guanine NH_2 group.

One anomaly of these experiments is the high GC acceptance of the dicationic netropsin on these footprints (174). The molecular formula given for netropsin in the Lown thiazole papers (173–175) shows a “truncated” netropsin with one methylene missing from the propylamidinium tail. The high GC acceptance can probably be attributed to the use of this “truncated” netropsin. Previous footprinting results with methidiumpropyl EDTA·FeII show little or no binding to GC base pairs by the naturally occurring netropsin. When GC binding does occur, it appears primarily at the end of the binding sequence (109). In the “truncated” netropsin footprints (174) the GC acceptance is not only quite high, but occurs in the middle of the footprint as well.

All of the thiazole compounds, whether sulfur is facing toward or away

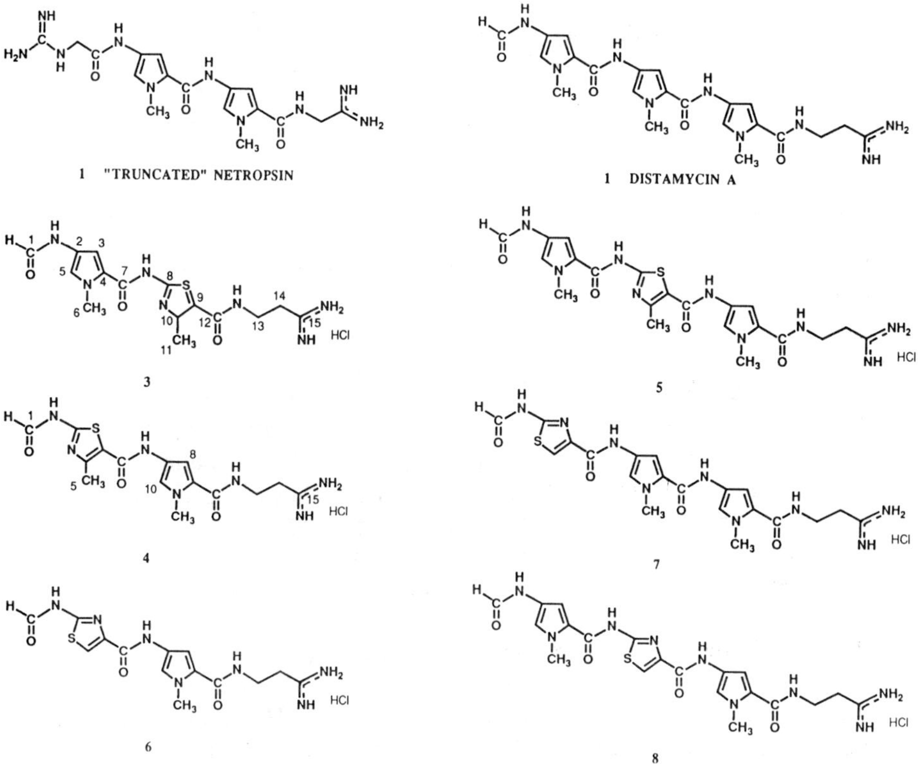

Figure 24 Molecular structures of thiazole lexitropsin analogs of **1** netropsin (truncated by removal of one methylene on the propylamidinium tail) and **2** distamycin A. Note that two types of thiazole rings have been synthesized with the sulfur atom in the ring at either carbon 3 or carbon 5 (see structures **3** and **6**). This allows the sulfur to face "in" or "out" of the groove, changing the binding characteristics of the lexitropsin.

from the groove floor, have weaker binding constants to poly(dA·dT) than netropsin or distamycin and stronger binding to poly(dG·dC) than these parent compounds (Table 5). The thiazole lexitropsins bind equally well to AT and GC sequences as evidenced by their similar binding constants to poly(dA·dT) and poly(dG·dC) of 10^6. Netropsin and distamycin, on the other hand, display a difference of three orders of magnitude in binding between poly(dA·dT) and poly(dG·dC), 10^7 versus 10^4, respectively. Based on these binding constants, although the GC specificity of the thiazole lexitropsins is better than netropsin and distamycin, the AT specificity has not been significantly reduced.

Table 5 Relative Binding Constants of Drug Ligands

	$K_{b\ \text{apparent}}$ (M^{-1})	
	poly (dA·dT)	poly (dG·dC)
Netropsin and distamycin	10^7	10^4
Thiazole lexitropsins	10^6	$\approx 0.5 \times 10^6$

Based on data from Ref. 174.

1. NMR Results: CHO-Th-Py-$CH_2CH_2Am^+$ + CGCAATTGCG

COSY, NOE and NOESY NMR methods were employed to define the binding of this groove-facing sulfur Th-Py lexitropsin (compound 4 of Fig. 24). Kumar and co-workers (175) interpret their NMR spectra as intercalation by the thiazole ring between the A4-T17 and A5-T18 base planes with the rest of the molecule binding in the minor groove, covering the remainder of the AATT region. Their basis for postulating intercalation is the absence of inter- and intranucleotide NOE cross-peaks in the C3, A4, A5 T6, and T7 region and three outer molecular NOEs in the A4-A5 region, namely: The *N*-formyl H1 proton cross-peaks with the adenine 5H8 major groove proton, and the cross peaks between pyrrole H8 and adenine 5H2, and methylene H13b and adenine 4H2. Also, the absence of certain expected NOE cross-peaks in the binding region supports the view that the duplex has been unwound as a result of drug intercalation. However, ^{31}P NMR did not wholly confirm the intercalation model, since the expected large downfield shift in the duplex from intercalation was not present; only a 0.2-ppm upfield shift was observed. An x-ray crystal analysis would prove useful in confirming this intercalation model.

D. Anthelvencin A: A Netropsin-like Antibiotic with a Chiral Center

Anthelvencin A is a naturally occurring dicationic antibiotic related to netropsin and distamycin, but containing a chiral center (Fig. 25). One end of the drug has the usual propylamidinium tail, while at the other end is the 2-aminopyrrolinium ring containing the chiral C-4 atom. Lee et al. (176) examined the binding in solution using biophysical means including ^1H NMR, footprinting, calorimetric and CD techniques. The natural (4*S*)-(+)-anthelvencin and its (4*R*)-(−)-enantiomer bind to CGCAATTGCG in the minor groove at -AATT-. Molecular modeling shows that the natural

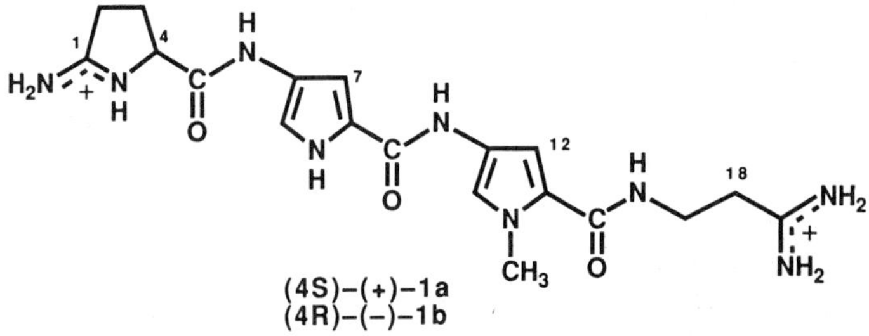

ANTHELVENCIN A

Figure 25 Molecular structure of **1a**—the naturally occurring antibiotic (4*S*)-(+)-anthelvencin A and **1b**—its (4*R*)-(−)-enantiomer. The chiral center is located at carbon 4 on the aminopyrrolinium ring. The guanadinium tail of the parent netropsin molecule has been replaced by a 2-aminopyrrolinium ring, which is followed by amide, pyrrole, *N*-methylpyrrole, amide and propylamidinium.

(4S)-enantiomer has the hydrogen on the chiral C-4 pointing out of the minor groove, whereas the (4*R*)-enantiomer points its charged 2-aminopyrrolinium group out of the groove. DNase I footprinting studies indicate little or no stereoselectivity in binding to DNA for the two enantiomers. Proton NMR substantiates this absence of selectivity.

The thermodynamic profiles of binding to poly(dAT)·poly(dAT) indicate lower binding affinity, as measured by $\Delta G°$, of the anthelvencin isomers when compared to netropsin. $\Delta G°$ for netropsin is -53.1 kJ mol^{-1} as compared to -33.5 kJ mol^{-1} and -33.0 kJ mol^{-1} for the (4*S*) and (4*R*) enantiomers, respectively. According to the authors, since all three drugs are bis-cations, the shape and conformation of the end groups affect the efficiency of binding, while chirality appears to have less of an effect. Since the anthelvencins are longer, containing three rings versus two for netropsin, this may also be a contributing factor to the weaker binding to the four-base AATT sequence as well. Distamycin, with three rings, prefers a five-base binding site while netropsin needs only four AT base pairs. However, comparison of binding constants of the three ligands to poly (dAT)·poly(dAT) shows that the strongest binding ligand netropsin has a binding constant 2×10^3 stronger than either anthelvencin enantiomer. The binding profiles also indicate that netropsin binding is enthalpy driven, while the binding of either isomer of anthelvencin is entropic in nature (Table II of Ref. 176).

E. Critique of Lexitropsins as Groove Binders

The reader may ask what has been accomplished by all the synthetic and investigative research in finding a GC-binding ligand. An assessment of the results reported reveals the following:

1. The initial proposal of making a GC-reading element by substituting a nitrogen or oxygen on the concave side of one or more of the five-membered rings of netropsin was not sufficient in itself to propel the ligand to GC-rich regions. The charged ends directed the ligand to the electrostatic minimum found in AT regions of the minor groove.
2. The presence of methylene groups on the charged ends of netropsin makes the ends AT "readers" because of close hydrophobic contacts between these methylenes and adenine bases on the DNA.
3. Replacement of the guanidinium group,

$$CH_2-CH\begin{matrix} \diagup NH_2 \\ \diagdown NH_2 \end{matrix}$$

   by the *N*-formyl, N—CHO, allows GC to be read because the guanidinium methylene is no longer present to sense the adenine C2H, and a positively charged group has been replaced by a neutral one.

4. The substitution of a lone pair nitrogen atom $\rangle$N: for $\rangle$CH in the pyrrole ring, while allowing a hydrogen bond to be made to guanine NH_2, does not take into account that the absence of hydrogen on the imidazole nitrogen gives it the capability of accepting either an adenine or bonding to a guanine. This factor alone may require an entirely different approach to making a semantophore completely specific for GC base pairs.

Although an exclusive GC-reading element has not been found in the lexitropsins studied to date, the ability of a groove-binding drug to recognize a $(GC)_3(AT)$ sequence is in itself an accomplishment. Lee et al. (170) in an assessment of the GC-reading ability, based on all the biophysical data (NMR, footprinting, binding constants, etc.), propose that the bis-imidazole is a better GC-reading element than the bis-furan which in turn is probably better than the thiazole lexitropsins. A comprehensive footprinting and binding experiment that allows direct comparison of all the various types of lexitropsins, i.e. imidazole, furan, thiazole and "truncated" *N*-formyl netropsin, would help clarify and rank the GC-reading propensity of each type.

F. Future Work

To complement the NMR experiments of Lown, an x-ray crystallography investigation of the binding properties of these lexitropsins and a number of DNA sequences is contemplated by the Dickerson laboratory at UCLA. The DNA oligomers, that have been synthesized for this purpose along with their putative binding sites, are shown in Table 6. Sequences 1, 2, 6, and 7 have been successfully crystallized without the drug ligand and data have been collected. The structures of sequences 2 and 7 have been solved with sequence 2 being solved independently by two groups (177,178). Sequence 6 is in the process of structure analysis (179), and sequence 1 needs higher-resolution data. The remaining sequences are still in preliminary stages of investigation. Another candidate for lexitropsin binding is sequence 7, CGATCGATCG, whose structure has been solved recently (180). It also may be worthwhile to add sequence 8, the self-complementary CATGGCCATG sequence, used by Lown and co-investigators for their NMR work, to the pool of oligomers slated for lexitropsin-DNA x-ray analysis.

VI. ALTERNATE STRATEGIES IN DRUG DESIGN

A. Isohelical Analysis

When Youngquist and Dervan (56) investigated the effect increasing the number of pyrrole-carboxamide units on the distamycin molecule had on DNA-binding ability, they found that a distamycin consisting of seven units

Table 6 Sequences for Lexitropsin Crystallographic Studies

Binding sites	Crystal structure of DNA without drug
1 CG**TG*TTA**A*CACG	x-ray data collected
2 CG**TG*AA**T*CACG	DNA crystal structure solved (Refs. 177, 178)
3 CGT**AGAT**CTACG	
4 CGC**ACAT**GTGCG	
5 CG**TGTA**CACG	
6 CG**TGAT**CACG	x-ray data collected
7 CGA**TCGA**TCG	DNA crystal structure solved (Ref. 180)
8 CA**TGGC**CATG	

*Sites are AT rich netropsin sites.
Bold type sites begin and end with A or T, indicative of dicationic netropsin-like binding.
Underlined sites begin with G or C, indicative of monocationic lexitropsin binding.

was the limit before a reduction in binding proficiency resulted. Goodsell and Dickerson (181) address this phenomenon of loss in binding ability with increased length of the drug in their "isohelical" analysis of groove-binding ligands. Longer netropsin and distamycin analogues lose their binding ability because they get out of phase with the base pairs along the minor groove floor of the helix. Both the repeating pyrrole-carboxamide unit and the benzimidazole of Hoechst are approximately 20% too long to synchronize with the base pairs along the groove of B-DNA. Goodsell calculates the degree of isohelicity for 14 different repeating ligand units, a few of which are found in known drugs. As an example, by shortening the carboxamide unit to either an amino or keto linkage, a drug more synchronous or "isohelical" with the DNA results (Fig. 26). This shorter tether links together the aromatic reading elements, either five-membered ring systems consisting of pyrroles and furans or six-membered ring systems of phenyls and pyridines. Zakrewska and Pullman (182,183) address the electrostatics of these isolexins and propose a new member, a vinylexin (183), in which the rings are *trans* across the C=C linkage, which replaces the amino NH group. The *trans* linkage permits a smooth curvature of the ligand and a "best" fit to the minor groove.

B. Flexible Tethers

Another approach to making a drug that can read a large number of base pairs, netropsin reads four base pairs and distamycin five, is to link together molecules that bind short stretches using a "linker." These linkers couple netropsin or distamycin yielding bis-netropsin or bis-distamycin. The linking elements used have been methylene chains (184–187), fumaride (188), succinamide (189), as well as short peptide chains (190). In another case, a hybrid molecule was prepared with netropsin joined with an intercalator, the phenoxazone chromophore ring system of actinomycin (186).

1. Polymethylene Linkers

The first linked molecules that were synthesized by Khorlin et al. (184) consisted of two netropsin-like molecules connected by methylene chains of various lengths, yielding bis-netropsin (Fig. 27). The use of tethers of the proper length (184,187) corrects alignment that would otherwise become "out of phase," a problem that occurs with the addition of too many carboxamide units. These tethers also allow an increase in the number of bases that can be bound by the drug ligand, increasing the specificity of the molecule. Youngquist and Dervan (56) found that the maximum number of pyrrole rings that could be joined together without loss of DNA binding was 7. A drug ligand consisting of seven pyrroles contains eight amides and, using the $n + 1$ rule for amides (see Section IIIA), a nine-base-pair

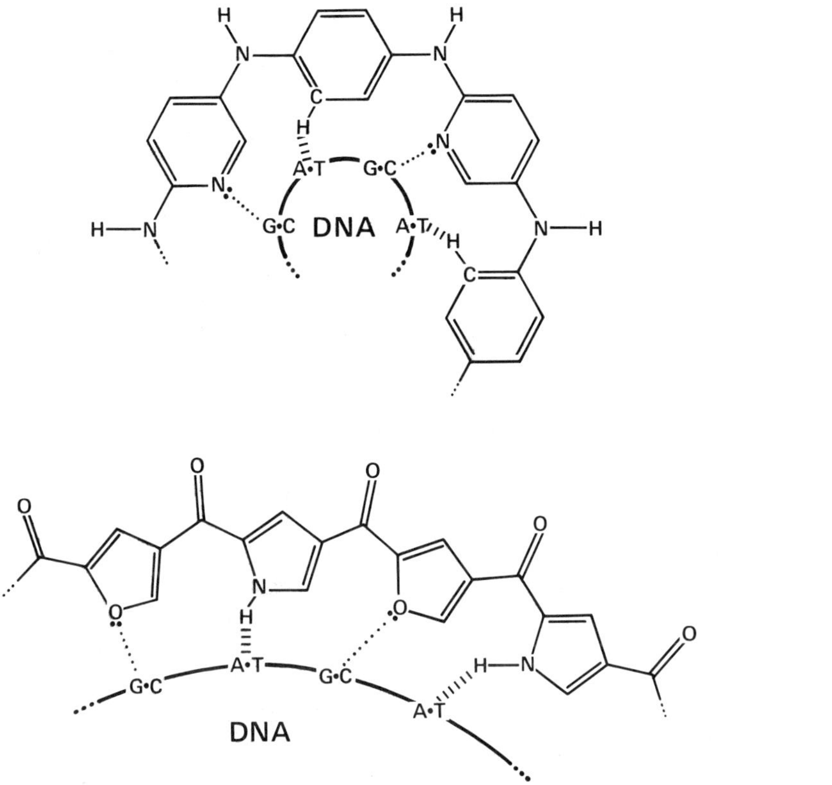

Figure 26 Two possible iso-lexins, or sequence-reading polymers that are geometrically isohelical with B-DNA. (top) pyridyl-amine polymer with phenyls in tight van der Waals contact with adenines at AT base pairs and pyridyls hydrogen bonded to guanine NH_2 at GC base pairs. (bottom) pyrrole-ketone polymer with pyrroles in tight contact with AT base pairs and furans hydrogen-bonded to guanine NH_2 at GC base pairs.

reading frame results. In the case of the polymethylene linkers, up to 11 base pairs have been recognized by these drugs (184), and 16-base-pair recognition has been reported using the amino acid alanine as the linking element between three distamycin molecules (201).

Increasing the length of the reading frame should increase drug efficiency and alleviate side effects by reducing the potential number of binding sites. For example, distamycin reads a five-base-pair sequence which would be expected to occur once in every $4^5 = 1024$ base pairs in double-stranded DNA of random sequence. A sequence-reading ability of 11 bases would occur once in $4^{11} = 4.2 \times 10^6$ base pairs. The human genome contains

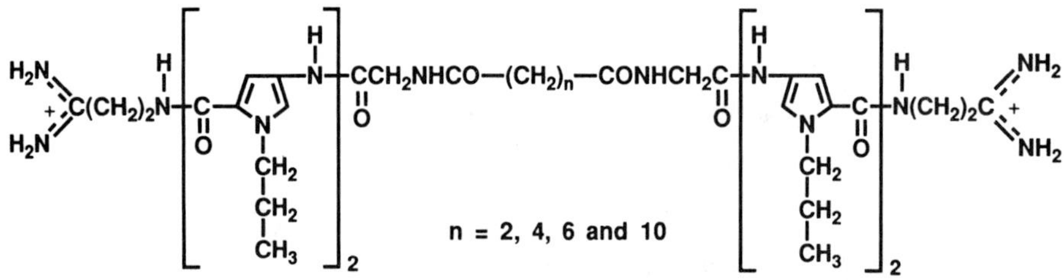

Figure 27 Netropsin-like molecules joined with polymethylene linkers of various lengths. Note the *N*-propylpyrrole rings have a *trans* linkage between pyrrole rings instead of the regular *cis* linkage found in the naturally occurring netropsin molecule.

approximately 3×10^9 base pairs. To select a single site on the genome a drug must have a 16-base recognition frame, i.e., $4^{16} = 4.3 \times 10^9$. Such specific targeting could eliminate many of the side effects of cancer chemotherapy where healthy cells are attacked as well as neoplastic ones.

Lown et al. (187) tested these polymethylene linked bis-netropsins and distamycins in vitro for their antiviral and cytostatic activity and found them to have increased antiviral function and improved tumor cell toxicity in a number of cases. They found that the rise and fall of cytotoxic and antiviral activity coincided with the number of methylene groups $(CH_2)_n$ used, reflecting maximal activity at $n = 1, 2, 6$, and 8. The same behaviour occurred when they measured the binding constants of the linked drugs, namely that the strongest binding coincided with drugs having 1, 2, 6, or 8 methylene units. This activity is attributed to two causes: (a) increased lypophilicity due to the presence of the polymethylene linkers promoting cellular uptake and (b) phase-dependent binding due to the number of linking elements employed.

DNA-Cleaving Molecules. Dervan, not only uses linkers to join two molecules together, but attaches an EDTA·Fe^{II} (185,191) to these DNA-binding ligands, creating a DNA-cleaving molecule (Fig. 28). These molecules are used for affinity cleavage footprinting experiments to delineate the exact binding site of the drug ligand on the DNA. Using a C-7 hydrocarbon linker (185,186) between two distamycin molecules (Fig. 28), Schultz and Dervan find an eight-base-pair reading frame with both distamycins bound, as well as a five-base-pair site with a single distamycin bound. This suggests that the C-7 tether is flexible enough to allow both dimeric and monomeric binding modes. In another case, a bis-distamycin joined by a fumaramide (C-4) linker (Fig. 28) binds to eight or nine base-pair sequences, binding only as the dimeric species (188).

To investigate the effect of a chiral linker on binding, Dervan synthesized

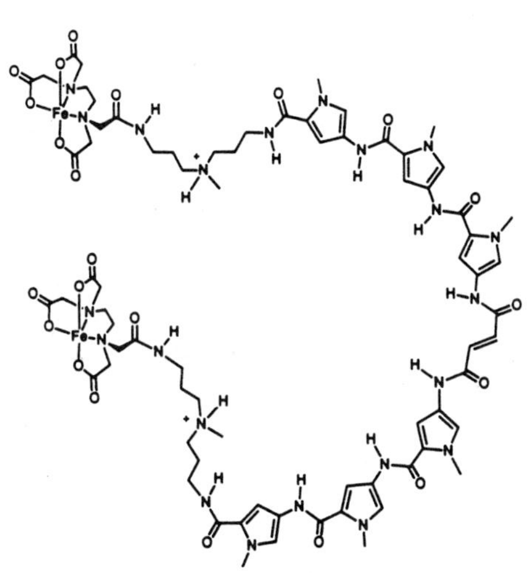

1

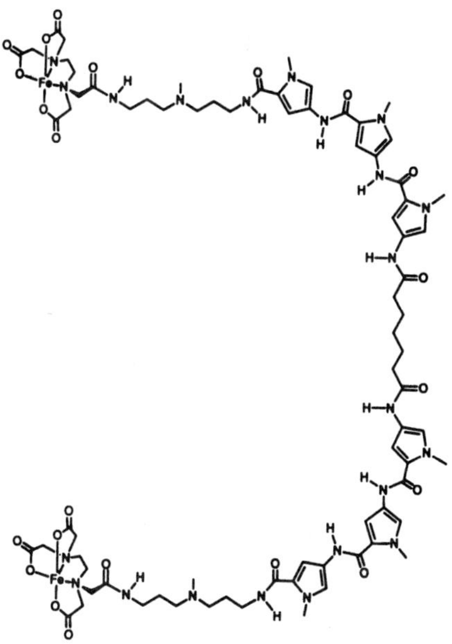

2

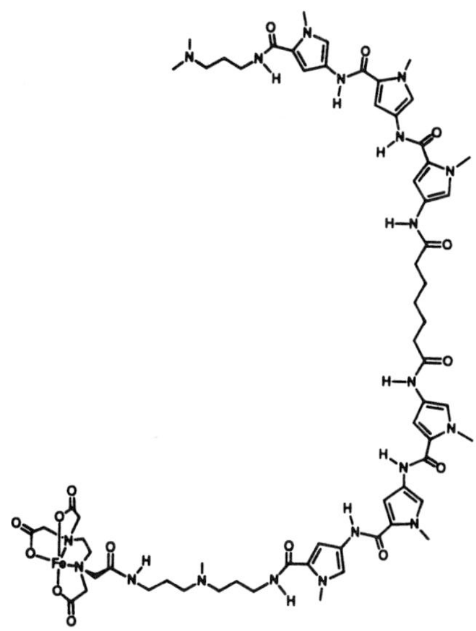

3

Figure 28 DNA cleaving molecules based on distamycin, containing iron-EDTA as the cleaving moiety. **1** bis(FE^{II}·EDTA-distamycin)fumaramide abbreviated BED-F·Fe(II). **2** bis(EDTA-distamycin)Fe^{II} or BED·Fe(II) and **3** EDTA-bis(distamycin)Fe^{II} or EBD·Fe(II). **2** and **3** are linked with a C-7 hydrocarbon linker.

bis(netropsin)succinamide and its two dihydroxybis-enantiomers, using the enantiomers of *dl*-tartaric acid as the linkers in preparing the optically active compounds (Fig. 29). These compounds are hexamides that bind seven base pairs (189). All three compounds exhibit dimeric binding with both netropsins binding to the DNA. The molecules are attached to EDTA·Fe^{II} to use in affinity cleavage experiments. The effect of chirality can be seen by comparing the concentrations required to achieve comparable cleaving efficiency. The concentration of the (*R,R*) and (*S,S*) forms vary as much as two orders of magnitude from bis(netropsin)succinamide (see Table 7). To explain the difference in binding, the authors state that in the crystal the (*S,S*)-enantiomer has a right-handed twist and the (*R,R*)-enantiomer a left-handed one. If these same conformations exist in solution, then the right-handed (*S,S*) matches the minor groove of B-DNA better than the (*R,R*)-enantiomer.

2. *Short Peptide Linkers*

Leinsoo et al. (190) constructed bis-netropsin analogs in which two netropsin-like molecules were attached to the peptide sequences gly-cys-gly-NH_2 and gly-cys-gly-gly-gly- and then linked together with the aid of a

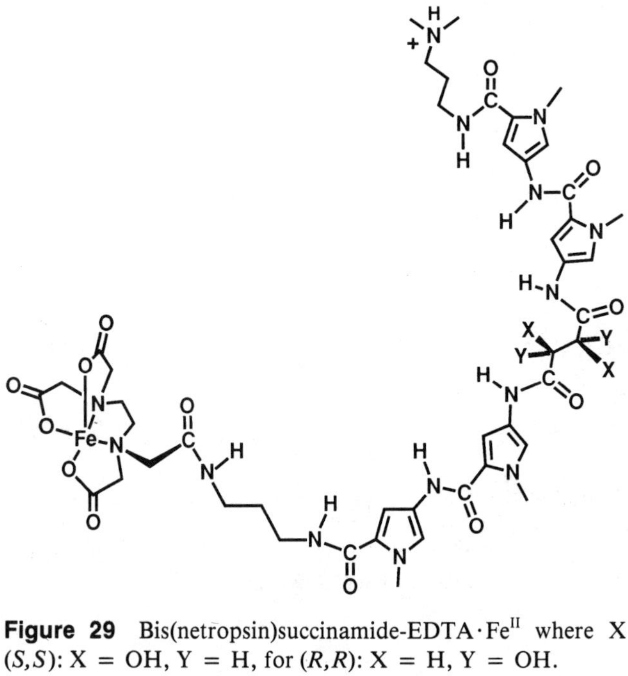

Figure 29 Bis(netropsin)succinamide-EDTA·Fe^{II} where X = H, Y = H, for (*S,S*): X = OH, Y = H, for (*R,R*): X = H, Y = OH.

Table 7 Effect of Chirality on Cleavage Efficiency

Concentration Required for Cleavage of 100-μM DNA	
Compound	Conc'n (μM)
Bis(netropsin)succinamide	0.5
(2R,3R)-Dihydroxybis(netropsin)succinamide	5.0
(2S,3S)-Dihydroxybis(netropsin)succinamide	50.0

Data from Ref. 189.

disulfide bond from one amino acid chain to a second amino acid chain of the same sequence containing the other cysteine residue (see Fig. 30). Footprinting experiments demonstrate that these two bis-netropsin analogs, joined by two glycine tethers of the same length containing a cystine bridge, preferentially bind an $(AT)_3(GC)_2(AT)_3$ sequence or $(AT)_4(GC)(AT)_3$ and not to straight runs of AT base pairs.

3. *Hybrid Molecules*

Another route to making an AT-GC reading molecule was the preparation of hybrid molecules containing GC-intercalating ligands coupled to groove binders (186,192). Krivtsova (192) and Dervan (186) speak of preparing netropsin and distamycin analogues attached to the phenoxazone ring of the GC intercalator, actinomycin. Krivtsova and co-workers attach from one to three pyrrolecarboxamides of netropsin directly to the 1,9-position of the phenoxazone chromophore (Fig. 31). The molecules with one and two pyrroles linked to phenoxazone appear to have both intercalation and groove-binding modes. The ligand with three pyrroles has only a groove-binding mode. Dervan links the distamycin via a glycine tether to the phenoxazone and also replaces the two cyclic pentapeptides of actinomycin with the tripeptide of distamycin, preparing a bis(distamycin)phenoxazone. Footprints of Dervan's bis(distamycin)phenoxazone reveal a 10-base-pair $(AT)_4(GC)_2(AT)_4$ site consistent with a groove binder–intercalator–groove binder mode of recognition. However, two other footprint sites indicated incomplete coverage by the entire molecule, with probably only one distamycin binding or a distamycin-phenoxazone binding. The combination of groove-binding and intercalating ligands in one molecule offers a different method of creating sequence specific reading elements.

C. Molecular Shape

Using a completely different approach, DesJarlais et al. (193,194) describe an algorithm (DOCK) for searching data bases to find ligands that match

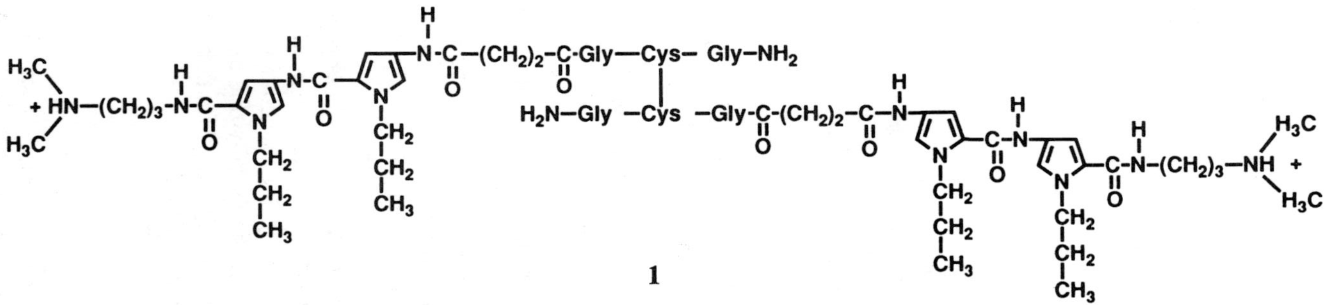

1

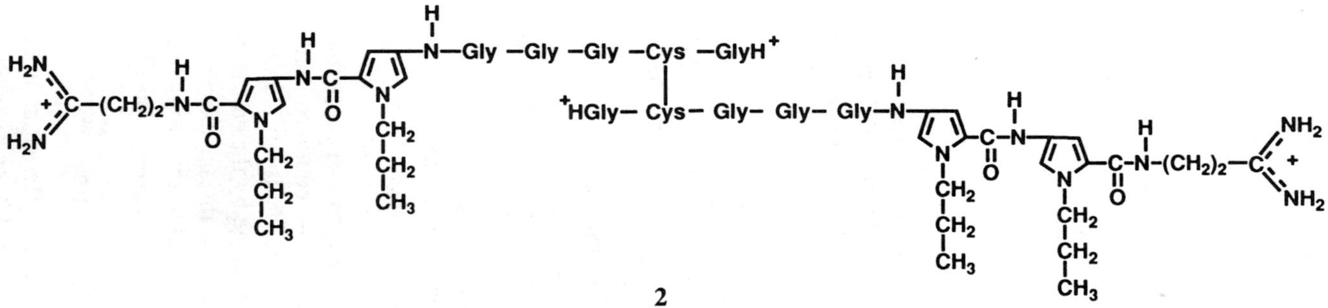

2

Figure 30 Netropsin analogs linked with short peptides. **1** gly-cys-gly chain cross-linked by a cysteine bridge with another gly-cys-gly chain. Note direction of amino acid chains. The peptide chains are joined to netropsin analog at carboxy group and end with amino group. **2** gly-gly-gly-cys-gly chain cross-linked by a cysteine bridge with a similar chain. The chain is joined in the opposite direction from **1**, i.e., the chains are attached at amino group and end with carboxy H^+.

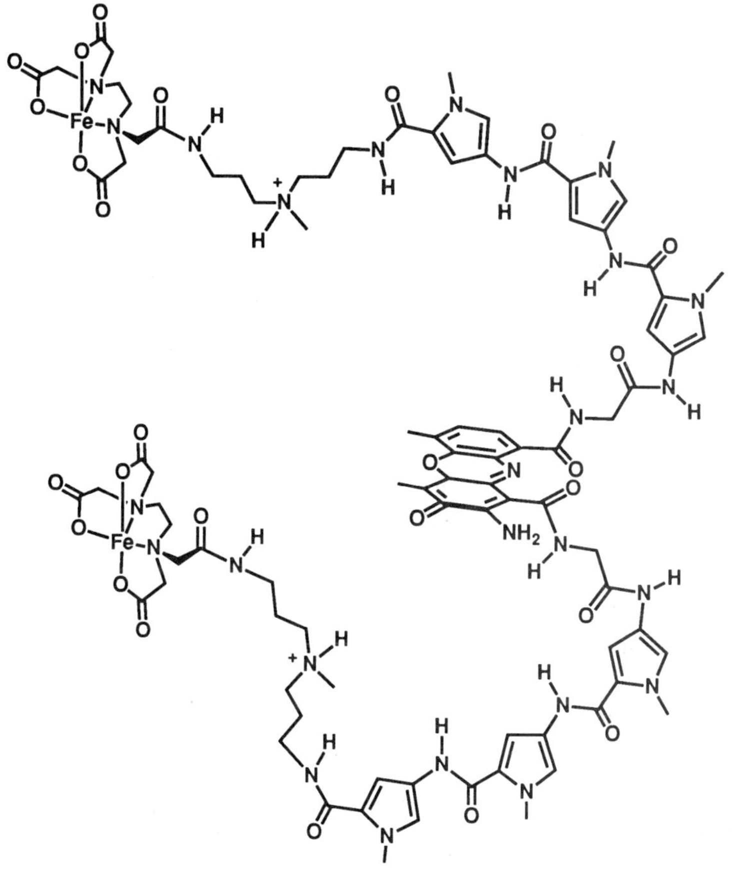

Figure 31 A hybrid sequence-reading molecule, bis(Fe^{II}·EDTA-distamycin)-phenoxazone abbreviated BEDP·Fe(II), consisting of groove binder–intercalator–groove binder.

the given shape of the receptor site of a known three-dimensional structure. They have tested this with proteins (194) and now with the DNA dodecamer sequence (195) used in a majority of the drug structures reported in this chapter, CGCGAATTCGCG. The ligand being investigated is fit into the shape of the binding site, which is characterized by a set of overlapping spheres that represent the receptor volume. The solvent accessible surface of the DNA is the starting point for generating the shape of this binding site. Using the computer graphics program MIDAS (196), each ligand is individually "docked" in a number of geometrically permissible orientations in this volume. Molecular mechanics calculations are then carried out using AMBER (197–199). In this study netropsin was compared to nine other compounds. On the basis of best fit to the shape of the binding site, the antitumor drug CC1065 was ranked first, followed by provitamin D and a biphenyl compound with netropsin sixth. However, they caution that this method doesn't take into account electrostatics or conformational adjustments. Nevertheless, it is an interesting screening method for known compounds found in a data base. Although these compounds may have been targeted for other uses originally, shapes complementary to the DNA target can be found in this way.

VII. CONCLUSION

A new generation of groove binders with the aptitude for reading both AT and GC base pairs of any combination is on its way. Since the mid-1980s organic chemists have been synthesizing new molecules based on the newly available biochemical and structural data on DNA and a variety of drug ligands to produce these sequence-reading drugs. The use of new computer programs to model and refine both the drug and drug-DNA interactions has and will be a tremendous aid in drug design to come. Hopefully one or more of these sequence-reading drugs, that are produced from this interweaving of biochemical, structural and graphical disciplines, will prove useful as a new antitumor or antiviral agent.

ACKNOWLEDGMENTS

This work was supported in part by American Cancer Society Grant CH-4720, National Institute of Health Grant GM-31299 and National Science Foundation Grant DMB-8916261. The authors wish to thank Professor Richard E. Dickerson for his support in providing the impetus for this research and his helpful comments and suggestions. Thanks also to Jordi Quintana for the useful discussions on drug binding and H. Bobby Zouvelos and Maria Grzeskowiak for help in library research and figure preparation.

REFERENCES

1. Mitsuya, H., and S. Broder, Strategies for antiviral therapy in AIDS. *Nature 325*:773–778 (1987).
2. Shakked, Z., D. Rabinovich, W.B.T. Cruse, E. Egert, O. Kennard, G. Sala, S.A. Salisbury, and M.A. Wiswamitra, Crystalline A-DNA: The x-ray analysis of the fragment d(G-G-T-A-T-A-C-C). *Proc. Roy. Soc. Lond. B 213*:479–487 (1981).

3. Conner, B.N., T. Takano, S. Tanaka, K. Itakura, and R.E. Dickerson, The molecular structure of d(^{i}CpCpGpG), a fragment of right handed double helical A-DNA. *Nature 295*:294–299 (1982).
4. Wing, R., H. Drew, T. Takano, C. Broka, S. Tanaka, K. Itakura, and R.E. Dickerson, Crystal structure analysis of a complete turn of B-DNA. *Nature 287*:755–758 (1980).
5. Wang, A.H. J., G.J. Quigley, F.J. Kolpak, J.L. Crawford, J.H. van Boom, G. van der Marel, and A. Rich, Molecular structure of a left-handed double helical DNA fragment at atomic resolution. *Nature 282*:680–686 (1979).
6. Drew, H., T. Takano, S. Tanaka, K. Itakura, and R.E. Dickerson, High-salt d(CpGpCpG), a left-handed Z-DNA double helix. *Nature 286*:567–573 (1980).
7. Hare, D.R., and B.R. Reid, Three dimensional structure of a DNA hairpin in solution: Two-dimensional NMR studies and distance geometry calculations on d(CGCGTTTTCGCG). *Biochemistry 25*:5341–5350 (1986).
8. Wolk, S., C.C. Hardin, M.W. Germann, H.J. van de Sande, and I. Tinoco, Comparison of the B- and Z-form hairpin loop structures formed by $d(CG)_5$–$T_4(CG)_5$. *Biochemistry 27*:6960–6967 (1988).
9. Chattopadhyaya, R., S. Ikuta, K. Grzeskowiak, and R.E. Dickerson, X-ray structure of a DNA hairpin molecule. *Nature 334*:175–179 (1988).
10. Chattopadhyaya, R., K. Grzeskowiak, and R.E. Dickerson, Structure of a T_4 hairpin loop on a Z-DNA stem and comparison with A-RNA and B-DNA loops. *J. Mol. Biol. 211*:180–210 (1990).
11. Rajagopal, P., and J. Feigon, NMR studies of triple-strand formation from the homopurine-homopyrimidine deoxyribonucleotides $d(GA)_4$ and $d(TC)_4$. *Biochemistry 28*:7859–7870 (1989).
12. Sklenar, V., and J. Feigon, Formation of a stable triplex from a single DNA strand. *Nature 345*:836–838 (1990).
13. Panyutin, I., O.I. Kovalsky, E.I. Budowsky, R.E. Dickerson, M.E. Rikhirev, and A.A. Lipanov, G-DNA: A twice folded DNA structure adopted by single-stranded oligo(dG) and its implications for telomeres. *Proc. Natl. Acad. Sci. (USA) 87*: 867–870 (1990).
14. Feigon, J., W.A. Denny, W. Leupin, and D.R. Kearns, Interactions of antitumor drugs with natural DNA: ^{1}H NMR study of binding mode and kinetics. *J. Med. Chem. 27*:450–465 (1984).
15. Zimmer, Ch., and U. Wahnert, Nonintercalating DNA-binding ligands: Specificity of the interaction and their use as tools in biophysical, biochemical and biological investigations of the genetic material. *Prog. Biophys. molec. Biol. 47*:31–112 (1986).
16. Rubin, J., and M. Sundaralingam, An unexpected major groove binding of netropsin and distamycin A to tRNA(phe). *J. Biomol. Struct. Dyn. 1*:165–174 (1984).
17. Sobell, H.M., S.C. Jain, T.D. Sakore, and C.E. Nordman, Stereochemistry of actinomycin-DNA binding. *Nature 231*: 200–205 (1971).
18. Jain, S.C., C-C. Tsai, and H.M. Sobell, Visualization of drug-nucleic acid interactions at atomic resolution. II. Structure of an ethidium/dinucleoside

monophosphate crystalline complex, Ethidium-Iodocytidylyl(3′-5′)guanosine. *J. Mol. Biol. 114*:317–331 (1977).

19. Wang, A.H.-J., G. Ughetto, G.J. Quigley, and A. Rich, Interactions between an anthracycline antibiotic and DNA: Molecular structure of daunomycin complexed to d(CpGpTpApCpG) at 1.2 Å resolution. *Biochemistry 26*:1152–1163 (1987).
20. Moore, M.H., W.N. Hunter, B.L. D'Estaintot, and O. Kennard, DNA-drug interactions. The crystal structure of d(CGATCG) complexed with daunomycin. *J. Mol. Biol. 206*:693–705 (1989).
21. Frederick, C.A., L.D. Williams, G. Ughetto, G.A. van der Marel, J.H. van Boom, A. Rich, and A.H.-J. Wang, Structural comparison of anti-cancer drug-DNA complexes: Adriamycin and daunomycin. *Biochemistry 29*:2538–2549 (1990).
22. Liaw, Y.-C., Y.-G. Gao, H. Robinson, G. van der Marel, J.H. van Boom, and A.H.-J. Wang, Antitumor drug nogalamycin binds DNA in both grooves simultaneously: Molecular structure of nogalamycin-DNA complex. *Biochemistry 28*:9913–9918 (1989).
23. Williams, L.D., M. Egli, Q. Gao, P. Bash, G.A. van der Marel, J.H. van Boom, A. Rich, and C.A. Frederick, Structure of nogalamycin bound to a DNA hexamer. *Proc. Natl. Acad. Sci. (USA) 87*:2225–2229 (1990).
24. Quigley, G.J., G. Ughetto, G.A. van der Marel, J.H. van Boom, A.H.-J. Wang, and A. Rich, Non-Watson-Crick G·C and A·T base pairs in a DNA-antibiotic complex. *Science 232*:1255–1258 (1986).
25. Wang, A.H.-J., G. Ughetto, G.J. Quigley, and A. Rich, Interactions of quinoxaline antibiotic and DNA: The molecular structure of a triostin A-d(GCGTACGC) complex. *J. Biomol. Struct. Dyn. 4*:319–342 (1986).
26. Ughetto, G., A.H.-J. Wang, G.J. Quigley, G.A. van der Marel, J.H. von Boom, and A. Rich, A comparison of the structures of echinomycin and triostin A complexed to a DNA fragment. *Nucl. Acids Res. 13*:2305–2323 (1985).
27. Gao, X.L., and D.J. Patel, Anti-tumor drug-DNA interactions: NMR studies of echinomycin and chromomycin complexes. *Quart. Rev. of Biophys. 22*: 93–138 (1989).
28. Gilbert, D.E., G.A. van der Marel, J.H. van Boom, and J. Feigon, Unstable Hoogsten base pairs adjacent to echinomycin binding sites within a DNA duplex. *Proc. Natl. Acad. Sci. (USA) 86*: 3006–3010 (1989).
29. Sherman, S.E., D. Gibson, A.H.-J. Wang, and S.J. Lippard, X-ray structure of the major adduct of the anticancer drug cisplatin with DNA:cis[$Pt(NH_3)_2$ (pGpG)]. *Science 230*:412–417 (1985).
30. Sherman, S.E., D. Gibson, A.H.-J. Wang, and S.J. Lippard, Crystal and molecular structure of cis-[$Pt(NH_3)_2$(pGpG)] the principal adduct formed by cis diammine-dichloroplatinum(II) with DNA. *J. Am. Chem. Soc. 110*:7368–7381 (1988).
31. Wing, R.M., P. Pjura, H.R. Drew, and R.E. Dickerson, The primary mode of binding of cisplatin to a B-DNA dodecamer: CGCGAATTCGCG. *EMBO J. 31*:1201–1206 (1984).

32. Krugh, T.R., D.E. Graves, and M.P. Stone, Two dimensional NMR studies on the anthramycin-d(ATGCAT)$_2$ adduct. *Biochemistry 28*:9988–9994 (1989).
33. Boyd, F.L., S.F. Cheatham, W. Remers, G.C. Hill, and L.H. Hurley, Characterization of the structure of the anthramycin-d(ATGCAT)$_2$ adduct by NMR and molecular modeling studies. Determination of the stereochemistry of the covalent linkage site, orientation in the minor groove of the DNA, and the effect on local DNA structure. *J. Am. Chem. Soc. 112*:3279–3289 (1990).
34. Jeffrey, A.M., K. Grzeskowiak, I.B. Weinstein, K. Nakanishi, P. Roller, and R.G. Harvey, Benzo(a)pyrene-7,8-dihydrodiol 9,10-oxide adenosine and deoxyadenosine adducts: Structure and stereochemistry. *Science 206*:1309–1311 (1979).
35. Kinoshita, T.M., H.M. Lee, R.G. Harvey, and A.M. Jeffrey, Structures of covalent adducts derived from the reactions of the 9,10-epoxides of 7,8,9,10-tetrahydrobenzo[*a*]pyrene and 9,10,11,12-tetrahydrobenzo[*e*]pyrene with DNA. *Carcinogenesis 3*:255–260 (1982).
36. Jeffrey, A.M., K.W. Jennette, S.H. Blobstein, I.B. Weinstein, F.A. Beland, R.G. Harvey, H. Kasai, I. Miura, and K. Nakanishi, Benzo[*a*]pyrene-nucleic acid derivative found in vivo: Structure of a benzo[*a*]pyrenetetrahydrodiol epoxide-guanosine adduct. *J. Am. Chem. Soc. 98*:5714–5715 (1976).
37. Watson, J.D., and F.H.C. Crick, Molecular structure of nucleic acids: A structure for deoxyribose nucleic acid. *Nature 171*:737–738 (1953).
38. Seeman, N.C., J.M. Rosenberg, and A. Rich, Sequence-specific recognition of double helical nucleic acids by proteins. *Proc. Natl. Acad. Sci. (USA) 73*: 804–808 (1976).
39. Dickerson, R.E., M.L. Kopka, and P. Pjura, Pathways of information readout in DNA. *Chemica Scripta 26B*:139–145 (1986).
40. Dickerson, R.E., M.L. Kopka, and P.E. Pjura, The major and minor grooves of the DNA helix as conduits for information transfer. In *DNA Ligand Interactions*, Guschlbauer, W., and Saenger, W. eds., 45–62 Plenum (1987).
41. Finlay, A.C., F.A. Hochstein, B.A. Sobin, and F.X. Murphy, Netropsin, a new antibiotic produced by a Streptomyces. *J. Am. Chem. Soc. 73*:341–343 (1951).
42. Thrum., H., Eine neue, von einer spezies der streptomyces-retucli-gruppe gebildete antibiotekakombination. *Naturweiss. 46*: 87 (1959).
43. Arcamone, F., S. Penco, P. Orezzi, V. Nicolella, and A. Pirelli, Structure and synthesis of distamycin A. *Nature 203*:1064–1065 (1964).
44. Wartell, R.M., J.E. Larson, and R.D. Wells, Netropsin: A specific probe for A-T regions of duplex deoxyribonucleic acid. *J. Biol. Chem. 249*:6719–6731 (1974).
45. Luck, G., and Ch. Zimmer, Interaction of netropsin with DNA in the course of the B-A transition. *Stud. Biophys. (Ber) 40*:9–12 (1973).
46. Ivanov, V.I., L.A. Minchenkova, E.E. Minyat, M.D. Frank-Kamenetskii, and A.K. Schyolkina, The B to A transition of DNA in solution. *J. Mol. Biol. 87*:817–833 (1974).

47. Wartell, R.M., J.E. Larson, and R.E. Wells, Netropsin: A specific probe for A-T regions of duplex deoxyribonucleic acid. *J. Biol. Chem. 249*:6719–6731 (1974).
48. Burckhardt, G., H. Votavova, J. Sponar, G. Luck, and Ch. Zimmer, Two binding modes of netropsin are involved in the complex formation with poly(dA-dT)·poly(dA-dT) and other alternating DNA duplex polymers. *J. Biomol. Struct. Dyn. 2*:721–736 (1985).
49. Wahnert, U., Ch. Zimmer, G. Luck, and C. Pitra, (dA·dT)-Dependent inactivation of the DNA template properties by interaction with netropsin and distamycin A. *Nucl. Acids Res. 2*:391–404 (1975).
50. Zimmer, Ch., G. Luck, E. Birch-Hirshfeld, R. Weiss, F. Arcamone, and W. Guschlbauer, Chain length dependent association of distamycin-type oligopeptides with A·T and G·C pairs in polydeoxynucleotide duplexes. *Biochim. Biophys. Acta 741*: 15–22 (1983).
51. Kolchinskii, A., A. Mirazabekov, A. Zasedatelev, G. Gurskii, S. Grokhovskii, A. Zhuze, and B. Gottikh, Structure of antibiotics of the distamycin type and actinomycin D with DNA: New experimental data on the location of the antibiotics in the narrow groove of DNA. *Mol. Biol. USSR 9*:14–20 (1975).
52. Kubwabara, M., C. Yoon, T. Goyne, T. Thederahn, and D.S. Sigman, Nuclease activity of 1,10-phenanthroline--Copper ion: Reaction with CGCGAATTCGCG and its complexes with netropsin and *Eco*RI. *Biochemistry 25*:7401–7408 (1986).
53. Krylov, A.S., S.L. Grokhovsky, A.S. Zasedatelev, A.L. Zhuse, G.V. Gursky, and B.P. Gottikh, Quantitative estimation of the contribution of pyrrolecarboxamide groups of the antibiotic distamycin A into specificity of its binding to DNA AT pairs. *Nucl. Acids. Res. 6*:289–304 (1979).
54. Van Dyke, M.W., R.P. Hertzberg, and P.B. Dervan, Map of distamycin, netropsin, and actinomycin binding sites on heterogeneous DNA: DNA cleavage inhibition patterns with methidium propyl-EDTA·Fe(II). *Proc. Natl. Acad. Sci. (USA) 79*:5470–5474 (1982).
55. Schultz, P.G., and P.B. Dervan, Distamycin and penta-*N*-methylcarboxamide binding sites on native DNA. A comparison of methidium propyl-EDTA-Fe(II) footprinting and DNA affinity cleaving. *J. Biomol. Struct. Dyn. 1*:1133–1147 (1984).
56. Youngquist, R.S., and P.B. Dervan, Sequence specific recognition of B-DNA by oligo-*N*-methylpyrrolecarboxamides. *Proc. Natl. Acad. Sci. (USA) 82*: 2565–2569 (1985).
57. Zimmer, Ch., B. Puschendorf, H. Grunicke, P. Chandra, and H. Venner, Influence of netropsin and distamycin A on the secondary structure and template activity of DNA. *Eur. J. Biochem. 21*:269–278 (1971).
58. Bruzik, J.P., D.T. Auble, and P.L. deHaseth, Specific activation of transcription initiation by the sequence-specific DNA-binding agents distamycin A and netropsin. *Biochemistry 26*:950–956 (1987).
59. Martello, P.A., J.P. Bruzik, P. deHaseth, R.S. Youngquist, and P.B. Dervan, Specific activation of open complex formation at an *Escherichia coli*

promoter by oligo(*N*-methylpyrrolecarboxamide)s: Effects of peptide length and identification of DNA target sites. *Biochemistry 28*:4455–4461 (1989).

60. Hahn, F.E., Distamycin A and netropsin. In *Antibiotics*, Corcoran, J.W. and Hahn, F.E., eds., Vol. 3:79–100 Springer-Verlag, New York (1975).
61. Bialer, M., Distamycin and derivatives: The pyrrole amidine antiviral antibiotics. In *Antiviral Drugs and Interferon: The Molecular Basis of Their Activity*, Becker, Y. ed., 143–156, Martinus Nijhoff, Boston (1984).
62. Zimmer, Ch., Effects of the antibiotics netropsin and distamycin A on the structure and function of nucleic acids. *Prog. Nucl. Acid Res. Mol. Biol. 15*: 285–318 (1975).
63. Patel, D.J., S.A. Kozlowski, J.A. Rice, C. Broka, and K. Itakura, Mutual interaction between adjacent dG·dC actinomycin binding sites and dA·dT netropsin binding sites on the self-complementary d(C-G-C-G-A-A-T-T-C-G-C-G) duplex in solution. *Proc. Natl. Acad. Sci. (USA) 78*:7281–7284 (1981).
64. Patel, D., Antibiotic-DNA interactions: Intermolecular nuclear overhauser effects in the netropsin-d(C-G-C-G-A-A-T-T-C-G-C-G) complex in solution. *Proc. Natl. Acad. Sci. (USA) 79*:6424–6428 (1982).
65. Patel, D.J., and L. Shapiro, Sequence-dependent recognition of DNA duplexes. Netropsin complexation to the AATT site of the d(G-G-A-A-T-T-C-C) duplex in aqueous solution. *J. Biol. Chem. 261*:1230–1240 (1986).
66. Patel, D.J., and L. Shapiro, Sequence dependent recognition of DNA duplexes: Netropsin complexation to the TATA site of the d(G-G-T-A-T-A-C-C) duplex in aqueous solution. *Biopolymers 25*:707–727 (1986).
67. Lee, M., R.T. Pon, K. Krowicki, and J.W. Lown, Structural and dynamic aspects of the sequence specific binding of netropsin and its bis-imidazole analogue on the decadeoxyribonucleotide d-$[CGCAATTGCG]_2$. *J. Biomol. Struct. Dyn. 4*:939–949 (1988).
68. Klevit, R.E., D.E. Wemmer, and B.R. Reid, ^1H NMR studies on the interaction between distamycin A and a symmetrical DNA dodecamer. *Biochemistry 25*:3296–3303 (1986).
69. Pelton, J.G., and D.E. Wemmer, Structural modeling of the distamycin A-$d(CGCGAATTCGCG)_2$ complex using 2D NMR and molecular mechanics. *Biochemistry 27*:8088–8096 (1988).
70. Pelton, J.G., and D.E. Wemmer, Structural characterization of a 2 : 1 distamycin A·d(CGCAAATTGGC) complex by two-dimensional NMR. *Proc. Natl. Acad. Sci. (USA) 86*:5723–5727 (1989).
71. Berman, H.M., S. Neidle, Ch. Zimmer, and H. Thrum, Netropsin, a DNA-binding oligopeptide: Structural and binding studies. *Biochem. Biophys. Acta 561*:124–131 (1979).
72. Kopka, M.L., C. Yoon, D. Goodsell, P. Pjura, and R.E. Dickerson, Binding of an anti-tumor drug to DNA: Netropsin and CGCGAATT^{Br}CGCG. *J. Mol. Biol. 183*:553–556 (1985).
73. Kopka, M.L., C. Yoon, D. Goodsell, P. Pjura, and R.E. Dickerson, The molecular origin of DNA-drug specificity in netropsin and distamycin. *Proc. Natl. Acad. Sci. (USA) 82*:1376–1380 (1985).

74. Coll, M., J. Aymami, G.A. van der Marel, J.H. van Boom, A. Rich, and A.H.-J. Wang, Molecular structure of the netropsin-d(CGCGATATCGCG) complex: DNA conformation in an alternating AT segment. *Biochemistry 28*:310–320 (1989).
75. Coll, M., C.A. Frederick, A.H.-J. Wang, and A. Rich, A bifurcated hydrogen-bonded conformation in the d(AT) base pairs of the DNA dodecamer d(CGCAAATTTGCG) and its complex with distamycin. *Proc. Natl. Acad. Sci. (USA) 84*:8385–8389 (1987).
76. Drew, H.R., and R.E. Dickerson, Structure of a B-DNA dodecamer. III. Geometry of hydration. *J. Mol. Biol. 151*:535–556 (1981).
77. Kopka, M.L., A.V. Fratini, H.R. Drew, and R.E. Dickerson, Ordered water structure around a B-DNA dodecamer. A quantitative study. *J. Mol. Biol. 163*:129–146 (1983).
78. Taylor, R., O. Kennard, and W. Versichel, Geometry of the N—H···O=C hydrogen bond. 2. Three-center ("bifurcated") and four-center ("trifurcated") bonds. *J. Am. Chem. Soc. 106*:244–248 (1984).
79. Jefferey, G.A., Hydrogen bonding in crystal structures of nucleic acid components: purines, pyrimidines, nucleosides and nucleotides. In *Landolt-Bornstein: Numerical Data and Functional Relationships in Science and Technology. Nucleic Acids*, Vol VII/1b:277–348, Saenger, W. ed., Springer-Verlag, New York (1989).
80. Marky, L.A., and K.J. Breslauer, Origins of netropsin binding affinity and specificity: Correlations of thermodynamic and structural data. *Proc. Natl. Acad. Sci. (USA) 84*:4359–4363 (1987).
81. Pullman, B., and A. Pullman, Structural factors involved in the binding of netropsin and distamycin A to nucleic acids. *Stud. Biophys. 86*:95–102 (1981).
82. Pullman, B., Electrostatics of polymorphic DNA. *J. Biomol. Struct. Dyn. 1*: 773–794 (1983).
83. Breslauer, K.J., R. Ferrante, L. Marky, P.B. Dervan, and R.S. Youngquist, The origin of the DNA binding affinity and specificity of minor groove directed ligands: Correlations of thermodynamic and structural data. In *Structure & Expression*, Sarma, R.H., and Sarma, M.H. eds., Vol. 2:273–290, Adenine Press, Schenectady (1988).
84. Breslauer, K.J., D.P. Remeta, W.-Y. Chow, R. Ferrante, J. Curry, D. Zaunczkowski, J.G. Snyder, and L.A. Marky, Enthalpy-entropy compensations in drug-DNA binding studies. *Proc. Natl. Acad. Sci. (USA) 84*:8922–8926 (1987).
85. Klug, A., A. Jack, M.A. Viswamitra, O. Kennard, Z. Shakked, and T.A. Steitz, Hypothesis on a specific sequence-dependent conformation of TA and its relation to the binding of the *lac*-repressor protein. *J. Mol. Biol. 131*:669–680 (1979).
86. Viswamitra, M.A., O. Kennard, Z. Shakked, P.G. Jones, G.M. Sheldrick, S. Salisbury, and L. Falvello, DNA double helical fragment at atomic resolution. *Nature 273*:687–688 (1978).
87. Viswamitra, M.A., Z. Shakked, P.G. Jones, G.M. Sheldrick, S.A. Salisbury, and O. Kennard, Structure of the deoxytetranucleotide d-pApTpApT and a

sequence-dependent model for poly (dA-dT). *Biopolymers 21*:513–533 (1982).

88. Yoon, C., G.P. Prive, D.S. Goodsell, and R.E. Dickerson, Structure of an alternating-B DNA helix and its relationship to A-tract DNA. *Proc. Natl. Acad. Sci. (USA) 85*:6332–6336 (1988).
89. Leupin, W., W.J. Chazin, S. Hyberts, W.A. Denny, and K. Wuthrich, NMR studies of the complex between the decadeoxynucleotide d-(GCATTAAT-GC)$_2$ and a minor groove binding drug. *Biochemistry 25*:5902–5910 (1986).
90. Taylor, J.S., P.G. Schultz, and P.B. Dervan, DNA affinity cleaving: Sequence specific cleavage of DNA by Distamycin-EDTA·Fe(II) and EDTA-Distamycin·Fe(II). *Tetrahedron 40*:457–465 (1984).
91. Ward, B., R. Rehfuss, J. Goodisman, and J.C. Dabrowiak, Determination of netropsin-DNA binding constants from footprinting data. *Biochemistry 27*:1198–1205 (1988).
92. Portugal, J., and M.J. Waring, Comparison of binding sites in DNA for berenil, netropsin and distamycin. A footprinting study. *Eur. J. Biochem. 167*:281–289 (1987).
93. Nelson, H.C.M., J.T. Finch, B.F. Luisi, and A. Klug, The structure of an oligo(dA)·oligo(dT) tract and its biological implications. *Nature 330*:221–226 (1987).
94. Dickerson, R.E., M.L. Kopka, and P. Pjura, A random-walk model for helix bending in B-DNA. *Proc. Natl. Acad. Sci. (USA) 80*:7099–7103 (1983).
95. Loewe, H., and J. Urbanietz, 2,6-Bis-benzimidazoles, a new class of anthelminitics. In *Abstracts and Reviews of Eighth International Congress on Tropical Medicine and Malaria*, Teheran, Iran, 139 (1968).
96. Lammler, G., and W. Raether, Chemotherapeutic experiments with a 2,6-bis-benzimidazole in cotton rats infected with *Litomosoides carinii*. In *Abstracts and Reviews of Eighth International Congress on Tropical Medicine and Malaria*, Teheran, Iran, 139–140 (1968).
97. Lammler, G., H. Herzog, E. Saupe, and H.R. Schutze, Chemotherapeutic studies on *Litomosoides carinii* infection of *Mastomys natalensis*. The filaricidal action of 2,6-bis-benzimidazoles. *Bull. Wld. Hlth. Org. 44*:751–756 (1971).
98. Lammler, G., and H.R. Schutze, Vital fluorescence staining of animal cell nuclei with a new fluorochrome. *Naturwiss. 56*:286 (1969).
99. Hilwig, I., and A. Gropp, Staining of constitutive heterochromatin in mammalian chromosomes with a new fluorochrome. *Exptl. Cell Res. 75*:122–126 (1972).
100. Holmquist, G., Hoechst 33258 fluorescent staining of *drosophilia* chromosomes. *Chromosoma 49*:333–356 (1975).
101. Kim, Y.-J., R.L.Y. Sah, J.-Y.H. Doong, and A.J. Grodzinsky, Fluorometric assay of DNA in cartilage explants using Hoechst 33258. *Anal. Biochem. 174*:168–176 (1988).
102. Yamomoto, A., T. Araki, K. Fujimori, M. Yamada, H. Yamaguchi, K. Izumi, and K. Matsumoto, NaCl-aided Hoechst 33258 staining method for DNA quantification and its application. *Histochemistry* 9265–9268 (1989).

103. Bontemps, J., C. Houssier, and E. Fredericq, Physico-chemical study of the complexes of '33258 Hoechst' with DNA and nucleohistone. *Nucl. Acids Res. 2*:971–984 (1975).
104. Stokke, T., and H.B. Steen, Multiple binding modes for Hoechst 33258 to DNA. *J. Histochem. Cytochem. 33*:333–338 (1985).
105. Stokke, T., and H.B. Steen, Binding of Hoechst 33258 to chromatin in situ. *Cytometry 7*:227–234 (1986).
106. Jorgenson, K.F., U. Varshney, and J.H. van de Sande, Interaction of Hoechst 33258 with repeating synthetic DNA polymers and natural DNA. *J. Biomol. Struct. Dyn. 5*:1005–1023 (1988).
107. Muller, W., and F. Gautier, Interactions of heteroaromatic compounds with nucleic acids: A·T-specific non-intercalating DNA ligands. *Eur. J. Biochem. 54*:385–394 (1975).
108. Mikhailov, M.V., A.S. Zasedatelev, A.S. Krylov, and G.V. Gurskii, Mechanism of the recognition of AT base pairs in DNA by molecules of the dye Hoechst 33258. *Molek. Biol. (USSR Eng. translation) 15*:541–554 (1981).
109. Harshman, K.D., and P.B. Dervan, Molecular recognition of B-DNA by Hoechst 33258. *Nucl. Acids Res. 13*:4825–4835 (1985).
110. Martin, R.F., and N. Holmes, Use of an ^{125}I-labeled DNA ligand to probe DNA structure. *Nature 302*:452–454 (1983).
111. Murray, V., and R.F. Martin, Sequence specificity of ^{125}I-labeled Hoechst 33258 damage in six closely related DNA sequences. *J. Mol. Biol. 203*:63–73 (1988).
112. Portugal, J., and M.J. Waring, Assignment of DNA binding sites for DAPI (4′,6-diamidine-2-phenylindole) and Hoechst 33258 (bisbenzimide). A comparative footprinting study. *Biochim. Biophys. Acta 949*:158–168 (1988).
113. Latt, S.A., Microfluorometric detection of deoxyribonucleic acid replication in human metaphase chromosomes. *Proc. Natl. Acad. Sci. (USA) 70*:3395–3399 (1973).
114. Latt, S.A., and J.C. Wohlleb, Optical studies of the interaction of 33258 Hoechst with DNA, chromatin, and metaphase chromsomes. *Chromsoma 52*:297–316 (1975).
115. Jin, R., and K.J. Breslauer, Characterization of the minor groove environment in a drug-DNA complex: Bisbenzimide bound to the poly [d(AT)]·poly-[d(AT)] duplex. *Proc. Natl. Acad. Sci. (USA) 85*:8939–8942 (1988).
116. Searle, M.S., and K.J. Embrey, Sequence-specific interaction of Hoechst 33258 with the minor groove of an adenine-tract DNA duplex studied in solution by ^{1}H NMR spectroscopy. *Nucl. Acids Res. 18*:3753–3762 (1990).
117. Pjura, P.E., K. Grzeskowiak, and R.E. Dickerson, Binding of Hoechst 33258 to the minor groove of B-DNA. *J. Mol. Biol. 197*:257–271 (1987).
118. Teng, M.-K., N. Usman, C.A. Frederick, and A.H.-J. Wang, The molecular structure of the complex of Hoechst 33258 and the DNA dodecamer d(CGC-GAATTCGCG). *Nucl. Acids Res. 16*:2671–2690 (1988).
119. Quintana, J.R., A.A. Lipanov, and R.E. Dickerson, Low temperature crystallographic analysis of the binding of Hoechst 33258 to the double-helical DNA dodecamer CGCGAATTCGCG. *Biochemistry 30*:10294–10306 (1991).

120. de C.T. Carrondo, M.A.A.F., M. Coll, J. Aymami, A.H.-J. Wang, G.A. van der Marel, J.H. van Boom, and A. Rich, Binding of a Hoechst dye to d(CGCGATATCGCG) and its influence on the conformation of the DNA fragment. *Biochemistry 28*:7849–7859 (1989).
121. Dann, O., G. Bergen, E. Demant, and G. Volz, Trypanocide diamidine des 2-phenyl-benzofurans, 2-phenyl-indens und 2-phenyl-indols. *Justus Liebigs Ann. Chem. 749*:68–89 (1971).
122. Chandra, P., B. Mildner, O. Dann, and A. Metz, Influence of 4′-6-diamidino-2-phenylindole on the secondary structure and template activities of DNA and polydeoxynucleotides. *Mol. Cell. Biochem. 18*:81–86 (1977).
123. Mildner, B., A. Metz, and P. Chandra, Interaction of 4′-6-diamidino-2-phenylindole to nucleic acids, and its implication to their template activity in RNA-polymerase reaction of *E. coli* bacteria and of Friend-virus infected mouse spleen. *Cancer Lett. 4*:89–98 (1978).
124. De Clerq, E., and O. Dann, Diaryl amidine derivatives as oncornaviral DNA polymerase inhibitors. *J. Med. Chem. 23*:787–795 (1980).
125. Kapuscinski, J., and W. Szer, Interactions of 4′,6-diamidine-2-phenylindole with synthetic polynucleotides. *Nucl. Acids Res. 6*:3519–3534 (1979).
126. Russell, W., C. Newman, and D.H. Williamson, A simple cytochemical technique for demonstration of DNA in cells infected with mycoplasmas and viruses. *Nature 253*:461–462 (1975).
127. Lin, M.S., D.E. Comings, and O.S. Alfi, Optical studies of the interaction of 4′6-diamidino-2-phenylindole with DNA and metaphase chromsomes. *Chromosoma 60*:15–25 (1977).
128. Schweizer, D., P. Ambros, and M. Andrle, Modification of DAPI banding on human chromosomes by pre-staining with a DNA-binding oligopeptide antibiotic, distamycin A. *Exptl. Cell Research 111*:327–332 (1978).
129. Kapuscinski, J., and K. Yanagi, Selective staining by 4′,6-diamidine-2-phenylindole of nanogram quantities of DNA in the presence of RNA on gels. *Nucl. Acids Res. 6*:3535–3542 (1979).
130. Macera, M.J., A. Babu, and R.S. Verma, Application of DA/DAPI technique in cancer cytogenetics. *Oncology 46*:61–62 (1989).
131. Takahama, M., and A. Kagaya, Hematoporphyrin/DAPI staining: Simplified simultaneous one-step staining of DNA and cell protein and trial application in automated cytological screening by flow cytometry. *J. Histochem. Cytochem. 36*:1061–1067 (1988).
132. Manzini, G., M.L. Barcellona, M. Avitabile, and F. Quadrifoglio, Interaction of diamidino-2-phenylindole (DAPI) with natural and synthetic nucleic acids. *Nucl. Acids Res. 11*:8861–8876 (1983).
133. Kubista, M., B. Akerman, and B. Norden, Characterization of interaction between DNA and 4′,6-diamidino-2-phenylindole by optical spectroscopy. *Biochemistry 26*:4545–4553 (1987).
134. Cavatorta, P., L. Masotti, and A.G. Szabo, A time-resolved fluorescence study of 4′,6′-diamidine-2-phenylindole dihydrochloride binding to polynucleotides. *Biophys. Chem. 22*:11–16 (1985).
135. Buys, C.H.C.M., J. Mesa, A.Y. van der Veen, and J.A. Aten, A comparison

of the effect of 5-bromodeoxyuridine substitution on 33258 Hoechst- and DAPI-fluorescence of isolated chromsomes by bivariate flow karyotyping. *Histochem. 84*:462–470 (1985).

136. Manzini, G., L. Xodo, M.L. Barcellona, and F. Quadrifoglio, Interaction of DAPI with double stranded ribonucleic acids. *Nucl. Acids Res. 24*:8955–8967 (1985).
137. Larsen, T., D.S. Goodsell, D. Cascio, K. Grzeskowiak, and R.E. Dickerson, The structure of DAPI bound to DNA. *J. Biomol. Struct. Dyn. 7*:477–491 (1989).
138. Jensch, H., Basic diazoaminobenzene compounds. U.S. Patent No. 2673197 (1954).
139. Jensch, H., 4,4′-Diamidino-diazoaminobenzol, ein neues mittel gegen trypanosomen- und babesien-infektionen. *Arzneimittel-Forsch. 5*:634–635 (1955).
140. Bauer, F., Uber den wirkungsmechanismus des berenil (4,4′-diamidino-diazoaminobenzol) bei *Trypanosoma congolense. Zbl. Bakt. Parasite., Abt. I, Orig. 172*:605 (1958).
141. Newton, B.A., and R.W.F. Le Page, Preferential inhibition of extranuclear deoxyribonucleic acid synthesis by the trypanocide berenil. *Proc. Biochem. Soc. 105*:50P (1967).
142. Bauer, F., The development of drug-resistance to berenil in *Trypanosoma congolense. Vet. Rec. 74*:265–266 (1962).
143. Gray, A.R., and C.J. Roberts, The cyclical transmission of strains of *Trypanosoma congolense* and *T. vivax* resistant to normal therapeutic doses of trypanocidal drugs. *Parasitology 63*:67 (1971).
144. Mbwambo, H.A., P.N.P. Mella, and K.A. Lekaki, Berenil (Diminazene aceturate)-resistant *Trypanosoma congolense* in cattle under natural tsetse challenge at Kibaha, Tanzania. *Acta Tropica 45*:239–244 (1988).
145. Newton, B.A., Berenil: A trypanocide with selective activity against extranuclear DNA. In *Antibiotics*, Corcoran, J.W. and Hahn, F.E. eds., Vol. 3, 34–47, Springer, New York (1975).
146. Festy, B., A.M. Lallemant, G. Riou, C. Brack, and E. Delain, Mecanisme d'action des diamidines trypanocides. Importance de la composition en bases dans l'association berenil-polynucleotides. *C. R. Acad. Sci. Paris Ser. D 271*: 684–687 (1970).
147. Newton, B.A., Interaction of berenil with deoxyribonucleic acid and some characteristics of the berenil deoxyribonucleic acid complex. *Proc. Biochem. Soc. 105*:50P–51P (1967).
148. Waring, M.J., Variation of the supercoils in closed circular DNA by binding antibiotics and drugs: Evidence for molecular models involving intercalation. *J. Mol. Biol. 54*:247–279 (1970).
149. Krey, A.K., Non-intercalative binding to DNA. *Prog. Molec. Subcell. Biol. 7*:43–87 (1980).
150. Festy, B., J. Strum, M. Daune, Interaction between hydroxystilbamidine and DNA. I. Binding isotherms and thermodynamics of the association. *Biochim. Biophys. Acta. 407*:24–42 (1975).

151. Gresh, N., and B. Pullman, A theoretical study of the nonintercalative binding of berenil and stilbamidine to double-stranded (dA-dT)$_n$ oligomers. *Molec. Pharm. 25*:452–458 (1984).
152. Pearl, L.H., J.V. Skelly, B.D. Hudson, and S. Neidle, The crystal structure of the DNA binding drug berenil: Molecular modelling studies of berenil-DNA complexes. *Nucl. Acids Res. 15*:3469–3477 (1987).
153. Brown, D.G., M.R. Sanderson, J.V. Skelly, T.C. Jenkins, T. Brown, E. Garman, D.I. Stuart, and S. Neidle, Crystal structure of a berenil-dodecanucleotide complex: The role of water in sequence-specific ligand binding. *EMBO J. 9*:1329–1334 (1990).
154. Yoshida, M., D.L. Banville, and R.H. Shafer, Structural analysis of d(GCAATTGC)$_2$ and its complex with berenil by nuclear magnetic resonance spectroscopy. *Biochemistry 29*:6585–6592 (1990).
155. Lown, J.W., K. Krowicki, U.G. Bhat, A. Skorobogaty, B. Ward, and J.C. Dabrowiak, Molecular recognition between oligopeptides and nucleic acids: Novel imidazole-containing oligopeptides related to netropsin that exhibit altered DNA sequence specificity. *Biochemistry 25*:7408–7416 (1986).
156. Lown, J.W., K. Krowicki, J. Balzarini, and E. De Clerq, Structure-activity relationship of novel oligopeptide antiviral and antitumor agents related to netropsin and distamycin. *J. Med. Chem. 29*:1210–1214 (1986).
157. Lown, J.W., and K. Krowicki, Efficient total syntheses of the oligopeptide antibiotics netropsin and distamycin. *J. Org. Chem. 50*:3774–3779 (1985).
158. Krowicki, K., M. Lee, J.A. Hartley, B. Ward, K. Kissinger, A. Skorobogaty, J.C. Dabrowiak, and J.W. Lown, Molecular recognition between oligopeptides and nucleic acids—Rational design of sequence specific DNA binding agents. In *Structure and Expression*, Sarma, R.H. and Sarma, M.H. eds., Vol. 2: 251–271, Adenine Press, Schenectady (1988).
159. Zakrzewska, K., R. Lavery, and B. Pullman, A theoretical study of the sequence specificity in binding of lexitropsins to B-DNA. *J. Biomol. Struct. Dyn. 4*:833–843 (1987).
160. Pullman, B., Sequence selectivity in DNA-antitumor interactions. In *Structure and Expression*, Sarma, R.H. and Sarma, M.H. eds., Vol. 2:237–249, Adenine Press, Schenectady (1988).
161. Burckhardt, G., G. Luck, C. Zimmer, J. Storl, K. Krowicki, and J.W. Lown, Variation of DNA sequence specificity of DNA-oligopeptide binding ligands related to netropsin: Imidazole-containing lexitropsins. *Biochim. Biophys. Acta 1009*:11–18 (1989).
162. Lee, M., D.-K. Chang, J.A. Hartley, R.T. Pon, K. Krowicki, and J.W. Lown, Structural and dynamic aspects of binding of a prototype lexitropsin to the decadeoxyribonucleotide d(CGCAATTGCG)$_2$ deduced from high-resolution ^{1}H NMR studies. *Biochemistry 27*:445–455 (1988).
163. Lee, M., R.T. Pon, K. Krowicki, and J.W. Lown, Structural and dynamic aspects of the sequence specific binding of netropsin and its bis-imidazole analogue on the decadeoxyribonucleotide d-[CGCAATTGCG]$_2$. *J. Biomol. Struct. Dyn. 5*:939–949 (1988).
164. Kissinger, K., K. Krowicki, J.C. Dabrowiak, and J.W. Lown, Molecular

recognition between oligopeptides and nucleic acids. Monocationic imidazole lexitropsins that display enhanced GC sequence dependent DNA binding. *Biochemistry 26*:5590–5595 (1987).
165. Randrianarivelo, M., K. Zakrzewska, and B. Pullman, Theoretical modeling of DNA-monocationic lexitropsin complexation: Influence of ligand binding on DNA curvature. *J. Biomol. Struct. Dyn. 4*:769–779 (1989).
166. Zimmer, Ch., G. Luck, G. Burckhardt, K. Krowicki, and J.W. Lown, Effect of ligand structure on the dA·dT-specific DNA groove binding of non-intercalative drugs. In *Structure and Expression*, Sarma, R.H. and Sarma, M.H. eds., Vol. 2:291–303, Adenine Press, Schenectady (1988).
167. Lee, M., J.A. Hartley, R.T. Pon, K. Krowicki, and J.W. Lown, Sequence specific molecular recognition by a monocationic lexitropsin of the decadeoxyribonucleotide d-[CATGGCCATG]$_2$: Structural and dynamic aspects deduced from high field ^{1}H-NMR studies. *Nucl. Acids Res. 16*:665–684 (1988).
168. Lee, M., D.M. Coulter, R.T. Pon, K. Krowicki, and J.W. Lown, Sequence specific molecular recognition and binding of a monocationic bis-imidazole lexitropsin to the decadeoxyribonucleotide d-[(GATCCGTATG)·(CATACGGATC)]: Structural and dynamic aspects of intermolecular exchange studied by ^{1}H-NMR. *J. Biomol. Struct. Dyn. 5*:1059–1087 (1988).
169. Lee, M., K. Krowicki, J.A. Hartley, R.T. Pon, and J.W. Lown, Molecular recognition between oligopeptides and nucleic acids: Influence of van der Waals contacts in determining the 3′-terminus of DNA sequences read by monocationic lexitropsins. *J. Am. Chem. Soc. 110*:3641–3649 (1988).
170. Lee, M., K. Krowicki, R.G. Shea, J.W. Lown, and R.T. Pon. Molecular recognition between oligopeptides and nucleic acids. Specificity of binding of a monocationic bis-furan lexitropsin to DNA deduced from footprinting and ^{1}H NMR studies. *J. Molec. Recog. 2*:84–93 (1989).
171. Borah, R., S. Roy, G. Zong, and J.S. Cohen, Unique pyrimidine 2D-COSY aromatic cross-peaks as monitors of pyrimidine environments and mobility in oligo- and polynucleotides. *Biochem. Biophys. Res. Comm. 133*:380–388 (1985).
172. Bailly, C., R. Houssin, J.-L. Bernier, and J.-P. Henichart, Depsipeptide analogs of the antitumor drug distamycin containing thiazole amino acids residues. *Tetrahedron 44*:5833–5843 (1988).
173. Rao, K.E., Y. Bathini, and J.W. Lown, Synthesis of novel thiazole-containing DNA minor groove binding oligopeptides related to the antibiotic distamycin. *J. Org. Chem. 55*:728–737 (1990).
174. Rao, K.E., R.G. Shea, B. Yadagiri, and J.W. Lown, Molecular recognition between oligopeptides and nucleic acids: DNA sequence specificity and binding properties of thiazole-lexitropsins incorporating the concepts of base site acceptance and avoidance. *Anti-Cancer Drug Design 5*:3–20 (1990).
175. Kumar, S., M. Jaseja, J. Zimmermann, B. Yadagiri, R.T. Pon, A.-M. Sapse, and J.W. Lown, Molecular recognition and binding of a GC site-avoiding thiazole-lexitropsin to the decadeoxyribonucleotide d-[CGCAATTGCG]$_2$: ^{1}H-NMR evidence for thiazole intercalation. *J. Biomol. Struct. Dyn. 8*:99–121 (1990).

176. Lee, M., R.G. Shea, J.A. Hartley, K. Kissinger, R.T. Pon, G. Vesnaver, K.J. Breslauer, J.C. Dabrowiak, and J.W. Lown, Molecular recognition between oligopeptides and nucleic acids: Sequence-specific binding of the naturally occurring antibiotic (4*S*)-(+)-anthelvencin A and its (4*R*)-(−)-enantiomer to deoxyribonucleic acids deduced from ^{1}H NMR, footprinting, and thermodynamic data. *J. Am. Chem. Soc. 111*:345–354 (1989).
177. Larsen, T.A., M.L. Kopka, and R.E. Dickerson, Crystal structure analysis of the B-DNA dodecamer CGTGAATTCACG. *Biochemistry 30*:4443–4449 (1991).
178. Narenda, N., S.L. Ginell, I.M. Russu, and H.M. Berman, Crystal and molecular structure of a DNA fragment d(CGTGAATTCACG). *Biochemistry 30*: 4449–4455 (1991).
179. Schultze, P., private communication.
180. Grzeskowiak, K., K. Yanagi, G.G. Prive, and R.E. Dickerson, The structure of B-helical CGATCGATCG, and comparison with CCAACGTTGG: The effect of base pair reversals. *J. Biol. Chem. 266*:8861–8883 (1991).
181. Goodsell, D., and R.E. Dickerson, Isohelical analysis of groove-binding drugs. *J. Med. Chem. 29*:727–733 (1986).
182. Zakrzewska, K., and B. Pullman, Theoretical study of the sequence selectivity of isolexins, isohelical DNA groove binding ligands. Proposal for the GC minor groove specific compounds. *J. Biomol. Struct. Dyn. 5*:1043–1058 (1988).
183. Zakrzewska, K., M. Randrianarivelo, and B. Pullman, Drug recognition of DNA. Proposal for GC minor groove specific ligands: Vinylexins. *J. Biomol. Struct. Dyn. 6*:331–344 (1988).
184. Khorlin, A.A., A.S. Krylov, S.L. Grokhovsky, A.L. Zhuze, A.S. Zasedatelev, G.V. Gursky, and B.P. Gottikh, A new type of AT-specific ligand constructed of two netropsin-like molecules. *FEBS Lett. 118*:311–314 (1980).
185. Schultz, P.G., and P.B. Dervan, Sequence-specific double-strand cleavage of DNA by bis(EDTA-distamycin·Fe^{II}) and EDTA-bis(distamycin)·Fe^{II}. *J. Am. Chem. Soc. 105*:7748–7750 (1983).
186. Dervan, P.B., Design of sequence-specific DNA-binding molecules. *Science 232*:464–471 (1986).
187. Lown, J.M., K. Krowicki, J. Balzarini, R.A. Newman, and E. De Clerq, Novel linked antiviral and antitumor agents related to netropsin and distamycin: Synthesis and biological evaluation. *J. Med. Chem. 32*:2368–2375 (1989).
188. Youngquist, R.S., and P.B. Dervan, Sequence-specific recognition of B DNA by bis(EDTA-distamycin)fumaramide. *J. Am. Chem. Soc. 107*:5528–5529 (1985).
189. Griffin, J.H., and P.B. Dervan, Sequence-specific chiral recognition of right-handed double-helical DNA by (2*S*,3*S*)- and (2*R*,3*R*)-dihydroxy-bis(netropsin) succinamide. *J. Am. Chem. Soc. 108*:5008–5009 (1986).
190. Leinsoo, T.A., V.A. Nikolaev, S.L. Grokhovskii, A.M. Surovaya, N. Yu. Sidorova, S.A. Strel'tsov, A.S. Zasedatelev, A.L. Zhuze, and G.V. Gurskii, Synthetic DNA-binding ligands containing reaction centers specific for AT and GC base pairs. *Molek. Biol. (USSR) 23*:1616–1637 (1989).

191. Hertzberg, R.P., and P.B. Dervan, Cleavage of double helical DNA by (methidiumpropyl-EDTA)iron(II). *J. Am. Chem. Soc. 104*:313–315 (1982).
192. Krivtsova, M.A., E.B. Moroshkina, E.N. Glibin, and E.V. Frisman, Interaction of DNA with low-molecular-weight ligands of differing structure. III. Complexes of DNA with Distactins. *Molek. Biol. (USSR) 18*:950–956 (1984).
193. DesJarlais, R.L., R.P. Sheridan, J.S. Dixon, I.D. Kuntz, and R. Venkataraghavan, Docking flexible ligands to macromolecular receptors by molecular shape. *J. Med. Chem. 29*:2149–2153 (1986).
194. DesJarlais, R.L., R.P. Sheridan, G.L. Seibel, J.S. Dixon, I.D. Kuntz, and R. Venkataraghavan, Using shape complementarity as an initial screen in designing ligands for a receptor binding site of known three-dimensional structure. *J. Med. Chem. 31*:722–729 (1988).
195. Grootenhuis, P.D. J., P.A. Kollman, G.L. Seibel, R.L. DesJarlais, and I.D. Kuntz, Computerized selection of potential DNA binding compounds. *Anti-Cancer Drug Design 5*:237–242 (1990).
196. Ferrin, T.E., C.C. Huang, L.E. Jarvis, and R. Langridge, The MIDAS display system. *J. Mol. Graphics 6*:13–27 (1988).
197. Weiner, P.K., and P.A. Kollman, AMBER: Assisted model building with energy refinement. A general program for modeling molecules and their interactions. *J. Comput. Chem. 2*:287–303 (1981).
198. Weiner, S.J., P.A. Kollman, D.A. Case, U.C. Singh, C. Ghio, G. Alagona, S. Profeta, Jr., and P. Weiner, A new force field for molecular mechanical simulation of nucleic acids and proteins. *J. Am. Chem. Soc. 106*:765–784 (1984).
199. Singh, U.C., P.K. Weiner, J. Caldwell, and D.A. Case, AMBER 3.0, University of California at San Francisco (1987).
200. Goodsell, D.S., RMS: programs for generating raster molecular surfaces. *J. Mol. Graphics 6*:39–44 (1988).
201. Youngquist, R.S., and P.B. Dervan, A synthetic peptide binds 16 base pairs of A, T double helical DNA. *J. Am. Chem. Soc. 109*:7564–7566 (1987).

8

Pyrrolo(1,4)benzodiazepines

Unraveling the Complexity of the Structures of the Tomaymycin-DNA Adducts in Various Sequences Using Fluorescence, ^{1}H-NMR, and Molecular Modeling

William A. Remers
University of Arizona, Tucson, Arizona

Mary D. Barkley
Louisiana State University, Baton Rouge, Louisiana

Laurence H. Hurley
The University of Texas at Austin, Austin, Texas

I. INTRODUCTION

The pyrrolo(1,4)benzodiazepine [P(1,4)Bs] antitumor antibiotics include anthramycin and tomaymycin (Fig. 1a) and as a class represent a unique group of drugs that react with DNA to form covalent adducts through the exocyclic amino group of guanine. The elucidation of the structures, strategies for synthesis, and biosynthesis by microorganisms of these antibiotics has been recently reviewed (1). Reviews on the mechanism of action of these potent DNA-reactive drugs and biological consequences have also been published (2,3).

While the P[1,4]Bs are highly potent cytotoxic agents that have significant efficacy as antitumor agents in animal model studies (3), their clinical application has been limited by various dose-limiting toxicities. For example, anthramycin and sibiromycin produce cardiotoxicity (3) and produce severe tissue necrosis at the site of injection. Nevertheless, they represent a structurally and mechanistically unique group of DNA reactive drugs that could provide the lead for the design of new antitumor agents that, while maintaining the antitumor activity of the parent compounds, would have reduced or negligible side effects. A particularly attractive possibility was that molecular modeling studies could be used to design new compounds, because the reaction of these drugs with DNA is very well characterized and these drugs are purported to be relatively nondistortive to DNA. Thus a combined NMR, fluorescence, and molecular modeling investigation was initiated to develop a firm basis for a subsequent design and synthesis proposal.

These drugs not only show a selectivity for covalent bonding through N_2 of guanine in DNA (4) but also exhibit a surprising sequence selectivity in

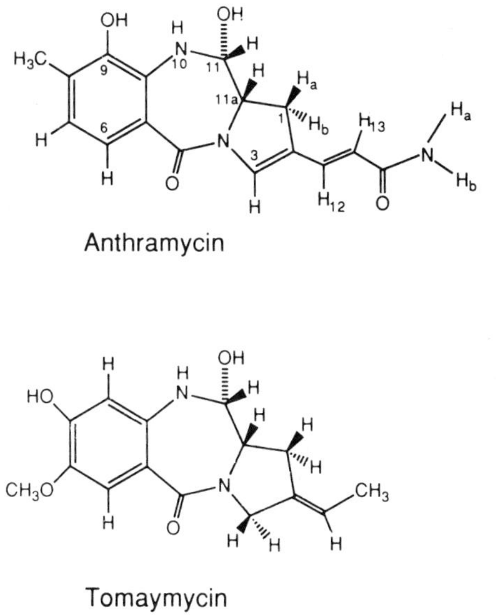

(a)

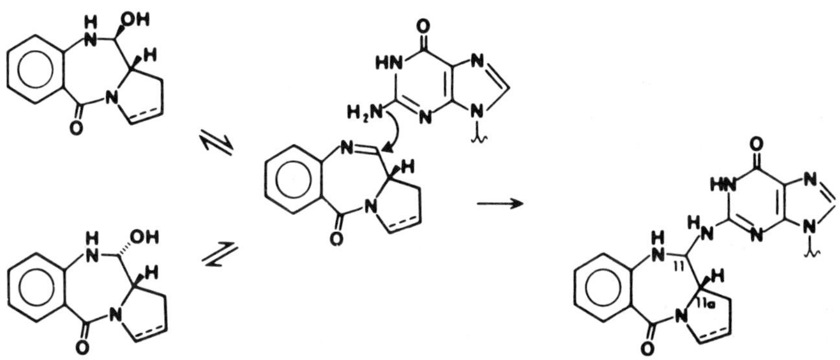

(b)

Figure 1 (a) Structures of anthramycin and tomaymycin. (b) Proposed mechanism for alkylation of DNA by the N10,C11 imine to form the P[1,4]B—N-2-guanine—DNA adduct.

view of their modest molecular weights. In this regard, DNA footprinting studies (5) and an exonuclease III stop assay (6) have shown that anthramycin, tomaymycin, and sibiromycin prefer to bond to 5′-PuGPu sequences, while 5′-PyGPu and 5′-PuGPy are of intermediate sequence preference and 5′-PyGPy sequences are least favored.

While the collaborative studies from Arizona, Louisiana, and Texas have examined both anthramycin (7) and tomaymycin (8–10), we have chosen to concentrate on just tomaymycin in this chapter in order to illustrate how techniques from molecular biophysics—namely fluorescence, high-field NMR, and molecular modeling—can be used in combination to provide considerable insight into the structures of the tomaymycin-DNA adducts. It is important to appreciate that no single technique or, indeed, even any two of the techniques could have provided anywhere near the insight that was afforded from using all three techniques side by side. So, for example, the fluorescence studies have allowed us to correlate the conformation variability of tomaymycin in solution with forms bound to oligomers and polymers, while high-field NMR has provided detailed structural and conformational information. Last, molecular modeling provides a means to rationalize the results and generate the graphical representations. It is therefore our intention to emphasize the advantages of a joint experimental/theoretical approach and provide a case analysis of how we built upon data obtained from free drug in solution to drug-oligomer studies, and how this may apply to drug-polymer systems. Finally, we propose how this information can be used to design a new generation of DNA reactive drugs of this class.

Studies from these laboratories provided unequivocal evidence that the P(1,4)Bs react with N_2 of guanine in DNA to form an animal linkage (Fig. 1b). We suspect that the imino or its imminium ion, resulting from loss of water from the carbinolamine, may well be the ultimate DNA-reactive species. In principle, electrophilic attack at C-11 of tomaymycin from N_2 of guanine could produce any one of four diastereomeric species shown in Figure 2. These differ in the sterochemistry at C-11 and orientation of the drug in the minor groove, i.e., aromatic ring to the 3′ or 5′ side of covalently modified guanine. In principle, in a particular sequence context (e.g., PuGPu or PyGPy) one species of adduct (e.g., 11*S*, 5′) or a mixture of adducts (e.g., 11*S*, 5′, and 11*R*, 5′) could be present. Thus, in addition to determining conformational aspects of the tomaymycin-DNA adducts, it was first important to define the complexity of the species as well as the stereochemistry at the covalent adduct site and orientation in the minor groove.

While the structural aspects were of sufficient merit alone to warrant these studies described here, there were other long-range objectives that

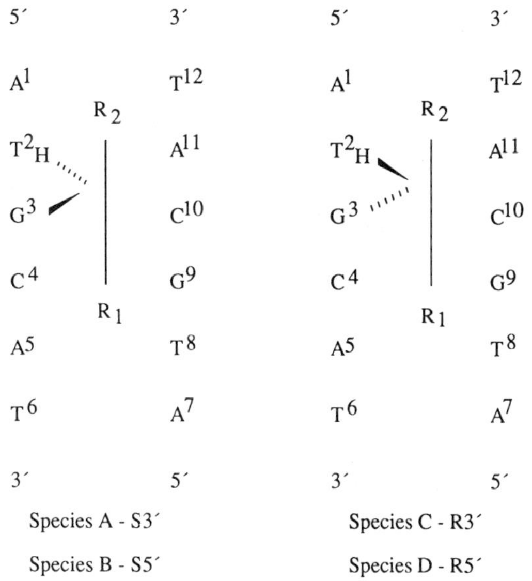

Figure 2 Four diastereomeric species.

also provided the motivation for these investigations. In response to the results of the earlier sequence selectivity studies, several theoretical investigations designed to rationalize these results had appeared (11–13). We were therefore anxious to provide experimental data that would test these theoretical studies. In addition, parallel studies from one of these laboratories and others had provided interesting data on the biochemical and biological consequences of DNA damage. The biochemical consequences included inhibition of replication (14), DNA repair recognition (15,16), and DNA strand breakage, which appeared to be repair-dependent (17), and the biological consequences included mutation and cell death (18,19). Rationalizing the biochemical and biological responses in accord with the structural modification of DNA was an exciting possibility.

II. TECHNIQUES

A. Fluorescence

Fluorescence decays were measured by the time-correlated single-photon counting technique. The sample decay curve was deconvolved by nonlinear least squares using a reference decay for the instrumental response. The data were fitted to a sum of exponentials with amplitudes α_i (λ_{ex}, λ_{em}) at excitation and emission wavelengths λ_{ex} and λ_{em}, respectively, and lifetimes

τ_i. For a mixture of fluorophores the relative amplitude α_i of the ith decay component depends on its absorption and emission spectra, radiative rate, and relative concentration. Decay-associated excitation and emission spectra are the steady-state spectra of the i components in a mixture. Decay-associated spectra are constructed from steady-state and time-resolved spectral data by scaling the amplitudes α_i to the steady-state intensity. All fluorescence experiments were done under magic angle conditions.

B. ^{1}H-NMR

The ^{1}H-NMR experiments described in this chapter were performed on a GN-500 NMR spectrometer at the University of Texas at Austin. The experimental parameters have been described before (7,9,10). In general, 10 to 15 mg of duplex or duplex adduct were used in NMR experiments. The assignments of exchangeable and nonexchangeable protons in the duplexes and corresponding duplex adduct were made by now well-established and routine procedures, as described in Chapter 2.

C. Molecular Modeling

The crystal structure of tomaymycin methyl ether (20) was used as the initial starting point. Partial atomic charges for tomaymycin with the methoxyl group removed (for covalent binding) were obtained from ab initio calculations (9). The resulting structure was minimized by molecular mechanics in AMBER 3.0 (21) using all atom force-field parameters derived from the crystal structure, under controls described in the literature (7,9). Polynucleotide duplex structures were constructed and minimized using AMBER 3.0 and parameters contained therein (22). The minimized tomaymycin structure was docked onto the duplexes in four different orientations near the guanine, to which it bonds covalently, using MIDAS interactive graphics (23), and the resulting adducts were minimized using molecular mechanics. They were then subjected to molecular dynamics with the duplexes constrained and tomaymycin free to move. Finally, the entire structures were relaxed in molecular mechanics minimization. Helix distortion energies were obtained by subtracting the energies of the separately minimized helices from their values in the adducts, and drug distortion energies were obtained in the same way. The different adducts were compared using net binding energies, which are the sum of distortion energies and total intermolecular binding energies. The effects of water and counterions were included in separate calculations to obtain qualitative estimates of entropy changes resulting from the disruption of ordered water molecules on adduct formation.

III. OLIGOMERS

A. Choice of Oligomers

The choice of oligomers for detailed experimental and theoretical work was made with consideration for size, bonding sequence, and preservation of a unique bonding site. For high-field NMR work, duplex molecules of between 6 and 12 base pairs appear to be optimal. The lower limit is established by the requirement for a duplex rather than single-stranded species (the duplex is stabilized by tomaymycin), while the upper limit is imposed by the complexity and overlapping signals that become too large and too many with duplexes of larger size. The choice of bonding sequence (i.e., flanking sequences one base pair each side of the covalently modified GC base pair) was made to sample sequences of various sequence preferences (i.e., PyGPy versus PyGPu).

The two oligomers described in this chapter were 6 and 12 base pairs in length and represented the bonding sequences 5′-TGC and 5′-CGA belonging to the PyGPy and PyGPu classes, respectively. The self-complementary 6-mer dATGCAT is sufficiently stabilized by tomaymycin that it forms a stable duplex at 23°C, although the 6-mer duplex alone shows considerable single stranded character under the same conditions (9).

The Dickerson dodecamer [d(C$\underline{G}$C$\underline{G}$AATTC$\underline{G}$CG)$_2$)] (24) was used as a starting point to design a self-complementary 12-mer sequence that would bond two tomaymycin molecules, each with the same defined orientation and stereochemistry. This dodecamer has three internal guanines (underlined in the 12-mer) that are potential tomaymycin covalent bonding sites (the terminal guanine is unlikely to be a bonding site). Two of these guanines were replaced with inosine, leaving only the guanine in the CGA sequence as a potential tomaymycin bonding site. Inosine differs from guanine only in that it lacks the exocyclic 2-amino group. In the inosine-disubstituted dodecamer [d(CICGAATTCICG)$_2$] the remaining tomaymycin-reactive guanine is contained within the sequence (5′-CGA) that is predicted from previous unpublished studies from our laboratories to bond predominantly one species of tomaymycin.

B. Synthesis and Purification

Oligomers for ^{1}H-NMR studies were prepared by solid-phase cyanoethyl phosphoramidite approach on an Applied Biosystems automated DNA synthesizer Model 381H. The oligomers were then deprotected separately with saturated ammonium hydroxide at 55°C overnight and purified in house on a Machery-Nagel Nucleogen-DEAE 60-7 HPLC column with an increasing gradient to 1 *M* NaCl in 15 mM sodium phosphate and 20% acetonitrile/aqueous buffer, pH 6.8.

IV. DETERMINATION OF THE STRUCTURAL ASPECTS OF THE TOMAYMYCIN-DNA ADDUCTS

A. Solution Studies on Tomaymycin

1. ^{1}H-NMR and Fluorescence of Tomaymycin in Aprotic Solvents

The crystalline form of tomaymycin 11-methyl ether (TME) is the 11*R*,11a*S* diastereomer (20), and when the crystals are dissolved in polar aprotic solvents, such as acetonitrile, one fluorescent ground-state species is observed. The partial ^{1}H-NMR of TME in CD_3CN is shown in Figure 3A.

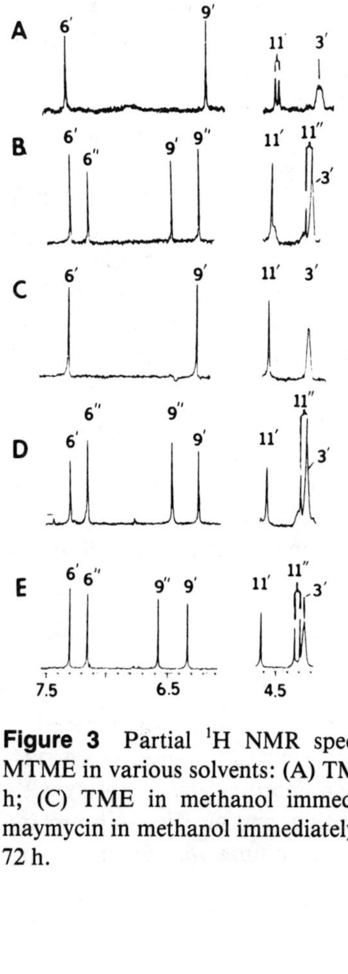

Figure 3 Partial ^{1}H NMR spectra of TME, 10,11-anhydrotomaymycin, and MTME in various solvents: (A) TME in acetonitrile; (B) TME in methanol after 72 h; (C) TME in methanol immediately after dissolution; (D) 10,11-anhydrotomaymycin in methanol immediately after dissolution; (E) MTME in methanol after 72 h.

The fluorescence decay gives a reasonable fit ($\chi_r^2 = 1.7$) to a single exponential with lifetime of 1.8 nsec (Table 1). In chloroform, tomaymycin is rapidly converted to 10,11-anhydrotomaymycin. This species was previously identified as an imine by ^{1}H-NMR (25), on the basis of the characteristic doublet found at $\delta 7.67$ ($J = 4.2$ Hz). The species in chloroform is nonfluorescent.

2. ^{1}H-NMR and Fluorescence of Tomaymycin in Methanol

^{1}H-NMR Studies. In protic solvents, such as alcohols and aqueous buffers, tomaymycin occurs in two fluorescent ground-state forms. The two forms are apparent in the ^{1}H-NMR of TME after equilibration in CD_3OD. As seen in Figure 3B, the H-11 protons of the two forms resonate between 4.20 and 4.70 ppm, whereas the aromatic H-6 and H-9 protons resonate between 6.20 and 7.40 ppm. Integration of these resonance signals indicates that the two species are present in approximate proportions of 65 : 35. In contrast, the ^{1}H-NMR of TME in methanol taken immediately after dissolving the crystalline sample shows only one species (Fig. 3C), which corresponds to the form found in acetonitrile (Fig. 3A). The ^{1}H-NMR of 10,11-anhydrotomaymycin was also examined immediately after dissolution in methanol. This spectrum (Fig. 3D) indicates that the imine has converted to the two forms of TME present in methanol, with a greater proportion of the species corresponding to the one present in lesser concentration at equilibrium (compare to Fig. 3B). However, if this sample is allowed to equilibrate, a spectrum identical with the spectrum of TME equilibrated in methanol (Fig. 3B) is obtained.

Table 1 Fluorescence Decay Data for TME, MTME, Tomaymycin, and 8-*O*-Methyltomaymycin[a] at 25°

Solvent	Compd	λ_{ex} (nm)	λ_{em} (nm)[b]	α_1[c]	τ_1 (ns)	α_2	τ_2 (ns)
MeOH	TMe	313	400	0.32	3.3	0.68	4.9
		337	400	0.15		0.85	
	MTME	313	400	0.17	2.7	0.86	5.4
		337	400	0.04		0.96	
Buffer, pH 5.1	Tomaymycin	313	410	0.999	1.1	0.001	10
	8-*O*-methyltomaymycin	313	410	0.99	1.3	0.01	8
		337	410	0.98		0.02	
MeCN	TME	313		1.00	1.8		
		337		1.00			
	MTME	313		1.00	2.3		
		337		1.00			

Although the species of tomaymycin in aprotic solvents have been identified previously by ^{1}H-NMR (25,26), the equilibrium mixture in methanol had not been reported. Conceivably, the two forms in protic solvents could be due to proton transfer, tautomers, conformers, or epimers at C-11. To test whether proton transfer or tautomerism might be responsible for the equilibrium mixture in methanol, the 8-*O*-methyl derivative (MTME) of TME was prepared. This derivative lacks the 8-phenolic proton of tomaymycin likely to participate in proton transfer to solvent or tautomerism with the C-5 carbonyl. Both possibilities are excluded by the ^{1}H-NMR, which still shows two species in methanol (Fig. 3E). If the two species in methanol represent two conformers of the drug molecule in protic solvents, then the ^{1}H-NMR signals of the two forms should begin to coalesce at elevated temperatures. However, heating the equilibrated sample of TME in methanol to 65°C did not simplify the ^{1}H-NMR spectrum (not shown), eliminating the possibility that the doubling of the proton signals in methanol is due to different conformational forms. A fourth and more likely possibility was epimerization at the C-11 position, a phenomenon first observed by Leimgruber et al. (27) for anthramycin. Since the crystalline form of TME is known to be the 11*R*,11a*S* diastereomer (20), the single species found in acetonitrile and also the predominance of the same species found immediately after dissolution in methanol are easily rationalized. Upon standing in methanol, the 11*R*,11a*S* diastereomer then equilibrates to a mixture of the two diastereomeric forms (11*S*,11a*S* and 11*R*,11a*S*), probably by conversion via the imine 10,11-anhydrotomaymycin (Fig. 1B). Finally, addition of methanol to 10,11-anhydrotomaymycin preferentially gives rise to the 11*S*,11a*S* diastereomer. This presumably reflects a sterically favored approach of methanol to one face of the molecule. The initial mixture is slowly converted to the same quilibrium mixture of the two diastereomeric forms found when the crystalline form (11*R*,11a*S*) is dissolved and equilibrated. In summary, we suggest that the two fluorescent ground-state forms of tomaymycin in methanol occur by epimerization at the C-11 position, which proceeds by way of attack by methanol of the intermediate 10,11-anhydrotomaymycin (Fig. B). In order to verify this interpretation, a ^{13}C-NMR experiment using $^{13}CH_3OH$ as the solvent for the equilibration was carried out (8). These results are consistent with the mechanism proposed in Figure 1B.

Fluorescence Studies. The fluorescence decay of tomaymycin is biexponential in protic solvents. The two emitting species represent two fluorescent ground-state forms with closely overlapped spectra and different excited-state lifetimes. This is manifest in the wavelength dependence of the steady-state spectra and fluorescence quantum yield. Figure 4A shows the absorption spectrum of TME equilibrated in methanol (broken line);

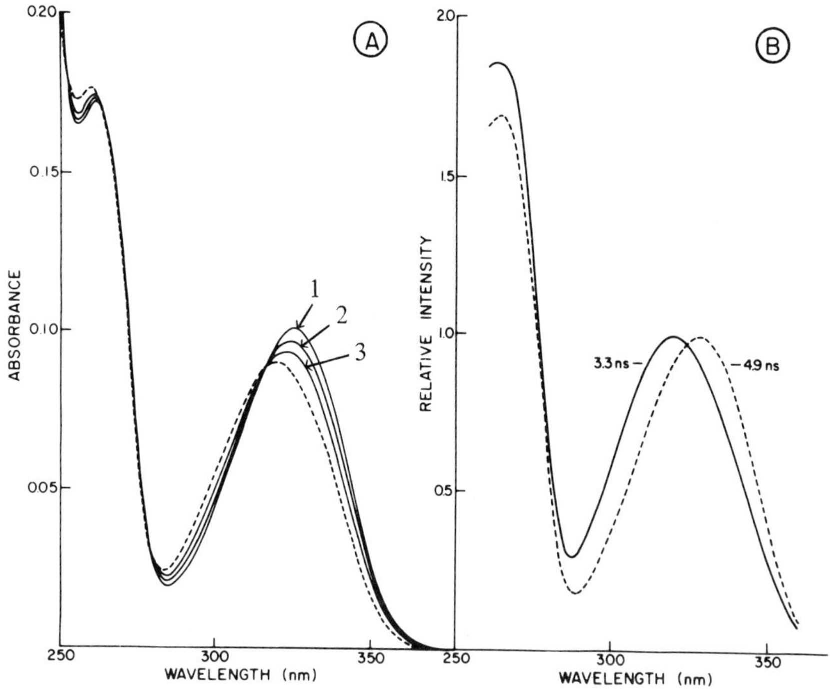

Figure 4 Absorption and fluorescence excitation spectra of TME in methanol. (A) Absorption spectra at various times after adding crystals to methanol: (1 , upper curve) 0.5 h; (2 , middle curve) 1.25 h; (3 , lower curve) 2.0 h; (---) equilibrium. Isosbestic points are at 316, 279, and 263 nm. (B) Decay-associated excitation spectra: (—) 3.3-ns component; (---) 4.9-ns component. Spectra are normalized to the long-wavelength peak.

the long-wavelength maximum is about 320 nm (28). The fluorescence emission maximum is around 400 nm. The emission and excitation maxima depend on excitation and emission wavelength. Figure 5A shows the emission spectra of TME in methanol corresponding to excitation at half-maximal absorption on either side of the 320-nm band. Excitation on the red (342 nm) shifts the emission spectra about 5 nm to the blue compared to excitation on the blue (296 nm). The excitation spectra show corresponding spectral shifts (data not shown). Thus, there are at least two absorbing and emitting species with different but overlapping spectra: the species that absorbs on the red emits on the blue, whereas the species that absorbs on the blue emits on the red. Furthermore, the apparent quantum yield of TME in methanol increases with increasing excitation wavelength (8). In principle, the quantum yield of a pure compound should be independent of

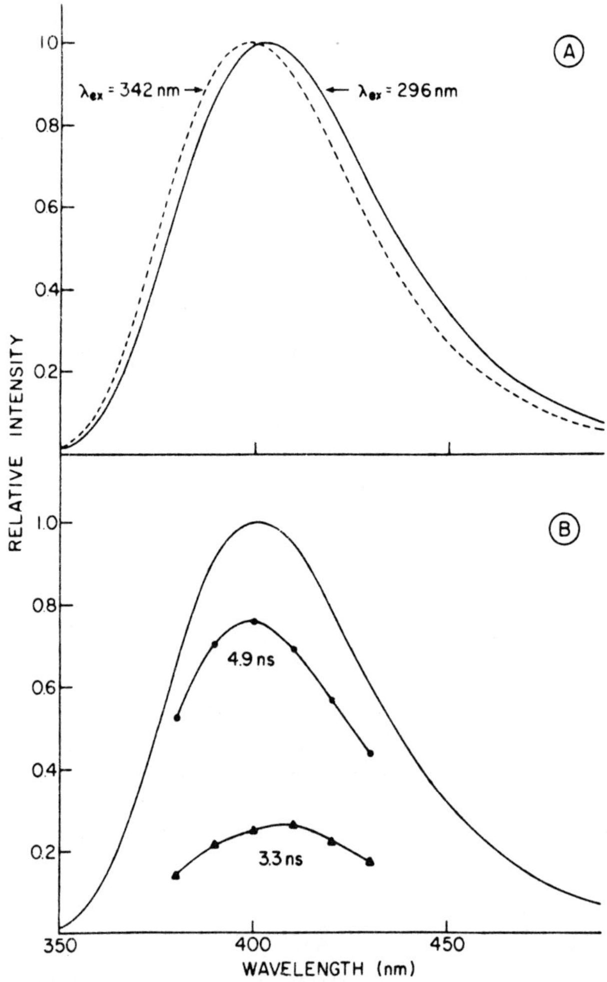

Figure 5 Fluorescence emission spectra of TME in methanol. (A) Corrected steady-state emission spectra at different excitation wavelengths: (—) λ_{ex} = 296 nm; (---) λ_{ex} = 342 nm. Spectra are normalized to the peak. (B) Decay-associated emission spectra: (—) corrected steady-state spectrum at λ_{ex} = 313 nm; (●) 4.9- and (▲) 3.3-ns components scaled to the steady-state spectrum. Lines were drawn to connect the points.

excitation wavelength. The wavelength dependence of the apparent quantum yield indicates that there is more than one absorbing species. If it were due entirely to the presence of a nonfluorescent compound, then the steady-state spectra would be independent of wavelength. The increase in quantum yield with excitation wavelength indicates that the species that absorbs on the red contributes an increasing fraction of the total emission.

The time-resolved fluorescence data extend the conclusions based on steady-state measurements. The fluorescence decay of TME in methanol is adequately described by a biexponential function with lifetimes of 3.3 and 4.9 ns at 25 °C and positive amplitudes that depend on excitation and emission wavelength (Table 1). Because accurate resolution of such closely spaced decays (lifetimes differing by less than a factor of 2) is generally not possible from analysis of single decay curves, the data presented in Table 1 were obtained from global analyses of multiple decay curves collected at different wavelengths (29). Fluorescence decays of TME in methanol were measured for six emission wavelengths (380–430 nm, 10-nm increments) at both 313- and 337-nm excitation. The 12 decay curves were analyzed by a global least-squares fitting program: the global $\chi_r^2 = 1.4$ for two exponentials compared to $\chi_r^2 = 3.5$ for a single exponential. No significant improvement in fit was obtained for three exponentials. As seen in Table 1, the amplitude α_1 of the 3.3-nsec decay decreases and the amplitude α_2 of the 4.9-nsec decay increases as the excitation wavelength is increased. The changes in the relative amplitudes with excitation wavelength at constant emission wavelength are due only to differences in the absorption spectra of the two lifetime components. Thus, the 4.9-nsec component absorbs to the red of the 3.3-nsec component. This is illustrated in Figure 4B, which shows the decay-associated excitation spectra of the two components normalized to the long-wavelength peak. The excitation maximum is at about 328 nm for the 4.9-nsec component compared to 320 nm for the 3.3-nsec component. The decay-associated emission spectra of the two components are depicted in Figure 5B. The emission maximum of the 4.9-nsec component lies about 10 nm to the blue of the peak of the 3.3-nsec component. The 4.9-nsec decay contributes about 75% of the fluorescence intensity at 313-nm excitation. The fluorescence properties of MTME in methanol are similar to those of TME, with minor quantitative differences (Table 1 and Ref. 8).

Correlation of ^{1}H-NMR and Fluorescence Results. The ^{1}H-NMR studies enable us to identify the two fluorescent ground-state forms of tomaymycin in methanol. According to the ^{1}H-NMR, the 11*R*,11a*S* diastereomer is present in about 65 : 35 excess of the 11*S*,11a*S* diastereomer at equilibrium. Moreover, the two diastereomers adopt different conformations in methanol, which accounts for the absorption spectral shifts. We can make a crude estimate of the relative concentrations of the two ground-state species from the amplitudes α_1 and α_2 of the fluorescence decay. Using the amplitudes in Table 1, we find the relative concentrations of the 4.9- and 3.3-nsec components to be roughly 75 : 25. Thus, the longer lifetime component present in excess must be the 11*R*,11a*S* diastereomer, and the shorter lifetime component must be the 11*S*,11a*S* diastereomer. Furthermore, the longer lifetime component has its absorption band on the red compared to the shorter lifetime component. The ^{1}H-NMR conformational analysis indicates that the C-5 carbonyl lies closer to the plane of the aromatic ring in the 11*R*,11a*S* diastereomer. The more extensive conjugation possible in the 11*R*,11a*S* diastereomer would shift the $\pi \rightarrow \pi^*$ transition to lower energy. The extinction coefficient of tomaymycin, which is typical of a weak $\pi \rightarrow \pi^*$ transition, is essentially independent of environment (8). This implies that the radiative rate does not vary much, so that the differences in observed lifetimes are due to nonradiative processes that compete with emission for deactivation of the excited state. Solvent isotope effects on the fluorescence lifetime and quantum yield for tomaymycin suggest that excited-state proton transfer reactions may account for the lifetime differences in various solvents and between diastereomers (Barkley, unpublished results).

The assignment of the longer lifetime species that absorbs on the red to the 11*R*,11a*S* diastereomer, and of the shorter lifetime species that absorbs on the blue to the 11*S*,11a*S* diastereomer, was confirmed by monitoring the time course of epimerization by ^{1}H-NMR, absorbance, and fluorescence. The ^{1}H-NMR shows only the 11*R*,11a*S* diastereomer immediately after dissolution of TME in methanol (Fig. 3C). The relative amounts of the 11*R*,11a*S* and 11*S*,11a*S* forms are about 8:1 after 0.5 h and 4:1 after 1 h. Figure 4A shows the absorption spectra of TME at various times after dissolving the crystals in methanol. The peak is at about 325 nm after 0.5 h and shifts about 5 nm to the blue in 3–4 h, with isosbestic points at 315, 279, and 263 nm. The isosbestic points signify two ground-state species. The absorption spectral changes correlated as expected with changes in the amplitudes of the fluorescence decay: the amplitude α_2 of the 4.9-nsec decay decreased while the amplitude of the 3.3-nsec decay increased with time. At 313-nm excitation and 405-nm emission, 89% of the fluorescence intensity arises from the 4.9-nsec component after about 1 h, compared to 73% at equilibrium.

B. Location of the Covalent Bonding Sites Between Tomaymycin and DNA

It has long been assumed, based upon indirect evidence, that the P(1,4)Bs covalently bond via C-11 to the exocyclic 2-amino group of guanine on

DNA, as shown in Figure 1b (30,31). However, only recently, using high-field NMR on a 12-mer duplex adduct, has this been unequivocally demonstrated. The proton NMR of the imino region of the bis(tomaymycin)-d(CICGAATTCICG)$_2$ adduct in 90% H_2O is plotted in Figure 6A. Six imino proton signals are found in the region between 12 and 16 ppm and are identified in Table 2. These were assigned by one-dimensional NOE experiments, and a representative NOE difference spectrum is shown in Figure 6B. The two protons upfield of the imino region in Figure 6A are the tomaymycin phenolic proton (10.5 ppm) and the ^{4}G hydrogen-bonded 2-amino proton (8.95 ppm). The assignment of the ^{4}G hydrogen-bonded 2-amino proton (NH_b) was made by irradiation of the ^{4}G imino proton and observation of the nonexchangeable base and amino proton region (7.0–9.0 ppm) (Fig. 6C). Irradiation of the ^{4}G-^{9}C imino proton results in NOEs into the ^{4}G NH_b, ^{9}C NH_b, and most likely the ^{9}C H6.

The ^{4}G NH_b proton identified through the one-dimensional NOE experiment (see Figure 6C) also appears in the one-dimensional proton spectrum run in D_2O. Normally guanine amino protons in duplex DNA are rapidly exchanged in D_2O, but in this case the exchange is relatively slow, suggesting tomaymycin modification of the duplex at this site. In the magnitude COSY (Fig. 7) scalar coupling is evident between ^{4}G NH_b and the proton at C-11 of tomaymycin (arrow A). This establishes the point of covalent attachment between tomaymycin and d(CICGAATTCICG)$_2$ as C-11 and the ^{4}G exocyclic 2-amino nitrogen, respectively.

C. Complexity of the Tomaymcyin (N_2-guanine) Adduct Species on Calf Thymus DNA, d(ATGCAT)$_2$ and d(CICGAATTCICG)$_2$, and Determination of Stereochemistry at the Covalent Linkage Site

1. Calf Thymus DNA

There are two fluorescent ground-state forms of tomaymycin in the calf thymus DNA adduct. The long-wavelength absorption band of the adduct is typical of tomaymycin in other solvents, but the maximum is red shifted to about 330 nm (8,32). The steady-state emission spectrum is almost independent of excitation wavelength with a maximum at about 405 nm, though the fluorescence quantum yield increases with increasing excitation wavelength. The fluorescence decay of the tomaymycin-DNA adduct gives a good fit (global $\chi_r^2 = 1.3$) to a biexponential function with lifetimes of 3.8 and 6.7 nsec at 5°C (Table 3). No improvement in the fit is obtained for three exponentials. Time-resolved emission spectroscopy shows that the emission spectra of the two components are almost identical (M. D. Barkley, F. N. Chowdhury, and K. Maskos, manuscript in preparation), in

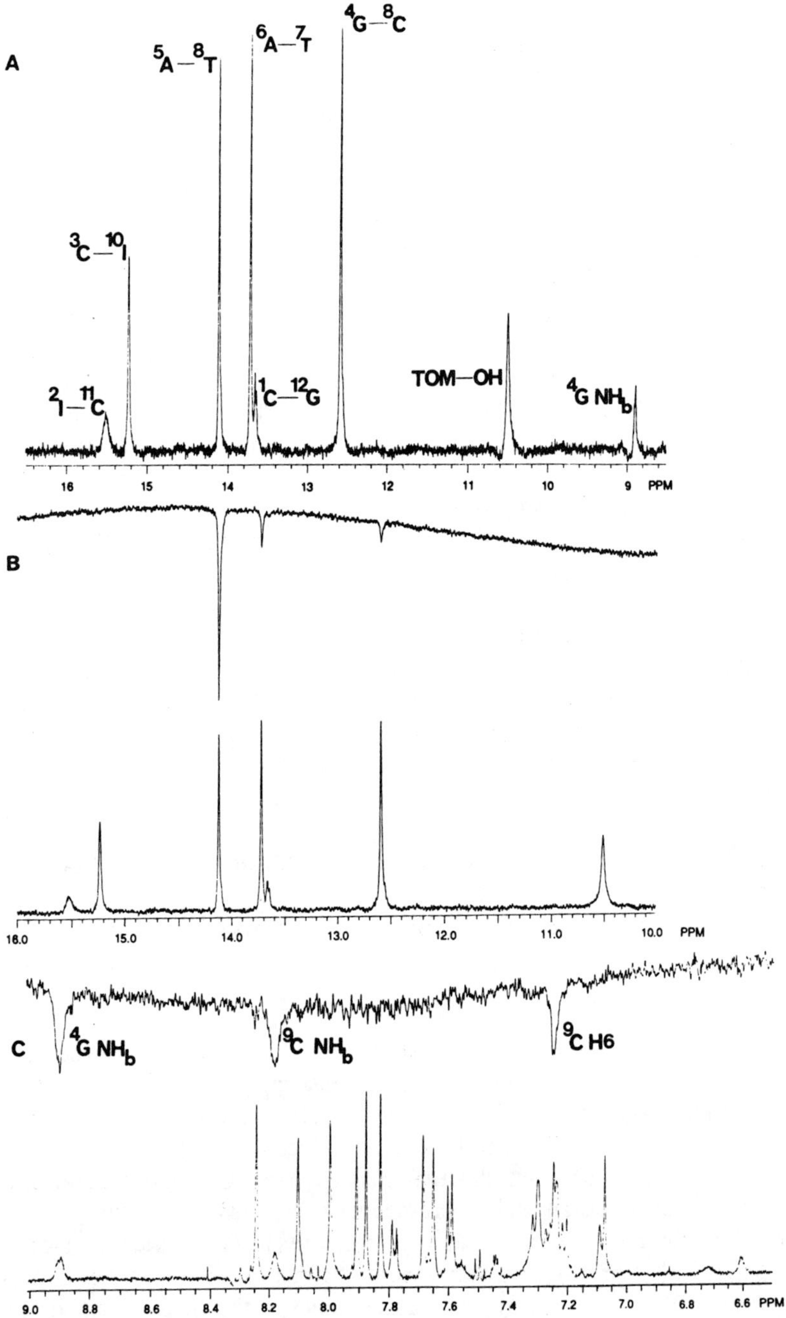

Figure 6 (A) 500-MHz proton NMR spectrum (12–16 ppm) of the bis(tomaymycin)-d(CICGAATTCICG)$_2$ adduct in buffered 90% H_2O/10% D_2O, pH 6.8 at 24°C. One-dimensional NOE difference spectra following saturation (2 s) of (B) the 14.13 ppm and (C) the 12.59 ppm exchangeable resonances in the complex. The assignments of the imino protons of the complex at 10°C are listed in Table 2.

Table 2

(a) Comparison of the Base Proton NMR Chemical Shift Assignments (ppm) and T_1 Relaxation Times of Nonexchangeable and Imino Protons of d(CICGAATTCICG)$_2$ and the bis(tomaymycin)-d(CICGAATTCICG)$_2$ Adduct in D_2O at 24°C

Assignment	d(CICGA-ATTCICG)$_2$, δ	bis(tomaymycin)-d(CICGAATT-CICG)$_2$ δ	bis(tomaymycin)-d(CICGAATT-CICG)$_2$ Δδ	Assignment	d(CICGAATTCICG)$_2$,δ	bis(tomaymycin)-d(CICGAATTCICG)$_2$ δ	bis(tomaymycin)-d(CICGAATTCICG)$_2$ Δδ
				Aromatic protons			
^{1}C H6	7.62	7.58	−0.04	^{12}G H8	7.89	7.84	−0.05
^{2}I H8	8.30	8.20	−0.10	^{1}C H5	5.90	5.82	−0.08
^{3}C H6	7.18	7.29	+0.11	^{2}I H2	7.85 (4.19 ± 0.08)	7.80 (5.06 ± 0.17)	−0.05 (+0.87)
^{4}G H8	7.81	7.62	−0.19	^{3}C H5	5.26	5.22	−0.04
^{5}A H8	8.10	7.86	−0.24	^{5}A H2	7.15 (3.10 ± 0.06)	7.24 (4.51 ± 0.15)	+0.09 (+1.41)
^{6}A H8	8.10	7.95	−0.25	^{6}A H2	7.57 (3.75 ± 0.09)	7.65 (5.19 ± 0.18)	+0.08 (+1.44)
^{7}T H6	7.07	7.27	+0.20	^{7}T Me	1.22	1.15	−0.07
^{8}T H6	7.33	7.07	−0.26	^{8}TMe	1.20	1.49	−0.01
^{9}C H6	7.47	7.24	−0.23	^{9}C H5	5.62	5.32	−0.30
^{10}I H8	8.25	8.06	−0.19	^{10}I H2	7.66 (3.66 ± 0.06)	7.22 (3.86 ± 0.09)	−0.44[a] (+0.20)
^{11}C H6	7.21	7.18	−0.03	^{11}C H5	5.35	5.22	−0.13
				Imino protons (10°C)			
^{1}C-^{12}G	12.70	12.80	+0.10	^{4}G-^{9}C	12.40	12.42	+0.02
^{2}I-^{11}C	15.38	15.41	+0.03	^{5}A-^{8}T	13.55	14.02	+0.47[a]
^{3}C-^{10}I	15.17	15.12	−0.05	^{6}A-^{7}T	13.60	13.60	−0.00

Table 2 (*Continued*)

(b) Comparison of the deoxyribofuranose proton chemical shift assignments (ppm) of d(CICGAATTCICG)$_2$ and the bis(tomaymycin)-d(CICGAATTCICG)$_2$ adduct

Deoxyribose protons

	H1′			H2′			H2″			H3′			H4′		
Assignment	12-mer	T-12-mer	Δδ	12-mer	T-12-mer	Δδ	12-mer	T-12-mer	Δδ	12-mer	T-12-mer	Δδ	12-mer	T-12-mer	Δδ
^{1}C	5.59	5.55	−0.04	1.96	1.90	−0.06	2.35	2.30	−0.05	4.61	4.62	+0.01	4.03	4.01	−0.02
^{2}I	6.15	6.18	+0.03	2.72	2.65	−0.07	2.85	2.82	−0.03	4.92	4.93	+0.01	4.42	4.38	−0.04
^{3}C	5.67	5.39	−0.28	1.58	2.06	<u>+0.48</u>	2.14	2.29	+0.15	4.67	4.75	+0.08	4.03	4.16	+0.13
^{4}G	5.26	6.00	<u>+0.74</u>[a]	2.63	2.43	−0.20	2.67	2.80	+0.13	4.88	4.92	+0.04	4.28	4.36	+0.08
^{5}A	5.94	5.75	−0.19	2.65	2.44	−0.21	2.89	2.55	<u>−0.34</u>	4.98	4.88	−0.10	4.40	4.24[b]	−0.16
^{6}A	6.08	5.80	−0.28	2.60	2.21	<u>−0.39</u>	2.88	2.79	−0.09	4.92	4.70	−0.22	4.42	3.92	<u>−0.50</u>
^{7}T	5.83	5.98	+0.15	1.95	1.95	+0.00	2.53	2.60	+0.07	4.71	4.85	+0.14	4.09	4.24[b]	+0.15
^{8}T	6.04	6.05	+0.01	2.12	2.15	+0.03	2.51	2.57	+0.06	4.82	4.89	+0.07	4.12	4.20[b]	+0.08
^{9}C	5.46	5.72	+0.26	2.05	1.80	−0.25	2.35	2.14	−0.21	4.79	4.84	+0.05	4.12	4.20[b]	+0.08
^{10}I	6.13	5.78	<u>−0.35</u>	2.70	2.55	−0.15	2.86	2.55	<u>−0.31</u>	4.94	4.81	−0.13	4.37	4.24	−0.13
^{11}C	5.78	5.83	+0.05	1.64	1.52	−0.12	2.20	2.15	−0.05	4.72	4.69	−0.03	4.18	3.81	−0.37
^{12}G	6.09	6.09	0.00	2.24	2.29	−0.05	2.61	2.55	−0.06	4.57	4.61	+0.04	4.15	4.00	−0.15

[a]Δδ underlined are > ±0.3 ppm. [b]The resonance of these H4′ protons in the bis(tomaymycin)-d(CICGAATTCICG)$_2$ adduct overlap.

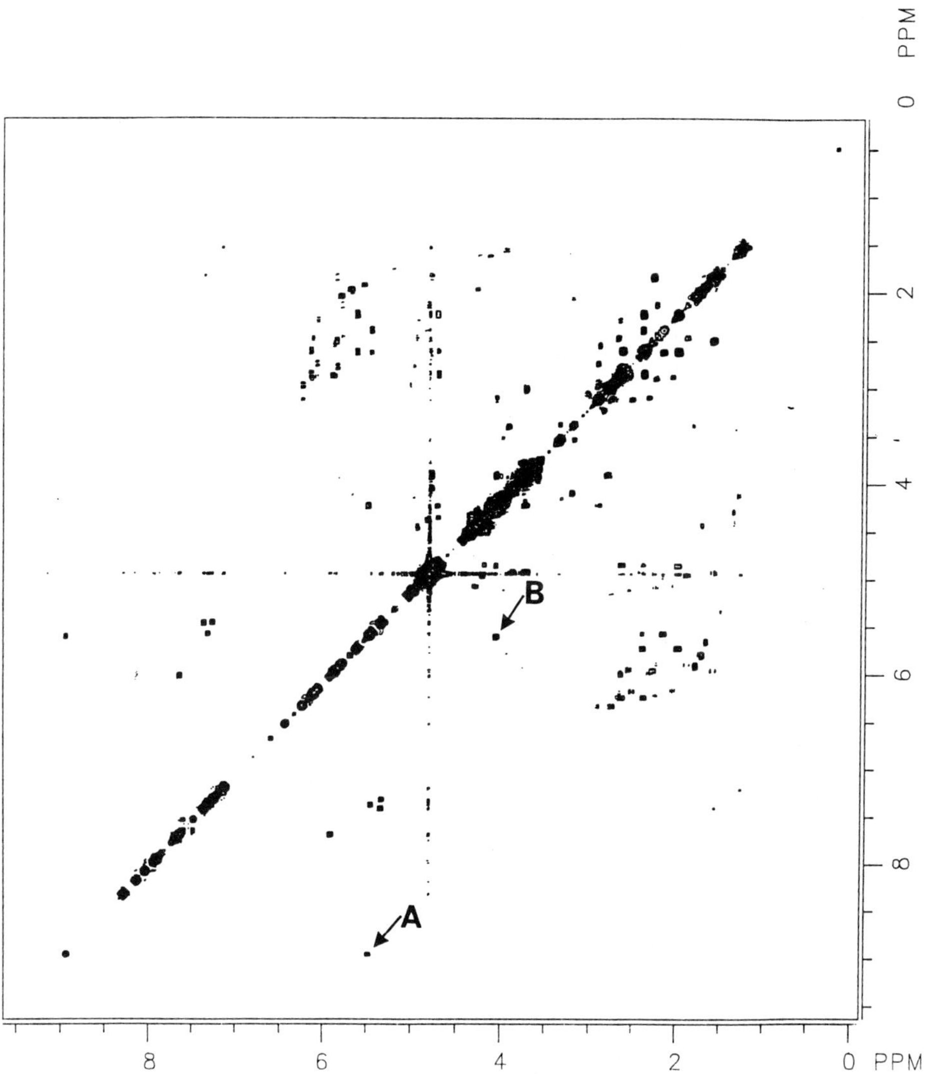

Figure 7 Contour plot of the magnitude COSY spectrum at 24°C. Arrow A shows the scalar coupling between the ^{4}G hydrogen-bonded amino proton and tomaymycin H11. Arrow B denotes the scalar coupling predicted for the 11*S* diastereomer of this adduct between the tomaymycin H11 and H11a protons.

Table 3 Fluorescence Decay Data for Tomaymycin Bound to Calf Thymus DNA, $d(ATGCAT)_2$ and $d(CIGCAATTCICG)_2$[a]

	λ_{ex} (nm)	α_1[b]	τ_1 (ns)	α_2	τ_2 (ns)
Calf thymus DNA[c]	313	0.35	3.8	0.65	6.7
	337	0.31		0.69	
	355	0.28		0.72	
$d(ATGCAT)_2$	313	0.43	2.3	0.57	5.7
	337	0.52		0.48	
	355	0.63		0.37	
$d(CIGCAATTCICG)_2$	313	0.11	1.6	0.89	6.4
	337	0.08		0.92	
	355	0.08		0.92	

[a]Decay parameters were obtained by global analysis of decay curves measured at the indicated excitation wavelengths and emission wavelengths between 390 and 490 nm.
[b]Fractional amplitudes $\Sigma\alpha_i = 1$.
[c]Data for pH 7.5, 5°C, from M.D. Barkley, F.N. Chowdhury, Q. Chen, and K. Maskos (manuscript in preparation). Amplitudes and lifetimes are independent of pH in the range 6.4–9.0.

agreement with the steady-state results. However, fluorescence decay measurements at three excitation wavelengths indicate that the absorption spectra of the two species are slightly different. As seen in Table 3, the amplitude of the 6.7-nsec component increases, whereas the amplitude of the 3.8-nsec component decreases, with increasing excitation wavelength. Thus for the tomaymycin-DNA adduct, as well as the free drug in protic solvents, the longer lifetime species is associated with an absorption band on the red, and the shorter lifetime species is associated with an absorption band on the blue. The longer lifetime species contributes about 80% of the fluorescence intensity.

The ^{1}H-NMR and fluorescence results for tomaymycin in protic solvents suggest a plausible explanation for the two fluorescent ground-state forms of the DNA adduct, namely, that the two emitting species represent the two diastereomeric forms of tomaymycin bound to DNA. The alternative explanation would be that the heterogeneous emission arises from a single diastereomer bound in different environments on DNA. The experiments described below on synthetic DNA duplexes confirm the former interpretation.

2. *Tomaymycin-d(ATGCAT)$_2$ Duplex Adduct*

The fluorescence decay of the $d(ATGCAT)_2$ adduct is also biexponential (Table 3). The lifetimes τ_i of the two species of tomaymycin-$d(ATGCAT)_2$

adducts are 2.3 and 5.7 nsec, compared to 3.8 and 6.7 nsec for the tomaymycin-DNA adduct. In both cases, the emission spectra associated with the two forms are identical, but the absorption spectra are slightly shifted. For the tomaymycin-d(ATGCAT)$_2$ adduct, the amplitude α_1 of the 2.3-nsec component increases, and the amplitude α_2 of the 5.7-nsec component decreases with increasing excitation wavelength, indicating that the shorter lifetime species has its absorption band on the red, compared to the longer lifetime species. This is opposite to tomaymycin in protic solvents and calf thymus DNA adducts, where the shorter lifetime component is associated with an absorption band on the blue, and the longer lifetime component is associated with an absorption band on the red. The greater excitation wavelength dependence of the amplitudes for the d(ATGCCAT)$_2$ adduct signifies a larger shift in the absorption spectra of the two species. We have argued elsewhere that the longer lifetime component represents the 11*S*,11a*S* diastereomeric adduct, while the shorter lifetime component represents the 11*R*,11a*S* adduct (8). Since the absorption spectra are not identical, the amplitudes are not identical to the relative amounts of the two diastereomers. Nevertheless, it appears that the two diastereomers occur in roughly equal proportions in the d(ATGCAT)$_2$ adduct, while the 11*S*,11a*S* diastereomer is the major species present in the DNA adduct. Calf thymus DNA will contain the full spectrum of binding sequences including the more favored 5′PuGPu, 5′PyGPu, and 5′PuGPy as well as the least favored 5′PyGPy sequence, which occurs in d(ATGCAT)$_2$. Consequently, we would predict that the more favored sequences would bind the 11*S*,11a*S* diastereomer with greater preference.

The d(ATGCAT)$_2$ duplex exhibits a two-fold symmetry axis resulting in equivalence of the corresponding base protons of each strand (33). In the anthramycin-d(ATGCAT)$_2$ duplex adduct, the symmetry of the duplex is destroyed by covalent association of drug (7). Consequently, signals from every base in the duplex are observed in the proton NMR spectrum, doubling the number of oligodeoxynucleotide signals relative to the duplex alone. Because of the signal doubling, COSY spectra of monodirectional duplex adducts show cross-peaks between four thymine methyl-H6 protons and two cytosine H5-H6 protons. In addition, one set of resonance signals from the drug molecule are present.

If tomaymycin were like anthramycin and bonded as just one species on d(ATGCAT)$_2$, then the loss of symmetry should result in the doubling of signals of the oligonucleotide. However, the fluorescence results presented above suggest that tomaymycin binds to d(ATGCAT)$_2$ with both diastereomeric linkages at C-11, which would result in two species of tomaymycin bound covalently to guanine. The COSY spectrum of two tomaymycin-d(ATGCAT)$_2$ adducts should then show connectivities between eight thymine methyl-H6 protons and four cytosine H5-H6 protons. In addition, the spectrum should also show two distinct sets of tomaymycin signals.

Examination of the ^{1}H-NMR spectrum of the tomaymycin-d(ATGCAT)$_2$ adduct reveals two distinct tomaymycin H9 protons at 5.98 and 6.13 ppm and a pair of tomaymycin H6 protons that occur close together, resonating at 6.99 and 7.03 ppm (Fig. 8). These tomaymycin H6 protons could be distinguished from the thymine H6 protons in the same region because all the thymine H6 protons exhibited cross-peaks to their respective methyl groups (Fig. 8, box A and expansion). The pattern of the tomaymycin aromatic protons in the d(ATGCAT)$_2$ adducts is very similar, though not identical with that observed for the 11*R*,11a*S* and 11*S*,11a*S* diastereomers of tomaymycin in D$_2$O buffer solution (8). In 10 mM phosphate buffer, the H9 and H6 protons of the 11*R*,11a*S* and 11*S*,11a*S* diastereomers of tomaymycin resonate at 6.26 and 6.51 ppm (H9) and at 7.11 and 7.19

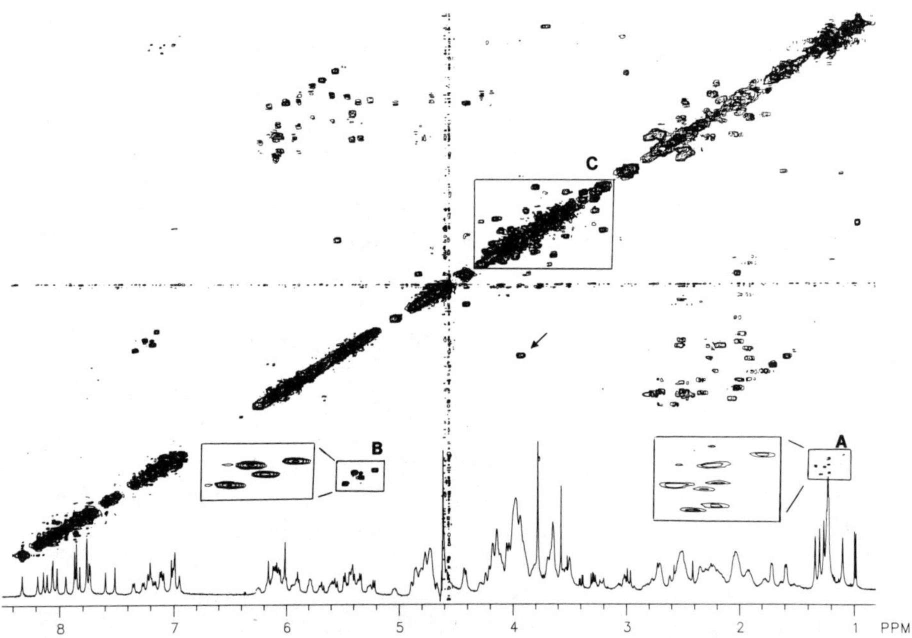

Figure 8 COSY spectrum of the tomaymycin–d(ATGCAT)$_2$ adducts in D$_2$O at 23°C. Box A shows the region containing the thymine methyl–H6 proton cross-peaks and its expansion, box B shows the cytosine H5–H6 proton cross-peaks and its expansion, and box C shows the region containing cross-peaks of the H3′, H4′, H5′, and H5′ protons of the duplex. The cross-peak denoted by the arrow is proposed to arise from a tomaymycin H11–H11a interaction (see text).

ppm (H6). The H13 ethylidene methyl signals of the two diastereomeric tomaymycin-d(ATGCAT)$_2$ adducts resonate downfield from the thymine methyl groups at 1.61 and 1.73 ppm (Fig. 8). Many of the remaining tomaymycin protons are obscured by signals originating from the oligomer. It is important to note that the chemical shifts of the tomaymycin signals in the d(ATGCAT)$_2$ adducts do not match the chemical shifts observed for the free drug in the same buffer. Also, extensive extraction of the duplex adduct with ethyl acetate, which removes free drug from aqueous buffer solutions, does not change the ^{1}H-NMR spectrum of the duplex adduct.

Four cytosine H5-H6 connectivities are present in the COSY spectrum of the tomaymycin-d(ATGCAT)$_2$ adduct (Fig. 8, panel B). COSY cross-peaks between seven of eight possible thymine methyl-H_6 protons are also observed (Fig. 8, box A). This pattern of signals for tomaymycin and the duplex molecule in the adduct is only consistent with the presence of two distinct tomaymycin-DNA adducts.

There are two strong cross-peaks in the COSY spectrum (arrows in Fig. 8); one of these is between a proton at 3.99 ppm and a doublet at 5.60 ppm. A similar cross-peak is also present in the spectrum of the anthramycin-d(ATGCAT)$_2$ adduct (7). In both cases, this COSY cross-peak is indicative of coupling between the protons at C-11 and C-11a of anthramycin or tomaymycin for the 11*R*,11a*S* diastereomer. The second cross-peak in the COSY spectrum between a proton at 4.78 ppm and a doublet at 4.42 ppm is presumably due to the 11*S*,11a*S* diastereomer. Molecular modeling using AMBER (Remers, unpublished results) predicts scalar coupling between the proton at C-11 and C-11a for both diastereomeric tomaymycin-d(ATGCAT)$_2$ adducts. Therefore the results of the COSY experiment on the tomaymycin-d(ATGCAT)$_2$ duplex adduct provides additional support for a mixture of the 11*S*,11a*S* and 11*R*,11a*S* diastereomeric species.

3. *Tomaymycin-d(CICGAATTCICG)$_2$ Duplex Adduct*

In contrast to the two fluorescent ground-state forms of tomaymycin in protic solvents and on calf thymus DNA and d(ATGCAT)$_2$ (8,9), we found essentially one fluorescent species of tomaymycin on d(CICGAATTCICG)$_2$ corresponding to the longer lifetime component or the 11*S*,11a*S* diastereomer (Table 3). The fluorescence decay of the tomaymycin-d(CICGAATTCICG)$_2$ adduct gave an acceptable fit to a monoexponential function with a χ_r^2 value of 1.4 and a lifetime of 6.4 nsec. A slightly better fit with χ_r^2 = 1.2 was obtained by assuming a biexponential function. Including the second exponential term in the data analysis had little effect on the lifetime of the predominant species. As seen in Table 3, the 6.4-nsec lifetime lies within the range of values associated with 11*S* diastereomeric adduct in calf thymus DNA and d(ATGCAT)$_2$. On the basis of the relative decay amplitudes

α_i in the bis(tomaymycin)-d(CICGAATTCICG)$_2$ adduct, the 11*S* diastereomer is present in at least 10-fold excess over the short-lifetime species. The 1.6-nsec lifetime may be due to small amounts of unbound tomaymycin, a second species of the tomaymycin-duplex adduct, or an unresolved mixture of the two.

Examination of the magnitude COSY spectrum of the bis(tomaymycin)-d(CICGAATTCICG)$_2$ adduct shows scalar coupling of the 11 and 11a protons of tomaymycin (arrow B in Fig. 7). In addition to firmly assigning the 5.48 ppm resonance signal to the proton at C-11 of tomaymycin (which is shifted downfield by 0.9 ppm relative to its chemical shift in free tomaymycin), the scalar coupling of protons at C-11 and C-11a potentially provides stereochemical information on the linkage site geometry. In TME only the 11*S*,11a*S* diastereomeric species shows scalar coupling between protons at C-11 and C-11a because the dihedral angle between these protons in the 11*R*,11a*S* diastereomer is close to 90° (8). Molecular modeling using AMBER was carried out on both diastereomeric species (11*R*,11a*S* and 11*S*,11a*S*) of tomaymycin bound covalently to the inosine-disubstituted dodecamer with each having a 3′ orientation of the aromatic ring of tomaymycin to ^{4}G. Only these two species were modeled since the NOE data narrowed down the four possible species to the 3′-oriented adducts (see later). The results showed that *only* the 11*S*,11a*S* species was predicted to have significant scalar coupling between protons at C-11 and C-11a; i.e., the dihedral angles between the protons at C-11 and C-11a in the 11*S* and 11*R* species are predicted to be 175.7° and 71.5°, respectively. The assignment of the species of tomaymycin bound to 12-mer duplex as the 11*S*,11a*S* diastereomer is also consistent with the fluorescence results (see Table 3).

D. Predictions for Complexity, Stereochemistry, and Orientation of Tomaymycin-Oligomer Duplex Adducts by Molecular Modeling

1. *Tomaymycin-d(ATGCAT)$_2$ Duplex Adduct*

For the molecular modeling studies, tomaymycin was covalently bound via the C-11 atom to the exocyclic amino group of guanine-3. All four cases illustrated in Figure 2 were used as the starting point for energy minimization using the all atom force-field parameters of AMBER. The net binding energy for each species (S3′; S5′; R3′; R5′) was calculated from the total intermolecular drug-DNA binding energies minus the helix and drug distortion energies that result from induced fits. In the initial study, which involved AMBER 2.0 parameters and molecular mechanics, species A and D in Figure 2 were found to be the preferred binding forms of tomaymycin-

D(ATGCAT)$_2$ (10). A more recent study, based on molecular dynamics and AMBER 3.0 parameters, indicates an order of preference S5′ > R5′ > S3′, with R3′ much less favored (32) (Table 4). The difference in net binding energies for the three best forms is only 7.3 kcal mol^{-1}, and it results from a combination of intermolecular binding energies, helix distortion energies, and drug distortion energies. To distinguish among these forms, we need to identify a unique tomaymycin-base interproton connectivity or connectivities that would define particular orientations (see later).

2. *Tomaymycin-d(CICGAATTCICG)$_2$ Duplex Adduct*

Likewise for the tomaymycin-12-mer duplex adduct, four different binding modes were modeled: the *R* and *S* configurations at C-11 of tomaymycin, each with the drug positioned in the minor groove so that its aromatic ring was pointing toward the 3′ end or the 5′ end of the covalently bound strand of d(CICGAAT·GCICTTA). To simplify the modeling, the 7-mer was used in place of the 12-mer. Following energy minimization in AMBER, the net binding energy for each species was calculated from its total intermolecular drug-DNA binding energy minus the helix and drug distortion energies that resulted from the induced fit (Table 5). There was a preference by 2.9 kcal mol^{-1} for the covalent adduct with 11*S* configuration and the benzene ring toward the 3′ direction over the 5′-oriented adduct with the same stereochemistry. While the molecular mechanics results clearly rule out both 11*R* diastereomers, the difference of 2.9 kcal mol^{-1} is considered insufficient to definitely differentiate between the two 11*S* diastereomers.

The explanation for the greater stability of the pair of 11*S* diastereomers lies not in differences in distortion energies but in intermolecular binding energies (Table 6). Through a combination of steric and electrostatic interactions, where steric binding energies predominate, the 11*S* diastereomers are favored equally over 11*R* diastereomers. According to these calculations the 3′ orientation is then marginally favored over the 5′ orientation on the basis of the relative distortion energies required for bonding in tomaymycin and DNA.

E. Determination of Orientation of Tomaymycin in the Minor Groove by ^{1}H-NMR Studies

1. *Tomaymycin-d(ATGCAT)$_2$ Duplex Adduct*

The NOESY spectra (Fig. 9A,B) of tomaymycin-d(ATGCAT)$_2$ showed clear evidence of cross-peaks between two pairs of drug-adenine H2 protons. These interproton through-space connectivities are between a tomaymycin H9 proton and an adenine H2 proton (Fig. 9A) and the tomaymycin ethylidene methyl and another adenine H2 proton (Fig. 9B). If the molecu-

Table 4 Energies (kcal mol^{-1}) for the Interactions between Species A-D in Figure 2

Species[a]	Configuration at C-11	Direction in minor groove[b]	Total tomaymycin-DNA binding	Helix distortion	Drug distortion	Net binding energy
A	S	3′	−34.3	12.5	0.2	−21.6
B	S	5′	−37.4	8.0	0.5	−28.9
C	R	3′	−31.5	15.8	8.5	−7.2
D	R	5′	−36.6	6.7	5.9	−24.0

[a]Refers to Figure 2.
[b]Orientation of the aromatic ring of tomaymycin relative to the covalently modified guanine.

Table 5 Energies (kcal mol^{-1}) of Covalent Complexes between Tomaymycin and d(CICGAAT·GCICTTA)[a]

			Intermolecular			Distortion[c]			
Confign at C-11	Direction[b]	Total	Steric	Elstat	Total	Drug	DNA	Total	Net binding[d]
S	3′	−443.5	−25.7	−13.4	−39.1	0.1	17.8	17.9	−21.2
S	5′	−440.6	−29.2	−9.9	−39.1	1.0	19.7	20.7	−18.3
R	3′	−420.9	−15.0	−8.4	−23.4	2.5	22.2	19.7	−3.7
R	5′	−432.3	−24.6	−7.0	−31.6	2.3	19.3	21.6	−10.0

[a]These energies are valid only for comparison within the table. They should not be compared with other drugs or other DNA segments.
[b]Direction refers to orientation in the minor groove. 3′ and 5′ are oriented with the aromatic ring to 3′ or 5′ sides, respectively.
[c]Drug distortion and helix distortion energies equal their values in the complex less their values when minimized alone (18.8 and -441.1 kcal mol^{-1}, respectively).
[d]Net binding energy equals the total intermolecular binding minus the total distortion energy.

Table 6 Energies (kcal mol^{-1}) for the Interactions of Tomaymycin with Individual Residues of d(CICGAAT·GCICTTA)[a]

Species		Residue					
Confign. at C-11	Direction	^{4}G	^{5}A	^{11}C	^{11,12}P	^{12}I^{12,13}P	^{13}S[b]
S	3′	−3.6	−4.7		−4.0		
S	5′	−3.1				−4.5	−4.7
R	3′	−3.1		−3.0			
R	5′	−3.9				−3.4	

[a]Only energies > −3.0 kcal mol^{-1} are listed.
[b]S stands for sugar.

lar modeling predictions of species S3′ and R5′ are correct, then these NOEs should represent interproton distances of less than 4.5 Å. The stereodiagrams of species A and D are shown in Figure 10. In Figure 10 (top), there is a close through-space interaction between the tomaymycin H9 and the ^{11}A H2 proton, which has a computed distance of 2.87 Å. This is in accord with the NOE shown in Figure 10. Likewise, in Figure 10 (bottom), there is a close through-space interaction between the tomaymycin ethylidene methyl and the ^{11}A H2 proton, which has a computed average interproton distance of 3.63 Å. This is in accord with the NOE found in Figure 9B. Significantly, the stronger NOE associated with the tomaymycin H9 compared to the ethylidene methyl is predicted by the shorter interproton distance found in Figure 10 (top) compared to that of Figure 10 (bottom). Thus the two best energy-minimized structures (A and D of Fig. 2) not only have the required opposite stereochemistries at C-11 and opposite orientations in the minor groove but also are predictive of the two interproton through-space connectivities identified in the NOESY spectra. Unfortunately, examination of the other stereodiagrams of the energy-minimized pairs of various diastereomeric tomaymycin-d(ATGCAT)$_2$ adducts have revealed short interproton distances between ^{11}A H2 protons and the same tomaymycin protons. Therefore, while the structures S5′ and R5′ in Figure 2 are predicted by molecular mechanics calculation to be the most likely forms of the tomaymycin-d(ATGCAT)$_2$, the NOEs demonstrated in this study suggests that two of the four possible sets (S3′/R3′ and S5′/R5′) can be eliminated, since these diastereomeric pairs have the same orientation in the minor groove. Clearly, this is at odds with the more recent molecular modeling studies. However, the difference in energy between the S3′ and

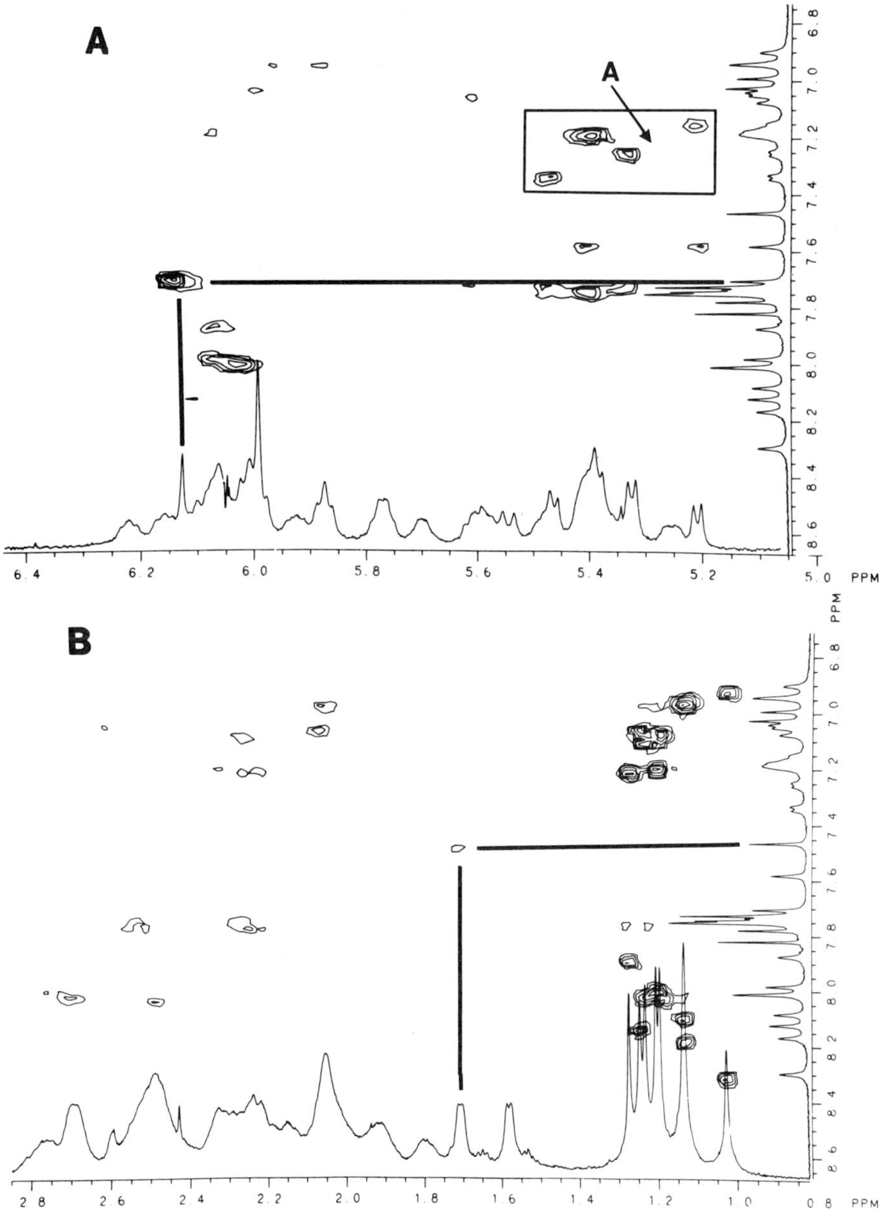

Figure 9 Variable-temperature analysis of the spectral region from 15.5 to 6.5 ppm of tomaymycin-d(ATGCAT)$_2$ duplex.

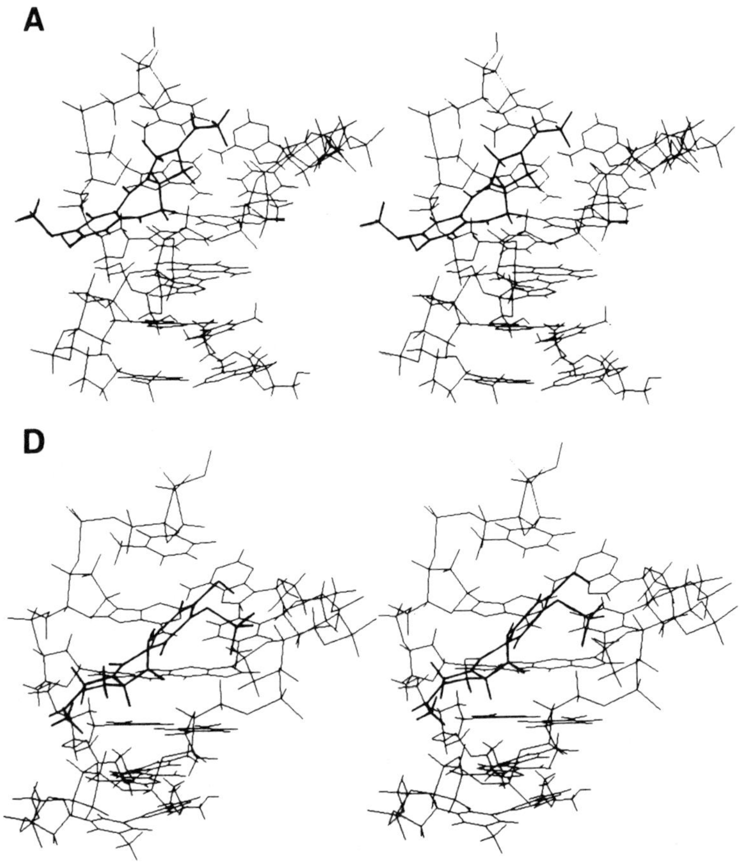

Figure 10 Stereodiagrams of the structures A and D from Figure 3.

S5′ is relatively small, and it is possible that a misassignment of an adenine H2 proton could also lead to this disagreement. At present we are unable to definitively distinguish between these possibilities. In conclusion, fluorescence, ^{1}H-NMR, and molecular modeling are in agreement that one of the species is the S3′; however, the second species may be either the R5′ or S5′.

2. *Tomaymycin-d(CICGAATTCICG)$_2$ Duplex Adduct*

In principle, the tomaymycin molecule may be oriented with the aromatic ring either to the 3′ side or the 5′ side of covalently modified guanine in the 12-mer adduct. To assign the orientation of tomaymycin in the minor groove of DNA, specific intermolecular drug-nucleotide NOEs need to be identified between unambiguously assigned tomaymycin and deoxynucleotide protons. Molecular modeling studies show that the H9 and ethylidene methyl protons of tomaymycin are in proximity to the floor of the minor groove of DNA (4). Adenine and inosine H2s that are located in the minor groove are potential "reporter atoms" to detect tomaymycin protons via NOE cross-peaks. In the expanded NOESY spectrum (Fig. 11), cross-peaks are evident between the tomaymycin H9 and both ^{5}A and ^{6}A H2 protons (box A). In addition, a NOE cross-peak is found between the tomaymycin ethylidine methyl and a ^{10}I H2 proton (10). Confirmation of these NOEs was obtained by one-dimensional NOE difference experiments, and the critical assignments of adenine, and inosine H2 protons were made by one-dimensional NOE difference experiments from imino protons (unpublished experiments).

V. EFFECT OF TOMAYMYCIN BONDING ON BACKBONE AND SUGAR GEOMETRY ON d(CICGAATTCICG)$_2$

A. Two-Dimensional NOESY Experiments

Parts A and B of Figure 11 show expanded contour plots of the phase-sensitive NOESY (250 ms) for the base of H1′ protons of d(CICGAATT-CICG)$_2$ in both the 12-mer duplex and bis(tomaymycin)-12-mer adduct. The one-dimensional spectra are drawn above the plots to illustrate the ample resolution between the individual nucleotide resonances, which simplified the task of resonance assignments. As observed in these figures, purine H8 and pyrimidine H6 protons exhibit NOEs to their own H1′ protons and to the H1′s of their 5′ neighbors, characteristic of a right-handed duplex (34–36). With these connectivities established, assignments of the majority of the oligodeoxynucleotide sugar protons (except H5′s) can likewise be made by using NOE cross-peaks between the base and other sugar protons (Table 2). These assignments are further confirmed with the complementary scalar coupling observed in the two-dimensional COSY spectrum.

The assignments of sugar H2″ and H2′ protons in the 12-mer and bis(tomaymycin)-12-mer adduct are made from their NOESY cross-peaks to inter- and intranucleotide base protons (Fig. 12). In addition, qualitative analysis of the intensity of NOESY cross-peaks from the base protons (H6, H8) to H2″ and H2′ protons at lower mixing times (50 and 100 msec) is a recognized means to evaluate the purine O_1'-C_1'-N_9-C_4 and pyrimidine O_1'-C_1'-N_1-C_2 glycosidic dihedral angle ranges syn, anti, and high anti (37–39). From the data in Tables 6 and 7 we tentatively conclude that a signifi-

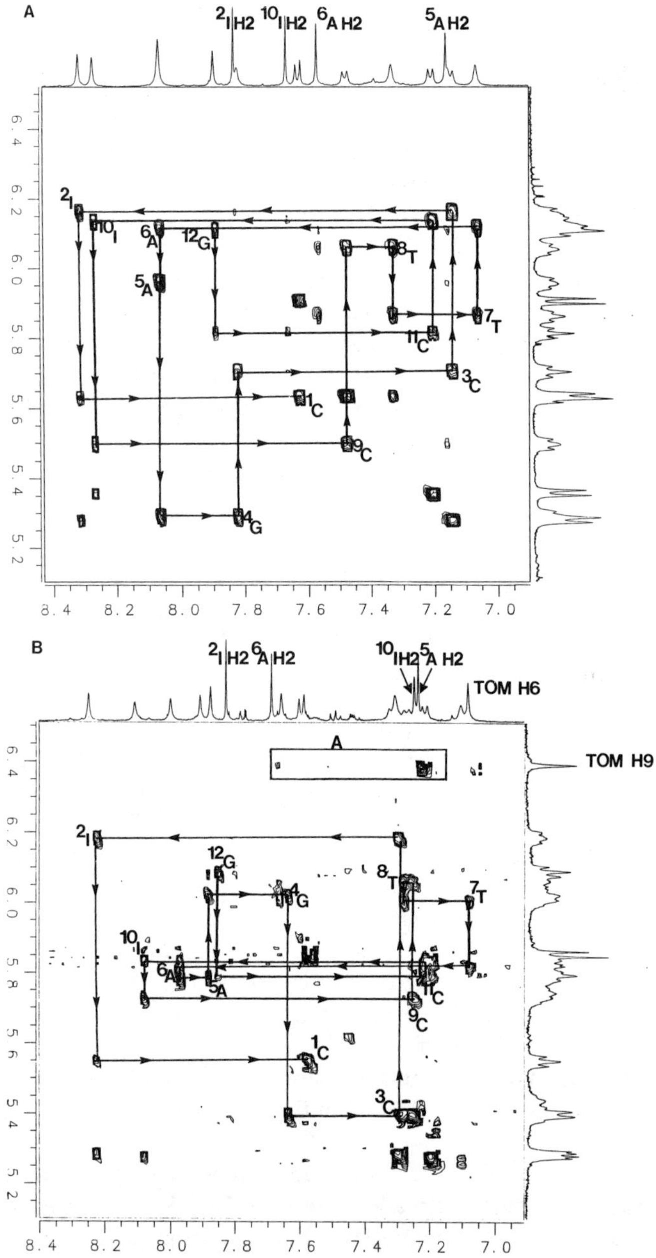

Figure 11 Phase-sensitive NOESY (250-ms mixing time) expanded contour plots of the PuH8 and the PyH6 to H1′ protons for the (A) duplex and (B) duplex adduct. Intranucleotide cross-peaks are indicated and the sequential 3′ to 5′ walks between the base and H1′ protons are traced. Box A in panel B shows the tomaymycin H9 cross-peaks to ^{5}A and ^{6}A H2s.

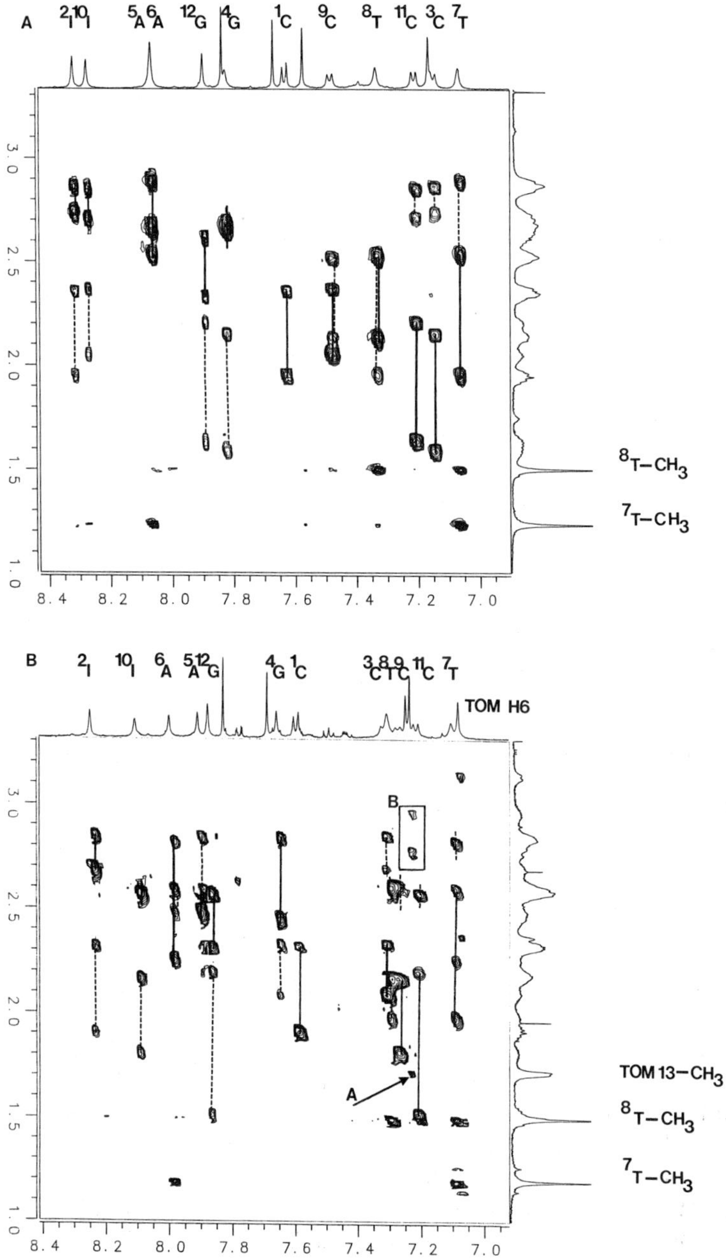

Figure 12 Phase-sensitive NOESY (250-msec mixing time) expanded contour plots of the PuH8 and PyH6 to H2′ and H2″ protons for the (A) duplex and (B) duplex adduct. Intranucleotide connectivities are indicated with solid lines. Internucleotide connectivities are indicated with broken lines. Arrow A and box B in panel B denote the cross peak(s) between ^{10}I H2 and the tomaymycin ethylidene methyl (A) and H1a and H1b (B), respectively.

Table 7 Structural Information from the Relative Intensities[a] of COSY[b] and NOESY[c] Cross-Peaks for d(CIGCAATTCICG)$_2$

Base	NOESY, NOE from H8/H6 to H2″	NOESY, NOE from H8/H6 to H2′	Inference, glycosidic angle	COSY, J coupling: H1′ with H2″	COSY, J coupling: H1′ with H2′	COSY, J coupling: H3′ with H2″	COSY, J coupling: H3′ with H2′	H3′ with H4′	Inference, sugar pucker
^{1}C	W	S	Anti	S	S	W	S	S	O-1′-*endo*
^{2}I	S	S	High anti	O (W)	M	A	O (M)	M	C-1′-*exo*
^{3}C	W	S	Anti	S	W	W	O (M)	O (S)	C-3′-*endo*
^{4}G	O	O		S	S	A	W	W	C-3′-*exo*
^{5}A	O	O		M	M	A	O (M)	S	O-1′-*endo*
^{6}A	O	O		O (W)	O (M)	A	M	O (S)	O-1′-*endo*
^{7}T	S	S	High anti	M	W	A	M	S	C-3′-*endo*
^{8}T	S	S	High anti	M	W	A	M	S	O-1′-*endo*
^{9}C	M	S	Anti	S	M	A	M	S	O-1′-*endo*
^{10}I	S	S	High anti	O (M)	O (M)	A	O (M)	O (S)	O-1′-*endo*
^{11}C	W	S	Anti	S	W	A	O (M)	S	C-3′-*endo*
^{12}G	S	M	High anti	O (M)	S	S	W	S	O-1′-*endo*

[a]A = absent; W = weak; M = moderate; S = strong; O = overlapping. Partially resolved cross-peak intensities are placed in parentheses.
[b]Magnitude COSY and two-dimensional phase-sensitive double quantum filtered COSY cross-peaks are examined.
[c]Data from phase-sensitive two-dimensional NOESY experiments with mixing times of 50 and 100 msec.

cant number of glycosidic dihedral angles in the unmodified duplex are in the high-anti domain (^{2}I, ^{7}T, ^{8}T, ^{10}I, ^{12}G), which suggests a DNA structure deviant from the classical B-form, although these results should be considered tentative since they are based upon qualitative trends in NOESY and COSY data sets. Two of the high-anti glycosidic angles are associated with substitution of deoxyinosine for deoxyguanosine into the dodecamer (^{2}I, ^{10}I). The high-anti glycosidic dihedral angle for nucleotide ^{12}G may be due to fraying at the oligomer termini, because this linkage is not as restricted as those more internal. In the bis(tomaymycin)-d(CICGAATTCICG)$_2$ adduct, only nucleotide ^{8}T continues to exhibit a glycosidic dihedral angle in the high-anti form, with the remainder of the nucleotides having glycosidic dihedral angles in the anti range, which is consistent with a B-form DNA.

The nonscalar cross-peaks between base H8/H6 and sugar H2″ and

H2′ protons have also been used to distinguish between A- and B-type helices (40). In the case of A-type DNA, the C-3′-*endo* sugar geometry predominates, and internucleotide cross-peaks between H8/H6 protons to the 5′-neighboring H2′ protons are stronger than those to the corresponding intranucleotide H2′ protons. The converse is true for B-type DNA, where C-2′-*endo* sugar geometry predominates and the intranucleotide distances are shorter. The intensities of these cross-peaks at lower mixing times (50–100 msec) were used to *qualitatively confirm* the sugar geometry assignments in the two-dimensional COSY (see below).

B. Two-Dimensional COSY Experiments

Since the vicinal coupling constants of DNA sugar ring protons vary as a function of pseudorotation angle or sugar pucker, it is possible to assign sugar geometries to each nucleotide on the basis of observed or absent scalar cross-peaks between the sugar protons listed in Table 2(b). As an example, one strong indicator of difference in sugar geometry is the comparison of cross-peak intensity between protons H1′ and H2″ and H1′ and H2′, as shown for the bis(tomaymycin)-d(CICGAATTCICG)$_2$ adduct (Fig. 13). In the case of C-2′-*endo* geometry for classical B-type DNA helices, strong scalar cross-peaks between protons of H1′ and H2″ and H1′ and H2′ are expected. Conversely for C-3′-*endo* geometries, or A-type DNA helices, the J couplings of H1′ to H2′ are expected to be weak or absent. With this concept, the results show that, for the most nucleotides in the 12-mer and 12-mer adduct, the sugar geometries are mostly around or near C-2′-*endo* or near O-1′-*endo* (Tables 7 and 8). In the duplex ^{3}C, ^{7}T, and ^{11}C are exceptions, where the average C-3′-*endo* sugar geometry of A-form helices predominates. Upon adduct formation of this duplex with tomaymycin, only at ^{11}C is the C-3′-*endo* geometry still present. For nucleotides ^{3}C and ^{7}T, there is a shift toward more B-like structure in the adduct.

C. Two-Dimensional Proton-Phosphorus J-Correlation Experiments

The majority of the phosphorus resonances in d(CICGAATTCICG)$_2$ only undergo small changes in chemical shift upon covalent modification with tomaymycin. However, one phosphorus resonance is significantly shifted downfield from the others (compare parts A and B of Fig. 14). This result is reminiscent of previous work in our laboratory with covalent bonding of anthramycin to d(ATGCAT)$_2$ (7). The observed downfield shift of the phosphorus resonance suggests a structural perturbation of the phosphate backbone as a result of covalent bonding. The contour plot of the phase-

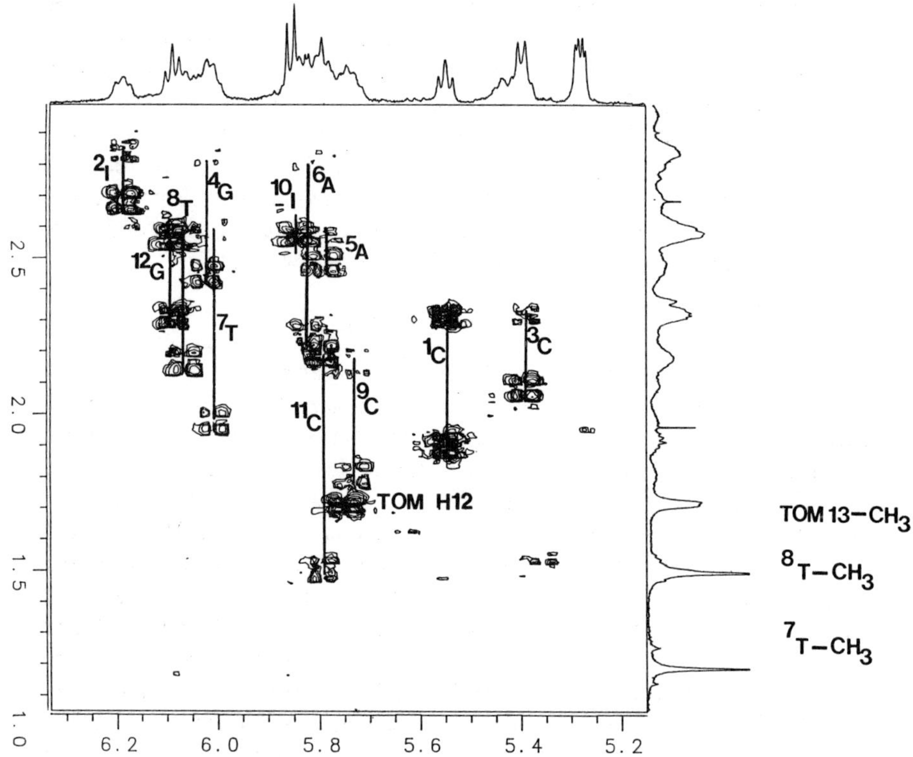

Figure 13 Phase-sensitive double quantum filtered COSY expanded contour plot of the H1′ to H2″, H2′ cross peaks for the bis(tomaymycin)–12-mer adduct. Solid lines connect the downfield H2″s to intranucleotide H2′s.

sensitive two-dimensional heteronuclear phosphorus-detected ^{31}P-^{1}H *J*-correlation experiment is shown in Figure 15. Cross-peaks connect the downfield-shifted phosphorus resonance to the ^{3}C H3′ and ^{4}G H4′, indicating that the perturbation of the phosphate backbone is localized to the 5′ side of the drug covalent bonding site.

D. Molecular Modeling

The energy-minimized structure of the bis(tomaymycin)-d(CICGAATTCICG)$_2$ adduct having the experimentally determined 11*S* stereochemistry and 3′ orientation in the minor groove is shown in Figure 16. An overlay of the d(CICGAAT·ATTCICG) duplex and the 11*S*, 3′-oriented tomaymycin 7-mer adduct is shown in Figure 17. Both figures predict minimal distortion

Table 8 Structural Information from the Relative Intensities[a] of COSY[b] and NOESY[c] Cross-Peaks for the Bis-(tomaymycin)-d(CICGAATTCICG)$_2$ Adduct

	NOESY, NOE from H8/H6 to			COSY, J coupling					
				H1′ with		H3′ with			
Base	H2″	H2′	Inference, glycosidic angle	H2″	H2′	H2″	H2′	H3′ with H4′	Inference sugar pucker
^{1}C	A	S	anti	S	S	A	S	S	O-1′-*endo*
^{2}I	W	S	anti	M	S	A	W	W	C-1′-*exo*–C-2′-*endo*
^{3}C	W	S	anti	M	S	M	W	O (S)	C-1′-*exo*
^{4}G	W	S	anti	W	M	A	A	A	C-1′*exo*–C-2′-*endo*
^{5}A	(W)	O (S)	anti	W	M	A	W	O (W)	C-1′-*exo*–O-1′-*endo*
^{6}A	W	S	anti	W	M	A	A	O (M)	C-1′-*exo*–O-1′-*endo*
^{7}T	W	S	anti	M	M	A	W	O (W)	C-1′-*exo*–O-1′-*endo*
^{8}T	S	S	high anti	O (M)	M	A	W	O (W)	C-1′-*exo*–O-1′-*endo*
^{9}C	(W)	M	anti	M	A	W	O (M)	O (W)	O-1′-*endo*
^{10}I	O	O		O	O	A		O	
^{11}C	W	S	anti	O (S)	M	A	O	O (S)	C-3′-*endo*
^{12}G	(S)	M	anti	O (S)	S	S	S	S	O-1′-*endo*

[a-c]For footnotes, see Table 7.

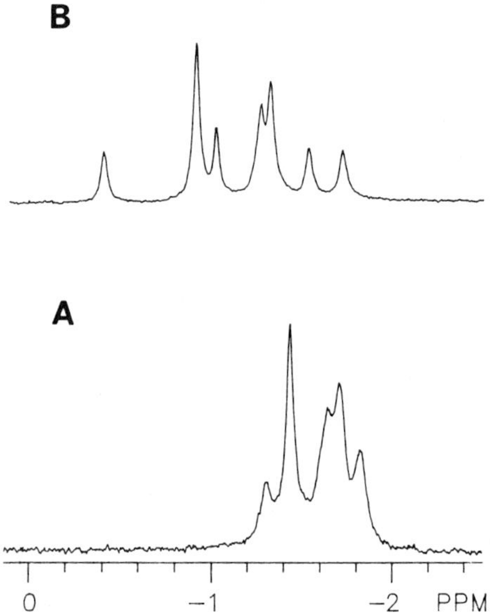

Figure 14 Proton-decoupled 202.44-MHz phosphorus NMR spectra of (A) the d(CICGAATTCICG)$_2$ duplex and (B) the tomaymycin–d(CICGAATTCICG)$_2$ adduct in buffered D_2O at 24°C.

of the duplex helix, which is in agreement with the analysis of the two-dimensional COSY and NOESY data (Tables 7 and 8).

A comparison was made of the pseudorotation angles of purine and pyrimidine nucleotides from the x-ray and ^{1}H-NMR data of d(CGCGAATTCGCG)$_2$ (24,41) and from our proton NMR data of d(CICGAATTCICG)$_2$ and its bis(tomaymycin) adduct (Fig. 18). A comparison of the x-ray-determined structure of the Dickerson dodecamer with the solution structure of the inosine-substituted oligomer show a pseudorotation angle shift from classical B-type toward O-1′-*endo*. This is not without precedent (35) and may also be influenced by the substitution of I for G with the accompanying loss of one hydrogen bond for each substitution. Upon formation of the bis(tomaymycin)-12-mer adduct, the population of sugar geometries for the inosine-disubstituted 12-mer shifts back toward C-2′-*endo*. Furthermore, from this analysis, it appears that tomaymycin stabilizes a B-type helical structure after formation of the covalent adduct. Molecular modeling results show less change in sugar geometries of the duplex upon adduct formation, but there is generally agreement between the experimental and theoretical studies.

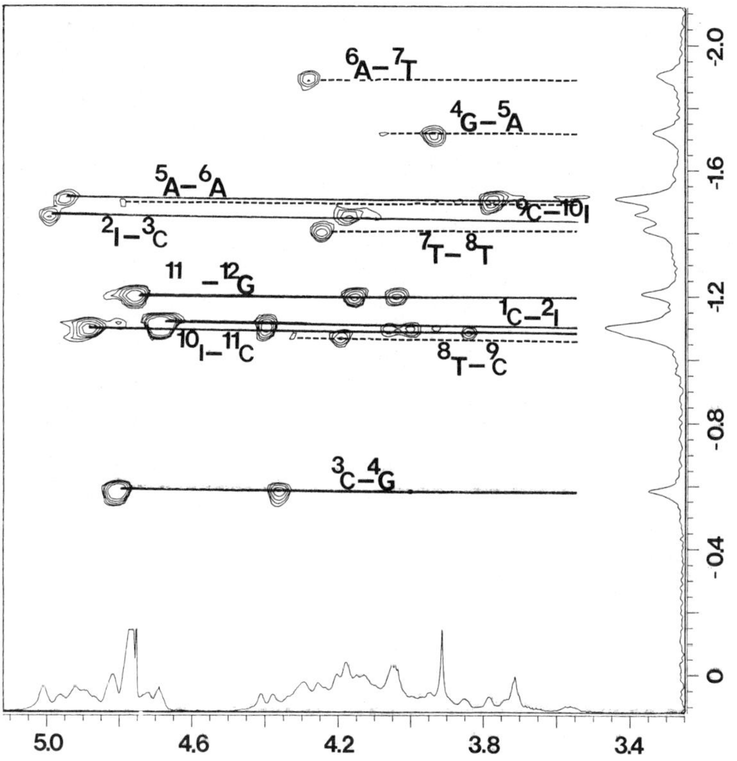

Figure 15 Expanded contour plot of the heteronuclear phosphorus-detected phase-sensitive two dimensional ^{31}P–^{1}H J correlation of the bis(tomaymycin)-12-mer adduct. Solid lines denote definitive assignments. Broken lines denote tentative assignments.

VI. INTERMOLECULAR INTERACTIONS DETERMINING THE ORIENTATION AND STEREOCHEMISTRY OF TOMAYMYCIN ON d(ATGCAT)$_2$ AND d(CICGAATTCICG)$_2$

The phosphorus and proton-NMR results show predominantly (i.e., > 90%) one species (11*S*,3′) of tomaymycin bound to the 12-mer. This is in contrast to the d(ATGCAT)$_2$-duplex where two species (11*S*, 3′ and 11*R*, 5′ or 11*R* 3′) of tomaymycin were bound in approximately equal amounts (9). Thus both the orientation in the minor groove and stereochemistry of tomaymycin bound to DNA is a function of the immediate flanking sequence. For the more preferred bonding sequence (5′-PyGPu) only one species is evident, while the less preferred sequence (5′-PyGPy) shows two

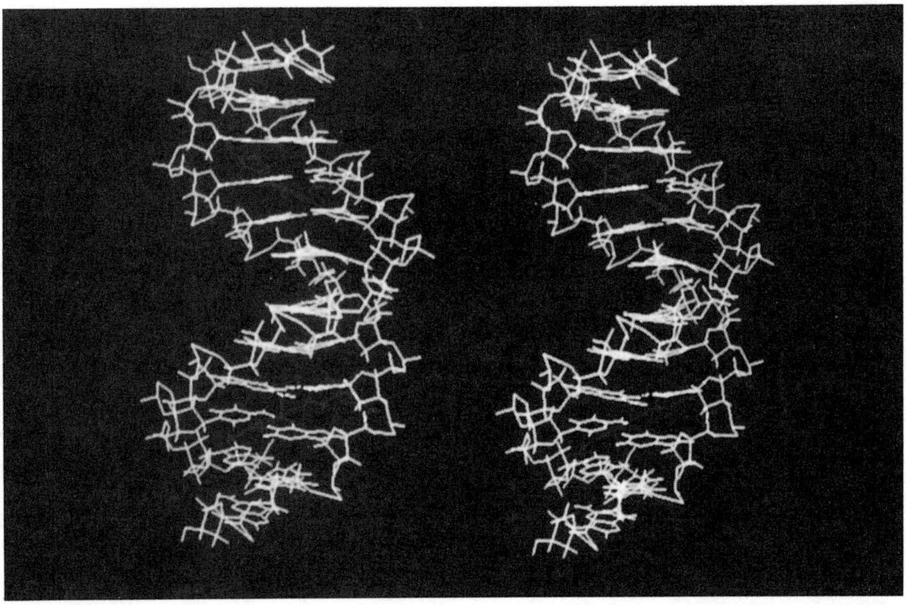

Figure 16 (see also Color Plate following Chapter 1) Stereo drawing of ^{4}G attached to tomaymycin showing the conformation obtained through minimization with AMBER (see Molecular Modeling).

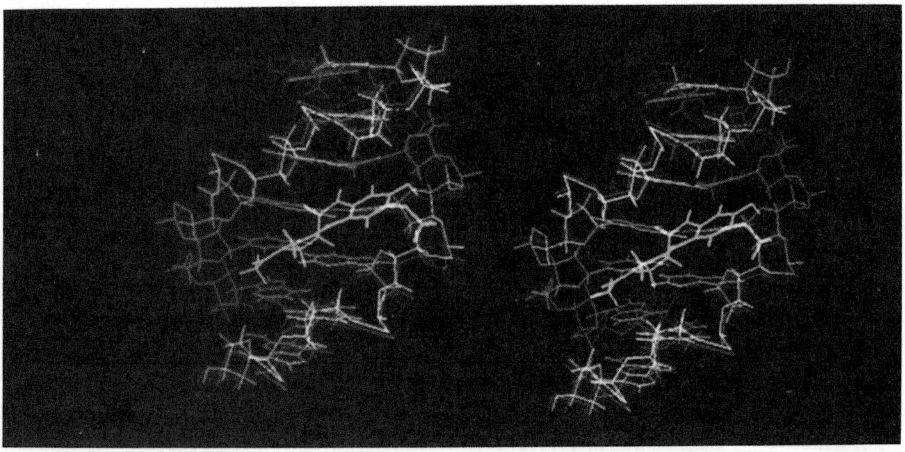

Figure 17 (see also Color Plate following Chapter 1) Stereo drawing of the superimposed d(CICGAATT·ATTCICG) in orange and tomaymycin–d(CICGAAT·ATTCICG) in magenta and yellow.

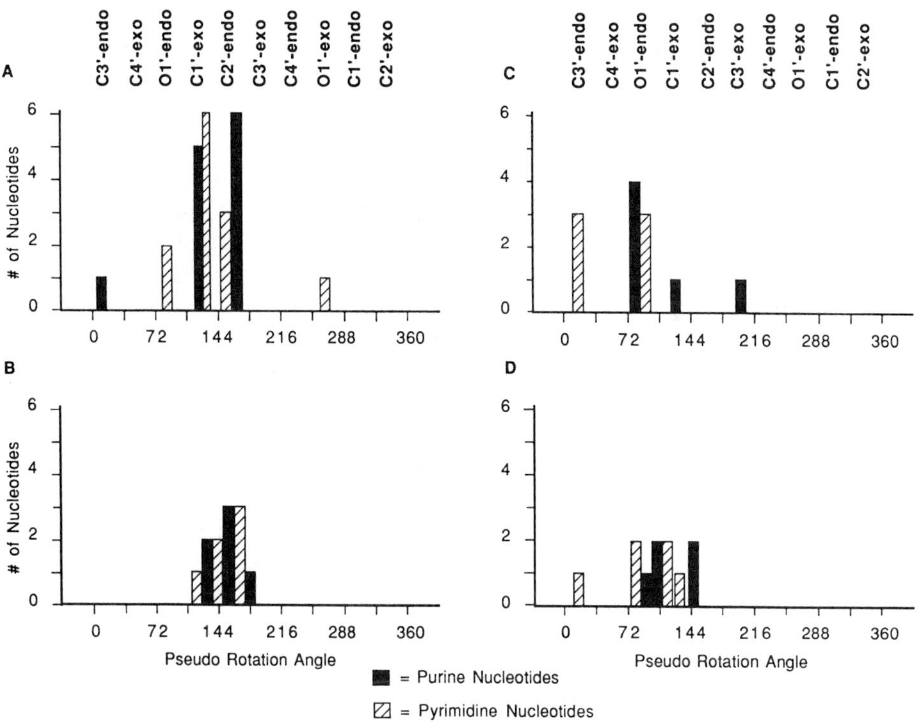

Figure 18 Comparison of the experimentally derived sugar pseudorotation angles for the purine and pyrimidine deoxynucleotides in $d(CGCGAATTCGCG)_2$ from (A) X-ray data (24), (B) ^1H NMR data (41) with the sugar pseudorotation angles from ^1H NMR data for (C) $d(CICGAATTCICG)_2$ (Table 7), and (D) *bis*-tomaymycin-$d(CICGAATTCICG)_2$ adduct (Table 8).

species. The species having an 11*S* configuration and 3′ orientation is found on both sequences examined, but the second species having the opposite stereochemistry at C-11 could not be detected on the more preferred bonding sequence. Molecular modeling using AMBER is not in complete accord with the experimentally determined species for both oligomers. The preferred species for $d(ATGCAT)_2$ binding is 11*S*, 5′, whereas the experientally found preference is 11*S*, 3′. There is agreement between the calculated and experimental evidence for the 11*R*, 5′ species as the second best one. For binding with the 12-mer, there is agreement that the 11*S*, 3′ species is favored; however, the small calculated difference between it and the 11*S*, 5′ species is probably within the error of the method. The most significant difference between the binding of tomaymycin to $d(ATGCAT)_2$ and the 12-mer is that the *R*-configuration is important only in the former. According to molecular modeling, this difference results from a combination of poorer intermolecular binding energies and increased distortion energies for models containing 12-mer (Tables 4 and 5).

The reason for this apparent discrepancy was addressed by examining the calculated interaction energies for tomaymycin and individual polynucleotide residues in each of the species listed in Table 9. Such analyses are complicated because the total binding energies reflect complex binding patterns and wide distribution of differences in energies between individual residues. However, insight into the effects of binding patterns can be obtained by focusing on the most important interactions, i.e., those of -2.0 kcal mol^{-1} or greater energy. Table 9 lists the binding patterns for the four species. The construction of this table reflects the increase of one in residue numbers in going from ATGCAT to CICGAAT, because of the location of the guanine residues (3G or 4C, respectively) to which the covalent bond is formed. It is evident from this table that the two patterns for the 11*S*, 3′ binding modes are identical, except for the pair 3G-4G, where the interaction is less than -2.0 kcal mol^{-1} with d(ATGCAT)$_2$. Even in this case, its value of -1.8 kcal mol^{-1} is, however, not much less than that of the corresponding interaction for d(CICGAAT·ATTCICG) (-3.6 kcal mol^{-1}). A greater contrast occurs in the binding patterns of the two 11*R*, 5′ modes. There are two more interactions between tomaymycin and the polynucleotide residues with the d(ATGCAT)$_2$ than with d(CICGAAT·ATTCICG). One of them involves a hydrogen bond between HN-10 of tomaymycin and N-3 of 11A of 1.09 Å length and -5.4 kcal mol^{-1} energy. The other is less significant (-2.1 kcal mol^{-1}). There is also an unfavorable interaction of $+1.7$ kcal mol^{-1} between O-4 of tomaymycin and O-3′ of

Table 9 Patterns for Binding of Tomaymycin with Individual Residues on Polyneucleotides[a]

Oligonucleotide	Confign. at C-11	Direction	Residues[b]											
d(ATGCAT)$_2$	*S*	3′	^{5}S		^{4}S	^{4}S	^{4}C		^{10}C	^{10,11}P	^{11}S	^{11}A	^{11,12}p	^{12}S
d(CICGAAT·ATTCICG)	*S*	3′	^{6}S		^{5}S	^{5}S	^{5}A	^{4}G	^{11}C	^{11,12}P	^{12}S	^{12}I	^{12,13}P	^{13}S
d(ATGCAT)$_2$	*R*	5′	^{5}S	^{4,5}P			^{4}C	^{3}G	^{10}C		^{11}S	^{11}A		^{12}S
d(CICGAAT·ATTCICG)	*R*	5′		^{5,6}p			^{5}A	^{4}G	^{11}C		^{12}S			^{13}S

[a]Only residues that bind to tomaymycin with an energy > -2.0 kcal mol^{-1} are included. Residues in CICGAAT·ATTCICG are shifted to one base higher than the corresponding ones in (ATGCAT)$_2$ because of the covalent bonding into ^{4}G in the former and ^{3}G in the latter.
[b]S = sugar.

12,13P in the d(CICGAAT·ATTCICG) sequence. (The O-3′ oxygen atom of a sugar is usually considered part of the phosphate to which it also is bonded in AMBER analysis.) This interaction does not occur in the adduct with $d(ATGCAT)_2$. These factors combine to disfavor the 11*R*, 5′ mode for the d(CICGAAT·ATTCICG) sequence as is evident in its relatively poor net binding energy (-10.0 kcal mol^{-1}) compared with that of the 11*S*, 3′ mode (-21.2 kcal mol^{-1}, Table 3). The total distortion energies of these two modes differ by only 3.7 kcal mol^{-1}, which means that binding energies are more important than distortion energies in favoring the 11*S*, 3′ mode on d(CICGAAT·ATTCICG). Comparable net binding energies for the tomaymycin-$d(ATGCAT)_2$ adducts are -17.9 kcal mol^{-1} for the 11*S*, 3′ mode and -21.9 kcal mol^{-1} for the 11*R*, 5′ mode. The resulting relatively small difference is consistent with the existence of both binding modes (8).

VII. BIOCHEMICAL AND BIOLOGICAL IMPLICATIONS

The relatively nondistortive nature of the P(1,4)B-DNA adducts shown in these studies, coupled with the absence of apparent significant changes in electronic or tautomeric characteristics of the covalently modified guanine, raises important questions on how the P(1,4)B-DNA adducts are recognized by repair enzymes. The UVRABC nuclease incision repair complex is able to recognize the anthramycin-DNA adduct and apparently incise both 3′ and 5′ sides of DNA flanking lesions (16). We know from the results reviewed here that the stereochemistry at the covalent linkage site and orientation in the minor groove is dependent upon the immediate bonding sequence. Such observations complicate the overall picture for repair recognition of P(1,4)B-DNA adducts, but these factors must be taken into account if we are to fully understand DNA repair recognition processes.

In both yeast and human cells anthramycin produces double-strand breaks (15,17). In human cells both the single- and double-strand breaks are persistent and are repair dependent, since in XP cells neither single- nor double-strand breaks are detected. This observation, coupled with the more recent data on the nondistortive nature of the P(1,4)B adducts, raises important questions about whether or not repair incision always takes place on the covalently modified strand.

Last but most importantly, from a drug design perspective the results summarized here provide an excellent basis for the design of new synthetic analogs of the P(1,4)Bs with improved therapeutic potential. Several possibilities can be outlined.

1. Replacement of the 9-Phenolic Group that Is Associated with Cardiotoxicity with Another Sterically Small Group

Anthramycin and sibiromycin are members of the P[1,4]Bs that both exhibit cardiotoxicity in animal and human studies (3). This dose-limiting toxicity has not been demonstrated with other members of the P(1,4)Bs such as tomaymycin and neothramycin. This has led us to postulate that a common structural feature (i.e., the 9-phenolic group) of anthramycin and sibiromycin absent in other noncardiotoxic members of the P[1,4]Bs is responsible for this effect. A chemical rationale involving the interconversion to an orthoquinone imine has been suggested to account for this selective toxicity (42). Although tomaymycin and neothramycin possess a phenolic substituent on the aromatic ring, it is meta to the imine and therefore not subject to the same chemistry. Synthetic analogs lacking the 9-phenolic group would appear to be desirable to avoid this anticipated problem.

2. Design of DNA-DNA Interstrand Cross-Linkers Based on the Structure of the Tomaymycin 12-mer Duplex

The bis-tomaymycin-12-mer duplex adduct described in this chapter has two tomaymycin molecules bound on opposite strands eight base pairs apart (10). In subsequent studies we have shown that the central two base pairs can be eliminated from the 12-mer while still maintaining the symmetry of a bis-tomaymycin-10-mer duplex adduct in which the two tomaymycin molecules are now just six base pairs apart in the same relative position as in the 12-mer duplex adduct (Wang and Hurley, unpublished results). The 10-mer duplex adduct serves as an ideal basis for drug design because molecular modeling permits us to construct linker units that will attach the two aromatic rings. In the 10- and 12-mer duplex molecules described above the aromatic rings of the drugs point toward each other. In principle, it is also possible to design DNA-DNA interstrand cross-linkers in which the side chains projecting from the pyrrolo rings could serve as linker attachment sites. The uncertainty remains whether such tail to tail projection still allows covalent attachment on guanine moieties six or even eight base pairs apart to the 3′ sides of a central point within the cross-linked site.

3. Attachment of Appendages at C-8 or on the Side Chain Will Point Along the Minor Groove of DNA

Molecular modeling studies show that substituents at C-8 on the side chain will project along the minor groove of DNA. Covalent attachment of the P(1,4)B at N-2 of guanine will fix the drug within the minor groove allowing for subsequent interactions to occur between appendages at C-8 or on the side chain and DNA at sites to the 3′ side (C-8 appendage) or 5′ side

(side chain appendage). In the simplest form, EDTA-Fe^{+4} substituents will produce strand breakage under appropriate conditions, while other more elaborate appendages such as intercalating moieties or even major groove recognition motives such as triple strands might be evaluated for binding and biological activity.

4. *Design of Intrastrand Head-to-Tail Cross-linkers Is also an Attractive Possibility*

Our NMR studies so far have concentrated on the design of bis-P[1,4]B-DNA-DNA interstrand cross-linkers. In principle, there is no obvious reason why intrastrand cross-linkers could not be designed based upon NMR characteristics of P[1,4]Bs bound to the same strand. In this case the chemistry would involve linking the 8-hydroxyl group to a side chain substituent on the second P[1,4]B molecule. Experiments to test such ideas are in progress.

VIII. CONCLUSIONS

High-field NMR, fluorescence, and molecular modeling have proved to be a powerful combination of techniques to study the structures of the tomaymycin-DNA adducts in various sequences. ^{1}H NMR provided the first unequivocal evidence for the covalent adduct between C-11 of tomaymycin and the exocyclic group of guanine. The species of tomaymycin bound to 5′-CGA has an 11*S* stereochemistry, with orientation of the aromatic ring to the 3′ side of guanine, while on 5′-TGC tomaymycin exists as a mixture of two species.

The minimal distortion produced upon covalent bonding is restricted to the deoxyribose at the covalently modified guanine and the immediate 3′-phosphate and deoxyribose. Tomaymycin appears to stabilize a B-type helical structure. The orientation of a P(1,4)B on DNA and the stereochemistry at the covalent linkage site are dependent on both the immediate surrounding sequence and the drug structure. Molecular modeling using AMBER is predictive of the orientation and stereochemistry of the P(1,4)Bs on oligodeoxynucleotides and is capable of defining the molecular interactions that favor the different species on various oligomers.

REFERENCES

1. Remers, W.A., In *The Chemistry of Antitumor Antibiotics*, Vol. 2:28–92, Wiley, New York (1988).
2. Hurley, L.H., Elucidation and formulation of novel biosynthetic pathways leading to the pyrrolo(1,4)benzodiazepine antibiotics anthramycin, tomaymycin, and sibiromycin. *Acc. Chem. Res. 13*:263 (1980).

3. Hurley, L.H., Pyrrolo(1,4)benzodiazepine antitumor antibiotics. Comparative aspects of anthramycin, tomaymycin, and sibiromycin. *J. Antibiotics 30*:349 (1977).
4. Petrusek, R.L., G.L. Anderson, T.F. Garner, Q.L. Fannin, D.J. Kaplan, S.G. Zimmer, and L.H. Hurley, Pyrrolo(1,4)benzodoazepine antibiotics. Proposed structures and characteristics of the in vitro DNA adducts of anthramycin, tomaymycin, sibiromycin, and neothramycins A and B. *Biochemistry 20*:1111 (1981).
5. Hertzberg, R., S. Hecht, V.L. Reynolds, I.J. Molineux, and L.H. Hurley, DNA sequence specificity of the pyrrolo(1,4)benzodiazepine antitumor antibiotics. MPE Fe(II) footprinting analysis of DNA binding sites for anthramycin and related drugs. *Biochemistry 25*:1249 (1986).
6. Hurley, L.H., T. Reck, D.E. Thurston, K.G. Holden, R.P. Hertzberg, J.R.E. Hoover, G. Gallagher, Jr., L.F. Faucette, S.-M. Mong, and R.K. Johnson, Pyrrolo(1,4)benzodiazepines antitumor antibiotics. Relationship of DNA binding and sequence specificity to biological activities of natural and synthetic compound. *Chem. Res. Toxicol. 1*:258 (1988).
7. Boyd, F.L., S.F. Cheatham, W. Remers, G.C. Hill, and L.H. Hurley, Characterization of structure of the anthramycin-d$(ATGCAT)_2$ adduct by NMR and molecular modeling studies. Determination of the stereochemistry at the covalent linkage site, orientation in the minor groove, and effect of drug binding on local DNA structure. *J. Amer. Chem. Soc. 112*:3279 (1990).
8. Barkley, M.D., S. Cheatham, D.E. Thurston, and L.H. Hurley, Pyrrolo(1,4)-benzodiazepine antitumor antibiotics. Evidence for two forms of tomaymycin bound to DNA. *Biochemistry 25*:3021 (1986).
9. Cheatham, S.F., A.M. Kook, L.H. Hurley, M. Barkley, and W. Remers. One- and two-dimensional ^{1}H-NMR fluorescence and molecular modeling studies on the tomaymycin-d$(ATGCAT)_2$ adduct. Evidence for two covalent adducts with opposite orientations and stereochemistry at the covalent linkage site. *J. Med. Chem. 31*:583 (1988).
10. Boyd, F.L., D. Stewart, W.A. Remers, M.D. Barkley, and L.H. Hurley, Characterization of a unique tomaymycin-d$(CICGAATTCICG)_2$ adduct containing two drug molecules per duplex by NMR, fluorescence, and molecular modeling studies. *Biochemistry 29*:2387 (1990).
11. Zakzewska, K., and B. Pullman, A theoretical investigation of the sequence specificity in the binding of the antitumor drug anthramycin to DNA. *J. Biomol. Struct. Dyn. 4*:127 (1986).
12. Remers, W.A., M. Mabilia, and A.J. Hopfinger, Conformations of complexes between pyrrolo[1,4]benzodiazepines and DNA segments. *J. Med. Chem. 29*: 2492 (1986).
13. Rao, S.N., U.C. Singh, and P.A. Kollman, Molecular mechanics simulations on covalent complexes between anthramycin and B DNA. *J. Med. Chem. 29*: 2484 (1986).
14. Nishioka, Y., T. Beppu, M. Kohsaka, and K. Arima, Mode of action of tomaymycin. *J. Antibiotics 25*:660 (1972).
15. Hannan, M.A., and L.H. Hurley, Pathways of DNA repair operating in yeast

treated with the pyrrolo(1,4)benzodiazepine antitumor antibiotics. *J. Antibiotics 31*:911 (1978).
16. Walter, R.B., J. Pierce, R. Case, and M.-S. Tang, Recognition of the DNA helix stabilizing anthramycin-N2 guanine adduct by UVRABC nuclease. *J. Mol. Biol. 203*:939 (1988).
17. Petrusek, R.L., E.L. Uhlenhopp, N. Duteau, and L.H. Hurley, Reaction of anthramycin with DNA. Biological consequences of DNA damage in normal and xeroderma pigmentosum cell lines. *J. Biol. Chem. 257*:6207 (1982).
18. Hannan, M.A., L.H. Hurley, and C. Gairola, Mutagenic and recombinogenic effects of the antitumor antibiotic anthramycin. *Cancer Research 38*:2795 (1978).
19. Horwitz, S.B., S.C. Chang, A.P. Grollman, and A.B. Borkovec, Chemosterilant action of anthramycin: A proposed mechanism. *Science 147*:159 (1971).
20. Arora, S., Structure of tomaymycin, a DNA binding antitumor antibiotic. *J. Antibiotics 34*:462 (1981).
21. Singh, V.C., P.K. Weiner, J. Caldwell, and P.A. Kollman, *AMBER (UCSF), version 3.0*, Dept. of Pharmaceutical Chemistry, University of California: San Francisco, 1986.
22. Weiner, S.J., P.A. Kollman, D.A. Case, U.C. Singh, C. Ghio, G. Alagona, S. Profeta, Jr., and P. Weiner, A new force field for molecular mechanical simulation of nucleic acids and proteins. *J. Am. Chem. Soc. 106*:765 (1984).
23. Ferrin, T.G., C.C. Huang, L.E. Jarvis, and R. Langridge, The MIDAS display system. *J. Mol. Graphics 6*:1,13 (1988).
24. Drew, H.R., R.M. Wing, T. Takano, C. Broka, S. Tanaka, K. Itakura, and R.E. Dickerson, Structure of a B-DNA dodecamer: Conformation and dynamics. *Proc. Natl. Acad. Sci. U.S.A. 78*:2179 (1981).
25. Tozuka, Z., and T. Takaya, Studies on tomaymycin. I: The structure determination of tomaymycin on the basis of NMR spectra. *J. Antibiotics 36*:142 (1983).
26. Tozuka, Z., H. Takasugi, and T. Takaya, Studies on tomaymycin. II: Total synthesis of the antitumor antibiotic E- and Z-tomaymycins. *J. Antibiotics 36*:276 (1983).
27. Leimgruber, W., V. Stefanovic, F. Schenker, A. Karr, and J. Berger, Isolation and characterization of anthramycin, a new antitumor antibiotic. *J. Am. Chem. Soc. 87*:5791 (1965).
28. Arima, K., M. Kohsaka, G. Tamura, H. Imanaka, and H. Sakai, Studies on tomaymycin, a new antibiotic. I: Isolation and properties of tomaymycin. *J. Antibiotics 25*:437 (1972).
29. Knutson, J.R., J.M. Beechem, and L. Brand, Simultaneous analysis of multiple fluorescence decay curves: A global approach. *Chem. Phys. Lett. 102*:501 (1983).
30. Kohn, K.W., D. Glaubiger, and C.L. Spears, The reaction of anthramycin with DNA II. Studies of kinetics and mechanism. *Biochim. Biophys. Acta 361*:288 (1974).
31. Hurley, L.H., and R. Petrusek, Proposed structure of the anthramycin-DNA adduct. *Nature 282*:529 (1979).

32. Barkley, M.D., T.J. Thomas, K. Maskos, and W.A. Remers, *Biochemistry 30*: 4421 (1991).
33. Patel, D.J., Peptide antibiotic-oligonucleotide interactions. Nuclear magnetic resonance investigation of complex formation between actinomycin D and d-ApTpGpCpApT in aqueous solution. *Biochemistry 13*:2396 (1974).
34. Hare, D.R., D.E. Wemmer, S.H. Chou, G. Drobay, and B.R. Reid, Assignment of the non-exchangeable proton resonances of d(C-G-C-G-A-A-T-T-C-G-C-G). Using two-dimensional nuclear magnetic resonance methods. *J. Mol. Biol. 171*:319 (1983).
35. Reid, B.R., Sequence-specific assignments and their use in NMR studies of DNA structure. *Q. Rev. Biophys. 20*:1 (1987).
36. Patel, D.J., L. Shapiro, and D. Hare, DNA and RNA: NMR studies of conformation and dynamics in solution. *Q. Rev. Biophys. 20*:35 (1987).
37. Hosur, R.V., M. Ravikumar, K.V.R. Chary, A. Sheth, G. Govil, T. Zu-Kun, and H.T. Miles, *FEBS Lett. 205*:71 (1986).
38. Kumar, R.M., R.V. Hosur, K.B. Roy, H.T. Miles, and G. Govil, Resonance assignment of the 500-MHz proton NMR spectrum of self-complementary dodecanucleotide d-GGATCCGGATCC: Altered conformations at *Bam*HI cleavage sites. *Biochemistry 24*:7703 (1985).
39. Chary, K.V.R., R.V. Hosur, G. Govil, T. Zu-Kun, and H.T. Miles, Novel solution conformation of DNA observed in d(GAATTCGAATTC) by two-dimensional NMR spectroscopy. *Biochemistry 26*:1315 (1987).
40. Haasnoot, C.A.G., H.P. Westerink, G.A. van der Marel, and J.H. van Boom, Discrimination between A-type and B-type conformations of double helical nucleic acid fragments in solution by means of two-dimensional nuclear overhauser experiments. *J. Biomol. Struct. Dyn. 2*:345 (1984).
41. Bax, A., and L. Lerner, Measurement of ^{1}H-^{1}H coupling constants in DNA fragments by 2D NMR. *J. Magn. Reson. 79*:429 (1988).
42. Thurston, D.E., and L.H. Hurley, A rational basis for development of antitumor agents in the pyrrolo(1,4)benzodiazepine group. *Drugs Future (CIPS) 8*: 957 (1984).

9

A Cooperative Quinolone-DNA Binding Model for DNA Gyrase Inhibition

Implications in Drug Design

Lester A. Mitscher
University of Kansas, Lawrence, Kansas

Linus L. Shen
Abbott Laboratories, Abbott Park, Illinois

I. INTRODUCTION

A. Role of Quinolones in Antibacterial Chemotherapy

The quinolones are an important new class of synthetic antimicrobial agents which are the subject of intense contemporary study in academic and industrial research laboratories (reviewed in Refs. 1–5). The newer members are broad spectrum in action, orally active, possess potencies equivalent to those of marketed fermentation-derived antibiotics and possess comparatively few serious side effects. More than a half dozen quinolones are commercially available in the United States, a larger number have been introduced overseas, and a great many more are competing for an eventual place in the $7 billion worldwide anti-infectives market. Their molecular mode of action is not yet completely established, but they have dramatic action against a fundamentally important and fairly recently discovered bacterial enzyme, DNA gyrase (reviewed in Refs. 6–22). This enzyme catalyzes changes in DNA topology, and its appropriate functioning is essential for orderly cellular life. The discovery of a novel mode of action is an all too rare event in antimicrobial chemotherapy and opens in principle new opportunities for the control of infectious disease. In the context of this book, the quinolones are particularly notable in that these agents are not genotoxic despite their pronounced potency against DNA-dependent processes. In this sense, they join the sulfonamides and trimethoprim as medically successful synthetic agents selectively toxic to bacteria because of interference with vital enzymes involved in DNA formation or processing unique to the bacterial cells. This chapter synopsizes some of the literature in this rapidly expanding field from the viewpoint of information of potential value in drug design.

Drugs whose primary function is to cause damage to DNA and to inhibit DNA-dependent functions are usually highly lethal to growing target cells. DNA function is too vital for cellular survival and reproduction for cells to accommodate to such poisoning if it is extensive or prolonged. Host cell death is not the usual purpose of drug therapy. Normally drugs are administered in order to return infected or transformed cells to normal functioning. It is only when this is not possible that cytotoxic drugs are employed to stop them from functioning at all. Examples can be found in cancer chemotherapy and in treatment of certain lethal viral infections where the patient will not likely recover or may be permanently disabled unless chemotherapy is instituted. In these cases the drug treatment may be sufficiently less harmful than the disease itself that successful drug therapy is contemplated even though many normal host cells will also die as a consequence. Examples of such agents are the alkylating agents and antimetabolites. That this stratagem can work at all without killing the patient is commonly attributed to kinetic factors. If all cells were proliferating (nonresting) and grew at precisely the same rate, successful therapy of this type would probably be impossible. Rapidly growing and actively dividing cells are often found in cancerous growths and in many viral diseases. Such cells have a greater appetite for metabolites and use their enzymic and genetic machinery to a greater extent than resting or mature host cells. This makes them exceptionally susceptible to cytotoxic agents and leaves a modest margin for safety that can be exploited for chemotherapy. Unfortunately, however, even this situation is not completely satisfactory. Many normal host cells are also rapidly dividing in healthy individuals. One can consider hair, GI epithelial cells, the immune system, hematopoietic cells, sperm, and so on, as examples. These rapidly growing normal host cells are also sensitive to kinetic poisons such as alkylating agents, mutagens, and intercalators. As a consequence, cancer patients, for example, often suffer hair loss, ulceration, loss of appetite, anemia, enhanced susceptibility to infections, and the like, because of the treatment of their disease by such agents. These side effects can be avoided only in degree by manipulation of the structure of the drugs employed because these side effects are the consequences of the molecular mode of action of the drugs against the disease itself and are not associated with some ancillary effect mediated in some other way. The chemist's task in these cases is to widen the therapeutic margin as much as possible, but it is not likely that molecular manipulation will ever produce an active agent of this type which has no significant host toxicity. As difficult as this task is seen to be, a much more difficult problem is to find drugs which will be effective against nonproliferating cells without unacceptable toxicity to normal host cells. Here no helpful kinetic effect can be invoked.

In principle, the problem is much more manageable if one is dealing with mixtures of host and infectious cells, such as bacteria, rather than mixtures of normal and rogue host cells. The major biochemical differences which often differentiate host (animal) and invader (microbial) cells can be exploited and the invasive cells killed without serious harm to host cells. The classic example of the β-lactam antibiotics comes readily to mind. These drugs inhibit enzymes responsible for bacterial cell wall formation. Humans have no counterpart structure and lack the enzymic targets, so these drugs are virtually without toxicity toward unallergic humans. When the intracellular target is DNA, success in achieving safe therapy is usually limited to agents attacking enzymes which synthesize or process DNA rather than attacking the DNA molecule itself. In some instances such enzymes are unique to the bacterial cells paralleling the β-lactam situation. A well-known example of this is dihydropteroate synthetase. This enzyme is essential for the biosynthesis of folic acid, itself essential for biosynthesis of DNA. Sulfonamide inhibition of this enzyme prevents formation of

DNA, and this inhibits and eventually kills susceptible bacterial cells. As we do not biosynthesize folic acid from simpler molecules but take it in intact from our diet as a vitamin, we have lost this enzyme somewhere along the evolutionary way and so escape this form of toxicity. This circumstance provides a readily understood and clinically useful margin of selective toxicity.

In other cases, both the host and the invasive cells must carry out the same biosynthetic transformation. Even here there are examples of successful selective toxicity. Inhibitors of dihydrofolate reductase are examples of this. Both infectious bacteria and humans possess enzymes capable of reducing dihydrofolate, and the enzymes responsible must be conservative at their active sites because they process the same substrate to the same product. They, however, need not be identical away from the active site and, fortunately, they are not. These slight differences in molecular architecture have been exploited successfully in the treatment of some bacterial infections. Trimethoprim inhibits the bacterial enzyme approximately 40,000 times more strongly than the corresponding mammalian enzyme and has found widespread safe clinical application.

These precedent findings provide the context for us to appreciate just why the discovery of the quinolones and their molecular mode of action has excited so much contemporary interest. The quinolones inhibit bacterial DNA gyrase (a type II DNA topoisomerase specifically associated with bacteria) while having only marginal inhibitory power against the analogous mammalian enzyme, DNA topoisomerase II. The function of these enzymes will be dealt with extensively in successive sections; however, put simply, these enzymes are essential for dictating the topological structure (conformation) of DNA (6,8,12). Many important cellular processes are sensitive to the shape of DNA. These include DNA replication, transcription, recombination, and repair. Both DNA gyrase and eukaryotic topoisomerase II catalyze change in the degree of supertwisting of a covalent closed circular DNA by introducing transient double-strand cuts, passing the molecule through the cut ends so altering the degree of DNA linking number, and resealing the double strands again. Failure to be able to shape DNA specifically and on demand presents great difficulties for cells and can be rapidly lethal. The detailed structural differences between bacterial DNA gyrase and mammalian DNA topoisomerase II are sufficient to provide selective toxicity for the treatment of bacterial infections. The quinolones exploit these differences by forming a ternary complex involving quinolone, substrate DNA and DNA gyrase (Fig. 1) (50). The ternary complex is apparently stuck part way along the catalytic cycle because artificial dissociation of the ternary complex by use of detergents liberates cut DNA. The stuck complex locks the DNA molecule in a nonfunctional shape. It is

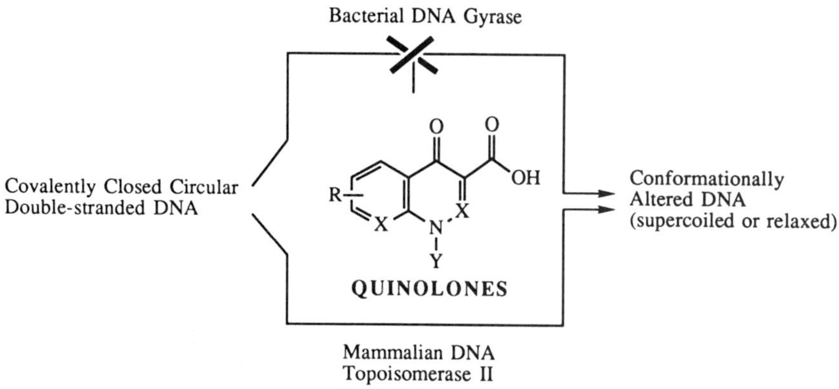

Figure 1 Selective inhibition of bacterial DNA gyrase by quinolones.

believed that the quinolones provide another example, then, of drugs which convert important biosynthetic host enzymes into rogue enzymes catalyzing lethal events. The action of the quinolones on DNA is also seen to induce the SOS response, an error-prone mechanism for repair of damaged DNA (11,24). If the SOS response is not controlled effectively, then severe lethality is experienced by the cell. Possibly these combined effects are of special relevance to the rapid cell killing quinolones cause in susceptible cells.

The first quinolone to find clinical use was nalidixic acid (25). Nalidixic acid has an antibacterial spectrum limited essentially to certain Gram-negative microorganisms, especially those causing urinary tract infections. Nalidixic acid remains on the market despite comparatively low potency, a pronounced tendency to elicit bacterial resistance, its comparatively narrow antimicrobial spectrum, its general poor performance in systemic infections and a toxicity spectrum including proconvulsant activity mediated at the CNS level. Other classic (first-generation) quinolones include oxolinic acid, piromidic acid, pipemidic acid, cinoxacin, miloxacin, and rosoxacin (Fig. 2) (26). These agents have somewhat broader antimicrobial spectra (including some Gram-positive bacteria), lesser toxicity, and produce higher blood levels than nalidixic acid, but have not become major drugs.

The fluorinated quinolone era began with norfloxacin (27). Norfloxacin has a much broader antimicrobial spectrum in that many Gram-positives are now susceptible in clinically achievable doses and this has not been achieved at the expense of loss of anti-Gram-negative activity. Indeed, activity against *Pseudomonas aeruginosa* is now seen. The potency levels are now roughly equivalent to those seen by many of the fermentation derived antibiotics well known to commerce, and the toxicity of these agents is

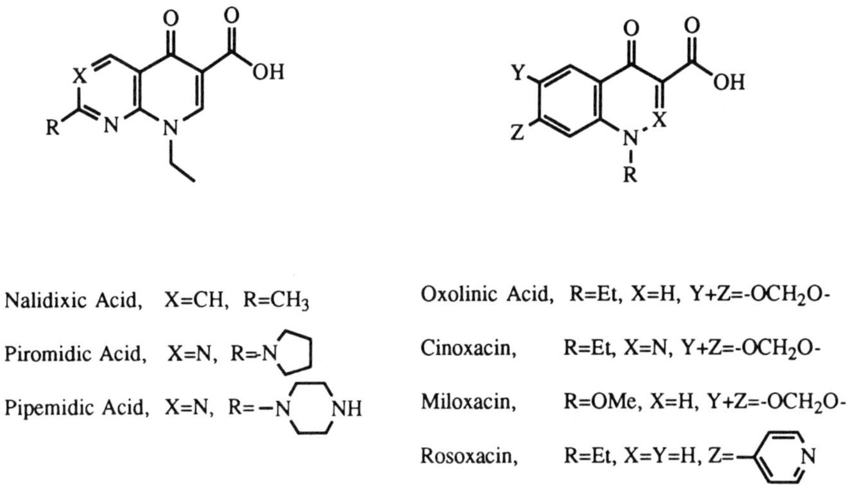

Figure 2 Chemical structures of some of the more important first generation quinolones.

generally more acceptable than that of the earlier quinolones. Other significant members of the second generation quinolone family include pefloxacin, ciprofloxacin, ofloxacin, enoxacin, lomefloxacin, fleroxacin, temafloxacin, and tosufloxacin (Fig. 3) (28). The antimicrobial activity of many of these agents against *Staphylococci, Pseudomonas aeruginosa, Enterobacteriaceae, Escherichia coli, Neisseria* sp., *Hemophilus* sp., *Chlamydia, Legionella,* etc., is good to excellent and is confirmed in a clinical setting. Individual agents, including some of the newest quinolones, show activity against certain *Streptococci*, anaerobes, mycobacteria and enhanced activity against pseudomonads, but these remain areas which require improvement. Nonetheless the newer quinolones find significant clinical application in important additional areas to the urinary tract and are rapidly becoming major drugs challenging the place of the aminoglycosides, the expanded spectrum penicillins, and the cephalosporins.

Quinolones are classified as bactericidal, as the "cidal" dose is only about twice the "static" dose, and these higher blood levels are often achievable in animal and clinical studies. There is a significant postantibiotic effect often seen, and these agents seriously intoxicate resting bacteria, so successful therapy of infections in immunocompromised patients is possible.

In contrast to the β-lactam antibiotics where the elaboration of drug metabolizing enzymes often is responsible for a significant inoculum size effect, this is usually a negligible factor with the quinolones. Indeed, there

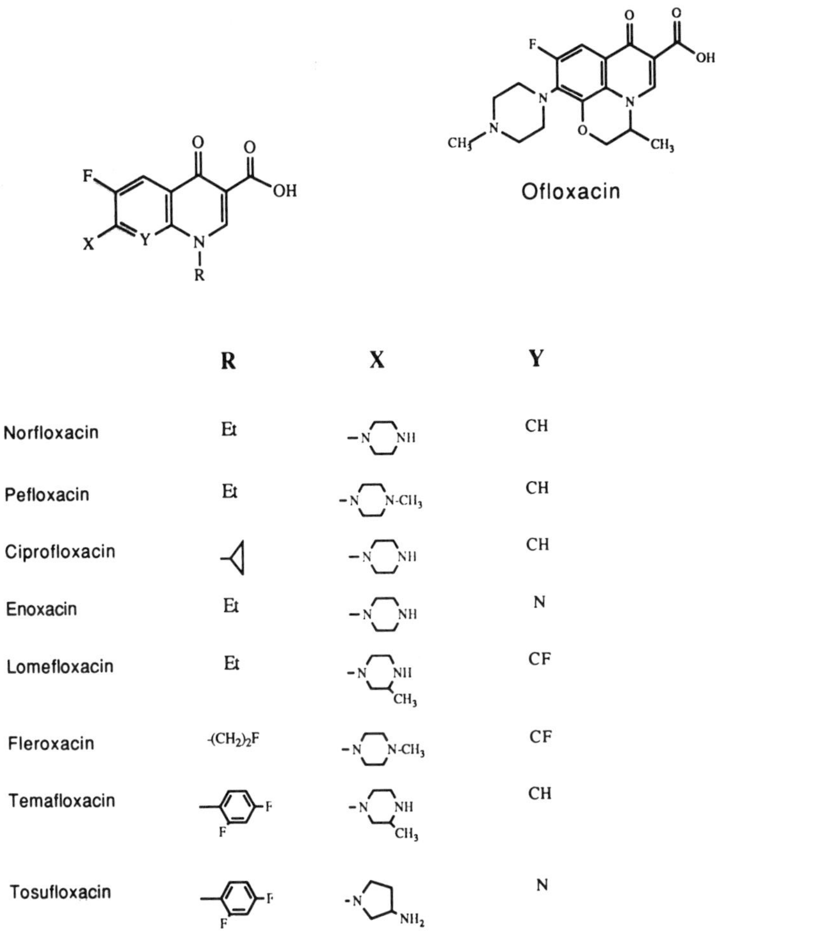

Figure 3 Chemical structures of some prominent second generation (fluorinated) quinolones.

is only a single, unconfirmed, report of R-factor resistance with quinolones in the literature (29). Resistance, rather, appears to be chromosomal and involves lack of cellular drug uptake or alterations in the DNA gyrase (30). Another favorable microbiological characteristic of the modern quinolones is that they can very often be combined with other antiinfective agents without producing significant antagonism. On the negative side, the potency of these agents is usually decreased by the presence of polyvalent metallic ions such as magnesium, aluminum, and iron, as well as by low pH's and lower (32°C) temperatures (31).

In the patient, toxicity is generally minimal although some of these agents are potentially neurotoxic. This is manifested by proconvulsant activity believed to be mediated through γ-aminobutyric acid receptors (32). There are also concerns about joint damage in immature animals, so these agents are generally not advised for use in adolescents or in sexually active females (33).

Because of these favorable characteristics and despite remaining drawbacks, this family of antimicrobial agents has become prominent and it appears that they have achieved a permanent place in human antibacterial therapy. More than 10,000 quinolones have now been synthesized and evaluated, nearly a dozen are in clinical use in one or more countries, and many more are in various stages of evaluation. This and their relatively unique molecular mode of action justify their study.

Next we consider the structures and functions of the DNA topoisomerases, their targets.

B. DNA Topoisomerases as Targets of Chemotherapeutic Agents

Since the discovery of the first topoisomerase (ω-protein) in *E. coli* in 1972 (34), extensive research efforts have been stimulated resulting in a better recognition of the important role of DNA topology in biological processes such as DNA replication, transcription and recombination (reviewed in Refs. 6–14). The subsequent discovery of this class of DNA-modifying enzymes in various living organisms and the finding that they are the target of various antibacterial and anticancer drugs make these enzymes even more intriguing. Consequently, many attempts have been made to use them as novel targets in designing new therapeutic agents (reviewed in Refs. 15–22).

There are two types of topoisomerases isolated from prokaryotic and eukaryotic cells. These enzymes catalyze their topological conversions (either relaxation or supercoiling) of covalently closed circular DNA substrates by introduction of transient protein-linked DNA breaks on one (type

I) or both DNA strands (type II), consequently changing the DNA linking number by steps of 1 or 2, respectively. Beside DNA supercoiling and relaxation, DNA topoisomerases can catenate/decatenate and knot/unknot DNA. All type II enzymes discovered so far require ATP for activity, while type I enzymes do not require energy cofactors for their catalytic function. The interconversion of circular DNA topoisomers catalyzed by DNA topoisomerases is illustrated in Figure 4.

Topoisomerases of DNA are ubiquitous enzymes, and both types of enzymes widely exist in various organisms, including bacteria, yeast, drosophila, T4-infected cells, HeLa cells, calf thymus, and *Saccharomyces cerevisiae* (Baker's yeast). Bacterial type II topoisomerase (also known as DNA gyrase) was isolated from *E. coli* (35), *M. luteus* (36), *Ps. aeruginosa* (37), *B. subtilis* (38), and, more recently, *Staphylococci* (39), and is the only topoisomerase discovered so far which negatively supercoils DNA and exists only in bacteria. The distinct DNA supercoiling mechanisms between eukaryotic and prokaryotic cells (Fig. 5) provide a basis of drug selectivity between bacterial and host systems. The biological role of the two topoisomerases is believed to be keeping an appropriate balance of the overall superhelicity in chromosomal DNA, as proper superhelicity is required in

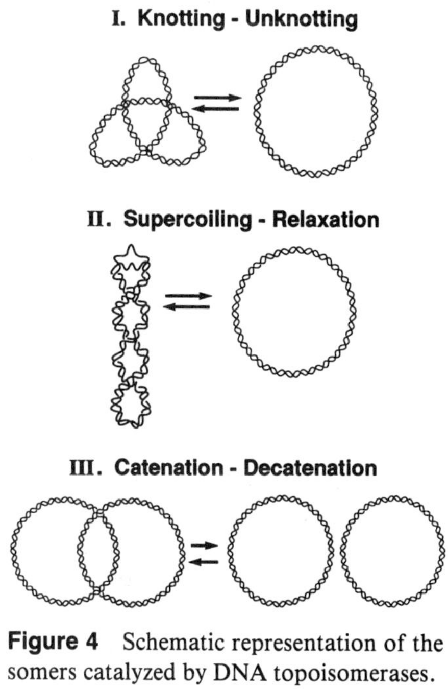

Figure 4 Schematic representation of the interconversion of circular DNA topoisomers catalyzed by DNA topoisomerases.

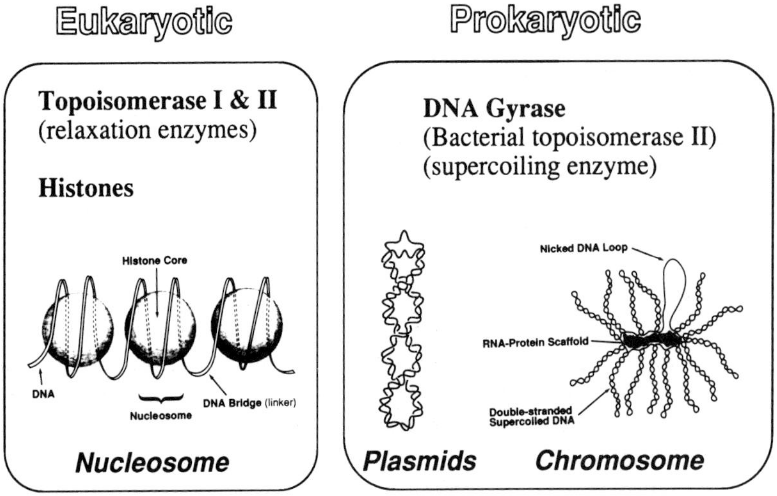

Figure 5 Illustration of DNA supercoiling differences in eukaryotic and prokaryotic cells.

living organisms for normal DNA processing procedures such as DNA replication and transcription (10,40). Both types of enzymes have been shown to be the targets of certain effective and potent therapeutic agents, although different drugs are required for optimum activity against the two different types. We shall discuss this topic in more detail in the following sections.

1. *Bacterial Type II DNA Topoisomerase (DNA Gyrase) as Target of Quinolone Antibacterial Agents*

Little systematic progress was made in the development of the quinolone antimicrobials until the discovery of DNA gyrase (35). Progress was made primarily through use of semiempirical medicinal chemical manipulations. Subsequent extensive studies on the mechanism of the gyrase-catalyzed DNA supercoiling process and on the genetic analysis of quinolone-resistant bacterial mutants led to the conclusion that the primary functional target of quinolone drugs is the A subunit of the enzyme (reviewed in Refs. 6–8), and this lays the groundwork for systematic drug development. DNA gyrase is a tetramer composed of two A subunits, the 105-kilodalton proteins encoded by the *gyrA* (formerly *nalA*) gene, and two B-subunits, the 95-kilodalton proteins encoded by the *gyrB* (formerly *cou*) gene. Electron microscopy of the DNA gyrase holoenzyme revealed that the enzyme adopts a scissors-like shape with the A subunits as the cutting blades and the

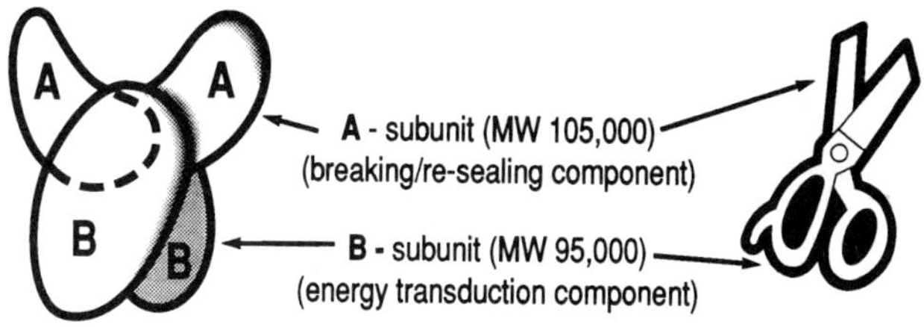

Figure 6 Structure and functions of the DNA gyrase subunits. Shape of *M. luteus* DNA gyrase revealed by electron microscopy studies of Kirchhausen, Wang and Harrison (*Cell 41*:933 (1985)).

B subunits as the handles (Fig. 6). Coincidentally, biochemical mechanistic studies have revealed that the functions of the two subunits are indeed responsible for the direct DNA cutting and for the ATP-driven energy-transduction process, respectively. The enzyme catalyzes the supercoiling reaction via a reduction of the DNA linking number of covalently closed circular DNA carried out by a concerted strand breaking-passing-resealing process (41). The end result of the sequence of enzyme-catalyzed reactions is the formation of an underwound DNA molecule which spontaneously adopts a negatively supercoiled form when the DNA strands resume the most energy-stable double-stranded configuration (Fig. 7). As shown in Figure 8, a complete run of supercoiling involves the following four steps: (i) binding of gyrase to DNA substrate to stabilize a positive DNA node; (ii) cleavage of DNA at 4-base pair staggered sites at the node, forming covalent linkages between a tyrosine group on gyrase subunit A and the

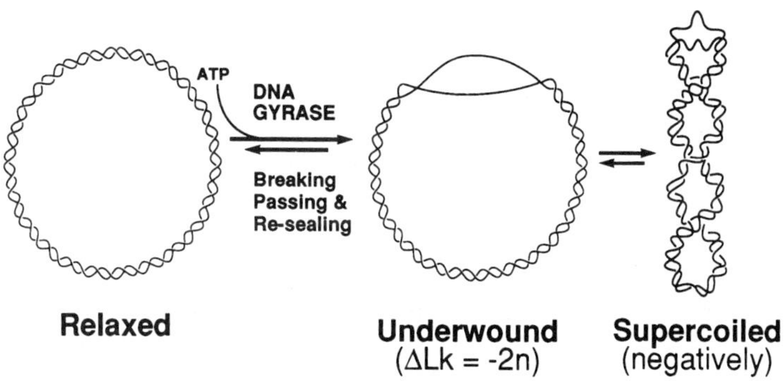

Figure 7 Negative supercoiling of circular duplex DNA brought about by DNA gyrase.

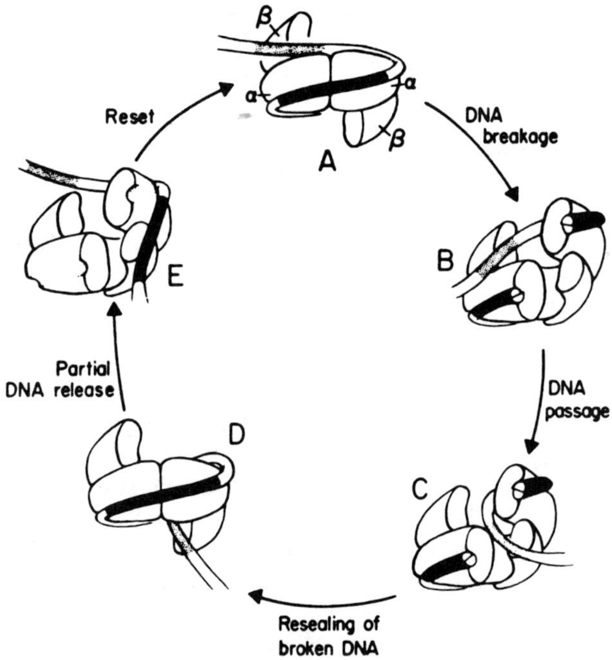

Figure 8 Schematic presentation of the stranded breaking-passing-resealing process in DNA gyrase supercoiling model. (from Ref. 44, with permission.)

5′-end of the DNA chain; (iii) passage of the intact DNA segment forming the node through the cleaved DNA gate, thus inverting the sign of the node; and (iv) resealing the DNA break and the release of enzyme for starting a new supercoiling run (41,44). Quinolone-type drugs such as oxolinic acid are shown to trap the cut intermediate (the DNA gate) by stabilizing the enzyme-DNA complex, thus preventing the enzyme's turnover (Fig. 9). Such a process leads to the formation of a cleavable complex. On addition of a protein denaturant the complex yields a double-stranded break in DNA with A subunit attached covalently to the revealed 5′-ends (42,43). The 4-base pair staggered cleavage site is in the center of the region bound by DNA gyrase (44). This observation provided key evidence that the target of the drug is the A subunit of DNA gyrase. Such drug-induced gyrase-dependent damage to genomic DNA would be lethal to bacteria. No definite explanation was given in these studies regarding exactly how the drug stops the enzyme reaction and what the structure of the stabilized enzyme-substrate complex is. The notion that *gyrA* is the exclusive target of quinolones is challenged by studies on quinolone resistance mutants that give

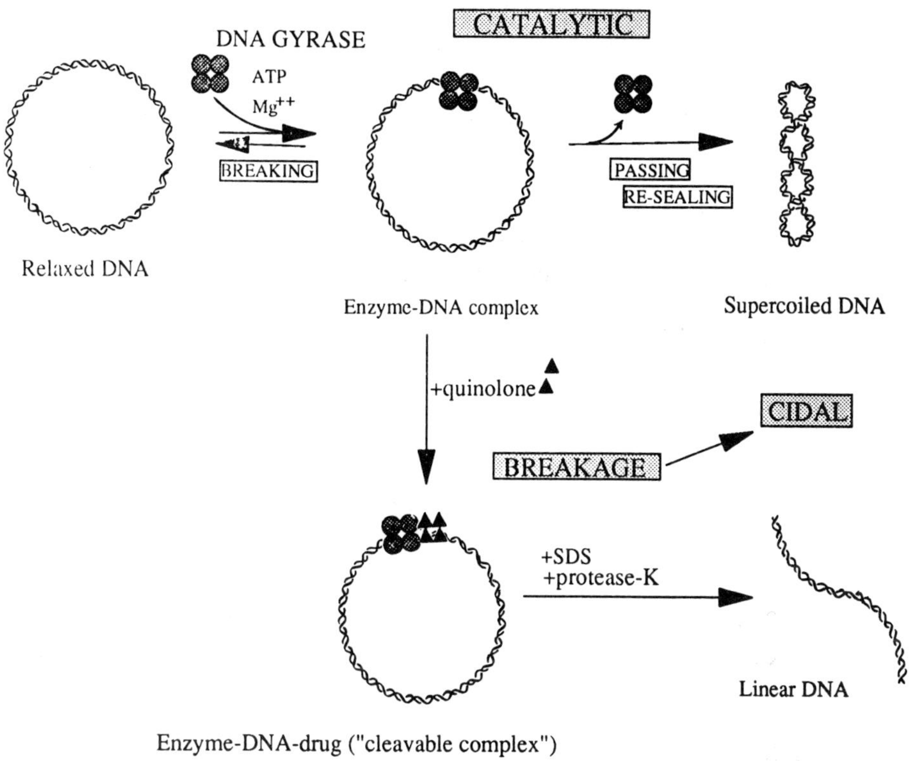

Figure 9 Catalytic function of DNA gyrase and the DNA breakage action of quinolone antibacterials.

somewhat contradicting results. Although the majority of the results indicate that mutations leading to high-level drug resistance are in *gyrA* (45), some mutations leading to resistance are also found in *gyrB* (46,47). No satisfactory explanation was available then. We started our investigation on the mechanism of quinolone inhibition by having ^{3}H-norfloxacin synthesized, and found that ^{3}H-norfloxacin binds not to DNA gyrase but to pure DNA (23). Several important features of the binding of quinolones to DNA are altered when the enzyme is also present. Later we proposed a model showing that the stabilization effect results from the formation of a ternary complex in which the drug binds to a DNA site induced by the enzyme in the presence of ATP (48–50). A more detailed description of the model is given in Section IV.

2. *Mammalian Type II Topoisomerase as Target of Certain Anticancer Agents*

Though the structure and function of the eukaryotic type II topoisomerases are quite different from their bacterial counterpart (DNA gyrase), these enzymes have a high degree of similarity in terms of their basic catalytic mechanism (double-stranded breaking and resealing) and the inhibition mechanism (cleavable complex formation) as well. The mammalian type II topoisomerases have been isolated from calf thymus (51) and from HeLa cells (52,53). The calf thymus enzyme exists in solution as a homodimer with a monomeric molecular weight of approximately 170,000. The first suggestion that DNA topoisomerase II might be involved in the action of anticancer agents came from the work of Kohn and his co-workers on adriamycin and ellipticine. These authors, using an alkaline elution technique, showed that protein-associated double-stranded DNA breaks were present in mouse leukemia cells when they were exposed to these agents (54–58). Subsequently, Liu and his co-workers, using highly purified calf-thymus topoisomerase II, demonstrated that these intercalating agents could promote DNA cleavage by this enzyme in a site-specific manner forming 5′-end links with DNA (59–62). Liu's group later showed that another class of nonintercalative anticancer agents, the epipodophyllotoxins (e.g., etoposide and teniposide), also produce double-stranded DNA breaks by a mechanism involving stabilization of a cleavable complex (63–64). The relationship between these two observations, one involving mammalian cells and perhaps relevant to cancer chemotherapy and one involving bacterial cells and relevant to anti-infective chemotherapy, is not completely understood at the molecular level as yet. Presently known quinolones do not have antitumor activity. The similarity in the two type II enzymes suggests, however, that suitable quinolones might possess useful antitumor properties.

3. *Mammalian Type I Topoisomerase as Target of the Anticancer Agent Camptothecin*

Though type I topoisomerases and eukaryotic type II topoisomerases share a similar catalytic function, i.e. DNA relaxation, the former enzyme has quite a different structure as well as different catalytic properties. There is no legitimate inhibitor for bacterial topoisomerase I reported in the literature; therefore the possible therapeutic value of this class of inhibitor as antibacterial agents waits to be demonstrated. However, efforts toward finding this class of inhibitor as antimicrobial agent have been encouraged by the finding that mammalian topoisomerase I is the target of camptothecin, a plant alkaloid with strong antitumor activity (68). This compound specifically interferes with the nicking-closing reaction of topoisomerase I and also traps the enzyme by forming a covalent linkage with DNA as demonstrated with pure enzyme (68) or in mammalian cells (69–70). Thus, there are analogous features linking the action of all these enzymes. The identification of the inhibition target and the discovery of the molecular action of camptothecin have triggered the synthesis of a large number of camptothecin derivatives. Studies have shown that a quantitative correlation may be demonstrated between the cytotoxicity of the drug, or the anticancer activity, and the potency of the compounds in producing the enzyme-dependent DNA breaking (71–72). Furthermore, recent results showed that this class of drugs is effective in the treatment of human colon cancer in xenografts (73). It is possible, therefore, that interesting findings lie ahead dealing with the antibacterial properties of inhibitors of bacterial type I topoisomerases.

C. Mechanism of Cell Killing and Drug Selectivity of Topoisomerase Inhibitors

The hallmark of the inhibition of both eukaryotic and prokaryotic topoisomerases is the resulting cell killing effect (Fig. 9). The quinolone-type DNA gyrase inhibitors are potent bactericidal agents. This class of compounds inhibits DNA replication and is a potent inducer of SOS repair (11,65–67). The exact mechanism of cell killing by quinolones is still unclear. However, the bactericidal effect is most likely due to DNA damage as a result of the stabilization of gyrase-DNA cleavable complex or the inhibition of DNA replication. Two possible mechanisms are proposed. In one of these, SOS-dependent *RecA* induction resulting from DNA damage triggers the expression of *sfiA*, a septum function inhibitor, which, in the absence of Lon protease, enhances cell killing. It seems true that, in the Lon^- strain, *sfiA*-dependent inhibition of cell division is the primary cause of cell killing by nalidixic acid (reviewed in Ref. 11). In the second, an SOS-independent cell division inhibition has been shown to be associated with cell death. The mechanism is modulated by a cAMP-cAMP receptor protein complex. The target of both mechanisms is the cell division protein FtsZ (reviewed in Ref. 22).

The precise mechanism of mammalian tumor cell killing by topoisomerase inhibitors is also unknown. Parallel with what has been observed with the bactericidal effect of bacterial topoisomerase II inhibitors (quinolones), it seems certain that the killing effect is due to the unusual DNA damage caused by the stabilization of the DNA-enzyme-inhibitor ternary complex. It has been speculated that cellular processes such as the interaction with a fast-moving replication fork, helicases in motion, or with other nuclear proteins may break the complex apart and trigger the killing process (60).

Evidence has been provided that suggests specific arrest of the replication fork by the cleavable complex to be one of the possible mechanisms of cell killing mediated by topoisomerase I (74).

Topoisomerases are ubiquitous enzymes which exist in various living organisms. Although this class of enzyme is highly conserved, it shows certain degrees of diversity in terms of molecular structure, properties and, most importantly, drug sensitivity. DNA gyrase, the only topoisomerase that negatively supercoils DNA, exists only in bacteria. A specific inhibitor of DNA gyrase is therefore unlikely to affect the host system (which has a distinct DNA supercoiling mechanism involving the formation of nucleosomes resulting from DNA-histone interaction). Drug selectivity of mammalian topoisomerase II inhibitors toward tumor cells is believed to be derived from the higher levels of the enzyme in rapidly proliferating tumor cells (75, 76). Topoisomerases from pathogenic fungi have not been reported in the literature. Figgitt et al. (77), using yeast topoisomerase II (from *Saccharomyces cerevisiae*) as a model, showed that there may be drug selectivity between mammalian topoisomerase II and its counterpart in the fungal system according to the differential response of etoposide and podophyllotoxins between these two enzymes. Interestingly, those podophyllotoxins that are active against yeast topoisomerase II are also active against some filamentous fungi, having minimum inhibitory concentration values in the range of 1 to 80 μg/mL. Fungi are presently outside the activity spectrum of quinolones. These findings suggest that suitable quinolones might have antifungal activity.

II. STRUCTURE-ACTIVITY RELATIONSHIPS OF QUINOLONES

Understanding the relationship between structure and activity of quinolones is not only of great importance in designing new analogs, but it is also important in devising credible theories of their molecular mode of action. The vast majority of the available information on the subject deals with comparative potencies in intact bacteria or in infected animal models.

The structure-activity relationships of quinolones have been the subject of extensive review (1–5). Our present understanding is somewhat different than that based upon the properties of the first 500 analogs (26). Preparation of several thousand further analogs, fortunately not based upon a slavish adherence to the initial understanding, has led to a somewhat different set of conclusions (78). These considerations are illustrated in Figure 10. Whereas N-1 still requires a comparatively compact substituent, it is now known that it can be aromatic, contain heteroatoms, or be cyclopropyl

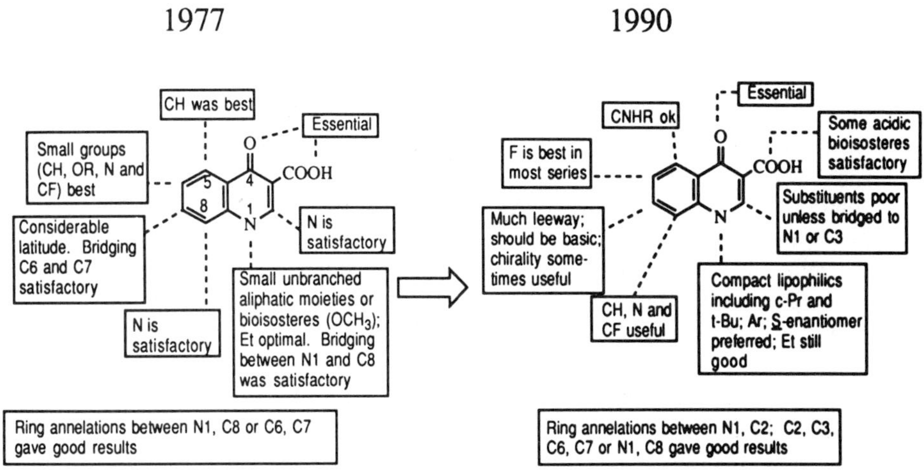

Figure 10 Comparison of structure-activity relationships in 1977 (26) and 1990 (78).

or even *tert*-butyl. It is also possible to bridge an N-1 substituent to C-2 or to C-8 with good effect. When this is done in such a way that a substituent must project above or below the plane of the molecule, it is optimal if it projects upward. The primary characteristic is that the group should be lipophilic and comparatively small. Groups attached to C-2 have been comparatively neglected. The CH group can be replaced by an N, but the replacement of the H by OH, OMe or other such groups has a negative influence on bioactivity unless the group is bridged to N-1 or to C-3. The carboxyl group at C-3 is nearly an absolute requirement. The single notorious exception to this is the presence of an isothiazolone group bridging from C-2 to C-3. This bioisosteric replacement for a carboxyl group produces a series of analogs with greatly enhanced in vitro activity. They are, however, without in vivo value. C-4 has not been studied. Groups at C-5 must be small, and an amino moiety appears to be the best of those which have been studied. C-6 has been consensually optimized at C-F. The quinolones are comparatively forgiving of substitution at C-7. Whereas in most series a piperazinyl or a 3-ethylaminomethylpyrrolidinyl moiety gives excellent results, a wide variety of five- and six-membered cycloaliphatic amines and bicycloderivatives give excellent results. Substitutions at C-8 are generally optimal with C-H, N, or C-F, and smaller groups are definitely better than large moieties. Bridging to N-1 has been mentioned previously as being acceptable. Bridging to C-7 also works, but is not as popular because the resulting biopotency is often decreased. When optical activity is present in quinolones, it is much more advantageously installed near N-1 than in the side chain at C-7.

These relationships have been confirmed in detail and reduced to equations using Hansch quantitative structure-activity techniques (79–80). Such equations putatively have predictive powers as well as retrospective correlative properties, and it will be interesting to see what novel structures will emerge from application of these equations. The main novel feature predicted is that unsaturation above the plane of the molecule connecting N-1 to a small terminal moiety should be especially suitable.

Pragmatically the structure-activity relationship information just reviewed is valuable for analog generation, but it is not entirely suitable for our purposes. Potency in intact bacteria is a complex composite sum of the ability of quinolones to penetrate into the cellular interior where their molecular targets are, their affinity for those targets, their ability to interfere with metabolic functioning following binding, and the lethality of this interference to the cell. Potency in infected animal models and in humans is even more difficult to use as it requires an understanding of pharmacokinetic factors influencing quinolone absorption, distribution, metabolism and excretion, etc.

It would seem, therefore, that the ability of quinolones to interfere with the functioning of purified DNA gyrase in cell-free systems would be the least complicated and most apt information for comparative structure-activity studies. Unfortunately, however, there is not enough of this sort of data available in the literature to enable us to construct a complete picture, and there are some reservations which must be observed in the thoughtful interpretation of the available data.

The first complication is that one cannot be absolutely certain that the action of quinolones can be completely explained by interference with the activity of DNA gyrase. Almost all antibacterial agents have multiple effects on bacterial cells and finding "the" mode of action is a complex enterprise. There is presently a consensus among investigators in the quinolone field that the prominent action of quinolones against DNA gyrase is of fundamental importance in their killing action and it may indeed be the single step of greatest importance, but one must keep an open mind. Indeed, certain lines of evidence demonstrate that functioning protein biosynthesis is essential for quinolone lethality (81,91). No killer protein has been demonstrated, but induction of a faulty SOS repair system is often implicated in the quinolone mode of action (24). This is not directly measured in any of the studies available for analysis. If there were additional important actions of quinolones, then data derived exclusively from free enzyme-binding studies would ultimately prove to be misleading.

A second complication which must be borne in mind is that the lethal action of quinolones against bacteria is most likely expressed against genomic DNA, whereas, for reasons of convenience and the fact that chromosomal DNA are essentially circular, the assays in use employ circular duplex DNA derived from plasmids. Whereas cells can survive quite nicely without plasmid DNA, they must have access to DNA in the genome, so interference with plasmid DNA is not as likely to be a lethal event as destruction of genomic DNA. Thus, the available biochemical data from which we can draw conclusions are derived from model reactions. Furthermore, the little data available relating to this question suggest a close parallel with respect to the relative potency of the drugs in the two DNA test systems (82).

A third complication is the necessity of choosing which bacterium or bacteria should serve as the source of the DNA gyrase to be used for the comparison. Whereas the modern quinolones are quite a broad spectrum, purified enzyme is available only from a relatively few strains of only a few microorganisms. To cover the waterfront adequately, one would prefer a side-by-side comparison of data derived from use of purified enzyme from a typical Gram-negative (*Escherichia coli*, for example), a typical *Gram-positive* (*Staphylococcus aureus*, for example), a difficult target organism (*Pseudomonas aeruginosa*, for example), and an anaerobe (*Bacteroides fragilis*, for example). Complete data do not exist.

A fourth complication is the necessity of choosing what aspect of the interaction of quinolones with DNA gyrase to measure. Affinity alone would not be satisfactory as this might not be the most relevant characteristic. Pharmacologists have long known that affinity and efficacy are not synonymous. Efficacy is a more useful measure for our purposes, but which kind of consequence of binding does one want to use? One can determine both interference with supercoiling action or strand cleavage as consequences, and the literature contains both kinds of data. We prefer to use strand cleavage as the appropriate end point because of our belief that this damage would be more difficult for the organism to survive. Crumplin points out, interestingly, that interference with supercoiling occurs at higher doses than strand cleavage, but that both effects parallel each other (82).

The difference between strand cleavage induction in vitro and action in whole cells is usually interpreted as representing the penetration factor and, with some reservations, we accept this convention. Unfortunately there are relatively few papers containing sufficient relevant data to evaluate these factors in structural terms. Domagala et al. (83) were among the first to note that agents which did not inhibit the purified enzyme were almost never effective in whole cells, but that a number of agents which had good antienzyme action were nonetheless comparatively ineffective in whole cell bacterial systems. The implication is that there is a penetration barrier toward uptake of some quinolones into bacterial cells and this idea is supported by the finding of numerous clinical strains of bacteria whose resistance to quinolones is attributable to the operation of such a barrier (30). Domagala's work and his subsequent papers are specially valuable for our purposes because they include side-by-side data from a single laboratory for significant numbers of classical and novel quinolones so that structural comparisons can be made. It is notoriously understood that quantitative comparisons of data from different laboratories is chancy.

Having warned the reader of some of the assumptions underlying the data, we will now proceed to make some generalizations. Looking at the cell-free *E. coli* enzyme inhibition data of Domagala et al. (83), it is comforting to note that the rank order of the classical quinolones and those introduced clinically more recently parallels reasonably their intact cell antibacterial and clinical efficacies. The divergences in rank order are taken to be a consequence of differences in penetration into intact *E. coli* (note pipemidic acid versus cinoxacin and rosoxacin versus amifloxacin, for examples) (Fig. 11).

From this sort of data one advances a tentative answer to the query, what is the best ring system to employ? The data set does not contain all of the possible ring systems and substitution patterns but in general the pyrimidinopyridones (such as piromidic and pipemidic acid) are rather less potent against *E. coli* DNA gyrase-induced strand cleavage than the 1,8-naphthyridines (such as enoxacin but *not* nalidixic acid) and the benzo-4-pyridones (such as ciprofloxacin) Data for intact bacteria are in general agreement with this conclusion. The *E. coli* strain chosen is the strain from which the DNA gyrase was isolated (83).

For fairer comparison in support of this proposition one can look at individual examples with otherwise identical substitution patterns (Fig. 11). It is difficult to compare the pyrimidinopyridones with the others as there are fewer opportunities to make analogous structural changes (C-6 does not exist and cannot be substituted in the same way as the others). The first three entries indicate that the benzo-4-pyridone is much more effective against the enzyme, whereas all three are nearly equivalent against the intact bacterium. The 1,8-naphthyridines with the same substitution patterns as the benzo-4-pyridones are roughly equivalent against the enzyme but are generally less active in the intact bacterium. The small sample size makes this generalization shaky. Tentatively one decides that the benzo-4-pyridones and the 1,8-naphthyridines are the better ring systems and that the most potent analog will be determined by the specific substitution pattern (83).

Within a given ring system, which substitution pattern is best? There is limited data available from which to attempt to answer this question (see Fig. 12). Nonetheless, for the benzo-4-pyridones it is better to have a cyclo-

Quinolone	DNA Gyrase Cleavage Assay	MIC (*E. coli* H560)
Nalidixic acid	50 μg/ml	6.3 μg/ml
Pipemidic Acid	50	3.1
Cinoxacin	43	50
Piromidic Acid	38	50
Oxolinic Acid	10	0.2
Miloxacin	10	0.4
Ofloxacin	5	0.1
Enoxacin	5	0.1
Rosoxacin	2.5	>10
Amifloxacin	2.5	0.025
Norfloxacin	1	0.1
Ciprofloxacin	0.5	0.025

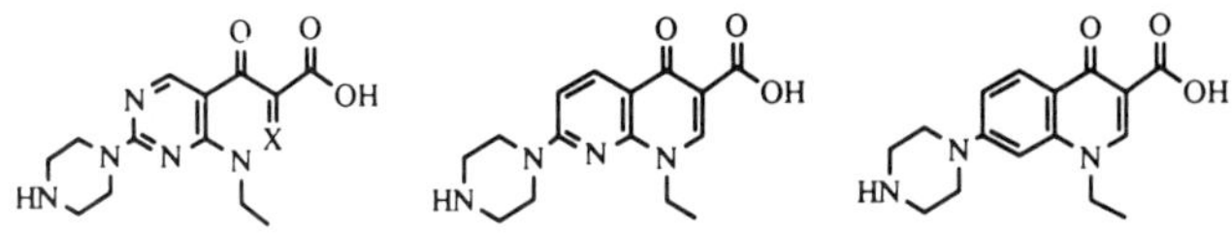

DNA gyrase cleavage:	50 μg/ml	75	18
MIC (E. coli H560):	3.1 μg/ml	3.1	6.3

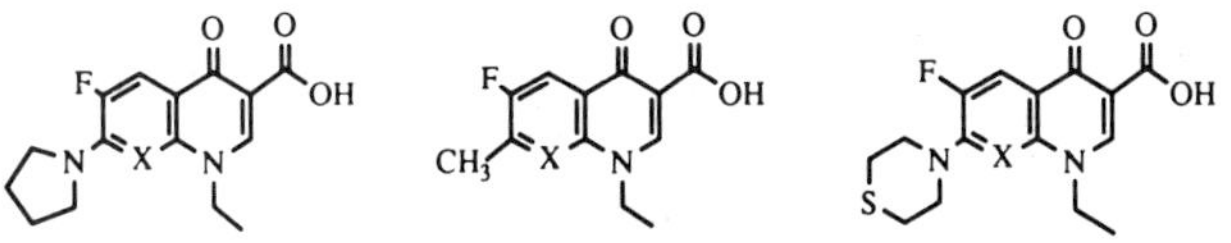

	X: N	X: CH	X: N	X: CH	X: N	X: CH
DNA gyrase cleavage:	2.5	5	25	25	18	5
MIC (E. coli H560):	0.4	1.6	3.1	0.4	1.6	0.1

Figure 11 A comparison of the *E. coli* DNA gyrase cleavage assay data and in vitro potency in intact *E. coli* of similarly substituted pyrimidopyridones, 1,8-naphthyridones and benzopyridones (83).

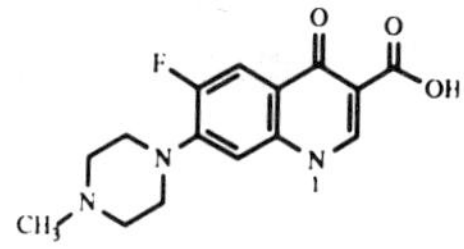

N-1	Et		c-Pr		NHCH3	
	1*	0.1**	0.5	0.025	2.5	0.025

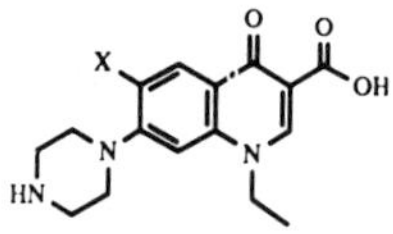

C-6 X	H		F	
	18*	6.3**	1	0.1

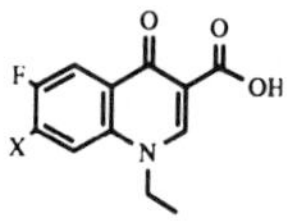

C-7 X	H						Cl		F		CH3	
	50*	25**	1	0.1	1	0.1	38	1.6	38	6.3	25	0.4

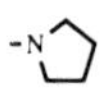		CH3NH-		NH2CH2CH2S-		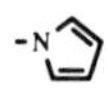		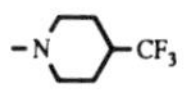		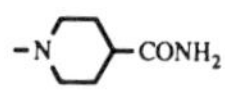		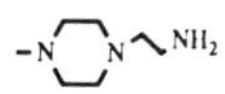	
5	1.6	18	3.1	5	3.1	2.5	1.6	5	6.3	7.5	25	10	25

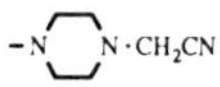		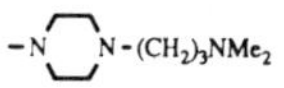		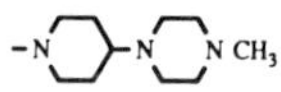		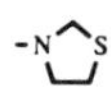		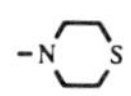	
10	0.8	50	3.1	50	12.5	5	0.2	5	0.1

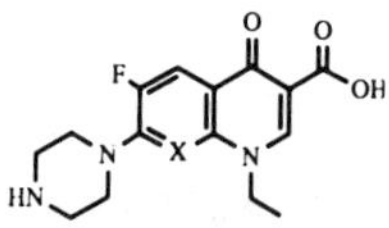

C-8 X	CH		CF		N	
	1*	0.1**	1	0.1	5	0.1

* mg/ml in *E. coli* gyrase cleavage.
** MIC (mg/ml) against *E. coli* H560.

Figure 12 A synopsis of the effect of ring substitution on DNA gyrase cleavage activity and on in vitro action against *E. coli* H560 (83).

propyl group at N-1 than an ethyl; a small, basic group at C-5 is satisfactory; at C-6, F is usually better than H; at C-7 there is substantial leeway for substituent variation although those short groups with a cyclic basic N atom are better than those with Me, F, or Cl; at C-8 F and H are comparable against the enzyme. Piperazine at C-7 is not an absolute requirement, but always gives excellent results (83).

The combination of C-6 F and C-7 piperazine is especially powerful in both the 1,8-naphthyridine and in the benzo-4-pyridone ring systems. The conclusion one can make from the data is that C-6 F gives particularly good enzyme inhibition and that the C-7 substituent contributes both to enzyme inhibition and to cellular uptake. One should not, however, always assume that a C-6 F will be the best possibility. There is data indicating that a C-6 F does not improve potency in either test with nalidixic acid (83). It could well be that in this, and in other ring systems, activity would optimize with something other than F at C-6.

Some data are available dealing with *M. luteus* (Gram-positive) gyrase (85). The data here involve supercoiling inhibition but the rank ordering (nalidixic acid ~ pipemidic acid ~ flumequin ~ oxolinic acid < cinoxacin ~ perfloxacin ~ amifloxacin < ofloxacin < rosoxacin ~ ciprofloxacin < norfloxacin) is in general agreement with the *E. coli* data just discussed. Cinoxacin, amifloxacin, and rosoxacin are in a different order, suggesting either the operation of a penetration barrier or a difference in the enzyme. One notes that the rank ordering is consistent with the *E. coli* data as long as weak binders are involved. For the more potent agents one begins to see some divergence.

Very little data are available using the *S. aureus* enzyme (86). From supercoiling inhibition data a nonsurprising rank ordering was seen (nalidixic acid ~ cinoxacin ≪ ofloxacin ~ norfloxacin < ciprofloxacin).

At the risk of overinterpretation from a limited data set, a common failing of medicinal chemists, it is comforting to note that the hard-earned structure-activity relationships emerging from the preparation of more than 7000 analogs and their examination in intact bacteria largely hold when compared with the limited cell-free antienzyme assay data. This can be confirmed by reference to Figure 10. Unfortunately, however, many potentially meaningful data points are unavoidably missing at present. For example, no enzyme cleavage data are available for some significant N-1 substituents (aryl, *t*-butyl), none at all for substituents at C-2, C-3, and C-4.

The enzyme and its substrate are both chiral so it would be expected that optical activity should play a role in enzyme inhibition. It does (87,88). The effect of optical asymmetry at N-1 is much greater than that seen at other centers investigated to date (particularly at C-7). For example, *S*-(*l*)-

ofloxacin is about 10-fold more potent at inhibiting supercoiling than is *R*-(*d*)-ofloxacin. The racemate is about half as potent as *S*-(*l*)-ofloxacin, as might be expected. The difference is even greater when intact bacteria are used and the ratio differs when different species are involved. This suggests strongly that penetration may also be a chiral process and therefore involve active processes rather than simple soluble passage through the cell wall. Optical asymmetry built into side chains at C-7 is variable in its effect. Sometimes a significant enantiopreference is seen and sometimes not. Where a preference is seen, it is not as pronounced as that seen at N-1.

A working molecular hypothesis of the mode of action of these agents must be consistent with these structures.

III. METHODS USED FOR STUDYING THE MODE OF ACTION OF TOPOISOMERASE INHIBITORS

We will briefly describe methods used for studies of the mode of action of topoisomerase inhibitors, and show the results obtained that demonstrate the usefulness of these techniques for derivation of an inhibition model.

A. Supercoiling Inhibition Assays

The supercoiling inhibition potency of a quinolone compound reflects the basic biological function of a DNA gyrase inhibitor. The assay is needed in order to guide the chemical synthesis for producing more potent DNA gyrase inhibitors. The procedures followed in supercoiling assay have been described in detail (89). The relaxed form of ColE1 or pBR322 DNA (prepared utilizing the relaxation activity of an eukaryotic DNA topoisomerase I) has been used commonly as the reaction substrate. The DNA topoisomers in the reaction mixture, after a certain incubation time, are separated by electrophoresis on a 1% agarose gel, and the amount of supercoiled DNA formed may be determined quantitatively from densitometer traces of the resulting bands. The inhibition of DNA gyrase by quinolone analogs is noncompetitive; the K_i value is calculated as the 50% inhibition concentration (IC_{50}) value from a Dixon plot of reciprocal velocity versus inhibitor concentration (90). This method has been used for the analysis of DNA gyrase inhibition by nalidixic acid, oxolinic acid (43), and various new fluoroquinolones (50). A Dixon plot using the newer fluoroquinolone norfloxacin is shown in Figure 13A. A simple arithmetic method (83) has been used for the determination of IC_{50} values of a variety of quinolones. In this method, the following formula was used: $IC_{50} = [IC_0 + IC_{100}]/2$, where IC_0 and IC_{100} represent the inhibitor concentrations that cause a minimum and near complete inhibition of the enzyme supercoiling activity, respec-

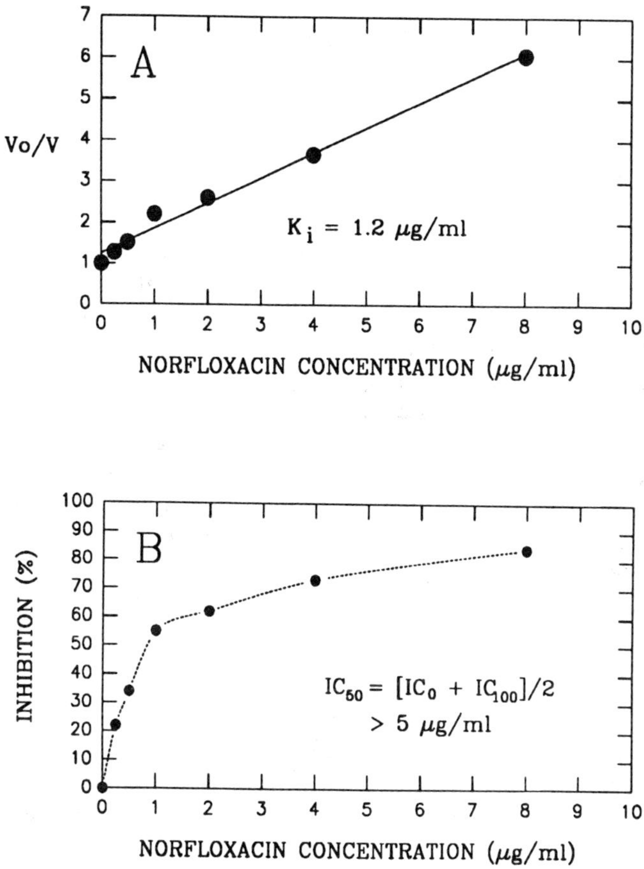

Figure 13 (A) a Dixon plot of the norfloxacin inhibition of DNA gyrase catalyzed supercoiling. (B) Nonlinearity of the inhibition of supercoiling with norfloxacin concentration.

tively. The IC_{50} values obtained this way are substantially higher than those obtained with the Dixon plot due to the fact that the response of supercoiling inhibition with drug concentration is nonlinear. This relationship is obvious from a comparison of Figures 13A,B.

B. Characterization of Topoisomerase-Dependent DNA Cleavage

As mentioned, the hallmark of the action of topoisomerase inhibitors is the induction of DNA cleavage mediated by the formation of the cleavable complex in which the enzyme is covalently attached to the DNA. The capability of the drug to induce such DNA breakage, a consequence of drug binding, has been correlated with the drug potency in killing the target cell (Fig. 14). Therefore a reliable and convenient DNA breakage assay is crucial for comparing for the biocidal effect and the DNA breaking properties of quinolones.

1. Unlabeled DNA Method

Topoisomerase-dependent single-strand DNA nicking and double-strand cleavage may be conveniently assessed by a "cold" method using unlabeled supercoiled plasmid DNA as the substrate. The conversion of supercoiled DNA to linear DNA resulting from a double-strand breakage may be determined on an agarose gel due to their differential electrophoretic mobilities. An example of the results one obtains by using this method is shown in Figure 15. The method is sensitive and has been used to detect the gyrase-dependent DNA breakage induced by quinolones (83), topoisomerase I-dependent DNA nicking induced by camptothecin (92,93), and the topoisomerase II-dependent DNA breakage induced by anticancer drugs (59,94). The formation of the nicked form needs to be determined on a gel containing either ethidium bromide or chloroquine. These intercalating drugs are used in the gel to separate the nicked circular from the relaxed circular DNA band since the intercalator can only change the topological supertwisting of the intact relaxed form but not the nicked form.

2. DNA Cleavage Mapping Using End-labeled DNA

A radioactive end-labeled DNA assay has been widely used for mapping the drug-induced DNA breakage associated with DNA gyrase (95) or eukaryotic DNA topoisomerase II in vitro (59,94). In this case, 3′-end labeled pBR322 DNA is used as the substrate of the breakage assay. The DNA is linearized by EcoRI and is labeled with [γ-^{32}P]dATP using the large fragment of *E. coli* DNA polymerase I and three other unlabeled deoxyribonucleotide triphosphates. Unreacted triphosphates may be removed by etha-

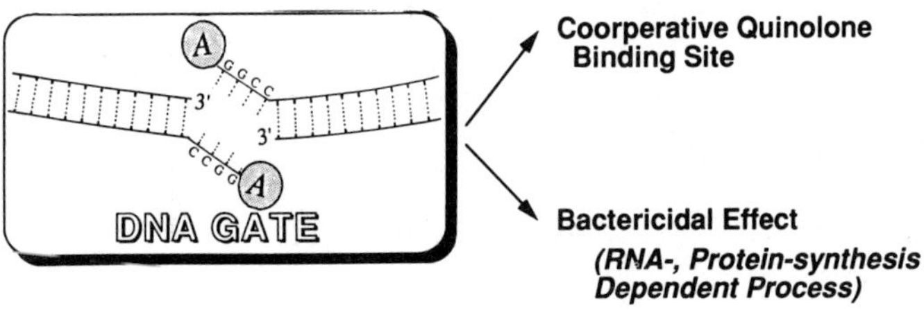

Figure 14 Illustration of the effects of quinolone-induced DNA cleavage highlighted with protein-linked DNA breaks.

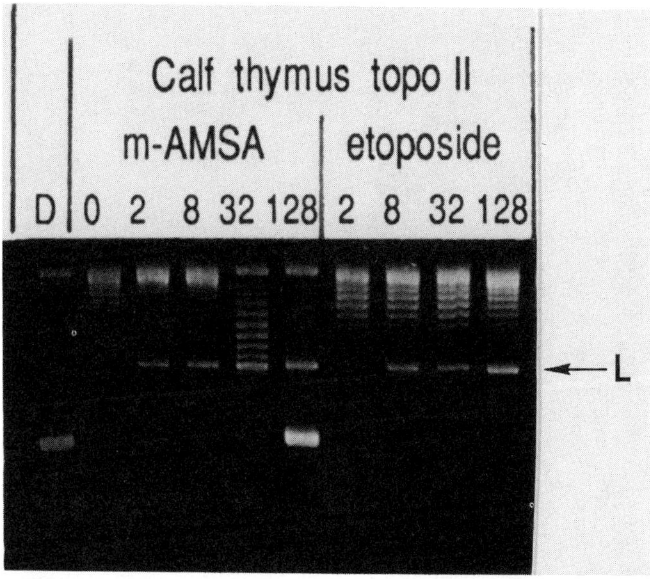

Figure 15 Topoisomerase-dependent DNA cleavage induced by anticancer drugs: *m*-AMSA and etoposide. D denotes reaction mixture containing DNA only. L indicates the linearized DNA band. Numbers on top of each lane represent the drug concentration in μg/mL.

nol precipitation in the presence of 2 *M* ammonium acetate. Topoisomerase II cleavage reactions in a mixture containing buffer, cofactors, DNA, and the drug, are carried out at 37°C for 30 min, and then the reaction is terminated by the addition of sodium dodecyl sulfate. Samples are treated with protease K to remove DNA-linked protein for sharper DNA bands before loading on to an agarose gel. Autoradiography is performed on the dried gel to visualize the cleaved DNA fragments. The fragments have greater electrophoretic mobility due to their smaller size than the original substrate. An example showing etoposide- and quinolone-promoted DNA cleavage by calf-thymus DNA topoisomerase II is shown in Figure 16. The above method may also be used for mapping single-stranded DNA breaks by the action of topoisomerase I. For this purpose, DNA samples are denatured in 0.15 *N* NaOH prior to electrophoresis in a 1% alkaline agarose gel with 30 *mM* NaOH (94).

C. Ultrafiltration Method—A Direct Binding Assay

This approach provides a powerful tool for the investigation of the mode of action of a drug, as it may be used to identify the direct target of the

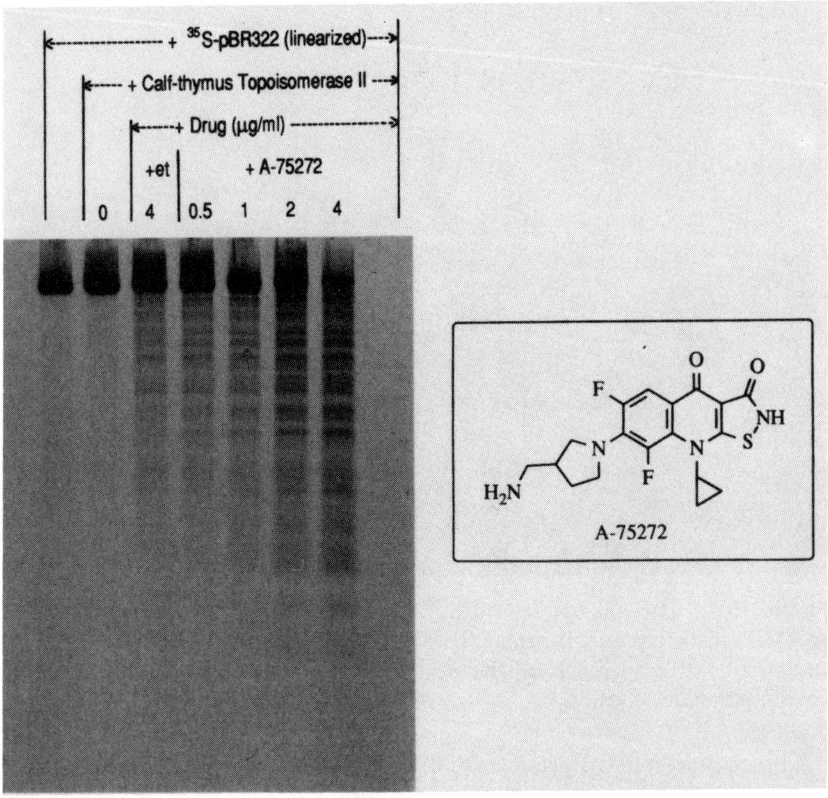

Figure 16 Topoisomerase-dependent DNA cleavage mapping using end-labeled DNA. Etoposide (Et) was used as a standard.

drug. The method described below is suitable for the study of ligand binding to macromolecules even if the drug has low binding affinity to the receptor site since the procedures allow maintenance of a ligand-receptor binding equilibrium throughout the entire process.

1. Methods of Binding Experiments

Quinolone binding may be determined by a membrane ultrafiltration method (23,49) using an ultrafiltration device, such as the Centrifree micropartition device (Amicon, no. 4103) which has membranes with a molecular weight cutoff of 30,000. Each sample consists of a 400 μL reaction mixture containing approximately 5 pmol of DNA and various amounts of radioactive ligand in a selected buffer system. After a certain incubation time, the mixture is transferred to the filtration device. After centrifugation, the filtrates which collected in the reservoir on the lower end of the device are removed and counted for radioactivity. The amount of ligand bound to the receptor is calculated by subtracting the free-ligand concentration in the reaction mixture from the initial ligand concentration, both ligand concentrations being determined from the radioactivity counts of the corresponding filtrates with and without the receptors. The binding of nonradiolabeled oxolinic acid, pefloxacin, and nalidixic acid to ColE1 DNA is measured with a fluorescence method. In these experiments, the same ultrafiltration procedures are used except that ligand concentrations are determined by measuring the drug's intrinsic fluorescence intensity instead of its radioactivity. The binding of ^{3}H-norfloxacin to oligo(dA) species is determined using Amicon YM-2 Centrifree Micropartition devices which have membranes with a molecular weight cutoff of 1000. When a thin, uniform membrane can be found, the amount of bound drug may be determined directly from the radioactivity retained on the membrane after subtracting the background counts.

2. *Data Analysis*

All the binding saturation data are preferably presented in the form of Klotz plots (96). A semi-empirical approach may be employed to analyze the binding cooperativity (97). A nonlinear least-square computer program, RS1 (Bolt, Beranek and Newman Software Products Corporation, Cambridge, MA) can be used to fit the binding data with the following basic binding equation:

$$R = \frac{R_m}{1 + (K_d/D)^n}$$

where R, R_m, K_d, and D represent the molar binding ratio, maximum molar binding ratio, apparent dissociation constant, and free-ligand concentration, respectively, while n is the Hill constant. When $n = 1$, the binding is purely noncooperative, and for large n values the binding is said to be increasingly cooperative. The R_m value, except as otherwise mentioned, is chosen manually as the R value at the apparent saturation plateau position, while other parameters are fitted by the computer program.

3. *Useful Information Obtained for Elucidating the Mode of Action*

Binding Affinity and Cooperativity. The Klotz plots for the binding of perfloxacin, oxolinic acid, and nalidixic acid to supercoiled DNA are shown in Figure 17. The binding cooperativity, expressed as the Hill constants (n) are also shown in the figure. These values indicate that the binding is highly cooperative to the supercoiled DNA; this means that the binding of the first drug molecule to the receptor site facilitates the binding of the second

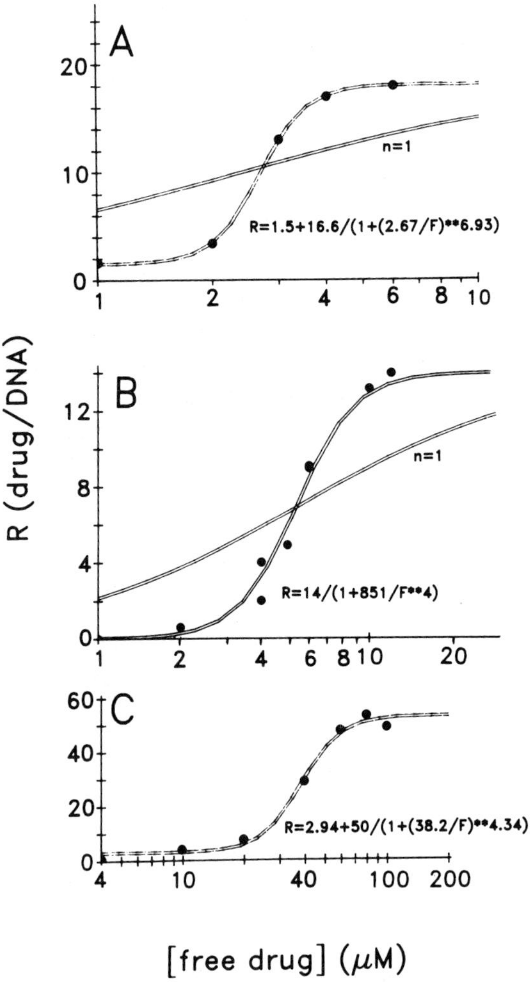

Figure 17 Klotz plots for the binding of pefloxacin (A), oxolinic acid (B), and nalidixic acid (C) to supercoiled DNA (49).

and subsequent drug molecules. Using the same approach, we were able to show that the binding of ^{3}H-norfloxacin to single-stranded (heat-denatured) DNA and to relaxed forms of DNA were nonspecific, since they lack binding cooperativity and demonstrate no trace of binding saturation. The binding of ^{3}H-norfloxacin to supercoiled ColE1 DNA at high drug concentrations (up to the drug's solubility of 1 mM) also fits to this type of binding (49).

Binding Specificity. In order to obtain a better understanding of the role of drug-DNA interactions in DNA gyrase inhibition and the molecular mode of the drug binding, we have extensively investigated the binding specificity of quinolone to various pure DNA forms (49). The binding affinity and specificity may be viewed at two levels. The first level of binding specificity is controlled by the topological structure of DNA. As our earlier studies indicated, ^{3}H-norfloxacin binds preferentially to single stranded rather than to double-stranded DNA (23). This fact is illustrated in Figure 18, which shows the binding of ^{3}H-norfloxacin to ColE1 DNA of different structural forms in a drug concentration range near its supercoiling inhibition constant. The data clearly demonstrate that the binding is greater with the heat-denatured single-stranded DNA rather than the double-stranded DNA of either linear or covalently closed relaxed form. The amount of the drug binding to the complex formed between relaxed DNA substrate and DNA gyrase (in the absence of ATP to prevent supercoiling) is low and is indistinguishable from that to the DNA alone, again demonstrating the absence of drug binding to DNA gyrase at this drug concentration range. We further studied the possible base binding preference to three native double-stranded DNA species of varying GC contents,

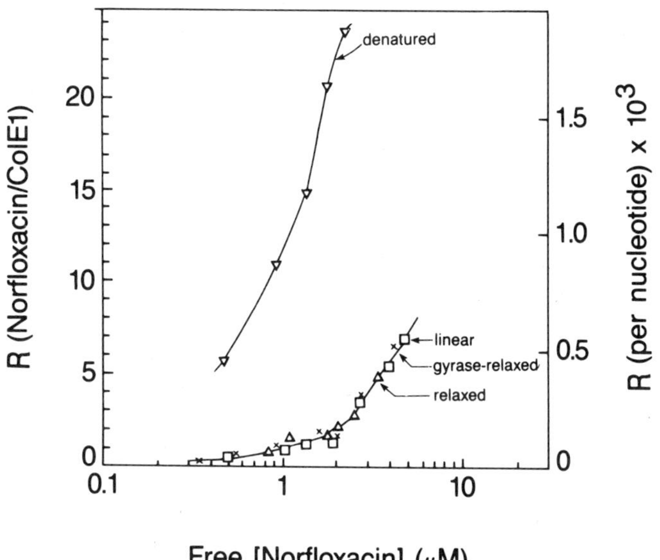

Figure 18 Binding of norfloxacin to ColE1 DNA of different structural forms at a concentration range near its supercoiling inhibition constant (data obtained from Ref. 23.)

even though the amount of the binding was expected to be low. The results showed that there was no correlation between the amount of drug binding and the GC content of the native DNA (49). These results suggest that when DNA strands are intact, the binding is limited and no trace of base preference can be demonstrated. When DNA strands are separated, the amount of drug binding increases and the second level of binding specificity, i.e., the base binding preference, is revealed. The emerging base preference on single-stranded DNA is illustrated in Figure 19A, which shows a preference to poly(dG), poly(dA), poly(dT), and poly(dC) in decreasing order, while the binding to the double-stranded poly(dA)-poly(dT) is virtually nondetectable. The binding to poly(dG) is distinctively greater than that to the other three polydeoxyribonucleotides. One unique structural feature of the guanine base is that it has two common hydrogen bond donors, while the other bases have only one. These results thus suggest that hydrogen bonds may be involved with the drug binding, possibly between the hydrogen bond donor groups on the base and the 4-keto and/or the 3-carboxyl group on the quinolone ring. Also shown in Figure 19A is the drug binding to poly(dI), which shows a level of drug binding roughly equal to 20% of that to poly(dG). The structural difference between guanosine and inosine is that the former has an extra amino group on the pyrimidine ring (Fig. 19A insert) which is an important hydrogen bond donor while pairing complementary strands. The result again suggests that the drug binds to unpaired DNA bases via hydrogen-bonding. Similar results supporting this conclusion were also obtained with synthetic polyribonucleotides (Fig. 19B). These results suggest that quinolones bind preferentially to an unpaired guanine base in a single-stranded DNA region, possibly through hydrogen bonds.

Binding Reversibility. The radioligand filter binding method described above may be used to test the reversibility of quinolone binding to DNA (109). In this way we demonstrated that binding of quinolones to DNA is noncovalent and readily reversible by a simple wash step. We also demonstrated that the binding of quinolones to DNA may be prevented or reversed when histones are added to DNA forming insoluble complexes. In the first set of experiments, we examined the binding of the radiolabeled norfloxacin and difloxacin to the complex formed by adding varying amounts of histones to DNA. It was seen that the levels of the drug bound to DNA are reduced proportionally with the amount of histones added to DNA. The amount of bound drug was reduced by 90% with 400-μg histones added per 100 μg DNA. In the second set of experiments, we showed that the drug bound to DNA may be rejected by the addition of histones. The bound drug was completely repulsed at a histone/DNA ratio (on a weight basis) of 4.

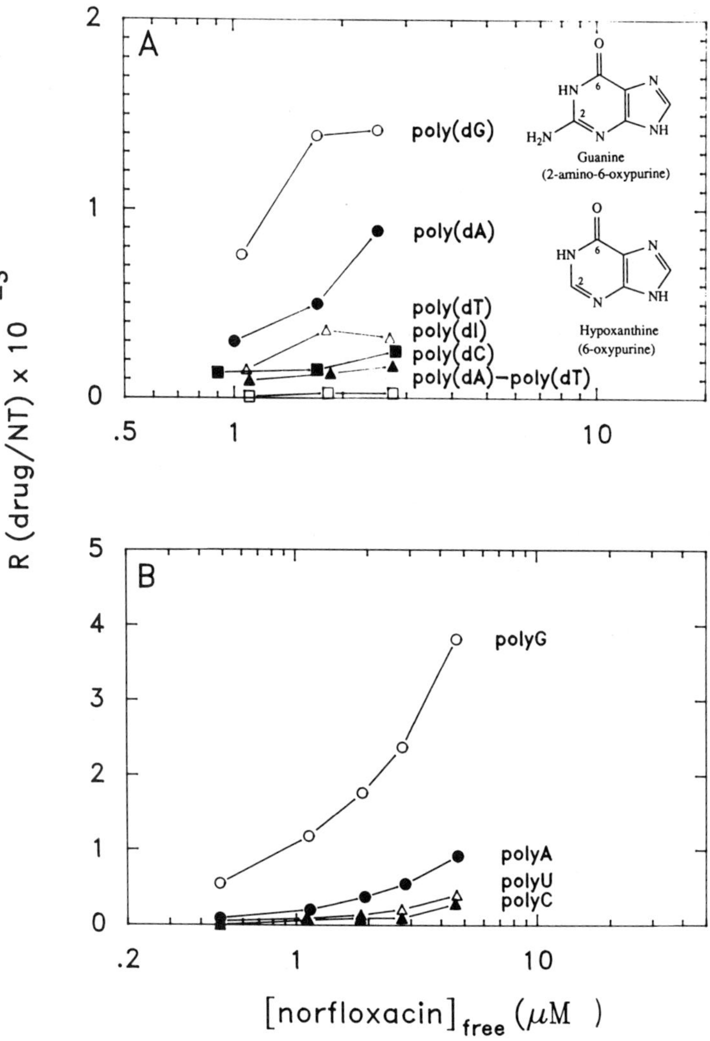

Figure 19 Base preference for norfloxacin binding to synthetic polydeoxyribonucleotide (A) and to polyribonucleotides (B) (49).

D. Spectroscopic Method—Indirect Observation of the Binding

Physicochemical methods normally are not practical for investigating quinolone-DNA binding, especially for the specific form of binding. The problem is mainly derived from the fact that only a very small fraction of the available drug molecules bind to the DNA receptor site under normal experimental conditions at the inhibitory drug concentration, thus giving small, not easily detectable signals. But the methods, however, can be used to demonstrate the nonspecific type of binding that takes place at relatively higher quinolone concentrations.

1. Quenching of Drug's Intrinsic Fluorescence

Quinolones possess a strongly fluorescent chromophore (99) while DNA molecules are essentially nonfluorescent. We capitalized on this fluorescence difference to investigate the interaction of quinolones with DNA. As shown in Figure 20A, we found that the drug's intrinsic fluorescence intensity was quenched when single-stranded DNA (a better receptor for quinolone binding) was added. The extent of quenching is proportional to the amount of drug bound to DNA (Fig. 20B), indicating a near complete fluorescence quenching of the DNA-bound drug. This phenomenon demonstrated that fluorescence intensity measurement cannot be used to detect

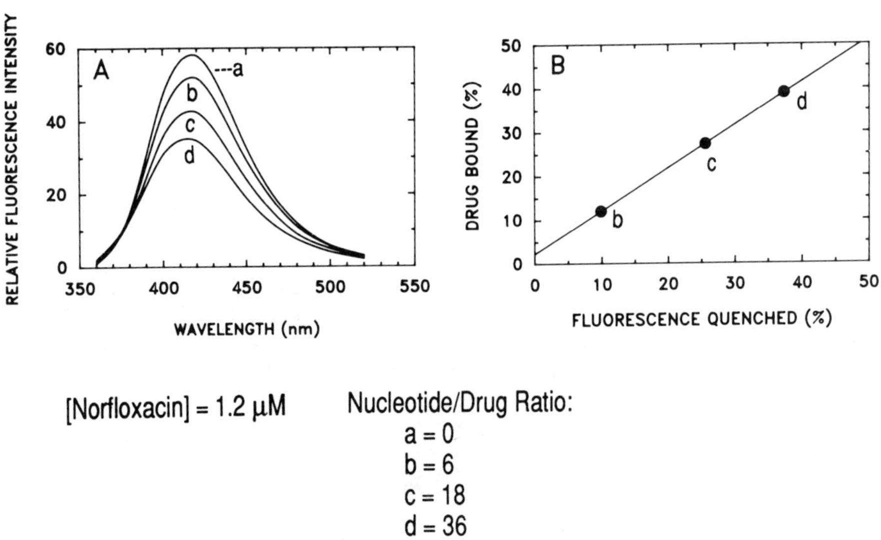

Figure 20 The quenching of quinolone fluorescence as an assay of DNA binding (for details, see Ref. 99).

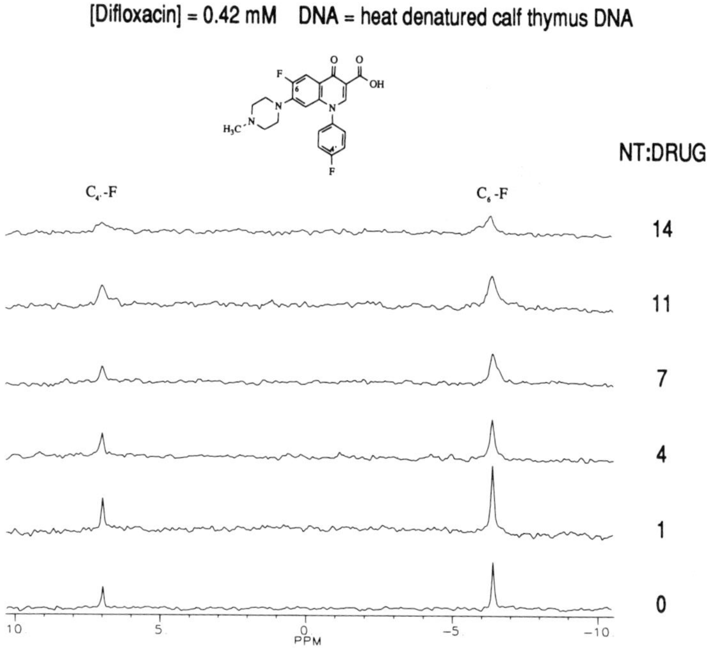

Figure 21 ^{19}F-NMR as a means of detecting quinolone binding to DNA (108).

the results of equilibrium dialysis experiments, since the DNA-bound drug molecules remain silent and cannot be detected.

2. *Fluorine-19 NMR Spectral Changes*

The fluorine-19 NMR spectrum of difloxacin in a phosphate buffer at pH 7 exhibits two sharp resonances corresponding to the two fluorine atoms of the compound (Fig. 21). When the compound is mixed with denatured DNA, a downfield shift is observed for both resonances as well as a dramatic broadening of both fluorine signals. Though it is not possible to determine the exact nature of the drug-DNA interaction from this experiment, results do indicate that the drug is indeed binding to the denatured DNA (108).

IV. MODEL OF QUINOLONE INHIBITION OF DNA GYRASE

Our findings on the binding of quinolones to DNA have led us to a more detailed investigation of the role of drug-DNA interaction in the enzyme inhibition. The binding of ^{3}H-norfloxacin to supercoiled ColE1 DNA (as illustrated in Fig. 17A) represents a specific type of drug binding based on the following three observations: (i) the binding is saturable and takes place at the drug's supercoiling inhibitory concentration, (ii) the binding affinities of some selected quinolones to the DNA site are directly proportional to their supercoiling inhibition constants (23); and (iii) the binding to this saturable site is highly cooperative (49). Such a specific binding pattern was observed only with the supercoiled form of covalently closed circular (ccc) DNA (such as ColE1 DNA), but not with the relaxed or the linear form. This observation can be explained by the fact that supercoiled DNA is the stabilized form of an underwound ccc DNA with reduced linking numbers (8). It is possible that a remnant single-stranded "bubble" may be retained in the supercoil, and it is to this denatured DNA pocket that the drug binds securely. Another possible candidate for the binding site is the cruciform (or hairpin) structure that forms in the region with palindromic sequence, and this structure is known to be promoted by negative supercoiling (100–102). The cruciform structural unit is characterized by single-stranded DNA bubbles that have been demonstrated by the cleavage with S-1 nuclease which has specificity to single-stranded DNA (103–104). Further studies with DNA having such structural features are needed to verify or choose between these possibilities.

The key experimental evidence leading to our proposed inhibition model was obtained from our studies on the norfloxacin binding to the enzyme-DNA complexes (48). We found that the binding of the drug to relaxed forms of DNA may be induced by the participation of DNA gyrase, and that the amount of the binding is far more under these conditions than the total amount of the binding to the DNA and to the enzyme when the experiments are performed separately. The possibility that the increased drug binding was due to the conversion of the substrate DNA to the supercoiled form was ruled out by the use of a nonhydrolyzable triphosphate nucleotide in the binding mixture. As shown in Figure 22, the induced binding to the gyrase-DNA complex showed a saturable phase that highly resembles the binding pattern to the supercoiled DNA in terms of the amount of the binding and the binding cooperativity (49). It was interesting to observe that the requirement of a nucleotide energy source for obtaining the specific form of binding may be eliminated if linearized DNA was used. It is conceivable that the open DNA ends do not restrict the DNA conformational change or DNA unwinding that is necessary for the creation of the drug binding site during the enzyme-wrapping around and gate-opening step following the DNA cleavage process. Moreover, there is a close relationship between the appearance of such characteristic binding and the level of drug-induced DNA breakage subject to different experimental conditions (103). These results suggested that the binding of DNA

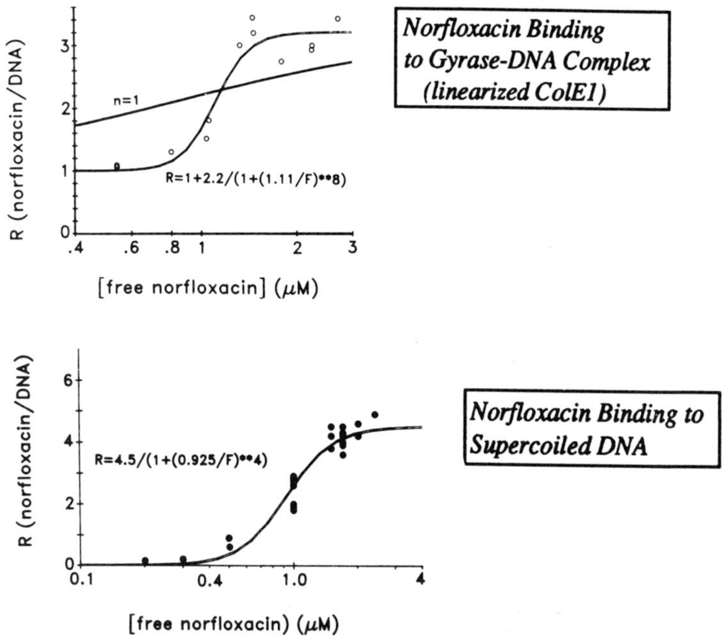

Figure 22 Comparison of ^{3}H-norfloxacin binding to linearized DNA-gyrase complex and supercoiled DNA (49).

gyrase to the DNA substrate creates a site that allows the drug to bind in a cooperative manner.

From all the experimental observations described above, a quinolone-DNA cooperative binding model for the inhibition of DNA gyrase has been proposed (50,98,105). All lines of experimental evidence favor the notion that the bound enzyme induces a drug binding site on the relaxed DNA substrate. We propose that the binding site is formed during the gate-opening step, which requires the binding of ATP. The separated short DNA strands between the 4-base pair staggered cuts form a simulated denatured DNA bubble that is an ideal site for the drug to bind (Fig. 23). As the high binding affinity and cooperativity is concerned, we envision this to occur as follows: drug molecules acquire high binding affinity through a cooperative binding mechanism achieved via self-association of the drug molecules. Two types of interactions, supported by observations in the structure of nalidixic acid crystal, are feasible: the π-π stacking between the quinolone rings and the tail-to-tail hydrophobic interactions between the N-1 substitution groups (Fig. 24). Such interactions result in the formation of a complex

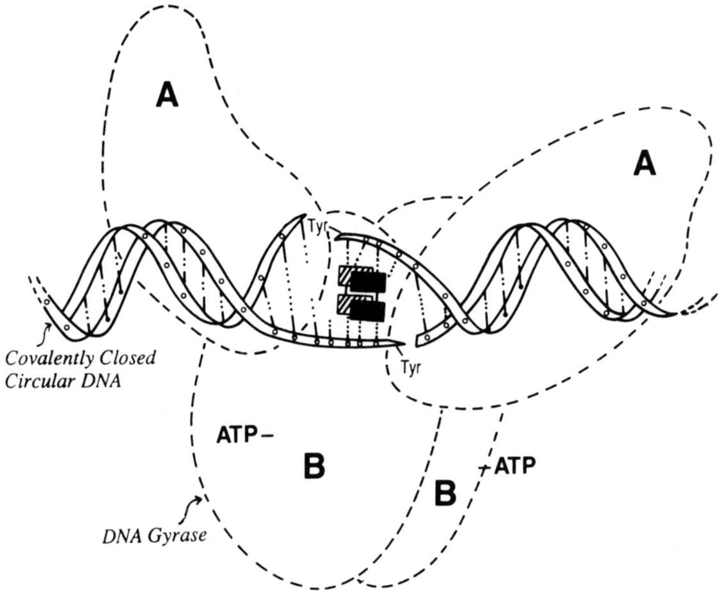

Figure 23 Cartoon of the proposed ternary complex between quinolone molecules, DNA gyrase and duplex DNA substrate (50).

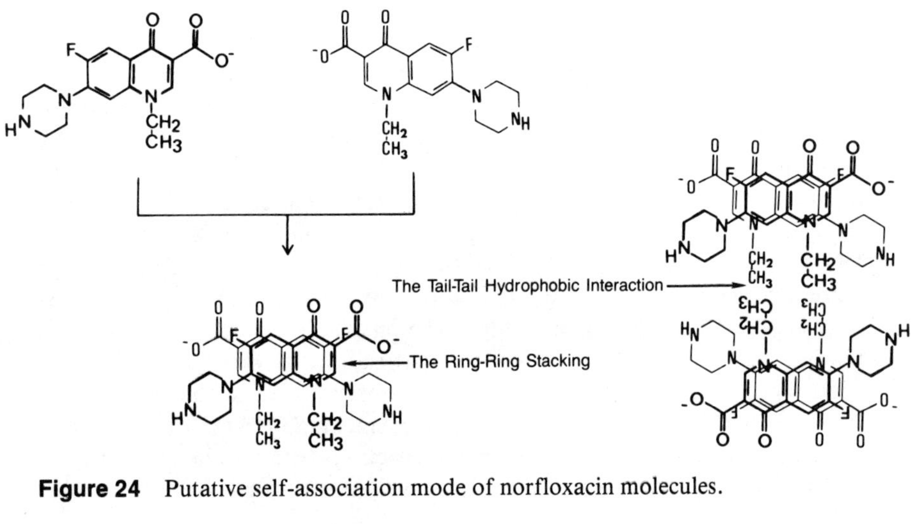

Figure 24 Putative self-association mode of norfloxacin molecules.

with multiple sets of hydrogen bond acceptors in a consolidated unit, thus extending the binding beyond a unidimensional domain; i.e., the assembled drug molecules can act together to occupy a site on DNA with binding groups distributed in a multidimensional space.

This working model has two unique features: (i) the unique configuration of a drug binding site on DNA induced by the action of DNA gyrase, and (ii) the unique ability of the drug molecules to occupy such a site. Both aspects are equally important to drug binding specificity. It is evident that the DNA binding specificity at the enzyme inhibition level is controlled by the binding of the enzyme that creates a favorable DNA site for the drug. This may be viewed as a third level of drug binding specificity.

This binding model offers an answer to questions concerning quinolone-resistance mutants. The model predicts that any mutations, either in *gyrA* or in *gyrB*, which affect the supercoiling process and which alters the configuration of the drug binding pocket on DNA, could lead to drug resistance. Such a mutation may be more likely to occur in *gyrA*, since the subunit A is directly involved in creating such a binding site and in stabilizing the drug bound to DNA. Gyrase subunit B, the energy-transduction machinery that provides a conformational change to assist the DNA gate opening and strand passing, is also important to the formation of such a site. Moreover, our binding and DNA cleavage experiments have also demonstrated the importance of the ATP component which has its receptor sites located on the subunit B.

V. IMPLICATIONS OF THE MODEL IN DRUG DESIGN

A. The Model Is Consistent with Quinolone Structure-Activity Relationships

The model assigns specific functional roles to the C-3 carboxyl and the C-4 keto moieties. These are the portions of the molecule believed to be in contact by strong hydrogen bonding interactions with the DNA bases in the single stranded regions of duplex DNA created by the action of the enzyme. This vital role is consistent with the empiric observation that this part of the quinolone molecule is its indispensable pharmacophore. The polar nature of these functional groups requires them to face outward toward polar environmental factors (solvent in solution; DNA bases in the active site). The relatively small and lipophilic requirement for the N-1 substituent is understood in light of the model as this is the central self-assembly part of the molecule. If the groups are the wrong size, then they will not fit into the space made available by the enzyme. If they are too polar, then they will not self-associate in the interior. The requirement that groups attached to C-5 and C-6 be small is understandable as they cannot be allowed to prevent interaction between the DNA bases and the upper portion of the molecule; but substitutions with hydrogen bond acceptor or donor groups are permissible. The wide structural latitude available at C-7 is understandable as this part of the molecule is distant from the two sites previously mentioned. The enhancement of activity seen when the group has a basic moiety follows as a natural consequence of the need of the molecules to self-associate cooperatively and in the right spatial orientation. Thus the basic group is attractive to the carboxyl moiety, and vice versa, and adjacent quinolone molecules will line up "head to tail" in the same plane and in vertical planes. It is possible that these groups find auxiliary binding sites on the protein. This would rationalize the finding that there are volume limits for the C-7 substituent that, while permissive, may not be exceeded. The requirement that C-8 be small is also consistent with the idea that this side of the molecule must be capable of close approach to another quinolone molecule. The helpful influence of bridging is also seen as a consequence of the need for the adjacent molecules to approach each other closely and form a compact virtual supermolecule. Bridging reduces the net volume requirements of the substituents. These inferences assign specific roles to the various regions of the quinolone pharmacophore as illustrated in Figure 25.

The strong influence of chirality at N-1 as compared with C-7 is also understandable. N-1 being the site of closest approach between quinolone molecules dictates the nature of the stacking interaction and must be com-

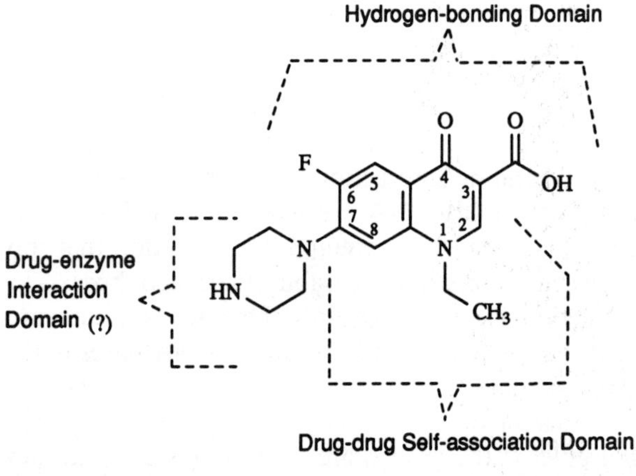

Figure 25 Functional domains of a quinolone antibacterial agent. Norfloxacin molecule is used as an example (50).

plementary in absolute stereochemistry to the pocket opened in the DNA by the action of the enzyme. This feature is discussed in greater detail in the next section. The relatively lesser effect of optical asymmetry at C-7 follows from its distance from either the pharmacophoric region or the self-assembly region.

There are some molecules, however, whose properties remain puzzling. The *N-tert*-butyl analogs are very bulky, and by necessity at least one of the methyl groups must lie below the plane of the molecule when associated in the manner proposed. One takes some comfort from the finding that these agents are more active against Gram-positive bacteria than against Gram-negative bacteria, where enzyme studies have shown there to be a smaller pocket (see below). They are, nonetheless, hard to rationalize.

The N-1 aryl analogs are not as troubling as might have been supposed. It is true that the aryl group is larger than an aliphatic *N*-ethyl moiety, but x-ray analysis has shown that these groups lie orthogonally to the quinolone ring system itself and can interact with each other by van der Waals forces (106). Thus what these analogs lose in bulk they regain to some extent by using binding factors not available to the aliphatic congeners.

A more direct test of the likelihood that self assembly forces play an important role in the action of the quinolones follows from experiments where such molecules have been tethered directly together to simulate the dimer (107). Following the observation that the methyl groups (of the N-1 ethyl moiety) in nalidixic acid were nearly within bonding distance from one another, a computer graphic simulation was performed in which these two moieties were bonded together with a carbon-carbon single bond. The atoms in the artificial molecule are nearly at the same locations in the artificial dimer. It was decided to synthesize such a molecule and subject it to the test. For this purpose, the nalidixic acid moiety was replaced by the more active norfloxacin ring system. This provided a molecule with two norfloxacin units tethered together by four methylenes. Gratifyingly, this molecule was easily synthesized and was very active against the DNA gyrase derived from *E. coli*, just as the hypothesis required. All of the precedent structure-activity relationship (SAR) considerations, whether whole cell or pure enzyme derived, predicted that this analog should be inactive unless self-association played an important role. To check this further, a series of analogs were synthesized in which the length of the tether arm was lengthened or shortened. Shorter arms would prevent coplanarity of the aromatic rings, and longer arms would fail to fit the space available in the unit crystal. Only the analog with four methylenes had significant enzyme inhibitory activity. Interestingly when this series of analogs was tested against the enzyme derived from Gram-positive *Micrococcus luteus*, the analog tethered by five methylenes was the active one. This suggests strongly that the melt zone created by the enzyme is of different dimensions in different microbes and that it is larger in at least some Gram-positives than in *E. coli*. This is in general agreement with experience (with the N-1 *t*-butyl analogs, for example), which suggests that it is very difficult to optimize a single structure against all possible classes of pathogens. It is also seen that Gram-positives are more tolerant of comparatively large substituents at N-1 (Fig. 26).

Another test of this hypothesis comes from the tethering of a simple aromatic ring at the end of the methylenes attached to N-1 of norfloxacin. These compounds were all inactive. Thus the quinolone structure is needed at both ends, as the model requires, and the result with the tethered "dimer" was not the consequence of accidentally reaching a previously unsuspected binding site for an aromatic ring.

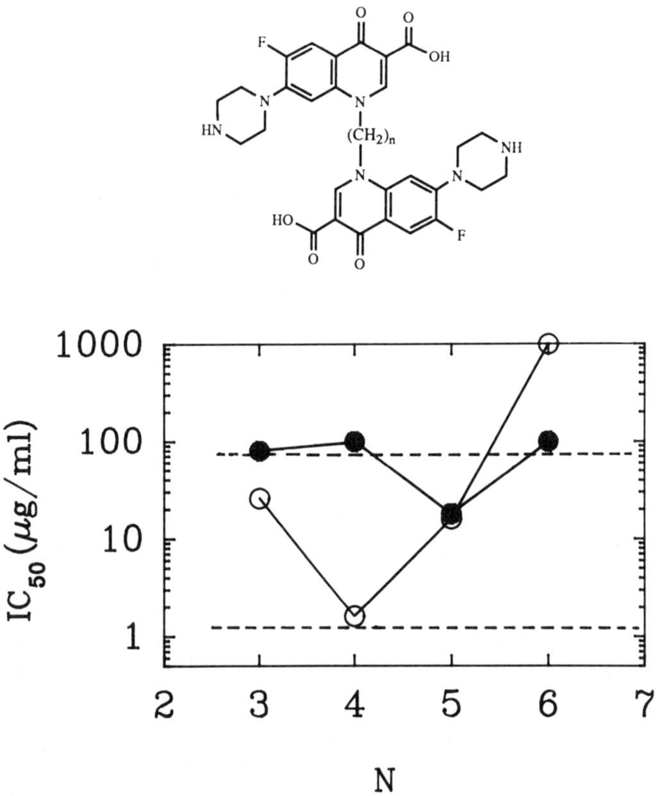

Figure 26 Dependence of the inhibition of DNA gyrase on length of linker arm with norfloxacin "dimers".

B. Interpretation of the Large Activity Difference of Ofloxacin Enantiomers

As suggested in the previous section, the groups attached to N-1 and C-8 represent points of close approach between adjacent quinolone monomers in the model. As the pocket opened in DNA by DNA gyrase is likely to be chiral, compatible chirality must be present in those drugs which possess chirality. Molecular modeling (108) shows clearly that there are two different ways in which *S*-ofloxacin can stack—methyls outside and methyls inside, and each puckers such that the methyls are either equatorial or axial (Fig. 27). When the methyls are outside, in the *endo* form, the stack is compact in the vertical dimension (the distance between the two rings is shortest). Further, the methyls are equatorial and nearly planar. In the *exo*-puckered form the methyls are axial and project significantly out of the plane of the molecule. The methyl groups, therefore, strongly affect the closeness of approach of molecules in the virtual supermolecule. In the *endo* conformers, when the methyls are inside, molecules adjacent in the vertical dimension cannot approach each other as closely due to the steric bulk of the methyl groups. The virtual supermolecule is intrinsically chiral. This means in particular that the pharmacophoric regions have a reverse screw sense when the stacked *R*- and *S*-monomers are compared with each other. In the more active *S*-analog the hydrogen-bond acceptors on the two quinolone rings are orientated "Seattle/Miami," whereas in the *R*-analog they are orientated "Boston/San Diego." Thus, the model predicts a particular complementary chirality to the enzyme-induced melt zone. It is not possible at present to verify or choose between these possibilities as the detailed structure of the ternary complex is unknown except by inference. Nonetheless, this provides an important and ultimately testable consequence of the model. The substituent at C-7 does not affect the closeness of approach nor the chirality of the stack, so optical asymmetry there should be less important. This is what is observed.

VI. CONCLUSIONS

In summary, the quinolones provide an important contemporary example of DNA-directed drugs of comparatively low toxicity based upon their action on the bacteria-specific DNA gyrase. Computer graphics simulations assisted in the design of important probe molecules in which two monomeric units were tethered together by an aliphatic linker chain. The activity of these novel agents in cell-free enzyme systems provides strong support for the model proposed to explain the molecular mode of action of these drugs. Presently this model is the only one available for understanding this

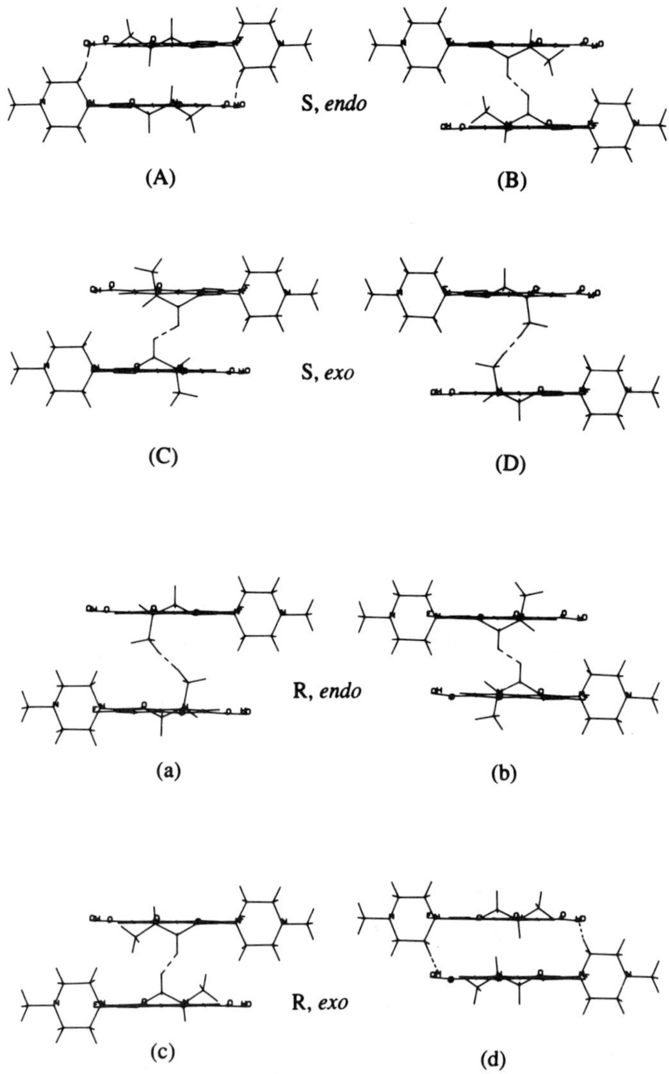

Figure 27 Eight possible stacking conformations of ofloxacin enantiomers. A and B represent *S, endo* methyls equatorial; they differ in the pucker conformation of the aliphatic ring bridging N-1–C-8. C and D represent *S, exo* methyls axial; they also differ in the pucker conformation of the aliphatic ring. Molecular pairs *a* to *d* represent the same conformers in the *R* series. The most compact orientations are *A* and *d* in which both the methyls and the ring pucker are away from adjacent stacked molecules (108).

interaction, and it correlates well the available information about the action of the drugs on the enzyme and in whole cells.

The most important characteristic of a useful model is its potential predictive power. Important inferences for future drug design along rational principles can be made from the model. For example, it is clear that the higher melting and comparatively poorly water-soluble analogs should be the more potent because these parameters are based upon the ability of the monomeric units to self-associate. One predicts also that placing a compatible purine or pyrimidine unit at one end of a tethered heterodimer should produce biochemical probes which would exhibit base pair specificity in the cleavage reaction. The present agents show a preference for guanine/cytosine because of the enhanced number of hydrogen bonding sites present, but in actuality this preference is significantly balanced by general base pair availability, so no consensus specificity is actually observed with natural DNA.

The mammalian and yeast topoisomerases II must bear some structural similarity to bacterial DNA gyrase because of their analogous actions on DNA. It should, then, in principle be possible to devise drugs for these cells based on analogous considerations. There are already in the literature indications that some quinolones interact with a mammalian topoisomerase II, and this suggests that agents based upon these leads could find a use in cancer chemotherapy. If this is so, then it is not far fetched to predict that agents active against fungal enzymes might also be found and belong to an analogous but idiosyncratically different structural type. Quite recently, quinolones have been found to react directly with DNA (that is, not requiring the enzyme). This raises the possibility that quinolone anticancer drugs can be found which will interact with DNA independently of the enzymes involved.

Thus the prospects for rational drug design based upon these considerations is bright in several important disease areas.

REFERENCES

1. Fernandes, P.B., *International Telesymposium of Quinolones*, J.R. Prous, Barcelona, Spain (1989).
2. Wolfson, J.S., and D.C. Hooper, *Quinolone Antimicrobial Agents*, American Society for Microbiology, Washington, DC (1989).
3. Andriole, V.T., *The Quinolones*, Academic Press, New York (1989).
4. Crumplin, G.C., *The 4-Quinolones: Antibacterial Agents* in Vitro, Springer-Verlag, New York (1990).
5. Siporin, C., C.L. Heifetz, and J.M. Domagala, *The New Generation of Quinolones*, M. Dekker, New York (1990).

6. Cozzarelli, N.R., DNA gyrase and the supercoiling of DNA. *Science 207*: 953–960 (1980).
7. Gellert, M., DNA topoisomerases, *Ann. Rev. Biochem. 50*: 879–910 (1981).
8. Wang, J.C., DNA topoisomerases. *Sci. Am. 247*: 94–109 (1982).
9. Liu, L.F., DNA topoisomerase—enzymes that catalyze the breaking and rejoining of DNA. *CRC Crit. Rev. Biochem. 15*: 1–24 (1983).
10. Wang, J.C., DNA topoisomerases. *Ann. Rev. Biochem. 54*: 665–697 (1985).
11. Drlica, K., Biology of bacterial deoxyribonucleic acid topoisomerases. *Microbiol. Rev. 48*: 273–289 (1984).
12. Wang, J.C., DNA topoisomerases: nature's solution to the topological ramifications of the double-helix structure of DNA. *The Harvey Lectures Series, 81*, 93–110 (1987).
13. Wang, J.C., Recent studies of DNA topoisomerases. *Biochem. Biophys. Acta 909*: 1–9 (1987).
14. Pruss, G.J., and K. Drlica, DNA supercoiling and prokaryotic transcription. *Cell 56*: 521–523 (1989).
15. Ross, W.E., DNA topoisomerases as target for cancer therapy. *Biochem. Pharmacol. 34*: 4191–4195 (1985).
16. Chen, G.L., and L.F. Liu, DNA topoisomerases as therapeutic targets in cancer chemotherapy. *Ann. Rep. Med. Chem. 21*: 257–262 (1986).
17. Epstein, R.J., Topoisomerases in human disease. *The Lancet 1*: 521–524 (1988).
18. Drlica, K., and R.J. Franco, Inhibitors of DNA topoisomerases. *Biochem. 27*: 2253–2259 (1988).
19. Hsiang, Y.-H., H.-Y. Wu, and L.F. Liu, Topoisomerases: novel therapeutic targets in cancer chemotherapy. *Biochem. Pharmacol. 37*: 1801–1802 (1988).
20. Bodley, A.L., and L.F. Liu, Topoisomerases as novel targets for cancer chemotherapy. *Biotechnology 6*: 1315—1319 (1988).
21. Long, B.H., and D.A. Stringfellow, Inhibitors topoisomerase. II: Structure-activity relationships and mechanism of action of podophyllin congeners. *Adv. Enz. Regul. 27*: 223–256 (1988).
22. Liu, L.F., DNA topoisomerase poisons as antitumor drugs. *Annu. Rev. Biochem. 58*: 351–375 (1988).
23. Shen, L.L., and A.G. Pernet, Mechanism of inhibition of DNA gyrase by analogues of nalidixic acid: the target of the drugs is DNA. *Proc. Nat. Acad. Sci. USA 82*, 307–311 (1985).
24. Drlica, K., S. Coughlin, and M.S. Gennaro, Mode of action of quinolones: biochemical aspects. In *The New Generation Quinolones*, Siporin, C., Heifetz, C.L., and Domagala, J.M. eds., Dekker, New York, 51–52 (1990).
25. Lesher, G.Y., E.J. Froelich, M.D. Gruett, J.H. Bailey, and R.P. Brundage, 1,8-Naphthyridine derivatives. A new class of chemotherapeutic agents. *J. Med. Chem. 5*, 1063–1065 (1962).
26. Albrecht, R., Development of antibacterial agents of the nalidixic acid type. *Prog. Drug Res. 21*: 9–104 (1977).
27. Koga, H., A. Itoh, S. Murayama, S. Suzue, and T. Irikura, Structure-activity relationships of antibacterial 6, 7- and 7, 8-disubstituted 1-aklyl-1,4-dihydro-4 oxoquinoline-3-carboxylic acids. *J. Med. Chem. 23*: 1358–1363 (1980).

28. Mitscher, L.A., R.M. Zavod, P.V. Devasthale, D.T.W. Chu, L.L. Shen, P.N. Sharma, and A.G. Pernet, Microbes beware: quinolones. I, II. *Chemtech. 21*: 50–56, 249–255 (1991).
29. Munshi, M.H., D.A. Sack, K. Haider, Z.U. Ahmed, M.M. Rahman, and M.G. Morshed, Plasmid-mediated resistance to nalidixic acid in *Shigella dysenteriae* type 1. *Lancet*, 419–421 (1987).
30. Wolfson, J.S., and D.C. Hooper, Bacterial resistance to quinolones: Mechanism and clinical importance. *Rev. Infect. Dis. 11*: S960–968 (1989).
31. Lewin, C.S., and J.T. Smith, 4-Quinolones and multivalent ions. *J. Antimicrob. Chemo. 26*: 149 (1990).
32. Tsuji, A., H. Sato, Y. Kume, I. Tamai, E. Okezaki, O. Nagata, and H. Kato, Inhibitory effects of quinolone antibacterial agents on γ-aminobutyric acid binding to receptor sites in rat brain membranes. *Antimicrob. Agts. Chemo. 32*: 190–194 (1988).
33. Stahlmann, R., Cartilage damaging effect of quinolones. *Infection 1951*: S38 (1991).
34. Wang, J.C., Interaction between DNA and *E. coli* protein ω. *J. Mol. Biol. 55*: 523–533 (1971).
35. Gellert, M., K. Mizuuchi, M.H. O'Dea, and H.A. Nash, DNA gyrase: an enzyme that introduces superhelical turns into DNA. *Proc. Nat. Acad. Sci. USA 73*: 3872–3876 (1976).
36. Liu, L.F., and J.C. Wang, Micrococcus luteus DNA gyrase: active components and a model for its supercoiling of DNA. *Proc. Nat. Acad. Sci. USA 75*: 2098–2102 (1978).
37. Miller, R.V., and T.R. Scrurlock, DNA gyrase (topoisomerase II) from *Pseudomonas aeruginosa. Biochem. Biophy. Res. Comm. 110*: 694–700 (1983).
38. Sugino, A., and K.F. Bott, Bacillus subtilis deoxyribonucleic acid gyrase. *J. Bacteriol. 141*: 1331–1339 (1980).
39. Takahata, M., and T. Nishino, DNA gyrase of *Staphylococcus aureus* and inhibitory effect of quinolones on its activity. *Antimicrob. Agents Chemo. 32*: 1192–1195 (1988).
40. Vosberg, H.-P., DNA topoisomerases: enzymes that control DNA conformation. *Curr. Top. Microbiol. Immunol. 114*: 19–102 (1985).
41. Brown, P.O., and N.R. Cozzarelli, A sign inversion mechanism for enzymatic supercoiling of DNA. *Science 206*: 1081–1083 (1979).
42. Gellert, M., K. Mizuuchi, M.H. O'Dea, T. Itoh, and J. Tomizawa, Nalidixic acid resistance: a second genetic character involved in DNA gyrase activity. *Proc. Nat. Acad. Sci. USA 74*: 4772–4776 (1977).
43. Sugino, A., C.L. Peebles, K.N. Kreuzer, and N.R. Cozzarelli, Mechanism of action of nalidixic acid: purification of *E. coli nalA* gene product and its relationship to DNA gyrase and a novel nicking-closing enzyme. *Proc. Nat. Acad. Sci. USA 74*: 4767–4771 (1977).
44. Morrison, A., and N.R. Cozzarelli, Contacts between DNA gyrase and its binding site on DNA: Features of symmetry revealed by protection from nucleases. *Proc. Nat. Acad. Sci. USA 78*: 1416–1420 (1981).
45. Higgins, N.P., C.L. Peebles, A. Sugino, and N.R. Cozzarelli, Purification

of subunits of *Escherichia coli* DNA gyrase and reconstitution of enzymatic activity. *Proc. Nat. Acad. Sci. USA 75*: 1773–1777 (1978).

46. Yamagishi, J., Y. Furutani, S. Inoue, T. Ohue, S. Nakamura, and M. Shimizu, New nalidixic acid resistance mutations related to deoxyribonucleic acid gyrase activity. *J. Bacteriol. 148*: 450–458 (1981).
47. Yamagishi, J., H. Yoshida, M. Yamayoshi, and S. Nakamura, Nalidixic acid-resistant mutations of the *gyrB* gene of *Escherichia coli. Mol. Gen. Genet. 204*: 367–373 (1986).
48. Shen, L.L., W.E. Kohlbrenner, D. Weigl, and J. Baranowski, Mechanism of quinolone inhibition of DNA gyrase. Appearance of unique norfloxacin binding sites in enzyme-DNA complexes. *J. Biol. Chem. 264*: 2973–2978 (1989).
49. Shen, L.L., J. Baranowski, and A.G. Pernet, Mechanism of inhibition of DNA gyrase by quinolone antibacterials. Specificity and cooperativity of drug binding to DNA. *Biochemistry 28*: 2879–2885 (1989).
50. Shen, L.L., L.A. Mitscher, P.N. Sharma, T.J. O'Donnell, D.W.T. Chu, C.S. Cooper, T. Rosen, and A.G. Pernet, Mechanism of inhibition of DNA gyrase by quinolone antibacterials. A cooperative drug-DNA binding model, *Biochemistry 28*: 2886–2894 (1989).
51. Halligan, B.D., K.A. Edwards, and L.F. Liu, Purification and characterization of a type II DNA topoisomerase from bovine calf thymus. *J. Biol Chem. 260*: 2475–2482 (1985).
52. Miller, K.G., L.F. Liu, and P.T. Englund, A homogeneous type II DNA topoisomerase from HeLa cell nuclei. *J. Biol. Chem. 256*: 9334–9339 (1981).
53. Liu, L.F., and K.G. Miller, Eukaryotic DNA topoisomerases: two forms of type I DNA topoisomerases from HeLa cell nuclei. *Proc. Nat. Acad. Sci. USA 78*: 3487–3491 (1981).
54. Ross, W.E., D.L. Glaubiger, and K.W. Kohn, Protein-associated DNA breaks in cells treated with adriamycin or ellipticine. *Biochem. Biophys. Acta 519*: 23–30 (1978).
55. Ross, W.E., D.L. Glaubiger, and K.W. Kohn, Qualitative and quantitative aspects of intercalator-induced DNA strand breaks. *Biochem. Biophys. Acta 562*: 41–50 (1979).
56. Ross, W.E., and M.O. Bradley, DNA double-strand breaks in mammalian cells after exposure to intercalating agents. *Biochem. Biophys. Acta 654*: 129–134 (1981).
57. Zwelling, L.A., S. Michaels, L.C. Erikson, R.S. Ungerleider, M. Nichols, and K.W. Kohn, Protein-associated DNA strand breaks in L1210 cells treated with DNA intercalating agents 4′-(9-acridinylamino)methanesulfon-*m*-anisidide and adriamycin. *Biochem. 20*: 6553–6563 (1981).
58. Zwelling, L.A., S. Michaels, D. Kerrigan, Y. Pommier, and K.W. Kohn, Protein-associated deoxyribonucleic acid strand breaks produced in mouse leukemia L1210 cells by ellipticine and 2-methyl-9-hydroxyellipticinium. *Biochem. Pharmacol. 31*: 3261–3267 (1982).
59. Nelson, E.M., K.M. Tewey, and L.F. Liu, Mechanism of antitumor drug action: poisoning of mammalian DNA topoisomerase II on DNA by 4′-(9-

acridinylamino)-methanesulfon-*m*-anisidide. *Proc. Nat. Acad. Sci. USA 81*: 1361–1365 (1984).

60. Tewey, K.M., G.L. Chen, E.M. Nelson, and L.F. Liu, Intercalative antitumor drugs interfere with the breakage-reunion reaction of mammalian DNA topoisomerase II. *J. Biol. Chem. 259*: 9182–9187 (1984).
61. Rowe, T.C., G.L. Chen, Y.-H. Hsiang, and L.F. Liu, DNA damage by antitumor acridines mediated by mammalian DNA topoisomerase II. *Cancer Res. 46*: 2021–2026 (1986).
62. Tewey, K.M., T.C. Rowe, L. Yang, B.D. Halligan, and L.F. Liu, Adriamycin-induced DNA damage mediated by mammalian DNA topoisomerase II. *Science 226*: 466–468 (1984).
63. Chen, G.L., L. Yang, T.C. Rowe, B.D. Halligan, K.M. Tewey, and L.F. Liu, Nonintercalative antitumor drugs interfere with the breakage-reunion reaction of mammalian DNA topoisomerase II. *J. Biol. Chem. 259*: 13560–13566 (1984).
64. Ross, W., T. Rowe, L. Yang, B. Glisson, J. Yalowich, and L.F. Liu, Role of topoisomerase II in mediating epipodophyllotoxin-induced DNA cleavage. *Cancer Res. 44*: 5857–5860 (1984).
65. Gudas, L.J. and A.B. Pardee, Model for regulation of *E. coli* DNA repair functions. *Proc. Nat. Acad. Sci. USA 72*: 2330–2334 (1975).
66. Piddock, L.J.V., and R. Wise, Induction of the SOS response in *E. coli* by 4-quinolone antimicrobial agents. *FEMS Microbiol. Lett. 41*: 289–294 (1987).
67. Phillips, I., E. Gulebras, F. Moreno, and F. Baquero, Induction of the SOS response by new 4-quinolones. *J. Antimicrob. Chemo. 20*: 631–638 (1987).
68. Hsiang, Y.-H., R. Hertzberg, S. Hecht, and L.F. Liu, Camptothecin induces protein-linked DNA breaks via mammalian DNA topoisomerase I. *J. Biol. Chem. 260*: 14873–14878 (1985).
69. Hsiang, Y.-H. and L.F. Liu, Identification of mammalian DNA topoisomerase I as an intracellular target of the anticancer drug camptothecin. *Cancer Res. 48*: 1722–1726 (1988).
70. Mattern, M.R., S.M. Mong, H.F. Bartus, C.K. Miravelli, S.T. Crooke, and R.K. Johnson, Relationship between the intracellular effects of camptothecin and the inhibition of DNA topoisomerase I in cultured L1210 cells. *Cancer Res. 47*: 1793–1798 (1987).
71. Hsiang, Y.-H., L.F. Liu, M.E. Wall, M.C. Wani, A.W. Nicholas, G. Manikumar, S. Kirschenbaum, R. Silber, and M. Potmesil, DNA topoisomerase I-mediated DNA cleavage and cytotoxicity of camptothecin analogues. *Cancer Res. 49*: 4385–4389 (1989).
72. Hertzberg, R.P., M.J. Caranfa, K.G. Holden, D.R. Jakas, G. Gallagher, M.R. Mattern, S.-M. Mong, J.O. Bartus, R.K. Johnson, and W.D. Kingsbury, Modification of the hydroxy lactone ring of camptothecin:Inhibition of mammalian topoisomerase I and biological activity. *J. Med. Chem. 32*: 715–720 (1989).
73. Giovanella, B.C., J.S. Stehlin, M.E. Wall, M.C. Wani, A.W. Nicholas, L.F. Liu, R. Silber, and M. Potmesil, DNA topoisomerase I-targeted chemotherapy of human colon cancer in xenografts. *Science 246*: 1046–1048 (1989).

74. Li, L.H., T.J. Fraser, E.J. Olin, and B.K. Bhuyab, Action of camptothecin on mammalian cells in culture. *Cancer Res. 32*: 2563–2570 (1972).
75. Heck, M.M.S., and W.C. Earnshaw, Topoisomerase II: A specific marker for cell proliferation. *J. Cell Biol. 103*: 2569–2581 (1986).
76. Hsiang, Y.-H., H.-Y. Wu, and L.F. Liu, Proliferation-dependent regulation of DNA topoisomerase II in cultured human cells. *Cancer Res. 48*: 3230–3235 (1988).
77. Figgitt, D.P., S.P. Denyer, P.M. Dewick, D.E. Jackson, and P. Williams, Topoisomerase II: A potential target for novel antifungal agents. *Biochem. Biophys. Res. Comm. 160*: 257–262 (1989).
78. Mitscher, L.A., P.V. Devasthale, and R.M. Zavod, Structure-activity relationships of fluoro-4-quinolones. In *The 4-Quinolones: Antibacterial Agents* in Vitro, Crumplin, G.C. ed., Springer-Verlag, New York, 115–146 (1990).
79. Ohta, M., and H. Koga, Three-dimensional structure-activity relationships and receptor mapping of N1-substitution of quinolone antibacterials. *J. Med. Chem. 34*: 131–139 (1991).
80. Domagala, J.M., C.L. Heifetz, M.P. Hutt, T.F. Mich, J.B. Nichols, M. Solomon, and D.F. Worth, 1-Substituted 7-[(ethylamino)methyl]-1-pyrrolidinyl]-6,8-difluoro-1,4-dihydro-4-oxo-3-quinolinecarboxylic acids. New quantitative structure-activity relationships at N1 for the quinolone antibacterials. *J. Med. Chem. 31*: 991–1001 (1988).
81. Zeiler, H., Evaluation of the in vitro bactericidal action of ciprofloxacin in cells of *E. coli* in the logarithmic and stationary phases of growth. *Antimicrob. Agents Chemother. 28*: 524–527 (1985).
82. Crumplin, G.C., Molecular effects of 4-quinolones upon DNA gyrase: DNA systems. In *The 4-Quinolones: Antibacterial Agents in* Vitro, Crumplin, G.C. ed., Springer-Verlag, New York, 53–68 (1990).
83. Domagala, J.M., L.D. Hanna, C.L. Heifetz, M.P. Hutt, T.F. Mich, J.P. Sanchez, and M. Solomon, New structure-activity relationships of the quinolone antibacterials using the target enzyme. The development and application of a DNA gyrase assay. *J. Med. Chem. 29*: 394–404 (1986).
84. Domagala, J.M., S.E. Hagen, C.L. Heifetz, M.P. Hutt, T.F. Mich, J.P. Sanchez, and A.K. Trehan, 7-Substituted 5-amino-1-cyclopropyl-6,8-difluoro-4-oxo-3-quinolinecarboxylic acids: Synthesis and biological activity of a new class of quinolone antibacterials. *J. Med. Chem. 31*: 503–506 (1988).
85. Zweerink, M.M., and A. Edison, Inhibition of *M. luteus* DNA gyrase by norfloxacin and 10 other quinolone carboxylic acids. *Antimicrob. Agts. Chemo. 29*: 598–601 (1986).
86. Takahata, M., and T. Nishino, DNA gyrase of Staphylococcus aureus and inhibitory effect of quinolones on its activity. *Antimicrob. Agts. Chemo. 32*: 1192–1195 (1988).
87. Mitscher, L.A., P.N. Sharma, and R. Zavod, The influence of optical isomerism on the biological properties of quinolone antimicrobial agents. In *International Telesymposium on Quinolones*, Fernandes, P.B. ed., J.R. Prous, Barcelona, Spain, 73–83 (1989).
88. Gerster, J.F., S.R. Rohlfling, N.J. Rustad, N.J. Reiter, S.E. Pecore, R.M.

Winandy, and J.E. Landmesser, The synthesis and pharmacological profile of the stereoisomers of a tricyclic quinolone antibacterial. In *International Telesymposium on Quinolones*, Fernandes, P.B. ed., J.R. Prous, Barcelona, Spain, 85–98 (1989).
89. Otter, R., and N.R. Cozzarelli, *E. coli* DNA gyrase. *Methods Enzymol. 100B*: 171–180 (1983).
90. Cozzarelli, N.R., The mechanism of action of inhibitors of DNA synthesis. *Ann. Rev. Biochem. 46*: 641–668 (1977).
91. Wolfson, J.S., and D.C. Hooper, Mechanisms of killing of bacteria by 4-quinolones. In *The 4-Quinolones: Antibacterial Agents* in Vitro, Crumplin, G.C. ed., Springer-Verlag, New York, 69–85 (1990).
92. Hertzberg, R.P., M.J. Caranfa, and S.M. Hecht, On the mechanism of topoisomerase I inhibition by camptothecin: evidence for binding to an enzyme-DNA complex. *Biochem. 28*: 4629–4638 (1989).
93. Jaxel, C.J., K.W. Kohn, M.C. Wani, M.E. Wall, and Y. Pommier, Structure-activity study of the actions of camptothecin derivatives on mammalian topoisomerase I: Evidence for a specific receptor site and a relation to antitumor activity. *Cancer Res. 49*: 1465–1469 (1989).
94. Liu, L.F., T.C. Rowe, L. Yang, K.M. Tewey, and G.L. Chen, Cleavage of DNA by mammalian DNA topoisomerase II. *J. Biol. Chem. 258*: 15365–15370 (1983).
95. Walton, L., and L.P. Elwell, In vitro cleavage-complex assay to monitor antimicrobial potency of quinolones. *Antimicrob. Agts. Chemo. 32*: 1086–1089 (1988).
96. Klots, I.M., Protein interactions with small molecules. *Acc. Chem. Res. 7*: 162–168 (1974).
97. Cantor, C.R., and P.R. Schimmel, *Biophys. Chem. Part III*, 863–864 (1980).
98. Shen, L.L., J. Baranowski, and T. Wai, Mechanism of inhibition of DNA gyrase by quinolone antibacterials: A cooperative drug-DNA binding model. *J. Cellular Biochemistry; Suppl. 10B*: abstract of UCLA Symposium on DNA replication and recombination, Park City, Utah (1986).
99. Shen, L.L., A reply: "do quinolones bind to DNA?"—yes. *Biochem. Pharmacol. 38*: 2042–2044 (1989).
100. Gellert, M., K. Mizuuchi, M.H. O'Dea, H. Ohmori, and J. Tomizawa, DNA gyrase and DNA supercoiling. *Cold Spring Harbor Symp. Quant. Biol. 43*: 35–40 (1978).
101. Lilley, D.M.J., The inverted repeat as a recognizable structural feature in supercoiled DNA molecules. *Proc. Nat. Acad. Sci. USA 77*: 6468–6472 (1980).
102. Mizuuchi, K., M. Mizuuchi, and M. Gellert, Cruciform structures in palindromic DNA are favored by DNA supercoiling. *J. Mol. Biol. 156*: 229–243 (1982).
103. Panayotatos, N., and R.D. Wells, Cruciform structures in supercoiled DNA. *Nature 289*: 466–470 (1981).
104. Singleton, C.K., and R.D. Wells, Relationship between superhelical density

and cruciform formation in plasmid pVH51. *J. Biol. Chem. 257*: 6292–6295 (1982).

105. Shen, L.L., Quinolone antibacterials: structure, activity and mechanism of inhibition of DNA gyrase, (abstract No. 47) Cold Spring Harbor Meetings on the Biological Effect of DNA Topology (1986).
106. Schlemperer, E., L.A. Mitscher, and S.R. Gollapudi, unpublished.
107. Sharma, P., R. Zavod, L. Shen, and L.A. Mitscher, unpublished.
108. Shen, L.L., M.G. Bures, D.T.W. Chu, and J.J. Plattner, Quinolone-DNA interaction: how a small drug molecule acquires high DNA binding affinity and specificity. In *Proc. 23rd Jerusalem Symp. on Molecular Basis of Specificity in Nucleic Acid-Drug Interaction*, Pullman, B. ed., Dordrecht, Holland, 495–512 (1990).
109. Shen, L.L., J. Baranowski, T. Wai, D.T.W. Chu, and G. Pernet, The binding of quinolones to DNA: Should we worry about it? In *International Telesymposium on Quinolones*, Fernandes, P.B. ed., J.R. Prous, Barcelona, Spain, 159–170 (1989).

10

Mechanism of DNA Damage by Neocarzinostatin and Other Bicyclic Enediyne Antibiotics

Peter C. Dedon* and Irving H. Goldberg
Harvard Medical School, Boston, Massachusetts

I. INTRODUCTION

Recent intense interest has focused on a relatively new group of naturally occurring antitumor antibiotics isolated from the eubacterial actinomy-

---

**Present affiliation:* Massachusetts Institute of Technology, Cambridge, Massachusetts.

cetes. The main feature of the family, which includes neocarzinostatin (NCS), calicheamicin, esperamicin, and dynemicin, is a large ring system containing an enediyne motif that undergoes electronic rearrangement to form a bicyclic diradical species, analogous to the Bergman rearrangement in the formation of 1,4-dehydroaromatic compounds (1). In spite of significant structural diversity, these agents appear to possess one common pathway that leads to mutagenesis and cytotoxicity: the activated diyl form of the drug abstracts hydrogens from the deoxyribose sugar backbone of DNA, leading to several specific types of DNA damage, including abasic sites and strand breaks.

The enediyne antibiotics provide a unique opportunity for study in the field of nucleic acid targeted drug design. The diversity of the structural embellishments within the enediyne family presents the drug designer with numerous examples of the elements needed for molecular recognition within the minor groove of DNA, while the commonality of their mechanisms of action and the chemistry of the DNA damage they produce attest to the importance of the conserved enediyne core structure. In this chapter, an attempt has been made to present the reader with the current knowledge about the DNA-damaging properties of enediyne antibiotics. The biochemical findings and methodologic details presented illustrate the important role of structure/function studies in not only the design of new drug molecules, the need for which is highlighted by the limited clinical utility of enediynes due to their toxicity, but also in the development of chemical probes of DNA and chromatin structure. Significant progress has been made toward the synthesis of enediyne analogs with the ultimate goal of greater selectivity of drug action.

The first portion of this chapter will deal with the chemistry, biochemistry, biology, and clinical utility of NCS, followed by similar coverage for the other naturally occurring members of the enediyne family. Later, the current theories concerning the chemical mechanisms controlling activation of enediyne molecules will be addressed along with recent progress in the synthesis of novel enediyne agents with DNA cleaving activity.

II. NATURALLY OCCURRING ENEDIYNES

A. Neocarzinostatin

1. *Structural Features of NCS*

Isolated in 1965 from *Streptomyces carzinostaticus* (2), neocarzinostatin is the most thoroughly studied of the enediyne antibiotics (for a recent review, see Ref. 3). The holoantibiotic consists of a protein that noncovalently binds to and protects a labile bicyclic chromophoric molecule (4). The acidic protein portion of neocarzinostatin (apo-NCS; M_r 11,000) consists of 113 amino acids for which the primary structure was originally determined using Edman degradation (5). Corrections of the primary structure have since been made by several groups utilizing enzymatic digestion and mass spectrometry in conjunction with standard Edman methods (6–8). The secondary structure of the protein consists of almost entirely β-sheets estimated to be made up of 8–10 total β-strands as determined by NMR, x-ray diffraction, circular dichroism, and Fourier transform-infrared (FT-IR) spectroscopy (9–11). The overall structure consists of a several-stranded antiparallel β-sheet that forms the main body of the protein with a 3–4-stranded region forming a binding cleft for the chromophore (10,11). Other *Streptomyces* antitumor antibiotics, such as auromomycin, possess similar protein-chromophore systems, and the protein components share significant sequence homology and structure (10,11). Crystal structure analysis of macromomycin, the apoprotein portion of auromomycin, reveals that, while the apoproteins of the various *Streptomyces* drugs share similar structures, the binding clefts possess unique amino acid side chains thought to be important for the binding and stabilization of the individual chromophores (12).

Neocarzinostatin chromophore (NCS; M_r 659), in which all the DNA damaging activity lies, is composed of four functional portions (Fig. 1): the highly strained chromophoric core containing the enediyne and epoxide moieties, a naphthoate moiety, an amino sugar, and a cyclic carbonate residue (13–17). The DNA-damaging activity was assigned to the chromophoric core of NCS in studies that assessed DNA fragmentation by acid solubility, gradient analysis, and the relaxation of supercoiled DNA (18–22), and, more recently, by spectroscopic examination of the chromophore and its degradation products (23). Using similar techniques, the damage was shown to be dependent on oxygen and added thiol or sodium borohydride (20–22,24). Changes in the viscometric and spectrophotometric properties of DNA in the presence of NCS indicated that the naphthoate residue was responsible for the intercalative activity demonstrated by NCS (18,25,26), while the positively charged amino sugar (D-fucosamine) was proposed to interact with the negatively charged sugar-phosphate backbone of DNA (18). The cyclic carbonate moiety appears to take part in the ability of the drug to enter cells, since analogs lacking this group show greatly reduced cytotoxic activity in vivo while retaining similar DNA-damaging activity in vitro (17,27). The epoxide moiety of the chromophoric core appears to have an important influence on the DNA cleavage activity of NCS (28). When the epoxide is opened as the diol monoethyl ether derivative,

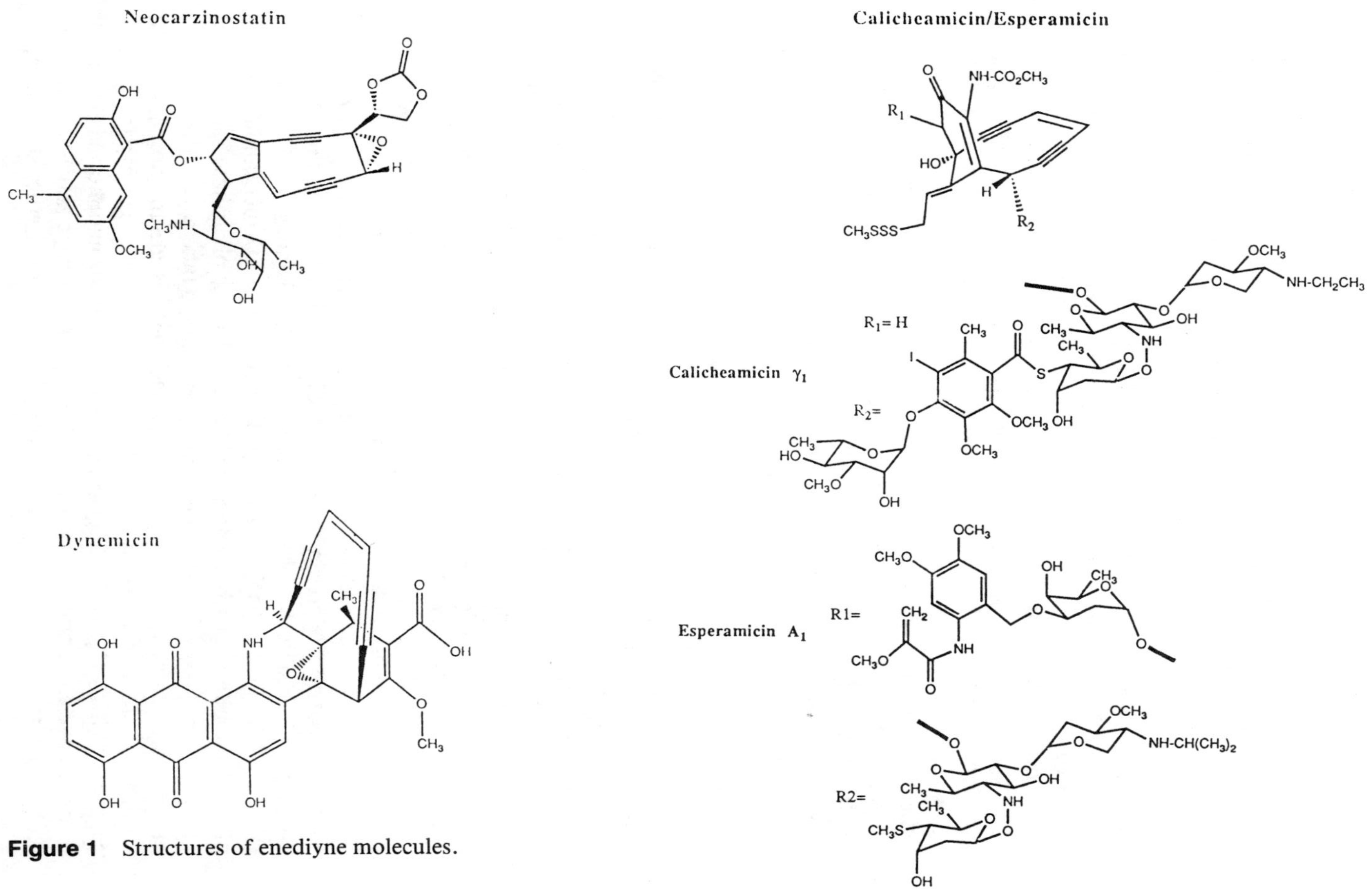

Figure 1 Structures of enediyne molecules.

the stability of the drug is increased, but its DNA cleavage activity is dramatically reduced. The chlorohydrin derivative possessed one-half the DNA-damaging activity of the epoxide. DNA binding properties of both derivatives and the parent NCS were similar, suggesting that the epoxide is important in the activation of NCS but not in its DNA binding activity.

The NCS chromophore is an unstable molecule under physiological conditions and undergoes degradation, in the absence of thiols, to form a product with different spectral characteristics than the thiol-inactivated chromophore (29). This product does not damage DNA and appears to have incorporated oxygen into its structure during the degradation reaction (29,30). The native chromophore is stabilized by the apoprotein (4,30), but undergoes inactivation upon exposure to heat, thiols, UV light and pH $>$ 4.8 (30). Chromophore inactivation is inversely related to methanol concentration, with stable, long-term storage achieved in 100% methanol at $-70°C$ (30).

The detailed picture of the structure of the NCS chromophore and its inactivated forms has come from several studies. The structures of the naphthoate, amino sugar, epoxide, and cyclic carbonate portions of NCS and their relationships to the undefined core were originally determined by NMR and mass spectral analysis (13–17). In the absence of more detailed structural data, it was originally proposed that a drug-based carbon-centered radical was the DNA-damaging agent rather than a molecular oxygen-mediated radical species (31). This conclusion was based on the biochemical findings that hydrogen from DNA became covalently bound to the NCS chromophore (31) and that hydrogen abstraction occurred in the absence of molecular oxygen (32). It was also shown that the 12 carbon center of the chromophore reacted with methyl thioglycolate to form a 1 : 1 adduct with the additional incorporation of two hydrogen atoms into the core (or deuterium atoms if sodium borodeuteride was employed) (14). Solution of the actual enediyne structure of the chromophoric core was later accomplished spectrometrically by Edo et al. (15). Myers related this structure of the chromophoric core to the structural information obtained by Hensens et al. (14) for the thiol- and borohydride-inactived forms and proposed a pathway leading to the reactive, DNA-damaging form of NCS as shown in Figure 2 (33–35). According to this model, attack by the thiol at C12 causes a Bergman-type rearrangement (1) of the enediyne system to form the radical centers at C-2 and C-6 (C in Fig. 2) via a cumulene intermediate (B of Fig. 2) (33–35). The structures of the thiol-conjugated inactive form and the borohydride-inactivated form were elucidated by spectroscopic methods (14,33,34,36). Further support for this cumulene-intermediate model comes from Myers' work with synthetic enediyne/cumulene systems that will be discussed later (37–39).

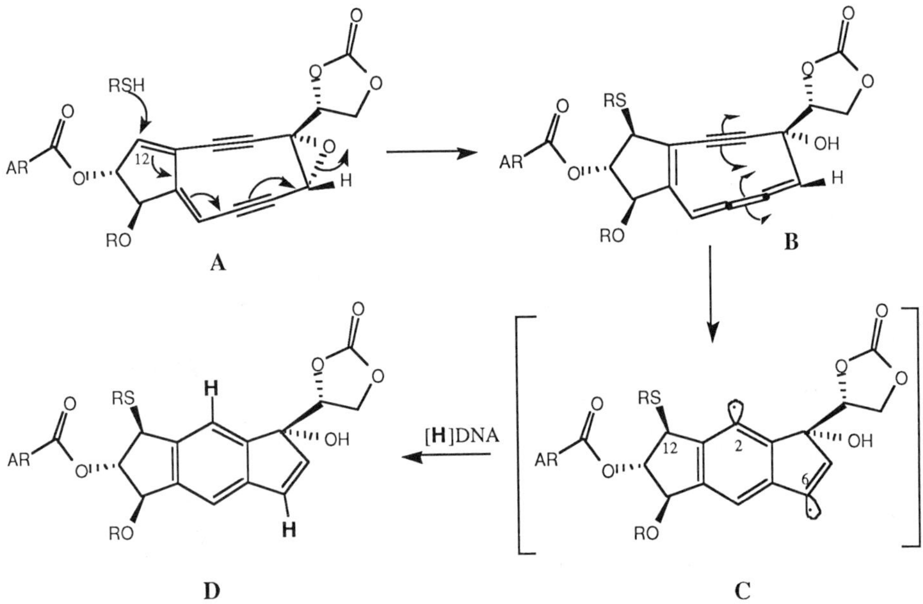

Figure 2 Proposed mechanism of activation of neocarzinostatin. [Based on Meyers (33).]

2. *Biochemistry of DNA Damage Caused by NCS*

DNA damage caused by NCS occurs by abstraction of hydrogen atoms from the ribose backbone by the radicals at positions C-2 and C-6 of the activated chromophore following intercalation of the naphthoate residue and positioning of the bicyclic ring system in the minor groove. Direct evidence for the incorporation of hydrogen atoms from DNA into positions C-2 and C-6 of the drug comes from two sources. Chin et al. (40) demonstrated that three deuterium atoms were incorporated into the chromophore at positions C-2, C-6, and C-12 when NCS was activated with sodium borodeuteride in the absence of DNA. When DNA was present, two of the deuterium atoms were replaced by hydrogen, at positions C-2 and C-6, thus implicating DNA as the source of the hydrogens present in the reacted drug molecule. Additionally, as will be discussed in the "NCS/DNA Interaction Models" section, Meschwitz and Goldberg showed that when the T of a GCT sequence is labeled with deuterium at the C-5′ position, the deuterium ends up in the drug molecule at position C-6 (41).

Additional evidence for this proposal comes from studies analyzing the chemistry of NCS-mediated DNA damage and molecular modeling studies.

The major method employed to study the biochemistry of the NCS-mediated DNA damage involves the use of polyacrylamide gel electrophoresis to compare the mobilities of DNA fragments with different sugar residues attached to the ends of the cleavage sites. The chemical identity of many of these breakdown products was originally accomplished by chemical, spectroscopic, and enzymatic analysis of the products. These defined chemical species, when present on the end of a DNA fragment, were then noted to produce variation in the mobility of the DNA fragment compared to phosphate-ended DNA fragments generated by standard Maxam-Gilbert chemistry (42). Thus, a nucleoside 5′-aldehyde present on the 3′-end of a DNA fragment generated by NCS-mediated cleavage (Fig. 3) will migrate two to four base pairs (depending on the length of the DNA fragment) more slowly than the phosphate-ended fragment, while a 3′-phosphoglycolate-ended fragment (Fig. 3) will migrate slightly faster than the same phosphate-ended DNA.

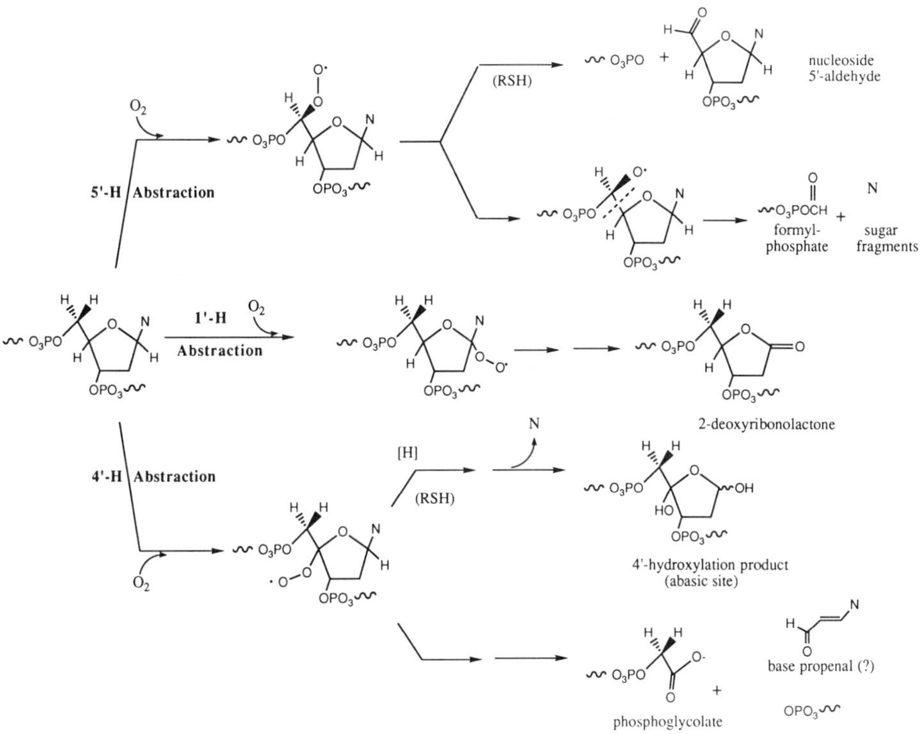

Figure 3 Proposed mechanisms of DNA damage produced by neocarzinostatin abstraction of deoxyribose hydrogen atoms.

Several pieces of evidence point to the minor groove as the location for the binding of the chromophore. First, it was demonstrated that bulky substituents in the major groove, such as halogen atoms and glucosyl moieties, did not interfere with the spectroscopically identified binding activity of NCS (43). However, occupation of the minor groove by methylation of the N-3 of adenine or competition with the minor-groove-binding agents netropsin and distamycin resulted in decreased binding of NCS to DNA (43). More recent DNA cleavage studies, to be discussed later, show a 3′-stagger to the cleavage sites on opposite strands of DNA in bistranded lesions (44,45), which indicates a minor groove location for the agent responsible for cutting (46).

Single-Strand Lesions. The chemistry of the DNA damage produced by NCS is similar to that of bleomycin (47) and iron-EDTA (46) in that the damage is caused by free-radical abstraction of hydrogens from the sugar phosphate backbone of DNA. Unlike these agents, however, the active DNA-damaging species is not a form of reactive oxygen (hydroxyl radical or its equivalent), but a radical form of NCS itself. The NCS damage also differs from that of x-rays, since damage to the nitrogen bases is minor, if it occurs at all (48). The major DNA lesion produced by NCS involves single-strand (SS) breaks that occur when the radical form of the drug abstracts hydrogen from the C-5′ position of the deoxyribose sugar of thymidine, and to a lesser extent adenosine, as shown in Figure 3 (49). The product resulting from C-5′ hydrogen abstraction is a nucleoside 5′-aldehyde (Fig. 3), which was characterized initially by mass spectroscopy and chemical derivatizations (50). The aldehyde is identified conveniently on sequencing gels by its breakdown to phosphate-ended DNA upon treatment with alkali, its characteristically slow migration, and the small increase in migration when the aldehyde is reduced with sodium borohydride (51). While there is a base preference for SS breaks ($T > A \gg C > G$), there is no clear-cut sequence specificity (52–54). Addition of molecular oxygen to the radical generated at the 5′ carbon (29,51) and subsequent reduction of the peroxide (29) results in formation of DNA fragments with predominantly (>80%) nucleoside 5′-aldehydes, with the remaining 20% composed of an unstable formylphosphate-ended DNA fragment (Fig. 3) (55,56).

Under anaerobic conditions, molecular oxygen is not available to participate in the sugar-damaging reactions so that strand scission is markedly reduced (57). Nitroaromatic radiation sensitizers such as misonidazole (Fig. 4), however, can substitute for oxygen and restore the DNA damage caused by NCS (57). The major DNA product under these conditions consists of a highly reactive 3′-formylphosphate species as shown in Figures 3 and 4 (55). Misonidazole adds to and transfers oxygen to the 5′ carbon radical as

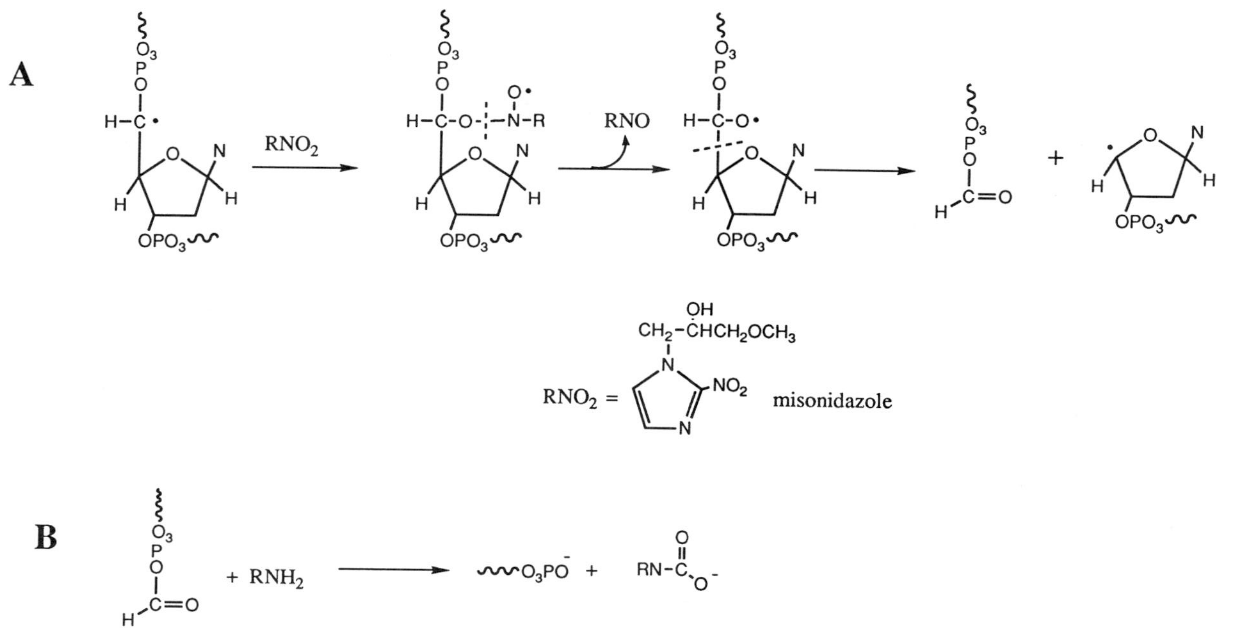

Figure 4 (A) Proposed mechanism of involvement of misonidazole (RNO_2) in neocarzinostatin-induced damage at the C5′ of deoxyribose. (B) Proposed transfer of the formyl residue from formylphosphate to amino nucleophiles.

demonstrated by ^{18}O-labeling of the misonidazole and mass spectrometric analysis of the reaction products (58). Subsequent chemical reactions result in the formation of a 3′-formylphosphate, which was detected by its reaction with nucleophilic amines, such as tris buffer, which generated formylamino bonds and a 5′-phosphate as shown in Figure 4B.

Another lesion enhanced under anaerobic conditions involves the formation of covalent NCS-deoxyribose adducts, presumably at the sites of radicals formed by NCS-mediated hydrogen abstraction (59–61). Nuclease digests of poly(dA·dT) treated with NCS under anaerobic conditions released dTAT trinucleotides that upon subsequent chromatographic analysis were shown to contain an NCS adduct at the C-5′ position of the A residue. The extent of adduct formation and the type of adduct formed varied with the thiol used to activate NCS, with dithiothreitol resulting in the highest level of adenosine deoxyribose adduct formation, and was inversely related to the amount of DNA strand breakage. Sites of adduct formation could be sequenced as exonuclease-resistant sites in defined sequence DNA (62).

Bistranded Lesions. Compared to SS lesions, bistranded DNA damage produced by NCS is minor, but it may be of greater significance with respect to cell lethality (63). Studies performed to date indicate two types of bistranded lesions: a direct double-strand (DS) break occurring predominantly at AGT·ACT sites (44) and a lesion involving an abasic site at the C of AGC accompanied by a direct strand break at the T of the complementary GCT (45,64). The abasic site at the C of AGC results from abstraction of the C-1′ hydrogen of the C residue (Fig. 3). Evidence supporting this conclusion comes from isotope labeling studies in which tritium used to label the C-1′ position of deoxyribose was found covalently bound to the NCS chromophore after reaction (65,66). There is also a significant deuterium isotope effect ($k_H/k_D \sim 4$) at this position as indicated by a reduced level of NCS-mediated DNA cleavage when deuterium substitutes for hydrogen at the C-1′ position (67) (Fig. 3). Abstraction of the C-1′ hydrogen results in the formation of an alkali-labile 2-deoxyribonolactone at the abasic site (Fig. 3) (65–67), which serves as a mutational hot spot for GC to AT transitions in *E. coli* (64) and the *cI* gene of lambda phage (68). The direct strand break at the opposing T residue results from removal of the C-5′ hydrogen, as shown in Figure 3 (45,65). The complete bistranded lesion at AGC·GCT is shown in Figure 5C.

The direct DS breaks produced by NCS, like the AGC·GCT bistranded lesion, occur with a trinucleotide specificity and involve predominantly GT steps. The identification of the direct DS breaks was accomplished by resolving products from NCS-treated, uniquely ^{32}P-end-labeled restriction fragments on nondenaturing polyacrylamide gels. The DS break fragments were eluted from gel pieces, and the base sequence at the cleavage site was

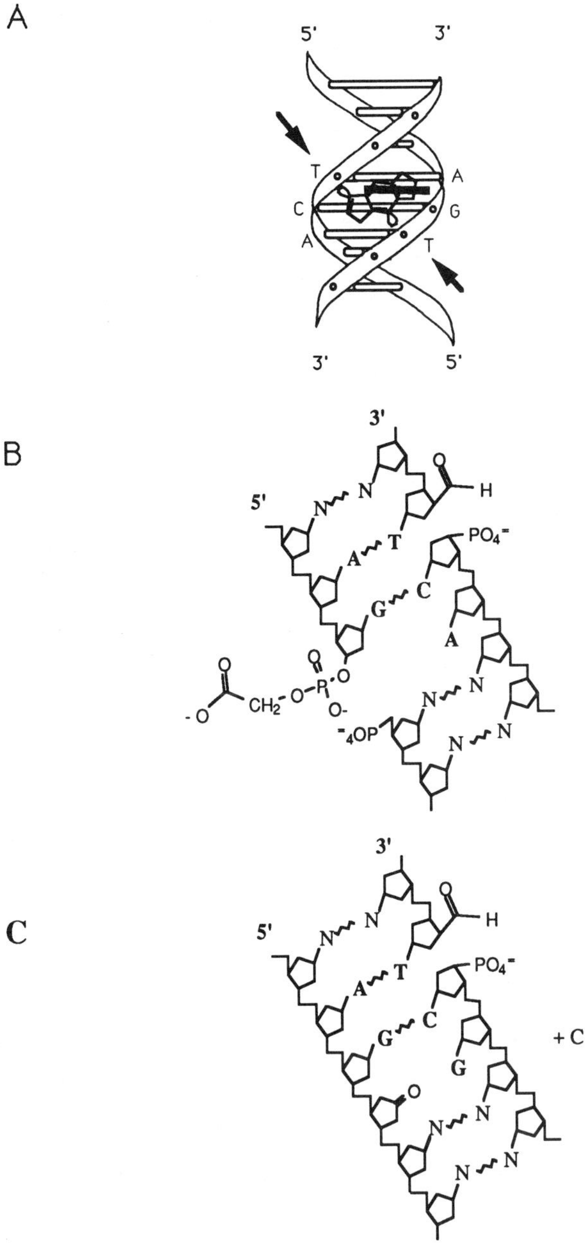

Figure 5 Bistranded lesions produced by neocarzinostatin. (A) Proposed binding of neocarzinostatin to the AGT·ACT sequence, based on the AGC·GCT modeling studies performed by Galat and Goldberg (75) and Meschwitz and Goldberg (41). (B) and (C) Stylized models of bistranded lesions at AGT·ACT and AGC·GCT, respectively, show examples of the biochemistry of the damage and demonstrate the 3′-directed 2-bp stagger of the cleavage sites.

identified by comparison to Maxam-Gilbert standards on a sequencing gel. The DS cleavages occur mainly at AGT·ACT sites (at the underlined residues), and to a lesser extent at TGT·ACA and TCT·AGA sites, when glutathione serves as activator (44). The chemical structures of the cleavage sites, again determined by comparing mobilities of DNA fragments on sequencing gels, revealed a predominance of C-4′ chemistry at the T of AGT and C-5′ chemistry at the T of ACT in the DS breaks. Removal of the C-4′ hydrogen is demonstrated by the presence of a phosphoglycolate residue at the 3′-end of the break (Fig. 3), as has been described for bleomycin-induced DNA damage (47), and by the deuterium isotope effect ($k_H/k_D \sim 4$) noted when the C-4′ hydrogen is replaced by deuterium at GT steps (69,70). In the original observation of C-4′ chemistry at GT steps, Saito and co-workers noted the predominance of 4′-hydroxylated abasic species rather than 3′-phosphoglycolate ($<3\%$) at the cleavage site (Fig. 3) (71). The C-4′ hydroxylation products were identified by their lability in alkali and by their reaction with hydrazine to form a stable pyridazine derivative (72). This altered distribution of C-4′-mediated breakdown products probably results from the use of a neutral thiol (4-hydroxythiophenol) as activator, which has been noted to shift the partitioning of the peroxyradical-containing sugar residue toward the 4′-hydroxylation product and away from the 3′-phosphoglycolate species (Fig. 3) (69,70).

Abstraction of the C-5′ hydrogen at the T of ACT in DS breaks is again confirmed by the presence of a nucleoside 5′-aldehyde at the 3′-end of the cleavage. The chemistry is different for SS lesions at either AGT or ACT, however, with removal of mainly the C-5′ hydrogen, perhaps as a result of a different drug binding mode as addressed below. As for the AGC·GCT site, the complete bistranded lesion at AGT·ACT is shown in Figure 5B.

The presence of two carbon-centered radicals on opposite sides of the NCS molecules begs the hypothesis that the bistranded DNA lesions are the result of hydrogen abstractions by a single molecule of NCS. Indirect support for this notion comes from two sources. First, in both types of bistranded lesions, the cleavage sites are staggered two base pairs in a 3′-direction from each other when the sequence is examined linearly, as shown in Figures 5B,C for the AGT·ACT and the AGC·GCT sites, respectively. This positioning, along with the one base pair spacing caused by intercalation of the naphthoate moiety, places the cleavage sites directly across the minor groove from each other in helical B DNA, as shown in Figure 5A, and raises the possibility that a single molecule of drug can have access to both of the abstracted hydrogen atoms.

Role of the Thiol in NCS Activity. The thiol used to convert NCS to its active form plays a complicated role in the NCS-mediated DNA damage by serving not only as activator, a role that can be substituted with $NaBH_4$, but also as reductant of the DNA damage intermediates. Nearly any thiol can serve to activate NCS, but the thiol of greatest biologic significance is glutathione (γ-glutamylcysteinylglycine). This tripeptide is found in most if not all living organisms and is present in mammalian cells at concentrations of 1–5 mM (see Ref. 73). As determined by structural analysis of the reaction products (14,34), the known dependence of NCS DNA damage on thiol concentration, and the stoichiometry of thiol consumption during drug activation (29), the formation of the reactive NCS intermediate probably depends on nucleophilic attack by the thiol at position C-12 of the drug (Fig. 2). Presumably, the more nucleophilic thiol will cause activation at a greater rate. This activity must be balanced by the ability of the thiol to quench the radical form of the drug by direct donation of hydrogen to the radical centers, or by nucleophilic attack at C-6 of the cumulene intermediate form of the drug (structure B in Fig. 2), which presumably results in the formation of a C-6′–C-12 bis-adduct of the thiol with inactivated NCS (35).

Additionally, the activating properties of the thiol are complicated by evidence that the thiols act as reductants of both the peroxy radical and hydroperoxy intermediates that form when molecular oxygen adds to the carbon-centered deoxyribose radicals produced by NCS-mediated hydrogen abstraction. Evidence to support this hypothesis comes from several sources, including studies of the stoichiometry of oxygen uptake and sulfhydryl consumption in the activation and DNA damage reactions of NCS (29). One mole of thiol is incorporated into the drug molecule while at least another appears to be consumed in the breakdown of the deoxyribose sugar (29).

Further support for the reductive role of the thiol comes from studies of the DNA damage product distribution that occurs with different thiols. Consider first the attack by NCS at the C-5′ of deoxyribose (Fig. 3). The peroxy radical is probably reduced to a hydroperoxy species and finally to a hydroxyl group by the thiol, with subsequent cleavage of the phosphodiester upon formation of the nucleoside 5′-aldehyde. Similarly, NCS-mediated C-4′ chemistry results in the formation of a C-4′ peroxy radical that may be reduced to the hydroperoxide by the thiol (Fig. 3). From this point, the reaction can proceed along two pathways, depending on the thiol: (1) formation of the 3′-phosphoglycolate-ended fragment; and (2) reduction to the C-4′ hydroxyl group with subsequent formation of an abasic site (47,69–71) (see Fig. 3). It has been suggested (70) that the thiol reduces the hydroperoxide to a C-4′ hydroxyl group, which results in the formation of the alkali-labile abasic site observed by Saito's group (71), while the 3′-phosphoglycolate pathway results from some other reaction of the hydroperoxide, conceivably 2′ intramolecular hydrogen abstraction

(70). Cleavage between C-3′ and C-4′ following oxyradical formation, as proposed in the misonidazole-dependent reaction (see above and Ref. 70), with the formation of a 3′-phosphoglycolate end also remains a possibility. The nature of the thiol has been shown to influence this product distribution. Saito and co-workers found a predominance of C-4′-hydroxylation product when 4-hydroxythiophenol was employed as activator (71), as was the case when other neutral thiols were used (70). However, acidic thiols cause a shift to the formation of 3′-phosphoglycolate-ended cleavage fragments (70). Preliminary evidence indicates that there is also an effect of the thiol on product distributions in DS breaks as well (P. C. Dedon, Z.-W. Jiang, and I. H. Goldberg, unpublished observations).

Finally, the thiol may also effect the drug's selection of C-4′ and C-5′ positions for hydrogen abstraction, although in an as yet unpredictable manner (70).

NCS/DNA Interaction Models. Molecular modeling and dynamics studies have been performed for the interaction of the postactivated form of NCS with several oligomeric DNA sequences. The studies were performed with the aid of several computer programs: the energy calculations for modeling, minimization, and dynamics were performed using the CHARMM program (74); coordinates for the oligonucleotide DNA were derived from Arnott and Hukins with the CORGEN program (see Ref. 75); and docking of the drug molecule with the DNA was performed with the aid of the HYDRA graphics system (see Ref. 75). The dynamics studies were performed to examine the stability of the optimized models and involved simulated heating of the structure and calculation of the fluctuation of interatomic distances about a mean value (75).

These modeling and dynamics studies support the concept of DS breaks mediated by a single drug molecule (75; Galat and Goldberg, unpublished observations). The oligomer 5′GAGCG·5′CGCTC contains the previously discussed trinucleotide consensus sequence for a bistranded lesion, AGC·GCT. Stable, low-energy structures can be rationalized that intercalate the naphthoate moiety between the A·T and G·C base pairs with the diradical core oriented toward the 3′-end of the AGC-containing strand (see similar model for AGT·ACT in Fig. 5A). With glutathione as the thiol activator, this orientation places the C-6 radical close to the C-4′ and C-5′ of thymidine (3.32 and 3.31 A, respectively) and the C-2 radical close to the C-1′ and C-4′ of cytosine (4.10 and 4.00 Å, respectively). In these modeling studies, glutathione was noted to push the C-2 radical closer to the C-1′ position than in the model without thiol present, a phenomenon attributed to the interaction of glutathione with the AGC-containing strand that forced the hydroindacene ring of the inactivated form of the drug (the form used for the modeling studies) deeper into the minor groove (75).

Examination of models of B-DNA suggests that the *S*-hydrogen of the C-5′ position, which points into the minor groove, is the one abstracted by NCS, rather than the *R*-hydrogen, which points toward the major groove.

Chemical evidence to support this model comes from the recent studies of Meschwitz and Goldberg (41), in which deuterium was substituted for both hydrogen atoms at the C-5′ of the T of AGC·GCT in a self-complementary oligonucleotide. Reaction of the oligonucleotide with NCS and subsequent proton nuclear magnetic resonance (^{1}H-NMR) analysis of the drug product revealed the presence of deuterium only at the C-6 position in the drug. This finding agrees with the molecular modeling studies.

Single-strand cleavage at the T residue, a lesion unaccompanied by the formation of an abasic site at the C of the complementary AGC, can be rationalized by two different mechanisms: different drug binding modes or subtle shifts in the positioning of the radical centers of the drug bound in one orientation in the minor groove. Molecular modeling predicts a stable binding mode in which the naphthoate intercalates between the G·C/C·G step, and the active core is oriented toward the 3′-end of the GCT-containing strand (75). This alternative places the C-2 of the drug close to the C-5′ hydrogen of the T of GCT. However, in the deuterium-labeling studies of Meschwitz and Goldberg described previously, SS breaks at the T of GCT were the predominant lesion that occurred in the AGC·GCT-containing oligonucleotide, and yet deuterium was incorporated only at the C-6 position of the drug (41). It is possible that the selection for SS breaks or bistranded lesions occurs as a result of minor-groove geometry placing the C-6 radical center closer to its target C-5′ hydrogen on one strand more frequently than the C-1′ on the other strand. Alternatively, the selectivity could result from differing chemical reactivities of the deoxyribose hydrogens, with the C-1′ hydrogen of the C of AGC less readily removed or the oxidation more easily repaired by thiols, for example, than the C-5′ hydrogen abstraction. The SS and DS lesions at AGT·ACT may be similarly explained.

A second stable model involves the same intercalation site at the GC·CG step, but places the active core perpendicular to the helix axis so that the C-6 radical now lies near the C-5′ hydrogen of the T of GCT at a distance of 3.47 Å (75). This could account for the deuterium incorporation noted previously. Another modeling study performed by Chen et al. (76), before the inactivated structure of NCS was known, also places the enediyne core perpendicular to the minor groove in a way that could account for SS breaks.

While the drug's selectivity for sugar hydrogens on opposite DNA strands is an important factor in the distribution of lesions induced by NCS, there is evidence for selectivity in the abstraction of different hydrogens from a deoxyribose sugar on one strand. If it is assumed that NCS interacts with GT steps, such as the AGT·ACT site, in a manner similar to the AGC·GCT model, then the radical centers of the drug are predicted to be in close proximity to both the C-4′ and C-5′ hydrogens of a single sugar residue (75). It is possible that the radical centers of the activated drug can shuttle between neighboring hydrogens of a single deoxyribose molecule without changes in the intercalation site. There should be sugar breakdown products representative of both sites of oxidation if this is true. Chemical evidence for this shuttling hypothesis has been derived again from deuterium labeling studies, in this case studies of deuterium isotope effects, in which NCS-mediated hydrogen abstraction at the T of a GT step in an oligonucleotide shifted from C-4′ to C-5′ when hydrogen was replaced by deuterium at the C-4′ position and vice versa (69,70). In these experiments, however, total cleavage at the GT step was less for C-4′ deuterium-labeled oligonucleotide than for C-5′ labeled sites, complicating the interpretation of the drug binding modes. As well, there is preliminary evidence for an influence by the thiol on the selection of the abstracted hydrogen (70).

The above explanations for the inter- and intrastrand reactivities of the radical centers all assume identical characteristics for the two radicals. Alternatively, the frequency of cleavage between strands and the selection of deoxyribose hydrogens on the same strand may reflect the different characteristics of the two radical centers. One of the radicals may be more reactive toward hydrogen abstraction than the other, biasing the frequency of cleavage toward a preponderance of SS breaks on one strand relative to the other. Additionally, formation of a carbon-hydrogen bond at one radical center may alter the reactivity of the other radical. To further complicate matters, it has been proposed that the radical center at C-6 (see Fig. 2C) may have significant carbene character by virtue of its relationship with the C-5—C-6 double bond, which would change the C-6 radical from an orientation of planarity with the ring system to one in which the radical is positioned in an orbital perpendicular to the plane of the rings and in the π-system (J. P. Snyder, personal communication). Such an altered orientation of the radical could effect its positioning relative to the deoxyribose hydrogens, in addition to the effect that carbene abstraction of a sugar proton could have on the NCS cleavage spectrum.

3. *Effects of NCS In Vivo*

While less thoroughly studied than the in vitro situation, experiments directed at NCS-mediated DNA damage and changes in DNA synthesis occurring in vivo have yielded several important findings. First, both NCS chromophore and the holoantibiotic cause both SS and DS DNA damage in intact bacterial and mammalian cells (21,63,77–82), isolated nuclei (82–

85), and purified chromatin (83; P. C. Dedon and I. H. Goldberg, unpublished observations). As measured by a variety of electrophoretic and gradient centrifugation techniques, the DNA damage consists primarily of internucleosomal cleavage of the chromatin as indicated by the generation of DNA fragments whose sizes are multiples of ~170 base pairs, the so-called nucleosome ladder. This is similar to the DNA cleavage induced in chromatin by the action of micrococcal nuclease and represents cutting in the linker region between nucleosome cores. Such a cleavage pattern is not apparent when purified DNA is treated with NCS (83,84).

The DNA damage in nuclei and purified chromatin is dependent on added thiol and drug (77,83,84), is complete within 30 sec of drug addition (P. C. Dedon and I. H. Goldberg, unpublished observations), and is not affected by the presence of 10 mM EDTA (P. C. Dedon and I. H. Goldberg, unpublished observations), which suggests that the production of the linker-specific cleavage is not dependent on endogenous cation-dependent endo- or exonucleases. The dependence of NCS-mediated DNA damage on thiols in vivo has been demonstrated by Kappen et al. (77), whereas the thiol dependence of NCS's cytotoxic activity has been demonstrated for intact cells by DeGraff et al., who demonstrated reduced NCS-mediated cytotoxicity and mutagenicity when glutathione levels are reduced in a Chinese hamster cell line (86,87). In addition, this group has demonstrated that it is possible to modulate NCS cytotoxicity by altering glutathione levels selectively in certain human tumor cell lines, cells in which there is a severalfold higher level of glutathione compared to normal cells (88). In the presence of a glutathione synthesis inhibitor, thiol levels drop more in the normal cells than in the tumor cells, which lowers NCS activity in the normal cells relative to the tumor cells (88). Cysteine does not play a major role in NCS activation in vivo (86).

The mechanism of action of NCS appears to be unaffected by the environment of the DNA in vivo, as suggested by the findings of Kappen et al. (77). HeLa cells were treated with NCS and the DNA isolated and analyzed for the presence of nucleoside 5′-aldehyde residues, presumptive evidence for the abstraction of the C-5′ hydrogen of deoxyribose sugar by NCS. The aldehydes were reduced to 5′-hydroxyl groups with sodium borohydride followed by polynucleotide kinase-mediated incorporation of ^{32}P-phosphate. Nucleoside 5′-aldehyde represented only 40% of the 5′-ends generated in HeLa cells versus 80% seen in vitro, a result attributed to incomplete detection of aldehydes rather than an alteration in NCS cutting mechanism. Furthermore, the base preference for cutting was identical to the in vitro situation: $T > A \gg C > G$ (77). This finding was based on HPLC end-group analysis of the ^{32}P-labeled DNA used for aldehyde reduction and on the results of DNA sequencing experiments utilizing the 340-

base-pair alphoid satellite DNA fragment isolated from drug-treated cells by hybridization to the (+) or (−) strand of cloned α-DNA inserted into M-13 (77). However, the quantity of NCS needed to generate similar levels of cutting in the α-DNA was significantly greater for the in vivo α-DNA fragment than for the purified DNA treated in vitro (77), a finding that can probably be attributed to the presence of nucleosomes in vivo and a multiplicity of nucleosome phasing in the α-DNA (89); as mentioned earlier, NCS appears to cut mainly in the linker region of chromatin.

The DNA damage produced by NCS precedes a reduction in DNA synthesis noted in both bacterial (90) and mammalian cells (21,82,91). Povirk and Goldberg (92) found that NCS selectively inhibited DNA replicon initiation but not chain elongation in CHO cells treated with low doses of NCS. The specific inhibition of initiation of new replication forks was noted to be identical to the situation with x-ray-induced DNA synthesis inhibition (92). It was hypothesized that, since the small quantity of DNA damage produced by the low levels of NCS precluded a break at each initiation site, the inhibition of replicon initiation may have been due to relaxation of DNA supercoiling, among other possibilities (92).

In apparent response to DNA damage caused by NCS, DNA repair systems are activated in cells treated with the drug. Neocarzinostatin has been shown to induce unscheduled DNA repair in lymphocytes (93) and to induce repair synthesis, as measured by ^{3}H-thymidine incorporation into the damaged parental DNA strand in HeLa cells and nuclei exposed to NCS (82). Repair of NCS-mediated SS and DS DNA cleavage has also been shown to occur as judged by the disappearance of DNA fragments in filter elution assays (78,79) and gradient centrifugation methods (63). An inability to repair DNA damage may be responsible for the hypersensitivity of ataxia-telangiectasia fibroblasts to the cytotoxic effects, i.e., DNA damage, of NCS, bleomycin and x-rays (78,79,92).

As alluded to earlier for the G : C→A : T transition caused by NCS at the apyrimidinic site at the C of AGC (64), NCS damage of DNA is highly mutagenic in both bacterial and eukaryotic cells (reviewed in Ref. 94). NCS caused extensive single-base substitution mutations in the lambda *cI* gene and in the *E. coli lacI* gene, with all possible types present. A : T→G : C (*cI*) and G : C→A : T (*lacI*) transitions were the most common (94). The lesions themselves are not mutagenic per se. The mutagenicity of NCS in *E. coli* has been shown to be SOS dependent, with requirements for both *recA* (95) and *umuCD* (96) loci in the mutagenic process.

Beerman and co-workers have demonstrated an interesting preference for NCS cutting at a DNase I sensitive site in a glucocorticoid-inducible gene in transformed mouse cells (97). Increased sensitivity toward NCS cleavage at this upstream regulatory region of the gene paralleled the change in DNase I sensitivity when the gene was induced to be transcriptionally active by steroid treatment. Such cleavage preference may reflect the loss of nucleosomal structure or disappearance of a DNA binding protein from this region upon transciptional activation, as indicated by the change in DNase I sensitivity.

B. Calicheamicin

The antitumor antibiotics esperamicin and calicheamicin are recently discovered members of a class of DNA cleaving molecules isolated from *Actinomadura verracosospora* and *Micromonospora echinospora*, respectively (109,110). The basic structures of calicheamicin and esperamicin have been solved by Lee et al. at Lederle Laboratories and Doyle and co-workers at Bristol-Myers, respectively, and both drugs have been shown to possess a novel bicyclo[7.3.1]tridecane structure containing a 3-ene-1,5-diyne system and a methyl trisulfide group (Fig. 1) (109–115). Like NCS, formation of the active intermediate is thought to involve generation of a diradical species initiated by a nucleophilic attack of a thiol as shown in Figure 6A. Unlike NCS, however, the drug molecules contain a trisulfide moiety that undergoes reduction to a single sulfhydryl and performs an intramolecular attack at the bridgehead carbon of the bicyclic ring system (114,116). As will be discussed in some detail later, this Michael-type addition to the double bond relieves ring strain and lowers the activation energy for the ring system to undergo a Bergman-type rearrangement (1). The resulting benzenoid diradical is thought to be the active DNA damaging species.

In analogy to NCS, the DNA-damaging activity of calicheamicin and esperamicin is hypothesized to involve abstraction of hydrogen atoms from the deoxyribose sugar rings when the drugs are situated in the minor groove. Similar to the results cited earlier for NCS (40), Ellestad and co-workers have used deuteriated DNA to demonstrate that the hydrogens present at positions C-1 and C-4 in the reacted drug (Fig. 6A, structure 4) are removed directly from DNA, which provides evidence for the 1,4-dehydrobenzene diradical species as an intermediate in the DNA damage reaction (117). Calicheamicin causes both SS and DS breaks in DNA. The predominant lesions in both types of strand scission involve removal on one strand of the 5′-hydrogen from the 5′-cytidine of TCCT, from which a nucleoside 5′-aldehyde (and possibly a 5′-carboxylate as was also noted with NCS (49)) was noted to form (118) (Fig. 6B). Direct support for this conclusion comes from studies similar to those cited for NCS, in which the 5′-hydrogens of the 5′-cytidine are replaced by deuterium (119). Caliche-

A

B

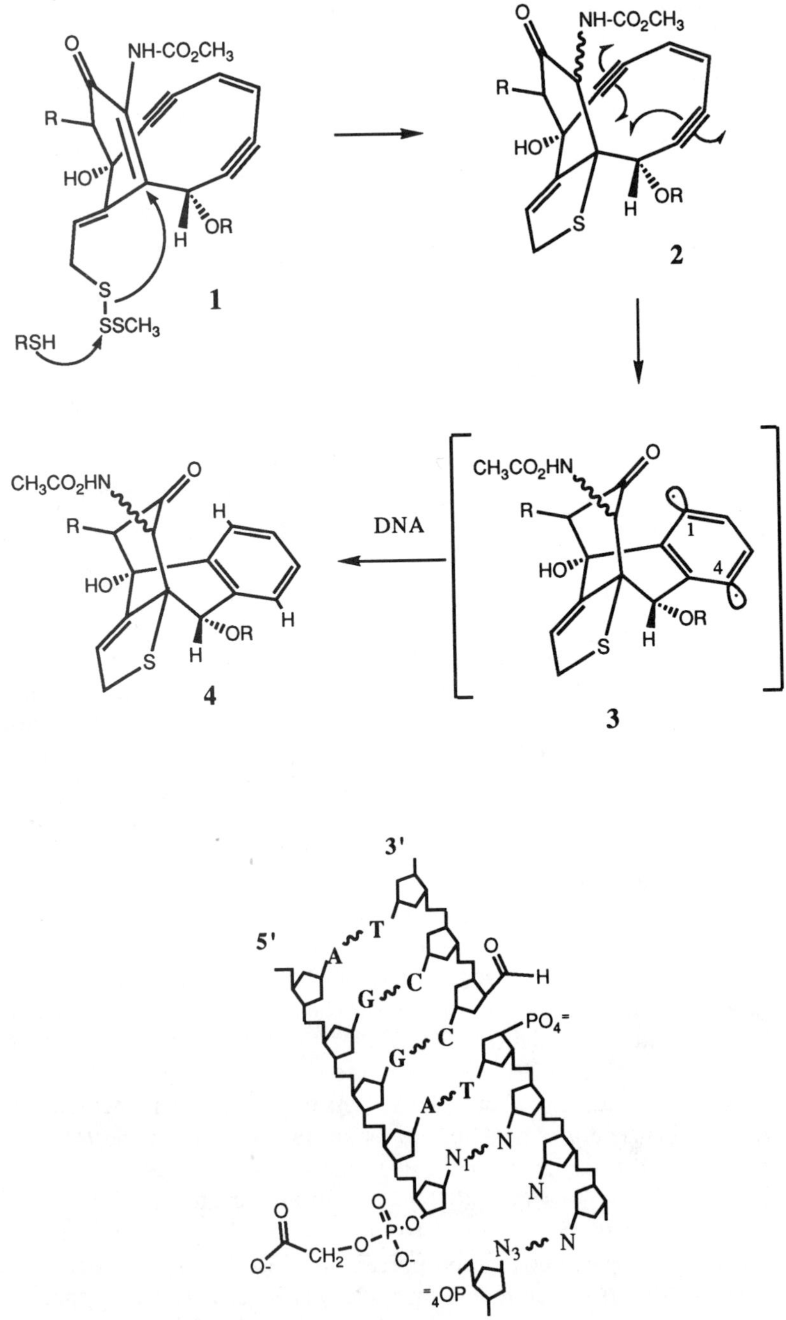

Figure 6 (A) Proposed mechanism of activation of calicheamicin and esperamicin. (B) Stylized model of DNA damage produced by calicheamicin.

amicin treatment of an oligonucleotide containing the deuteriated cytidine results in incorporation of deuterium at only the C-4 position of the inactivated drug, one of the presumed radical centers in the activated form. Cleavage of the other strand results in the formation of a band moving slightly faster than the marker phosphate-ended fragment, suggesting formation of a 3′-phosphoglycolate residue by abstraction of the 4′-hydrogen three nucleotides from the 3′-guanosine of AGGA (120) (Fig. 6B). Other minor sites for DS breaks include short runs of pyrimidines such as GCCT, TCCG, TCCC, TCTC, ACCT, TCCA, and their complementary strands. Again, since cleavage sites on opposite strands are staggered three base pairs in a 3′-direction (Fig. 6B) and netropsin interferes with the cutting activity of calicheamicin, the drugs appear to act via the minor groove of DNA (118). Like NCS, the lesions may also be produced by a single drug molecule abstracting hydrogens across the minor groove from each other. Since NCS binds by intercalation of its naphthoate moiety, it essentially adds the equivalent of one base pair to the interaction site, resulting in lesions separated by two actual pairs rather than the three found with calicheamicin.

Structure-function studies have revealed that both the thiobenzoate-rhamnose and the 4-ethylamino sugar moieties of calicheamicin are important for cutting intensity but not sequence specificity (120), a finding that runs contrary to the prediction that the carbohydrate tails serve as AGGA recognition elements (121). Ellestad's group has proposed a model in which the carbohydrate tail guides the drug into the minor groove, while the carbohydrate-thiobenzoate moiety positions the drug in the groove by hydrogen bonding and quasi-intercalation (120). In this model, the carbohydrate tail is oriented in the minor groove in a 5′-direction to the TCCT site, which positions the C-4 radical center near the cleavage site in the AGGA strand (Fig. 6) (120). Modeling studies performed by Schreiber and co-workers (121), however, suggest that the carbohydrate tail is directed to the 3′-side of the TCCT site, a conclusion supported by the previously mentioned deuterium incorporation from the cytidine deoxyribose into the C-4 radical position (119). Their modeling studies also suggest that the iodine atom on the thiobenzoate ring interacts with the N-2 of guanine residues of the AGGA sequence, an interaction for which there is ample precedent in the literature (121).

C. Esperamicin

Esperamicin and calicheamicin differ in structure only in the carbohydrate tails present at positions R1 and R2 in Figure 1. Instead of a combined thiotrisaccharide-monosaccharide/thiobenzoate moiety at position R1, this group has been split in esperamicin to a 2-deoxy-L-fucose/aromatic ring moiety at position R1 and a thiotrisaccharide at R2. The sequence requirements for DNA damage caused by esperamicin compounds are less clear than for calicheamicin. The drug causes both SS and DS breaks in DNA, with a base preference order of T > C > A > G and a strong preference for pyrimidines (122). These findings are in contrast to the weak base specificity reported in earlier work and the finding that the most intense cleavage occurs at pyrimidines located two base pairs in a 3′-direction from a guanine residue (123). Oligopyrimidine sequences such as CTC, TTC, and TTT are most frequently cleavaged, but there is also moderate cleavage occurring at the pyrimidines of TG and CG sequences (122,124).

Preliminary structure-function studies indicate that the trisaccharide moiety (Fig. 1) is important in the cleavage activity of the drug. Successive removal of sugar residues from the trisaccharide caused a progressive loss of DNA cleaving activity (123). Removal of only the thiomethylhexapyranose portion of the trisaccharide results in diminished cleavage frequency but similar sequence specificity compared to the parent compound (122). The role of the deoxyfucose-anthranilate residue, however, remains unclear. Long et al. argue that this moiety increases the quantity of SS breaks at the expense of DS breaks, but no rigorous quantitative assessment was undertaken in the experiments (123). Sugiura et al. argue that the deoxyfucose-anthranilate group, while reducing cleavage frequency, does not alter sequence specificity (122). The deoxyfucose-anthranilate residue does, however, appear to enhance the uptake of the drug into cells, making esperamicin A the most potent of the analogs in vivo in spite of similar in vitro levels of DNA damage (123).

Again, a minor-groove location for the DNA cleavage activity of esperamicin has been proposed based on two findings (122,124). First, minor-groove agents such as netropsin and distamycin A significantly reduce the DNA damage caused by esperamicin (122,124), while the major-groove-binding endonuclease, *Eco*RI, does not interfere (122). In addition, there is weak evidence for a three base pair stagger in a 3′-direction for cleavages on opposite strands, which suggests a minor-groove location for the cleavage.

D. Dynemicin

Like the esperamicins and calicheamicins, dynemicins, isolated from *Micromonospora chersina*, possess a 1,5-diyne-3-ene core (125,126). However, the trisulfide moiety is absent, and the enediyne core is fused to an anthraquinone molecule (Fig. 1). The mechanism of activation of the drug to the diradical intermediate is proposed to involve reduction of the quninone nucleus (Fig. 7A) to the hydroquinone (Fig. 7B), followed by formation of a quinone methide and opening of the epoxide (Fig. 7C) (127,128). Subsequent addition of a nucleophile (R- in Fig. 7) to regenerate the hydroquinone (Fig. 7D), or tautomerization of the quinone methide to give the

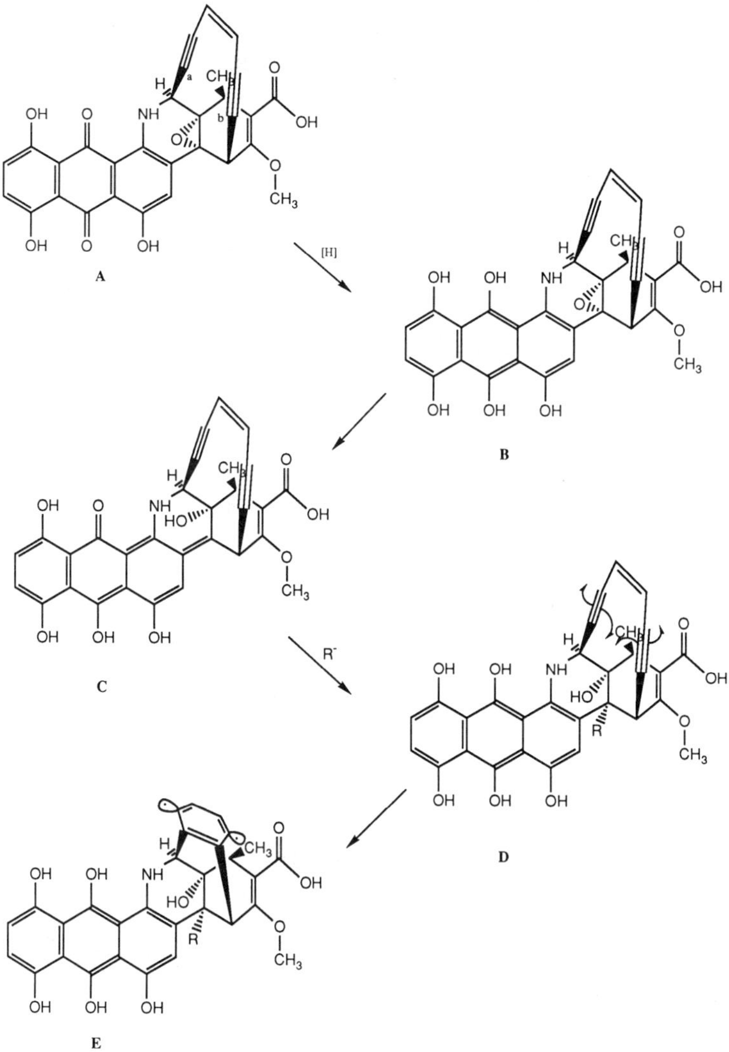

Figure 7 Proposed mechanism of reductive activation of dynemicin.

quinone form, results in a structure that would readily cyclize to form a *p*-phenylene diradical (Fig. 7E).

The epoxide appears to act as a barrier to cyclization by providing enough strain energy to perturb the geometry of the enediyne system and prevent progression to the biradical form (Fig. 7E) (126–128). Molecular mechanics calculations reveal that the change in proximity of the alkyne termini (a and b in structure A, Fig. 7) upon opening of the epoxide ring, from 3.35 to 3.16 Å, allows the Bergman rearrangement to occur readily under physiologic conditions, while the parent molecule is stable to cyclization (128). This dependence of the cyclization reaction on acetylene promixity and transition state energy barriers will be discussed more fully in the next section.

In the presence of reducing agents such as NADPH and thiols, dynemicin has been shown to cleave DNA with a weak sequence preference for the 3′-side of purine bases, although no information is available concerning the frequency of SS and DS breaks or the identification of DS break sites (124,129). A minor-groove location for the enediyne moiety was inferred from the DNA damage interference produced by the known minor-groove binders, netropsin (124,129), distamycin A (124,129), anthramycin (129), while questionable evidence for intercalation of the anthracycline moiety was claimed when actinomycin and adriamycin, two intercalating agents the latter of which contains an anthraquinone nucleus like dynemicin, were shown to interfere with dynemicin-mediated DNA damage (129).

An interesting feature of the mechanism of dynemicin-induced DNA damage involves the ability of low-energy visible light to activate the drug (130). Light appears to be more effective than either NADPH or thiols at activating the molecule and does not change its cleavage specificity. The photoproduct generated is identical to that produced by the reducing agents, namely the aromatized form of dynemicin (structure E in Fig. 7 with hydrogens added at the radical centers) (130).

E. Clinical Utility of Enediynes

Neocarzinostatin has shown some clinical utility against several tumors, but it remains an experimental drug for reasons of significant toxicity. Early work showed that NCS possessed antitumor activity against several animal tumor models (2,98,99), and subsequent studies in humans demonstrated some effectiveness against acute leukemia but limited efficacy against solid tumors of the stomach, colon, kidney, and bladder (100–103). The drug is administered by intravenous or intraperitoneal injection and, in a rat model, is distributed mainly to the kidney, bladder, ovary/uterus, bone marrow, intestine, lung, skin, stomach, and muscle (104). It is excreted mainly in the urine (104). In humans as with many other antineoplastic agents, the dose-limiting toxicity involves bone marrow suppression, along with nausea and vomiting, fever, and, in patients given intermittent therapy, severe allergic reactions (105).

Recent attention has been focused on two new methods for the delivery of NCS: conjugation of the drug to other molecules and the administration of the drug with antidotal compounds in another body compartment. The latter method involves intraperitoneal injection of the NCS along with intravenous, systemic administration of tiopronin, a thiol compound known to inactivate NCS and reduce the drug's toxic side effects (106). Such a treatment protocol produced significant improvement in the antitumor properties by allowing larger doses of the drug to be administered, while producing no increase in the toxicities associated with NCS (106).

Another approach to increasing the selectivity of an enediyne drug for cancer cells involves conjugating the drug to antibodies or polymers. Neocarzinostatin has been linked to a copolymer of styrene and maleic anhydride, yielding a compound with the acronym SMANCS (107). This formulation of NCS resulted in increased activity against a murine tumor model, prolonged plasma half-life, and a reduction in the lethal toxicity of the drug by fourfold (107). Kohgo et al. (108) conjugated NCS to transferrin, an iron transport protein that binds to cell receptors and is internalized, with the effect of increasing the cytotoxicity of the drug ninefold. Much of the conjugate was recycled back to the cell surface unchanged, while about 25% was degraded in lysosomes of the cell (108).

Calicheamicin and esperamicin are among the most potent antitumor agents known, with calicheamicin approximately 4000-fold more active against murine tumor models than adriamicin (115,131). Esperamicin A_1^2 is currently undergoing phase I clinical trials (132). Calicheamicin was recently linked to monoclonal antibodies raised against several tumor antigens (133). This technique allows concentration of the drug in the area of the tumor and reduces toxicity to nontumorous tissue. The drug-antibody conjugates showed superior antitumor activity compared to the drug or antibody alone in nude mice bearing human breast, lung, ovarian, and melanoma cancer cells, with either complete or partial regression of the tumors (133).

While these techniques show promise in improving the therapeutic index of NCS as an antitumor agent, the major research effort toward the preparation of effective enediyne antitumor agents involves rational design based on a thorough knowledge of the chemical and biochemical properties of the drugs. The remainder of this text will address these properties and the efforts to synthesize new and better drug molecules.

III. DESIGN AND SYNTHESIS OF DIRADICAL DNA-CLEAVING MOLECULES

A. Synthesis of Enediyne Compounds

1. Neocarzinostatin-Based Molecules

The full chemical synthesis of an enediyne agent has not yet been achieved, but significant progress has been made toward this end. Two groups have focused on synthesizing the carbocyclic core of neocarzinostatin. Wender and co-workers were the first group to synthesize the unembellished carbocyclic core of NCS, bicyclo[7.3.0] dodeca-1,8-dien-2,6-diyne (structure A in Fig. 8), and employed a ring contraction strategy to prepare the highly

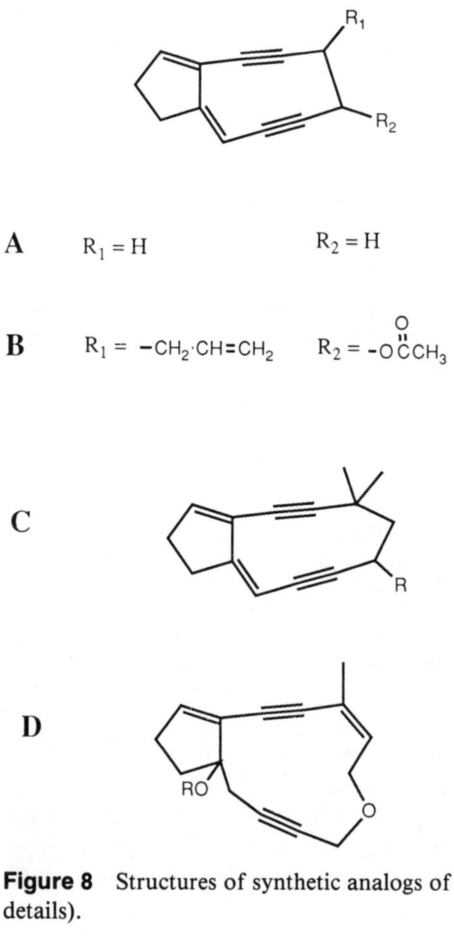

Figure 8 Structures of synthetic analogs of the neocarzinostatin core (see text for details).

strained nine-membered enediyne ring (134). Hirama and co-workers synthesized the less strained bicyclo[8.3.0]trideca-1,9-dien-2,7-diyne by a simple ring closure (structure C in Fig. 8) (135). Wender and co-workers later succeeded in preparing a partially functionalized NCS core (structure B, Fig. 8) that would allow the preparation of the 4,5-epoxide moiety thought to be important for the stability of the parent enediyne (136). An oxacycloalkenediyne related to NCS (structure D in Fig. 8) has been synthesized by Krebs et al. (137).

Several groups have succeeded in the synthesis of acyclic enediyne and eneyne-allene systems related to NCS. Myers and coworkers have developed synthetic strategies for the preparation of noncyclic (Z)-enediynes useful as synthetic precursors and as model compounds to study the formation of diradical agents (37–39). These cumulene compounds represent models of the intermediate proposed for the activation of NCS (33) and undergo cyclization reactions to form diradical species as shown in Figure 9A (38,39). Myers and Dragovich first demonstrated that (Z)-1,2,4-heptatriene-6-yne was converted to α,3-dehydrotoluene and subsequently into various products dependent on the reaction conditions (Fig. 9A) (38). When the allene terminus of (Z)-1,2,4-heptatriene-6-yne is substituted with a methyl group, which should stabilize the radical at the benzylic carbon of the dehydrotoluene, the cyclization reaction was shown to occur six times faster than the parent compound (38). It was similarly hypothesized that a sulfur group should additionally stabilize the radical intermediate. This was verified by the synthesis of the thiol shown in Figure 9B, which cyclized via a putative radical intermediate with an estimated half-life of 2 min at 37°C, significantly faster than the previous compounds (39).

Nakatani et al. (138) have also synthesized two acyclic analogs of NCS differing only in the E- or Z-stereochemistry about the C8-C9 double bond (Fig. 9C). These compounds were tested for in vitro cytotoxicity against P388 murine leukemia cells, and it was determined that the Z-isomer was more toxic than the E-isomer, as one would expect if there was no rotation about the C-8—C-9 bond to interconvert the isomers during formation of the active intermediate. No evidence for the formation of a radical intermediate nor for any DNA-damaging activity was presented.

Two groups have synthesized allenylphosphine oxides as model acyclic enediyne compounds hypothesized to form diradical intermediates. The work of Nicolaou and co-workers and Nagat et al. will be discussed in the next section.

In all these experiments, evidence for the presence of free-radical centers is obtained by performing the cyclization reactions in deuteriated solvents and assessing the location of incorporated deuterium in the product molecules. Similarly, the radical intermediates can undergo dismutation to cre-

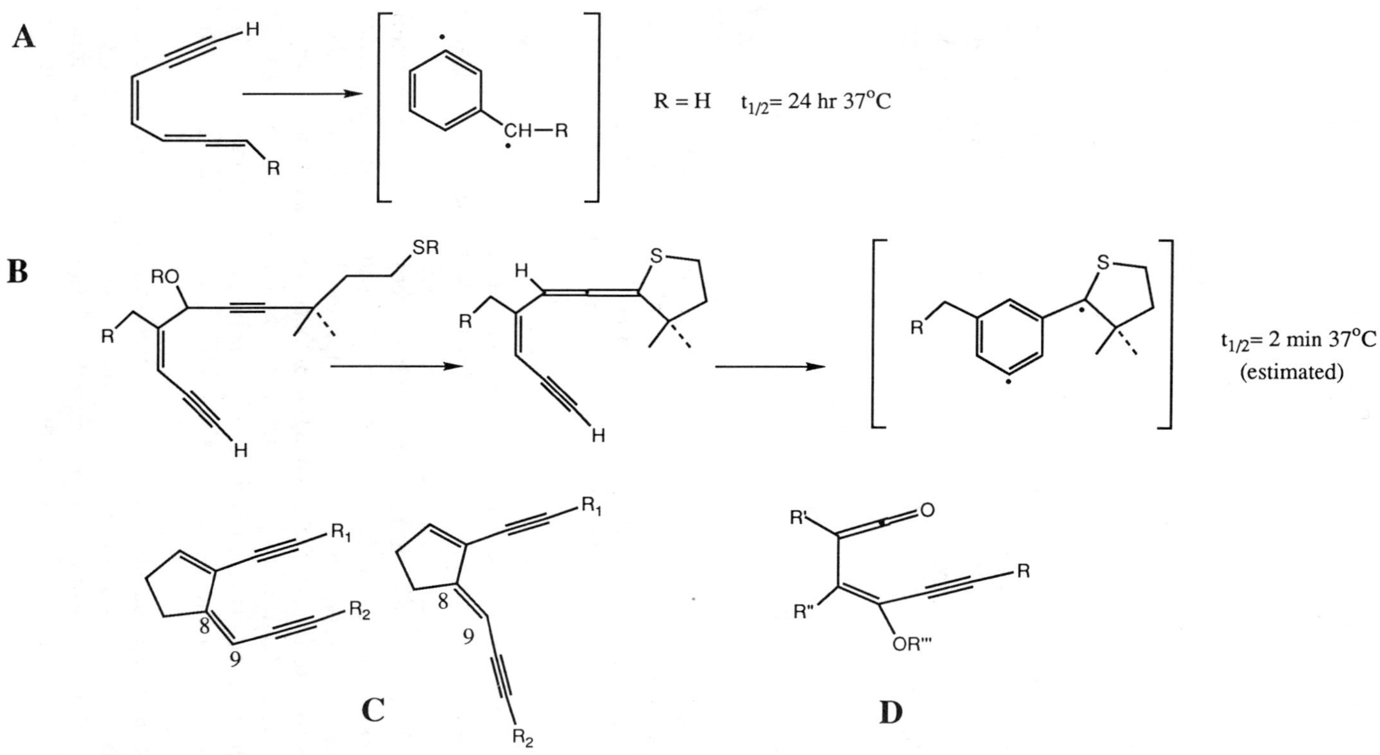

Figure 9 Structures of synthetic acyclic enediyne and eneyne-allene molecules and the proposed mechanisms of formation of their diradical intermediates (see text for details).

ate dimeric compounds and other characteristic products. In all these cases, however, the cyclization reactions take place in nonaqueous, organic solvents, suggesting that considerable work remains to be done before synthetic diradical-intermediate drugs are of practical import.

It should be noted that Foland *et al.* (139) have made a thorough study of analogous compounds in their work on the rearrangement of 4-alkynylcyclobutenones to form (2-alkynylethenyl)ketenes, compounds similar to the (*Z*)-1,2,4-heptatriene-6-yne of Myers except for the presence of a ketene instead of an allene (Fig. 9D). These compounds rearrange to form benzoquinones or cyclopentenediones.

2. *Calicheamicin/Esperamicin-Type Compounds*

The first successful synthesis of the bicyclic core of the calicheamicin/esperamicin class was achieved by Schreiber and Kiessling (140). Utilizing an intramolecular type II Diels-Alder strategy, they prepared a molecule containing an enediyne motif and a bridgehead olefin, but which lacked a functionality for placement of the ethylene trisulfide moiety present in the natural compounds (structure A in Fig. 10). Danishefsky and co-workers employed a different synthetic strategy and started with a benzenoid system to which the enediyne ring was attached by an intramolecular cyclization (structure B, Fig. 10) (141). The core subunit they synthesized possessed derivatizable functionalities at all positions modified in the natural compounds. Danishefsky's group proceeded to synthesize the calicheamicin/esperamicin core containing an intramolecular sulfur nucleophile as in the natural compounds (Fig. 10C) and demonstrated that the molecule underwent Michael-type addition of the sulfur to the bridgehead double bond with subsequent aromatization of the ring system (142).

Less functionalized versions of the calicheamicin/esperamicin core have been prepared using a similar strategy by Kende and Smith (143), Magnus and co-workers (144–146), and Tomioka and Koga (147) (structures D, E, and F, respectively, Fig. 10). Using one of these synthetic cores that lacked the trisulfide moiety necessary for activation, Magnus and Lewis demonstrated that attack by a thiol at C-2, the other end of the carbon-carbon double bond at the bridgehead, could lead to formation of the active diradical intermediate (Fig. 10E). In the other synthetic work of Magnus' group (144,145) and that of Tomioka and Koga (147), evidence is also presented for the intermediacy of a biradical molecule.

Efforts have also been directed to the synthesis of the carbohydrate side chains of the calicheamicin/esperamicin class of ene-diynes, the portion of the molecules that appears to be responsible for much of the DNA sequence specificity of the drugs. Nicolaou and Groneberg (148) have succeeded in synthesizing an analog of the oligosaccharide moiety that contains the

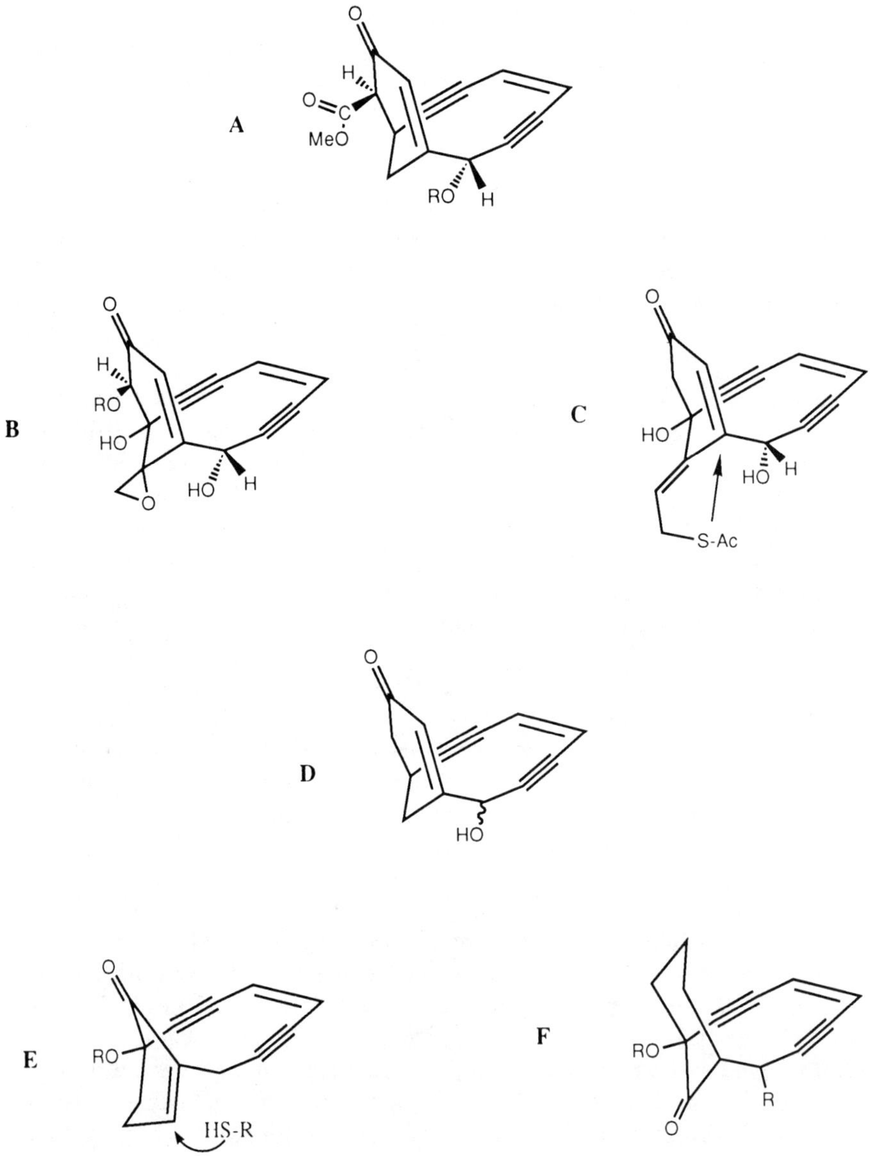

Figure 10 Structures of synthetic calicheamicin/esperamicin-type molecules (see text for details).

necessary stereochemistry as well as the unusual alkoxylamine and thioester linkages present in the natural product, but substitutes a trimethylbenzoate for the 4-hydroxy-5-iodo-2,3-dimethoxy-6-methylbenzoate species normally present. Nicolaou's group has synthesized and characterized this natural aromatic moiety (149). van Laak and Scharf have succeeded in synthesizing the 2,6-dideoxy-4-thio-D-ribohexopyranose moiety present in calicheamicin (150).

B. Energetics of Formation of the Diradical Intermediates

While there remains much to be characterized in the naturally occurring enediyne antibiotics, significant progress has been made in elaborating the features necessary for the conversion of the enediyne molecules into the radical species under physiological conditions. The enediyne starting molecule, a relatively unstable structure, must undergo electronic rearrangement under physiological conditions to form the diyl species. Two hypotheses have arisen to explain the driving force behind diradical formation: one involving ground state energy factors, specifically the conformation of the acetylene groups, and another focusing on the energetics of the biradicaloid transition state.

1. Ground State Energy Factors

Nicolaou and co-workers originally proposed that the major factor in the cyclization to form the diradical intermediate is the distance between the two acetylene groups (151). Support for this proposal came from molecular mechanics calculations and synthesis of model monocyclic enediyne systems. The calculations, performed with the MacroModel program using the MM2 force-field parameters, showed a good correlation between the interacetylenic distance and the tendency of the ring system to cyclize. These calculations predicted that the enediyne system shown in Figure 11A with $n \leq 2$ would have interacetylenic distances (distance r in Fig. 11A) equal to or less than 3.2–3.3 Å and would cyclize spontaneously. With $n \geq 3$, the distance between acetylenes would be greater than 3.3 Å and the molecules should be stable at ambient temperatures. Synthesis of rings with $n = 2$–8 confirmed the prediction. The $n = 2$ ring cyclized spontaneously at room temperature ($r \sim 3.25$ Å), while the molecules with $n = 3$–8 were stable to cyclization. At first glance, these studies support the hypothesis that the ground state energy of the ring system, as represented by the spatial relationship of the acetylene groups, correlates with the tendency for spontaneous electronic rearrangement. However, the models chosen for study, unsubstituted monocyclic ring systems, represent a very small subset of the whole population of enediynes, for which a better correlation exists

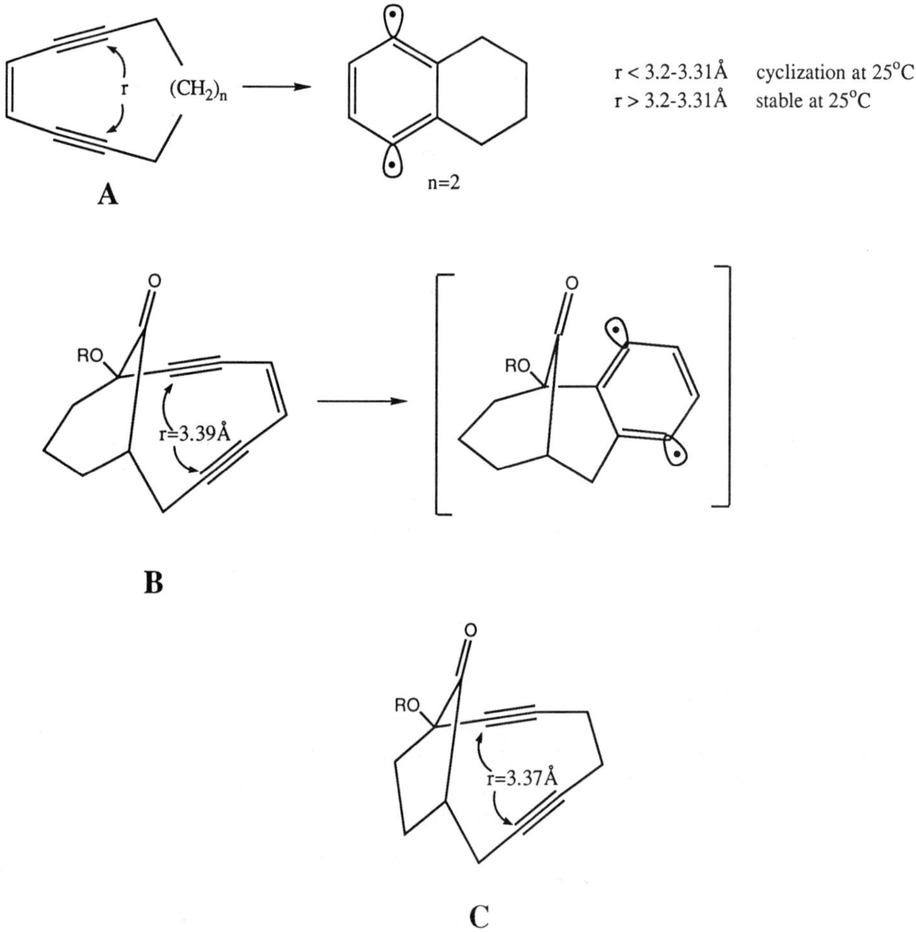

Figure 11 Mono- and bicyclic enediynes employed to study the energetics of formation of diradical intermediates (see text for details).

between the energetics of the biradicaloid transition state and the rate of cyclization.

2. *Transition State Energy Considerations*

Independent studies performed by the Magnus and Snyder groups made it apparent that, with ring systems more complicated than the monocyclic ene-diynes, factors other than acetylene proximity played a role in the cyclization reaction (144,145,152–154). Theoretical studies performed by Snyder

(152) reproduced the findings of Nicolaou for proximity control in the unsubstituted monocyclic ring systems: a shorter distance between the termini of the acetylene groups results in additional ground state strain energy, thus reducing the difference in energy between the ground and proposed biradicaloid transition states. However, in fused ring systems, such as the esperamicin/calicheamicin core, and in substituted monocyclic systems, new strain components appear that are independent of the group state interacetylenic distance (144,145,153,154). In one set of experiments, it was determined that a cyclohexanone-containing enediyne molecule (Fig. 11B), with an interacetylenic distance r = 3.39 Å, underwent ring closure two to three orders of magnitude faster than the cyclopentanone form (Fig. 11C) in spite of the smaller distance between acetylenes of the latter (r = 3.37 Å). Quantum mechanics calculations (the PRDDO program) in conjunction with molecular mechanical studies (MacroModel with MM2) revealed that a difference in activation energies for the two molecules accounted for the difference in cyclization rates. The most significant contribution to this activation energy difference is made by conformational changes in the five- or six-membered rings upon moving into the partially cyclized, biradicaloid transition state. As the molecule moves from ground state to the biradicaloid structure, the cyclohexanone ring can undergo a conformational change from boat to chair form with a release of 6 kcal of strain energy, while the energy of the cyclopentanone group increases by 1.5 kcal. No energy-relieving conformational change is available to the pentanone ring, and there is no relief of the energy barrier imposed by the cycloalkanone ring system. Thus, even though the acetylenes in the cyclopentanone form are close to the Nicolaou spontaneous cyclization range (r = 3.2–3.31 Å), the molecule is unable to surmount the alkanone-imposed energy barrier on moving from the enediyne configuration to the diradical intermediate, as is the cyclohexanone-containing molecule. Reduction of the carbonyl moiety to an alcohol in both of these ring systems resulted in a large decrease in this energy barrier, and the cyclization for each proceeded significantly faster in both theory and experiment (152,153).

Snyder makes similar arguments for substituted monocyclic enediynes (154). Again, quantum mechanics calculations coupled with molecular mechanics calculations were used to predict that, for both nine- and ten-membered monocyclic enediyne ring systems, nonbonded ring substituent interactions contribute to both ground and transition state strain energies (154). As with the bicyclic systems, it is proposed that differences in the rates of biradical formation are best correlated with differences in the ground or transition state strain energies and not more simply with the spatial relationship of the acetylene groups. It appears, then, that nature has selected ring systems whose acetylenes are placed in close enough proximity to cyclize spontaneously (under physiological conditions), but which are held in check by structural embellishments of the ring system. For example, the bridgehead double bond of the calicheamicin/esperamicin core prevents cyclization of the ring system, and it is nucleophilic addition of the thiol to this double bond that results in a lowering of the activation energy for the cyclization reaction. Other structural features of NCS, including the epoxide, may also play a role in the stabilization of the drug.

C. DNA Cleavage by Synthetic Enediyne Molecules

Recently, Nicolaou's group made the first demonstration of DNA cleavage by synthetic enediyne-type molecules with the cyclodecaenediyne diol shown in Figure 12A (155). Except for the placement of the two hydroxyl groups presumably for reasons of solubility, this molecule is nearly identi-

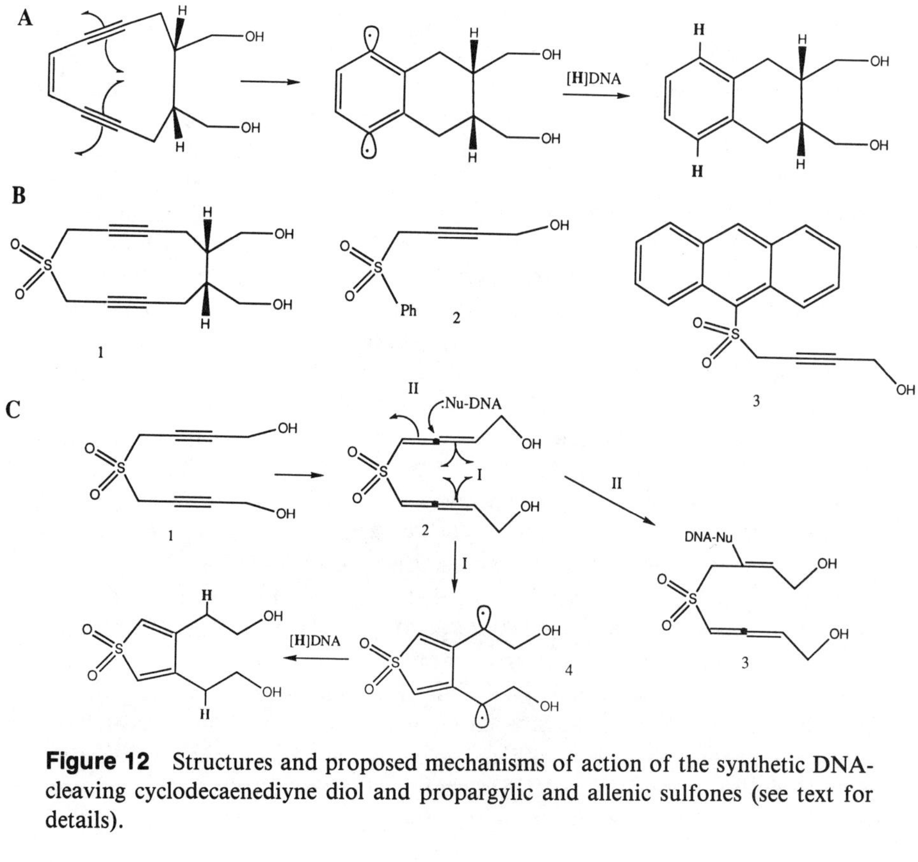

Figure 12 Structures and proposed mechanisms of action of the synthetic DNA-cleaving cyclodecaenediyne diol and propargylic and allenic sulfones (see text for details).

cal to the cyclodecenediyne molecule from the series described earlier (Fig. 11A), which was hypothesized to form a benzenoid diradical as the active intermediate. It was selected on the basis of its acetylene proximity (r = 3.2 Å) and its activation energy of 24 kcal/mol, both of which suggested that the ring would undergo closure at a reasonable rate. Indeed, the molecule cyclized with a half-life of 11.8 hr at 37°C and caused cleavage of closed circular plasmid DNA under aqueous, aerobic conditions in a time frame similar to its cyclization kinetics.

Nicolaou and co-workers have also synthesized a series of compounds that, while originally thought to undergo rearrangement to a free-radical-containing intermediate, probably cause DNA strand cleavage by a mechanism different than the natural enediyne compounds. One set of such molecules consists of mono- and bis-propargylic (structures 1–3, Fig. 12B; and structure 1, Fig. 12C) and -allenic sulfones (structure 2, Fig. 12C) hypothesized to proceed through the reaction sequence shown in Figure 12C (156). Not only did the bis compounds cleave closed circular DNA at 25°C, but there was also DNA damage caused by the monopropargylic and -allenic sulfone molecules (Fig. 12B, structure 2), which presumably do not form a radical species. In addition, these compounds also exhibited a pH dependence for the DNA damage, with ideal cleavage occurring at pH = 8.5. These findings suggested that DNA damage was occurring by the alternate pathway (path II in Fig. 12C), involving reaction of the cumulene intermediate (structure 2 in Fig. 12C) with nucleophilic nitrogens of DNA bases in analogy to Maxam-Gilbert chemistry (42). This would cause DNA strand scission without proceeding through a radical intermediate. Whatever mechanism was responsible for the DNA cleavage, the importance of a DNA binding moiety to the activity of the molecule was born out by attaching the monopropargylic sulfone to an anthracene ring system (Fig. 12B, structure 3). The DNA cleavage activity of this conjugate increased by two orders of magnitude over the unconjugated molecules (156).

Similarly, Nicolaou's group prepared two compounds with structures analogous to the asymmetric, *cis*-enediyne core of NCS (Figs. 13A,B), both of which were hypothesized to cause DNA damage by either a radical intermediate or by alkylation of DNA (157,158). The radical forms were hypothesized to form with the intermediacy of the cumulene-ene-yne structure proposed by Myers for NCS (34,35), following nucleophilic attack at one of the triply unsaturated groups (see pathways outlined in Fig. 13). Golfomycin A (Fig. 13A) caused cleavage of DNA under physiologic conditions, but the cleavage was inhibited by methyl thioglycolate, the nucleophilic "activator" for the radical-mediated pathway (157). This finding and other features of golfomycin chemistry led to the conclusion that the alkylation pathway (path II in Fig. 13A) was the likely cause of the DNA

A

B

C

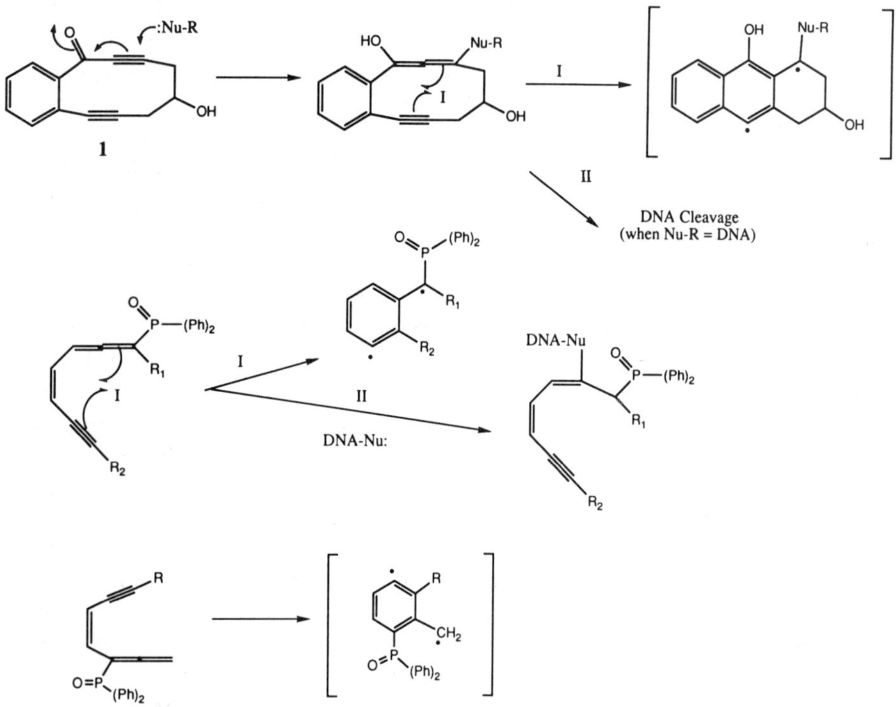

Figure 13 Structures and proposed mechanisms of action of synthetic DNA-cleaving agents: (A) golfomycin A; (B) phosphine oxide-allene-*cis*-eneyne; (C) allenyl phosphine oxide (see text for details).

damage. The same conclusion was reached for the phosphine oxide-allene-*cis*-ene-yne compound (Fig. 13B), which cyclized to the expected aromatic forms in the absence of nucleophiles and cleaved DNA (158). Nucleophiles readily added to the central allene carbon, in support of the alkylation pathway (158). These compounds and the propargylic and allenic sulfones all showed enhanced DNA cleavage properties under alkaline conditions, which supports the alkylation pathway as the major route to DNA damage.

Nagata et al. reasoned that, in the case of an unstrained, acyclic 1,5-diyne-3-ene, the interacetylenic distance would be too great (>4 Å) for efficient cyclization to occur, and that by introducing an acetylene-equivalent allene group, the distance could be shortened and the cyclization enhanced (159). They introduced an allene group into an acyclic, symmetric 1,5-diyn-3-ene in much the same way as Myers' group, except that the driving force to form the allene group was a sigmatropic rearrangement of a propargylic phosphinite, which yielded the allenylphosphine oxide shown in Figure 13C (159). This compound was shown to cyclize under moderate conditions (37°C) to products consistent with the formation of a diradical intermediate. Nagata et al. also tested these compounds for DNA cleavage activity in vitro (160). The phosphine analog (Fig. 13C) and a sulfoxide analog both caused strand breaks in supercoiled DNA, but again only at high concentrations (100–500 μM) and long incubation times (2–12 hr), raising the possibility that the molecules are cleaving DNA by a mechanism other than radical formation.

D. DNA Binding Modalities in Synthetic Enediynes

Little attention has been paid to the factors responsible for DNA binding properties and sequence-specificity of the enediyne drugs. The importance of a DNA binding modality is evident from Nicolaou's experiments with the propargylic sulfones discussed earlier, in which attachment of an anthracycline moiety enhanced DNA cleavage by 100-fold (156). The attachments to the core molecules appear to fall into two major groups: intercalating moieties (the anthracycline of dynemicin, the naphthoate of NCS, and possibly the iodobenzene of calicheamicin) and carbohydrate chains. As suggested by the structure-function studies with the enediynes, the embellishments impart sequence specificity for the DNA cleavage in addition to their function of enhancing the binding of the molecules to DNA. As Schreiber and co-workers point out, common properties of glycoconjugate antibiotics that interact with DNA, such as NCS, calicheamicin, and esperamicin, are the *trans* geometries of the carbohydrate and alkoxy moieties attached to the aglycone core and the 1,4-orientation of the sugar linkages, a bonding that maximizes linearity of the carbohydrate tails (121). For these antibiotics, such properties give the molecule a linear structure with a right-handed twist that fits into right-handed minor groove of DNA. Zein et al. point out that, while the amino sugars may be involved in electrostatic binding to DNA, the high hydrophobicity of the enediyne molecules may also be crucial to the drug-DNA interaction, with the water insoluble drug probably finding suitable environment in the relatively hydrophobic minor groove of DNA (161). Optimum minor-groove width then places abstractable deoxyribose hydrogens in close proximity to the radicals of the activated form of the drug.

IV. CONCLUSIONS

Our understanding of the enediyne class of DNA cleaving agents is obviously limited, but the findings presented here illustrate the extent to which this family of drugs has progressed along the paths of nucleic acid targeted drug design. While the major focus has been placed upon the enediyne core of the molecules, the field of enediyne-mediated DNA cleavage is still in a relative infancy, so it is anticipated that the DNA binding properties and determinants of sequence specificity will be studied more extensively in the future. Ultimately, the complete molecule will be understood to the point that analogs can be synthesized that are capable of efficient and sequence-specific DNA strand scission closely mimicking the natural products, but with lessened toxic side effects or more specific targeting to the DNA of diseased cells.

ACKNOWLEDGMENTS

The authors wish to thank Dr. James Snyder at Searle Research and Development for critical review of this manuscript and for communicating results prior to publication. This work was supported by a Bristol-Myers Squibb Fellowship (P.C.D.) and Grants CA44257 and GM12573 from the National Institutes of Health (I.H.G.).

REFERENCES

1. Bergman, R.G., Reactive, 1,4-dehydroaromatics. *Acc. Chem. Res. 6*:25–31 (1973).
2. Ishida, N., K. Miyazaki, K. Kumagai, and M. Rikimaru, Neocarzinostatin: an antitumor antibiotic of high molecular weight. *J. Antiobiot. (Tokyo) Ser.A. 18*:63–76 (1965).
3. Goldberg, I. H., Mechanism of neocarzinostatin action: role of DNA microstructure in determination of chemistry of bistranded oxidative damage. *Accts. Chem. Res. 24:*191–196 (1991).
4. Napier, M.A., B. Holmquist, D.J. Strydom, and I.H. Goldberg, Neocarzinostatin: spectral characterization and separation of the nonprotein chromophore. *Biochem. Biophys. Res. Commun. 89*:635–642 (1979).
5. Meienhofer, J., H. Maeda, C.B. Glaser, J. Czombos, and K. Kuromizu, Primary structure of neocarzinostatin, an antitumor protein. *Science 178*: 875–876 (1972).
6. Biemann, K., Sequencing of proteins. *Int. J. Mass Spectrom. Ion. Phys. 45*: 183–194 (1982).
7. Gibson, B.W., W.C. Herlihy, T.S.A. Samy, K. Hahm, H. Maeda, J. Meienhofer, and K. Biemann, A revised primary structure for neocarzinostatin based on fast atom bombardment and gas chromatographic-mass spectrometry. *J. Biol. Chem. 259*:10801–10806 (1984).
8. Hirayama, K., T. Ando, R. Takahashi, and A. Murai, A revised primary structure of neocarzinostatin using the combination of endopeptidase, carboxypeptidase Y, and fast atom bombardment mass spectrometry. *Bull. Chem. Soc. Jpn. 59*:1371–1378 (1986).

9. Saito, K., Y. Sato, K. Edo, Y. Akiyama-Murai, Y. Koide, N. Ishida, and M. Mizugaki, Characterization of secondary structure of neocarzinostatin apoprotein. *Chem. Pharm. Bull. 37*:3078–3082 (1989).
10. Adjadj, E., J. Mispelter, E. Quiniou, J.-L. Dimicoli, V. Favaudon, and J.-M. Lhoste, Proton NMR studies of apo-neocarzinostatin from *Streptomyces carzinostaticus*. Sequence-specific assignment and secondary structure. *Eur. J. Biochem. 190*:263–271 (1990).
11. Remerowski, M.L., S.J. Glaser, L.C. Sieker, T.S.A. Samy, and G.P. Drobny, Sequential ^{1}H NMR assignments and secondary structure of apo-neocarzinostatin in solution. *Biochemistry 29*:8401–8409 (1990).
12. Van Roey, P., and T.A. Beerman, Crystal structure analysis of auromomycin apoprotein (macromomycin) shows importance of protein side chains to chromophore binding selectivity. *Proc. Nat. Acad. Sci. USA 86*:6587–6591 (1989).
13. Albers-Schonberg, G., R.S. Dewey, O.D. Hensens, J.M. Liesch, M.A. Napier, and I.H. Goldberg, Neocarzinostatin: chemical characterization and partial structure of the nonprotein chromophore. *Biochem. Biophys. Res. Commun. 95*:1351–1356 (1980).
14. Hensens, O.D., R.S. Dewey, J.M. Liesch, M.A. Napier, R.A. Reamer, J.L. Smith, G. Albers-Schonberg, and I.H. Goldberg, Neocarzinostatin chromophore: presence of a highly strained ether ring and its reaction with mercaptan and sodium borohydride. *Biochem. Biophys. Res. Commun. 113*:538–547 (1983).
15. Edo, K., Y. Mizugaki, H. Koide, K. Seto, N. Furihata, N. Otake, and N. Ishida, The structure of neocarzinostatin chromophore possessing a novel bicyclo[7.3.0]dodecadiyne system. *Tetrahedron Lett. 26*:331–334 (1985).
16. Edo, K., Y. Akiyama, K. Saito, M. Mizugaki, Y. Koide, and N. Ishida, Absolute configuration of the amino sugar moiety of the neocarzinostatin chromophore. *J. Antibiot. 39*:1615–1619 (1986).
17. Napier, M.A., I.H. Goldberg, O.D. Hensens, R.S. Dewey, J.M. Liesch, and G. Albers-Schonberg, Neocarzinostatin chromophore: presence of a cyclic carbonate subunit and its modification in the structure of other biologically active forms. *Biochem. Biophys. Res. Commun. 100*:1703–1712 (1981).
18. Napier, M.A., and I.H. Goldberg, Neocarzinostatin chromophore: assignment of spectral properties and structural requirements for binding to DNA. *Molec. Pharmacol. 23*:500–510 (1983).
19. Kappen, L.S., M.A. Napier, and I.H. Goldberg, Roles of chromophore and apo-protein in neocarzinostatin action. *Proc. Nat. Acad. Sci. USA 77*:1970–1974 (1980).
20. Kappen, L.S., and I.H. Goldberg, Activation and inactivation of neocarzinostatin-induced cleavage of DNA. *Nucleic Acids Res. 5*:2959–2967 (1978).
21. Beerman, T.A., and I.H. Goldberg, The relationship between DNA strand-scission and DNA synthesis inhibition in HeLa cells treated with neocarzinostatin. *Biochem. Biophys. Acta 475*:281–293 (1977).
22. Poon, R., T.A. Beerman, and I.H. Goldberg, Characterization of DNA strand breakage *in vitro* by the antitumor protein neocarzinostatin. *Biochemistry 16*:486–493 (1977).

23. Koide, Y., A. Ito, K. Edo, and N. Ishida, The biologically active site of neocarzinostatin-chromophore. *Chem. Pharm. Bull. 34*:4425–4428 (1986).
24. Burger, R.M., J. Peisach, and S.B. Horwitz, Effect of light and oxygen on neocarzinostatin stability and DNA-cleaving activity. *J. Biol. Chem. 253*: 4830–4832 (1978).
25. Povirk, L.F., N. Dattagupta, B.C. Warf, and I.H. Goldberg, Neocarzinostatin chromophore binds to deoxyribonucleic acid by intercalation. *Biochemistry 20*:4007–4014 (1981).
26. Dasgupta, D., D.S. Auld, and I.H. Goldberg, Cryoscopic evidence for the mode of reversible binding of NCS chromophore to poly(deoxyadenylic-thymidylic acid). *Biochemistry 24*:7049–7054 (1985).
27. Napier, M.A., B. Holmquist, D.J. Strydom, and I.H. Goldberg, Neocarzinostatin chromophore: purification of the major active form and characterization of its spectral and biological properties. *Biochemistry 20*:5602–5608 (1981).
28. Lee, S.H., and I.H. Goldberg, Role of epoxide in NCS chromophore stability and action. *Molecular Pharmacology 33*:396–401 (1988).
29. Povirk, L.F., and I.H. Goldberg, Stoichiometric uptake of molecular oxygen and consumption of sulfhydryl groups by neocarzinostatin chromophore bound to DNA. *J. Biol. Chem. 19*:11763–11767 (1983).
30. Kappen, L.S., and I.H. Goldberg, Stabilization of neocarzinostatin nonprotein chromophore activity by interaction with apoprotein and with HeLa cells. *Biochemistry 19*:4786–4790 (1980).
31. Charnas, R.L., and I.H. Goldberg, NCS abstracts hydrogen during formation of nucleotide 5′-aldehyde on DNA. *Biochem. Biophys. Res. Comm. 122*:642–648 (1984).
32. Kappen, L.S., and I.H. Goldberg, Activation of neocarzinostatin chromophore and formation of nascent DNA damage do not require molecular oxygen. *Nucleic Acids Res. 13*:1637–1648 (1985).
33. Meyers, A.G., Proposed structure of the neocarzinostatin chromophore-methyl thioglycolate adduct: a mechanism for the nucleophilic activation of neocarzinostatin. *Tetrahedron Lett. 28*:4493–4496 (1987).
34. Meyers, A.G., P.J. Proteau, and T.M. Handel, Stereochemical assignment of neocarzinostatin chromophore. Structures of neocarzinostatin chromophore-methyl thioglycolate adducts. *J. Am. Chem. Soc. 110*:7212–7214 (1988).
35. Myers, A.G., and P.J. Proteau, Evidence for spontaneous, low-temperature biradical formation from a highly reactive neocarzinostatin chromophore-thiol conjugate. *J. Am. Chem. Soc. 111*:1146–1147 (1989).
36. Hensens, O.D., and I.H. Goldberg, Mechanisms of activation of the antitumor antibiotics neocarzinostatin by mercaptan and sodium borohydride. *J. Antibiot. 42*:761–768 (1989).
37. Meyers, A.G., M.M. Alauddin, M.A.M. Fuhry, P.S. Dragovich, N.S. Finney, and P.M. Harrington, Versatile precursors for the synthesis of enynes and enediynes. *Tetrahedron Lett. 30*:6997–7000 (1989).
38. Myers, A.G., E.Y. Kuo, and N.S. Finney, Thermal generation of α,3-

dehydrotoluene from (*Z*)-1,2,4-heptatriene-6-yne. *J. Am. Chem. Soc. 111*: 8057–8059 (1989).

39. Myers, A.G., and P.S. Dragovich, Design and dynamics of a chemically triggered reaction cascade leading to biradical formation at subambient temperature. *J. Am. Chem. Soc. 111*:9130–9133 (1989).
40. Chin, D.-H., C.-h. Zeng, C.E. Costello, and I.H. Goldberg, Sites in the diyne-ene bicyclic core of neocarzinostatin chromophore responsible for hydrogen abstraction from DNA. *Biochemistry 27*:8106–8114 (1988).
41. Meschwitz, S.M., and I.H. Goldberg, Selective abstraction of ^{2}H from C5′ of thymidylate in oligodeoxynucleotide by radical center at C6 of diradical species of neocarzinostatin: chemical evidence for stucture of activated drug-DNA complex. *Proc. Nat. Acad. Sci. USA 88*:3047–3051 (1991).
42. Maxam, A., and W. Gilbert, Sequencing end-labeled DNA with base-specific chemical cleavages. *Methods Enz. 65*:499–560 (1980).
43. Dasgupta, D., and I.H. Goldberg, Mode of reversible binding of neocarzinostatin chromophore to DNA: evidence for binding via the minor groove. *Biochemistry 24*:6913–6920 (1985).
44. Dedon, P.C., and I.H. Goldberg, Sequence-specific double-strand breakage of DNA by neocarzinostatin involves different chemical mechanisms within a staggered cleavage site. *J. Biol. Chem. 265*:14713–14716 (1990).
45. Povirk, L.F., C.W. Houlgrave, and Y.-H. Han, Neocarzinostatin-induced DNA base release accompanied by staggered oxidative cleavage of the complementary strand. *J. Biol. Chem. 263*:19263–19266.
46. Dervan, P.B., Design of sequence-specific DNA-binding molecules. *Science 232*:464–471.
47. Stubbe, J., and J.W. Kozarich, Mechanisms of bleomycin-induced DNA degradation. *Chem. Rev. 87*:1107–1136.
48. von Sonntag, D., in *The Chemical Basis of Radiation Biology*, Taylor and Francis, London (1987).
49. Kappen, L.S., and I.H. Goldberg, Deoxyribonucleic acid damage by neocarzinostatin chromophore: strand breaks generated by selective oxidation of C-5′ of deoxyribose. *Biochemistry 22*:4872–4878 (1983).
50. Kappen, L.S., I.H. Goldberg, and J.M. Liesch, Identification of thymidine-5′-aldehyde at DNA strand breaks induced by neocarzinostatin chromophore. *Proc. Nat. Acad. Sci. 79*:744–748 (1982).
51. Chin, D.-H., S.A. Carr, and I.H. Goldberg, Incorporation of $^{18}O_2$ into thymidine 5′-aldehyde in neocarzinostatin chromophore-damaged DNA. *J. Biol. Chem. 259*:9975–9978 (1984).
52. Hatayama, T., I.H. Goldberg, M. Takeshita, and A.P. Grollman, Nucleotide specificity in DNA scission by neocarzinostatin. *Proc. Nat. Acad. Sci. USA 75*:3603–3607 (1978).
53. Takeshita, M., L.S. Kappen, A.P. Grollman, M. Eisenberg, and I.H. Goldberg, Strand scission of deoxyribonucleic acid by neocarzinostatin, auromomycin, and bleomycin: studies on base release and nucleotide sequence specificity. *Biochemistry 20*:7599–7606 (1981).

54. Lee, S.H., and I.H. Goldberg, Sequence-specific, strand-selective, and directional binding of neocarzinostatin chromophore to oligodeoxyribonucleotides. *Biochemistry 28*:1019–1026 (1989).
55. Chin, D.-H., L.S. Kappen, and I.H. Goldberg, 3′-Formyl phosphate-ended DNA: high energy intermediate in antibiotic-induced DNA sugar damage. *Proc. Nat. Acad. Sci. USA 84*:7070–7073 (1987).
56. Kawabata, H., H. Takeshita, T. Fujiwara, H. Sugiyama, T. Matsuura, and I. Saito, Chemistry of neocarzinostatin-mediated degradation of d(GCATGC). Mechanism of spontaneous thymine release. *Tetrahedron Lett. 30*:4263–4266 (1989).
57. Kappen, L.S., and I.H. Goldberg, Nitroaromatic radiation sensitizers substitute for oxygen in NCS-induced DNA damage. *Proc. Nat. Acad. Sci. USA 81*:3312–3316.
58. Kappen, L.S., T.R. Lee, C.-c. Yang, and I.H. Goldberg, Oxygen transfer from the nitro group of a nitroaromatic radiosensitizer to a DNA sugar damage product. *Biochemistry 28*:4540–4542 (1989).
59. Povirk, L.F., and I.H. Goldberg, Covalent adducts of DNA and the nonprotein chromophore of NCS contain a modified deoxyribose. *Proc. Nat. Acad. Sci. USA 79*:369–373 (1982).
60. Povirk, L.F., and I.H. Goldberg, NCS chromophore-DNA adducts: evidence for a covalent linkage to the oxidized C-5′ of deoxyribose. *Nucleic Acids Res. 10*:6255–6264 (1982b).
61. Povirk, L.F., and I.H. Goldberg, Competition between anaerobic covalent linkage of NCS chromophore to deoxyribose in DNA and oxygen-dependent strand breakage and base release. *Biochemistry 23*:6304–6311 (1984).
62. Povirk, L.F., and I.H. Goldberg, Detection of neocarzinostatin chromophore-deoxyribose adducts as exonuclease-resistant sites in defined-sequence DNA. *Biochemistry 24*:4035–4040 (1985).
63. Hatayama, T., and I.H. Goldberg, DNA damage and repair in relation to cell killing in neocarzinostatin-treated HeLa cells. *Biochim Biophys. Acta 563*:59–71 (1979).
64. Povirk, L.F., and I.H. Goldberg, Endonuclease-resistant apyrimidinic sites formed by neocarzinostatin at cytosine residues in DNA: evidence for a possible role in mutagensis. *Proc. Nat. Acad. Sci. USA 82*:3182–3186 (1985).
65. Kappen, L.S., C.-q. Chen, and I.H. Goldberg, Atypical abasic sites generated by neocarzinostatin at sequence-specific cytidylate residues in oligodeoxynucleotides. *Biochemistry 17*:4331–4340 (1988).
66. Kappen, L.S., and I.H. Goldberg, Identification of 2-deoxyribonolactone at the site of neocarzinostatin-induced cytosine release in the sequence d(AGC). *Biochemistry 28*:1027–1032 (1989).
67. Kappen, LS., I.H. Goldberg, S.H. Wu, J.A. Stubbe, L. Worth, and J.W. Kozarich, Isotope effects on the sequence-specific cleavage of dC in d(AGC) sequences by neocarzinostatin: elucidation of chemistry of minor lesions. *J. Am. Chem. Soc. 112*:2797–2798 (1990).
68. Povirk, L.F., and I.H. Goldberg, Base substitution mutations in the *cI* gene

of lambda phage by neocarzinostatin chromophore: correlation with depyrimidination hotspots at the sequence AGC. *Nucleic Acids Res. 14*:1417–1426 (1986).

69. Frank, B.L., L. Worth, Jr., D.F. Christner, J.W. Kozarich, J. Stubbe, L.S. Kappen, and I.H. Goldberg, Isotope effects on the sequence-specific cleavage of DNA by neocarzinostatin: kinetic partitioning between 4′- and 5′-hydrogen abstraction at unique thymidine sites. *J. Am. Chem. Soc. 113*:2271–2275 (1991).
70. Kappen, L.S, I.H. Goldberg, B.L. Frank, L. Worth, D.F. Christner, J.W. Kozarich, and J. Stubbe, Necarzinostatin-induced hydrogen atom abstraction from C-4′ and C-5′ of the T residue at a d(GT) step in oligonucleotides: shuttling between deoxyribose attack sites based on isotope selection effects. *Biochemistry 30*:2034–2042 (1991).
71. Saito, I., H. Kawabata, T. Fujiwara, H. Sugiyama, and T. Matsuura, A novel ribose C-4′ hydroxylation pathway in neocarzinostatin-mediated degradation of oligonucleotides. *J. Am. Chem. Soc. 111*:8302–8303 (1989).
72. Sugiyama, H., C. Xu, N. Murugesan, S.M. Hecht, G.A. van der Marel, and J.H. van Boom, Chemistry of the alkali-labile lesion formed from iron(II) bleomycin and d(CGCTTTAAAGCG). *Biochemistry 27*:58–67 (1988).
73. Metzler, D.E. *Biochemistry: The Chemical Reactions of Living Organisms*, Academic Press, New York, 421 (1977).
74. Brooks, B.R., R.E. Bruccoleri, B.D. Olafson, D.J. States, S. Swaminathian, and M. Karplus, CHARMM—a program for macromolecular energy, minimization, and dynamics calculations. *J. Comput. Chem. 4*:187–217 (1983).
75. Galat, A., and I.H. Goldberg, Molecular models of neocarzinostatin damage of DNA: analysis of sequence dependence in 5′GAGCG : 5′CGCTC. *Nucleic Acids Res. 18*:2093–2099 (1990).
76. Chen, K.-X., N. Gresh, and B. Pullman, A tentative model of the intercalative binding of the neocarzinostatin chromophore to double-stranded tetranucleotides. *Nucleic Acids Res. 15*:2175–2189 (1987).
77. Kappen, L.S., T.E. Ellenberger, and I.H. Goldberg, Mechanism and base specificity of DNA breakage in intact cells by neocarzinostatin. *Biochemistry 26*:384–390 (1987).
78. Shiloh, Y., G.P. van der Schans, P.H.M. Lohman, and Y. Becker, Induction and repair of DNA damage in normal and ataxia telangiectasia skin fibroblasts treated with noecarzinostatin. *Carcinogenesis 4*:917–921 (1983).
79. Kuo, W.-L., R.E. Meyn, and C.W. Haidle, Neocarzinostatin-mediated DNA damage and repair in wild-type and repair-deficient Chinese hamster ovary cells. *Cancer Res. 44*:1748–1751 (1984).
80. Boye, E., W. Kohnlein, and K. Skarstad, Characterization of intracellular DNA strand breaks induced by neocarzinostatin in *Escherichia coli* cells. *Nucleic Acids Res. 12*:8281–8291 (1984).
81. Grimwade, J.E., and T.A. Beerman, Measurement of bleomycin, neocarzinostatin, and auromomycin cleavage of cell-free and intracellular simian virus 40 DNA and chromatin. *Mol. Pharmacol. 30*:358–363 (1986).

82. Kappen, L.S., and I.H. Goldberg, Neocarzinostatin induction of DNA repair synthesis in HeLa cells and isolated nuclei. *Biochim. Biophys. Acta 520*:481–489 (1978).
83. McHugh, M.M., J. Woynarowski, and T. Beerman, Degradation of HeLa cell chromatin by neocarzinostatin and its chromophore. *Biochim. Biophys. Acta 696*:7–14 (1982).
84. Beerman, T., G. Mueller, and H. Grimmond, Effects of neocarzinostatin on chromatin in HeLa S_3 nuclei. *Mol. Pharmacol. 23*:493–499 (1983).
85. Kuo, M.T., and T.S.A. Samy, Effects of neocarzinostatin on mammalian nuclei: release of nucleosomes. *Biochim. Biophys. Acta 518*:186–190 (1978).
86. DeGraff, W.G., A. Russo, and J.B. Mitchell, Glutathione depletion greatly reduces neocarzinostatin cytotoxicity in Chinese hamster V79 cells. *J. Biol. Chem. 260*:8312–8315 (1985).
87. DeGraff, W.G., and J.B. Mitchell, Glutathione dependence of neocarzinostatin cytotoxicity and mutagenicity in Chinese hamster V-79 cells. *Cancer Res. 45*:4760–4762 (1985).
88. Russo, A., W. DeGraff, N. Friedman, and J.B. Mitchell, Selective modulation of glutathione levels in human normal *versus* tumor cells and subsequent differential response to chemotherapy drugs. *Cancer Res. 46*:2845–2848 (1986).
89. van Holde, K.E., *Chromatin*, Springer-Verlag, New York, 310–314 (1989).
90. Ono, Y., Y. Watanabe, and N. Ishida, Mode of action of neocarzinostatin: inhibition of DNA synthesis and degradation of DNA in *Sarcina lutea. Biochim. Biophys. Acta 119*:46–58 (1966).
91. Swada, H., K. Tatsumi, M. Sasada, S. Shirakawa, T. Nakamura, and G. Wakisaka, Effect of neocarzinostatin on DNA synthesis in L1210 cells. *Cancer Res. 34*:3341–3346 (1974).
92. Povirk, L.F., and I.H. Goldberg, Inhibition of mammalian deoxyribonucleic acid synthesis by neocarzinostatin: selective effect on replicon initiation in CHO cells and resistant synthesis in ataxia telangiectasia fibroblasts. *Biochemistry 21*:5857–5862 (1982).
93. Tatsumi, K., T. Sakane, H. Swada, S. Shirakawa, T. Nakamura, and G. Wakisaka, Unscheduled DNA synthesis in human lymphocytes treated with neocarzinostatin. *Gann 66*:441–444 (1975).
94. Povirk, L.F., and I.H. Goldberg, A role for oxidative DNA sugar damage in mutagensis by neocarzinostatin and bleomycin. *Biochimie 69*:815–823 (1987).
95. Tatsumi, K., and H. Nishioka, Effect of DNA repair systems on antibacterial and mutagenic activity of an antitumor protein, neocarzinostatin. *Mutat. Res. 48*:195–204 (1977).
96. Eisenstadt, E., M. Wolf, and I.H. Goldberg, Mutagensis by neocarzinostatin in *Escherichia coli* and *Salmonella typhimurium*: Requirement for *umuC*+ or plasmid pKM101. *J. Bacteriol. 144*:656–660 (1980).
97. Beckmann, R.P., M.J. Agostino, M.M. McHugh, R.D. Sigmund, and T.A. Beerman, Assessment of preferential cleavage of an actively transcribed retroviral hybrid gene in murine cells by deoxyribonuclease I, bleomycin, neocarzinostatin, or ionizing radiation. *Biochemistry 26*:5409–5415 (1987).
98. Bradner, W.T., and D.J. Hutchison, Neocarzinostatin (NSC-69856): An an-

titumor antibiotic effective against ascitic leukemia L1210 in mice. *Cancer Chemother. Rep.* *50*:79–84 (1966).
99. Kumagi, K. Antitumor activity of carzinostatin. *J. Antibiot.* Ser. A. 15:53–59 (Tokyo) (1962).
100. Lagha, S.S., D.D. Von Hoff, M. Rozencweig, D. Abraham, M. Slavik, and M. Muggia, Neocarzinostatin (NSC 157365): A new carcinostatic compound. *Oncology* *33*:265–270 (1976).
101. Maeda, H., S. Sakamoto, and J. Ogata, Mechanism of accumulation of the antitumor protein antibiotic neocarzinostatin in bladder tissue: intravenous administration, urinary excretion, and absorption into bladder tissue. *Antimicrob. Agents Chemother.* *11*:941–945 (1977).
102. Kinami, Y., H. Miyazaki, F. Koyama, M. Noguchi, K. Konishi, N. Furukawa, and Y. Nishida, Treatment of pancreas cancer with neocarzinostatin. *J. Jpn. Soc. Cancer Ther.* *10*:502–508 (1975).
103. Satake, I., K. Tari, M. Yamamoto, and H. Nishimura, Neocarzinostatin-induced complete regression of metastatic renal cell carcinoma. *J. Urology* *133*:87–89 (1985).
104. Maeda, H., N. Yamamoto, and A. Yamashita, Fate and distribution of [^{14}C] succinyl neocarzinostatin in rats. *Europ. J. Cancer* *12*:865–870 (1976).
105. McKelvey, E.M., M.A. Burgess, K.B. McCredie, W.K. Murphy, and G.P. Bodey, Neocarzinostatin: A phase I clinical trial with five-day intermittent and continuous infusions. *Cancer* *44*:1182–1188 (1979).
106. Hasuda, K., H. Kobayashi, T. Kuroiwa, K. Aoki, S. Taniguchi, and T. Baba, Efficacy of two-route chemotherapy using intraperitoneal neocarzinostatin and its antidote, intravenous tiopronin, for peritoneally disseminated tumors in mice. *Jpn. J. Cancer Res.* *80*:283–289 (1989).
107. Maeda, H., M. Ueda, T. Morinaga, and T. Matsumoto, Conjugation of poly(styrene-*co*-maliec acid) derivatives to the antitumor protein neocarzinostatin: Pronounced improvements in pharmacologic properties. *J. Med. Chem.* *28*:455–461 (1985).
108. Kohgo, Y., H. Kondo, J. Kato, K. Sasaki, N. Tsushima, T. Nishisato, M. Hirayama, K. Fujikawa, N. Shintani, Y. Mogi, and Y. Niitsu, Kinetics of internalization and cytotoxicity of transferrin-neocarzinostatin conjugate in human leukemia cell line, K562. *Jpn. J. Cancer Res.* *81*:91–99 (1990).
109. Konishi, M., H. Ohkuma, K.-i. Saitoh, H. Kawaguchi, J. Golik, G. Dubay, G. Groenewald, B. Krishnan, and T. Doyle, Esperamicins, a novel class of potent antitumor antibiotics. I: Physico-chemical data and partial structure. *J. Antibiot.* *38*:1605–1609 (1985).
110. Lee, M.D., T.S. Dunne, M.S. Marshall, C.C. Chang, G.O. Morton, and D.B. Borders, Calicheamicins, a novel family of antitumor antibiotics. I: Chemistry and partial structure of calicheamicin γ_1I. *J. Am. Chem. Soc.* *109*:3464–3466 (1987).
111. Golik, J., H. Wong, D.M. Vyas, and T. Doyle, Stereochemical studies on esperamicins: determination of the absolute configuration of isopropylamino sugar moiety. *Tetrahedron Lett.* *30*:2497–2500 (1989).
112. Lee, M.D., T.S. Dunne, C.C. Chang, G.A. Ellestad, M.M. Siegel, G.O.

Morton, W.J. McGahren, and D.B. Borders, Calichemicins, a novel family of antitumor antibiotics. 2. Chemistry and structure of calichemicin $\gamma 1^{I}$. *J. Am. Chem. Soc. 109*:3466–3468 (1987).

113. Kahne, D., D. Yang, and M.D. Lee, Preparation of the 4-ethylamino sugar of calicheamicin: Assignment of absolute configuration. *Tetrahedron Lett. 31*:21–22 (1990).
114. Golik, J., J. Clardy, G. Dubay, G. Groenewald, H. Kawaguchi, M. Konishi, B. Krishnan, H. Ohkuma, K.-i. Saitoh, and T.W. Doyle, Esperamicins, a novel class of potent antitumor antibiotics. 2. Structure of esperamicin X. *J. Am. Chem. Soc. 109*:3461–3462 (1987).
115. Ibid., 3462–3464.
116. Ellestad, G.A., P.R. Hamann, N. Zein, G.O. Morton, M.M. Siegel, M. Pastel, D.B. Borders, and W.J. McGahren, Reactions of the trisulfide moiety in calicheamicin. *Tetrahedron Lett. 30*:3033–3036 (1989).
117. Zein, N., W.J. McGahren, G.O. Morton, J. Ashcroft, and G.A. Ellestad, Exclusive abstraction of nonexchangeable hydrogens from DNA by calicheamicin $\gamma 1^{I}$. *J. Am. Chem. Soc. 111*:6888–6890 (1989).
118. Zein, N., A.M. Sinha, W.J. McGahren, and G.A. Ellestad, Calicheamicin $\gamma 1^{I}$: an antitumor antibiotic that cleaves double-stranded DNA site specifically. *Science 240*:1198–1201 (1988).
119. DeVoss, J.J., C.A. Townsend, W.-D. Ding, G.O. Morton, G.A. Ellestad, N. Zein, A.B. Tabor, and S.L. Schreiber, Site-specific atom transfer fron DNA to a bound ligand defines the geometry of a DNA-calicheamicin $\gamma 1^{I}$ complex. *J. Am. Chem. Soc. 112*:9669–9670 (1990).
120. Zein, N., M. Poncin, R. Nilakantan, and G.A. Ellestad, Calicheamicin $\gamma 1^{I}$ and DNA: molecular recognition process responsible for site-specificity. *Science 244*:697–699 (1989).
121. Hawley, R.C., L.L. Kiessling, and S.L. Schreiber, Model of the interactions of calicheamicin $\gamma 1^{I}$ with a DNA fragment from pBR322. *Proc. Natl. Acad. Sci. USA 86*:1105–1109 (1989).
122. Sugiura, Y., Y. Uesawa, Y. Takahashi, J. Kuwahara, J. Golik, and T.W. Doyle, Nucleotide-specific cleavage and minor-groove interaction of DNA with esperamicin antitumor antibiotics. *Proc. Nat. Acad. Sci. USA 86*:7672–7676 (1989).
123. Long, B.H., J. Golik, S. Forenza, B. Ward, R. Rehfuss, J.C. Dabrowiak, J.J. Catino, S.T. Musial, K.W. Brookshire, and T.W. Doyle, Esperamicins, a class of potent antitumor antibiotics: mechanism of action. *Proc. Nat. Acad. Sci. USA 86*:2–6 (1989).
124. Shiraki, T., and Y. Sugiura, DNA binding and cleavage by enediyne antitumor antibiotics, esperamicin and dynemicin. *Nucleic Acids Res. Symp. Ser. 21*:53–54 (1989).
125. Konishi, M., H. Ohkuma, K. Matsumoto, T. Tsuno, H. Kamei, T. Miyaki, T. Oki, and H. Kawaguchi, Dynemicin A, a novel antibiotic with the anthraquinone and 1,5-diyne-3-ene subunit. *J. Antiobiot. 42*:1449–1452 (1989).
126. Konishi, M., H. Ohkuma, T. Tsuno, T. Oki, G.D. VanDuyne, and J. Clardy,

Crystal and molecular structure of dynemicin A: a novel 1,5-diyne-3-ene antitumor antibiotic. *J. Am. Chem. Soc. 112*:3715–3716 (1990).
127. Snyder, J.P., and G.E. Tipsword, Proposal for blending classical and biradical mechanisms in antitumor antibiotics: dynemicin A. *J. Am. Chem. Soc. 112*:4040–4042 (1990).
128. Semmelhack, M.F., J. Gallagher, and D. Cohen, Bioreductive alkylation as a trigger for toxic effects of dynemicin. *Tetrahedron Lett. 31*:1521–1522 (1990).
129. Sugiura, Y., T. Shiraki, M. Konishi, and T. Oki, DNA intercalation and cleavage of an antitumor antibiotic dynemicin that contains anthracycline and enediyne cores. *Proc. Nat. Acad. Sci. USA 87*:3831–3835 (1990).
130. Shiraki, T., and Y. Sugiura, Visible light induced DNA cleavage by the hybrid antitumor antibiotic dynemicin A. *Biochemistry 29*:9795–9798 (1990).
131. Lee, M.D., T.S. Dunne, M.M. Siegel, C.C. Chang, G.O. Morton, and D.B. Borders, Calicheamicins, a novel family of antitumor antibiotics. 1. Chemistry and partial structure of calicheamicin $\gamma 1^I$. *J. Am. Chem. Soc. 109*:3464–3466 (1987).
132. Golik, J., T.W. Doyle, G. Van Duyne, and J. Clardy, Stereochemical studies on esperamicin A1: A single crystal x-ray structure of thiosugar moiety. *Tetrahedron Lett. 31*:6149–6150 (1990).
133. Wallace, R.E., L.H. Hinman, P. Hamann, J. Upeslacis, and F.E. Durr, Calicheamicin immunoconjugates: activity against human tumors in athymic mice. *Proc. Amer. Assoc. Cancer Res. 31*:285 (1990).
134. Wender, P.A., M. Harmata, D. Jeffrey, C. Mukai, and J. Suffert, Studies on DNA-active agents: the synthesis of the parent carbocyclic subunit of neocarzinostatin chromophore A. *Tetrahedron Lett. 29*:909–912 (1988).
135. Hirama, M., K. Fujiwara, K. Shigemata, and Y. Fukazawa, The 10-membered ring analogues of neocarzinostatin chromophore: design, synthesis, and mode of decomposition. *J. Am. Chem. Soc. 111*:4120–4122 (1989).
136. Wender, P.A., J.A. McKinney, and C. Mukai, General methodology for the synthesis of neocarzinostatin chromophore analogues: intramolecular chromium-mediated closures for strained-ring systems. *J. Am. Chem. Soc. 112*: 5369–5370 (1990).
137. Krebs, A., T. Wehlage, and C.-P. Kramer, Synthesis of a bicyclic oxacycloalkene diyne system related to neocarzinostatin chromophore A. *Tetrahedron Lett. 31*:3533–3536 (1990).
138. Nakatani, K., K. Arai, N. Hirayama, F. Matsuda, and S. Terashima, Synthesis and cytotoxicity of the acyclic (*E*)- and (*Z*)-dienediyne systems related to neocarzinostatine chromophore. *Tetrahedron Lett. 31*:2323–2326 (1990).
139. Foland, L.D., J.O. Karlsson, S.T. Perri, R. Schwabe, S.L. Xu, S. Patil, and H.W. Moore, Rearrangement of 4-alkynylcyclobutenones. A new synthesis of 1,4-benzoquinones. *J. Am. Chem. Soc. 111*:975–989 (1989).
140. Schreiber, S.L., and L.L. Kiessling, Synthesis of the bicyclic core of the esperamicin/calicheamicin class of antitumor agents. *J. Am. Chem. Soc. 110*:631–633 (1988).

141. Danishefsky, S.J., N.B. Mantlo, and D.S. Yamashita, A concise route to the calichemicin-esperamicin series: the crystal structure of a core subunit. *J. Am. Chem. Soc. 110*:6890–6891 (1988).
142. Haseltine, J.N., S.J. Danishefsky, and G. Schulte, Experimental modeling of the priming mechanism of the calicheamicin/esperamicin antibiotics: actuation by the addition of intramolecular nucleophiles to the bridgehead double bond. *J. Am. Chem. Soc. 111*:7638–7640 (1989).
143. Kende, A.S., and C.A. Smith, Synthesis of a calicheamicin deoxyaglycone model by an intramolecular acetylide cyclization. *Tetrahedron Lett. 29*:4217–4220 (1988).
144. Magnus, P., and P.A. Carter, A model for the proposed mechanism of action of the potent antitumor antibiotic esperamicin A_1. *J. Am. Chem. Soc. 110*:1626–1628 (1988).
145. Magnus, P., R.T. Lewis, and J.C. Huffman, Synthesis of a remarkably stable bicyclo[7.3.1]diynene esperamicin A_1/calicheamicin γ system. Structural requirements for facile formation of a 1,4-diyl. *J. Am. Chem. Soc. 110*:6921–6923 (1988).
146. Magnus, P., and R.T. Lewis, Synthetic studies on the esperamicin/calicheamicin antitumor antibiotics. Conjugate addition of thiol to initiate 1,4-diyl formation. *Tetrahedron Lett. 30*:1905–1906 (1989).
147. Tomioka, K., H. Fujita, and K. Koga, Synthesis of the core enediyne structure of esperamicin-calicheamicin class of antitumor antibiotics. *Tetrahedron Lett. 30*:851–854 (1989).
148. Nicolau, K.C., and R.D. Groneberg, Novel strategy for the construction of the oligosaccharide fragment of calicheamicin $\gamma 1\alpha^I$. Synthesis of the ABC skeleton. *J. Am. Chem. Soc. 112*:4085–4086 (1990).
149. Nicolau, K.C., T. Ebata, N.A. Stylianides, R.D. Groneberg, and P.J. Carrol, Methyl 4-hydroxy-5-iodo-2,3-dimethoxy-6-methylbenzoate: The aromatic fragment of calicheamicin $\gamma 1\alpha^I$. Synthesis, x-ray crystallographic analysis, and properties. *Angew. Chem. Int. Ed. Engl. 27*:1097–1099 (1989).
150. van Laak, K., and H.D. Scharf, Synthesis of methyl 2,6-dideoxy-4-thio-α-*D-ribo*-hexo-pyranoside, a new thio sugar found in calicheamicins. *Tetrahedron Lett. 30*:4505–4506 (1989).
151. Nicolaou, K.C., G. Zuccarello, Y. Owaga, E.J. Schweiger, and T. Kumazawa, Cyclic conjugated enediynes related to calicheamicins and esperamicins: calculations, synthesis, and properties. *J. Am. Chem. Soc. 110*:4866–4868 (1988).
152. Snyder, J.P. The cyclization of calicheamicin-esperamicin analogues: a predictive biradicaloid transition state. *J. Am. Chem. Soc. 111*:7630–7632 (1989).
153. Magnus, P., S. Fortt, T. Pitterna, and J.P. Snyder, Synthetic and mechanistic studies on esperamicin A_1 and calicheamicin $\gamma 1$. Molecular strain rather than π-bond proximity determines the cycloaromatization rates of bicyclo[7.3.1]enediynes. *J. Am. Chem. Soc. 112*:4986–4987 (1990).
154. Snyder, J. Monocyclic enediyne collapse to 1,4-diyl biradicals: a pathway under strain control. *J. Am. Chem. Soc. 112*:5367–5369 (1990).
155. Nicolaou, K.C., Y. Ogawa, G. Zucarello, and H. Kataoka, DNA cleavage

by a synthetic mimic of the calicheamicin-esperamicin class of antibiotics. *J. Am. Chem. Soc. 110*:7247–7248 (1988).

156. Nicolaou, K.C., G. Skokotas, P. Maligres, G. Zuccarello, E.J. Schweiger, K. Toshima, and S. Wendeborn, A new class of DNA-cleaving molecules: pH-dependent DNA cleavage by propargylic and allenic sulfones. *Angew. Chem. Int. Ed. Engl. 28*:1272–1275 (1989).
157. Nicolaou, K.C., G. Skokotas, S. Furuya, H. Suemune, and D.C. Nicolaou, Golfomycin A, a novel designed molecule with DNA-cleaving properties and antitumor activity. *Angew. Chem. Int. Ed. Engl. 29*:1064–1066 (1990).
158. Nicolaou, K.C., P. Maligres, J. Shin, E. de Leon, and D. Rideout, DNA cleavage and antitumor activity of designed molecules with conjugated phosphine oxide-allene-ene-yne functionalities. *J. Am. Chem. Soc. 112*:7825–7826 (1990).
159. Nagata, R., H. Yamanaka, E. Okazaki, and I. Saito, Biradical formation from acyclic conjugated eneyne-allene system related to neocarzinostatin and esperamicin-calichemicin. *Tetrahedron Lett. 30*:4995–4998 (1989).
160. Nagata, R., H. Yamanaka, E. Murahashi, and I. Saito, DNA cleavage by acyclic eneyne-allene systems related to neocarzinostatin and esperamicin-calicheamicin. *Tetrahedron Lett. 31*:2907–2910 (1990).
161. Zein, N., W.-D. Ding, and G.A. Ellestad, Interaction of calicheamicin with DNA. In *Molecular Basis of Specificity in Nucleic Acid-Drug Interactions* Pullman, B. and Jortner J., eds., Kluwer, Netherlands, 323–330 (1990).

11
Comparative Studies of Intercalative DNA-Drug Interactions Using Molecular Modeling and Biophysical Techniques

Mavis Agbandje and Robert McKenna
Purdue University, West Lafayette, Indiana

I. INTRODUCTION

Cancer chemotherapeutic drugs may be classified according to their mechanism of action into several classes: hormones (1), immunological agents (2), radiosensitizers (3), antimetabolites (4), mitotic spindle inhibitors (5), and (6) those which interact with deoxyribose nucleic acid (DNA). Drugs which interact with DNA include the aklylating agents (1,6,7), groove binders (7), and intercalators (1,6,7). In this chapter two types of intercalating compounds with potential anticancer activity are discussed, namely, anthraquinone-based and phenylquinoline-8-carboxamide-based derivatives. Discussions will range from the rationale of molecular design to experimental results of their interactions with DNA.

This chapter studies the interaction of these intercalating compounds with DNA using computerized molecular modeling and energy minimization techniques, and biophysical methods such as spectroscopic binding, thermal denaturation, and unwinding of closed circular supercoiled plasmid DNA. In the search for new anticancer drugs, potential agents are critically evaluated in animal tumor models in which human neoplasia has been simulated, following preliminary testing in tumor cells in vitro. These types of evaluation were also applied to the compounds discussed.

II. INTERCALATION

Intercalators are the most significant form of noncovalent DNA binders. The theory of DNA intercalation, proposed by Lerman in 1961 (8) (Fig. 1), explained fiber diffraction and solution studies observations of DNA-acridine complexes. These observations were the continued presence of the 3.4 Å meridional repeat in diffraction patterns, a decrease in mass per unit length of the DNA, and the DNA helix extension seen using hydrodynamic techniques (9).

This class of drugs includes daunomycin, actinomycin D and echinomycin (Fig. 2). These drugs possess a planar aromatic chromophore which is able to stack between adjacent base pairs within the DNA helix. This causes the base pairs at the intercalation site to become separated by 6.8 Å instead of the usual 3.4 Å so as to accommodate the drug. Intercalators stabilize the DNA by stacking interactions which are essentially van der Waals forces. The side chains on the chromophore in the DNA grooves can also participate in hydrogen bonding with the bases or the phosphodiester backbone to further stabilize the complex (10).

Detailed studies (11) of the conformational features of the crystal structures of different intercalated dinucleotides have shown that the backbone torsion angles do not fundamentally differ. The torsion angles about the

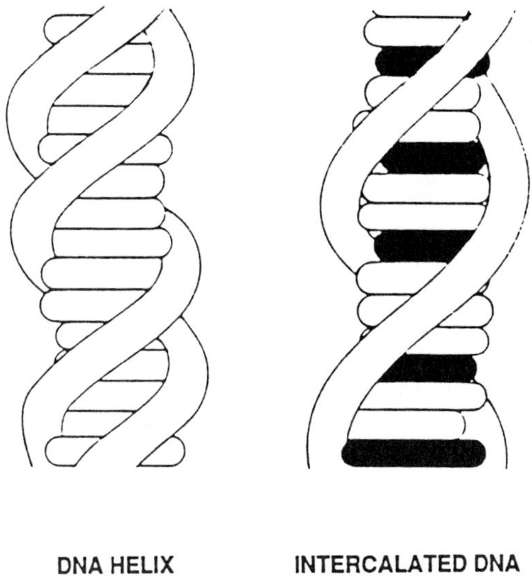

Figure 1 Lerman's intercalation model. The black disks indicate intercalated drug.

O5′-C5′ bond in each strand are greater by about 50° than the value of 175° in A-DNA, and the glycosidic angles at the 3′-end of each strand are greater than these corresponding polynucleotide angles by about 60–70°. The fundamental changes therefore are simple expansions about the above two torsion angles per strand.

III. RATIONALE FOR DRUG DESIGN

A. Anthraquinone Derivatives

The development of anthraquinone-based anticancer agents has been dominated by reports of the anthracycline antibiotics daunomycin and adriamycin (12,13). Daunomycin and adriamycin presently rank among the most important chemotherapeutic drugs in clinical use. Daunomycin is used in the treatment of acute lymphocytic and myelogenous leukemias (14). Adriamycin is an active antileukemic drug, but is also effective against a broad spectrum of solid tumors (15). The clinical efficacy of both daunomycin and adriamycin is, however, limited by dose-related specific toxicity to the heart (16,17), which occurs in addition to the usual toxic effects of anticancer agents, for example, leukocytopenia, lymphocytopenia, immunosupression and alopecia. Consequently, much effort has gone into the elucidation

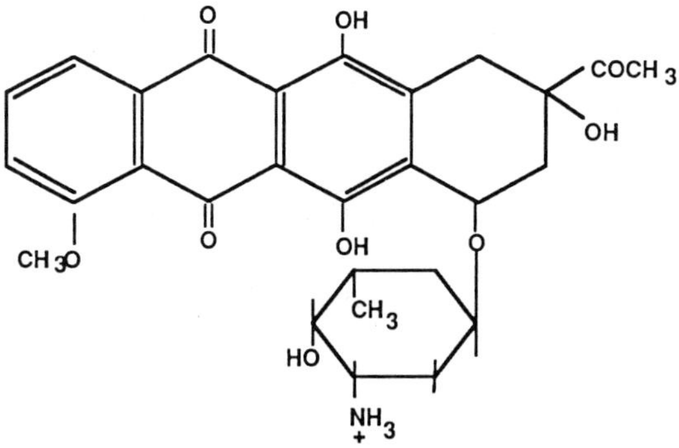

DAUNOMYCIN

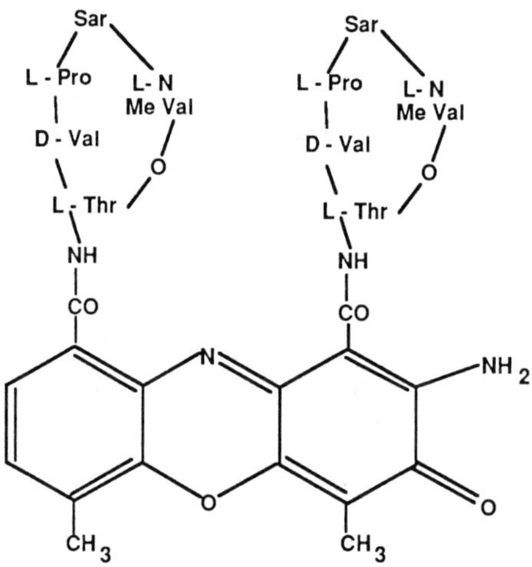

ACTINOMYCIN D

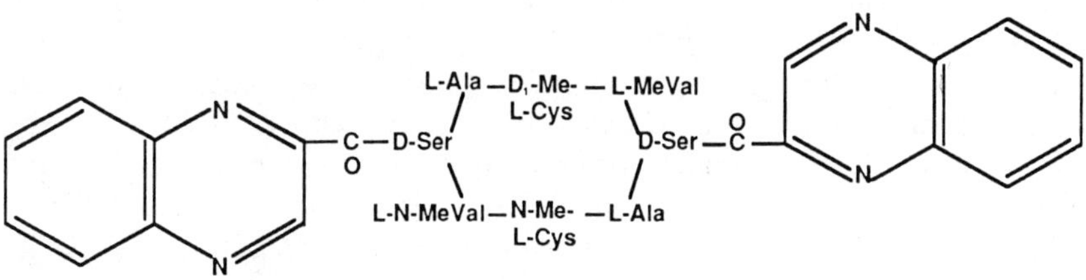

ECHINOMYCIN

Figure 2 Classic DNA intercalators.

of the underlying molecular origin of this cardiotoxicity (18,19,20), so that attempts can be made to alleviate or minimize this effect, either by structural modification (21) or by the design and de novo synthesis of less toxic agents (22,23). Although the mechanism of the cardiotoxicity of these anthracyclines remains incompletely understood, studies of its molecular basis suggest that the effects may be related to redox reduction of the quinone moiety, leading to the formation of free radical intermediates in vivo (18,24,25). The free-radical species formed are thought to induce destructive cellular events such as enzyme inactivation, DNA strand cleavage and membrane lipid peroxidation.

The anthraquinones have a lower tendency to undergo one-electron reduction than the anthracyclines (26). Thus synthetic efforts in the quest for less cardiotoxic anticancer agents have recently focussed attention upon the development of aminoalkyl-functionalized anthraquinones, particularly following the demonstration of clinical activity with mitoxantrone, 1,4-dihydroxy-5,8-bis((2-(2-hydroxyethyl)amino)ethylamino)-9,10-anthracenedione (27,28,29).

The anthraquinones discussed here, aminoalkanamido-substituted anthracene-9,10-diones (Fig. 3), were designed with the aid of computer modeling to intercalate into DNA. The compounds thus feature a planar fused aromatic chromophore and amine functionalized side chains attached via the two- and six-ring positions, which modeling suggests may interact electrostatically with the phosphodiester backbone of DNA. A similar group

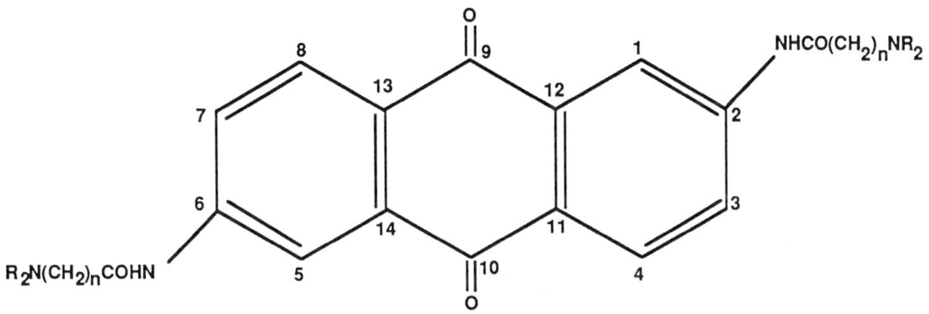

where n=1-3, NR_2=secondary amine

Figure 3 The general structural formula for the 2,6-bis(ω-aminoalkanamido)-9,10-anthracenediones.

of compounds functionalized at positions 1 and 1,4 on the anthracene-9,10-dione chromophore has been developed (30). The molecular modeling studies of the interaction of these compounds with DNA indicated that the usual 1,4-ring disubstitution positions may not be optimal, and suggested that the 2,6-positions may yield compounds whose side chains better interact with the DNA backbone atoms (D. Collier and S. Neidle, personal communication).

B. Phenylquinoline-8-Carboxamides

The DNA-intercalating abilities of acridines and related compounds in relation to cytotoxic and antineoplastic activities, have been extensively studied, with particular reference to the anticancer drug amsacrine (Fig. 4), a 9-aminoacridine derivative (31,32,33,34).

In general, for many analogues of amsacrine it has been found that DNA intercalation is a necessary, though by itself insufficient, property for high in vivo and in vitro anticancer activity to be shown. Analogues with reduced levels of DNA binding, such as 1-methyl amsacrine (35) which has a nonplanar chromophore hindering full intercalation of the molecule, have lower biological activity. Over the past few years several new series of carboxamide-substituted acridines and 9-aminoacridines have been developed (36,37,38,39). The 9-aminoacridine-4-carboxamide derivative (Fig. 4) (acridine $pK_a = 8.30$) is a highly active antileukemic agent but shows no activity against solid tumors, e.g., the Lewis lung carcinoma, whereas the more weakly basic acridine-4-carboxamide compound (Fig. 4) (acridine $pK_a = 3.54$) shows good activity in both tumor models (38). This observation is seen in other weakly basic tricyclic compounds, for example in phenazines (40) and phenanthridines (41) (Fig. 4). Therefore an uncharged chromophore appears to be critical for an intercalating acridine drug to have activity in solid tumors.

The compounds discussed here are a series of substituted quinolines (42) that have a carboxamide side chain at the 8-position and a phenyl ring at differing places on the quinoline chromophore (Fig. 5). These compounds have an acridine pK_a of less than 3.2 and thus should have the same activity against solid tumors as other weakly basic acridine derivatives. While it is generally accepted that for efficient intercalative binding three fused rings are required, this in no way implies that smaller ring systems cannot intercalate (43). The reduction of the size of the potential intercalating chromophore of these compounds to a bicyclic system was rationalized on the basis of possible improved cellular distribution as well as reduction of nonspecific binding to other cell macromolecules such as proteins and lipids.

AMSACRINE

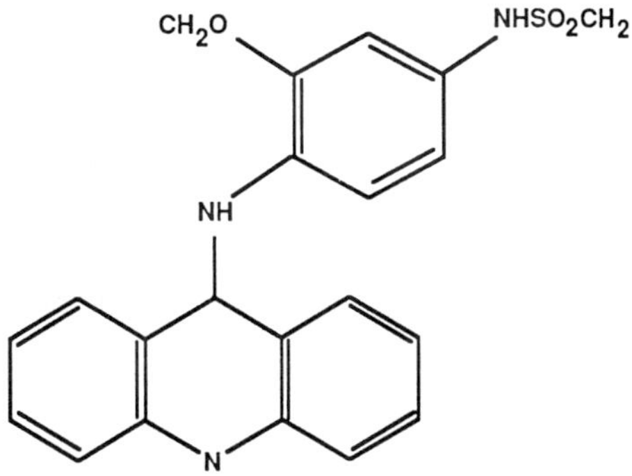

9-AMINOACRIDINE-4-CARBOAMIDE $R = NH_2$

ACRIDINE-4-CARBOXAMIDE $R = H$

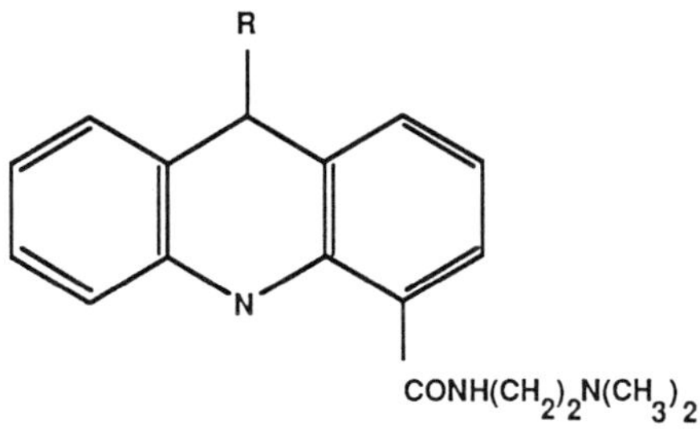

PHENAZINE

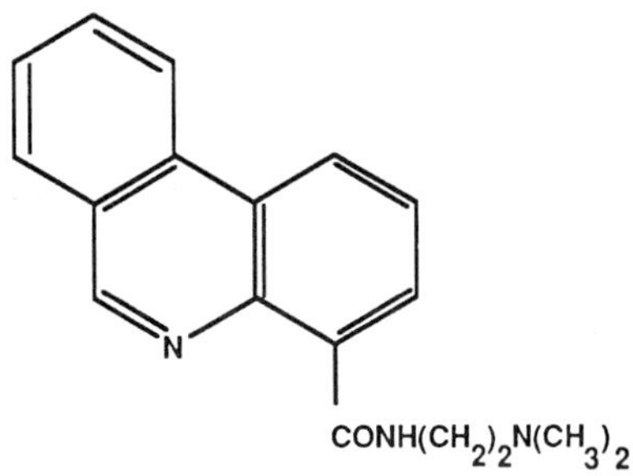

PHENANTHRIDINE

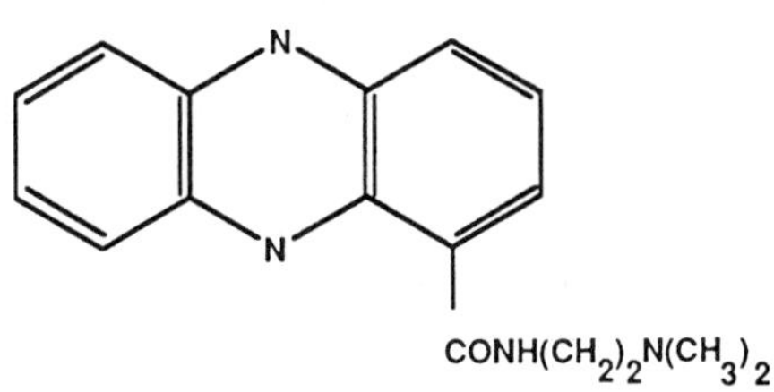

Figure 4 Structural formulae for some anticancer intercalating acridine drugs.

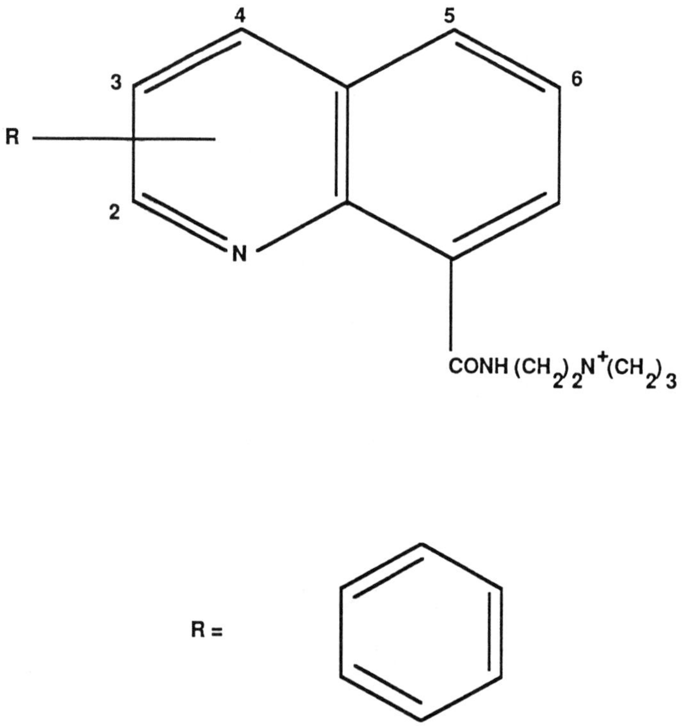

For B1 R= H

B2 -B6 Phe at positions 2,3,4,5 & 6

Figure 5 The structural formulae of the phenylquinoline-8-carboxamides.

IV. MOLECULAR MODELING

Computerized graphics technology developed in the last two decades has become a very powerful tool in the field of drug design. This is because it allows theoretical studies of the interaction between a substrate and receptor of known structure to be modeled visually (44). It has made possible the investigation of both potential and established anticancer agents with DNA on a theoretical basis (45). In many cases this type of investigation has been used as the first step in the rational design of new anticancer agents (46). This is because these studies result in the understanding and verification of the modes of interaction and the determination of features of the DNA-drug interaction that may be relevant for the design of new drugs. There are however limitations in the use of molecular modeling results in rational drug design as it is focused only on the structural and electronic aspects of the actual DNA-drug interaction, whereas additional factors such as drug absorption, transport, and metabolism should also be considered. Therefore, extrapolation from these types of theoretical studies have to be done with a great deal of caution.

The computer-assisted molecular modeling studies of the DNA-drug interactions discussed in this chapter used a combination of interactive computer graphics and semiemperical energy calculations. The molecular modeling procedure involved four main steps:

1. Acquisition of a predetermined intercalation geometry in a hexamer sequence (47)
2. Building of three-dimensional models of compounds to be modeled
3. Study of intercalative DNA-drug interactions using interactive computer graphics, followed by conformational energy calculations of the complexes formed
4. Visualization of the geometries of the intercalation complex, using computer graphics

The molecular modeling procedures involved the use of a consensus dinucleoside intercalation site geometry d(CpG) derived from the x-ray crystal structure of a proflavine-d(CpG) intercalation complex (48). The hexamer sequence d(TACGTA) (Fig. 6) was used as a model for intercalation into duplex DNA. This hexamer sequence was constructed from a deca-

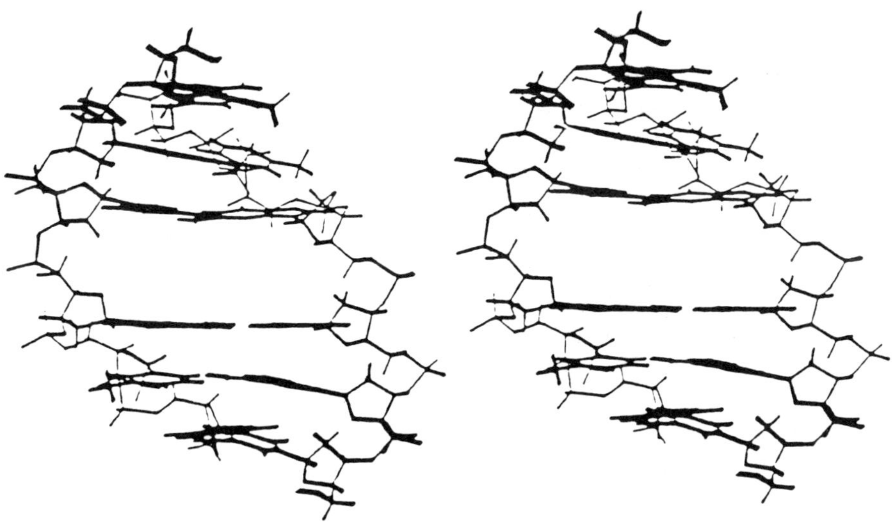

Figure 6 The hexanucleotide sequence d(TACGTA).

nucleotide complex by means of least-squared fitting of B-form DNA residues to both ends of the d(CpG) dinucleotide intercalation geometry, followed by energy minimization (47).

The coordinates of the chromophore and side chains for the molecules were taken from X-ray crystal structures of representative parent compounds. Partial atomic charge densities for the molecules, in their intercalated conformation, were calculated on a VAX 11/750 computer using the quantum methods CNDO/2 (49) and MNDO as part of the AMPAC package (50).

Interactive computer graphics, on a Silicon Graphics IRIS workstation, with the molecular modeling program GEMINI (51) was used to dock the compounds into the intercalation site of the hexamer sequence. The dinucleotides at the intercalation site were held fixed throughout the docking maneuvers.

A. Anthraquinone Derivatives

The molecular modeling studies on the anthraquinone derivatives (Fig. 7), were carried out in an attempt to predict the effect of the different structural features present in the molecules, i.e., the number of methylene groups (n) in the side chains, the steric bulk of the terminal amines, and the presence of hydrophilic hydroxyl groups in the side chain, on their interactions with DNA. These compounds have been given the numbers A1–A7 for simplicity, their serial numbers (Ref. 74) are BSU-1002(A1), BSU-1015(A2), BSU-1045(A3), BSU-1038(A4), BSU-1006(A5), BSU-1017 (A6), and BSU-1041(A7). The compounds were protonated on the amine nitrogens to simulate physiological conditions (pH 7.0–7.4) as they have pK_a values greater than 7. The initial objective was to intercalate the planar chromophore at an average distance of 3.4 Å from the adjacent intercalation site base pairs. The positioning of the side chains in the major and minor grooves were then studied. First, the compounds were modeled with both side chains in the major groove. Next both side chains were modeled in the minor groove. Finally, they were modeled with one side chain in the major groove and the other in the minor groove. The docking maneuvers showed that intercalation of the drugs, leading to maximum base stacking interactions with the chromophore can only be achieved with one side chain in each groove. The successful docking of these chromophores were followed by a search for possible hydrogen bond interactions between the hydrogen atoms attached to the nitrogens in the side chains and the phosphate oxygens on the backbone. The possibility of hydrogen bond formation between the side chain amide oxygen and base hydrogens was also explored alongside the possibility of hydrophobic van der Waals contacts.

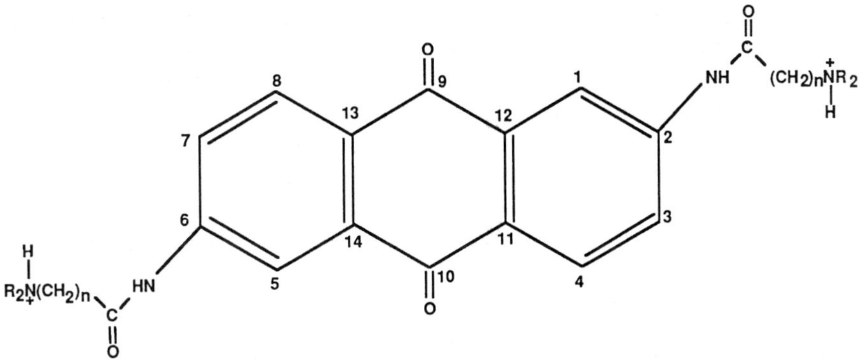

COMPOUND NO.	n	$-NR_2$	COMPOUND NO.	n	$-NR_2$
A1	1	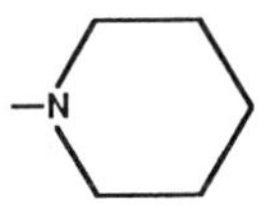	A5	1	$-N(CH_2CH_3)_2$
A2	2	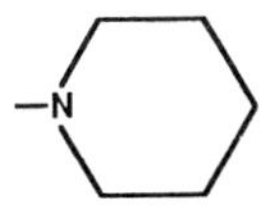	A6	2	$-N(CH_2CH_3)_2$
A3	1	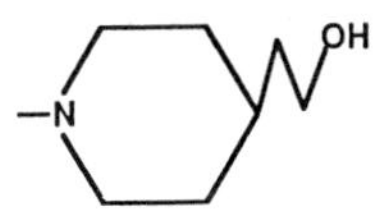	A7	2	$-N(CH_2CH_2OH)_2$
A4	2	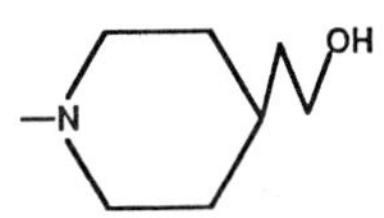			

Figure 7 The structural formulae of the compounds used in the molecular modeling of the interactions of the 2,6-bis(ω-aminoalkanamido)-9,10-anthracenediones with DNA.

These procedures were carried out by rotating different torsion angles along the side chains to alter their geometry, without altering the geometry of the amide group. A position of low potential energy for the intercalated complex was decided interactively, the criteria for which were maximal overlap between the intercalation site base pairs and the anthraquinone chromophore while maximizing possible hydrogen bonding between the side-chain NH groups and the backbone. Repulsive contacts of the side chains with the backbone were minimized, and van der Waals interactions maximized.

B. Phenylquinoline-8-Carboxamides

The molecular modeling studies on the phenylquinolines were conducted in an attempt to determine the effect of the position of the phenyl ring substituent on their intercalative properties. The pK_a of the chromophores for all six phenylquinolines compounds (Fig. 5) is less than 3.2 (40), hence the hydrogen atom on the N1 atom (as found in the crystal structure of compound B2) was removed in the compounds to be modeled (the converse situation of the anthraquinones). This gave rise to the possibility of a hydrogen bond between N-1 and the N-18 atom H-181 of the carboxamide, which was achieved by a change of the C-7–C-8–C-17–N-18 torsion angle by 180°. A similar situation has been described for *N*-(2-dimethylaminoethyl)- and *N*-(2-dimethylaminobutyl)-9-aminoacridine-4-carboxamides (52). The structures of the five other compounds were generated using this basic quinoline-8-carboxamide, by relocating the phenyl substituent at the 3-,4-, 5-, or 6-position on the quinoline ring to produce structures for B3–B6 and also by completely removing the phenyl group in the case of compound B1 (Fig. 5). The possibility of occupation of either the major or minor DNA groove by the carboxamide side chain was considered for all six compounds. In the case of compounds B2 and B6, the positioning of both the phenyl group and the side chain simultaneously in the groove was sterically possible and was thus studied. The criteria for visual interactive docking to obtain an initial starting model for energy refinement was as previously mentioned for the anthraquinone derivatives.

V. ENERGY CALCULATIONS

The coordinates of the visualized intercalated drug-DNA complexes were subjected to molecular mechanics full-geometry energy minimization, using the program EMPMDS (53) implemented on a VAX 11/750 computer, to determine the total energy (enthalpy) for the low energy state. The total energy computed consists of a number of discrete noninteracting energy contributions (54). The force-field computation of the total energy as

$$E_{\text{TOTAL}} = \Sigma k_r(r - r_0)^2 + \Sigma k\Theta(\Theta - \Theta_0)^2 + \Sigma \frac{V_n}{2[1 + \cos(n\Phi - \gamma)]} + \Sigma \left[\frac{A_{ij}}{R_{ij}^{12}} - \frac{B_{ij}}{R_{ij}^{6}}\right] + \Sigma \frac{q_i q_j}{\epsilon R_{ij}} \quad (1)$$

where the terms represent bond length and angle deformation, torsion angle barriers, van der Waals nonbonded interactions, and electrostatic energies. The force field parameters used for the DNA were an all-atom representation, which were essentially identical to those derived by Kollman and his associates (55). The force-field parameters for the drug molecules were those which had been developed from previous studies of intercalating compounds (56). A distance-dependent dielectric constant of the form $\epsilon = 4r_{ij}$ was employed (47). Previous use of this formalism suggests that it gives results for nucleic acids that approximate solvent and counterion shielding (47,57). The DNA-drug complexes were judged to have reached convergence in total energy refinement when the root-mean-square gradient was less than 0.2 kcal mol^{-1} Å^{-1}, which was achieved within approximately 2000 cycles of energy refinement. The binding energies (ΔE), for the DNA-drug interactions, were calculated from minimization on the complexes, the drugs in isolation and on the hexamer sequence, d(TACGTA) in a B-DNA form helix conformation. This energy was defined as the energy of the complex minus the sum of the B-DNA and drug energies:

$$\Delta E = E_{\text{complex}} - (E_{\text{drug}} + E_{\text{DNA}}) \quad (2)$$

VI. BIOPHYSICAL METHODS USED FOR THE STUDY OF DNA-DRUG INTERACTIONS

There are several methods available for the study of the intercalative process, which allow both the nature and the affinity of the binding to be determined. When a compound intercalates into nucleic acids, there are changes which occur in both the DNA and the compound during the complexation which can be used for studying the binding process. The binding of the compounds under discussion with DNA were studied using methods that monitor changes in the spectroscopic properties of the drug and in the physical properties of the DNA.

The intercalation of a drug into DNA is an equilibrium binding process. Thus the determination of the equilibrium binding constant for the intercalation process should give an insight into the nature and strength of the DNA-drug interaction involved. To determine the association constant for this process, the concentration of free and bound drug in the DNA-drug mixture must be determined. This can be done by spectrophotometric methods (58,59), equilibrium dialysis (60), ultracentrifugation (61), or the solute-enhanced partition analysis method (62,63). The interaction of the compounds discussed in this chapter with DNA resulting in changes in their spectral properties were studied using spectrophotometric binding and equilibrium dialysis methods.

The interaction of intercalators with DNA cause changes in its physical properties (7,64). These changes include the stabilization of the DNA double helix against thermal denaturation and the unwinding of the double helix of the DNA. Hydrodynamic methods used in monitoring physical changes in DNA resulting from intercalation show an increase in viscosity and a decrease in the sedimentation coefficient of the DNA (65–68). In the body of work discussed here stabilization of DNA against thermal denaturation and unwinding of the double helix were used to observe the physical changes in DNA induced following binding by the compounds.

VII. EVALUATION OF COMPOUNDS IN TUMOR MODELS

The preclinical screening of potential anticancer agents in tumor models involves in vitro and in vivo systems. This screening is based on the assumption that a correlation exists between the activity of a compound against transplantable tumors in animals and its therapeutic effectiveness in human cancer (69). The anthraquinones and the phenylquinolines were screened for anticancer activity in vitro and in vivo tumor models. Both series of derivatives were screened for their inhibition of cellular proliferation in the L1210 leukemia tumor cell line. The animal tumor model systems used for the anthraquinones and the phenylquinolines were L1210 (70) and P388 (71) leukemia, respectively.

VIII. RESULTS OF THE MOLECULAR MODELING

The results of the energy minimization calculations for the simulated DNA-drug complexes are presented in Tables 1 and 2. The energy-minimized complexes were visualized on a Silicon Graphics IRIS workstation using the program GEMINI (51). The intercalated models, plotted using the BONZO (72) plotting routine, are shown in Figures 8 and 9, for the anthraquinones and phenylquinolines, respectively.

A. Anthraquinone Derivatives

The high binding energies calculated for the anthraquinone derivatives (Table 1) indicate that these compounds are strong DNA binders when modeled as intercalators. In general, the $n = 2$ anthraquinone derivatives were ob-

Table 1 Energies Calculated for the Interaction(s) of the 2,6-Bis(ω-aminoalkanamido)-9,10-Anthraquinones with the Intercalation Site in d(TACGTA), in kcal mol^{-1}

Compound	Drug-hexamer total energy	Drug energy	Binding energy (ΔE)[a]	Model number
A1	−310.39	−3.20	−35.34	1
A2	−324.98	−3.40	−49.73	2
A3	−316.84	−4.37	−40.62	3
A4	−331.63	−6.19	−53.59	4
A5	−299.82	1.79	−29.76	5
A6	−302.36	4.70	−35.21	6
A7	−301.51	9.49	−39.13	7

[a]The total energy of the d(TACGTA) energy-refined B-DNA double helix is −271.85 kcal mol^{-1}.

Table 2 Energies Calculated for the Interaction(s) of the Phenylquinoline-8-Carboxamide Derivatives with the Intercalation Site in d(TACGTA), in kcal mol^{-1}

Compound	Groove for side chain	Drug-hexamer total energy	Drug energy	Binding energy (ΔE)[a]	Model number
B1	Minor	−287.9	−0.2	−15.9	1
	Major	−285.1	−0.2	−13.1	2
B2	Minor	−287.3	−0.6	−14.9	3
	Major	−288.1	−0.6	−15.7	4
B2*	Minor	−294.0	−0.6	−21.6	5
	Major	−271.1	−0.6	1.2	6
B3	Minor	−290.3	1.7	−20.2	7
	Major	−285.3	1.7	−15.2	8
B4	Minor	−286.4	5.0	−19.4	9
	Major	−283.9	5.0	−17.1	10
B5	Minor	−285.2	4.9	−18.3	11
	Major	−282.3	4.9	−15.4	12
B6	Minor	−299.1	1.8	−25.7	13
	Major	−283.9	1.8	−13.9	14
B6*	Minor	−282.0	1.8	−12.0	15
	Major	−265.2	1.8	4.8	16

The compounds marked (*) have their chromophore-carboxamide torsion angle around C8–C17, approximately at the crystallographically observed value.

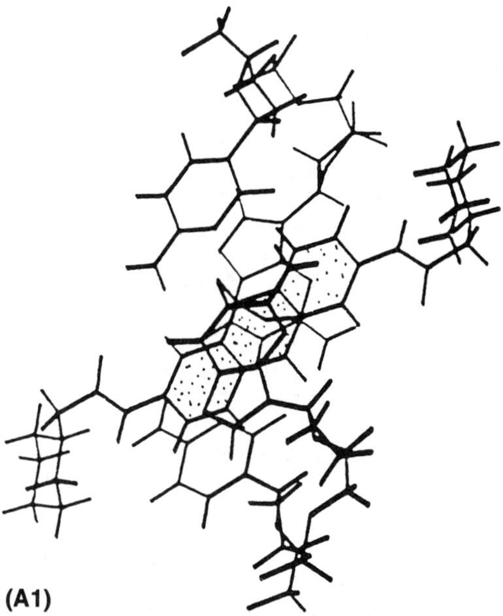

(A1)

Figure 8 Views of the intercalated models for compounds A1–A7 in the d(TACGTA) hexanucleotide, looking down the helical axis. The intercalated drugs are stippled for clarity.

served to have better overlap between their chromophores and the base pairs in the intercalation site. Their chromophores were positioned almost parallel to the long axis of the base pairs. With the $n = 1$ derivatives, parallel positioning of the chromophore resulted in repulsive contacts between the side chains and the DNA backbone, which could not be relieved due to the inflexibility of the shorter chain. In all the models, the binding energy of the $n = 2$ compound is greater than that for the $n = 1$. This difference in binding energy of the $n = 1$ and $n = 2$ derivatives can be clearly seen between models 1 and 2, where both compounds have a piperidine as their terminal amine, models 3 and 4, where the terminal amine is 4-(2-hydroxy)ethylpiperidine and models 5 and 6, where the terminal amine is ethylamine. The different DNA binding energies obtained for all the compounds indicated that the formation of an intercalation complex with DNA is influenced by structural factors. The better overlap of the chromophore of the $n = 2$ derivatives in combination with formation of intermolecular hydrophobic contacts between their more flexible methylene hydrogens and the bases, as well as the hydrogens of the sugars in the two DNA

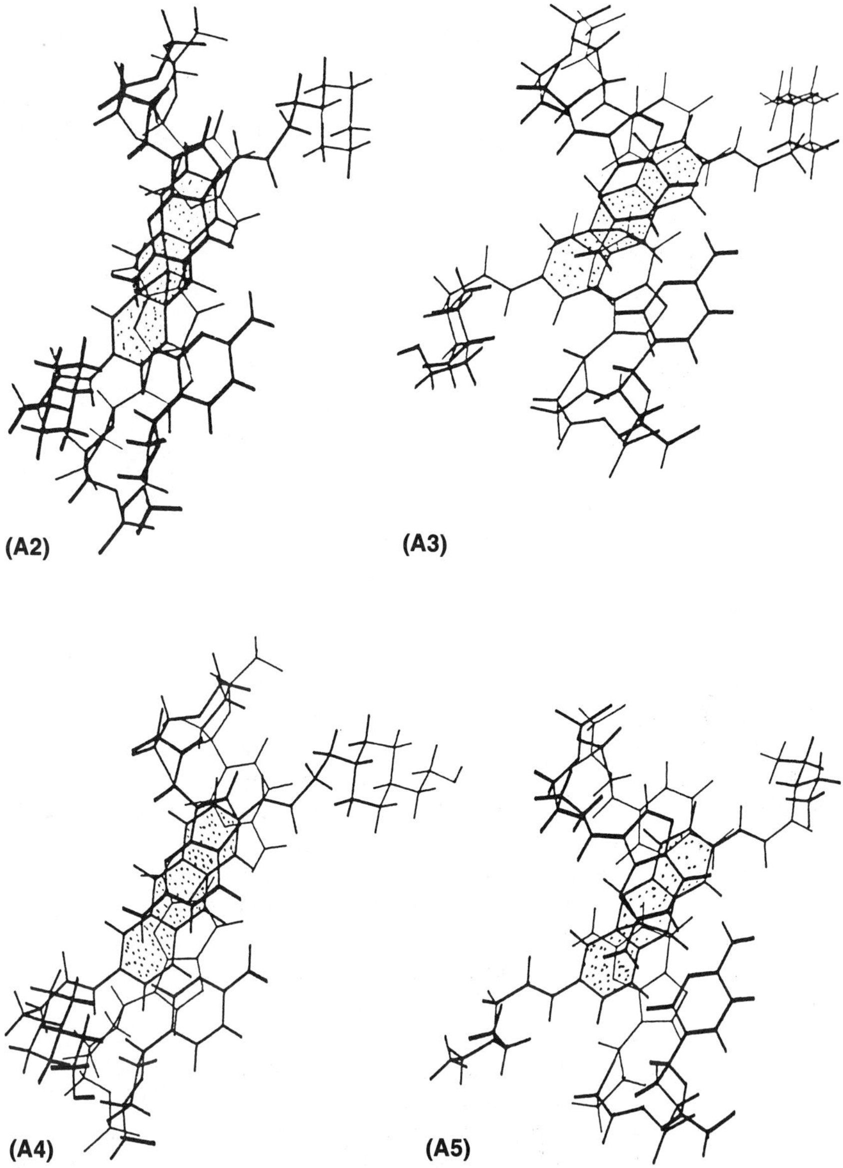

Figure 8 continued

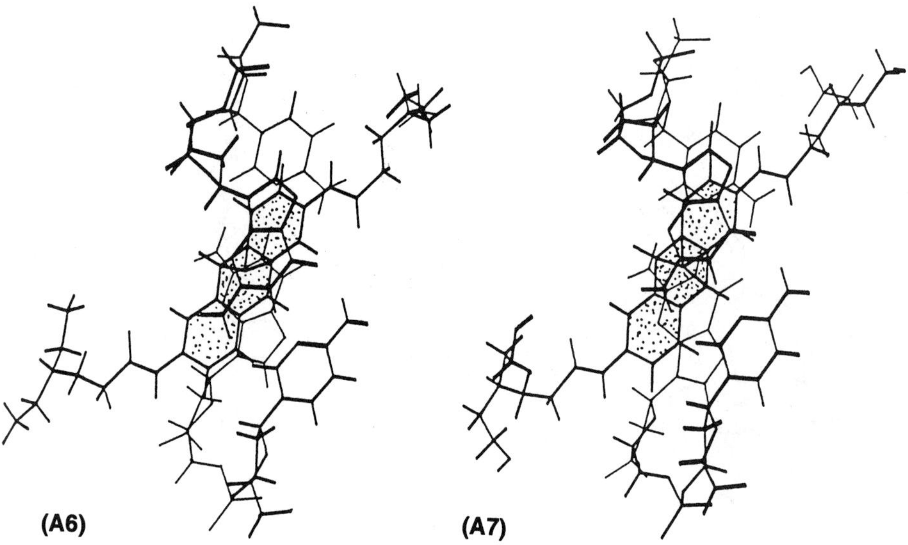

(A6) (A7)

Figure 8 continued

strands, are probably the determinant factors in their improved binding energies. The observed differences in the binding energies calculated for the compounds with similar side-chain lengths indicated that major roles are played by the differences in the intermolecular contacts of the different side chains with the DNA strands. The examination of the effect of bulkiness of the terminal amine substituents of the side chains on complex formation with DNA showed that the compounds with the bulkier amine had better binding energies. This was exemplified by the observed binding order of A4 > A2 > A6 for the $n = 2$ derivatives and A3 > A1 > A5 for the $n = 1$ derivatives. This order of calculated binding energy was probably a result of the greater number of close van der Waals contacts observed between the bulkier amine substituents and the hydrophobic walls of the two DNA strands. The presence of hydroxyl groups in the side chain was also observed to affect the binding energy of the DNA-drug complex. Hydroxyl groups in the side chain increased binding energies by approximately 4 kcal mol^{-1} as observed when comparing A3 and A1, A2, and A4, and A6 and A7. The increase in binding energy is probably due to the possibility of hydrogen bond formation between these side-chain hydroxyl group hydrogens and the oxygens in the sugar-phosphate backbone brought about by close hydrophobic interactions.

The complexes obtained with the anthraquinone derivatives demon-

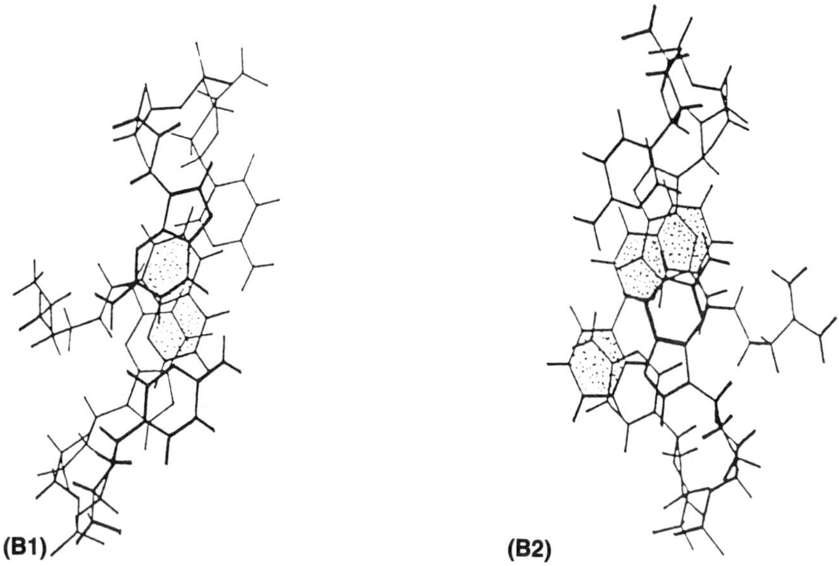

Figure 9 Views of the intercalated "best" models for compounds B1–B6 in the d(TACGTA) hexanucleotide, looking down the helical axis. The intercalated drugs are stippled for clarity.

strated neighbor exclusion (73) in their binding with DNA. The neighbor exclusion principle of drug intercalation states that maximal drug binding occurs every other site. The models indicated that binding at every site was impossible, on steric grounds. The side chains were observed to span up to three base bases in each groove. The number of base pairs occluded by the compounds during the modeling studies are in good agreement with the results for the binding site size observed during the spectrophotometric determination of the binding constants for this group of compounds, with calculated values of approximately three to four base pairs (74) (Table 3).

B. Phenylquinoline-8-Carboxamides

The molecular modeling of intercalative-type interactions for compounds B1–B6 to the CpG site demonstrated that there are both quantitative (Table 2) and qualitative differences between the complexes for each compound. The most stable models had their side chain lying in the DNA minor groove. From a quantitative view it was observed that compounds B1, B4, and B5 are poor intercalators characterized by having only their quinoline chromophore stacked between base pairs. This was in contrast to the best (minor

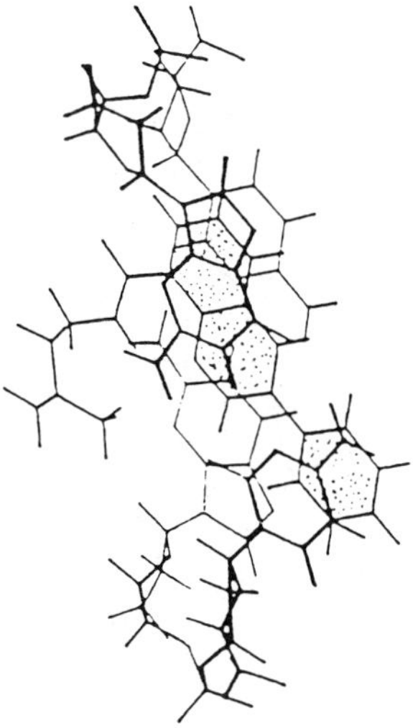

(B3)

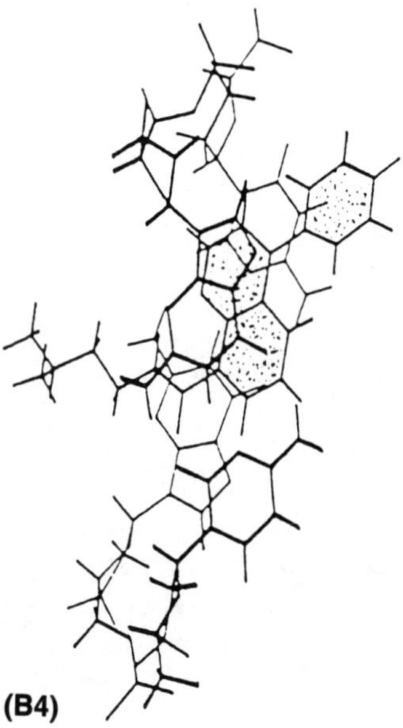

(B4)

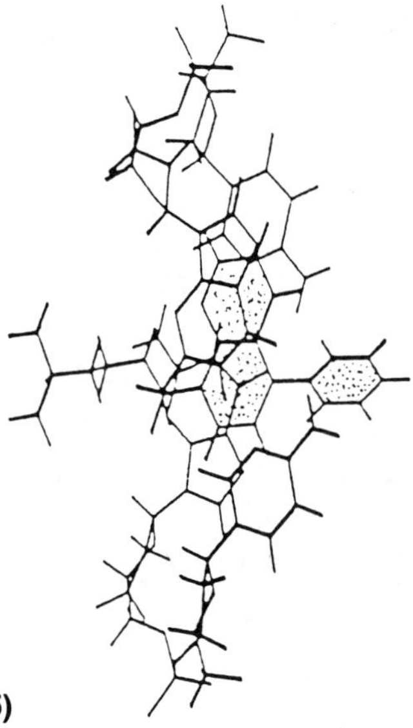

(B5)

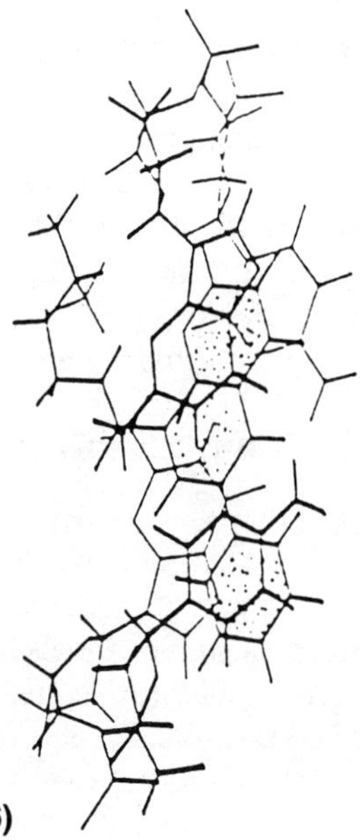

(B6)

Figure 9 continued

Table 3 Experimental, Physiochemical, and Biological Data of the 2,6-Bis(ω-aminoalkanamido)-9,10-Anthraquinones

Compound no.	Unwinding[a] angle	logK^b	n^c	IC_{50}[d]	L1210[e] ILS	ΔE^f	ΔG^g
A1	ND	ND	ND	76.00	NA	−35.34	ND
A2	21	7.89	3.65	0.19	NA	−49.73	−41.3
A3	ND	ND	ND	ND	ND	−40.62	ND
A4	ND	ND	ND	2.30	36.8	−53.59	ND
A5	ND	ND	ND	27.00	ND	−29.76	ND
A6	26	6.14	4.38	0.34	NA	−35.21	−32.1
A7	23	6.59	4.10	4.20	ND	−39.13	−34.4

[a]The degree of local unwinding of the plasmid DNA helix due to each molecule of drug bound to DNA, relative to a value of 26° determined for ethidium bromide, estimated values of ±2°.
[b]Affinity constants from the McGhee-von Hippel extended model (76).
[c]The binding site size.
[d]IC_{50} = the concentration of compound (μmol dm^{-3}) required to kill 50% of cell population.
[e]The percentage increase in life-span of mice with L1210, NA indicates inactive compound.
[f]The calculated binding energy.
[g]Free energies of binding, in kcal mol^{-1} from K values using $\Delta G = RT\,lnK$.
ND = not determined.

groove) models for B2, B3 and B6 which had their phenyl groups twisted so as to be involved in the stacking interactions with a cytosine base as well as having their chromophore stacked parallel to other bases in the intercalation site. The position of the attached phenyl ring is of major importance in allowing these interactions to be sterically achieved. Crucial roles were also played by differences in close intermolecular contacts involving the side chain, with its relatively tight fit in the DNA minor groove being between the hydrophobic walls of the two strands. There were numerous attractive close contacts between these strands and atoms of compounds B1 and B6. In the case of the minor-groove model 13 for compound B6, its superiority over all the other models was probably largely due to its very effective phenyl group stacking. By contrast, the major-groove model 16 for B6 (not shown), although it has some quinoline-guanine stacking, had no close contacts between the side chain and phenyl ring, and the DNA groove atoms. With model 6 for compound B2 (not shown), the relative disposition of the side chain and phenyl group produced too bulky an arrangement for comfortable positioning even in the major groove, resulting in a model with some close repulsive contacts with distances less than van der Waals separations.

IX. RESULTS OF THE DNA BINDING STUDIES

A. Anthraquinone Derivatives

The equilibrium binding constants determination for calf thymus DNA binding of the anthraquinone derivatives, using a spectrophotometric method (74) resulted in the values shown in Table 3. The data required for the calculation of the equilibrium binding constant, the binding site size, and the cooperativity factor (K, n, and w, respectively), were calculated by the method of Peacocke and Skerret (58). The experimental data were plotted according to the method of Scatchard (75) and analyzed using the extended neighbor exclusion model of McGhee and von Hippel (76). Parameters K, n, and w, were estimated by an iterative procedure designed to satisfy Eq. (3), given experimentally determined values of r and c, together with initial starting estimates of K, n, and w.

$$\frac{r}{c} = K(1 - nr)\left[\frac{(2w - 1)(1 - nr) + r - R}{2(w - 1)(1 - nr)}\right]^{n-1}\left[\frac{1 - (n + 1)r + R}{2(1 - nr)}\right]^2 \quad (3)$$

where r = number of sites occupied by the drug per DNAp
c = concentration of unbound drug and
$R = \{(1 - (n + 1)r)^2 + 4wr(1 - nr)\}^{1/2}$

Table 3 shows that these anthraquinone derivatives interact strongly with DNA. The equilibrium binding constants calculated ($K > 10^5$ dm^3 mol^{-1}) are of a magnitude commonly associated with classical DNA intercalating ligands (59). These results must, however, be treated with caution as the binding curve fitting is normally poor, due to the requirement for precise measurements of extremely small changes in optical absorption as the DNA approaches saturation by the drug during the experiment. The calculations resulted in low values of w (not given), indicating that cooperativity binding is unimportant in the interaction of these compounds with DNA.

The results from the determination of unwinding angle produced by the interaction of some of the modeled anthraquinone derivatives with closed-circular PM-2 plasmid DNA (74) are shown in Table 3. These compounds were able to unwind DNA, with A6 having a value as good as ethidium bromide. With the exception of A7, all the other compounds had the ability to reverse the supercoils of the PM-2 plasmid DNA.

Table 3 also contains the IC_{50} (in L1210 tumor cells) and increase in life span (of mice passaged with L1210 leukemia tumor cells) results for the anthraquinone derivatives under consideration (74). The results show that most of these compounds are cytotoxic to L1210 cells in vitro. These compounds were however at least 100-fold less toxic than mitoxantrone (IC_{50} of 0.002) which was used as a control. It was also observed that the in vitro toxicities of the $n = 1$ compounds (A1 and A5) were less than those for the $n = 2$ derivatives (A4, A6, and A7), suggesting that the number of methylene groups in the side chain is an important factor in the determination of in vitro activity. The in vivo antitumor activity results for these anthraquinone derivatives showed that they were inactive and toxic to the animals, with the exception of A4.

The effects of the anthraquinone derivatives on the helix-to-coil thermal denaturation transition of different DNAs (*Micrococcus lysodiektius*, calf thymus, *Clostridium perfringens* and poly[d(A-T)]·poly[d(A-T)] caused by complexation (74) are given in Table 4. Attempts to determine the effects of these compounds on the thermal denaturation of poly[d(G-C)]·poly[d(G-C)] were unsuccessful due to the very high melting temperatures of this polymer when complexed, even in buffer with no stabilizing salt and at low drug : DNA ratios. These studies with different DNAs were carried out in an attempt to determine possible base pair or sequence preferences for the anthraquinones. In all the studies the acetate salts of the anthraquinone derivatives were used and mitoxantrone as its hydrochloride salt was the control. Table 4 shows that all the compounds examined induced posi-

Table 4 Increase in Melting Temperature (ΔT_m) for Various Polynucleotides Following Complexation with Acetate Salts of the 2,6-Bis(ω-aminoalkanamido)-9,10-Anthraquinones

Compound no. ($-NR_2 \cdot HOAc$)	ΔT_m(°C)[e] of Different DNAs (DNA : drug ratio of 10 : 1)			
	AT[a]	CP[b]	CT[c]	ML[d]
A1	31.7	19.0	14.3	>18.4
A2	33.7	>18.9	>15.3	>14.5
A3	ND	ND	ND	ND
A4	ND	ND	ND	ND
A5	31.2	16.6	13.5	>16.6
A6	32.9	>20.5	>15.5	>18.1
A7	ND	11.7	10.9	>15.7
Mitoxantrone	ND	>26.4	>15.0	>11.7

[a]poly[d(A-T)]·poly[d(A-T)](T_m of native was 44.1 ± 0.2°C).
[b]*Clostridium perfringens* (CP, 72% A-T, T_m of native was 61.1 ± 0.2).
[c]Calf thymus (CT, 58% A-T, T_m of native was 67.0 ± 0.2°C).
[d]*Micrococcus lysodiekticus* (ML, 28% A-T, T_m of native was 72.0°C ± 0.2).
[e]The ΔT_m values were determined at the DNA : DRUG ratio shown for the CP, CT, and MT DNAs, but at a ratio of 5 : 1 for AT; values represent means from at least three determinations with esd of ±0.1–0.3°C.

tive shifts in the melting temperature of the different DNAs examined. The shifts produced by the majority of these compounds are quite large, showing that they are strong DNA binders and are comparable to those for mitoxantrone, adriamycin (77), and daunomycin (78). The length of the side chain of these compounds appeared to have little influence upon the increase in the T_m of the nucleic acids as did the presence of hydroxyl groups. There was no significant effect of steric bulk of the side-chain amine upon thermal stabilization of the DNAs. There was no positive evidence of sequence preferences observed with these anthraquinone derivatives with the different DNAs; however, the $n = 2$ compound A2 showed indications of AT preference and A7 greatly stabilized GC-rich *Micrococcus lysodiekticus* DNA. Mitoxantrone was observed to have AT-rich sequence binding preference, in contrast to modeling studies by other workers who suggest a GC site preference for this intercalator.

B. Phenylquinoline-8-Carboxamides

The experimental data on the DNA binding affinities and in vitro activity for the phenylquinoline-8-carboxamides were obtained from Dr. W. A. Denny and are published in Ref. 42; hence, only a summary of this work is described here. The equilibrium binding constants, K, given in Table 5, are

Table 5 Experimental, Physicochemical and Biological Data for the Phenylquinoline-8-carboxamides

Compound	Unwinding angle[a]	log K[b]	IC_{50}[c]	P388 ILS[d]	ΔE[e] (kcal mol^{-1})	ΔG[f] (kcal mol^{-1})
B1	0°	5.15	725.0	NA	−15.9	−12.7
B2*	14°	5.95	1.3	91	−21.6	−14.7
B3	20°	5.69	2.6	27	−20.2	−14.0
B4	0°	4.75	14.7	NA	−19.4	−11.8
B5	0°	4.76	13.7	NA	−18.3	−11.8
B6	20°	6.04	0.91	29	−25.7	−14.9

[a]As for Table 3.
[b]Equilibrium binding constant determined by means of the ethidium displacement method. Mean of values found for poly[d(AT)] and poly[d(GC)].
[c]As for Table 3.
[d]Percentage increase in life span of treated animals with intraperitoneally injected P388 leukemia cells. NA indicates inactive compounds.
[e, f]As for Table 3.
Source: From Ref. 42.

the mean for poly[d(A-T)]·poly[d(A-T)] and poly[d(G-C)]·poly[d(G-C)] as there was no experimentally significant difference between the values for a given compound. The high *K* values demonstrated strong association to DNA for compounds B2, B3, and B6, whereas the other three compounds showed weaker affinities by approximately one order of magnitude. The anthraquinone derivatives showed greater DNA association than the 'best' phenylquinoline B6 (also by approximately one order of magnitude). The results of the unwinding of the closed circular pNZ116 plasmid DNA, showed that only compounds B2, B3, and B6, were able to unwind DNA, but again to a lesser extent than the anthraquinones. The IC_{50} for the compounds against L1210 leukemia cells showed that the three compounds with strongest DNA association also inhibit cell growth in vitro. When the compounds were tested for in vivo activity they were shown to be effective and prolonged the life span of animal injected with P388 leukemia cells. Compound B2 was outstanding as the best anticancer compound in the series extending the life span of the injected animals by 91%, the average life span of the control animals being 11 days.

X. DISCUSSION

The experimental data on the DNA interactions for the two series of compounds discussed enables comparisons between the modeling and the experimental results to be made. The binding energies calculated here do not directly take into account differences in solvation between different complexes or between drug-bound and drug-free DNA (79). Thus, the calculated ΔE values cannot be formally compared with experimentally derived free energies, ΔG. Nevertheless the ranking orders of the calculated ΔE values and ΔG for the two types of compounds treated separately do correspond moderately well. This suggests that the scaling of the relatively large electrostatic component of the force field by a $4r_{ij}$ term was a satisfactory procedure.

A. Anthraquinone Derivatives

The theoretical molecular modeling and the biophysical studies conducted on this series of potential anticancer agents were carried out in an attempt to verify their postulated mode of intercalative interaction with nucleic acids and to determine the structural features required for their interaction and possible antitumor activity. The results obtained indicate that there are definite structural considerations involved in the determination of the activity of a series of compounds towards DNA and also their antitumor capability, at least in vitro. All the compounds were shown to interact with DNA, with intercalation inferred to be their most likely mode, based on the experimental results obtained from unwinding studies, DNA binding constants ($>10^5$ dm^3 mol^{-1}) and thermal stabilization values in the range expected for intercalating compounds. The strong interaction of these compounds with DNA observed with the biophysical techniques were supported by the molecular modeling studies. The observation that the binding site size for the compounds were three to four base pairs using spectrophotometric techniques was also supported by the observation of neighbor exclusion with the modeling studies. It is evident from all the techniques used for the study of these compounds that the length of the side chain attached to the anthraquinone chromophore has a critical influence upon both the in vitro cytotoxicity and DNA binding, with the $n = 2$ derivatives having greater activity. A comparison of the in vitro cytotoxicity results for these 2,6-disubstituted anthracene-9,10-diones (with $n = 2$) with the 1,4-derivatives (also $n = 2$) (Ref. 30, previously mentioned in Sec. III.A), where the terminal amines are the same, indicated that the position of the side chains on the anthracene-9,10-dione chromophore may also be important in determining toxicity in vitro. It was found that the 2,6-compounds were more cytotoxic than the 1,4. This result supported the observation, during the molecular modeling of the 1,4-compounds, that this position may not be the optimal for disubstitution on the anthracene-9,10-dione chromophore. There was no notable correlation between the improved calculated binding energies of the 2,6-compounds with hydroxyl groups in their side chains and the results from the biophysical studies, however the only compound which showed in vivo biological activity of the compounds tested, A4, contains a hydroxyl group in its side chain. These result tentatively suggests that the side-chain length of two methylene groups, together with the presence of polar hydroxylalkyl substituents, may be important factors for in vivo activity, particularly in view of the clinical activity of mitoxantrone (27,28,29).

B. Phenylquinoline-8-Carboxamides

Table 5 shows that compounds B1, B4, and B5 are poor intercalators, with the other three compounds having significantly higher association constants, in the order B6 > B2 > B3, as well as being classical intercalators, unwinding covalently closed circular supercoiled DNA (80). These results were reflected in their ranking order in calculated binding energies. The poorest DNA binders of the series show very low potency, which extends to their lack of in vivo activity, whereas the other three intercalating compounds are all active (to varying extents) against P388 leukemia.

XI. CONCLUSION

This study has provided a quantitative molecular basis for the DNA binding behavior and biological activity of the two types of intercalating compounds. The molecular modeling has correctly indicated the overall order of DNA affinity in both series, which also suggests that the intercalation model used here has relevance for the conformation of the binding site in solution, although detailed confirmation must await the results of X-ray or NMR analyses.

As demonstrated here, with the availability of detailed crystallographic structures of DNA-drug complexes, "drug design" can be employed to produce and/or understand analogs of drugs which best utilized a particular feature of the DNA structure (in this case the intercalation site), to maximize their interactions. This however does not always lead to clinically useful drugs. Other factors such as metabolism, cellular uptake, drug persistency, cytotoxicity, and possible reactions with other cellular components must also be examined. These additional factors can result in the design of biologically limited molecules as has been so far observed for the anthraquinone derivatives (members of this series of compounds are still undergoing biological evaluation). In other instances potentially useful molecules are produced. The phenylquinolines, for example, showed reasonable correlation between the in vitro and in vivo results, suggesting that the design goals of improved cellular distribution and reduction of nonspecific binding to cell macromolecules may have been achieved. The above points underline both the usefulness and the dangers of using only structural knowledge in the search for active anticancer drugs.

ACKNOWLEDGMENTS

We would like to thank Stephen Neidle, Terry Jenkins, William A. Denny, Phylis Goddard, and Melanie Valenti for their involvement in this work. Tom Smith for his help in the preparation of this manuscript. This work was supported by the Cancer Research Campaign (U.K.) and The Institute of Cancer Research (London, U.K.).

REFERENCES

1. Farmer, P.B., Cancer chemotherapy. I: design and mechanism of action of cytotoxic drugs. In *The Molecular Basis of Cancer*, Farmer, P.B. and Walker, J.M. eds., Croom Helm, U.K., 259–285 (1985).
2. Powis, G., Cancer chemotherapy. II: laboratory to clinic. In *The Molecular Basis of Cancer*, Farmer, P.B. and Walker, J.M. eds., Croom Helm, U.K. 287–324 (1985).

3. Workman, P., Radiation therapy. In *The Molecular Basis of Cancer*, Farmer, P.B. and Walker, J.M. eds., Croom Helm U.K. 237-258 (1985).
4. Farber, S., L.K. Diamond, R.D. Mercer, R.F. Sylvester, and J.A. Wolff, *New Eng. J. Med. 238*:787- (1948).
5. Johnson, I.S., J.G. Armstrong, M. Gorman, and J.P. Burnett, The vinca alkaloids: a new class of oncolytic agents. *Cancer Res. 23*:1390–1427 (1963).
6. Pratt, W.B., and R.W. Ruddon, eds., *The Anticancer Drugs*, Oxford University Press, New York (1979).
7. Gale, E.F., E. Cundliffe, P.E. Reynolds, M.H. Richmond, and M.J. Waring, eds., *The Molecular Basis of Antibiotic Action*, 2nd ed., Wiley, London (1981).
8. Lerman, L.S., Structural considerations in the interaction of DNA and acridines. *J. Mol. Biol. 3*:18–30 (1961).
9. Cantor, C.R. and P.R. Schimmel, *Biophysical Chemistry*, Part II, *Techniques for the Study of Biological Structure and Function*, Freeman, San Francisco (1980).
10. Neidle, S., and Z. Abraham. Structural and sequence-dependent aspects of drug intercalation into nucleic acids. *CRC Crit. Rev. Biochem. 17*:73–121 (1984).
11. Berman, H.M., S. Neidle, and R.K. Stodola, Drug-nucleic acid interactions. Conformational flexibility at the intercalation site. *Proc. Nat. Acad. Sci. USA 75*:828–832 (1978).
12. Arcamone, F. *Doxorubicin: Anticancer Antibiotics*, Academic Press, New York (1981).
13. Wiernik, P.H., In *Anthracyclines: Current Status and New Developments*, S.T. Crooke, and S.D. Reich, eds., Academic Press, New York (1980).
14. Bernard, J., R. Paul, M. Boiron, U. Macquillat, and R.D. Maral, eds., *Recent Results in Cancer Research: Rubidomycin*, Springer-Verlag, New York (1969).
15. Blum, R.H., and S.K. Carter, Adriamycin. A new anticancer drug with significant clinical activity. *Ann. Inter. Med. 80*:249–259 (1974).
16. Tritton, T.R., S.A. Murphree, and A.C. Sartorelli, Adriamycin: A proposal on the specificity of drug action. *Biochem. Biophys. Res. Comm. 84*:802–808 (1978).
17. Young, R.C., R.F. Ozols, and C.E. Myers, The anthracycline antineoplastic drugs. *New Eng. J. Med. 305*:139–153 (1981).
18. Lown, J.W., The mechanism of action of quinone antibiotics. *Mol. Cell. Biochem. 55*:17–40 (1983).
19. Doroshow, J.H., G.Y. Locker, and C.E. Myers, Enzymatic defenses of the mouse heart against reactive oxygen metabolites: alterations produced by doxorubicin. *J. Clin. Invest. 65*:128–135 (1980).
20. Goodman, J., and P. Hochstein, Generation of free radicals and lipid peroxidation by redox cycling of adriamycin and daunomycin. *Biochem. Biophys. Res. Commun. 77*:797–803 (1977).
21. Tong, G.L., D.W. Henry, and E.M. Acton, 5-Iminodaunorubicin. Reduced cardiotoxic properties in an antitumor anthracycline. *J. Med. Chem. 22*:36–39 (1979).
22. Lown, J.W., S.M. Sondhi, S.B. Mandal, and J. Murphy, Synthesis and redox

properties of chromophore modified glycosides related to anthracyclines. *J. Org. Chem. 47*:4304–4310 (1982).

23. Showalter, H.D.H., D.W. Fry, W.R. Leopold, J.W. Lown, J.A. Plambeck, and K. Reszka, Design, biochemical pharmacology, electrochemistry and tumor biology of antitumor anthrapyrazoles. *Anti-Cancer Drug Design 1*:73 (1986).
24. Bachur, N.R., S.L. Gordon, and M.V. Gee, A general mechanism for microsomal activation of quinone anticancer agents to free radicals. *Cancer Res. 38*:1745–1750 (1978).
25. Myers, C.E., In *Anthracycline Antibiotics in Cancer Therapy*, Muggia, F.M., Young, C.N., and Carter, S.K. eds., Martinus Nijhoff, Boston (1982).
26. Kharash, E.D., and R.F. Novak, Anthracenedione activation by NADPH-cytochrome P-450 reductase; comparison with anthracyclines. *Biochem. Pharm. 30*:2881–2884 (1981).
27. Cheng, C.C., and R.K.-Y. Zee-Cheng, The design, synthesis and development of a new class of potential antineoplastic anthraquinones. *Prog. Med. Chem. 20*:83–118 (1983).
28. Stuart-Harris, R.C., and I.E. Smith, Mitoxantrone: a phase II study in the treatment of patients with advanced breast carcinoma and other solid tumours. *Cancer Chemother. Pharm. 8*:179–182 (1982).
29. Estley, E.H., M.J. Keating, K.B. McCredie, G.P. Bodey, and E.J. Freidreich, Phase II trials of mitoxantrone in refractory acute leukemia. *Cancer Treat. Rep. 67*:389–390 (1983).
30. Collier, D.A., and S.J. Neidle, *J. Med. Chem. 31*:847–857 (1988).
31. Waring, M.J., DNA-binding characteristics of acridinylmethane-sulphonalide drugs: comparison with antitumour properties. *Europ. J. Cancer 12*:995–1001 (1976).
32. Baguley, B.C., W.A. Denny, G.F. Atwell, and B.F. Cain, Potential antitumor agents. 34. Quantitative relationships between DNA binding and molecular structure for 9-anilinoacridines substituted in the anilino ring. *J. Med. Chem. 24*:170–177 (1981).
33. Baguley, B.C., W.D. Denny, J.G. Atwell, and B.F. Cain, Potential antitumor agents. 35. Quantitative relationships between antitumor (L1210) potency and DNA binding for 4′-(9-acridinylamino)methanesulfon-*m*-anisidide analogues. *J. Med. Chem. 24*:520–525 (1981).
34. Wilson, W.R., B.C. Baguley, L.P.G. Wakelin, and M.J. Waring, Interactions of the antitumor drug 4′-(9-acridinylamino)methane-sulfon-*m*-anisidide and related acridines with nucleic acids. *Mol. Pharmacol. 20*:404–414 (1981).
35. Neidle, S., G.D. Webster, B.C. Baguley, and W.A. Denny, The crystal structure of 1-methylamsacrine hydrochloride; relationships to DNA-binding ability and anti-tumor activity. *Biochem. Pharmacol. 35*:3915–3921 (1986).
36. Atwell, G.J., B.F. Cain, B.C. Baguley, and W.A. Denny, Potential antitumor agents. 43. Synthesis and biological activity of dibasic 9-aminoacridine-4-carboxamides, a new class of antitumor agents. *J. Med. Chem. 27*:1481–1485 (1984).
37. Rewcastle, G.W., G.J. Atwell, D. Chambers, B.C. Baguley, and W.A. Denny,

Potential antitumor agents. 46. Structure-activity relationships for acridine monosubstituted derivatives of antitumor agent *N*-[2-(dimethylamino)ethyl]-9-aminoacridine-4-carboxamide. *J. Med. Chem. 29*:472–477 (1986).
38. Atwell, G.L., G.W. Rewcastle, B.C. Baguley, and W.A. Denny, Potential antitumor agents. 50. *In vivo* solid tumor activity of derivatives of *N*-[2-(dimethylamino)ethyl]acridine-4-carboxamide. *J. Med. Chem. 30*:664–669 (1987).
39. Denny, W.A., I.A.G. Roos, and L.P.G. Wakelin, Interactions between antitumor activity, DNA breakage, and DNA binding kinetics for 9-aminoacridinecarboxamide anti-tumor agents. *Anti-Cancer Drug Design 1*:141–147 (1986).
40. Rewcastle, G.W., W.A. Denny, and B.C. Baguley, Potential antitumor agents. 51. Synthesis and antitumor activity of substituted phenazine-1-carboxamides. *J. Med. Chem. 30*:843–851 (1987).
41. Palmer, B.D., G.W. Rewcastle, G.J. Atwell, B.C. Baguley, and W.A. Denny, Potential antitumor agents. 54. Chromophore requirements for *in vivo* antitumor activity among the general class of linear tricyclic carboxamides. *J. Med. Chem. 31*:707–712 (1988).
42. Atwell, G.L., B.C. Bos, B.C. Baguley, and W.A. Denny, Potential antitumor agents. 56. "Minimal" DNA-intercalating ligands as antitumor drugs: phenylquinoline-8-carboxamides. *J. Med. Chem. 31*:1048–1052 (1988).
43. Gabbay, E.J., and A.J. DePaolis, Topography of nucleic acid helices in solutions. Steric requirements for intercalation. *J. Am. Chem. Soc. 93*:562–564 (1971).
44. Cohen, C., *Advances in Drug Design 14*:41 (1985).
45. Neidle, S., L.H. Pearl, and J.V. Skelly, DNA structure and pertubation by drug binding. *Biochem. J. 243*:1–13 (1987).
46. Miller, K.J., and D.D. Nelvin, Interactions of molecules with nucleic acids. VI: Computer design of chromophoric intercalating agents. *Biopolymers 21*: 633–652 (1982).
47. Neidle, S., L.H. Pearl, P. Herzyk, and H.M. Berman, A molecular model for proflavine-DNA intercalation. *Nucleic Acids Res. 16*:8999–9016 (1988).
48. Shieh, H.S., H.M. Berman, M. Dabrow, and S. Neidle, The structure of drug-deoxydinucleoside phosphate complex: generalised conformational behaviour of intercalation complexes with RNA and DNA fragments. *Nucleic Acids Res. 8*:85–97 (1980).
49. Pople, J., and D. Beveridge, *Approximate Molecular Orbital Theory*, McGraw-Hill, New York (1978).
50. AMPAC, *Semi-Emperical Molecular-Orbital Package*, Quantum Chemistry Program Exchange No. 506, Dept. of Chemistry, Indiana University, Bloomington, IN.
51. Beveridge, A., GEMINI (*G*raphics *E*mulation of *M*olecules by *IN*teractive *I*nstruction), The Institute of Cancer Research, Sutton, Surrey, U.K. (1988).
52. Hudson, B.D., R. Kuroda, W.A. Denny, and S. Neidle, Crystallographic and molecular mechanics calculations on the antitumor drugs *N*-[(2-dimethylamino)butyl]-9-aminoacridine-4-carboxamides and their dications: Implications for models of DNA-binding. *J. Biomol. Struct. Dyn. 5*:145–158 (1987).

53. Haneef, I., Ph.D. thesis, Birbeck College, London (1985).
54. Kollman, P.A., In *X-ray Crystallography and Drug Action*, A.S. Horn, and C.J. De Ranter, eds., Oxford University Press, Oxford, U.K. (1984).
55. Weiner, J., P.A. Kollman, D.T. Nuggen, and D.A. Case, An all atom force field for simulations of protein and nucleic acids. *J. Computational Chem. 7*: 230 (1986).
56. Abraham, Z.H.L., M., Agbandje, S. Neidle, and R.M. Acheson, Experimental DNA binding and computer modeling studies on an analogue of the antitumour drug amsacrine. *J. Biomol. Struct. Dyn. 6*:471–488 (1988).
57. Whitlow, M., and M.M. Teeter, An empirical examination of potential energy minimisation using the well determined structure of the protein crambin. *J. Amer. Chem. Soc. 108*:7163–7172 (1986).
58. Peacocke, A.R., and J.N.H. Skerret, The interactions of aminoacridines with nucleic acids. *Trans. Faraday. Soc. 52*:261 (1956).
59. Dougherty, G., A single-beam microsphectrophotometer structure for investigating the linear dichroism of DNA-drug fibers. *Anal Biochem. 115*:52–57 (1981).
60. Muller, W., and D.M. Crothers, Interactions of heteroaromatic compounds with nucleic acids. 1. The influence of heteroatoms and polarizability on the base specificity of intercalating agents. *Eur. J. Biochem. 54*:267 (1975).
61. Lloyd, P.H., R.N. Prutton, and A.R. Peacocke, Sedimentation studies on the interactions of Proflavine with deoxyribonucleic acid. *Biochem. J. 107*:353–359 (1968).
62. Wakelin, L.P.G., and M.J. Waring, The unwinding of circular deoxyribonucleic acid by phenanthridium drugs: Structure-activity relations for the intercalation reaction. *Molec. Pharmacol. 10*:544–561 (1974).
63. Krugh, T.R., S.A. Winkle, and D.E. Graves, Solute enhanced partition analysis—a novel method for measuring the binding of drugs to DNA. *Biochem. Biophys. Res. Commun. 98*:317–323 (1981).
64. Neidle, S., The molecular basis for the action of some DNA-binding drugs. *Prog. Med. Chem. 16*:151–221 (1979).
65. Zunino, F., Studies on the mode of interaction of daunomycin with DNA. *Febs. Letts. 18*(2):249–253 (1971).
66. Kersten, W., H. Kersten, and W. Szybalski, Physiochemical properties of complexes between deoxyribonucleic acid and antibiotics which affect ribonucleic acid synthesis (actinomycin, daunomycin, cinerubin, nogalomycin, chromomycin, mithramycin, and olivomycin). *Biochemistry 5*:236–244 (1966).
67. Muller, W., and D.M. Crothers, Studies of the binding of Actinomycin and related compounds to DNA. *J. Mol. Biol. 35*:251–290 (1968).
68. O'Brien, R.L., J.L. Allison, and F.E. Hahn, Evidence for intercalation of Chloroquine into DNA. *Biochim. Biophys. Acta. 129*:622–624 (1966).
69. Farmer, P.B., and J.M. Walker, eds., *The Molecular Basis of Cancer*, Croom Helm, London (1985).
70. Zubrod, C.G., Quantitative concepts in the clinical study of drugs, In *Advances in Chemotherapy*, Goldin, A. and Hawkins, F. eds., Academic Press, New York 9–34 (1964).

71. Vendetti, J.M., and B.J. Abbott, Studies on oncolytic agents from natural sources: correlation of activity against tumors and clinical experiments. *Lloydia 30*:332–348 (1967).
72. Webster, G.D., *BONZO, Molecular Stick Plotting Program*. The Institute of Cancer Research, Sutton Surrey, U.K. (1988).
73. Crothers, D.M., Calculation of binding isotherms for heterogenous polymers. *Biopolymers 6*:575–584 (1968).
74. Agbandje, M., T.C. Jenkins, R. McKenna, A. P. Reszka, P. Goddard, M. Valenti, and S. Neidle, (to be published).
75. Scatchard, G., *Ann. N.Y. Acad. Sci. 51*:660 (1949).
76. McGhee, J.D., and P.H. von Hippel, Theoretical aspects of DNA-protein interactions: Co-operative and non-co-operative binding of large ligands to a one-dimensional homogenous lattice. *J. Mol. Biol. 86*:469–489 (1974).
77. Zunino, F., R. Gambetta, A. Di Marco, and A. Zaccara, *Biochim. Biophys. Acta. 277*:489 (1972).
78. Henry, D.W., *Am. Chem. Soc. Symp. Ser. 30*:15 (1976).
79. Lybrand, T., and P.A. Kollman, Molecular mechanical calculations on the interaction of ethidium cation with double-helical DNA. *Biopolymers 24*: 1863–1879 (1985).
80. McKenna, R., A. Beveridge, T.C. Jenkins, S. Neidle, and W.A. Denny, Molecular modeling of DNA-Antitumour drug intercalation interactions: correlation of structural and energetic features with biological properties for a series of phenylquinoline-8-carboxamide compounds. *Molec. Pharmacol. 35*:720–728 (1989).

12

Catalytic Antisense RNAs

Principles and Design

James O. Deshler and John J. Rossi
Beckman Research Institute of the City of Hope, Duarte, California

I. INTRODUCTION

Catalytic RNAs (ribozymes) were simultaneously discovered in a group I self-splicing intron of the tetrahymena pre-rRNA (1) and in the RNAse P enzyme purified from *E. coli* (2). The evidence provided by analyzing these two molecules revolutionized views on the origins of life, because it indicated that RNA can create not only a genotype (in many RNA viruses) but a phenotype. RNA is the only known molecule which can do this, and, therefore, is thought to have been an early self replicating molecule in the prebiotic world. Further evidence supporting this notion comes from the discovery and characterization of plant viroids and virusoids (3) which are small (250–350 nts), single-stranded, circular RNA molecules possessing pathogenic capabilities in plants. The replication of these pathogenic RNAs is thought to involve their self-cleaving capabilities (4).

In this chapter, we will focus on several aspects of catalytic RNAs including their potential use in the treatment of human disease. First, a background to naturally occurring ribozymes and parameters which affect the rates of their catalysis will be discussed. Secondly, a summary of what is known about gene regulation by naturally occurring and artificial antisense RNAs will be presented to give a rational foundation for designing ribozymes capable of inhibiting gene expression. Finally, ways in which RNA catalytic centers can be engineered to target specific genes of interest will be addressed. We will examine the delivery methods available for ribozyme therapies and practical problems which still have to be overcome before ribozyme therapy becomes a realistic form of medical treatment.

II. NATURAL RNA CATALYSIS

Examples of catalytic RNAs have now been found in the plant, eubacterial, protista, animal, and fungal kingdoms. We can, therefore, conclude that ribozymes represent a common molecular phenomenon throughout nature.

A. Group I Intron

One of the first examples of ribozymes is the group I intron of the pre-rRNA of tetrahymena. Group I introns are defined by a complex, but phylogenetically conserved secondary structure (5). Cech and co-workers showed that the group I intron from the tetrahymena pre-rRNA can self splice in the absence of protein (1). Additional work with this and other group I intron systems has led to the following general conclusions which are important for ribozyme design:

1. It is possible to engineer a form of the intron which can repeatedly perform the first chemical step of the splicing process (cleavage at the 5′ splice site) on substrate molecules in trans, indicating that it is a true enzyme (6).
2. Base-pairing interactions between the substrate and an internal guide sequence within the intron are required for cleavage, and changing the sequence of the guide sequence alters the cleavage specificity of the ribozyme (7,8).
3. Tertiary structures as well as secondary structures are required within the intron for proper ribozyme conformation (9).
4. The core ribozyme associates with 1000-fold higher affinity with the substrate than predicted by base pairing alone between the internal guide sequence and substrate molecule (10).
5. The cleavage reaction shows a requirement for Mg^{2+} ions.

B. RNAse P

RNAse P is a ubiquitous enzyme required for the maturation of tRNAs. Specifically, this enzyme is an endonuclease which cleaves, thus defining the 5′-ends of tRNAs. RNAse P has been well studied in bacterial systems where it has been shown to consist of both an RNA and protein component. The RNA component of the *E. coli* enzyme (referred to as the M1 RNA) is 377 nucleotides long (MW 132,000) and the protein is 14 kd. In vitro analyses have demonstrated that the M1 RNA can perform the 5′ cleavage of pre-tRNAs in the absence of the protein subunit (11). However, the protein-independent reaction only occurs under conditions of high ionic strength. It is thought, therefore, that the protein moiety creates a local vicinity of positively charged ions under physiological conditions (lower ionic strength) which disperse the negative charge repulsions between the M1 RNA and pre-tRNA substrates (12).

The cleavage of artificial RNAse P substrates in vitro has shown that a minimal substrate consists of a 5′ single-stranded region (containing the cleavage site), followed by a stem loop which has the sequence CCA at its 3′-end (13). The loop of the stem, however, is not required for cleavage, implying that any target RNA can be cleaved by endogenous RNAse P provided a complementary piece of RNA containing the sequence NCCA (where N is any nucleotide) is annealed to the desired cleavage site via RNA-RNA base pairing (14) (Fig. 1). As with the group I intron, a higher order of tertiary structure is thought to form the catalytic RNA subunits found in all known RNAse P enzymes (15).

Both the RNAse P and the group I intron cleavage reactions require

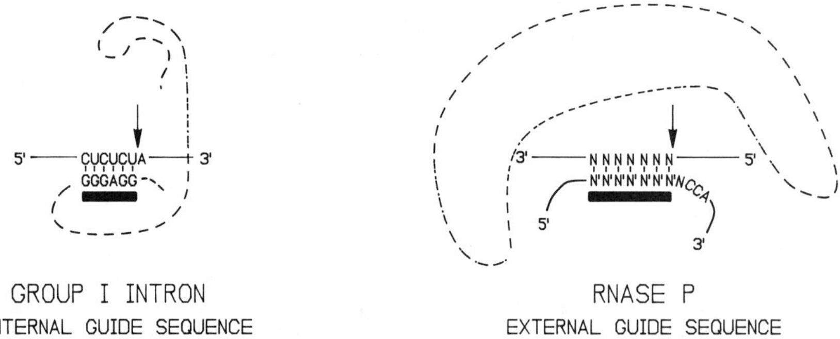

Figure 1 Two types of guide sequences are used by two different ribozymes. The cleavage site on the target RNA is indicated with an arrow, and the guide sequences are underlined with a thick line. As can be seen, the group I intron contains the guide sequence making it "internal" to the ribozyme, whereas, the guide sequence utilized by RNAse P is "external" to the ribozyme and can be supplied additionally to the reaction or as part of the substrate as with pre-tRNAs. Both ribozymes are indicated schematically with a dotted line drawing.

RNA-RNA base pairing interactions between nucleotides near the cleavage site and a "guide" sequence. The guide sequence for RNAse P can either be part of the substrate molecule (as in unprocessed tRNAs), or it can be provided by an unattached short ribonucleotide complementary to nucleotides adjacent to the cleavage site. This guide sequence is external to the RNAse P M1 RNA. In contrast, the group I catalytic intron contains the guide sequence and it is referred to as an "internal" guide sequence. Therefore, each ribozyme can theoretically be designed to recognize and cleave any target RNA molecule by making the "guide" sequence complementary to the target site. The length of the guide sequence is 6 bases for the intron (7) and 6–15 for RNAse P (14). The two guide sequences are schematically shown in Figure 1. When considering human nucleic acid sequences it is important to remember that a target sequence of 17 nucleotides ($4^{17} \approx 10^{10}$ bp) or greater is required to render it unique in the human genome. Therefore, the ribozymes just described would have the potential to cut at unwanted sites in human cells. The effect of long guide sequences is not known. In addition, cleavage by these ribozymes to newly targeted sites has not yet been observed to occur inside a living cell.

C. Viroids and Virusoids

A more recent addition to the list of natural ribozymes came from the analysis of certain plant diseases. It was found that single stranded, circular

RNA molecules termed *viroids* are the pathogenic agents in many commercial crops (16). One viroid and several virusoids (similar to viroids, but found in association with plant viruses as satellite circular RNAs) were shown to self-cleave their RNA genomes. The self cleavage reaction is thought to be critical for their replication cycles within host cells (17). Sequence analysis of a variety of the self-cleaving RNAs led to the identification of a highly conserved nucleotide sequence and secondary structure in the vicinity of the cleavage sites. The consensus structure, shown in Fig. 2a, is referred to as a hammerhead (17). The hammerhead structure is the smallest catalytic RNA structure known to date and consists of three stem-loop structures and ten highly conserved single stranded ribonucleotides thought to form the catalytic center or enzymatic "core."

While these structures are found in nature residing within a single strand of RNA, it has been shown that an active hammerhead can be formed between two separate RNA molecules. For example, Uhlenbech synthesized a separate top (consisting of half of stems I and III, and all of stem II) and bottom strand (containing the complementary halves of stems I and III and a portion of the catalytic core) which could form the secondary structures depicted in Figure 2b. The top strand contains the cleavage site. It was shown that the bottom strand could repetitively cleave upper strands in vitro (18). These results are exciting because they imply that any 5′-NNNGAAAC–GUCUXXXXXX-3′ can be the target site for ribozyme cleavage by simply designing a bottom strand 5′X′X′X′X′X′X′-CUGANGAGN′N′N′-3′, where N and X can base-pair with N′ and X′, respectively. A variation of this type of analysis, done by Haseloff and Gerlach (19), indicates that any GUC trinucleotide can be a cleavage site by providing half of stems II and III, and all of stem II on the ribozyme (Fig. 2c). In fact, CAT and HIV specific messages have been cleaved at GUC sequences by hammerhead ribozymes in vitro (19,20). In addition, the total number of flanking complementary nucleotides can be greater than 17 making the cleavage site unique in the human genome.

D. Enzymology

Kinetic analyses of the ribozymes described above all indicate that ribozyme mediated cleavage is one of the slowest enzymatic processes in nature. The turnover numbers (k_{cat} or V_{max}) of different ribozymes show little variability in value and average approximately one cleavage event per molecule per minute. This is similar to the turnover number of the DNA restriction enzyme *Eco* RI (21). These low turnover numbers are sufficient for DNA restriction enzymes to protect their hosts from infectious viral DNAs (22). While the catalytic efficiencies of all the ribozymes described are quite

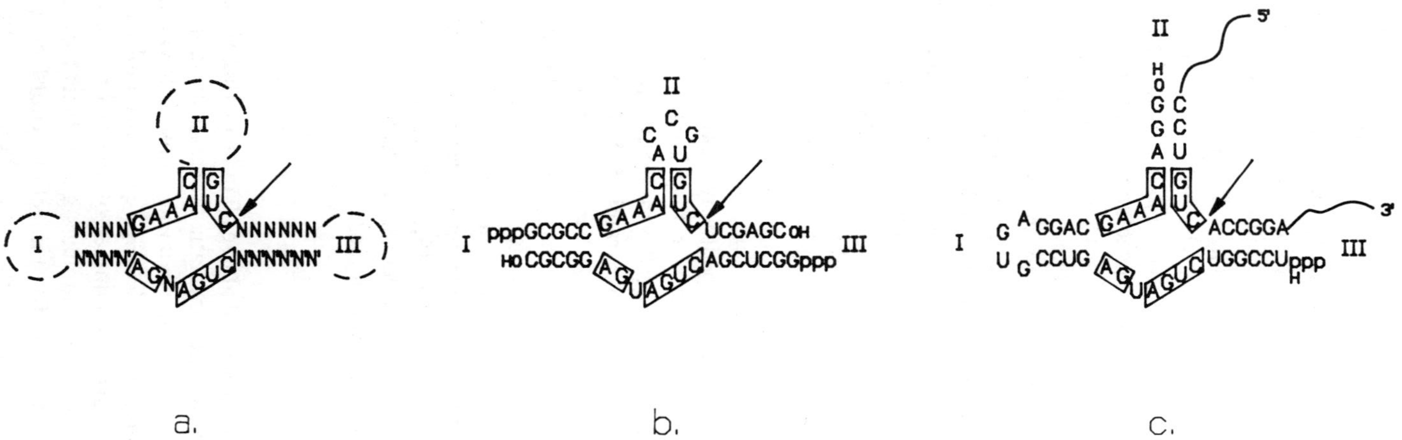

Figure 2 Development of the in vitro hammerhead ribozyme. (a) The consensus secondary structure derived by phylogenetically comparing the potential base-pairing structures of several self-cleaving RNA sequences (17). Note that three stem loops form the consensus structure. A self-cleaving molecule can be discontinuous in any of the loops. (b) By synthesizing two short RNA molecules in vitro, Uhlenbeck, demonstrated that ribozyme cleavage could be an intermolecular event as well as the intramolecular cleavage event found in natural virusoid transcripts (18). (c) Haseloff and Gerlach synthesized a long and a short RNA in vitro showing that any 5′-GUC-3′ could be used as a cleavage site (19). Again the cleavage sites are indicated with an arrow.

comparable (see below), the hammerhead has been favored for use as a therapeutic agent because it appears to be the simplest (smallest), making it easy to manipulate. Also, the length and composition of the flanking secondary structures are not critical for cleavage and hammerhead ribozymes can be designed which are capable of specifically cleaving target molecules within a complex milieu of RNA molecules (20) present inside a living human cell.

The K_m's of ribozymes are in the nanomolar range, and for the hammerhead alone, can vary over about two orders of magnitude in vitro (23). It appears that the highly conserved catalytic core of the hammerhead ribozyme maintains a three dimensional structure and mutations within this core seriously decrease the cleavage rate. Variation of the stem regions of the hammerhead, on the other hand do not significantly affect k_{cat}, but can affect the K_m to various degrees. Since the catalytic efficiencies of enzymes (k_{cat}/K_m) depend on both values, parameters affecting the k_{cat} and K_m are discussed below.

Enzymatic analyses of various mutant forms of the hammerhead ribozyme indicate that the single-stranded core sequences and one of the two phylogenetically conserved base pairs (the A-U base pair adjacent to the cleavage site in stem II) are absolutely required for cleavage in vitro (24). These bases define the catalytic center of the ribozyme. Alteration of nucleotides within the stems influence the cleavage efficiencies to a much lesser degree. While base stacking in the stems might affect the k_{cat} to some extent, it appears that the most problematic variable affecting cleavage rates is the potential of the target or ribozyme to form alternative secondary structures which are incompatible with ribozyme mediated cleavage. Such competing structures are thought to increase the K_m, slowing the rate of hammerhead formation (24). This is an important feature likely to render a target accessible or inaccessible to gene inhibition by artificial ribozymes. Another factor which affects the rate of dissociation of product from ribozyme is the length of the intermolecularly base-paired stems (19). Longer stems slow dissociation rates making it difficult for the ribozyme to recycle for another round of cleavage. Recycling of the ribozyme in vivo, however, may not be a critical factor in determining its therapeutic effectiveness if a molar excess of ribozyme over target can be achieved.

E. Gene Regulation

Before discussing the use of ribozymes as an applied science in the treatment of disease, it is first necessary to consider what has been learned about gene regulation by naturally occurring antisense RNA molecules which have no known catalytic activity. Gene regulation by antisense RNA was first discovered in *E. coli* and shown to regulate both ColE1 plasmid copy numbers and transposition frequencies of the Tn10 transposable element. The most extensive analysis of antisense-sense RNA interaction has been carried out by Tomizawa for the ColE1 plasmid replication system. Using a combination of genetic and biochemical approaches Tomizawa derived a model to describe RNA-RNA intermolecular base pairing which indicates that pairing is initiated through single-stranded (looped) regions of the complementary RNAs which have inherent secondary structure. This initial rate-limiting step is reversible (25) and facilitates subsequent intermolecular pairing events until a complete RNA-RNA hybrid is formed which completely disrupts the intramolecular secondary structures of the individual complementary RNAs. A similar mechanism of RNA-RNA base pairing has been proposed to regulate translation of the Tn10 transposase mRNA. In this case, nucleotides within the loop domain of a stem-loop structure of the antisense RNA are thought to initially pair with the ribosome binding site of the mRNA (26), thus preventing the association of ribosomes (27). The initial, transient molecular event in both systems has been termed "kissing" to distinguish it from stable pairing (28) and is consistent with the idea that intermolecular base-pairing reactions are optimized when the interacting regions of RNA are in a single-stranded conformation. Recently, many other bacterial regulatory mechanisms have been shown to involve naturally transcribed antisense RNA molecules (29,30). It has been shown that transcription, translation, DNA replication, and RNA stability are controlled by antisense RNAs in the various regulatory systems. The mechanisms of antisense action include sequestration of ribosomal binding sites (27,31,32), cleavage of RNA-RNA hybrids by double-stranded ribonucleases (33,34), and alteration of target conformation (35,36), Finally, it should be mentioned that the analysis of various mutant antisense RNAs indicate that their stability, which in many instances is a result of their inherent structure, and the levels of expression of an antisense RNA, can significantly affect their ability to inhibit gene expression (37).

Even though ribozyme-mediated gene inhibition involves a different mechanism (target cleavage) than the bacterial antisense RNAs, many of the essential steps required for gene inhibition by ribozymes and bacterial antisense RNAs are identical. For example, the antisense RNA and the ribozyme must exist long enough (long $t_{1/2}$) to find their target sites inside a cell. This is not a trivial point because RNA is known to be a relatively unstable molecule in vivo unless it is protected from degradation. Also, the antisense RNA and the ribozyme must locate their targets through intermolecular base pairing. This implies that a favorable conformation of target and ribozyme must exist to optimize the base-pairing interaction (at least some of the base pairing regions of the target and ribozyme should be single stranded to allow the "kissing" reaction to occur). Finally, the relative concentrations of target and antisense or ribozyme affects the kinetics of target localization which can lead to increased gene inhibition.

No naturally occurring antisense RNAs have yet been observed to control gene expression in eucaryotic cells. However, many successful attempts at inhibiting specific gene expression using artificial antisense RNAs and DNAs have been reported. Early studies involved the inhibition of Rous sarcoma virus replication and cell transformation with antisense DNA (38). It was shown that translation of viral RNAs were inhibited in vitro by the addition of antisense DNA oligonucleotides (39). It was subsequently discovered that antisense DNA biding to RNA causes the site-specific degradation of the RNA by endogenous RNAse H activity (RNAse H recognizes heteroduplex molecules consisting of one strand of DNA and one strand of RNA) (40). Inhibition of Herpes thymidine kinase gene expression demonstrated the first use of antisense RNA (41). The use of artificial antisense RNAs in eucaryotic cells has demonstrated various degrees of effectiveness, and again, as in bacterial systems, many different mechanisms explain the inhibition of gene expression by different antisense RNAs. Transcription has been inhibited by antisense RNA (42), mRNA degraded by ribonucleases upon hybridization to antisense RNAs (43), viral replication inhibited (44), ribosomal binding to mRNAs prevented (45), and RNA has been covalently modified by cellular enzymes responding to the double-stranded RNA formed between the target and antisense RNAs (46). Why these different mechanisms act in the various systems is not clear.

One might ask, why use ribozymes when antisense DNAs and RNAs have been shown to inhibit gene expression in so many cases? An answer lies in looking at the drawbacks of these two antisense approaches. A disadvantage of using antisense DNA oligonucleotides is that their effects are transient making the constant and repetitive addition of oligonucleotides mandatory. Therefore, the amount of chemically synthesized DNA required for practical therapies is large and costly. Also, it is thought that this method generally requires the presence of endogenous RNAse H activity which may or may not be present in the desired cell type. In addition, there is presently no way to genetically engineer a cell to synthesize short DNA oligonucleotides for use in gene therapy. Gene therapy technology, on the other hand, is applicable to antisense RNA approaches in eucaryotes. Plasmids can be engineered to produce high levels of antisense RNAs when introduced into cells. The effectiveness of an antisense RNA at inhibiting gene expression, however, is unpredictable. Several reports indicate that stable and abundant expression of antisense RNAs have no notable effect on gene expression of their target genes (47,48,49). Therefore, the reliability of antisense RNAs is not 100%. The reasons for this are unknown.

The fact that ribozymes cleave their substrate molecules simplifies concerns of therapeutic approaches inherent in antisense RNA schemes because the mechanism (cleavage of the target molecule) is simple and effective at inactivating any target RNA. When applying ribozyme technology for use as an antiviral or anticarcinogenic agent, the important considerations are the stability, conformation, cellular location, and steady-state levels of ribozyme within the host cell. Also, ribozymes can be engineered onto plasmids to take advantage of gene therapy technologies which, of course, is one of the definitive goals in the field of antisense RNA and ribozyme drug technologies.

III. DESIGN OF RIBOZYMES

Considering the facts learned about the action of naturally occurring antisense RNAs and the in vitro kinetics of ribozymes, we will discuss the design of optimal ribozymes directed to inactivate any target molecule. In this section we only discuss the hammerhead design because it has been well characterized and is small, making it easy to manipulate in the laboratory.

A. First Step: Selection of a Cleavage Site in the Target RNA

As discussed previously, any 5′-GUC-3′ should be susceptible to cleavage by a synthetic ribozyme (Fig. 2), and while GUC is the most conserved and probably most efficiently cleaved sequence, it is possible to target any UX (X=U,C, or A) (24,50). The target RNA will contain more that one GUC and evaluating which GUC sequences will be targeted should be considered carefully.

When choosing the cleavage site on the RNA target molecule, several aspects of mRNA biosynthesis in eucaryotic cells should be considered. First, pre-mRNAs are complexed with heteronuclear ribonucleoproteins (hnRNPs) during transcription in the nucleus. It is thought that this process condenses and packages the RNAs similar to the function of histone proteins in DNA packaging (51,52). It has been shown that certain antisense cleavage sites can be protected by associating with nuclear RNA binding proteins (snRNPs) (53,54). However, all regions are not protected by hnRNPs because many artificial antisense RNAs and DNAs are successful at decreasing gene expression. To determine which regions of the RNA are bound to hnRNPs is labor intensive, so accessible regions of the target should be found empirically by designing separate ribozymes to cleave several potential sites and testing their relative effectiveness. One study which directly tested this issue found that only some regions of the HIV genome, including an AUG initiation codon and splices sites, were accessible to degradation by oligonucleotides (55).

Another factor which can affect the availability of a cleavage site is the structure of the target RNA. Intuition tells us that if a sequence of RNA is sequestered by a folded secondary or tertiary structure, the accessibility of this sequence to degradation by a ribozyme or any other antisense molecule will be impaired. Experimental data concerning the structure of pre-mRNAs in vivo is limited, but there is one in vitro example in which a 500 molar excess of an oligonucleotide complementary to a stem region of an RNA secondary structure only weakly competed for an intermolecularly hybridized sequence (56). Because the biophysical principles governing RNA folding in vivo have not been determined, one can only check the region surrounding the cleavage site using a computer program which analyzes RNA secondary folding to see if the target site is buried in a thermodynamically stable secondary structure. This technique is unreliable, however, and empirical testing will, again, give the only meaningful answer as to the accessibility of the cleavage site.

B. Second Step: Sequence of the Ribozyme

After choosing the cleavage site(s) in the target RNA, it is relatively simple to design an RNA molecule similar to the Haseloff and Gerlach ribozyme shown in Figure 1(c). The sequence of the ribozyme will be identical to that shown in Figure 1(c) except the ribozyme portions of stems II and III should be complementary to sequences flanking either side of the cleavage site (GUC). In order to make the ribozyme cleave the target site and no other unwanted sites in the human genome, the total number of flanking nucleotides should be at least 17. As mentioned earlier, longer flanking sequences not only increase target specificity, but can also shorten the association time of the ribozyme with its target. However, it is thought that longer flanking sequences usually decrease the cycling efficiency of the ribozyme by slowing the dissociation rate after the cleavage step (19). This may or may not be important depending on whether the intracellular concentration of the ribozyme exceeds that of the target. In short, there is a trade-off between specificity and cycling efficiency which may need to be empirically optimized for each system.

C. Third Step: Ribozyme Delivery to Cells

The therapeutic use of ribozymes will require efficient delivery of active ribozymes to the intracellular environment. Two general approaches exist for introducing catalytic RNA molecules to cells. One involves the use of exogenously added presynthesized ribozymes, and the other employs the use of genetically engineered plasmids which direct cellular enzymes to transcribe the catalytic RNAs (gene therapy).

Many experiments involving the use of antisense molecules to inhibit gene expression in eucaryotes have utilized exogenous delivery systems. In other words, the antisense DNAs or RNAs were synthesized in vitro and subsequently added to cell cultures. Several types of exogenous delivery systems have been used. The simplest is the direct addition of nucleic acids to the culture medium (57). Some factors limit this approach, however, the most important of which is the susceptibility of the DNA or RNA to degradation by nucleases and poor penetration of the nucleic acids into cells. Other delivery methods attempt to overcome these limitations by packaging the nucleic acids into liposomes (58) or polycationic glycoprotein conjugates (59). An advantage of packaging is that a specific subpopulation of cells can be targeted by incorporating monoclonal antibodies or receptor ligands to the liposomal or polycationic glycoprotein conjugates, respectively.

Another way of protecting nucleic acids from degradation is done using chemical modifications which render the DNA and RNA more stable within cells. For example, phosphorothioates are more resistant to nuclease degradation than unmodified oligonucleotides (60), and are effective inhibitors of HIV-1 replication in infected cells (61). A phosphorothioate ribonucleoside analog (J. Nielsen and K.H. Peterson, unpublished) and a modified ribose group (2′*O*-methyl) (62,63) have been synthesized and RNA oligomers containing these modifications can be made using conventional chemistry. Also, chimeric molecules containing a ribonucleotide catalytic center and deoxyribonucleotide flanking sequences may be another strategy to create more stable ribozymes (J. Rossi, B. Kaplan, and P. Chang, unpublished). And recently, certain conserved nucleotides in the hammerhead catalytic center were changed to the deoxy form without preventing the catalytic activity (64). It is therefore conceivable that deoxyribonucleotide substitutions confined to the sequences flanking the RNA catalytic center will enhance the stability without impairing activity. Additional studies are required to ascertain the feasibility of using RNA-DNA chimeric ribozymes. These and other modifications that protect the RNA molecule from enzymatic degradation, coupled to lipophilic or other transport vehicles, may play important roles in the early phase of ribozyme applications to human diseases.

While exogenous delivery of nucleic acids is feasible (65) in the treatment of disease, the effects of nucleic acids delivered in this fashion are transient and require repeated inoculations. Therefore, exogenous delivery would require frequent infusions of preformed ribozymes to maintain an intracellular concentration sufficient to effect a significant reduction in disease causing RNAs. An alternative approach is gene therapy, which involves the endogenous expression of a ribozyme from a DNA template permanently maintained within the cell. Endogenous expression should theoretically eliminate most of the difficulties associated with the delivery of preformed ribozyme molecules.

IV. EXAMPLES OF RIBOZYMES USED TO INHIBIT SPECIFIC GENES

Only a few examples of endogenously expressed ribozymes exist in which specific gene inhibition was obtained in vivo. A major consideration in all projects of this nature is the stability and expression levels of the ribozyme RNAs. Different expression systems have been tested in order to attain stable and high level production of ribozymes. These systems employ the use of a strong promoter to drive transcription of the ribozyme in the desired cell type, and also involve cloning the ribozyme into some kind of cassette which incorporates flanking sequences on both the 5′- and 3′-sides of the ribozymes. It is thought that certain flanking sequences might prolong the half life of the ribozyme giving it more time to locate its target RNA within the cell. A general schematic view of a ribozyme vector is shown in Figure 3. The ribozyme-expressing gene contains a strong promotor, a transcription terminator which provides a defined and stable 3′-end, and both 5′ and 3′ flanking sequence which provide stability to the ribozyme without disrupting its activity. Below are described specific cases in which in vivo ribozyme activity has been pursued.

A. Xenopus tRNA

Cotten and Birnstiel (66) inserted a 50-nucleotide ribozyme into the anticodon loop of a Xenopus tRNA gene. They reasoned that tRNA genes would be good cassettes for expressing ribozymes for the following reasons:

1. tRNA genes are transcribed at high rates by RNA polymerase III in almost all tissue types.
2. Both the tRNA promoter and its transcriptional terminator are not affected by small insertions within the anticodon loop.
3. Higher stability of the ribozyme might be provided by the normal folding pattern of the tRNA.
4. The chimeric genes would not be expected to affect expression of nearby genes upon chromosomal insertion.
5. tRNAs are predominantly localized in the cytoplasm where mRNAs are translated.

Incorporation of the ribozyme into the tRNA stabilized the ribozyme RNA, but had several effects on the tRNA worth noting. First, the transcription

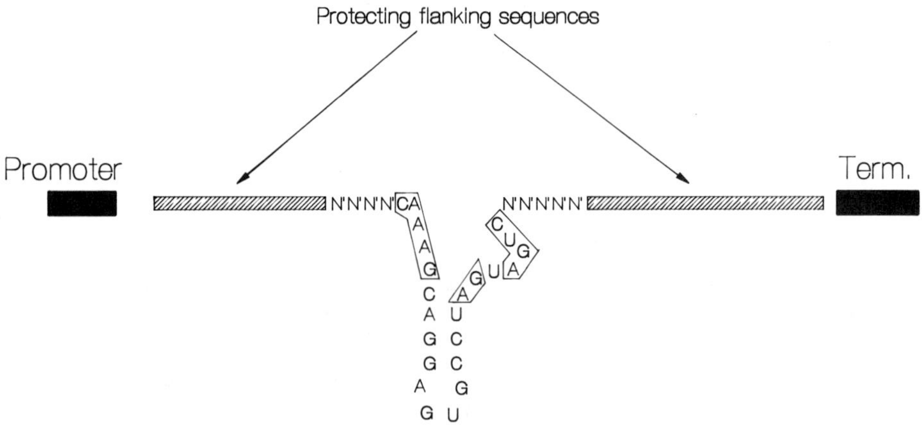

Figure 3 Schematic representation of a ribozyme expression vector. As described in the text, the chimeric genes which have been used to successfully express ribozymes in vivo contain a strong promoter, an appropriate transcriptional terminator, flanking sequences which provide stability to the ribozyme (hatched lines), and the ribozyme containing the catalytic core (outlined nucleotides) and nucleotides complementary to both sides of the desired cleavage site (N's). The partial ribozyme here is shown containing stem I of the Haseloff and Gerlach model (Fig. 2(c)) where the complementary sequences of stems II and III are provided by the target molecule.

levels and stability of the chimeric tRNA were only slightly lower than the uninterrupted tRNA. However, while the correct transcriptional start and stop points were maintained, posttranscriptional processing of the 3′-end was disrupted. In addition, transport of the tRNA ribozyme to the cytoplasm was blocked and only a small fraction was localized in the cytoplasm. These effects were attributed to a possible disruption of the normal tRNA structure by the ribozyme.

The amount of tRNA ribozyme which was transported to the cytoplasm, though, appeared to be sufficient to mediate specific and complete degradation of the coinjected U7 snRNA target molecule located in the cytoplasm when the ratio of ribozyme to target was estimated at 1000 : 1 in the cytoplasm. It was not formally proven, however, that ribozyme cleavage was the mechanism by which the U7 snRNA was degraded. In short, the chimeric tRNA ribozyme was highly expressed, quite stable (half-life of many hours), and active, while cellular transport and posttranscriptional processing were adversely affected.

B. CAT mRNA

Another example of in vivo gene inhibition was seen by Cameron and Jennings using an entirely different system (67). The target was the bacterial chloramphenicol acetyltransferase (CAT) mRNA expressed from the weak thymidine kinase promoter in COS1 tissue culture cells. This RNA is known not to be sensitive to antisense RNAs. Three different ribozymes were expressed from a large cassette consisting of the firefly luciferase gene transcribed to high levels from the SV40 early promoter. The luciferase cassette used for expressing ribozymes provided a simple assay for analyzing the quantitative effects on RNA stability of forming a chimeric ribozyme containing RNA. Each ribozyme was inserted into the 3′ untranslated region of the luciferase gene and was designed to cleave a different GUC cleavage site in the target CAT RNA. Surprisingly, all three of the ribozymes had similar effects on CAT activity by reducing its level to approximately 70% of that obtained with simple antisense or reversely oriented ribozyme control constructs.

The luceriferase message was destabilized somewhat by all ribozyme and control insertions to equal extents (about 30%) suggesting that ribozymes, per se, do not decrease the stability of cassette RNAs. The ratio of ribozyme to target was again estimated to be approximately 1000 : 1, and decreasing this ratio eliminated the effectiveness of the chimeric ribozyme. Another factor increasing the ribozyme's effectiveness was increasing the flanking sequences from 8 nucleotides to 12 and 13 nucleotides for the 5′- and 3′-sides, respectively. From these studies it was shown that incorporation of ribozymes into large RNA polymerase II transcripts can lead to a reduction of gene expression in vivo.

C. HIV RNA

Our laboratory, in collaboration with others, has attempted to inhibit HIV viral replication in tissue culture using ribozymes. Throughout our work several characteristics of ribozymes have been analyzed. A substantial amount of information has been gained by analyzing ribozyme cleavage reactions in vitro, before attempting to inhibit viral replication in vivo. It was found that the Hasaloff and Gerlach model shown in Figure 2 was more effective than the Uhlenbeck design for a given HIV target molecule in vitro (20). This is advantageous because there are more GUC cleavage sites in target RNAs than the $GAAAX_1$ and X_2GUC (when X_1 and X_2 nucleotides can form Watson-Crick base pairs) required for the Uhlenbeck design. In addition, most of the work characterizing the kinetics of ribozymes in vitro utilizes a purified system consisting of the ribozyme, the substrate, and buffer. We reasoned that in vivo, competition with nonspecific sequences within a cell might affect ribozyme activity. An in vitro experiment was done to address this issue in which total cellular RNA from either HIV infected or uninfected cells was added to an in vitro cleavage reaction. Only the RNA from infected cells inhibited the cleavage reaction suggesting that nonspecific sequences within the cell do not interfere with ribozyme activity. It was also shown that the 5′ GpppG cap structure found on most Pol II transcripts did not interfere with ribozyme activity in vitro (68).

Subsequent to the in vitro analyses, an expression vector was chosen to express the anti HIV ribozyme. This vector contains the highly active human β-actin promoter and an SV40 late transcriptional termination and polyadenylation signals. Insertion of an anti-HIV ribozyme DNA sequence into this expression vector resulted in stable expression of the ribozyme in transfected CD4+ HeLa cells. To our surprise, total RNA purified from transfected clones was capable of cleaving in vitro transcribed *gag* target RNA indicating that the endogenously expressed ribozyme containing approximately 400 nucleotides of flanking sequences was still active (20). Further experiments demonstrated that upon exposure to HIV, clones transfected with the anti-HIV ribozyme demonstrated decreased expression of the HIV *gag* RNA, 100-fold decrease in proviral DNA sequences, and 20–40-fold reduction in GAG p24 antigen 7 days postinfection when compared to cells which did not express the ribozyme. In addition, several growth studies have demonstrated that ribozyme expression had no notable effect on the viability of ribozyme transformed cells, suggesting that the anti-HIV ribozyme does not cleave nonspecifically within the cell as was demonstrated by in vitro experiments with competing total cellular RNA (68).

For gene therapy, the stability and intracellular localization of catalytic RNAs will be critical. A number of naturally occurring RNA modifications, such as 5′-capping and 3′-polyadenylation, may prolong the half-life and facilitate intracellular compartmentalization of catalytic RNAs. For example, a catalytic RNA derived from a Pol II transcriptional cassette will acquire a 5′-m^7GpppG cap typical of mRNAs, which may enhance its cytoplasmic stability. In contrast, an snRNA catalytic RNA transcript will acquire a 5′- $m_3^{2,2,7}$GpppG cap, and this, combined with an Sm antigen binding site found in most snRNAs (69) should localize the ribozyme to a nuclear compartment where it could interact with unspliced pre-mRNAs. Much work is required to determine which parameters are most important for the efficient design of ribozymes which efficiently inhibit gene expression in vivo. However, the application of the ideas summarized in this chapter should lead to successful ribozyme technology.

V. CONCLUSION

Review of the literature summarized in this chapter shows that RNA-RNA interactions are complex, and that antisense RNAs can block different steps required for gene expression. Also, it shows that a number of RNAs are catalytic in nature. All of the catalytic RNAs can cleave RNA molecules and require Watson-Crick base-pairing interactions adjacent to the cleavage site. By combining the knowledge gained from the analysis of ribozyme and antisense RNA mechanisms, it is hoped that therapeutic nucleic acid drug technologies will be developed into practical treatments for viral and genetic diseases. Most exciting is the potential combined application of gene therapy and ribozyme technologies. Theoretically, it is possible to inhibit the expression of any gene by introducing a DNA vector which produces an effective ribozyme specific for a disease-causing RNA. In fact, this approach has been successful in phenotypically altering the characteristics of transgenic mice using antisense RNA (70). The development of ribozyme technologies should help to establish their usefulness in the field of disease prevention and treatment.

ACKNOWLEDGMENT

The authors acknowledge the support of NIH Grants AI29329 and AI25959 to JJR.

REFERENCES

1. Kruger, K., P.J. Grabowski, A.J. Zaug, J. Sands, D.E. Gottschling, and T.R. Cech, Self-splicing RNA: autoexcision and autocyclization of the ribosomal RNA intervening sequence of tetrahymena. *Cell 31*:147 (1982).
2. Takada, C.G., K. Gardiner, T. Marsh, N. Pace, and S. Altman, The RNA moiety of ribonuclease P is the catalytic subunit of the enzyme. *Cell 35*:849 (1983).
3. Riesner, D., and H.J. Gross, Viroids. *Ann. Rev. Biochem. 54*:531 (1985).
4. Forster, A., and R.H. Symons, Self-cleavage of plus and minus RNAs of a virusoid and a structural model for the active site. *Cell 49*:211 (1987).
5. Michel, F., and B. Dujon, Conservation of RNA secondary structures in two intron families including mitochondrial-, chloroplast- and nuclear-encoded members. *EMBO J. 2*:33 (1983).
6. Zaug, A.J., and T.R. Cech, The intervening sequence RNA of *tetrahymena* is an enzyme. *Science 231*:470 (1986).
7. Zuag, A.J., M.D. Been, and T.R. Cech, The *tetrahymena* ribozyme acts like an RNA restriction endonuclease. *Nature 324*:429 (1986).

8. Murphy, F.L., and T.R. Cech, Alteration of substrate specificity for the endoribonucleolytic cleavage of RNA by the *tetrahymena* ribozyme. *Proc. Nat. Acad. Sci. USA. 86*:9218 (1989).
9. Michel, F., A.D. Ellington, S. Couture, and J.W. Szostak, Phylogenetic and genetic evidence for base-triples in the catalytic domain of group I introns. *Nature 347*:578 (1990).
10. Pyle, A.M., J.A. McSwiggen, and T.R. Cech, Direct measurement of oligonucleotide substrate binding to wild-type and mutant ribozymes from *tetrahymena. Proc. Nat. Acad. Sci. USA., 87*:8187 (1990).
11. Guerrier-Takada, C., K. Gardiner, T. Marsh, N. Pace, and S. Altman, The RNA moiety of ribonuclease P is the catalytic subunit of the enzyme. *Cell 35*: 849 (1983).
12. Reich, C., G.J. Olsen, B. Pace, and N.R. Pace, Role of the protein moiety of ribonuclease P, a ribonucleoprotein enzyme. *Science 239*:178 (1988).
13. McClain, W.H., C.G. Takada, and S. Altman, Model substrates for an RNA enzyme. *Science 238*:527 (1987).
14. Forster, A.C., and S. Altman, External guide sequences for an RNA enzyme. *Science 249*:783 (1990).
15. Forster, A.C., and S. Altman, Similar cage-shaped structures for the RNA components of all ribonuclease P and ribonuclease MRP enzymes. *Cell 62*: 407 (1990).
16. Riesner, D., and H.J. Gross, Viroids. *Ann. Rev. Biochem. 54*:531 (1985).
17. Forster, A.C., and R.H. Symons, Self-cleavage of plus and minus RNAs of a virusoid and a structural model for the active sites. *Cell 49*:211 (1987).
18. Uhlenbeck, O.C., A small catalytic oligoribonucleotide. *Nature 328*:596 (1987).
19. Haseloff, J., and W.L. Gerlach, Simple RNA enzymes with new and highly specific endoribonuclease activity. *Nature 334*:585 (1988).
20. Chang, P.S., E.M. Cantin, J.A. Zaia, P.A. Ladne, D.A. Stephens, N. Sarver, and J.J. Rossi, Ribozyme-mediated site-specific cleavage of the HIV-1 genome. *Clin. Biotechnol. 2*:23 (1990).
21. Modrich, P., and D. Zabel, *Eco*RI endonuclease: physical and catalytic properties of the homogeneous enzyme. *J. Biol. Chem. 251*:5866 (1976).
22. Lederberg and Meselson, Degradation of non-replication bacteriophage DNA in non-accepting cells. *J. Mol. Biol. 8*:623 (1964).
23. Fedor, M.J., and O.C. Uhlenbeck, Substrate sequence effects on "hammerhead" RNA catalytic efficiency. *Proc. Nat. Acad. Sci. USA 87*:1668 (1990).
24. Ruffner, D.E., G.D. Stormo, and O.C. Uhlenbeck, Sequence requirements of the hammerhead RNA self-cleavage reaction. *Biochemistry 29*:10695 (1990).
25. Tomizawa, J., Control of ColE1 plasmid replication: initial interaction of RNA I and the primer transcript is reversible. *Cell 40*:527 (1985).
26. Kittle, J.D., R.W. Simons, and N. Kleckner, Insertion sequence IS*10* anti-sense pairing initiates by an interaction between the 5′ end of the target RNA and a loop in the anti-sense RNA. *J. Mol. Biol. 210*:561 (1989).
27. Ma, C., and R.W. Simons, The IS*10* antisense RNA blocks ribosome binding at the transposase translation initiation site. *EMBO J. 9*:1267 (1990).

28. Tomizawa, J., Control of ColE1 plasmid replication: the process of binding of RNA I to the primer transcript. *Cell 38*:861 (1984).
29. Simons, R.W., Naturally occurring antisense RNA control—a brief review. *Gene 72*:35 (1988).
30. Inouye, M., Antisense RNA: its functions and applications in gene regulation—a review. *Gene 72*:25 (1988).
31. Liao, S., T. Wu, C.H. Chiang, M.M. Susskind, and W.R. McClure, Control of gene expression in bacteriophage P22 by a small antisense RNA. I: Characterization in vitro of the P_{sar} promoter and the *sar* RNA transcript. *Genes Dev. 1*:197 (1987).
32. Wu, T., S. Liao, W.R. McClure, and M.M. Susskind, Control of gene expression in bacteriophage by a small antisense RNA. II: Characterization of mutants defective in repression. *Genes Dev. 1*:204 (1987).
33. Krinke, L., and D.L. Wulff, RNase III-dependent hydrolysis of lambda cII-*O* gene mRNA mediated by lambda OOP antisense RNA. *Genes Dev. 4*:2223 (1990).
34. Krinke, L., and D.L. Wulff, OOP RNA, produced from multicopy plasmids, inhibits lambda cII gene expression through an RNase III-dependent mechanism. *Genes Dev. 1*:1005 (1987).
35. Okamoto, K., and M. Freundlich, Mechanism for the autogenous control of the *crp* operon: transcriptional inhibition by a divergent RNA transcript. *Proc. Nat. Acad. Sci. USA 83*:5000 (1986).
36. Tomizawa, J., Control of ColE1 plasmid replication: binding of RNA I to RNA II and inhibition of primer formation. *Cell 47*:89 (1986).
37. Case, C.C., S.M. Roels, P.D. Jensen, J. Lee, N. Kleckner, and R.W. Simons, The unusual stability of the IS10 anti-sense RNA is critical for its function and is determined by the structure of its stem-domain. *EMBO J. 13*:4297 (1990).
38. Zamecnik, P.C., and M.L. Stephenson, Inhibition of Rous sarcoma virus replication and cell transformation by a specific oligodeoxynucleotide. *Proc. Nat. Acad. Sci. USA 75*:280 (1978).
39. Stephenson, M.L., and P.C. Zamecnik, Inhibition of Rous sarcoma viral RNA translation by a specific oligodeoxyribonucleotide. *Proc. Nat. Acad. Sci. USA 75*:285 (1978).
40. Dash, P., I. Lotan, M. Knapp, E.R. Kandel, and P. Goelet, Selective elimination of mRNAs *in vivo*: complementary oligodeoxynucleotides promote RNA degradation by an RNAse-H-like activity. *Proc. Nat. Acad. Sci. USA 84*:7896 (1987).
41. Izant, J.G., and H. Weintraub, Inhibition of thymidine kinase gene expression by anti-sense RNA: a molecular approach to genetic analysis. *Cell 36*:1007 (1984).
42. Yokoyama, K., and F. Imamoto, Transcriptional control of the endogenous *MYC* protooncogene by antisense RNA. *Proc. Nat. Acad. Sci. USA 84*:7363 (1987).
43. Wormington, W.M., Stable repression of ribosomal protein L1 synthesis in *Xenopus* oocytes by microinjection of antisense RNA. *Proc. Nat Acad. Sci. USA 83*:8639 (1986).

44. Jennings, P.A., and P.L. Molloy, Inhibition of SV40 replicon function by engineered antisense RNA transcribed by RNA polymerase III. *EMBO J. 6*: 3043 (1987).
45. Melton, D.A., Injected anti-sense RNAs specifically block messenger RNA translation *in vivo. Proc. Nat. Acad. Sci. USA 82*:144 (1985).
46. Bass, B.L., and H. Weintraub, An unwinding activity that covalently modifies its double-stranded RNA substrate. *Cell 55*, 1089 (1988).
47. Kerr, M.S., G.R. Stark, and I.M. Kerr, Excess antisense RNA from infectious recombinant SV40 fails to inhibit expression of a transfected, interferon-inducible gene. *Eur. J. Biochem. 175*:65 (1988).
48. Gunning, P., J. Leavitt, G. Muscat, S. Ng, and L. Kedes, A human β-actin expression vector system directs high-level accumulation of antisense transcripts. *Proc. Nat. Acad. Sci. USA 84*:4831 (1987).
49. Salmons, B., B. Groner, R. Friis, D. Muellener, and R. Jaggi, Expression of anti-sense mRNA in H-*ras* transfected NIH/3T3 cells does not suppress the transformed phenotype. *Gene 45*:215 (1986).
50. Koizumi, M., Y. Hayase, S. Iwai, H. Kamiya, H. Inoue, and E. Ohtsuka, Design of RNA enzymes distinguishing a single base mutation in RNA. *Nucl. Acids Res., 17*:7059 (1989).
51. Conway, G., J. Wooley, T. Bibring, and W.M. Lestrourgeon, Ribonucleoproteins package 700 nucleotides of pre-mRNA into a repeating array of regular particles. *Mol. Cell. Bio. 8*:2884 (1988).
52. Dreyfuss, G., M.S. Swanson, and S. Pinol-Roma, Heterogeneous nuclear ribonucleoprotein particles and the pathway of mRNA formation. *Trends Biochem. Sci. 13*:86 (1988).
53. Rinke, J., A. Bernd, H. Blocker, R. Frank, and R. Lurhmann, The 5′-terminal sequence of U1 RNA complementary to the consensus 5′ splice site of hnRNA is single-stranded in intact U1 snRNP particles. *Nucleic Acids Res. 12*:4111 (1984).
54. Black, D.L., B. Chabot, and J.A. Steitz, U2 as well as U1 small nuclear ribonucleoproteins are involved in premessenger RNA splicing. *Cell 42*:737 (1985).
55. Goodchild, J., S. Agrawal, M.P. Civeira, P.S. Sarin, D. Sun, and P.C. Zamecnik, Inhibition of human immunodeficiency virus replication by antisense oligodeoxynucleotides. *Proc. Nat. Acad. Sci. USA 85*:5507 (1988).
56. Winkler, M.E., K. Mullis, J. Barnett, I. Stroynowski, and C. Yanofsky, Transcription at the tryptophan operon attenuator is decreased *in vitro* by an oligomer complementary to a segment of the leader transcript. *Proc. Nat. Acad. Sci. USA 79*:2181 (1982).
57. Yakubov, L.A., E.A. Deeva, V.F. Zarytova, E.M. Ivanova, A.S. Ryte, L.V. Yurchenko, and V.V. Vlassov, Mechanism of oligonucleotide uptake by cells: Involvement of specific receptors? *Proc. Nat. Acad. Sci. USA 86*:6454 (1989).
58. Leonetti, J., P. Machy, G. Degols, B. Lebleu, and L. Leserman, Antibody-targeted liposomes containing oligodeoxyribonucleotides complementary to viral RNA selectively inhibit viral replication. *Proc. Nat. Acad. Sci. USA 87*: 2448 (1990).

59. Wu, C.H., J.M. Wilson, and G.Y. Wu, Targeting gene delivery and persistent expression of a foreign gene driven by mammalian regulatory elements in vivo. *J. Biol. Chem. 264*:16985 (1989).
60. Stein, C.A. and J.S. Cohen, *Oligonucleotide Antisense Inhibitors of Gene Expression*, Cohen, J.S. ed., CRC Press, Boca Raton, FL, 97 (1990).
61. Matsukura, M., G. Zon, K. Shinozuka, M. Robert-Guroff, R. Shimada, C.A. Stein, H. Mitsuya, G. Wong-Staal, J.S. Cohen, and S. Broder, Regulation of viral expression of human immunodeficiency virus in vitro by an antisense phosphorothioate oligodeoxynucleotide against *rev (art/trs)* in chronically infected cells. *Proc. Nat. Acad. Sci. USA 86*:4244 (1989).
62. Inoue, H., Y. Hayase, A. Imura, S. Iwai, K. Miura, and E. Ohtsuka, Synthesis and hybridization studies on two complementary nona(2′-*o*-methyl)ribonucleotides. *Nucl. Acid. Res. 15*:6131 (1987).
63. Shibahara, S., S. Mukai, H. Morisawa, H. Nakashima, S. Kobayashi, and N. Yamamoto, Inhibition of human immunodeficiency virus (HIV-I) replication by synthetic oligo-RNA derivatives. *Nucl. Acid. Res. 17*:239 (1989).
64. Perreault, J.P., W. Talfeng, B. Cousineau, K.K. Ogilive, and R. Cedergren, Mixed deoxyribo- and ribooligonucleotides with catalytic activity. *Nature 344*: 565 (1990).
65. Miller, P.S., and P.O. Ts'O, A new approach to chemotherapy based on molecular biology and nucleic acid chemistry: Matagen (masking tape for gene expression). *Anticancer Drug Des. 2*:117 (1987).
66. Cotten, M., and M.L. Birnstiel, Ribozyme mediated destruction of RNA in vivo. *EMBO J. 8*:3861 (1989).
67. Cameron, F.H., and P.A. Jennings, Specific gene suppression by engineered ribozymes in monkey cells. *Proc. Nat. Acad. Sci. USA 86*:9139 (1989).
68. Sarver, N., E. Cantin, P. Chang, P. Ladne, D. Stephens, J. Zaia, and J. Rossi, Ribozymes as potential anti-HIV-1 therapeutic agents. *Science 247*:1222 (1990).
69. Guthrie, C., and B. Patterson, Spliceosomal snRNAs. *Ann. Rev. Gene. 22*: 387 (1989).
70. Katsuki, M., M. Sato, M. Kimura, M. Yokoyama, K. Kobayashi, and T. Nomura, Conversion of normal behavior to shiverer by myelin basic protein antisense cDNA in transgenic mice. *Science 241*:593 (1988).

13
Oligonucleotide-Based Therapeutics

Norbert Bischofberger and Regan G. Shea
Gilead Sciences, Inc., Foster City, California

I. INTRODUCTION

A. Rationale, Design, and Specificity of Targeting DNA/RNA

The majority of drugs currently on the market or in development target proteins. Such proteins can be host proteins or receptors where agonist or antagonist activity leads to a therapeutic effect or they can be foreign proteins (e.g., of viral origin) that one wishes to inhibit. The discovery and design of drugs interacting with proteins is a lengthy process, in part because of the dearth of structural information that is usually available on the target protein. Drug candidates arise generally by synthesizing analogs of natural ligands or from random screening. Frequently specificity is a problem; a drug should act on the target protein but not on any similar, closely related proteins. The difficulty in designing specific inhibitors is especially apparent with viruses that utilize the host's cellular machinery for multiplication and therefore offer few specific targets for therapeutic intervention.

A potentially general strategy for drug design involves targeting oligonucleotides against complementary base sequences in double-stranded DNA (dsDNA) or in RNA (see Fig. 1). When oligonucleotides are added extracellularly, they are expected to enter the cell and specifically bind to their respective targets interfering with transcription, RNA processing and transport, or translation. Targeting dsDNA with oligonucleotide leads to triple-helix formation (see Fig. 1); targeting RNA will lead to double-helix formation. This latter approach of recognizing intracellular sense RNA is widely referred to as antisense DNA. Both approaches have been termed code blockers.

This strategy of targeting nucleotide sequences potentially offers significant advantages over traditional drugs that target proteins. By identifying the sequences one wants to inhibit, one can in principle readily design the corresponding complementary sequence that will bind to and inactivate the target sequence. Also, code blocking agents can be used against many disease states such as cancers, viral, bacterial, and yeast infections, and others as long as a unique target sequence exists.

The premise behind the design of any drug is its ability to show efficacy without causing toxicity. A common measure for this selectivity is the therapeutic index of a drug. For antisense oligonucleotides this therapeutic index is dependent on the specificity of hybridization, i.e., that the oligonucleotide can discriminate target sequences from host endogenous sequences. Statistically, any sequence of 17 bases long should occur only once in the human genome. Thus, a 17-mer or longer oligonucleotide should theoretically recognize only its unique target sequence without hybridizing to similar sequences containing one or more mismatched bases. Another promising mechanism for developing oligonucleotide-based drugs uses selection and amplification of random-sequence DNA or RNA to produce binding-competent "aptamers," specific nucleic acid ligands, for any molecule (see Fig. 1).

A number of reviews are available on various aspects of code blockers

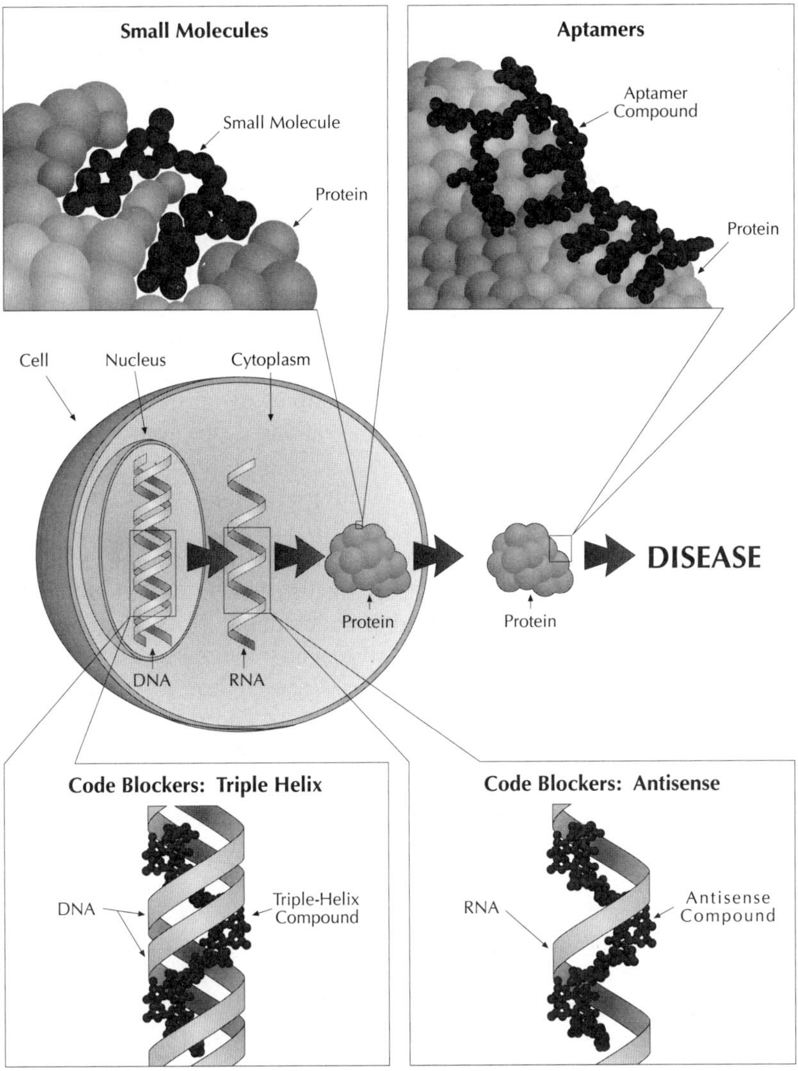

Figure 1 Some possible molecular sites of action for classic small-molecule drugs and nucleotide-based compounds. Small molecules as a class can be targetd against all types of intra- and extracellular molecules, but they are difficult to design. Aptamer nucleic acids can potentially also be designed to interact with any chosen class; extracellular or membrane surface proteins represent attractive early targets. Code blockers are designed to bind to the DNA double helix (triple-helix compounds) or to RNA (antisense compounds) within the cell. (From Gilead Sciences, Inc.)

and can be consulted for more in-depth information (1–6). The present account is intended to give a broad overview of the subject, introducing the concept and discussing some specific problems that have to be overcome to make code blockers feasible as therapeutics.

B. Antisense RNA

Distinct from antisense DNA is the strategy called antisense RNA (for reviews see 2,7). Antisense RNA refers to the intracellular transcription of an RNA sequence that selectively binds to a target gene sequence to modulate its expression. The antisense RNA is produced from either a genetically engineered plasmid or retroviral vector that has been introduced into the cell or from naturally occurring genes. Examples of antisense RNA from both nature and experimental studies have validated this approach to gene regulation. Naturally occurring antisense RNA molecules play an important role in regulating protein synthesis in certain plasmids, bacteriophages, and prokaryotes (7–9).

Antisense RNA has been utilized successfully to inhibit viral genes in in vitro tissue culture assays for pathogens such as Rous sarcoma virus (10), avian retroviruses (11), and human immunodeficiency virus (12,13). In one case, a plasmid expressing a functional ribozyme targeted against the HIV-1 *gag* transcripts was shown to reduce HIV p24 protein levels in tissue culture (14). Recently, an in vivo result has been reported for antisense RNA (15). Transgenic mice containing genes that expressed antisense proviral packaging sequences of Moloney murine leukemia virus (M-MuLV) were demonstrated to be completely resistant to developing leukemia symptoms following M-MuLV infection. The antisense RNA strategy has also been successfully applied in plants where expression of antisense RNA to the rate-limiting enzyme in the biosynthetic pathway of ethylene was found to inhibit fruit ripening in tomato plants (16). This approach might be valuable to extend the life span of plant tissues, thereby preventing spoilage.

The exact mechanism(s) for antisense RNA inhibition has not been fully characterized. However, one potential mechanism that has been described utilizes a double-stranded RNA base modification activity that is found ubiquitously in mammalian somatic cells (17). This activity converts up to 50–60% of adenosine residues in the duplex to inosine, creating RNase-sensitive or mutational "hot-spot" I-U mismatches throughout the duplexed RNA region (18,19). Although antisense RNA offers an attractive method for gene inhibition, the problems faced by other gene therapy schemes, such as efficient transfection, stable integration, and high expression of the (anti-) gene, each apply to antisense RNA.

C. Chemical Modifications

For code blockers to be of practical value for the treatment of various diseases, a number of hurdles have to be overcome. The externally added oligonucleotide has to find the target cell, cross the cell membrane, find its target RNA or dsDNA in the cytoplasm or the nucleus, bind with high enough affinity and specificity, and exert the desired biological effect for a specified period of time. A number of problems exist, most notably stability of the oligonucleotide in serum and intracellularly, inefficient permeation across membranes, affinity, and pharmacokinetic considerations. These problems will be discussed in more detail later.

Some of these problems can be addressed by chemically modifying oligonucleotides resulting in derivatives with enhanced properties. Chemical modifications of oligonucleotides have included attachment of a variety of linkers at both the 5′ and the 3′ end to improve cellular uptake, stability, and affinity (see Fig. 2, structure 1) (4). Chemically or photochemically reactive groups have been attached to form stable cross-links with or catalytically degrade the target sequence. The phosphodiester linkage has been replaced with other phosphorous linkages, most notably thioates (X = S) and methyl phosphonates (X = methyl; structure 2 in Fig. 2), or with non-phosphorous-containing linkages, such as formacetal (structure 3) and peptides (structure 6). Modifications in the sugar part have included 2′-*O*-alkylated species (R = methyl, allyl; structure 4) (20) or α-anomeric nucleosides (structure 5) to increase affinity and stability. Some of the individual modifications will be discussed in more detail later.

II. APPROACHES

A. Antisense DNA: RNA as a Target

The first reported use of a synthetic antisense DNA oligomer came in 1978 (21). The apparently straightforward mechanism and potential power of the approach led to a rapid proliferation of research. It is now obvious, however, that there is not a singular mechanism for antisense inhibition. Indeed much research that has been called antisense might not actually involve recognition of a target mRNA by an oligonucleotide. Proposed mechanisms that are consistent with antisense action will be described here. Some alternative modes will be presented in the practical applications section (Section III.E).

Recognition of the target RNA by the antisense oligonucleotide leads to duplex formation through normal Watson-Crick base pairing. The orienta-

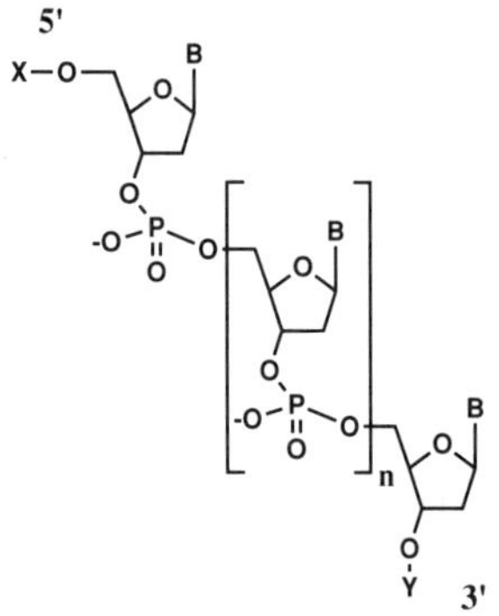

1

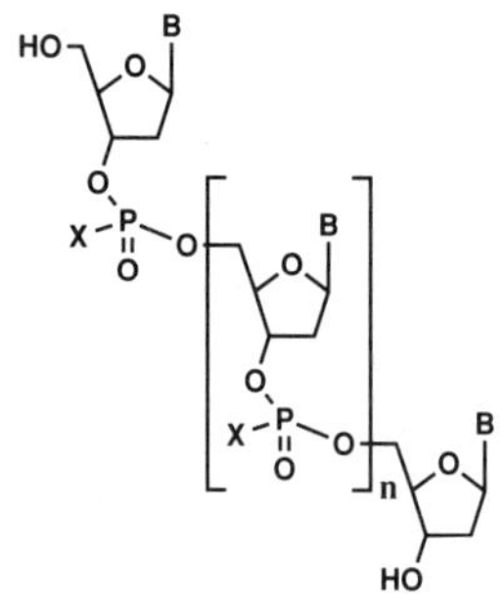

2

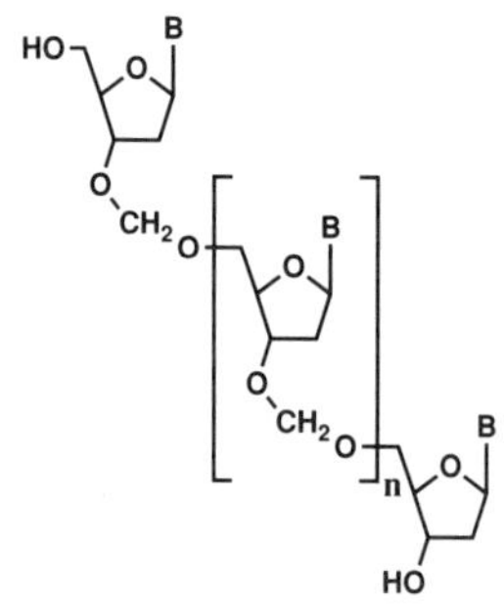

3

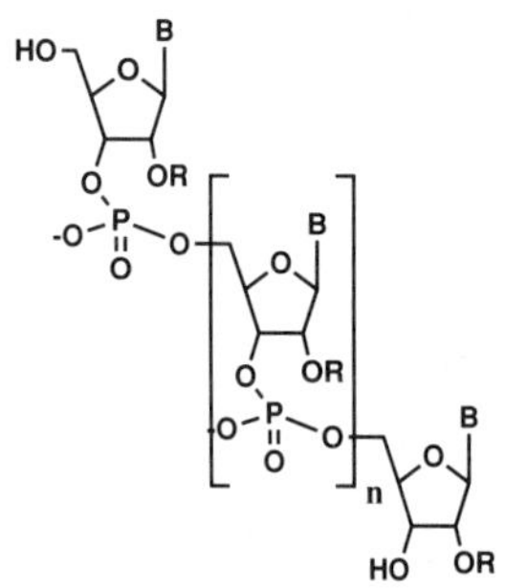

4

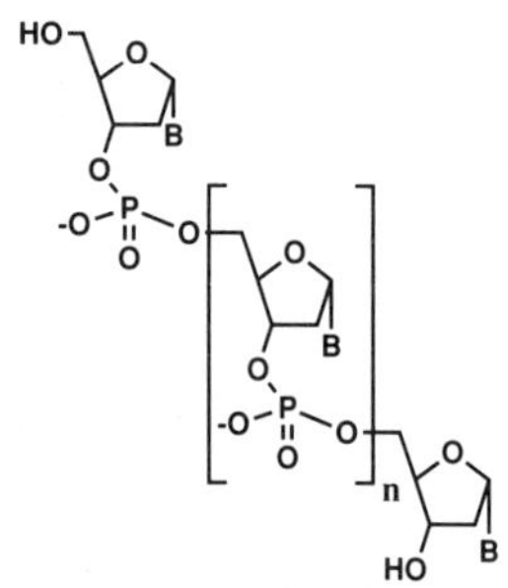

5

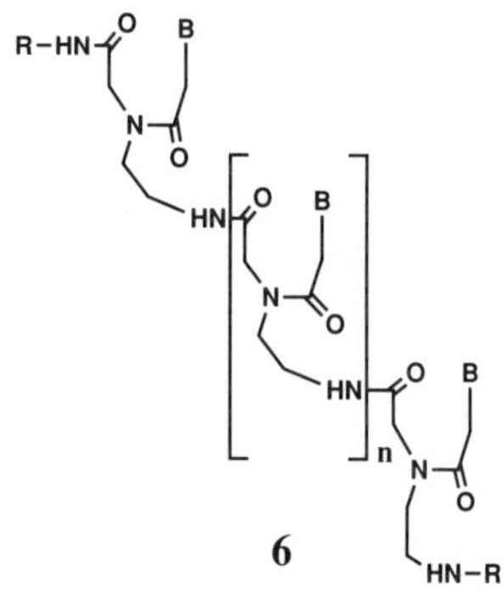

6

Figure 2 Nucleic acids with modified backbones. Structure 1 shows normal diester DNA with 5′ and 3′ appended groups. Structure 2 had modified phosphates; X = S- in phosphorothioates and X = Me in methylphosphonates. Structure 3 shows a formacetal backbone. Structure 4 depicts α-anomer DNA. Structure 6 shows a peptide nucleic acid (PNA). In all structures, B is a nucleobase.

tion of the two strands is normally antiparallel, but some modified oligonucleotides (e.g., α-anomers) will form parallel duplexes (22). The kinetics and thermodynamics of duplex formation have been studied by a variety of techniques, most notably thermal denaturation experiments where the UV absorbance of the duplex is measured at various temperatures. The melting curves obtained yield information on the stability of the duplex.

The use of target mRNA in a cell-free translation system allows for assessment of the influence of an antisense oligonucleotide on protein expression. Using this system thermal denaturation studies of the antisense oligonucleotide can be correlated with biological activity. Oligonucleotides, acridine-derivatized oligonucleotides, and oligonucleotide methylphosphonates complementary to the 5′-cap site, the initiation codon region, or the coding regions of mRNA have been tested (23–26). In rabbit reticulocyte lysates and wheat germ extracts, generally, these agents inhibit mRNA translation in a selective manner at low (0.1–10 μM) oligomer concentration. Oligomers targeted against the 5′-cap or the initiation codon region appear to be more effective inhibitors than those targeted against the coding region (23). The efficiency of translation inhibition by antisense oligonucleotides may also be a function of the mRNA secondary structure that influences oligomer accessibility and binding.

Antisense oligonucleotides have also shown biological activity in tissue culture in a variety of cell lines and assay systems. For references the reader is referred to the reviews (1–4). In a typical example, antisense oligonucleotides exhibit a biological effect at relatively high concentration, i.e., 5-200 μM, where corresponding control sequences show decreased or no activity. A wide variety of chemically modified oligonucleotides have been employed for the antisense approach. These will be discussed in more detail later.

One mechanism by which antisense oligonucleotides exhibit biological effects involves the interaction of the oligonucleotide/RNA duplex with the ribosome, to prevent protein expression. Another possible mechanism is the enzymatic hydrolysis of the RNA part of the oligonucleotide/RNA duplex by RNase H, provided the chemical form of the oligonucleotide allows the duplex to be an RNase H substrate (27). Such a catalytic mechanism of antisense action is very appealing because it potentially reduces the amount of oligonucleotide required to target multiple copies of RNA. Attempts to utilize a catalytic antisense mechanism have also included the attachment of various functional groups to oligonucleotides that degrade RNA by base catalysis (28), by a redox mechanism (29,30), or enzymatically (31). A special example of oligonucleotides that enzymatically degrade the target RNA are ribozymes, and attempts to make stable, small synthetic ribozymes have been described (32). So far, however, none of these catalytically acting antisense oligonucleotides have proven more efficacious in tissue culture assays than unmodified oligonucleotides.

Another site of action for oligonucleotide intervention is mRNA processing such as blockage of the splicing reaction or mRNA transport. For example, oligonucleotides targeted to the acceptor splice junction of HSV-I immediate early pre-mRNA have been reported to inhibit viral replication (33).

One consideration in targeting genetic sequences by antisense oligonucleotides is the abundance of target RNA. In a cell, the number of copies of an RNA ranges from a few hundred to a few million (34). In a virally infected cell, the total viral mRNA concentration may be as high as 750 nM [assuming a cell volume of 4×10^{-14} L (35)]. The intracellular concentration of antisense oligonucleotide required to efficiently hybridize to the target RNAs should be in excess of the concentration of target. Moreover, this effective concentration of antisense oligonucleotide has to be maintained at a steady state because RNA is being continuously transcribed. Therefore, both efficient permeation and intracellular stability of antisense oligonucleotides are important parameters for potent biological activity.

B. Triple-Helix Formation: Duplex DNA as a Target

A short time after the structure of duplex DNA was proven, DNA was found to also be capable of forming a triple-helical structure (36). Only polymers containing a homopurine strand and a complementary homopyrimidine strand were observed to adopt the triple-helix form (37–39). Most commonly, the triple helix consisted of two polypyrimidine strands bound to one polypurine strand. An early proposal that one polypyrimidine strand forms Watson-Crick hydrogen bonds with the polypurine strand while the other makes Hoogsteen hydrogen bonds with the polypurine strand in the duplex major groove has been confirmed now by many assay techniques. Early on it was discovered that under certain conditions, the strand stoichiometry observed was reversed: two polypurine strands complexed with a single polypyrimidine strand. The biological relevance of triple-helix nucleic acids was disputed, and the structure was seen as merely a curiosity by most for 30 years after its discovery.

Subsequently several developments have heightened interest in triplex DNA as a viable drug design goal. One was the discovery that three-stranded complexes are found in nature and may play a role in transcriptional regulation (40–45). Another was the observation that short, defined-sequence oligonucleotides can form triplexes (46–48). This raised the possibility that synthetic DNA or analogs could be used to form triplex in vivo, potentially providing a highly specific means to interfere with dsDNA function. A significant advantage of the triple-helix approach to code blocking over antisense inhibition is that the target binding sites for triplex formation are present at much lower concentration (perhaps one or two copies per cell) than are the mRNA targets of antisense.

Best characterized among triplexes are those in which the major groove-bound third strand contains only T hydrogen bonded to dA of the duplex and dC (protonated at N3) hydrogen bonded to dG (49) (see Fig. 3). In this "CT motif" the third strand maintains parallel polarity relative to the duplex homopurine strand (see Fig. 4). There is room for variation in this motif, however, since isolated dG in the third strand has been observed to tolerate a T within the polypurine tract of the duplex (50). The structural details of this variant have not yet been elucidated.

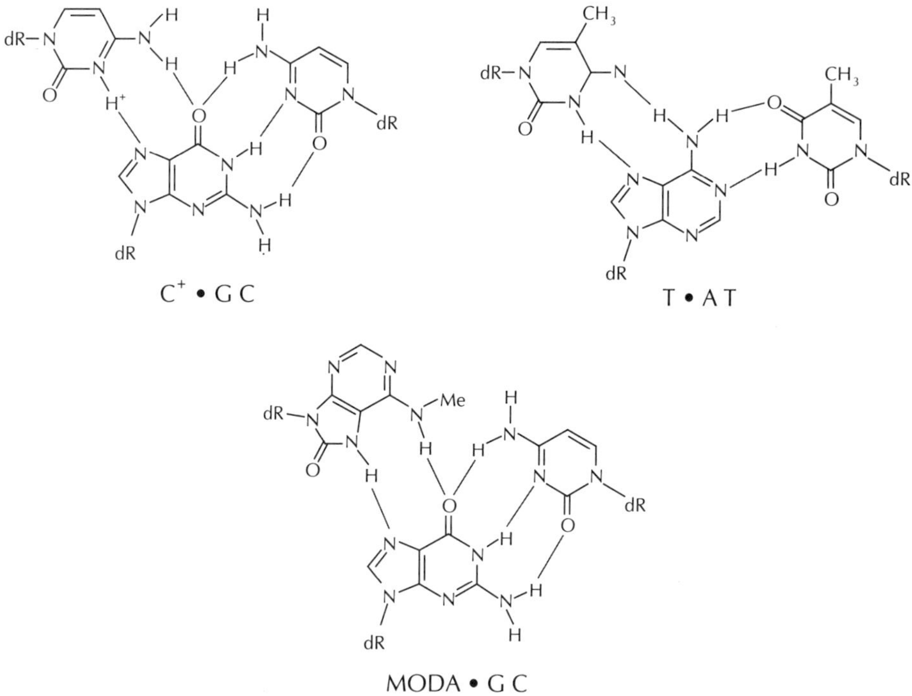

Figure 3 Details of the base-base hydrogen bond interactions in the standard CT motif triple helix (C+·GC and T:AT) and a pH-insensitive variant using a protonated C mimic (MODA·GC).

The so-called "G driven motif" is not as well understood. Reports indicate that this triplex forms G-GC, A-AT, and T-AT triplets. Third-strand orientation can be parallel or antiparallel to the duplex polypurine strand, depending on the sequence studied (51–54). Since protonated dC is not involved, this motif is insensitive to pH.

A third motif has recently been described where RecA can catalyze the formation of three-stranded complexes in which the incoming third strand has the same sequence as one of the duplex strands. Although RecA is required for complex formation, when the protein is removed from the system the triplex remains intact (55–57). The most intriguing feature is that all four bases are allowed in each of the three strands. It has been proposed that the third strand lies in the center of the duplex major groove and that the bases of this strand form hydrogen bonds to the adjacent duplex purine regardless of strand (55). However, duplexes containing dea-

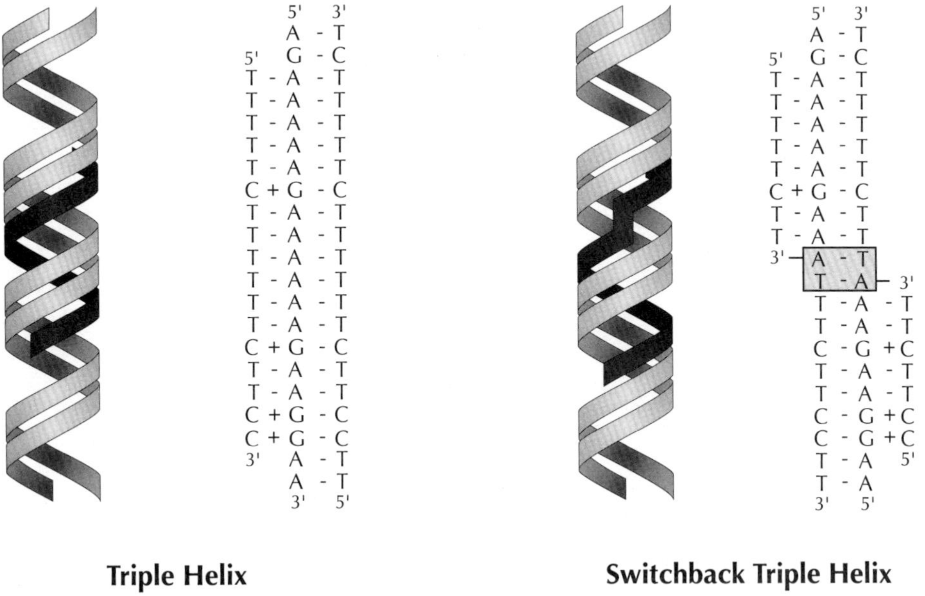

Figure 4 Oligonucleotide third strand binding in parallel fashion to an oligopurine region within the major groove of duplex DNA via the CT motif (left). Analogous representation of alternate-strand triplex formation by a 3′-3′ linked oligonucleotide, also via the CT motif (right).

zapurines can still take part in strand exchange, a fact that appears to be incompatible with the proposed model (57). Although the details of the RecA-mediated triplex remain uncertain, the very existence of mixed-sequence, three-stranded structures raises the possibility that a general recognition code may be attainable.

Another approach to sequence specific recognition of dsDNA involves the recently published use of peptide nucleic acids (PNAs), which are DNA analogs where the sugar-phosphate backbone has been replaced by a peptide (see Fig. 2, structure 6) (58). It was found that PNA oligomers containing thymine bases recognize their complementary oligo-dA target in dsDNA by strand displacement. This displacement seems to arise from the extraordinarily high stability of the PNA-DNA duplex. This approach, if applicable to mixed sequences, may present a general strategy to target dsDNA. At present, however, only cell-free assay systems have been studied, and it is unknown how PNAs would work in tissue culture assays or in vivo biological systems. Likewise, RecA-mediated recognition of dsDNA has only been studied in cell-free assay systems.

One method used to elucidate details of triplex formation is thermal denaturation, which has confirmed the parallel orientation of the third strand and has provided qualitative and quantitative information regarding the stability of mixed-sequence triplexes (54,59–63). A great deal of information has also been obtained through a variety of DNA-cleavage–based techniques. In these studies an agent able to sever the backbone of the radiolabeled duplex target is applied either independently [e.g., DNase I (64), copper phenanthroline (65), restriction enzymes (66,67), light (68,69)] or covalently linked to the third strand [e.g., iron EDTA, staphylococcal nuclease (70)]. Visualized by denaturing PAGE and autoradiography, the products of the partial digestion reflect the location of the third-strand binding site.

Additional characterization of triple helices has come from gel shift experiments similar to those frequently used to study DNA-protein binding. Thus the shift in mobility of a radiolabeled DNA duplex upon addition of the third strand can be seen by nondenaturing PAGE (54,60,64). By performing a titration of duplex vs. third strand, it is possible to obtain triplex binding constants.

The most detailed view of triplex DNA has been afforded by 2D NMR experiments. Important contributions include confirmation of the Hoogsteen base pairing of the third strand and protonation of N3 of dC (71–75). Although early fiber diffraction studies were useful for the study of long sequence-regular homopolymers (76), no x-ray crystal structures of triple helix have yet been published.

While duplex recognition offers an attractive alternative to the single-strand targeting of antisense, there are two fundamental problems unique to triplex. One is that, for the CT motif, acidic pH is required for triplex formation (46). The second is that the recognition sequence is limited to oligopurines. The first problem can be ameliorated by replacing third-strand dC with 5-methyl-deoxycytosine (77,78). Although this substitution allows triplex formation at neutral pH, it does not render the complex pH insensitive. The continuing pH dependence of triplexes containing 5-Me-dC is expected since this base too must be protonated to bind and its pKa value is comparable to that of dC. Other reports describe the replacement of dC by nonnatural bases that bear proper hydrogen bonding functionality at neutral pH. Successful mimics of protonated dC include 6-methyl-8-oxo-2′-deoxyadenosine [MODA; see Fig. 3 (79)] and 2′-*O*-methylpseudoisocytosine (80). However, aside from the occasional acceptance of a G-TA triplet noted previously, there has been little success in extending the basic triple-helix recognition code. The greatest success has come from linking together two standard triplex-forming oligonucleotides with opposing polarity (i.e., joined at their 3′ ends through a linker). These dimeric "switchback" molecules are capable of recognizing an oligopurine stretch in the usual way, crossing the major groove via the linker, and recognizing an adjacent run of purines on the opposite strand of the duplex (see Fig. 4) (81–83). This approach provides access to a greater number of target sequences but still does not address the problem of recognizing duplexed pyrimidines. There has been a report on the use of a random screening method (discussed in greater detail in the following section) to generate third-strand sequences without any bias for known triple helices (84). Although some unexpected base triples were tolerated in the final structures obtained, no new rules for allowable triplets were discovered.

Triple-helix formation by oligonucleotides on dsDNA has been found to interfere with the action of methylases and restriction enzymes (66,67,85). More recently, a triple-helix–forming oligonucleotide targeted to the coding region of dsDNA was found to inhibit RNA polymerase-mediated transcription in a cell-free system (66).

An experiment suggestive of intracellular triple-helix formation was the tissue culture inhibition of SV40 replication in COS cells by acridine-linked T_8 (86). It is supposed that the antiviral effect of the oligonucleotide was due to specific interaction of this molecule with an A_8 segment within the SV40 replication origin, but it was not conclusively shown that the biological event was due to intracellular triple-helix formation. In another example, an oligonucleotide targeted against the IL2Rα promoter region was shown to cause less accumulation of IL2Rα mRNA in lymphocytes relative to cells treated with a control oligomer of similar composition (53).

C. Aptamers: Nucleic Acid Ligands for Any Molecule

Perhaps the most promising opportunity for developing therapeutic oligonucleotides lies in the "aptamer" (from the Greek "aptos" = to fit) approach. The method is potentially useful for the identification of RNA, ssDNA, or dsDNA ligands for any molecule (84,87–93). In these experiments large sequence-random pools of oligonucleotides are screened for binding against target small molecules, proteins, or nucleic acids. A basic premise is that the pool nucleic acids populate a large number of classes of structures, some of which may possess desirable binding or catalytic activities. Affinity chromatography, immunoprecipitation, electrophoretic mobility shift, or filter binding techniques allow separation of the highest-affinity molecules from the pools. Amplification of the selected oligonucleotides by PCR provides a new pool enriched in sequences with improved affinity for the target relative to the starting pool. Successive iterations of the selection and amplification cycle "evolve" a product pool containing only those oligonucleotides ("aptamers") having the greatest affinity (see Fig. 5). Once the pool has converged, individual nucleic acids sequences are accessible through cloning.

The selection of high-affinity ligands for a wide variety of targets has been reported. An RNA enzyme, or ribozyme, specific for cleaving a target ssDNA was isolated in one case (87,88). In another, a mutant RNA was found that bound the bacteriophage T4 DNA polymerase as well as the natural RNA ligand bound (89). Other experiments have produced consensus dsDNA ligands for the transcription factor SRF, the Fos oncoprotein (93), and homo- and hetero-oligomers of the helix-loop-helix proteins MyoD and E2A (90). As mentioned above, selection and amplification has yielded novel DNA sequences capable of binding duplex DNA (84). Perhaps the most intriguing studies in the context of drug design are those in which molecules having no known specificity for binding nucleic acids have been used to select oligomers. One example was the generation of a limited pool of RNA able to discriminate between two structurally similar dyes (92). In the most powerful use of the technology to date, ssDNA was screened against human thrombin to produce a 15-mer aptamer capable of nanomolar inhibition of thrombin activity in vitro (94). Thus it appears that this method of "random drug design" will occupy an important place in the future of nucleic acids–based therapeutics.

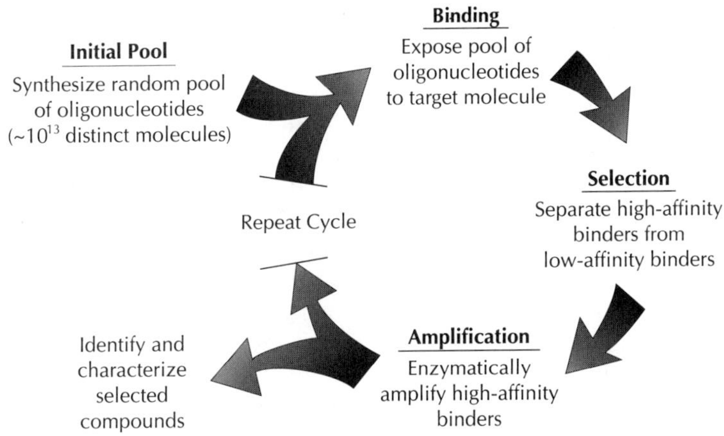

Figure 5 Schematic representation of the selection and amplification cycle used for the creation of aptamer nucleic acid ligands.

III. PRACTICAL CONSIDERATIONS

Although a large number of examples of sequence-specific effects of oligonucleotides have been published, major problems for the development of these agents into therapeutics remain. Generally, the extracellular concentration of oligonucleotides required for a biological effect in tissue culture is high, i.e., about 10–200 μM. One of the main reasons for the low potency of oligonucleotides in tissue culture assays is the low efficiency of cellular permeation. Oligonucleotides must be present at sufficiently high concentrations in the cytoplasm or the nucleus of the cell for antisense or triple-helix action. Another potential reason is the susceptibility of unmodified oligonucleotides to degradative enzymes. While chemical modifications exist that make oligonucleotides more stable to nucleases, they often also lead to decreased binding to the target RNA or DNA. Another concern is specificity, i.e., that oligonucleotides having a sequence that is not expected to hybridize to any target sequence still show biological activity. Such sequence unspecific effects could arise from the presence of complementary sequences other than those which are being targeted, hybridization to mismatched sequences, or binding to other cellular targets such as proteins.

A. Stability

Nuclease activity in serum or cellular extracts rapidly degrades unmodified diester DNA and RNA. In human serum, the predominant nuclease activity against oligodeoxynucleotides is a 3′-exonuclease. Substantial stability to 3′-exonuclease degradation is conferred by certain modifications to the DNA chemical structure. It is now generally accepted that the addition of just two backbone substitutions such as methylphosphonates (95), phosphoroamidates (96), or phosphorothioates (96) at the 3′ end of an oligodeoxynucleotide is sufficient to protect it from serum nucleases for the length of many antisense experiments. Such modifications were found to maintain the ability of the oligonucleotide to bind to its complementary sequence. Additional modifications beyond the first two do not substantially improve stability to 3′ nucleases. A single reversed linkage (3′-3′ diester or amidate) also protected the 3′ end from degradation (96). DNA conjugates with acridine (4) or anthraquinone (97) are also more stable than the parent oligomers. Such 3′-exonuclease-resistant oligonucleotides have not yet been proven more effective than unmodified oligonucleotides in tissue culture assays. Another method to stabilize oligonucleotides to degradation in serum is by attachment of a cholesterol. Such conjugated oligonucleotides were found to associate with plasma lipoproteins, resulting in a prolonged half-life (98).

While serum degradation of oligonucleotides is caused by a 3′-exonuclease, the intracellular pathway of degradation has not been characterized. Recent studies have suggested that intracellularly oligonucleotides are at least in part degraded through an endonucleolytic process and that 3′- and 5′-capped oligonucleotide diesters might not be stable enough to degradation (99). The earliest studies on enhanced oligomer stability used DNA derivatized at every phosphate as, e.g., phosphorothioates, amidates, and methylphosphonates. These oligonucleotide backbone analogs are stable to various exo- and endonucleases in vitro, and they have exhibited activities in tissue culture where the unmodified oligonucleotides were inactive. Of these, the phosphorothioates are especially attractive because they still maintain their ability to hybridize to their complementary sequence (although somewhat diminished) while being more resistant to nuclease degradation.

A number of hydrolysis-resistant DNA analogs have been prepared in which the backbone contains no phosphorous. Examples include carbamates (100), siloxanes (101), thioethers (102), PNAs (58), and formacetals (103–105). While they appear to generally be completely nuclease resistant, these molecules for the most part do not hybridize well with normal nucleic acids. They also tend to be poorly soluble in water. Exceptional among this class of derivatives are the formacetals and PNAs. Formacetals not only confer nuclease stability, but they are also competent for hybridization to ssDNA, ssRNA, and dsDNA. PNAs have solubility problems, and their mode of binding to normal nucleic acids has not been fully characterized, but their complexes with DNA are exceptionally stable. Of course, PNAs might be degraded by peptidases.

B. Permeation

DNA rapidly associates with cells, and its entrance into cells is a naturally occurring, albeit inefficient, phenomenon (106,107). There is evidence that certain plasma membrane proteins may function as specific receptors for oligonucleotide uptake (108–110). This oligonucleotide uptake appears to take place via endocytosis. Because of the poor permeation efficiency, transfection experiments requiring significant cellular penetration by DNA have commonly included microinjection (111), liposome (112), calcium phosphate (113), virus-mediated transfection (114), or electroporation methods (115). It has been observed that a cationic lipid can enhance transfection efficiency when premixed with DNA (116). The subject of cellular uptake of oligonucleotides has recently been reviewed (117).

DNA permeation may be assayed by measuring the extent of cell association of radiolabeled oligonucleotides. Two higher-resolution techniques have been applied successfully as well: cell fractionation and confocal microscopy (using fluorescently tagged oligomers). The two latter methods allow the determination of the intracellular localization of oligonucleotides, which is important for antisense and triple-helix action.

Covalent linkage of a variety of cell receptor ligands to otherwise unmodified DNA has been tested for transport via receptor-mediated endocytosis (e.g., epidermal growth factor (118), sugar lysine complexes (119), mitochondrial signal peptide (120), transferrin polycation conjugates (121)]. Polylysine-mediated internalization follows a classic endocytic pathway (122). Along with enhanced antiviral activity against both VSV (123,124) and HIV (125), polylysine conjugation also greatly increases the uptake of fluorescently labeled oligomers. However, under the fluorescent microscope the intracellular fluorescence was punctate, an indication of endosomal localization (126). Connecting lipophilic groups to oligonucleotides produced activity against both VSV (127) and influenza virus (128) under conditions where the unconjugated oligonucleotides were inactive. Similarly, antibody-targeted liposomes were used for the intracellular delivery of oligonucleotides targeted against VSV. These liposome-encapsulated oligomers were active in amounts 10- to 100-fold lower than unencapsulated materials (129).

The importance of good permeation along with appropriate intracellular localization for successful application of code blocking was striking in one recent publication. Phosphorothioate antisense oligomers exhibited sequence-specific inhibition of an HIV LTR/human placental alkaline phosphatase gene fusion only when transfected using calcium phosphate, in effect bypassing the plasma membrane (130). Simple application of the same oligonucleotides to the cells was ineffective at inhibiting production of the gene product. There is evidence that fluorescently labeled oligonucleotides concentrate in the nucleus following microinjection into the cytoplasm of cells (35,131). However, these compounds do not accumulate in the nucleus to an appreciable extent when added exogenously to cells in the culture medium, as shown both by fluorescent microscopy (35,109) and by cell fractionation using radiolabeled oligonucleotides (132). Rather, these compounds accumulate in cytoplasmic compartments, presumably lysosomes. It thus appears that diffusion of oligonucleotides across the cellular membrane is an extremely inefficient process.

That endocytosed DNA remains trapped is consistent with the lack of leakage of other macromolecules from intracellular compartments. Polyvinylpyrrolidone, for example, accumulates in lysosomes, where it remains for days (133). Likewise, the rate of diffusion of fluorescein-tagged dextrans from lipid-bilayer vesicles occurs only very slowly (134). Hence, one would predict that oligonucleotide escape from lysosomes would be a very slow process unless the oligonucleotide could follow a specialized macromolecule-sorting pathway, as seen with ricin A or diphtheria toxin (135) or the HIV tat protein (136).

Both the anionic character and large size of DNA makes it a poor candidate for entering cells. Even at the level of monomers, it is well established that the negatively charged phosphate group of mononucleotides makes them very poor cell penetrators (137). Uncharged mononucleosides, on the other hand, can enter cells either through specific transport or by passive diffusion (137). One obvious way to enhance the cellular permeation of oligonucleotides is to replace the polyanionic phosphate diester backbone by an uncharged analog. Accordingly, a number of oligonucleotides containing methylphosphonate or phorphoroamidate linkages have been reported (for a review see 4). Of these backbone-modified oligonucleotides only methylphosphonates have been studied more extensively for activity in biological systems. They were found to be taken up by cells more efficiently than diesters in a cell association experiment (138), although fluorescence microscopy revealed a vesicular subcellular distribution suggestive of an endosomal localization. This observation indicates that the pathway of cellular uptake of oligonucleotide methylphosphonates involves fluid-phase or absorptive endocytosis (139). Methylphosphonate oligomers have proven efficient in a number of biological assays (33,140–143), but despite enhanced access to cells, the effective concentration at which methylphosphonate oligonucleotides show efficacy in tissue culture remains high (generally 100–200 μM). Methylphosphonates also exhibit compromised binding to the target DNA/RNA. This is thought to be at least partly due to the presence of diastereomers less competent for hybridization. This makes the synthesis of stereoregular methylphosphonate oligonucleotides an attractive goal (144). Selective synthesis may not be a panacea though, because methylphosphonate/RNA duplexes are incompetent as substrates for the RNase H degradation of the target RNA (145).

Cellular uptake of oligonucleotides is currently an inefficient process because both unmodified and modified oligonucleotides localize mainly in intracellular endosomal compartments, if they enter cells at all. So permeation remains a major roadblock to the development of code blockers as drugs. Only with a clearer understanding of how oligonucleotides interact with cells will it be possible to produce more potent molecules.

C. Affinity and Specificity

Both the binding affinity and sequence specificity of an oligonucleotide to its complementary target increase with increasing length. On the other hand, shorter oligomers are easier to make and more likely to be cell permeant than longer fragments. Attempts to increase the affinity of short oligonucleotides have focused on the attachment of intercalators, e.g., acridines, ellipticine (for a review see 4). Such conjugates have exhibited biological effects in assay systems where unmodified oligonucleotides did not (146). Affinity of an oligonucleotide to its target can also be improved by covalently conjugating the DNA with chemically or photochemically reactive functionalities. The reactive groups can then form stable cross-links to the target sequences (for a review see 4). An example of a specific triple-helix-mediated cross-linking to dsDNA has recently been published (147). However, the intercalators or cross-linkers may alter the specificity and the kinetics of binding to the complementary sequences. It has been suggested that neutral or cationic backbones might improve affinity of code blockers for the anionic target. On the whole, backbone-modified oligomers exhibit lower Tms than the corresponding unmodified DNA duplexes and triplexes. The diastereomer problem mentioned above may be critical in this regard. A cationic backbone did show favorable properties, but only under very specific salt and pH conditions (148). Another way of increasing the affinity of an oligonucleotide to its target is by using circular oligonucleotides. Such oligos were found to bind to their complementary targets with high affinity owing to their prearranged structure (149).

The degree of specificity, i.e., discrimination of an oligomer between its precisely complementary sequence and a given closely related sequence, can be estimated from thermal denaturation studies of oligonucleotides. For example, the self-complementary oligonucleotide GGGAAGCTTCC was found to have a value for the free energy of hybridization $\Delta G°_{298}$ of -16.8 kcal/mol (100 mM NaCl, pH 7.0), from which one can calculate an equilibrium constant $K = 2 \times 10^{12}$ (150). For the imperfectly self-complementary sequence GGGAAGATTCC, which contains two central mismatched bases, $\Delta G°$ was found to be -11.3 kcal/mol and $K = 2 \times 10^{8}$. Therefore, the presence of the two mismatched bases causes a 10,000-fold decrease in the equilibrium constant. It must be noted, however, that this difference in K applies only if the hybridization of an antisense DNA to its target can be carried out under equilibrium conditions. Since hybridization inside a cell will compete with other processes (e.g., metabolism, sequestration), it seems likely that kinetic parameters will be important and have a major influence on the specificity and the mechanism of code-blocking oligonucleotides. Recent studies of antisense RNA mechanism have indeed shown that hybridization inside a cell occurs under nonequilibrium conditions and that K is of minor importance for the association rate (151).

It is worth noting that individual domains or subunits of sequence-specific dsDNA binding proteins tend to recognize only a few base pairs. Greater specificity is attained through cooperative interaction of multiple binding components. A recent application of this concept to triple-helix formation has been published (152). The subunits in this case are two DNA strands, each made up of both a "recognition domain" for triplex formation at one of two closely spaced oligopurine sites on a 2.3-kbp target and a "dimerization domain" for hybridizing to one another. Thus, duplex formation by the complementary dimerization domains engenders cooperativity in the triplex binding of the recognition domains. This motif for dsDNA recognition is reminiscent of the leucine zipper dimeric transcription factors (153).

In an absolute sense, the specificity goal of code blockers is unique targeting of a single human, bacterial, or viral sequence with negligible interaction elsewhere. One of the most impressive results to date was the demonstration of binding and cleavage at a single site within a 340-kb yeast chromosome by a triplex-forming artificial nuclease (154). The main goal for the future is to maintain this high level of specificity with molecules showing enhanced uptake and stronger binding affinity.

D. Large-Scale Manufacturing/Cost

Synthesis of sequence-defined oligonucleotides has become feasible with the development of solid-phase synthesis methodologies and efficient coupling techniques. Current DNA synthesis and purification technologies provide oligonucleotides on a microgram-to-milligram scale mainly for application in molecular biology and genetic engineering. For the therapeutic application of oligonucleotides, multigram quantities will be needed for toxicological and pharmacological evaluation. Currently, the cost of a 21-mer oligonucleotide produced on a 2500-μmol scale is estimated to be about \$3400/g (155). Through improved and optimized synthesis and purification conditions, this cost could be reduced to ~\$1000/g. Even so, oligonucleotide synthesis needs to become more cost-efficient in order to be practically useful.

The present status of large-scale oligonucleotide synthesis can be best assessed when compared to large-scale peptide synthesis. Examples of peptides synthesized in solution range from 2 to 70 amino acids in length, and in amounts from milligrams to tons. Representative of large-scale, solid-phase peptide synthesis are the production of LHRH analogs and salmon calcitonin, a 32-amino-acid peptide that is produced in multihundred-gram quantities per year (156).

Thus, for peptides the problems of efficient large-scale production has largely been solved. For oligonucleotides this is just a matter of time once the proper commercial incentive exists for the development of more efficient techniques.

E. Mechanism of Action

To maximize the powerful drug design promise of code blockers, detailed knowledge will be required of all mechanisms by which oligonucleotides exert biological activity. A variety of possible antisense DNA modes of action have been elucidated. Although sequence-specific inhibitory activity by antisense oligonucleotides is well established, many reports also describe nonspecific interactions of oligonucleotides. The issue of specificity is most apparent with respect to antiviral activities of oligonucleotides. This will be discussed for HSV and HIV, along with the proposed mechanism of action.

Antisense oligonucleotides have been extensively explored as potential anti-HIV agents. In a representative example, modified and unmodified oligonucleotides were targeted against various regions of the HIV genome. Various 14-mer diester and methylphosphonate oligomers failed to show any anti-HIV activity; phosphorothioate oligomers did exhibit anti-HIV activity. In all cases sequence-nonspecific effects were noted. Upon further investigation it was found that a 28-base homo-oligomer of dC was the most active among a number of phosphorothioates, implying that the effects was non-sequence-specific (157). When a series of compounds was assayed in chronically infected H9 cells, however, neither the sense, random, dC_{28}, nor any other control sequences exhibited any activity, whereas the antisense sequence showed antiviral activity at 10 μM (158). In another investigation, external application of oligonucleotide phosphorothioates inhibited HIV replication in both acute and chronically infected viral assays, but without sequence specificity. When these thioates were delivered into cells by transfection, sequence-specific inhibition of transactivation was observed in response to oligonucleotides targeted against TAR. Oligonucleotide diesters showed no effect in this system (130).

Attachment of cholesterol to phosphorothioate DNA increased the potency of the compounds (159) even though the binding to the complementary target sequences was diminished. In the most favorable case, a 20-mer cholesterol phosphorothioate at a concentration of 0.2 μM completely inhibited HIV p24 production. Decamers were active and some activity was observed with a heptanucleotide that binds poorly to complementary sequences. The finding that the activity of the phosphorothioate decamers did not correlate with a specific sequence suggests that a mechanism other than antisense played a role in these systems. One possible mechanism is inhibition of the viral reverse transcriptase (RT). It has indeed been found that phosphorothioates are potent inhibitors of RT; e.g., a phosphorothioate dC_{28} is a competitive inhibitor of RT with respect to template primer with a K_i of 2.8 nM (160). Oligonucleotide phosphorothioates also partially block the binding of gp120 to CD4 in a solid-phase ELISA, suggesting this is a second possible mechanism of action (161).

The dC_{28} phosphorothioate described above is also a potent inhibitor of HSV-II. The compound was relatively nontoxic while showing 90% viral inhibition at 1 μM. Although the effect was not driven by complementary binding to a specific sequence, it was base-specific. The order of activity was deoxycytidine = thymidine > deoxyadenosine. Interestingly, the oligo dC compound was not as active against HSV-I. Previous studies had shown that HSV-II DNA polymerase inhibition by phosphorothioate oligomers was greater than inhibition of HSV-I polymerase (or than human polymerases α, β, and γ). This observation may indicate that the phosphorothioates are acting at the level of HSV-II polymerase inhibition, but this mechanism has not been proven. Another possible clue to the basis for activity was that the phosphorothioate was most active when present during the 1-hr viral adsorption period. Therefore, it may be that the oligonucleotide interferes with interactions between the virus and the plasma membrane by binding either the viral coat proteins or membrane receptor proteins (162).

In summary, both sequence-specific and nonspecific activities of oligonucleotides have been characterized in in vitro tissue culture assays. The exact mechanism for the nonspecific inhibition, especially by the phosphorothioate oligomers, is not known, although several theories are possible. One is that the oligonucleotide enters the cell and acts as a viral polymerase inhibitor. Another mechanism is that the oligomer binds to proteins on the virus or on the plasma membrane, interfering with viral adsorption and/or entry into the cell. This mechanism has been demonstrated for the blocking of HSV adsorption by another sulfated polyanion, heparan sulfate (163). Another possibility is that the oligonucleotide enters the cell in the same compartment with the virus and thereby disturbs cellular sorting pathways essential for viral release into the cytoplasm. Endocytosed polyanion disruption of vesicle trafficking has been well documented (164).

F. In Vivo Applications

To date, there are only a few reports dealing with pharmacokinetics, toxicology, and efficacy in animal models of oligonucleotides. The distribution and excretion of methylphosphonates was studied by injecting them into the tail vein of mice. The half-life of distribution was 6 min and the excretion half-life of the DNA was 17 min. About 70% of the total dose was excreted in urine after 60–120 min (165). Oligonucleotide phosphorothioates have been found to be relatively nontoxic when injected into mice up to 40 mg/kg. At a larger dose of 160 mg/kg, toxicity was observed (166). After i.v. or i.p. administration of a single dose, the labeled oligonucleotide thioate was found in most tissues for up to 48 hr. About 30% of the dose was excreted in urine and the oligonucleotide was highly degraded (167).

In a recent demonstration of in vivo efficacy, a phosphorothioate 21-mer targeted against the UL-13 gene of HSV-I was used in an animal model of HSV-induced keratitis (168). Mice infected with HSV were treated with various concentrations of the oligomer thioate topically on the keratitic lesion. Treatment with a thioate reduced blepharitis, vascularization, and stromal disease but did not change viral titer in the eyelids, eyes, or trigeminal ganglia. This experiment included no control oligomer, so it is not certain that the observed effects were due to an antisense mechanism. Nevertheless, the reduction in disease severity seen with these oligonucleotide thioates is encouraging and suggests and antisense oligomers may be useful in treating HSV keratitis.

HSV was the target of another promising experiment. Methylphosphonate analogs prevented expression of herpetic lesions when applied topically to the HSV infected ear of a mouse (169). In another experiment, an antisense oligonucleotide containing a cross-linking agent was effective against tick-borne encephalitis virus in mice (170).

Very recently an Investigational New Drug application was filed with the FDA for the first use of code blockers in humans. This clinical trial proposes to use oligonucleotide methylphosphonates as a systemic therapy for chronic myelogenous leukemia in an ex vivo treatment (171).

IV. SUMMARY AND CONCLUSIONS

The prospect of using sequence-specific nucleic acids and their analogs to target gene sequences associated with disease presents an exciting opportunity in the field of drug design. Code blockers may conceivably be applied to any target, virus, or disease as long as a unique sequence exists. Oligonucleotides have been successfully used for the inhibition of a number of targets in tissue culture and more recently against important viruses in animal experiments. However, a number of questions must be answered before sequence-specific targeting will be feasible for commercial development of new drugs. More information needs to be assembled to determine the exact mechanism of antisense action of oligonucleotides and to distinguish thermodynamic and kinetic influences. This would allow the design of oligonucleotide analogs that show much higher specificity than phosphorothioates, which show a high degree of non–sequence-specific effects.

A major challenge that needs to be solved is permeation. Efficient permeation of oligonucleotides into cells would allow these agents to exhibit therapeutic effects at much lower concentrations. This is necessary from both a cost and manufacturing standpoint. Future efforts should also focus on oligonucleotide absorption, distribution, metabolism, and elimination in vivo.

In the triplex arm of code blockers all the same questions apply, as well as the unique problem of generating a "dictionary" of interactions that will allow one to write down the third-strand sequence correct for any given duplex sequence. Though still in its infancy, the aptamer field may rapidly become a ripe ground for drug discovery and development. Aptamers targeted against extracellular or membrane surface components circumvent the permeation problem. Another advantage is that a general "recognition code" need not be known a priori. For these reasons and since the approach offers a general method to target any biologically relevant molecule, the first successful nucleic acids-based drugs may well be aptamers.

REFERENCES

1. Helene, C., and J-J. Toulme, Specific regulation of gene expression by antisense, sense and antigene nucleic acids. *Biochim. Biophys. Acta 1049*: 99–125 (1990).
2. van der Krol, A.R., J.N.M. Mol, and A.R. Stuitje, Modulation of eukaryotic gene expression by complementary RNA or DNA sequences. *BioTechniques 6*: 958–976 (1988).
3. Cohen, J.S., ed., *Oligodeoxynucleotides: Antisense Inhibitors of Gene Expression*. CRC Press, Boca Raton, FL (1987).
4. Goodchild, J., Conjugates of oligonucleotides and modified oligonucleotides: a review of their synthesis and properties. *Bioconjugate Chem. 1*: 165–187 (1990).
5. Riordan, M.L., and J.C. Martin, Oligonucleotide-based therapeutics. *Nature 350*: 442–443 (1991).
6. Uhlmann, E., and A. Peyman, Antisense oligonucleotides: a new therapeutic principle. *Chem. Rev. 90*: 544–584 (1990).
7. Green, P.J., O. Pines, and M. Inouye, The role of antisense RNA in gene regulation. *Annu. Rev. Biochem. 55*: 569–597 (1986).
8. Simons, R.W., and N. Kleckner, Biological regulation by antisense RNA in procaryotes. *Annu. Rev. Genet 72*: 567–600 (1988).
9. Simons, R.W., Naturally occurring antisense RNA control-α. *Gene 72*: 35–44 (1988).
10. Chang, L-J., and C.M. Stoltzfus, Gene expression from both intronless and intron-containing Rous sarcoma virus clones is specifically inhibited by antisense RNA. *Mol. and Cell. Biol. 5*: 2341–2348 (1985).

11. To, RY-L., S.C. Booth, and P.E. Neiman, Inhibition of retroviral replication by anti-sense RNA. *Mol. Cell. Biol. 6*: 4758–4762 (1986).
12. Rittner, K., and G. Sczakiel, Identification and analysis of antisense RNA target regions of the human immunodeficiency virus type 1. *Nucleic Acids Res. 19*: 1421–1426 (1991).
13. Rhodes, A., and W. James, Inhibition of human immunodeficiency virus replication in cell culture by endogenously synthesized antisense RNA. *J. Gen. Virol. 71*: 1965–1974 (1990).
14. Sarver, N., E.M. Cantin, P.S. Chang, J.A. Zaia, P.A. Ladne, D.A. Stephens, and J.J. Rossi, Ribozymes as potential anti-HIV-1 therapeutic agents. *Science 247*: 1222–1225 (1989).
15. Han, L., J.S. Yun, and T.W. Wagner, Inhibition of Moloney murine leukemia virus-induced leukemia in transgenic mice expressing antisense RNA complementary to the retroviral packaging sequences. *Proc. Natl. Acad. Sci. USA 88*: 4313–4317 (1991).
16. Oeller, P.W., L. Min-Wong, L.P. Taylor, D.A. Pike, and A. Theologis, Reversible inhibition of tomato fruit senescence by antisense RNA. *Science 254*: 437–439 (1991).
17. Wagner, R.W., C. Yoo, L. Wrabetz, J. Kamholz, J. Buchhalter, N.F. Hassan, K. Khalili, S.U. Kim, B. Perussia, F.A. McMorris, and K. Nishikura, Double-stranded RNA unwinding and modifying activity is detected ubiquitously in primary tissues and cell lines. *Mol. Cell. Biol. 10*: 5586–5590 (1990).
18. Bass, B.L., and H. Weintraub, An unwinding activity that covalently modifies its double-stranded RNA substrate. *Cell 55*: 1089–1098 (1988).
19. Wagner, R.W., J.E. Smith, B.S. Cooperman, and K. Nishikura, A double-stranded RNA unwinding activity introduces structural alterations by means of adenosine to inosine conversions in mammalian cells and *Xenopus* eggs. *Proc. Natl. Acad. Sci. USA 86*: 2647–2651 (1989).
20. Iribarren, A.M., B.S. Sproat, P. Neuner, I. Sulston, U. Ryder, and A.I. Lamond, 2′-*O*′alkyl oligoribonucleotides as antisense probes. *Proc. Natl. Acad. Sci. USA 87*: 7747–7751 (1990).
21. Zamecnik, P.C., and M.L. Stephenson, Inhibition of Rous sarcoma virus replication and cell transformation by a specific oligodeoxynucleotide. *Proc. Natl. Acad. Sci. USA 75*: 280–284 (1978).
22. Sun, J-s., U. Asseline, D. Rousaud, T. Montenay-Garestier, N.T. Thuong, and C. Helene, Oligo-(α)-deoxynucleotides covalently linked to an intercalating agent. Double helices with parallel strands are formed with complementary oligo-(β)-deoxynucleotides. *Nucleic Acids Res. 15*: 6149–6159 (1987).
23. Blake, K.R., A. Murakami, S.A. Spitz, S.A. Glave, M.P. Reddy, P.O.P. Ts'o, and P.S. Miller, Hybridization arrest of globin synthesis in rabbit reticulocyte lysates and cells by oligodeoxyribonucleoside methylphosphonates. *Biochemistry 24*: 6139–6145 (1985).
24. Cazenave, C., N. Loreau, N.T. Thuong, J-J. Toulme, and C. Helene, Enzymatic amplification of translation inhibition of rabbit b-globin mRNA mediated by antimessenger oligodeoxynucleotides covalently linked to intercalating agents. *Nucleic Acids Res. 15*: 4717–4736 (1987).

25. Cazenave, C., C.A. Stein, N. Loreau, N.T. Thuong, L.M. Neckers, C. Subasinghe, C. Helene, J.S. Cohen, and J-J. Toulme, Comparative inhibition of rabbit globin mRNA translation by modified antisense oligodeoxynucleotides. *Nucleic Acids Res. 17*: 4255–4273 (1989).
26. Verspieren, P., N. Loreau, N.T. Thuong, D. Shire, and J-J. Toulme, Effect of RNA secondary structure and modified bases on the inhibition of trypanosomatid protein synthesis in cell free extracts by antisense oligodeoxynucleotides. *Nucleic Acids Res. 18*: 4711–4717 (1990).
27. Cazenave, C., M. Chevrier, N.T. Thuong, and C. Helene, Rate of degradation of [α]- and [β]-oligodeoxynucleotides in *Xenopus* oocytes. Implications for anti-messenger strategies. *Nucleic Acids Res. 15*: 10507–10521 (1987).
28. Bashkin, J.K., J.K. Gard, and A.S. Modak, Synthesis and characterization of nucleoside peptides: toward chemical ribonucleases. *J. Org. Chem. 55*: 5125–5132 (1990).
29. Dreyer, G.B., and P.B. Dervan, Sequence-specific cleavage of single-stranded DNA: oligodeoxynucleotide-EDTA.Fe(II). *Proc. Natl. Acad. Sci. USA 82*: 968–972 (1985).
30. Chen, C.B., and D.S. Sigman, Sequence-specific scission of RNA by 1,10-phenanthroline-copper linked to deoxyoligonucleotides. *J. Am. Chem. Soc. 110*: 6570–6572 (1988).
31. Corey, D.R., and P.G. Schultz, Generation of a hybrid sequence-specific single-stranded deoxyribonuclease. *Science 238*: 1401–1403 (1987).
32. Pieken, W.A., D.B. Olsen, F. Benseler, H. Aurup, and F. Eckstein, Kinetic characterization of ribonuclease-resistant 2′-modified hammerhead ribozymes. *Science 253*: 314–317 (1991).
33. Kulka, M., C.C. Smith, L. Aurelian, R. Fishelevich, K. Meade, P. Miller, and P.O.P. Ts'o, Site specificity of the inhibitory effects of oligo(nucleoside methylphosphonate)s complementary to the acceptor splice junction of herpes simplex virus type 1 immediate early mRNA 4. *Proc. Natl. Acad. Sci. USA 86*: 6868–6872 (1989).
34. Lewin, B., *Gene Expression*, Vol. 2: *Eucaryotic Chromosomes*, 2nd ed., Wiley, New York (1980).
35. Chin, D.J., G.A. Green, G. Zon, F.C. Szoka, Jr., and R.M. Straubinger, Rapid nuclear accumulation of injected oligodeoxyribonucleotides. *New Biologist 2*: 1091–1100 (1990).
36. Felsenfeld, G., D.R. Davies, and A. Rich, Formation of a three-stranded polynucleotide molecule. *J. Am. Chem. Soc. 79*: 2023–2024 (1957).
37. Wells, R.D., D.A. Collier, J.C. Hanvey, M. Shimizu, and F. Wohlrab, The chemistry and biology of unusual DNA structures adopted by oligopurine. oligopyrimidine sequences. *FASEB J. 2*: 2939–2949 (1988).
38. Michelson, A.M., J. Massoulie, and W. Guschlbauer, Synthetic polynucleotides. *Progr. Nucleic Acid Res. Mol. Biol. 6*: 83–141 (1967).
39. Felsenfeld, G., and H.T. Miles, The physical and chemical properties of nucleic acids. *Annu. Rev. Biochem. 36*: 407–448 (1967).
40. Johnston, B.H., The S1-sensitive form of d(C-T)n.d(A-G)n: chemical evidence for a three-stranded structure in plasmids. *Science 241*: 1800–1804 (1988).

41. Kinniburgh, A.J., A *cis*-acting transcription element of the *c-myc* gene can assume an H-DNA conformation. *Nucleic Acids Res. 17*: 7771–7778 (1989).
42. Bacolla, A., and Y-H. Wu, Mung bean nuclease cleavage pattern at a polypurine.polypyrimidine sequence upstream from the mouse metallothionein-I gene. *Nucleic Acids Res. 19*: 1639–1647 (1991).
43. Roy, C., and B. LeBleu, DNA protein interactions at the interferon-responsive promoter elements: potential for an H-DNA conformation. *Nucleic Acids Res. 19*: 517–524 (1991).
44. Lee, J.S., L.J.P. Latimer, B.L. Haug, D.E. Pulleyblank, D.M. Skinner, and G.D. Burkholder, Triplex DNA in plasmids and chromosomes. *Gene 82*: 191–199 (1989).
45. Lyamichev, V.I., S.M. Mirkin, O.N. Danilevskaya, O.N. Voloshin, S.V. Balatskaya, V.N. Dobrynin, S.A. Filippov, and M.D. Frank-Kamenetskii, An unusual DNA structure detected in a telomeric sequence under superhelical stress and at low pH. *Nature 339*: 634–637 (1989).
46. Moser, H.E, and P.B. Dervan, Sequence-specific cleavage of double helical DNA by triple helix formation. *Science 238*: 645–650 (1987).
47. Lyamichev, V.I., S.M. Mirkin, M.D. Frank-Kamenetskii, and C.R. Cantor, A stable complex between homopyrimidine oligomers and the homologous regions of duplex DNAs. *Nucleic Acids Res. 16*: 2165–2178 (1988).
48. Fedorova, O.S., D.G. Knorre, L.M. Podust, and V.F. Zarytova, Complementary addressed modification of double-stranded DNA within a ternary complex. *FEBS Lett. 228*: 273–276 (1988).
49. Matteucci, M.D., and N. Bischofberger, Sequence-defined oligonucleotides as potential therapeutics. *Annu. Rep. Med. Chem. 26*: 287–296 (1991).
50. Griffin, L.C., and P.B. Dervan, Recognition of thymine.adenine base pairs by guanine in a pyrimidine triple helix motif. *Science 245*: 967–971 (1989).
51. Beal, P.A., and P.B. Dervan, Second structural motif for recognition of DNA by oligonucleotide-directed triple-helix formation. *Science 251*: 1360–1361 (1991).
52. Durland, R.H., D.J. Kessler, S. Gunnell, M. Duvic, B.M. Pettitt, and M.E. Hogan, Binding of triple helix forming oligonucleotides to sites in gene promoters. *Biochemistry 30*: 9246–9255 (1991).
53. Orson, F.M., D.W. Thomas, W.M. McShan, D.J. Kessler, and M.E. Hogan, Oligonucleotide inhibition of IL2Rα mRNA transcription by promoter region collinear triplex formation in lymphocytes. *Nucleic Acids Res. 19*: 3435–3440 (1991).
54. Pilch, D.S., C. Levenson, and R.H. Shafer, Structure, stability, and thermodynamics of a short intermolecular purine-purine-pyrimidine triple helix. *Biochemistry 30*: 6081–6087 (1991).
55. Hsieh, P., C.S. Camerini-Otero, and R.D. Camerini-Otero, Pairing of homologous DNA sequences by proteins: evidence for three-stranded DNA. *Genes Dev. 4*: 1951–1963 (1990).
56. Rao, B.J., M. Dutreix, and C.M. Radding, Stable three-stranded DNA made by RecA protein. *Proc. Natl. Acad. Sci. USA 88*: 2984–2988 (1991).
57. Fox, J.L., Case for DNA triple helix strengthens. *ASM News 57*: 509–513 (1991).

58. Nielsen, P.E., M. Egholm, R.H. Berg, and O. Buchardt, Sequence-selective recognition of DNA by strand displacement with a thymine-substituted polyamide. *Science 254*: 1497–1500 (1991).
59. Lee, J.S., D.A. Johnson, and A.R. Morgan, Complexes formed by (pyrimidine)n.(purine)n DNAs on lowering the pH are three-stranded. *Nucleic Acids Res. 6*: 3073–3091 (1979).
60. Shea, R.G., P. Ng, and N. Bischofberger, Thermal denaturation profiles and gel mobility shift analysis of oligodeoxynucleotide triplexes. *Nucleic Acids Res. 18*: 4859–4866 (1990).
61. Xodo, L.E., G. Manzini, and F. Quadrifoglio, Spectroscopic and calorimetric investigation on the DNA triplex formed by d(CTCTTCTTTCTTTTCTTTCTTCTC) and d(GAGAAGAAAGA) at acidic pH. *Nucleic Acids Res. 18*: 3557–3564 (1990).
62. Plum, G.E., Y-W. Park, S.F. Singleton, P.B. Dervan, and K.J. Breslauer, Thermodynamic characterization of the stability and the melting behavior of a DNA triplex: a spectroscopic and calorimetric study. *Proc. Natl. Acad. Sci. USA 87*: 9436–9440 (1990).
63. Pilch, D.S., R. Brousseau, and R.H. Shafer, Thermodynamics of triple helix formation: spectrophotometric studies on the d(A)10.2d(T)10 and d(C+3T4-C+3).d(G3A4G3).d(C3T4C3) triple helices. *Nucleic Acids Res. 18*: 5743–5750 (1990).
64. Cooney, M., G. Czernuszewicz, E.H. Postel, S.J. Flint, and M.E. Hogan, Site-specific oligonucleotide binding represses transcription of the human *c-myc* gene in vitro. *Science 241*: 456–459 (1988).
65. Francois, J-C., T. Saison-Behmoaras, and C. Helene, Sequence-specific recognition of the major groove of DNA by oligodeoxynucleotides via triple helix formation. Footprinting studies. *Nucleic Acids Res. 16*: 11431–11440 (1988).
66. Mayer, L.J., III, B. Wold, and P.B. Dervan, Inhibition of DNA binding proteins by oligonucleotide-directed triple helix formation. *Science 245*: 725–730 (1989).
67. Hanvey, J.C., M. Shimizu, and R.D. Wells, Site-specific inhibition of EcoRI restriction/modification enzymes by a DNA triple helix. *Nucleic Acids Res. 18*: 157–161 (1990).
68. Lyamichev, V.I., O.N. Voloshin, M.D. Frank-Kamenetskii, and V.N. Soyfer, Photofootprinting of DNA triplexes. *Nucleic Acids Res. 19*: 1633–1638 (1991).
69. Perrouault, L., U. Asseline, C. Rivalle, N.T. Thuong, E. Bisagni, C. Giovannangeli, T. LeDoan, and C. Helene, Sequence-specific artificial photo-induced endonucleases based on triple helix-forming oligonucleotides. *Nature 344*: 358–360 (1990).
70. Pei, D., D.R. Corey, and P.G. Schultz, Site-specific cleavage of duplex DNA by a semisynthetic nuclease via triple-helix formation. *Proc. Natl. Acad. Sci. USA 87*: 9858–9862 (1990).
71. Rajagopal, P., and J. Feigon, Triple-strand formation in the homopurine: homopyrimidine DNA oligonucleotides d(G-A)4 and d(T-C)4. *Nature 339*: 637–640 (1989).

72. de los Santos, C., M. Rosen, and D. Patel, NMR studies of DNA (R+)n.(Y+)n triple helices in solution: imino and amino proton markers of T.A.T and C.G.C+ base-triple formation. *Biochemistry 28*: 7282–7289 (1989).
73. Radhakrishnan, I., X. Gao, C. de los Santos, D. Live, and D.J. Patel, NMR structural studies of intramolecular (Y+)n.(R+)n(Y−)n DNA triplexes in solution: imino and amino proton and nitrogen markers of G.TA base triple formation. *Biochemistry 30*: 9022–9030 (1991).
74. Radhakrishnan, I., D. Patel, and X. Gao, NMR assignment strategy for DNA protons through three-dimensional proton-proton connectivities. Application to an intramolecular DNA triplex. *J. Am. Chem. Soc. 113*: 8542–8544 (1991).
75. Macaya, R.F., D.E. Gilbert, S. Malek, J.S. Sinsheimer, and J. Feigon, Structure and stability of X.G.C mismatches in the third strand of intramolecular triplexes. *Science 254*: 270–274 (1991).
76. Arnott, S., and E. Selsing, Structures for the polynucleotide complexes poly (dA).poly(dT) and poly(dT).poly(dA).poly(dT). *J. Mol. Biol. 88*: 509–521 (1974).
77. Lee, J.S., M.L. Woodsworth, L.J.P. Latimer, and A.R. Morgan, Poly-(pyrimidine).poly(purine) synthetic DNAs containing 5-methylcytosine form stable triplexes at neutral pH. *Nucleic Acids Res. 12*: 6603–6614 (1984).
78. Povsic, T.J., and P.B. Dervan, Triple helix formation by oligonucleotides on DNA extended to the physiological pH range. *J. Am. Chem. Soc. 111*: 3059–3061 (1989).
79. Young, S., S. Krawczyk, M. Matteucci, and J. Toole, Triple helix formation inhibits in vitro transcription. *Proc. Natl. Acad. Sci. USA 88*: 10023–10026 (1991).
80. Ono, A., P.O.P. Ts'o, and L. Kan, Triplex formation of oliogonucleotides containing 2′-O′methylpseudoisocytidine in substitution for 2′-deoxycytidine. *J. Am. Chem. Soc. 113*: 4032–4033 (1991).
81. Horne, D.A., and P.B. Dervan, Recognition of mixed-sequence duplex DNA by alternate-strand triple-helix formation. *J. Am. Chem. Soc. 112*: 2435–2437 (1990).
82. McCurdy, S., C. Moulds, and B. Froehler, Deoxyoligonucleotides with inverted polarity: synthesis and use in triple-helix formation. *Nucleosides Nucleotides 10*: 287–290 (1991).
83. Ono, A., C-N. Chen, and L. Kan, DNA triplex formation of oligonucleotide analogues consisting of linker groups and octamer segments that have opposite sugar-phosphate backbone polarities. *Biochemistry 30*: 9914–9921 (1991).
84. Pei, D., H.D. Ulrich, and P.G. Schultz, A combinatorial approach toward DNA recognition. *Science 253*: 1408–1411 (1991).
85. Francois, J-C., T. Saison-Behmoaras, N.T. Thuong, and C. Helene, Inhibition of restriction endonuclease cleavage via triple helix formation by homopyrimidine oligonucleotides. *Biochemistry 28*: 9617–9619 (1989).
86. Birg, F., D. Praseuth, A. Zerial, N.T. Thoung, U. Asseline, T. LeDoan, and C. Helene, Inhibition of Simian virus 40 DNA replication in CV-1 cells by an oligodeoxynucleotide covalently linked to an intercalating agent. *Nucleic Acids Research 18*: 2901–2907 (1990).

87. Joyce, G.F., Amplification, mutation and selection of catalytic RNA. *Gene 82*: 83–87 (1989).
88. Robertson, D.L., and G.F. Joyce, Selection in vitro of an RNA enzyme that specifically cleaves single-stranded DNA. *Nature 344*: 467–468 (1990).
89. Tuerk, C., and L. Gold, Systematic evolution of ligands by exponential enrichment: RNA ligands to bacteriophage T4 DNA polymerase. *Science 249*: 505–510 (1990).
90. Blackwell, T.K., and H. Weintraub, Differences and similarities in DNA-binding preferences of MyoD and E2A protein complexes revealed by binding site selection. *Science 250*: 1104–1110 (1990).
91. Blackwell, T.K., L. Kretzner, E.M. Blackwood, R.N. Eisenman, and H. Weintraub, Sequence-specific DNA binding by the *c-myc* protein. *Science 250*: 1149–1151 (1990).
92. Ellington, A.D., and J.W. Szostak, In vitro selection of RNA molecules that bind specific ligands. *Nature 346*: 818–822 (1990).
93. Pollock, R., and R. Treisman, A sensitive method for the determination of protein-DNA binding specificities. *Nucleic Acids Res. 18*: 6197–6204 (1990).
94. Bock, L.C., L.C. Griffin, J.A. Latham, E.H. Vermaas, and J.J. Toole, Selection of single-stranded DNA molecules that bind and inhibit human thrombin. *Nature 355*: 564 (1992).
95. Tidd, D.M., and H.M. Warenius, Partial protection of oncogene, antisense oligodeoxynucleotides against serum nuclease degradation using terminal methylphosphonate groups. *Br. J. Cancer 60*: 343–350 (1989).
96. Shaw, J-P., K. Kent, J. Bird, J. Fishback, and B. Froehler, Modified deoxyoligonucleotides stable to exonuclease degradation in serum. *Nucleic Acids Res. 19*: 747–750 (1991).
97. Lin, K-Y., and M. Matteucci, Hybridization properties of deoxyoligonucleotides containing anthraquinone pseudonucleosides. *Nucleic Acids Res. 19*: 3111–3114 (1991).
98. de Smidt, P.C., T. Le Doan, S. de Falco, and T.J.C. van Berkel, Association of antisense oligonucleotides with lipoproteins prolongs the plasma half-life and modifies the tissue distribution. *Nucleic Acids Res. 19*: 4695–4700 (1991).
99. Hoke, G.D., K. Draper, S.M. Freier, C. Gonzalez, V.B. Driver, M.C. Zounes, and D.J. Ecker, Effects of phosphorothioate capping on antisense oligonucleotide stability, hybridization and antiviral efficacy versus herpes simplex virus infection. *Nucleic Acids Res. 19*: 5743–5748 (1991).
100. Stirchak, E.P., J.E. Summerton, and D.D. Weller, Uncharged stereoregular nucleic acid analogues. 1. Synthesis of a cytosine-containing oligomer with carbamate internucleoside linkages. *J. Org. Chem. 52*: 4202–4206 (1987).
101. Cormier, J.F., and K.K. Ogilvie, Synthesis of hexanucleotide analogues containing diisopropylsilyl internucleotide linkages. *Nucleic Acids Res. 16*: 4583–4594 (1988).
102. Kawai, S.H., and G. Just, Synthesis of a non-hydrolyzable dinucleoside analogue. *Nucleosides Nucleotides 10*: 1485–1498 (1991).
103. Matteucci, M., Deoxyoligonucleotide analogs based on formacetal linkages. *Tetrahedron Lett. 31*: 2385–2388 (1990).

104. Matteucci, M., Hybridization properties of a deoxyoligonucleotide containing four formacetal linkages. *Nucleosides Nucleotides 10*: 231–234 (1991).
105. Veeneman, G.H., G.A. Van Der Marel, H. Van Den Elst, and J.H. Van Boom, An efficient approach to the synthesis of thymidine derivatives containing phosphate-isosteric methylene acetal linkages. *Tetrahedron 47*: 1547–1562 (1991).
106. Bennett, R.M., G.T. Gabor, and M.M. Merritt, DNA binding to human leukocytes: evidence for a receptor-mediated association, internalization, and degradation of DNA. *J. Clin. Invest. 76*: 2182–2190 (1985).
107. Paffenholz, V., H. Van Le, Y-k. Ho, and T.J. Bardos, Uptake of partially thiolated DNA by ascites tumor cells. *Cancer Res. 36*: 1445–1451 (1976).
108. Stein, C.A., K. Mori, S.L. Loke, C. Subasinghe, K. Shinozuka, J.S. Cohen, and L.M. Neckers, Phosphorothioate and normal oligodeoxyribonucleotides with 5′-linked acridine: characterization and preliminary kinetics of cellular uptake. *Gene 72*: 333–341 (1988).
109. Loke, S.L., C.A. Stein, X.H. Zhang, K. Mori, M. Nakanishi, C. Subasinghe, J.S. Cohen, and L.M. Neckers, Characterization of oligonucleotide transport into living cells. *Proc. Natl. Acad. Sci. USA 86*: 3474–3478 (1989).
110. Yakubov, L.A., E.A. Deeva, F. Zarytova, E.M. Ivanova, A.S. Ryte, L.V. Yurchenko, and V.V. Vlassov, Mechanism of oligonucleotide uptake by cells: involvement of specific receptors? *Proc. Natl. Acad. Sci. USA 86*: 6454–6458 (1989).
111. DePamphilis, M.L., S.A. Herman, E. Martinez-Salas, L.E. Chalifour, D.O. Wirak, D.Y. Cupo, and M. Miranda, Microinjecting DNA into mouse ova to study DNA replication and gene expression and to produce transgenic animals. *BioTechniques 6*: 662–680 (1988).
112. Mannino, R.J., and S. Gould-Fogerite, Liposome mediated gene transfer. *BioTechniques 6*: 682–690 (1988).
113. Chen, C.A., and H. Okayama, Calcium phosphate–mediated gene transfer: a highly efficient transfection system for stably transforming cells with plasmid DNA. *BioTechniques 6*: 632–638 (1988).
114. Eglitis, M.A., and W.F. Anderson, Retroviral vectors for introduction of genes into mammalian cells. *BioTechniques 6*: 608–614 (1988).
115. Andreason, G.L., and G.A. Evans, Introduction and expression of DNA molecules in eukaryotic cells by electroporation. *BioTechniques 6*: 650–660 (1988).
116. Felgner, P.L., T.R. Gadek, M. Holm, R. Roman, H.W. Chan, M. Wenz, J.P. Northrup, G.M. Ringold, and M. Danielsen, Lipofection: a highly efficient, lipid-mediated DNA-transfection procedure. *Proc. Natl. Acad. Sci. USA 84*: 7413–7417 (1987).
117. Jaroszewski, J.W., and J.S. Cohen, Cellular uptake of antisense oligodeoxynucleotides. *Adv. Drug Delivery Rev. 6*: 235–250 (1991).
118. Myers, A.E., and B. Dousse, A method for internalizing nucleic acids into eukaryotic cells. Eur. Pat 273,085 (1988).
119. Wu, G.Y., and C.H. Wu, Receptor-mediated gene delivery and expression in vivo. *J. Biol. Chem. 263*: 14621–14624 (1988).

120. Vesterweber, D., and G. Schatz, DNA-protein conjugates can enter mitochondria via the protein import pathway. *Nature 338*: 170–172 (1989).
121. Wagner, E., M. Zenke, M. Cotten, H. Beug, and M.L. Birnstiel, Transferrin-polycation conjugates as carriers for DNA uptake into cells. *Proc. Natl. Acad. Sci. USA 87*: 3410–3414 (1990).
122. Leonetti, J-P., G. Degols, and B. Lebleu, Biological activity of oligonucleotide-poly(L-lysine) conjugates: mechanism of cell uptake. *Bioconjugate Chem. 1*: 149–153 (1990).
123. Degols, G., J-P. Leonetti, C. Gagnor, M. Lemaitre, and B. Lebleu, Antiviral activity and possible mechanism of action of oligonucleotides-poly(L-lysine) conjugates targeted to vesicular stomatitis virus mRNA and genomic RNA. *Nucleic Acids Res. 17*: 9341–9350 (1989).
124. Lemaitre, M., B. Bayard, and B. Lebleu, Specific antiviral activity of a poly(L-lysine)-conjugated oligodeoxyribonucleotide sequence complementary to vesicular stomatitis virus N protein mRNA initiation site. *Proc. Natl. Acad. Sci. USA 84*: 648–652 (1987).
125. Stevenson, M., and P. Iversen, Inhibition of human immunodeficiency virus type 1-mediated cytopathic effects by poly(L-lysine)-conjugated synthetic antisense oligodeoxyribonucleotides. *J. Gen. Virol. 70*: 2673–2682 (1989).
126. Leonetti, J.P., N. Mechti, G. Degols, C. Gagnor, and B. Lebleu, Intracellular distribution of microinjected antisense oligonucleotides. *Proc. Natl. Acad. Sci. USA 88*: 2702–2706 (1991).
127. Shea, R.G., J.C. Marsters, and N. Bischofberger, Synthesis, hybridization properties and antiviral activity of lipid-oligodeoxynucleotide conjugates. *Nucleic Acids Res. 18*: 3777–3783 (1990).
128. Kabanov, A.V., S.V. Vinogradov, A.V. Ovcharenko, A.V. Krivonos, N.S. Melik-Nubarov, V.I. Kiselev, and E.S. Severin, A new class of antivirals: antisense oligonucleotides combined with a hydrophobic substituent effectively inhibit influenza virus reproduction and synthesis of virus-specific proteins in MDCK cells. *Febs Lett. 259*: 327–330 (1990).
129. Leonetti, J-P., P. Machy, G. Degols, B. Lebleu, and L. Leserman, Antibody-targeted liposomes containing oligodeoxyribonucleotides complementary to viral RNA selectively inhibit viral replication. *Proc. Natl. Acad. Sci. USA 87*: 2448–2451 (1990).
130. Vickers, T., B.F. Baker, P.D. Cook, M. Zounes, R.W. Buckheit, Jr., J. Germany, and D.J. Ecker, Inhibition of HIV-LTR gene expression by oligonucleotides targeted to the TAR element. *Nucleic Acids Res. 19*: 3359–3368 (1991).
131. Leonetti, J.P., N. Mecht, G. Degols, C. Gagnor, and B. Lebleu, Intracellular distribution of microinjected antisense oligonucleotides. *Proc. Natl. Acad. Sci. USA 88*: 2702–2706 (1991).
132. Ceruzzi, M., K. Draper, and J. Schwartz, Natural and phosphorothioate-modified oligodeoxyribonucleotides exhibit a non-random cellular distribution. *Nucleosides Nucleotides 9*: 679–695 (1990).
133. Steinman, R.M., I.S. Mellman, W.A. Muller, and Z.A. Cohn, Endocytosis and the recycling of plasma membrane. *J. Cell Biol. 96*: 1–27 (1983).

134. Järnefelt, J., T. Laurent, and R. Rigler, Diffusion of fluorescein-labelled molecules in suspensions of erythrocyte ghosts. *FEBS 242*: 129–133 (1988).
135. Olsnes, S., K. Sandvig, O.W. Peterson, and B. van Deurs, Immunotoxins: entry into cells and mechanisms of action. *Immunol. Today 10*: 291–295 (1989).
136. Mann, D.A., and A.D. Frankel, Endocytosis and targeting of exogenous HIV-1 Tat protein. *EMBO J. 10*: 1733–1739 (1991).
137. Plagemann, P.G.W., and R.M. Wohlhueter, Permeation of nucleosides, nucleic acid bases, and nucleotides in animal cells. *Current Topics Membrane Transport 14*: 225–330 (1980).
138. Miller, P.S., K.B. McParland, K. Jayaraman, and P.O.P. Ts'o, Biochemical and biological effects of nonionic nucleic acid methylphosphonates. *Biochemistry 20*: 1874–1880 (1981).
139. Shoji, Y., S. Akhtar, A. Periasamy, B. Herman, and R.L. Juliano, Mechanism of cellular uptake of modified oligodeoxynucleotides containing methylphosphonate linkages. *Nucleic Acids Res. 19*: 5543–5550 (1991).
140. Agris, C.H., K.R. Blake, P.S. Miller, M.P. Reddy, and P.O.P. Ts'o, Inhibition of vesicular stomatitis virus protein synthesis and infection by sequence-specific oligodeoxyribonucleoside methylphosphonates. *Biochemistry 25*: 6268–6275 (1986).
141. Zaia, J.A., J.J. Rossi, G.J. Murakawa, P.A. Spallone, D.A. Stephens, B.E. Kaplan, R. Eritja, R.B. Wallace, and E.M. Cantin, Inhibition of human immunodeficiency virus by using an oligonucleoside methylphosphonate targeted to the *tat-3* gene. *J. Virol. 62*: 3914–3917 (1988).
142. Sarin, P.S., S. Agrawal, M.P. Civeira, J. Goodchild, T. Ikeuchi, and P.C. Zamecnik, Inhibition of acquired immunodeficiency syndrome virus by oligodeoxynucleoside methylphosphonates. *Proc. Natl. Acad. Sci. USA 85*: 7448–7451 (1988).
143. Chang, E.H., P.S. Miller, C. Cushman, K. Devadas, K.F. Pirollo, P.O.P. Ts'o, and Z.P. Yu, Antisense inhibition of ras p21 expression that is sensitive to a point mutation. *Biochemistry 30*: 8283–8286 (1991).
144. Lesnikowski, Z.J., M. Jaworska, and W.J. Stec, Octa(thymidine methanephosphonates) of partially defined stereochemistry: synthesis and effect of chirality at phosphorus on binding to pentadecadeoxyriboadenylic acid. *Nucleic Acids Res. 18*: 2109–2115 (1990).
145. Quartin, R.S., C.L. Brakel, and J.G. Wetmur, Number and distribution of methylphosphonate linkages in oligodeoxynucleotides affect exo- and endonuclease sensitivity and ability to form RNase H substrates. *Nucleic Acids Res. 17*: 7253–7262 (1989).
146. Zerial, A., N.T. Thuong, and C. Helene, Selective inhibition of the cytopathic effect of type A influenza viruses by oligodeoxynucleotides covalently linked to an intercalating agent. *Nucleic Acids Res. 15*: 9909–9919 (1987).
147. Shaw, J-P., J. Milligan, S. Krawczyk, and M. Matteucci, Specific, high efficiency, triple helix mediated crosslinking to duplex DNA. *J. Am. Chem. Soc., 113*: 7765–7766 (1991).
148. Letsinger, R.L., C.N. Singman, G. Histand, and M. Salunkhe, Cationic oligonucleotides. *J. Am. Chem. Soc. 110*: 4470–4471 (1988).

149. Kool, E.T., Molecular recognition by circular oligonucleotides: increasing the selectivity of DNA binding. *J. Am. Chem. Soc. 113*: 6265–6266 (1991).
150. Kawase, Y., S. Iwai, H. Inoue, K. Miura, and E. Ohtsuka, Studies on nucleic acid interactions I. Stabilities of mini-duplexes ($dG_2A_4XA_4G_2 \cdot dC_2T_4YT_4C_2$) and self-complementary d(GGGAAXYTTCCC) containing deoxyinosine and other mismatched bases. *Nucleic Acids Res. 14*: 7727–7736 (1986).
151. Persson, C., E.G.H. Wagner, and K. Norström, Control of replication of plasmid R1: formation of an initial transient complex is rate-limiting for antisense RNA - target RNA pairing. *EMBO J. 9*: 3777–3785 (1990).
152. Distefano, M.D., J.A. Shin, and P.B. Dervan, Cooperative binding of oligonucleotides to DNA by triple helix formation: dimerisation via Watson-Crick hydrogen bonds. *J. Am. Chem. Soc. 113*: 5901–5902 (1991).
153. Harrioson, S.C., A structural taxonomy of DNA-binding domains. *Nature 353*: 715–719 (1991).
154. Strobel, S.A., and P.B. Dervan, Site-specific cleavage of a yeast chromosome by oligonucleotide-directed triple helix formation. *Science 249*: 73–75 (1990).
155. Geiser, T., Large-scale economic synthesis of antisense phosphorothioate analogues of DNA for preclinical investigations. *Ann. NY Acad. Sci. 616*: 173–183 (1990).
156. Maccecchini, M., Techniques for the process scale-up of commercially important peptides. *Genetic Engineering News*, p. 12 (Nov./Dec. 1990).
157. Matsukura, M., K. Shinozuka, G. Zon, H. Mitsuya, M. Reitz, J.S. Cohen, and S. Broder, Phosphorothioate analogs of oligodeoxynucleotides: Inhibitors of replication and cytopathic effects of human immunodeficiency virus. *Proc. Natl. Acad. Sci. USA 84*: 7706–7710 (1987).
158. Matsukura, M., G. Zon, K. Shinozuka, M. Robert-Guroff, T. Shimada, C.A. Stein, H. Mitsuya, F. Wong-Staal, J.S. Cohen, and S. Broder, Regulation of viral expression of human immunodeficiency virus in vitro by an antisense phosphorothioate oligodeoxynucleotide against *rev (art/trs)* in chronically infected cells. *Proc. Natl. Acad. Sci. USA 86*: 4244–4248 (1989).
159. Letsinger, R.L., G. Zhang, D.K. Sun, T. Ikeuchi, and P.S. Sarin, Cholesteryl-conjugated oligonucleotides: synthesis, properties, and activity as inhibitors of replication of human immunodeficiency virus in cell culture. *Proc. Natl. Acad. Sci. USA 86*: 6553–6556 (1989).
160. Majumdar, C., Stein, C.A., J.S. Cohen, S. Broder, and S.H. Wilson, Stepwise mechanism of HIV reverse transcriptase: primer function of phosphorothioate oligodeoxynucleotide. *Biochemistry 28*: 1340–1346 (1989).
161. Zon, G., Innovations in the use of antisense oligonucleotides. *Ann. NY Acad. Sci. 616*: 161–172 (1990).
162. Gao, W-Y., R.N. Hanes, M.A. Vazquez-Padua, C.A. Stein, J.S. Cohen, and Y.C. Cheng, Inhibition of herpex simplex virus type 2 growth by phosphorothioate oligonucleotides. *Antimicrob. Agents Chemother. 34*: 808–812 (1990).
163. Wu Dunn, D., and P.G. Spear, Initial interaction of herpes simplex virus with cells is binding to heparan sulfate. *J. Virol. 63*: 52–58 (1989).
164. Goren, M.B., A.E. Vatter, and J. Fiscus, Polyanionic agents as inhibitors

of phagosome-lysosome fusion in cultured macrophages: evolution of an alternative interpretation. *J. Leukocyte Biol. 41*: 111–121 (1987).

165. Colvin, O.M., Disposition and metabolism of oligodeoxynucleoside methylphosphonate following a single iv injection in mice. *Drug Metabol. Disposition 18*: 815–818 (1990).
166. Goodchild, J., S. Agrawal, M.P. Civeira, P.S. Sarin, D. Sun, and P.C. Zamecnik, Inhibition of human immunodeficiency virus replication by antisense oligodeoxynucleotides. *Proc. Natl. Acad. Sci. USA 85*: 5507–5511 (1988).
167. Agrawal, S., J. Temsamani, and J.Y. Tang, Pharmacokinetics, biodistribution, and stability of oligodeoxynucleotide phosphorothioates in mice. *Proc. Natl. Acad. Sci. USA 88*: 7595–7599 (1991).
168. Draper, K.G., and V. Brown-Driver, Reduction of herpes simplex virus infection using phosphorothioate oligonucleotides complementary to viral mRNA. *J. Antiviral Res. Suppl 1*, Program Abstracts from 4th International Conference on Antiviral Research, New Orleans, LA, April 1991.
169. Ts'o, P.O.P., P.S. Miller, L. Aurelian, A. Murakami, C. Agris, K.R. Blake, S.B. Lin, B.L. Lee, and C.C. Smith, An approach to chemotherapy based on base sequence information and nucleic acid chemistry. *Ann. NY Acad. Sci. 507*: 220–241 (1987).
170. Vlassov, V.V., Inhibition of tick-borne viral encephalitis expression using covalently linked oligonucleotide analogs. Conference on Oligodeoxynucleotides as Antisense Inhibitors of Gene Expression: Therapeutic Implications, Rockville, MD, June 1989.
171. Genta developing non-oligonucleotide dermatologicals as adjunct to antisense—Nov. 25 IPO filing. *FDC Rep. Prescription OTC Pharmaceuticals 53*: T&G-5 (Dec. 9, 1991).

Index